T0191570

Modular Forms and Fermat's Last Theorem

Springer
New York
Berlin
Heidelberg
Barcelona
Hong Kong
London
Milan
Paris
Singapore
Tokyo

Gary Cornell Joseph H. Silverman
Glenn Stevens
Editors

Modular Forms and Fermat's Last Theorem

Springer

Gary Cornell
Department of Mathematics
University of Connecticut
Storrs, CT 06268
USA

Joseph H. Silverman
Department of Mathematics
Brown University
Providence, RI 02912
USA

Glenn Stevens
Department of Mathematics
Boston University
Boston, MA 02215
USA

Mathematics Subject Classification (1991): 11D41, 11G18, 14Hxx, 11-03

Library of Congress Cataloging-in-Publication Data
Modular forms and Fermat's last theorem / edited by Gary Cornell,
Joseph H. Silverman, Glenn Stevens ; with contributions by B. Conrad
[et al.].
p. cm.
Papers from a conference held Aug. 9-18, 1995, at Boston
University
Includes bibliographical references and index.
ISBN 0-387-98998-6 (alk. paper)
1. Curves, Elliptic—Congresses. 2. Forms, Modular—Congresses.
3. Fermat's last theorem—Congresses. I. Cornell Gary.
II. Silverman, Joseph H., 1955- . III. Stevens, Glenn, 1953- .
QA567.2.E44M63 1997
512'.74—dc21 97-10930

Printed on acid-free paper.

First softcover printing, 2000.

Production managed by Natalie Johnson; manufacturing supervised by Johanna Tschebull.
Photocomposed copy prepared from the authors' AMS-TEX, LaTEX, and TEX files.
Printed and bound by Maple-Vail Book Manufacturing Group, York, PA.
Printed in the United States of America.

9 8 7 6 5 4 3 2 1

ISBN 0-387-98998-6 Springer-Verlag New York Berlin Heidelberg SPIN 10755699

Preface

This volume is the record of an instructional conference on number theory and arithmetic geometry held from August 9 through 18, 1995 at Boston University. It contains expanded versions of all of the major lectures given during the conference. We want to thank all of the speakers, all of the writers whose contributions make up this volume, and all of the "behind-the-scenes" folks whose assistance was indispensable in running the conference. We would especially like to express our appreciation to Patricia Pacelli, who coordinated most of the details of the conference while in the midst of writing her PhD thesis, to Jaap Top and Jerry Tunnell, who stepped into the breach on short notice when two of the invited speakers were unavoidably unable to attend, and to Stephen Gelbart, whose courage and enthusiasm in the face of adversity has been an inspiration to us.

Finally, the conference was only made possible through the generous support of Boston University, the Vaughn Foundation, the National Security Agency and the National Science Foundation. In particular, their generosity allowed us to invite a multitude of young mathematicians, making the BU conference one of the largest and liveliest number theory conferences ever held.

January 13, 1997

G. Cornell
J.H. Silverman
G. Stevens

Preface

[illegible]

Contents

Preface v

Contributors xiii

Schedule of Lectures xvii

Introduction xix

CHAPTER I 1

An Overview of the Proof of Fermat's Last Theorem

GLENN STEVENS

§1. A remarkable elliptic curve 2
§2. Galois representations 3
§3. A remarkable Galois representation 7
§4. Modular Galois representations 7
§5. The Modularity Conjecture and Wiles's Theorem 9
§6. The proof of Fermat's Last Theorem 10
§7. The proof of Wiles's Theorem 10
References 15

CHAPTER II 17

A Survey of the Arithmetic Theory of Elliptic Curves

JOSEPH H. SILVERMAN

§1. Basic definitions 17
§2. The group law 18
§3. Singular cubics 18
§4. Isogenies 19
§5. The endomorphism ring 19
§6. Torsion points 20
§7. Galois representations attached to E 20
§8. The Weil pairing 21
§9. Elliptic curves over finite fields 22
§10. Elliptic curves over $\mathbb{C}$ and elliptic functions 24
§11. The formal group of an elliptic curve 26
§12. Elliptic curves over local fields 27
§13. The Selmer and Shafarevich-Tate groups 29
§14. Discriminants, conductors, and L-series 31
§15. Duality theory 33

§16. Rational torsion and the image of Galois 34
§17. Tate curves 34
§18. Heights and descent 35
§19. The conjecture of Birch and Swinnerton-Dyer 37
§20. Complex multiplication 37
§21. Integral points 39
References 40

CHAPTER III 41
Modular Curves, Hecke Correspondences, and L-Functions
DAVID E. ROHRLICH
§1. Modular curves 41
§2. The Hecke correspondences 61
§3. L-functions 73
References 99

CHAPTER IV 101
Galois Cohomology
LAWRENCE C. WASHINGTON
§1. H^0, H^1, and H^2 101
§2. Preliminary results 105
§3. Local Tate duality 107
§4. Extensions and deformations 108
§5. Generalized Selmer groups 111
§6. Local conditions 113
§7. Conditions at p 114
§8. Proof of theorem 2 117
References 120

CHAPTER V 121
Finite Flat Group Schemes
JOHN TATE
Introduction 121
§1. Group objects in a category 122
§2. Group schemes. Examples 125
§3. Finite flat group schemes; passage to quotient 132
§4. Raynaud's results on commutative p-group schemes 146
References 154

CHAPTER VI 155
Three Lectures on the Modularity of $\bar{\rho}_{E,3}$
and the Langlands Reciprocity Conjecture
STEPHEN GELBART
Lecture I. The modularity of $\bar{\rho}_{E,3}$ and automorphic representations of weight one 156
§1. The modularity of $\bar{\rho}_{E,3}$ 157
§2. Automorphic representations of weight one 164
Lecture II. The Langlands program: Some results and methods
§3. The local Langlands correspondence for $GL(2)$ 176
§4. The Langlands reciprocity conjecture (LRC) 179
§5. The Langlands functoriality principle theory and results 182

Lecture III. Proof of the Langlands-Tunnell theorem 192
§6. Base change theory 192
§7. Application to Artin's conjecture 197
References 204

CHAPTER VII 209
Serre's Conjectures
BAS EDIXHOVEN
§1. Serre's conjecture: statement and results 209
§2. The cases we need 222
§3. Weight two, trivial character and square free level 224
§4. Dealing with the Langlands–Tunnell form 230
References 239

CHAPTER VIII 243
An Introduction to the Deformation Theory of Galois Representations
BARRY MAZUR
Chapter I. Galois representations 246
Chapter II. Group representations 251
Chapter III. The deformation theory for Galois representations 259
Chapter IV. Functors and representability 267
Chapter V. Zariski tangent spaces and deformation problems subject to "conditions" 284
Chapter VI. Back to Galois representations 294
References 309

CHAPTER IX 313
Explicit Construction of Universal Deformation Rings
BART DE SMIT AND HENDRIK W. LENSTRA, JR.
§1. Introduction 313
§2. Main results 314
§3. Lifting homomorphisms to matrix groups 317
§4. The condition of absolute irreducibility 318
§5. Projective limits 320
§6. Restrictions on deformations 323
§7. Relaxing the absolute irreducibility condition 324
References 326

CHAPTER X 327
Hecke Algebras and the Gorenstein Property
JACQUES TILOUINE
§1. The Gorenstein property 328
§2. Hecke algebras 330
§3. The main theorem 331
§4. Strategy of the proof of theorem 3.4 334
§5. Sketch of the proof 335
Appendix 340
References 341

CHAPTER XI 343

Criteria for Complete Intersections

BART DE SMIT, KARL RUBIN, AND RENÉ SCHOOF

Introduction 343
§1. Preliminaries 345
§2. Complete intersections 347
§3. Proof of Criterion I 350
§4. Proof of Criterion II 353
Bibliography 355

CHAPTER XII 357

ℓ-adic Modular Deformations and Wiles's "Main Conjecture"

FRED DIAMOND AND KENNETH A. RIBET

§1. Introduction 357
§2. Strategy 358
§3. The "Main Conjecture" 359
§4. Reduction to the case $\Sigma = \emptyset$ 363
§5. Epilogue 370
Bibliography 370

CHAPTER XIII 373

The Flat Deformation Functor

BRIAN CONRAD

Introduction 373
§0. Notation 374
§1. Motivation and flat representations 375
§2. Defining the functor 394
§3. Local Galois cohomology and deformation theory 397
§4. Fontaine's approach to finite flat group schemes 406
§5. Applications to flat deformations 412
References 418

CHAPTER XIV 421

Hecke Rings and Universal Deformation Rings

EHUD DE SHALIT

§1. Introduction 421
§2. An outline of the proof 424
§3. Proof of proposition 10 – On the structure of the Hecke algebra 432
§4. Proof of proposition 11 – On the structure of the universal deformation ring 436
§5. Conclusion of the proof: Some group theory 442
Bibliography 444

CHAPTER XV 447

Explicit Families of Elliptic Curves with Prescribed Mod N Representations

ALICE SILVERBERG

Introduction 447
Part 1. Elliptic curves with the same mod N representation 448
§1. Modular curves and elliptic modular surfaces of level N 448
§2. Twists of Y_N and W_N 449
§3. Model for W when $N = 3$, 4, or 5 450
§4. Level 4 451

Part 2. Explicit families of modular elliptic curves 454
§5. Modular j invariants 454
§6. Semistable reduction 455
§7. Mod 4 representations 456
§8. Torsion subgroups 457
References 461

CHAPTER XVI 463
Modularity of Mod 5 Representations
KARL RUBIN

Introduction 463
§1. Preliminaries: Group theory 465
§2. Preliminaries: Modular curves 466
§3. Proof of the irreducibility theorem (Theorem 1) 470
§4. Proof of the modularity theorem (Theorem 2) 470
§5. Mod 5 representations and elliptic curves 471
References 473

CHAPTER XVII 475
An Extension of Wiles' Results
FRED DIAMOND

§1. Introduction 475
§2. Local representations mod ℓ 476
§3. Minimally ramified liftings 480
§4. Universal deformation rings 481
§5. Hecke algebras 482
§6. The main results 483
§7. Sketch of proof 484
References 488

APPENDIX TO CHAPTER XVII 491
Classification of $\bar{\rho}_{E,\ell}$ by the j Invariant of E
FRED DIAMOND AND KENNETH KRAMER

CHAPTER XVIII 499
Class Field Theory and the First Case of Fermat's Last Theorem
HENDRIK W. LENSTRA, JR. AND PETER STEVENHAGEN

CHAPTER XIX 505
Remarks on the History of Fermat's Last Theorem 1844 to 1984
MICHAEL ROSEN

Introduction 507
§1. Fermat's last theorem for polynomials 507
§2. Kummer's work on cyclotomic fields 508
§3. Fermat's last theorem for regular primes and certain other cases 513
§4. The structure of the p-class group 517
§5. Suggested readings 521
Appendix A: Kummer congruence and Hilbert's theorem 94 522
Bibliography 524

CHAPTER XX 527

On Ternary Equations of Fermat Type and Relations with Elliptic Curves

GERHARD FREY

§1. Conjectures 527
§2. The generic case 540
§3. $K = \mathbf{Q}$ 542
References 548

CHAPTER XXI 549

Wiles' Theorem and the Arithmetic of Elliptic Curves

HENRI DARMON

§1. Prelude: plane conics, Fermat and Gauss 549
§2. Elliptic curves and Wiles' theorem 552
§3. The special values of $L(E/\mathbf{Q}, s)$ at $s = 1$ 557
§4. The Birch and Swinnerton-Dyer conjecture 563
References 566

Index 573

Contributors and Speakers

BRIAN CONRAD
Department of Mathematics, Harvard University, One Oxford Street, Cambridge, MA 02138 USA.
(bconrad@math.harvard.edu)

GARY CORNELL
Department of Mathematics, University of Connecticut at Storrs, Storrs, CT 06269 USA.
(gcornell@nsf.gov)

HENRI DARMON
Department of Mathematics, McGill University, Montréal, Québec, H3A-2K6 Canada.
(darmon@math.mcgill.ca, www.math.mcgill.ca/~darmon)

EHUD DE SHALIT
Institute of Mathematics, Hebrew University, Giv'at-Ram, 91904 Jerusalem Israel.
(deshalit@math.huji.ac.il)

BART DE SMIT
Vakgroep Wiskunde, Universiteit van Amsterdam, Plantage Muidergracht 24, 1018 TV Amsterdam, The Netherlands.
(bds@wins.uva.nl)

FRED DIAMOND
Department of Mathematics, Massachusetts Institute of Technology, 77 Massachusetts Avenue, Cambridge, MA 02139 USA.
(fdiamond@math.mit.edu)

BAS EDIXHOVEN
Institut Mathématique, Université de Rennes 1, Campus de Beaulieu, 35042 Rennes cedex France.
(edix@univ-rennes1.fr)

GERHARD FREY
Institute for Experimental Mathematics, University of Essen, 29, Ellernstrasse, 45326 Essen Germany.
(frey@exp-math.uni-essen.de)

STEPHEN GELBART
Department of Mathematics, Weizmann Institute of Science, Rehovot 76100 Israel.
(gelbar@wisdom.weizmann.ac.il)

BENEDICT H. GROSS
Department of Mathematics, Harvard University, One Oxford Street, Cambridge, MA 02138 USA.
(gross@math.harvard.edu)

KENNETH KRAMER
Department of Mathematics, Queens College, City University of New York, 65-30 Kissena Boulevard, Flushing, NY 11367 USA.
(kramer@qcvaxa.acc.qc.edu)

HENDRIK W. LENSTRA, JR.
Department of Mathematics 3840, University of California, Berkeley, CA 94720–3840 USA.
(hwl@math.berkeley.edu)

BARRY MAZUR
Department of Mathematics, 1 Oxford Street, 325 Science Center, Harvard University, Cambridge, MA 02138 USA.
(mazur@math.harvard.edu)

KENNETH A. RIBET
Department of Mathematics 3840, University of California, Berkeley, CA 94720 USA.
(ribet@math.berkeley.edu)

DAVID E. ROHRLICH
Department of Mathematics, Boston University, 111 Cummington Street, Boston, MA 02215 USA.
(rohrlich@math.bu.edu)

MICHAEL ROSEN
Department of Mathematics, Box 1917, Brown University, Providence, RI 02912 USA.
(michael_rosen@brown.edu)

KARL RUBIN
Department of Mathematics, Ohio State University, 231 W. 18th Avenue, Columbus, OH 43210 USA.
(rubin@math.ohio-state.edu, www.math.ohio-state.edu/~rubin)

RENÉ SCHOOF
2ª Università di Roma "Tor Vergata", Dipartimento di Matematica, I-00133 Roma Italy.
(schoof@fwi.uva.nl)

ALICE SILVERBERG
Department of Mathematics, Ohio State University, 231 W. 18 Avenue, Columbus, OH 43210 USA.
(silver@math.ohio-state.edu)

JOSEPH H. SILVERMAN
Department of Mathematics, Box 1917, Brown University, Providence, RI 02912 USA.
(jhs@gauss.math.brown.edu, www.math.brown.edu/~jhs)

PETER STEVENHAGEN
Faculteit WINS, Universiteit van Amsterdam, Plantage Muidergracht 24, 1018 TV Amsterdam, The Netherlands.
(psh@wins.uva.nl)

GLENN STEVENS
Department of Mathematics, Boston University, 111 Cummington Street, Boston, MA 02215 USA.
(ghs@math.bu.edu)

JOHN TATE
Department of Mathematics, University of Texas at Austin, Austin, TX 78712 USA.
(tate@math.utexas.edu)

JACQUES TILOUINE
Départment de Mathématiques, UA742, Université de Paris-Nord, 93430 Villetaneuse France.
(tilouine@math.univ-paris13.fr)

JAAP TOP
Vakgroep Wiskunde RuG, P.O. Box 800, 9700 AV Groningen, The Netherlands.
(top@math.rug.nl)

JERRY TUNNELL
Department of Mathematics, Rutgers University, New Brunswick, NJ 08903 USA.
(tunnell@math.rutgers.edu)

LAWRENCE C. WASHINGTON
Department of Mathematics, University of Maryland, College Park, MD 20742 USA.
(lcw@math.umd.edu)

ANDREW WILES
Department of Mathematics, Princeton University, Princeton, NJ 08544 USA.
(wiles@math.princeton.edu)

Schedule of Lectures

Wednesday, August 9, 1995

9:00–10:00	Glenn Stevens, *Overview of the proof of Fermat's Last Theorem*
10:30–11:30	Joseph Silverman, *Geometry of elliptic curves*
1:30–2:30	Jaap Top, *Modular curves*
3:00–4:00	Larry Washington, *Galois cohomology and Tate duality*

Thursday, August 10, 1995

9:00–10:00	Joseph Silverman, *Arithmetic of elliptic curves*
10:30–11:30	Jaap Top, *The Eichler-Shimura relations*
1:30–2:30	John Tate, *Finite group schemes*
3:00–4:00	Jerry Tunnell, *Modularity of* $\overline{\rho}_{E,3}$

Friday, August 11, 1995

9:00–10:00	Dick Gross, *Serre's Conjectures*
10:30–11:30	Barry Mazur, *Deformations of Galois representations: Introduction*
1:30–2:30	Hendrik Lenstra, Jr., *Explicit construction of deformation rings*
3:00–4:00	Jerry Tunnell, *On the Langlands Program*

Saturday, August 12, 1995

9:00–10:00	Jerry Tunnell, *Proof of certain cases of Artin's Conjecture*
10:30–11:30	Barry Mazur, *Deformations of Galois representations: Examples*
1:30–2:30	Dick Gross, *Ribet's Theorem*
3:00–4:00	Gerhard Frey, *Fermat's Last Theorem and elliptic curves*

Monday, August 14, 1995

9:00–10:00	Jacques Tilouine, *Hecke algebras and the Gorenstein property*
10:30–11:30	René Schoof, *The Wiles-Lenstra criterion for complete intersections*
1:30–2:30	Barry Mazur, *The tangent space and the module of Kähler differentials of the universal deformation ring*
3:00–4:00	Ken Ribet, *p-adic modular deformations of mod p modular representations*

Tuesday, August 15, 1995

9:00–10:00	René Schoof, *The Wiles-Faltings criterion for complete intersections*
10:30–11:30	Brian Conrad, *The flat deformation functor*
1:30–2:30	Larry Washington, *Computations of Galois cohomology*
3:00–4:00	Gary Cornell, *Sociology, history and the first case of Fermat*

Wednesday, August 16, 1995

9:00–10:00	Ken Ribet, *Wiles' "Main Conjecture"*
10:30–11:30	Ehud de Shalit, *Modularity of the universal deformation ring (the minimal case)*

Thursday, August 17, 1995

9:00–10:00	Alice Silverberg, *Explicit families of elliptic curves with prescribed mod n representations*
10:30–11:30	Ehud de Shalit, *Estimating Selmer groups*
1:30–2:30	Ken Ribet, *Non-minimal deformations (the "induction step")*
3:00–4:00	Michael Rosen, *Remarks on the history of Fermat's Last Theorem: 1844 to 1984*

Friday, August 18, 1995

9:00–10:00	Fred Diamond, *An extension of Wiles' results*
10:30–11:30	Karl Rubin, *Modularity of mod 5 representations*
1:30–2:30	Henri Darmon, *Consequences and applications of Wiles' theorem on modular elliptic curves*
3:00–4:00	Andrew Wiles, *Modularity of semistable elliptic curves: Overview of the proof*

Introduction

The chapters of this book are expanded versions of the lectures given at the BU conference. They are intended to introduce the many ideas and techniques used by Wiles in his proof that every (semi-stable) elliptic curve over **Q** is modular, and to explain how Wiles' result combined with Ribet's theorem implies the validity of Fermat's Last Theorem.

The first chapter contains an overview of the complete proof, and it is followed by introductory chapters surveying the basic theory of elliptic curves (Chapter II), modular functions and curves (Chapter III), Galois cohomology (Chapter IV), and finite group schemes (Chapter V). Next we turn to the representation theory which lies at the core of Wiles' proof. Chapter VI gives an introduction to automorphic representations and the Langlands-Tunnell theorem, which provides the crucial first step that a certain mod 3 representation is modular. Chapter VII describes Serre's conjectures and the known cases which give the link between modularity of elliptic curves and Fermat's Last Theorem. After this come chapters on deformations of Galois representations (Chapter VIII) and universal deformation rings (Chapter IX), followed by chapters on Hecke algebras (Chapter X) and complete intersections (Chapter XI). Chapters XII and XIV contain the heart of Wiles' proof, with a brief interlude (Chapter XIII) devoted to representability of the flat deformation functor. The final step in Wiles' proof, the so-called "3-5 shift," is discussed in Chapters XV and XVI, and Diamond's relaxation of the semi-stability condition is described in Chapter XVII. The volume concludes by looking both backward and forward in time, with two chapters (Chapters XVIII and XIX) describing some of the "pre-modular" history of Fermat's Last Theorem, and two chapters (Chapters XX and XXI) placing Wiles' theorem into a more general Diophantine context and giving some ideas of possible future applications.

As the preceding brief summary will have made clear, the proof of Wiles' theorem is extremely intricate and draws on tools from many areas of mathematics. The editors hope that this volume will help everyone, student and professional mathematician alike, who wants to study the details of what is surely one of the most memorable mathematical achievements of this century.

Introduction

[illegible]

AN OVERVIEW OF THE PROOF OF FERMAT'S LAST THEOREM

GLENN STEVENS

The principal aim of this article is to sketch the proof of the following famous assertion.

Fermat's Last Theorem. *For $n > 2$, we have*

$$\mathbf{FLT}(n): \quad \left.\begin{matrix} a^n + b^n = c^n \\ a, b, c \in \mathbf{Z} \end{matrix}\right\} \Longrightarrow abc = 0.$$

Many special cases of Fermat's Last Theorem were proved from the 17th through the 19th centuries. The first known case is due to Fermat himself, who proved FLT(4) around 1640. FLT(3) was proved by Euler between 1758 and 1770. Since FLT(d) $\Longrightarrow$ FLT(n) whenever $d|n$, the results of Euler and Fermat immediately reduce our theorem to the following assertion.

Theorem. *If $p \geq 5$ is prime, and $a, b, c \in \mathbf{Z}$, then $a^p + b^p + c^p = 0 \Longrightarrow abc = 0$.*

The proof of this theorem is the result of the combined efforts of innumerable mathematicians who have worked over the last century (and more!) to develop a rich and powerful arithmetic theory of elliptic curves, modular forms, and galois representations. It seems appropriate to emphasize the names of five individuals who had the insight to see how this theory could be used to prove Fermat's Last Theorem and to supply the final crucial ingredients of the proof:

Gerhart Frey (1985), who first suggested that the existence of a solution of the Fermat equation might contradict the Modularity Conjecture of Taniyama, Shimura, and Weil;

Jean-Pierre Serre (1985-6), who formulated and (with J.-F. Mestre) tested numerically a precise conjecture about modular forms and galois representations mod p and who showed how a small piece of this conjecture — the so-called *epsilon conjecture* — together with the Modularity Conjecture would imply Fermat's Last Theorem;

Ken Ribet (1986), who proved Serre's *epsilon conjecture*, thus reducing the proof of Fermat's Last Theorem to a proof of the Modularity Conjecture for semistable elliptic curves;

Richard Taylor (1994), who collaborated with Wiles to complete the proof of Wiles's numerical criterion in the *minimal case*;

Andrew Wiles (1994), who had the vision to identify the crucial numerical criterion from which the Modularity Conjecture for semistable elliptic curves would follow, and who finally supplied a proof of this criterion, thus completing the proof of Fermat's Last Theorem.

To prove the theorem we follow the program outlined by Serre in [16]. Fix a prime $p \geq 5$ and suppose $a, b, c \in \mathbf{Z}$ satisfy $a^p + b^p + c^p = 0$ but $abc \neq 0$. The triple (a^p, b^p, c^p) is what Gerhard Frey has called a "remarkable" triple of integers, so remarkable in fact, that we suspect it does not exist. To derive a contradiction, we will transform this triple into another object with remarkable properties, namely a very special modular form f_{a^p,b^p,c^p}, something firmly rooted in the fertile grounds of modern number theory. The construction of this modular form is a two-step process. First, by a simple but insightful construction due independently to Yves Hellegouarch and Gerhard Frey, we obtain a certain semistable elliptic curve E_{a^p,b^p,c^p} defined over $\mathbf{Q}$. Then, by Wiles's semistable modularity theorem, we deduce the existence of a modular form f_{a^p,b^p,c^p} associated to E_{a^p,b^p,c^p} by the correspondence of Eichler and Shimura.

With f_{a^p,b^p,c^p} in hand, we seek a contradiction within the realm of modular forms. The crucial ingredients that finally lead to a contradiction are encoded in a certain irreducible galois representation $\overline{\rho}_{a^p,b^p,c^p} : G \longrightarrow \mathrm{GL}_2(\mathbf{F}_p)$ associated to f_{a^p,b^p,c^p}. As noted by Frey and Serre, the remarkableness of the triple (a^p, b^p, c^p) is reflected by some remarkable local properties of $\overline{\rho}_{a^p,b^p,c^p}$. Indeed, they noted that $\overline{\rho}_{a^p,b^p,c^p}$ can ramify only at 2 and p, and that the ramification at 2 and p is rather mild (semistable at 2 and what Serre called *peu ramifiée*). But experience with galois representations shows that it is difficult to make large galois representations with so little ramification. As Serre conjectured and Ribet proved, the existence of such a *modular* galois representation has untenable consequences in the theory of modular forms. Fermat's Last Theorem follows.

§1. A Remarkable Elliptic Curve

In this section we describe the crucial construction of an elliptic curve E_{a^p,b^p,c^p} out of a hypothetical solution of the Fermat equation $a^p+b^p+c^p = 0$. For any triple (A, B, C) of coprime integers satisfying $A + B + C = 0$, Gerhart Frey [8] considered the elliptic curve $E_{A,B,C}$ defined by the Weierstrass equation

$$E_{A,B,C} : y^2 = x(x - A)(x + B)$$

and explained some of the ways in which the arithmetic properties of $E_{A,B,C}$ are related to the diophantine properties of the triple (A, B, C). Especially interesting are the connections with the Masser-Oesterle A-B-C conjecture and its generalizations. For a discussion of this line of thought including connections with modular curves, we refer the reader to [7] and to Frey's article in this volume (chapter XX).

For our purposes it suffices to consider only the special case where $(A, B, C) = (a^p, b^p, c^p)$ corresponds to a hypothetical solution of the Fermat equation. Without loss of generality, we may assume $a \equiv -1$ modulo 4 and $2|b$. It is not hard to calculate both the minimal discriminant Δ_{a^p,b^p,c^p} and the conductor N_{a^p,b^p,c^p} of the elliptic curve E_{a^p,b^p,c^p}.

(1.1) Proposition. *Let $p \geq 5$ be prime and let a, b, c be coprime integers satisfying $abc \neq 0$, $a \equiv -1$ modulo 4, $2|b$, and $a^p + b^p + c^p = 0$. Then E_{a^p,b^p,c^p} is a semistable elliptic curve whose minimal discriminant and conductor are given by the formulas*
(a) $\Delta_{a^p,b^p,c^p} = 2^{-8} \cdot (abc)^{2p}$, *and*
(b) $N_{a^p,b^p,c^p} = \prod_{\ell | abc} \ell$.

For definitions of semistability and of the conductor and minimal discriminant see Silverman's article in this volume (chapter II, especially §14 and §17). In general the primes dividing the minimal discriminant of an elliptic curve over $\mathbf{Q}$ are the same as those dividing the conductor and this might lead us to suspect that the discriminant and conductor should be close to one another. Indeed, Szpiro has formulated the following conjecture (see [19] where a slightly stronger form of the conjecture is formulated).

Conjecture. (Szpiro) *For any $\epsilon > 0$ there is a constant $C > 0$ such that the minimal discriminant Δ_E and conductor N_E of any elliptic curve $E_{/\mathbf{Q}}$ satisfy the inequality*

$$|\Delta_E| < C \cdot N_E^{6+\epsilon}.$$

On the other hand, proposition 1.1 shows that a counterexample to $FLT(p)$ for sufficiently large p gives rise to an elliptic curve whose minimal discriminant and conductor are so far apart that they would contradict Szpiro's conjecture. We might thus hope to uncover a contradiction within the field of diophantine geometry. We will follow a different but related path and examine certain galois representations attached to E_{a^p,b^p,c^p}.

The idea of using elliptic curves to study Fermat's Last Theorem and vice versa goes back at least to the work of Y. Hellegouarch [9] (1972) who studied connections between the Fermat equation and torsion points on elliptic curves. Gerhart Frey seems to have been the first to suspect that a counterexample to Fermat's Last Theorem might contradict the Modularity Conjecture and to investigate various approaches based on this idea.

§2. Galois Representations

In this section we collect the basic definitions and conventions from the theory of galois representations that we will need later. For more details we refer the reader to the article by Mazur in this volume (chapter VIII).

Let $\overline{\mathbf{Q}}$ be the algebraic closure of $\mathbf{Q}$ in $\mathbf{C}$. We endow the galois group $G_{\mathbf{Q}} := \mathrm{Gal}(\overline{\mathbf{Q}}/\mathbf{Q})$ with the Krull topology in which a basis of neighborhoods of the origin is given by the collection of subgroups $H \subseteq G_{\mathbf{Q}}$ of finite index in $G_{\mathbf{Q}}$. With this topology, $G_{\mathbf{Q}}$ is a profinite group and in particular is a compact topological group.

By a two dimensional galois representation over a topological ring A we mean a continuous group homomorphism

$$\rho : G_{\mathbf{Q}} \longrightarrow \mathrm{GL}_2(A).$$

In this paper, the topological ring A will always be what Mazur calls a *coefficient ring* (in chapter VIII). Since these rings will play an important role in what follows, we make a formal definition.

(2.1) Definition. A *coefficient ring* is a complete noetherian local ring with finite residue field of characteristic p (our fixed prime).

Whenever we write that $\rho : G_{\mathbf{Q}} \longrightarrow \mathrm{GL}_2(A)$ is a galois representation, it is understood that A is a coefficient ring and that ρ is continuous.

(2.2) Residual representations and deformations. Let A be a coefficient ring with maximal ideal m_A and let $k_A := A/m_A$ be the residual field. We define the *residual representation* of a galois representation $\rho : G_{\mathbf{Q}} \longrightarrow \mathrm{GL}_2(A)$ to be the representation

$$\overline{\rho} : G_{\mathbf{Q}} \longrightarrow \mathrm{GL}_2(k_A)$$

obtained by composing ρ with the reduction map $\mathrm{GL}_2(A) \longrightarrow \mathrm{GL}_2(k_A)$. Conversely, if $\rho_0 : G_{\mathbf{Q}} \longrightarrow \mathrm{GL}_2(k)$ is a two dimensional galois representation over a finite field k, then we say that ρ is a *lifting* of ρ_0 to A if $k = k_A$ and $\overline{\rho} = \rho_0$. Two liftings ρ, ρ' of ρ_0 to A are said to be *equivalent* if ρ' can be conjugated to ρ by a matrix in $\mathrm{GL}_2(A)$ that is congruent to the identity matrix modulo m_A.

A *deformation* of ρ_0 to A is an equivalence class of liftings of ρ_0 to A. For a given lifting ρ of ρ_0, we will abuse notation and also write ρ to denote the deformation to which it belongs. This should not cause confusion in our discussion.

(2.3) The determinant of a galois representation. If ρ is a two dimensional galois representation over A then

$$\det(\rho) : G_{\mathbf{Q}} \longrightarrow A^{\times}$$

will denote the composition of ρ with the determinant homomorphism

$$\det : \mathrm{GL}_2(A) \longrightarrow A^{\times}.$$

In the applications it is sometimes convenient to restrict our attention to representations with prescribed determinant.

For example, let $\chi_p : G_{\mathbf{Q}} \longrightarrow \mathbf{Z}_p^\times$ denote the *cyclotomic character*, which is characterized by the property $\sigma(\zeta) = \zeta^{\chi_p(\sigma)}$ for any p-power root of unity ζ and any $\sigma \in G_{\mathbf{Q}}$. Any coefficient ring A admits a unique continuous ring homomorphism $\mathbf{Z}_p \longrightarrow A$ and we therefore have a canonical group homomorphism $\mathbf{Z}_p^\times \longrightarrow A^\times$. We say that ρ has determinant χ_p if $\det(\rho)$ is the composition of χ_p with the canonical homomorphism $\mathbf{Z}_p^\times \longrightarrow A^\times$.

(2.4) Local galois groups. For each prime ℓ, we let $\mathbf{Q}_\ell$ denote the field of ℓ-adic rationals, i.e., the completion of $\mathbf{Q}$ with respect to the ℓ-adic absolute value $|\cdot|_\ell$. We fix once and for all an algebraic closure $\overline{\mathbf{Q}}_\ell$ of $\mathbf{Q}_\ell$ as well as an embedding of $\overline{\mathbf{Q}}$ into $\overline{\mathbf{Q}}_\ell$. For $\ell = \infty$ we let $\mathbf{Q}_\infty := \mathbf{R}$, the completion of $\mathbf{Q}$ with respect to the usual absolute value $|\cdot|_\infty$, and we take $\overline{\mathbf{Q}}_\infty := \mathbf{C}$. For each ℓ (ℓ prime, or $\ell = \infty$), the *local galois group* at ℓ is the group

$$G_{\mathbf{Q}_\ell} := \mathrm{Gal}(\overline{\mathbf{Q}}_\ell/\mathbf{Q}_\ell).$$

For $\ell = \infty$, we have

$$G_{\mathbf{Q}_\infty} := \mathrm{Gal}(\mathbf{C}/\mathbf{R}) = \langle c \rangle,$$

the cyclic group of order 2 generated by complex conjugation c. It is well-known that for each ℓ there is a unique absolute value $|\cdot|_\ell$ on $\overline{\mathbf{Q}}_\ell$ extending the given absolute value on $\mathbf{Q}_\ell$. From this it follows easily that the elements of $G_{\mathbf{Q}_\ell}$ are *continuous* automorphisms of $\overline{\mathbf{Q}}_\ell$.

Using our fixed embeddings $\overline{\mathbf{Q}} \subseteq \overline{\mathbf{Q}}_\ell$, we may restrict any automorphism of $\overline{\mathbf{Q}}_\ell$ to obtain an automorphism of $\overline{\mathbf{Q}}$. Since $\overline{\mathbf{Q}}$ is dense in $\overline{\mathbf{Q}}_\ell$, the induced homomorphisms $G_{\mathbf{Q}_\ell} \to G_{\mathbf{Q}}$ are injective and we will regard them as inclusions:

$$G_{\mathbf{Q}_\ell} \subseteq G_{\mathbf{Q}}.$$

These subgroups are often called the decomposition subgroups of $G_{\mathbf{Q}}$. Of course, strictly speaking, they are not well-defined since their definition depends on our choice of the fixed embeddings of $\overline{\mathbf{Q}}$ into $\overline{\mathbf{Q}}_\ell$. However, changing any one of these embeddings has the effect of conjugating the corresponding decomposition subgroup by an element of $G_{\mathbf{Q}}$. This ambiguity will not be important to us.

(2.5) Inertia groups. For $\ell \neq \infty$, $G_{\mathbf{Q}_\ell}$ preserves the ring $\overline{\mathbf{Z}}_\ell$ of integers in $\overline{\mathbf{Q}}_\ell$ and also preserves the maximal ideal $\lambda \subseteq \overline{\mathbf{Z}}_\ell$. Thus, $G_{\mathbf{Q}_\ell}$ acts naturally on the residual field $\overline{\mathbf{F}}_\ell = \overline{\mathbf{Z}}_\ell/\lambda$ and we obtain a natural map $G_{\mathbf{Q}_\ell} \longrightarrow \mathrm{Gal}(\overline{\mathbf{F}}_\ell/\mathbf{F}_\ell)$, which is easily seen to be surjective. Its kernel I_ℓ is called the inertia group at ℓ. Thus for each $\ell \neq \infty$, we have an exact sequence

$$1 \longrightarrow I_\ell \longrightarrow G_{\mathbf{Q}_\ell} \longrightarrow \mathrm{Gal}(\overline{\mathbf{F}}_\ell/\mathbf{F}_\ell) \longrightarrow 1.$$

(2.6) Local properties of galois representations. Given a *global* galois representation $\rho : G_{\mathbf{Q}} \longrightarrow \mathrm{GL}_2(A)$, we may restrict ρ to the decomposition groups $G_{\mathbf{Q}_\ell}$ and obtain the family $\{\rho|G_{\mathbf{Q}_\ell}\}$ of *local* galois representations

$$\rho|G_{\mathbf{Q}_\ell} : G_{\mathbf{Q}_\ell} \longrightarrow \mathrm{GL}_2(A).$$

In many important examples from number theory one knows that the global representation ρ is determined up to isomorphism by the family of local representations $\{\rho|G_{\mathbf{Q}_\ell}\}_{\ell \notin S}$, where ℓ ranges over the complement of any finite set S of primes. By the local properties at ℓ of a galois representation ρ we mean the properties of the local representation $\rho|G_{\mathbf{Q}_\ell}$. The next three definitions describe three local properties that play a special role in what follows.

(2.7) Definition. We say that ρ is *odd* if $\det \rho(c) = -1$, where c is the complex conjugation generating $G_{\mathbf{Q}_\infty}$.

(2.8) Definition. We say that ρ is *unramified* at a prime ℓ if $I_\ell \subseteq \ker \rho|G_{\mathbf{Q}_\ell}$.

Since the galois group $\mathrm{Gal}(\overline{\mathbf{F}}_\ell/\mathbf{F}_\ell)$ is a topologically cyclic group generated by the ℓth power Frobenius automorphism Frob_ℓ, when ρ is unramified at ℓ, $\rho|G_{\mathbf{Q}_\ell}$ may be viewed as a homomorphism $\mathrm{Gal}(\overline{\mathbf{F}}_\ell/\mathbf{F}_\ell) \longrightarrow \mathrm{GL}_2(A)$ and is thus determined by its value on any representative of Frob_ℓ in $G_{\mathbf{Q}_\ell}$.

When $\ell = p$ we need the following weaker condition.

(2.9) Definition. We say that ρ is *flat* at p if, for every ideal $I \subseteq A$ for which A/I is finite, the representation $G_{\mathbf{Q}_p} \longrightarrow \mathrm{GL}_2(A/I)$, obtained by reducing $\rho|G_{\mathbf{Q}_p}$ mod I, extends to a finite flat group scheme over $\mathbf{Z}_p$ (see Tate's article in this volume (chapter V)).

(2.10) Examples from number theory. The galois representations that arise naturally in number theory have the especially nice property of being unramified almost everywhere, that is, they are unramified at all but finitely many primes ℓ. For example, let $E_{/\mathbf{Q}}$ be an elliptic curve. Then for each $n \geq 0$ the galois group $G_\mathbf{Q}$ acts on the group $E[p^n] \cong (\mathbf{Z}/p^n\mathbf{Z})^2$ of p^n-torsion points on E. Since the action of $G_\mathbf{Q}$ commutes with multiplication by p on E, $G_\mathbf{Q}$ acts naturally on the *Tate module*

$$\mathrm{Ta}_p(E) := \varprojlim E[p^n] \cong \mathbf{Z}_p^2$$

and we obtain the p-adic galois representation

$$\rho_{E,p} : G_\mathbf{Q} \longrightarrow \mathrm{GL}_2(\mathbf{Z}_p)$$

associated to E. The residual representation $\overline{\rho}_{E,p} : G_\mathbf{Q} \longrightarrow \mathrm{GL}_2(\mathbf{F}_p)$. describes the action of $G_\mathbf{Q}$ on $E[p] \cong \mathbf{F}_p^2$. We have the following basic result concerning the properties of these representations.

(2.11) Theorem. *Let $\rho_{E,p}$ be the p-adic galois representation associated to an elliptic curve $E_{/\mathbf{Q}}$ and let N_E be the conductor of E. Then*

- *the determinant of $\rho_{E,p}$ is χ_p, and*
- *$\rho_{E,p}$ is unramified outside of pN_E.*

In particular, $\rho_{E,p}$ is odd. If E is semistable with minimal discriminant Δ_E, then the residual representation $\overline{\rho}_{E,p}$ has the following local properties.

- *If $\ell \neq p$, then $\overline{\rho}_{E,p}$ is unramified at $\ell \iff p | ord_\ell(\Delta_E)$.*
- *$\overline{\rho}_{E,p}$ is flat at $p \iff p | ord_p(\Delta_E)$.*

§3. A Remarkable Galois Representation

Let $E := E_{a^p,b^p,c^p}$ be as in §1 and consider the galois representation

$$\overline{\rho}_{a^p,b^p,c^p} : G_{\mathbf{Q}} \longrightarrow \mathrm{GL}_2(\mathbf{F}_p)$$

given by $\overline{\rho}_{a^p,b^p,c^p} = \overline{\rho}_{E,p}$. Gerhart Frey [7,8] and Jean-Pierre Serre [16] noted that this representation has some remarkable local properties. More precisely they proved the following theorem.

(3.1) Theorem *Let $p \geq 5$ be prime and $a, b, c \in \mathbf{Z}$ satisfy $a^p + b^p + c^p = 0$ and $abc \neq 0$. Assume further that $a \equiv -1$ modulo 4 and $2|b$. Then*

(a) *$\overline{\rho}_{a^p,b^p,c^p}$ is absolutely irreducible;*

(b) *$\overline{\rho}_{a^p,b^p,c^p}$ is odd;*

(c) *$\overline{\rho}_{a^p,b^p,c^p}$ is unramified outside $2p$, flat at p, and semistable at 2.*

(See §7.1 for the definition of semistability of galois representations.) One suspects that there are no galois representations $\rho_0 : G_{\mathbf{Q}} \longrightarrow \mathrm{GL}_2(\mathbf{F}_p)$ satisfying properties (a), (b) and (c), but this suspicion remains unproven. On the other hand, by a theorem of Ribet, we *do* know that no such galois representation lives in the world of modular forms, in a sense that we will make precise in the next section.

§4. Modular Galois Representations

The theory of modular forms offers a rich source of galois representations. Using the Hecke operators, these "modular" galois representations can be constructed out of the torsion groups on the modular jacobians $J_1(N)$, $N > 0$ by the method of Eichler and Shimura. For an introduction to the theory of modular forms and the Eichler-Shimura theory, see David Rohrlich's article in this volume (chapter III).

(4.1) Galois representations associated to newforms. Fix, once and for all, a prime $\mathfrak{p}$ of $\overline{\mathbf{Q}}$ lying over p. Let $f = \sum_{n\geq 1} a_n q^n$ be a weight two (normalized) newform of conductor N and character ϵ (in (3.5) of chapter III, newforms are called primitive forms). We let K_f be the completion at $\mathfrak{p}$ of the number field generated by the values of ϵ and the fourier coefficients a_n ($n \geq 1$), and we let $\mathcal{O}_f \subseteq K_f$ be the ring of integers in K_f. The theory of Eichler and Shimura associates to f an odd two dimensional galois representation

$$\rho_f : G_{\mathbf{Q}} \longrightarrow \mathrm{GL}_2(\mathcal{O}_f)$$

such that for all sufficiently large primes ℓ, ρ_f is unramified at ℓ and

$$\mathrm{Trace}\big(\rho_f(\mathrm{Frob}_\ell)\big) = a_\ell \quad \text{and} \quad \det\big(\rho_f(\mathrm{Frob}_\ell)\big) = \epsilon(\ell)\ell.$$

For the details of the Eichler-Shimura construction, we refer to section 3.7 of Rohrlich's chapter III in this volume, where ρ_f appears as ρ_λ. By the work of Carayol and others, we now have a good understanding of the local structure of ρ_f at all primes. In particular we know that ρ_f is unramified outside pN and that the above conditions on the trace and determinant of $\rho_f(\mathrm{Frob}_\ell)$ are satisfied for these primes.

By the work of Deligne [3] and Deligne-Serre [4], we know that similar assertions hold for newforms of any weight $w \geq 1$. Indeed, if f is a weight w newform of conductor N then Deligne has constructed an odd two dimensional p-adic galois representation ρ_f, which is unramified outside pN and satisfies $\mathrm{Trace}(\rho_f(\mathrm{Frob}_\ell)) = a_\ell$ and $\det(\rho_f(\mathrm{Frob}_\ell)) = \epsilon(\ell)\ell^{w-1}$ for all $\ell \nmid pN$. In this paper, we will be concerned almost exclusively with the case $w = 2$.

(4.2) Hecke algebras. Let $N > 0$ be an integer and let $S_2(N)$ denote the space of weight 2 cusp forms for $\Gamma_1(N)$ (see (3.2) of chapter III). We let

$$\mathbf{T}'(N) := \mathbf{Z}[T_\ell, \langle d\rangle] \subseteq \mathrm{End}\big(S_2(N)\big)$$

be the $\mathbf{Z}$-subalgebra of $\mathrm{End}(S_2(N))$ generated by the Hecke operators T_ℓ and the diamond operators $\langle d\rangle$ where ℓ runs over all primes not dividing pN, and d runs over $(\mathbf{Z}/N\mathbf{Z})^\times$ (see (3.3) of chapter III).

(4.3) Modularity of galois representations. Motivated by (4.1) we say that a galois representation

$$\rho : G_{\mathbf{Q}} \longrightarrow \mathrm{GL}_2(A)$$

over a coefficient ring A is *modular* if there exists an integer $N > 0$ and a homomorphism $\pi : \mathbf{T}'(N) \longrightarrow A$ such that ρ is unramified outside Np and for every prime $\ell \nmid pN$ we have

$$\mathrm{Trace}(\rho(\mathrm{Frob}_\ell)) = \pi(T_\ell) \quad \text{and} \quad \det(\rho(\mathrm{Frob}_\ell)) = \pi(\langle \ell\rangle)\ell.$$

Remark: In view of the above restriction on the determinant it might be more appropriate to call these modular representations of weight 2. However, since all of our representations will have weight 2, we will drop that modifier from our language.

(4.4) Serre's Conjectures. In the special case where $A = k$ is a finite field, Serre [16] has formulated some precise conjectures about modularity of galois representations over k. One consequence of Serre's conjectures is the following conjecture.

Conjecture. *Every odd absolutely irreducible galois representation*

$$\rho_0 : G_{\mathbf{Q}} \longrightarrow \mathrm{GL}_2(k)$$

is modular (in the sense of (4.3)).

In fact, Serre's conjectures are much more precise. They predict — in terms of the local structure of ρ — the optimal weight, conductor and character of a newform f for which $\overline{\rho}_f = \rho_0$. For precise statements of Serre's conjectures and an account of what is known about them today, see the article by Edixhoven in this volume (chapter VII). An important special case of these conjectures, which Serre called the *epsilon conjecture* in [16], is the following theorem of Ribet [13] (see §3 of chapter VII for a sketch of the proof).

(4.5) Ribet's Theorem. *Let f be a weight two newform of conductor $N\ell$ where $\ell \nmid N$ is a prime. Suppose $\overline{\rho}_f$ is absolutely irreducible and that one of the following is true:*

- *$\overline{\rho}_f$ is unramified at ℓ; or*
- *$\ell = p$ and $\overline{\rho}_f$ is flat at p.*

Then there is a weight two newform g of conductor N such that $\overline{\rho}_f \cong \overline{\rho}_g$.

§5. The Modularity Conjecture and Wiles's Theorem

We say that an elliptic curve $E_{/\mathbf{Q}}$ is modular if there is a weight two newform f of conductor N_E and trivial character for which

$$L(f, s) = L(E, s).$$

There are a number of equivalent ways of defining modularity of elliptic curves. Here are a few.

(5.1) Theorem. *The following assertions are equivalent for an elliptic curve $E_{/\mathbf{Q}}$.*

(a) *E is modular;*
(b) *for some prime p, $\rho_{E,p}$ is modular;*
(c) *for every prime p, $\rho_{E,p}$ is modular;*
(d) *there is a non-constant morphism $\pi : X_0(N_E) \longrightarrow E$ of algebraic curves defined over $\mathbf{Q}$;*
(e) *E is isogenous to the modular abelian variety A_f associated to some weight two newform f of conductor N_E.*

We have the following profound conjecture developed between 1957 and 1967 by Shimura, Taniyama, and Weil.

(5.2) The Modularity Conjecture. *Every elliptic curve over $\mathbf{Q}$ is modular.*

The Modularity Conjecture is still open in general, but thanks to the work of Wiles [20] and Taylor–Wiles [18], we know at least that it is true for a large and important class of elliptic curves, namely the semistable ones.

(5.3) Wiles's Theorem. *Every semistable elliptic curve over* $\mathbf{Q}$ *is modular.*

We will sketch the proof in §7. In fact, by improving Wiles's methods, Fred Diamond [5] has proven the much stronger result that every elliptic curve $E_{/\mathbf{Q}}$ that is semistable at 3 and 5 is modular. The proof is outlined in chapter XVII by Diamond.

§6. The proof of Fermat's Last Theorem

Returning to the situation of §1 and §3 we suppose $p \geq 5$ and assume $a, b, c \in \mathbf{Z}$ satisfy $a^p + b^p + c^p = 0$ but $abc \neq 0$. We derive a contradiction by the method described in [16] (see also [8]). Without loss of generality, we may assume $a \equiv -1 \pmod 4$ and $2|b$. Let E_{a^p,b^p,c^p} be the elliptic curve $y^2 = x(x - a^p)(x + b^p)$ and let ρ_{a^p,b^p,c^p} be the associated p-adic galois representation.

By proposition 1.1, E_{a^p,b^p,c^p} is semistable and has conductor

$$N_{a^p,b^p,c^p} = \prod_{\ell|abc} \ell.$$

Hence, by Wiles's theorem, E_{a^p,b^p,c^p} is modular and there is a weight two newform f_{a^p,b^p,c^p} of conductor N_{a^p,b^p,c^p} associated to E_{a^p,b^p,c^p}. In particular, we have $\rho_{a^p,b^p,c^p} \cong \rho_{f_{a^p,b^p,c^p}}$. But according to theorem 2.11 $\overline{\rho}_{a^p,b^p,c^p}$ is absolutely irreducible and is unramified outside $2p$ and flat at p. Applying Ribet's Theorem we conclude that there is a weight two newform g of conductor 2 such that $\overline{\rho}_g \cong \overline{\rho}_{a^p,b^p,c^p}$. But the dimension of $S_2(\Gamma_0(2))$ is equal to the genus of $X_0(2)$, which is easily seen to be zero. Thus there are no weight two newforms of conductor 2. This is a contradiction and Fermat's Last Theorem is proved.

§7. The proof of Wiles's Theorem

In this final section, we describe the structure of the proof of Wiles's Theorem [18,20]. For other surveys of the proof, we recommend [2,12,14,17]. Here we assume that the distinguished prime p is ≥ 3. Let k be a finite field of characteristic p and let

$$\rho_0 : G_{\mathbf{Q}} \longrightarrow \mathrm{GL}_2(k)$$

be a galois representation. As we move through this section we will impose a number of cumulative hypotheses on ρ_0. The first of these is the following.

Hypothesis A. ρ_0 has determinant χ_p.

(7.1) Semistable galois representations. We say that a galois representation

$$\rho : G_{\mathbf{Q}} \longrightarrow \mathrm{GL}_2(A)$$

is *ordinary* at p if the restriction of ρ to the inertia group I_p at p has the form $\rho|_{I_p} = \begin{pmatrix} \chi_p & * \\ 0 & 1 \end{pmatrix}$ for a suitable choice of basis. We say that ρ is *semistable* at a prime ℓ if one of the following two conditions is satisfied.

- $\ell = p$ and ρ is either flat at p or ordinary at p (or both).
- $\ell \neq p$ and $\rho|_{I_\ell} = \begin{pmatrix} 1 & * \\ 0 & 1 \end{pmatrix}$ for a suitable choice of basis.

We say that a two dimensional galois representation ρ is semistable if it is semistable at every prime. From now on, we impose the following additional hypothesis on ρ_0.

Hypothesis B. ρ_0 is semistable.

The use of the word *semistable* in this context is motivated by the simple fact that if $E_{/\mathbf{Q}}$ is a semistable elliptic curve, then the p-adic galois representation $\rho_{E,p} : G_{\mathbf{Q}} \longrightarrow \mathrm{GL}_2(\mathbf{Z}_p)$ is semistable in the above sense.

(7.2) Deformation types. A deformation type $\mathcal{D}$ is a list of conditions to be imposed on deformations of a residual representation

$$\rho_0 : G_{\mathbf{Q}} \longrightarrow \mathrm{GL}_2(k).$$

Using more sophisticated terminology, a deformation type may be regarded as a functor from the category of coefficient rings with residue field k to the category of sets, where, for a given coefficient ring A, $\mathcal{D}(A)$ is the set of deformations of ρ_0 to A that satisfy the conditions of $\mathcal{D}$. For more discussion of deformation types we refer the reader to Mazur's chapter VIII in this volume.

Wiles considers a variety of different deformation types, but for the application to the semistable modularity conjecture it suffices to restrict to the following special cases. Let $S := \{\, \ell \neq p \mid \rho_0 \text{ is ramified at } \ell \,\}$. A deformation type $\mathcal{D}$ is associated to a finite set of primes $\Sigma_{\mathcal{D}}$ disjoint from S. We say that a deformation ρ of ρ_0 is of type $\mathcal{D}$ if the following conditions are satisfied.

- ρ has determinant χ_p,
- ρ is unramified outside $S \cup \{\, p \,\} \cup \Sigma_{\mathcal{D}}$,
- ρ is semistable outside $\Sigma_{\mathcal{D}}$, and
- if $p \notin \Sigma_{\mathcal{D}}$ and if ρ_0 is flat at p, then ρ is also flat at p.

Roughly speaking, the last three conditions say that ρ has the same local properties as ρ_0 at primes not in $\Sigma_{\mathcal{D}}$. We remark that in any case, if ρ_0 is ordinary at p then ρ is also ordinary at p.

(7.3) Universal deformation rings and Hecke rings. In addition to hypotheses A and B above we suppose ρ_0 satisfies the following hypothesis.

Hypothesis C. ρ_0 is absolutely irreducible.

Using Mazur's theory of deformations of galois representations [10], Wiles associates to each deformation type $\mathcal{D}$ a *universal deformation ring* $R_{\mathcal{D}}$ (which is, in particular, a coefficient ring) and a *universal deformation*

$$\rho_{\mathcal{D}} : G_{\mathbf{Q}} \longrightarrow \mathrm{GL}_2(R_{\mathcal{D}})$$

of ρ_0 of type $\mathcal{D}$. The representation $\rho_{\mathcal{D}}$ satisfies the following universal property: for every deformation $\rho : G_{\mathbf{Q}} \longrightarrow \mathrm{GL}_2(A)$ of ρ_0 of type $\mathcal{D}$ there is a unique homomorphism $\pi_A : R_{\mathcal{D}} \longrightarrow A$ such that the diagram

$$\begin{array}{ccc} G_{\mathbf{Q}} & \xrightarrow{\rho_{\mathcal{D}}} & \mathrm{GL}_2(R_{\mathcal{D}}) \\ & \rho \searrow \quad \swarrow \pi_A & \\ & \mathrm{GL}_2(A) & \end{array}$$

is commutative. For details on the properties and construction of $R_{\mathcal{D}}$ see chapter VIII by Mazur and chapter XIII by Brian Conrad. An explicit approach to constructing deformation rings is given in chapter IX by de Smit and Lenstra.

Hypothesis D. ρ_0 is modular, and $\rho_0|_{G_{\mathbf{Q}(\sqrt{-3})}}$ is absolutely irreducible.

Under this hypothesis, Wiles defines another coefficient ring $\mathbf{T}_{\mathcal{D}}$, the *universal modular deformation ring* and a *universal modular deformation*

$$\rho_{\mathcal{D},\mathrm{mod}} : G_{\mathbf{Q}} \longrightarrow \mathrm{GL}_2(\mathbf{T}_{\mathcal{D}})$$

of ρ_0 of type $\mathcal{D}$. The representation $\rho_{\mathcal{D},\mathrm{mod}}$ satisfies the analogous universal property for modular deformations of type $\mathcal{D}$. Namely, for every *modular* deformation $\rho : G_{\mathbf{Q}} \longrightarrow \mathrm{GL}_2(A)$ of ρ_0 of type $\mathcal{D}$ there is a unique homomorphism $\pi_A : \mathbf{T}_{\mathcal{D}} \longrightarrow A$ such that the obvious diagram commutes.

The constructions of $\mathbf{T}_{\mathcal{D}}$ and $\rho_{\mathcal{D},\mathrm{mod}}$ are quite difficult. The algebra $\mathbf{T}_{\mathcal{D}}$ is defined in chapter XII by Diamond and Ribet. It's existence depends on the highly non-trivial fact (described in chapter VII by Edixhoven) that there exists a weight two newform f such that ρ_f is a deformation of ρ_0 of type $\mathcal{D}$. The representation $\rho_{\mathcal{D},\mathrm{mod}}$ is cut out of the Tate module of a modular Jacobian using the Hecke operators. Wiles's proof that this representation is a free rank two $\mathbf{T}_{\mathcal{D}}$-module depends on the Gorenstein property of $\mathbf{T}_{\mathcal{D}}$ (see Tilouine's chapter X in this volume). Later, other proofs of this fact were given that do not make explicit use of the Gorenstein property, but rather have the Gorenstein property as a by-product (for example, see [6]).

(7.4) The main theorem. By the universal property of $\rho_{\mathcal{D}}$ there is a unique homomorphism $\varphi_{\mathcal{D}} : R_{\mathcal{D}} \longrightarrow \mathbf{T}_{\mathcal{D}}$ such that $\rho_{\mathcal{D},\mathrm{mod}} = \varphi_{\mathcal{D}} \circ \rho_{\mathcal{D}}$. The following theorem is a special case of the main theorem of Wiles [20].

Theorem. *Suppose ρ_0 satisfies hypotheses* A-D. *Then the canonical map $\varphi_{\mathcal{D}} : R_{\mathcal{D}} \longrightarrow \mathbf{T}_{\mathcal{D}}$ is an isomorphism of complete intersection rings.*

For the definition of complete intersection rings, we refer to chapter XI by de Smit, Rubin, Schoof and in this volume. For our purposes what matters is the conclusion that $\varphi_{\mathcal{D}}$ is an isomorphism. The proof of the theorem is based on the numerical criterion of Wiles described in the next section, which reduces the proof to an inequality between two numbers. The theorem has the following important corollary as an immediate consequence.

Corollary. *Suppose ρ_0 satisfies hypotheses* A-D. *Then every deformation of ρ_0 of type $\mathcal{D}$ is modular.*

(7.5) Wiles's numerical criterion. *Let R and T be coefficient rings and suppose we have a commutative diagram*

$$\begin{array}{ccc} R & \xrightarrow{\varphi} & T \\ & \pi_R \searrow \quad \swarrow \pi_T & \\ & \mathcal{O} & \end{array}$$

in which $\mathcal{O}$ is a complete discrete valuation ring and all the arrows are surjective. Let $I_R := \ker \pi_R$, $I_T := \ker \pi_T$, and let $\eta_T := \pi_T(\mathrm{Ann}_T(I_T))$. Then the following three assertions are equivalent.

- *φ is an isomorphism of complete intersection rings;*
- *I_R/I_R^2 is finite and $\#(I_R/I_R^2) \le \#(\mathcal{O}/\eta_T)$;*
- *I_R/I_R^2 is finite and $\#(I_R/I_R^2) = \#(\mathcal{O}/\eta_T)$.*

This is a special case of Criterion I given in chapter IX by Schoof, Rubin, and de Smit.

(7.6) Selmer groups and congruence modules. Now let f be a weight two newform and suppose $\rho_f : G_{\mathbf{Q}} \longrightarrow \mathrm{GL}_2(\mathcal{O}_f)$ is a deformation of ρ_0 of type $\mathcal{D}$. By the universality of $\mathbf{T}_{\mathcal{D}}$ there is a unique homomorphism $\pi_{\mathbf{T}_{\mathcal{D}}} : \mathbf{T}_{\mathcal{D}} \longrightarrow \mathcal{O}_f$ such that $\rho_f = \pi_{\mathbf{T}_{\mathcal{D}}} \circ \rho_{\mathcal{D},\mathrm{mod}}$. Let $\pi_{R_{\mathcal{D}}} := \pi_{\mathbf{T}_{\mathcal{D}}} \circ \varphi_{\mathcal{D}}$ so that we have the following commutative diagram:

$$\begin{array}{ccc} R_{\mathcal{D}} & \xrightarrow{\varphi_{\mathcal{D}}} & \mathbf{T}_{\mathcal{D}} \\ & \pi_{R_{\mathcal{D}}} \searrow \quad \swarrow \pi_{\mathbf{T}_{\mathcal{D}}} & \\ & \mathcal{O}_f. & \end{array}$$

To prove that $\varphi_{\mathcal{D}}$ is an isomorphism, Wiles establishes the middle inequality in the above criterion. For this, he first interprets the two sides of the inequality in terms of other objects that have been studied in some detail in the literature. More precisely, Wiles interprets the "tangent space" $\mathrm{Hom}_{\mathcal{O}}(I_{R_{\mathcal{D}}}/I_{R_{\mathcal{D}}}^2, K/\mathcal{O})$ as a *Selmer group* $H^1_{\mathcal{D}}(G_{\mathbf{Q}}, \mathrm{ad}^0(\rho_f) \otimes K/\mathcal{O})$, i.e., as a certain subgroup of the galois cohomology group $H^1(G_{\mathbf{Q}}, \mathrm{ad}^0(\rho_f) \otimes K/\mathcal{O})$ determined by local conditions associated to $\mathcal{D}$, and he interprets $\mathcal{O}/\eta_{\mathbf{T}_\mathcal{D}}$ as a *congruence module* classifying congruences between f and other newforms of type $\mathcal{D}$. For precise definitions, see sections 4.2 and 4.3 of chapter XII by Diamond and Ribet, chapter VIII by Mazur, and chapter IV by Washington. The isomorphism between tangent spaces and Selmer groups is described in chapter VIII.

The proof of the crucial numerical inequality divides into two parts. The case where $\Sigma_\mathcal{D} = \emptyset$, which is called the *minimal case*, is proved by Wiles with Taylor in [18]. Their original proof has been simplified by making use of another criterion due to Faltings, a generalization of which is given as criterion II in chapter XI. This is the method followed by de Shalit in chapter XIV. The *non-minimal case* is proved by induction on the number of primes in $\Sigma_\mathcal{D}$. The proof is accomplished by analyzing how the Selmer groups and congruence modules grow as $\Sigma_\mathcal{D}$ is enlarged to conclude that if the numerical inequality is satisfied for one $\mathcal{D}$ then it is also satisfied when more primes are included in $\Sigma_\mathcal{D}$. See chapter XII by Diamond and Ribet for more details.

(7.7) The Proof of Wiles's Theorem. We prepare for the proof by noting that hypotheses A and B are satisfied by $\overline{\rho}_{E,p}$ for every prime p. Indeed hypothesis A is contained in theorem 2.11 and hypothesis B is a consequence of the semistability of E.

Moreover, by a theorem of Serre ([15], prop. 21, and [17], §3.1), the semistability of E guarantees that $\overline{\rho}_{E,p}$ is either surjective or reducible for every prime $p \geq 3$. Hence for $p \geq 3$, absolute irreducibility of $\overline{\rho}_{E,p}$ is equivalent to irreducibility of $\overline{\rho}_{E,p}$, and if $p = 3$ this is equivalent to absolute irreducibility of $\overline{\rho}_{E,3}|G_{\mathbf{Q}(\sqrt{-3})}$. Thus the following lemma is a consequence of corollary 7.4.

(7.8) Lemma. *Let $E_{/\mathbf{Q}}$ be a semistable elliptic curve and suppose $\overline{\rho}_{E,p}$ is both modular and irreducible for some prime $p \geq 3$. Then E is modular.*

Wiles gave an ingenious argument to show that for E semistable, the hypotheses of the lemma are satisfied by either $p = 3$ or $p = 5$. The proof is based on the following three theorems.

(7.9) Theorem. *Let E be an arbitrary elliptic curve and suppose $\overline{\rho}_{E,3}$ is irreducible. Then $\overline{\rho}_{E,3}$ is modular.*

This follows from a deep theorem of Langlands and Tunnell and depends in a crucial way on the theory of Langlands for GL_2. For an exposition of the Langlands theory and the proof of Theorem 7.9, see chapter VI by Stephen Gelbart in this volume.

(7.10) Theorem. *Let $E_{/\mathbf{Q}}$ be a semistable elliptic curve and suppose $\overline{\rho}_{E,5}$ is irreducible. Then there is another semistable elliptic curve $E'_{/\mathbf{Q}}$ for which*

(a) *$\overline{\rho}_{E',3}$ is irreducible, and*
(b) *$\overline{\rho}_{E',5} \cong \overline{\rho}_{E,5}$.*

Indeed, proposition 11 and the argument in section 4 of Rubin's chapter XVI in this volume provide us with a family of elliptic curves $E'_{/\mathbf{Q}}$ satisfying conditions (a) and (b). All of these curves are semistable away from 5. By taking E' in this family sufficiently close 5-adically to E, we obtain the desired semistable curve.

(7.11) Theorem. *Let $E_{/\mathbf{Q}}$ be a semistable elliptic curve. Then at least one of the representations $\overline{\rho}_{E,3}$ or $\overline{\rho}_{E,5}$ is irreducible.*

Indeed, if both $\overline{\rho}_{E,3}$ and $\overline{\rho}_{E,5}$ were reducible, then $E[15]$ would contain a galois invariant subgroup of order 15. This contradicts Lemma 9 (iv) of chapter XVI by Karl Rubin (see also [11]).

(7.12) Conclusion of the proof. Let $E_{/\mathbf{Q}}$ be a semistable elliptic curve. If $\overline{\rho}_{E,3}$ is irreducible then, according to theorem 7.9, $\overline{\rho}_{E,3}$ is also modular, so E is modular by lemma 7.8. If $\overline{\rho}_{E,3}$ is not irreducible, then by theorem 7.11, $\overline{\rho}_{E,5}$ is irreducible. Then there is another semistable elliptic curve $E'_{/\mathbf{Q}}$ satisfying (a) and (b) of theorem 7.10. In particular, $\overline{\rho}_{E',3}$ is irreducible. Repeating the above argument we see that E' is modular. Hence $\overline{\rho}_{E',5}$ is modular and by (b) of 7.10, $\overline{\rho}_{E,5}$ is modular. Once again we use lemma 7.8 to conclude E is modular.

References

[1] Carayol, H.: Sur les représentations galoisiennes modulo ℓ attachées aux formes modulaires. *Duke Math. J.* **59** (1989), 785-801.

[2] Darmon, H., Diamond, F., Taylor, R. L.: Fermat's Last Theorem. In *Current Developments in Mathematics, 1995*, International Press. To appear.

[3] Deligne, P.: Formes modulaires et représentation ℓ-adiques. Sém. Bourbaki, 1968/69, Exposé 355. *Lect. Notes in Math.* **179** (1971), 139-172.

[4] Deligne, P., Serre, J.-P.: Formes modulaires de poids 1. *Ann. Sci. E.N.S.* **7** (1974), 507-530.

[5] Diamond, F.: On deformations rings and Hecke rings. *Ann. of math.*. To appear.

[6] Diamond, F.: The Taylor-Wiles construction and multiplicity one. *Invent. Math.*. To appear.

[7] Frey, G.: Links between solutions of $A - B = C$ and elliptic curves. In *Number Theory, proceedings of the Journees arithmetiques, held in Ulm, 1987*, H.P. Schlickewei, E. Wirsing, editors. Lecture notes in mathematics **1380**. Springer-Verlag, Berlin, New York, 1989.

[8] Frey, G.: Links between stable elliptic curves and certain Diophantine equations. *Ann. Univ. Saraviensis, Ser. Math.* **1** (1986), 1-40.

[9] Hellegouarch, Y.: Points d'ordre $2p^h$ sur les courbes elliptiques. *Acta. Arith.* **26** (1974/75), 253-263.

[10] Mazur, B.: Deforming Galois representations. In *Galois groups over* **Q**: proceedings of a workshop held March 23-27, 1987, Y. Ihara, K. Ribet, J.-P. Serre, editors. Mathematical Sciences Research Institute publications **16**. Springer-Verlag, New York,1989, pp. 385-437.

[11] Mazur, B.: Modular curves and the Eisenstein ideal. *Publ. Math. I.H.E.S.* **47** (1977), 33-186.

[12] Murty, V.K.: Modular elliptic curves. in *Seminar on Fermat's Last Theorem.* Canadian Math. Soc. Conf. Proc. **17**, 1995.

[13] Ribet, K.A.: On modular representations of $\mathrm{Gal}(\overline{\mathbf{Q}}/\mathbf{Q})$ arising from modular forms. *Invent. math.* **100** (1990), 431-476.

[14] Oesterlé, J.: Travaux de Wiles (et Taylor, ...), Partie II. *Asterisque* **237** (1996), 333-355.

[15] Serre, J.-P.: Propriétés galoisiennes des points d'ordre fini des courbes elliptiques. *Invent. Math.* **15** (1972), 259-331.

[16] Serre, J.-P.: Sur les représentations modulaires de degré 2 de $\mathrm{Gal}(\overline{\mathbf{Q}}/\mathbf{Q})$, *Duke Math. J.* **54** (1987), 179-230.

[17] Serre, J.-P.: Travaux de Wiles (et Taylor, ...), Partie I. *Asterisque* **237** (1996), 319-332.

[18] Taylor, R. L., Wiles, A.: Ring theoretic properties of certain Hecke algebras. *Annals of Math.* **141** (1995), 553-572.

[19] Vojta, P.: *Diophantine Approximations and Value Distribution Theory.* Lect. Notes in Math. **1239**, 1987

[20] Wiles, A.: Modular elliptic curves and Fermat's Last Theorem. *Annals of Math.* **141** (1995), 443-551.

A SURVEY OF THE ARITHMETIC THEORY OF ELLIPTIC CURVES

JOSEPH H. SILVERMAN

§1. BASIC DEFINITIONS

An *elliptic curve* is a pair (E, O), where E is a smooth projective curve of genus one and O is a point of E. The elliptic curve is said to be *defined over the field K* if the underlying curve is defined over K and the point O is defined over K.

Every elliptic curve can be embedding as a smooth cubic curve in $\mathbb{P}^2$ given by an equation of the form

$$E: y^2 + a_1xy + a_3y = x^3 + a_2x^2 + a_4x + a_6. \tag{1}$$

Such an equation is called a *Weierstrass equation for E*. The point O is the point $[0,1,0]$ at infinity. If E is defined over K, then the a_i's can be chosen in K. If in addition $\operatorname{char}(K) \neq 2,3$, then E has a Weierstrass equation of the form

$$E: y^2 = x^3 + Ax + B. \tag{2}$$

The non-singularity assumption on E implies that the *discriminant*

$$\Delta = -16(4A^3 + 27B^2) \neq 0.$$

We also define the *j-invariant of E* to be the quantity

$$j(E) = -1728\frac{64A^3}{\Delta} = 1728\frac{4A^3}{4A^3 + 27B^2}.$$

(When using the general Weierstrass equation (1), the formulas for Δ and j are more complicated, see [10] or [8].)

Theorem. *Let E and E' be elliptic curves defined over an algebraically closed field K. Then E is K-isomorphic to E' if and only if $j(E) = j(E')$.*

Two special types of elliptic curves are those with j-invariant 0 and 1728. These curves are given by equations of the form

$$E: y^2 = x^3 + Ax \qquad j = 1728,$$
$$E: y^2 = x^3 + B \qquad j = 0.$$

This survey summarizes, without proof, some of the basic theory of elliptic curves. Proofs for most of the theorems can be found in the references listed at the end, see especially [3], [8], and [9].

§2. THE GROUP LAW

The points on an elliptic curve form a group. The group law can be characterized in a number of equivalent ways. Let E be an elliptic curve and $P, Q \in E$. The sum $P+Q$ is the (unique) point R satisfying

$$(P)+(Q) \sim (R)+(O),$$

where $\sim$ denotes linear equivalence of divisors. Geometrically, three points sum to zero if and only if they are collinear. Using this geometric characterization, one can write down explicit formulas. For example, if $P=(x,y)$ and $P'=(x',y')$ are on the curve given by the equation (2), then

$$x(P+P') = \left(\frac{y'-y}{x'-x}\right)^2 - x - x' \quad \text{and} \quad x(2P) = \frac{x^4-2Ax^2-8Bx+A^2}{4x^3+4Ax+4B}.$$

Similarly, the additive inverse of $P=(x,y)$ is $-P=(x,-y)$.

Repeated addition gives *multiplication maps*

$$[m] : E \to E, \qquad [m]P = \begin{cases} P+P+\cdots+P & \text{if } m>0, \\ O & \text{if } m=0, \\ -(P+P+\cdots+P) & \text{if } m<0. \end{cases}$$

Further, for any point $Q \in E$, there is the *translation-by-Q* map

$$\tau_Q : E \to E, \qquad \tau_Q(P) = P+Q.$$

Riemann-Roch tells us that an elliptic curve has a unique holomorphic differential (up to scalar). On the Weierstrass equations (1) and (2) it is given by

$$\omega_E = \frac{dx}{2y+a_1x+a_3} \qquad \text{and} \qquad \omega_E = \frac{dx}{2y} \qquad \text{respectively.}$$

The uniqueness of ω_E implies that it is translation invariant,

$$\tau_Q^*(\omega_E) = \omega_E \qquad \text{for all } Q \in E.$$

§3. SINGULAR CUBICS

If the discriminant of a Weierstrass equation (1) or (2) vanishes, then the curve is singular, with exactly one singular point. There are two possible behaviors. Either the singular point has two distinct tangent directions (a *node*), or it has only a single tangent direction (a *cusp*). The non-singular locus is denoted

$$E^{\mathrm{ns}} = \{P \in E : P \text{ is a non-singular point of } E\}.$$

The group law described above makes the non-singular locus into a group:

$$E^{\mathrm{ns}} \cong \begin{cases} \text{the multiplicative group } \mathbb{G}_m \text{ if } E \text{ has a node,} \\ \text{the additive group } \mathbb{G}_a \text{ if } E \text{ has a cusp.} \end{cases}$$

§4. ISOGENIES

A non-constant morphism $\phi : E_1 \longrightarrow E_2$ between elliptic curves which satisfies $\phi(O) = O$ is called an *isogeny.*

Proposition. *An isogeny $\phi : E_1 \longrightarrow E_2$ is always a group homomorphism. That is, $\phi(P+Q) = \phi(P) + \phi(Q)$.*

It follows that the kernel of an isogeny $\phi : E_1 \longrightarrow E_2$ is a finite subgroup of E_1. The *degree of* ϕ is its degree as a finite map of curves. (The constant map sending E_1 to O is defined to have degree zero.)

Associated to an isogeny $\phi : E_1 \longrightarrow E_2$ of degree n is a *dual isogeny*

$$\hat{\phi} : E_2 \longrightarrow E_1$$

characterized by the property that

$$\hat{\phi} \circ \phi = [n]_{E_1} \qquad \text{and} \qquad \phi \circ \hat{\phi} = [n]_{E_2}.$$

The dual isogeny has the following additional properties:

$$\hat{\hat{\phi}} = \phi, \qquad \widehat{\phi + \psi} = \hat{\phi} + \hat{\psi}, \qquad \widehat{\phi \circ \lambda} = \hat{\lambda} \circ \hat{\phi}, \qquad \widehat{[m]} = [m].$$

§5. THE ENDOMORPHISM RING

The set of isogenies from E to itself, together with the zero map, form a ring which we denote by $\mathrm{End}(E)$ and call the *endomorphism ring of* E. We make $\mathrm{End}(E)$ into a ring via the rules

$$(\phi + \psi)(P) = \phi(P) + \psi(P) \qquad \text{and} \qquad (\phi\psi)(P) = \phi(\psi(P)).$$

The unit group of $\mathrm{End}(E)$ consists of the isomorphisms from E to itself. It is called the *automorphism group of* E and is denoted $\mathrm{Aut}(E)$.

Theorem. *Let E be an elliptic curve defined over a field K.*
(a) *The endomorphism ring of E is one of the following three sorts of rings:*

$$\mathrm{End}(E) = \begin{cases} \mathbb{Z}, \\ \textit{an order in a quadratic imaginary field,} \\ \textit{a maximal order in a quaternion algebra.} \end{cases}$$

The third possibility can only occur if $\mathrm{char}(K) > 0$.
(b) *Assume* $\mathrm{char}(K) \neq 2, 3$. *Then the automorphism group of E is given by*

$$\mathrm{Aut}(E) = \begin{cases} \boldsymbol{\mu}_2 & \textit{if } j(E) \neq 0, 1728, \\ \boldsymbol{\mu}_4 & \textit{if } j(E) = 1728, \\ \boldsymbol{\mu}_6 & \textit{if } j(E) = 0. \end{cases}$$

(Here $\boldsymbol{\mu}_n$ is the group of nth roots of unity.)

An elliptic curve whose endomorphism ring is strictly larger than $\mathbb{Z}$ is said to have *complex multiplication* (or CM for short). For example, the curves with $j = 0$ and $j = 1728$ have CM.

§6. Torsion Points

The kernel of the multiplication-by-m map consists of the points whose order divides m. This subgroup is denoted

$$E[m] = \ker[m] = \{P \in E : [m]P = O\}.$$

The *torsion subgroup of* E is the set of all points of finite order,

$$E_{\text{tors}} = \{P \in E : [m]P = O \text{ for some } m \geq 1\} = \bigcup_{m \geq 1} E[m].$$

Remark. When we write E, $E[m]$, E_{tors}, etc., we are always referring to geometric points, that is, to points defined over an algebraically closed field. If E is defined over K and we want to discuss only the points defined over K, we will write $E(K)$, $E(K)[m]$, and $E_{\text{tors}}(K)$.

Proposition. *Let E/K be an elliptic curve.*
(a) *If* $\text{char}(K) = 0$ *or if* $\text{char}(K) = p$ *with* $p \nmid m$, *then*

$$E[m] \cong \mathbb{Z}/m\mathbb{Z} \times \mathbb{Z}/m\mathbb{Z}.$$

(b) *If* $\text{char}(K) = p > 0$, *then*

$$E[p^r] \cong \mathbb{Z}/p^r\mathbb{Z} \quad \textit{or} \quad 0.$$

For a fixed prime ℓ, consider the inverse system of ℓ-power torsion points via the maps $[\ell] : E[\ell^{n+1}] \to E[\ell^n]$. The inverse limit is called the (ℓ*-adic*) *Tate module of* E and denoted

$$T_\ell(E) = \varprojlim E[\ell^n].$$

If $\text{char}(K) \neq \ell$, then $T_\ell(E)$ is a free $\mathbb{Z}_\ell$-module of rank 2,

$$T_\ell(E) \cong \mathbb{Z}_\ell \times \mathbb{Z}_\ell.$$

It is often more convenient to work with the $\mathbb{Q}_\ell$-vector space

$$V_\ell(E) = T_\ell(E) \otimes \mathbb{Q} \cong \mathbb{Q}_\ell \times \mathbb{Q}_\ell.$$

§7. Galois Representations Attached to E

If E is defined over K, then its torsion points are defined over the algebraic closure of K, and we can look at the associated Galois action. To simplify our exposition, we will always assume that

$$K \text{ is a perfect field.}$$

We also fix an algebraic closure $\bar{K}$ of K.

The action of Galois commutes with the group law on E, so if $\text{char}(K) = 0$ or if $\text{char}(K) = p$ with $p \nmid m$, then we obtain a two-dimensional representation

$$\bar{\rho}_m : G_{\bar{K}/K} \longrightarrow \text{Aut}\big(E[m]\big) \cong \text{GL}_2(\mathbb{Z}/m\mathbb{Z}).$$

These representations are extremely important in studying the arithmetic properties of E.

Proposition. *The determinant* $\det(\bar{\rho}_m)$ *of the representation* $\bar{\rho}_m$ *is equal to the cyclotomic character*

$$\chi_m : G_{\bar{K}/K} \longrightarrow \mathrm{Aut}(\boldsymbol{\mu}_m) \cong (\mathbb{Z}/m\mathbb{Z})^*.$$

The ℓ-power representations $\bar{\rho}_{\ell^n}$ fit together to give the *ℓ-adic representation of E*,

$$\rho_\ell : G_{\bar{K}/K} \longrightarrow \mathrm{Aut}(T_\ell(E)) \cong \mathrm{GL}_2(\mathbb{Z}_\ell).$$

The associated vector space representation is also denoted ρ_ℓ,

$$\rho_\ell : G_{\bar{K}/K} \longrightarrow \mathrm{Aut}(V_\ell(E)) \cong \mathrm{GL}_2(\mathbb{Q}_\ell).$$

Remark. The Tate module $V_\ell(E)$ is dual to the étale cohomology group $H^1_{\text{ét}}(E, \mathbb{Q}_\ell)$, so the associated representation can equally well be defined using cohomology.

§8. THE WEIL PAIRING

Let E/K be an elliptic curve, and fix an integer $m \geq 2$. If $\mathrm{char}(K) > 0$, we assume that it does not divide m. The *Weil pairing* is a pairing

$$e_m : E[m] \times E[m] \longrightarrow \boldsymbol{\mu}_m$$

defined as follows: Let $S, T \in E[m]$. Choose a function g on E whose divisor satisfies

$$\mathrm{div}(g) = [m]^*(T) - [m]^*(O).$$

Then

$$e_m(S,T) = \frac{g(X+S)}{g(X)}$$

for any point $X \in E$ such that g is defined at X and at $X+S$.

Proposition. *The Weil pairing is*

Bilinear: $e_m(S_1 + S_2, T) = e_m(S_1, T)e_m(S_2, T)$.
$e_m(S, T_1 + T_2) = e_m(S, T_1)e_m(S, T_2)$.

Alternating: $e_m(T,T) = 1$.

Non-degenerate: $e_m(S,T) = 1$ *for all* $S \iff T = O$.

Galois Equivariant: $e_m(S^\sigma, T^\sigma) = e_m(S,T)^\sigma$ *for all* $\sigma \in G_{\bar{K}/K}$.

Thus e_m induces an isomorphism

$$E[m] \wedge E[m] \xrightarrow{\sim} \boldsymbol{\mu}_m$$

of Galois modules. Let $\bar{\rho} : G_{\bar{K}/K} \to \mathrm{Aut}(E[m])$ be the Galois representation attached to E, and let $\chi : G_{\bar{K}/K} \to \mathrm{Aut}(\boldsymbol{\mu}_m)$ be the cyclotomic character. Then with this identification, we have for any $\sigma \in G_{\bar{K}/K}$,

$$\chi(\sigma)(S \wedge T) = (S \wedge T)^\sigma = S^\sigma \wedge T^\sigma = \bar{\rho}_m(\sigma)S \wedge \bar{\rho}_m(\sigma)T,$$

which verifies the formula $\det(\bar{\rho}_m) = \chi_m$ as stated in §7.

Let $\phi : E_1 \to E_2$ be an isogeny. Then the dual isogeny $\hat{\phi} : E_2 \to E_1$ is dual (i.e., adjoint) with respect to the Weil pairing:

$$e_m\bigl(S, \hat{\phi}(T)\bigr) = e_m\bigl(\phi(S), T\bigr) \qquad \text{for all } S \in E_1[m] \text{ and } T \in E_2[m].$$

The ℓ-power Weil pairings e_{ℓ^n} fit together to define a bilinear, alternating, non-degenerate, Galois equivariant pairing

$$e_\ell : T_\ell(E) \times T_\ell(E) \longrightarrow T_\ell(\boldsymbol{\mu}),$$

where $T_\ell(\boldsymbol{\mu}) = \varprojlim \boldsymbol{\mu}_{\ell^n}$ is the Tate module of the multiplicative group $\mathbb{G}_m$.

§9. Elliptic Curves Over Finite Fields

Let $E/\mathbb{F}_q$ be an elliptic curve defined over a field with q elements. Then the group of rational points $E(\mathbb{F}_q)$ is a finite group.

Theorem. (Hasse)

$$\bigl|q + 1 - \#E(\mathbb{F}_q)\bigr| \le 2\sqrt{q}.$$

Proof sketch. Let $\phi : E \to E$ be the Frobenius morphism given on Weierstrass coordinates by $\phi(x, y) = (x^q, y^q)$. Then $E(\mathbb{F}_q) = \ker(1 - \phi)$. Further, one can show that the map $1 - \phi$ is separable by looking at its action on the invariant differential, so

$$\#E(\mathbb{F}_q) = \#\ker(1 - \phi) = \deg(1 - \phi).$$

We know that

$$\phi \circ \hat{\phi} = \deg \phi = q \in \mathbb{Z} \subset \operatorname{End}(E),$$

and we let

$$a = \phi + \hat{\phi} \in \mathbb{Z} \subset \operatorname{End}(E).$$

Then for any $m, n \in \mathbb{Z}$ we have

$$0 \le \deg(m + n\phi) = (m + n\phi) \circ (m + n\hat{\phi}) = m^2 + amn + qn^2.$$

The non-negativity implies that the quadratic form is positive semi-definite, so its discriminant is non-positive, $a^2 - 4q \le 0$. In particular, putting $m = 1$ and $n = -1$ yields

$$\#E(\mathbb{F}_q) = \deg(1 - \phi) = 1 - a + q,$$

which combined with $|a| \le 2\sqrt{q}$ gives the desired result.

Remark. Examining the above proof, we see that we have proven the following fundamental formula for the sum of the q-power Frobenius map and its dual:

$$\phi + \hat{\phi} = \begin{pmatrix} \text{multiplication} \\ \text{by } q+1-\#E(\mathbb{F}_q) \\ \text{on } E \end{pmatrix}.$$

Hasse's theorem says that the *trace of Frobenius*, that is $\phi + \hat{\phi}$, is an integer in $\mathrm{End}(E)$ of magnitude at most $2\sqrt{q}$.

The *zeta function of an elliptic curve* $E/\mathbb{F}_q$ is defined by the formal power series

$$Z(E/\mathbb{F}_q, T) = \exp\Bigl(\sum_{n=1}^{\infty} \#E(\mathbb{F}_{q^n}) \cdot \frac{T^n}{n}\Bigr).$$

Theorem. *Let $E/\mathbb{F}_q$ be an elliptic curve. The zeta function of E is a rational function of the form*

$$Z(E/\mathbb{F}_q, T) = \frac{1 - aT + qT^2}{(1-T)(1-qT)},$$

where a is the trace of Frobenius,

$$a = q + 1 - \#E(\mathbb{F}_q) = \phi + \hat{\phi}.$$

Further,

$$1 - aT + qT^2 = (1-\alpha T)(1-\beta T) \qquad \textit{with } |\alpha| = |\beta| = \sqrt{q}.$$

Isogenous elliptic curves have the same number of points, since if $\psi : E \to E'$ is an isogeny defined over $\mathbb{F}_q$, then

$$\begin{aligned} \deg(\psi)\bigl(q+1-\#E(\mathbb{F}_q)\bigr) &= \deg(\psi)\deg(1-\phi_E) \\ &= \deg(\psi - \psi \circ \phi_E) \\ &= \deg(\psi - \phi_{E'} \circ \psi) \\ &= \deg(1-\phi_{E'})\deg(\psi) \\ &= \bigl(q+1-\#E'(\mathbb{F}_q)\bigr)\deg(\psi). \end{aligned}$$

The converse is also true, but harder to prove:

Theorem. *Two elliptic curves $E/\mathbb{F}_q$ and $E'/\mathbb{F}_q$ are isogenous over $\mathbb{F}_q$ if and only if $Z(E/\mathbb{F}_q, T) = Z(E'/\mathbb{F}_q, T)$.*

For an elliptic curve over a finite field, the p-torsion, the Frobenius map, and the endomorphism ring are closely related.

Theorem. *Let $E/\mathbb{F}_q$ be an elliptic curve, let* $p = \mathrm{char}(\mathbb{F}_q)$ *and let* $\phi : E \to E$ *be the* q^{th}*-power Frobenius map. The following are equivalent:*
(i) $E[p] = 0$.
(ii) *The dual* $\hat{\phi}$ *of Frobenius is purely inseparable.*
(iii) *The map* $[p] : E \to E$ *is purely inseparable.*
(iv) $\mathrm{End}(E)$ *is an order in a quaternion algebra.*
If these conditions hold, we say that E *is* supersingular, *otherwise we say that* E *is* ordinary. *If* E *is ordinary, then* $E[p] \cong \mathbb{Z}/p\mathbb{Z}$ *and* $\mathrm{End}(E)$ *is an order in a quadratic imaginary field.*

The supersingular elliptic curves in characteristic p all have j-invariants lying in $\mathbb{F}_{p^2}$. Up to $\bar{\mathbb{F}}_p$-isomorphism, there are approximately $p/12$ of them.

§10. Elliptic Curves Over $\mathbb{C}$ and Elliptic Functions

The complex analytic theory of elliptic curves is vast, so we will only hit on a few highlights. Let $L \subset \mathbb{C}$ be a lattice. An *elliptic function* is an L-periodic meromorphic function $f(z)$, that is, $f(z+\omega) = f(z)$ for all $z \in \mathbb{C}$ and all $\omega \in L$. The collection of all elliptic functions for L forms a field, denoted $\mathbb{C}(L)$.

The *Weierstrass* $\wp$*-function*

$$\wp(z) = \wp(z, L) = \frac{1}{z^2} + \sum_{\omega \in L,\, \omega \neq 0} \left(\frac{1}{(z-\omega)^2} - \frac{1}{\omega^2} \right)$$

is an elliptic function with a double pole at each point of L and no other poles. Also associated to the lattice L are the *Eisenstein series*

$$G_{2k}(L) = \sum_{\omega \in L,\, \omega \neq 0} \frac{1}{\omega^{2k}}.$$

These series are absolutely convergent for all integers $k \geq 2$. Notice that G_{2k} has the property the $G_{2k}(\lambda L) = \lambda^{-2k} G_{2k}(L)$ for any $\lambda \in \mathbb{C}^*$. It is standard to set

$$g_2(L) = 60 G_4(L) \qquad \text{and} \qquad g_3(L) = 140 G_6(L).$$

Theorem. (a) $\mathbb{C}(L) = \mathbb{C}\big(\wp(z), \wp'(z)\big)$.
(b) *The Weierstrass* $\wp$*-function and its derivative satisfy the identity*

$$\wp'(z)^2 = 4\wp(z)^3 - g_2(L)\wp(z) - g_3(L).$$

Further, the discriminant

$$\Delta(L) = g_2(L)^3 - 27 g_3(L)^2$$

of the cubic polynomial is non-zero, so the equation

$$E_L : y^2 = 4x^3 - g_2(L)x - g_3(L)$$

defines an elliptic curve over $\mathbb{C}$.

(c) *The map*

$$\phi_L : \mathbb{C}/L \longrightarrow E_L(\mathbb{C}), \qquad z \longrightarrow (\wp(z), \wp'(z)),$$

is a complex analytic isomorphism of complex Lie groups.

(d) *Conversely, given any elliptic curve* $E/\mathbb{C}$, *there exists a lattice* L, *unique up to homothety, such that* $E_L \cong E$.

Corollary. *Let* $E/\mathbb{C}$ *be an elliptic curve and let* $m \geq 1$ *be an integer. Then as an abstract group,* $E[m] \cong \mathbb{Z}/m\mathbb{Z} \times \mathbb{Z}/m\mathbb{Z}$.

Proof. $E[m] = \ker(\mathbb{C}/L \xrightarrow{z \mapsto mz} \mathbb{C}/L) = (1/m)L/L \cong (\mathbb{Z}/m\mathbb{Z})^2$.

Another useful function is the Weierstrass σ-function

$$\sigma(z) = \sigma(z, L) = z \prod_{\omega \in L,\, \omega \neq 0} \left(1 - \frac{z}{\omega}\right) e^{z/\omega + (1/2)(z/\omega)^2}.$$

It is a theta function and can be used to construct elliptic functions. For example,

$$\wp(z) - \wp(a) = -\frac{\sigma(z+a)\sigma(z-a)}{\sigma(z)^2\sigma(a)^2} \qquad \text{and} \qquad \wp'(z) = -\frac{\sigma(2z)}{\sigma(z)^4}.$$

If E_1 and E_2 are associated to the lattices L_1 and L_2 respectively, then one can show that

$$\mathrm{Hom}(E_1, E_2) \cong \{\alpha \in \mathbb{C} : \alpha L_1 \subset L_2\},$$

where the isogeny associated to α is given analytically by

$$\mathbb{C}/L_1 \longrightarrow \mathbb{C}/L_2, \qquad z \longmapsto \alpha z.$$

Using this, it is not hard to show that if $L = \omega_1\mathbb{Z} + \omega_2\mathbb{Z}$, then either

(i) $\mathrm{End}(E_L) = \mathbb{Z}$, or

(ii) $\mathbb{Q}(\omega_1/\omega_2)$ is a quadratic imaginary field, and $\mathrm{End}(E_L)$ is isomorphic to an order in $\mathbb{Q}(\omega_1/\omega_2)$.

Homothetic lattices correspond to isomorphic elliptic curves, so it is common practice to use the normalized lattices

$$L_\tau = \tau\mathbb{Z} + \mathbb{Z} \qquad \text{with } \mathrm{Im}(\tau) > 0.$$

One then writes $\wp(z, \tau)$, $\sigma(z, \tau)$, $G_{2k}(\tau)$, etc. An elliptic function for L_τ is $\mathbb{Z}$-periodic, and thus may be written as a function of

$$u = e^{2\pi i z} \qquad \text{and} \qquad q = e^{2\pi i \tau}.$$

This is equivalent to using the natural isomorphism

$$\mathbb{C}/L_\tau \xrightarrow{\sim} \mathbb{C}^*/q^{\mathbb{Z}}, \qquad z \longmapsto u = e^{2\pi i z}.$$

Theorem.

$$\frac{1}{(2\pi i)^2}\wp(z,\tau) = \sum_{n\in\mathbb{Z}} \frac{q^n u}{(1-q^n u)^2} + \frac{1}{12} - 2\sum_{n\geq 1}\frac{q^n}{(1-q^n)^2}.$$

$$\sigma(z,\tau) = -\frac{1}{2\pi i} e^{\eta z^2/2} e^{-\pi i z}(1-u)\prod_{n\geq 1}\frac{(1-q^n u)(1-q^n u^{-1})}{(1-q^n)^2}.$$

(Here $\eta = \eta(\tau)$ is a complex number called a quasi-period of the lattice L_τ.)

Finally, I want to mention the q-expansions for Δ and j and the Eisenstein series G_{2k}, and also to state Jacobi's beautiful product formula for the discriminant function.

Theorem. *As functions of $q = e^{2\pi i\tau}$, the Eisenstein series G_{2k}, the discriminant function $\Delta(\tau)$ and the j-invariant $j(\tau)$ have the following expansions in $\mathbb{Z}[\![q]\!]$:*

$$G_{2k}(\tau) = 2\zeta(2k) + 2\frac{(2\pi i)^{2k}}{(2k-1)!}\sum_{n\geq 1}\Big(\sum_{d\mid n} d^{2k-1}\Big)q^n.$$

$$\Delta(\tau) = (2\pi)^{12}\sum_{n\geq 1}\tau(n)q^n = q - 24q^2 + 252q^3 - 1472q^4 + \cdots.$$

$$j(\tau) = q^{-1} + \sum_{n\geq 0} c(n)q^n = q^{-1} + 744 + 196884q + 21493760q^2 + \cdots.$$

(Here $\zeta(s)$ is the Riemann zeta function.)

The discriminant function also has the following product expansion:

$$\Delta(\tau) = \Delta(L_\tau) = (2\pi)^{12} q \prod_{n\geq 1}(1-q^n)^{24}. \quad \text{(Jacobi's formula)}$$

The integer coefficients $\tau(n)$ and $c(n)$ of Δ and j have many wonderful arithmetic properties.

§11. The Formal Group of an Elliptic Curve

Substituting $x = z/w$ and $y = -1/w$ into a Weierstrass equation for E gives

$$w = z^3 + a_1 zw + a_2 z^2 w + a_3 w^2 + a_4 z w^2 + a_6 w^3,$$

and then repeated substitution (or Hensel's lemma) can be used to express w as a formal power series $w(z) \in \mathbb{Z}[a_1,\ldots,a_6][\![z]\!]$. This in turn can be used to express x, y, and the invariant differential ω_E as formal series in z, and then the group law is given by a power series $F_E(z_1,z_2)$ in two variables. The first few terms of these series are:

$$\begin{aligned} w(z) &= z^3 + a_1 z^4 + (a_1^2 + a_2)z^5 + (a_1^3 + 2a_1a_2 + a_3)z^6 + \cdots, \\ x(z) &= z^{-2} - a_1 z^{-1} - a_2 - a_3 z - (a_4 + a_1a_3)z^2 - \cdots, \\ y(z) &= -z^{-3} + a_1 z^{-2} + a_2 z^{-1} + a_3 + (a_4 + a_1 a_3)z - \cdots, \\ \omega_E(z) &= \left(1 + a_1 z + (a_1^2 + a_2)z^2 + (a_1^3 + 2a_1a_2 + 2a_3)z^3 + \cdots\right)dz, \\ F_E(z_1, z_2) &= z_1 + z_2 - a_1 z_1 z_2 - a_2(z_1^2 z_2 + z_1 z_2^2) + \cdots. \end{aligned}$$

The *formal group* $\hat{E}$ *associated to* E is the formal group defined by the formal group law $F_E(z_1, z_2) \in \mathbb{Z}[a_1, \ldots, a_6][\![z_1, z_2]\!]$.

Let R be a complete local ring with maximal ideal $\mathfrak{p}$, and suppose that the a_i's are in R. Then F_E converges for $z_1, z_2 \in \mathfrak{p}$ and gives $\mathfrak{p}$ a group structure which we denote by $\hat{E}(\mathfrak{p})$. The series F_E also induces a group structure on the powers $\mathfrak{p}^r$, which gives $\hat{E}(\mathfrak{p})$ a natural filtration $\hat{E}(\mathfrak{p}^r)$. The following is a general property of formal groups.

Proposition. *The group* $\hat{E}(\mathfrak{p})$ *has no prime-to-*$\mathfrak{p}$ *torsion. In other words, if* $m \not\equiv 0 \pmod{\mathfrak{p}}$*, then* $\hat{E}(\mathfrak{p})$ *has no non-trivial points of order* m*.*

§12. ELLIPTIC CURVES OVER LOCAL FIELDS

For this section we set the following notation:

- K a complete local field with normalized valuation $v : K^* \to \mathbb{Z}$.
- R the ring of integers of K.
- $\mathfrak{p}$ the maximal ideal of R.
- k the residue field $k = R/\mathfrak{p}$.

A *minimal Weierstrass equation* for an elliptic curve E/K is a Weierstrass equation

$$E : y^2 + a_1xy + a_3y = x^3 + a_2x^2 + a_4x + a_6$$

with $a_i \in R$ and $v(\Delta)$ minimized. If $\operatorname{char}(k) \neq 2, 3$, then E always has a minimal equation with $a_1 = a_2 = a_3 = 0$.

The *reduction of* E *modulo* $\mathfrak{p}$, denoted $\tilde{E}$, is then the curve over k defined by the equation

$$\tilde{E} : y^2 + \tilde{a}_1xy + \tilde{a}_3y = x^3 + \tilde{a}_2x^2 + \tilde{a}_4x + \tilde{a}_6,$$

where the tilde denotes reduction modulo $\mathfrak{p}$. The curve $\tilde{E}$ may be singular; its non-singular part is denoted $\tilde{E}^{\text{ns}}$. We say that

E has *good* (or *stable*) *reduction* if $\tilde{E}$ is non-singular.

E has *multiplicative* (or *semi-stable*) *reduction* if $\tilde{E}$ has a node. The reduction is called *split* if the tangent directions are defined over k, otherwise it is *non-split*.

E has *additive* (or *unstable*) *reduction* if $\tilde{E}$ has a cusp.

Remark. It is becoming common to use the term "semi-stable" to refer to an elliptic curve which has either good or multiplicative reduction, while "unstable" retains its meaning of additive reduction.

Proposition. *Let E/K be an elliptic curve. Then there is a finite extension K'/K such that E has either good or split multiplicative reduction over K'.*

We define a filtration on $E(K)$ by

$$\begin{aligned} E_0(K) &= \{P \in E(K) : \tilde{P} \in \tilde{E}^{\text{ns}}(k)\} \\ E_1(K) &= \{P \in E(K) : \tilde{P} = \tilde{O}\} \\ E_r(K) &= \{P \in E(K) : v(x(P)) \le -2r\} \quad (\text{for } r \ge 1). \end{aligned}$$

Proposition. (a) *There is an exact sequence*

$$0 \longrightarrow E_1(K) \longrightarrow E_0(K) \longrightarrow \tilde{E}^{\text{ns}}(k) \longrightarrow 0.$$

(b) *There is an isomorphism $E_1(K) \cong \hat{E}(\mathfrak{p})$. This isomorphism identifies $E_r(K)$ with $\hat{E}(\mathfrak{p}^r)$.*
(c) *The quotient group $E(K)/E_0(K)$ is finite. More precisely, it has order 1, 2, 3, or 4 unless $\tilde{E}$ has split multiplicative reduction, in which case it is a cyclic group of order $v(\Delta)$.*

Remark. Another description of the group $E(K)/E_0(K)$ is that it is isomorphic to the group of components of the Néron model of E over R.

The following corollary is of fundamental importance.

Corollary. *If E has good reduction at $\mathfrak{p}$ and m is relatively prime to* char(k), *then the reduction map*

$$E(K)[m] \longrightarrow \tilde{E}(k)$$

is injective. Equivalently, the extension $K\big(E[m]\big)$ generated by the m-torsion points is an unramified extension of K.

The following converse is often useful. Let $I_{\bar{K}/K}$ denote the inertia subgroup of $G_{\bar{K}/K}$, and recall that a $G_{\bar{K}/K}$-module M is said to be *unramified* if $I_{\bar{K}/K}$ acts trivially on M.

Theorem. (Criterion of Néron-Ogg-Shafarevich) *The following are equivalent:*
(i) *E has good reduction.*
(ii) *$E[m]$ is unramified for infinitely many m prime to* char(k).
(iii) *$T_\ell(E)$ is unramified for some $\ell \ne$* char(k).

Corollary. *If E_1/K and E_2/K are isogenous over K, then they either both have good reduction, or neither has good reduction.*

An elliptic curve E/K is said to have *potential good reduction* if it acquires good reduction over a finite extension of K.

Proposition. *An elliptic curve E/K has potential good reduction if and only if $j(E) \in R$.*

§13. THE SELMER AND SHAFAREVICH-TATE GROUPS

For this section we fix the following notation:

K a number field.

R the ring of integers of K.

For any place v of K, we write K_v for the completion of K with respect to v. If v is non-archimedean, we write R_v, $\mathfrak{p}_v$, and k_v for the ring of integers of K_v, maximal ideal of R_v, and residue field of R_v respectively.

Mordell-Weil Theorem. *Let E/K be an elliptic curve. Then the group of rational points $E(K)$ is a finitely generated abelian group.*

In this section we will consider a weak form of the Mordell-Weil theorem which asserts that the quotient group $E(K)/mE(K)$ is finite. This assertion is one of the main ingredients in the proof of the full theorem.

Fix an integer $m \geq 2$ and consider the exact sequence

$$0 \longrightarrow E[m] \longrightarrow E(\bar{K}) \xrightarrow{[m]} E(\bar{K}) \longrightarrow 0.$$

Taking Galois cohomology gives the long exact sequence

$$\to E(K) \xrightarrow{[m]} E(K) \to H^1(G_{\bar{K}/K}, E[m]) \to H^1(G_{\bar{K}/K}, E(\bar{K})) \xrightarrow{m} H^1(G_{\bar{K}/K}, E(\bar{K})) \to,$$

and this in turn gives the *Kummer sequence for E/K*,

$$0 \to E(K)/mE(K) \to H^1(G_{\bar{K}/K}, E[m]) \to H^1(G_{\bar{K}/K}, E(\bar{K}))[m] \to 0.$$

Unfortunately, the group $H^1(G_{\bar{K}/K}, E[m])$ need not be finite. However, any element of $H^1(G_{\bar{K}/K}, E[m])$ which comes from a point of $E(K)$ will necessarily come from a point in $E(K_v)$ for every completion of K. In other words, if we consider the Kummer sequence for E/K_v and restriction maps on cohomology, we get a commutative diagram

$$\begin{array}{ccccccc}
0\to & \dfrac{E(K)}{mE(K)} & \to & H^1(G_{\bar{K}/K}, E[m]) & \to & H^1(G_{\bar{K}/K}, E(\bar{K}))[m] & \to 0 \\
 & \downarrow & & \downarrow & & \downarrow & \\
0\to & \prod_v \dfrac{E(K_v)}{mE(K_v)} & \to & \prod_v H^1(G_{\bar{K}_v/K_v}, E[m]) & \to & \prod_v H^1(G_{\bar{K}_v/K_v}, E(\bar{K}_v))[m] & \to 0
\end{array}$$

This suggests the following definitions: The *m-Selmer group of E/K* is the group

$$S^{(m)}(E/K) = \ker\Big\{H^1(G_{\bar{K}/K}, E[m]) \longrightarrow \prod_v H^1(G_{\bar{K}_v/K_v}, E(\bar{K}_v))\Big\}.$$

The *Shafarevich-Tate group of E/K* is the group

$$Ш(E/K) = \ker\Big\{H^1(G_{\bar{K}/K}, E(\bar{K})) \longrightarrow \prod_v H^1(G_{\bar{K}_v/K_v}, E(\bar{K}_v))\Big\}.$$

It is immediate from these definitions that there is an exact sequence

$$0 \to E(K)/mE(K) \to S^{(m)}(E/K) \to Ш(E/K)[m] \to 0.$$

Theorem. *The Selmer group $S^{(m)}(E/K)$ is finite. Hence $E(K)/mE(K)$ and $Ш(E/K)[m]$ are also finite.*

Proof sketch. Let $\mathfrak{p}$ be a prime of K not dividing m for which E has good reduction. Then $E[m] \hookrightarrow \tilde{E}(k_\mathfrak{p})$ (i.e., the m-torsion injects into the reduction modulo $\mathfrak{p}$). This implies that any cocycle in $S^{(m)}(E/K)$ is unramified at $\mathfrak{p}$, so $S^{(m)}(E/K)$ consists of cocycles which are unramified outside a finite set of primes, specifically outside the set

$$\{\mathfrak{p} : E \text{ has bad reduction at } \mathfrak{p}\} \cup \{\mathfrak{p} : \mathfrak{p} \text{ divides } m\}.$$

Finally, it is an elementary consequence of Dirichlet's unit theorem and the finiteness of the class group that for any finite $G_{\bar{K}/K}$-module M and any finite set of places S, the set of cocycles in $H^1(G_{\bar{K}/K}, M)$ unramified outside S is finite.

Remark. More generally, if $\phi : E \to E'$ is an isogeny of elliptic curves defined over K, there is an associated Kummer sequence

$$0 \to E'(K)/\phi(E(K)) \to H^1(G_{\bar{K}/K}, E[\phi]) \to H^1(G_{\bar{K}/K}, E(\bar{K}))[\phi] \to 0.$$

Using this, one defines in an analogous fashion the *ϕ-Selmer group*, denoted $S^{(\phi)}(E/K)$, which can be shown to be finite, and an associated exact sequence

$$0 \to E'(K)/\phi(E(K)) \to S^{(\phi)}(E/K) \to Ш(E/K)[\phi] \to 0.$$

The group $H^1(G_{\bar{K}/K}, E(\bar{K}))$ can also be interpreted as the collection of homogeneous spaces of E/K. Generally, one defines the *Weil-Châtelet group of E/K* to be[†]

$$\mathrm{WC}(E/K) = \left\{\begin{matrix} K\text{-isomorphism classes of smooth projective curves} \\ C/K \text{ such that } C \text{ is isomorphic to } E \text{ over } \bar{K} \end{matrix}\right\}.$$

[†]This is cheating a little bit. The Weil-Châtelet group is actually the group of principal homogeneous spaces for E/K. That is, an element of $\mathrm{WC}(E/K)$ consists of a curve C/K and a simply transitive algebraic group action of E on C defined over K. Further, in defining the associated cocycle, we need to choose an isomorphism $f : C \to E$ with the property that $f^\sigma \circ f^{-1}$ is a pure translation map on E.

A homogeneous space C/K represents the zero element of $\mathrm{WC}(E/K)$ if and only if $C(K)$ is non-empty.

There is a natural isomorphism $\mathrm{WC}(E/K) \cong H^1(G_{\bar{K}/K}, E(\bar{K}))$ defined in the following way. Let $[C/K] \in \mathrm{WC}(E/K)$ and choose an isomorphism $f : C \to E$ defined over $\bar{K}$ and a point $P \in C(\bar{K})$. Then the cocycle

$$G_{\bar{K}/K} \longrightarrow E(\bar{K}), \qquad \sigma \longmapsto f(P^\sigma) - f(P),$$

represents the cohomology class in $H^1(G_{\bar{K}/K}, E(\bar{K}))$ associated to C/K. With this identification, the subgroup $\text{Ш}(E/K)$ in $\mathrm{WC}(E/K)$ consists of all homogeneous spaces C/K such that $C(K_v)$ is non-empty for all places v of K.

Remark. Each Selmer group $S^{(m)}(E/K)$ is effectively computable in theory, and frequently computable in practice. At present, there is no proven effective method for determining which part of $S^{(m)}(E/K)$ comes from $E(K)/mE(K)$ and which part comes from $\text{Ш}(E/K)$.

§14. DISCRIMINANTS, CONDUCTORS, AND L-SERIES

Let K be a number field and E/K an elliptic curve. For each prime $\mathfrak{p}$ of K we can consider a minimal Weierstrass equation for the local field $K_\mathfrak{p}$ and the discriminant $\Delta_\mathfrak{p}$ of this minimal equation. The *minimal discriminant of E/K* is the integral ideal

$$D_{E/K} = \prod_{\mathfrak{p}} \mathfrak{p}^{v_\mathfrak{p}(\Delta_\mathfrak{p})}.$$

If K has class number one (e.g., $K = \mathbb{Q}$), it is possible to find a Weierstrass equation

$$E : y^2 + a_1xy + a_3y = x^3 + a_2x^2 + a_4x + a_6$$

which is simultaneously minimal at all primes of K. The discriminant Δ of this *global minimal Weierstrass equation* is then equal to the discriminant of E/K (and is uniquely determined up to multiplication by the 12th-power of a unit.)

The minimal discriminant is a measure of the bad reduction of E. Another such measure is the *conductor of E/K*. This is an ideal

$$N_{E/K} = \prod_{\mathfrak{p}} \mathfrak{p}^{f_\mathfrak{p}(E/K)},$$

where the exponents $f_\mathfrak{p}(E/K)$ are given by

$$f_\mathfrak{p}(E/K) = \begin{cases} 0 & \text{if } E \text{ has good reduction at } \mathfrak{p}, \\ 1 & \text{if } E \text{ has multiplicative reduction at } \mathfrak{p}, \\ 2 & \text{if } E \text{ has additive reduction at } \mathfrak{p} \text{ and } \mathfrak{p} \nmid 6. \end{cases}$$

If $\mathfrak{p}$ has residue characteristic 2 or 3 and E has additive reduction at $\mathfrak{p}$, then the exponent of the conductor is equal to $2+\delta_\mathfrak{p}$, where $\delta_\mathfrak{p}$ is a measure of the wild ramification in the extensions $K_\mathfrak{p}(E[m])/K_\mathfrak{p}$. Over $\mathbb{Q}$, for example, the conductor exponents are bounded by

$$f_3 \le 5 \qquad \text{and} \qquad f_2 \le 8.$$

Even in characteristics 2 and 3, the conductor can easily be computed using an algorithm of Tate and a formula of Ogg and Saito.

Remark. If E has everywhere semi-stable (i.e., good or multiplicative) reduction, then its conductor is simply the product of its primes of bad reduction.

For each prime $\mathfrak{p}$ of K, let $q_\mathfrak{p}$ be the norm of $\mathfrak{p}$. If E has good reduction at $\mathfrak{p}$, we also let

$$a_\mathfrak{p} = q_\mathfrak{p} + 1 - \#\tilde{E}(k_\mathfrak{p}).$$

The *local factor of the L-series of E at* $\mathfrak{p}$ is the polynomial

$$L_\mathfrak{p}(T) = \begin{cases} 1 - a_\mathfrak{p}T + q_\mathfrak{p}T^2 & \text{if } E \text{ has good reduction at } \mathfrak{p}, \\ 1 - T & \text{if } E \text{ has split multiplicative reduction at } \mathfrak{p}, \\ 1 + T & \text{if } E \text{ has non-split multiplicative reduction at } \mathfrak{p}, \\ 1 & \text{if } E \text{ has additive reduction at } \mathfrak{p}. \end{cases}$$

In all cases the relation

$$L_\mathfrak{p}(1/q_\mathfrak{p}) = \#\tilde{E}^{\mathrm{ns}}(k_\mathfrak{p})/q_\mathfrak{p}$$

holds. The *global* (or *Hasse-Weil*) *L-series of* E/K is then defined by the Euler product

$$L(E/K, s) = \prod_\mathfrak{p} L_\mathfrak{p}(q_\mathfrak{p}^{-s})^{-1}.$$

It is not hard to prove that isogenous curves have the same L-series. The following converse is a consequence of (and in fact equivalent to) Faltings' isogeny theorem.

Theorem. *Two elliptic curves E/K and E'/K are isogenous over K if and only if $a_\mathfrak{p}(E) = a_\mathfrak{p}(E')$ for all (or all but finitely many, or even all but a set of density zero) primes $\mathfrak{p}$ of K.*

Remark. Over $\mathbb{Q}$ it is even true that $E/\mathbb{Q}$ and $E'/\mathbb{Q}$ are isogeneous if and only if $L(E/\mathbb{Q}, s) = L(E'/\mathbb{Q}, s)$, but this need not be true over number fields. An example, given in [7, remark 3.4], is $K = \mathbb{Q}(i)$ and

$$E^\pm : y^2 = x^3 \pm ix + 3.$$

In this example, E^+ and E^- are not isogeneous, but $L(E^+, s) = L(E^-, s)$, since if we write $G_{K/\mathbb{Q}} = \{1, \sigma\}$, then $a_\mathfrak{p}(E^+) = a_{\mathfrak{p}^\sigma}(E^-)$.

The estimate $|a_\mathfrak{p}| \le 2q_\mathfrak{p}^{1/2}$ implies that the Euler product converges and gives an analytic function in the half-plane $\mathrm{Re}(s) > 3/2$.

Conjecture. *The L-series $L(E/K,s)$ has an analytic continuation to the entire complex plane and satisfies a functional equation relating its values at s and $2-s$.*

Over $\mathbb{Q}$, the conjecture asserts that the function

$$\xi(E/\mathbb{Q},s) = N_{E/K}^{s/2}(2\pi)^{-s}\Gamma(s)L(E/\mathbb{Q},s)$$

has an analytic continuation and satisfies the functional equation

$$\xi(E/\mathbb{Q},2-s) = \pm\xi(E/\mathbb{Q},s).$$

This is known to be true for modular elliptic curves.

§15. DUALITY THEORY

There are both local and global duality theorems for the cohomology of an elliptic curve.

Local Duality Theorem. (Tate) *Let K be a complete local field and let E/K be an elliptic curve. There is a bilinear, non-degenerate pairing*

$$E(K)\times H^1(G_{\bar{K}/K},E(\bar{K})) \longrightarrow \mathbb{Q}/\mathbb{Z}.$$

More precisely, the pairing induces a duality of locally compact groups, where $E(K)$ is given the topology induced by the topology on K, and where the cohomology group $H^1(G_{\bar{K}/K},E(\bar{K}))$ is given the discrete topology.

Here is one of the many equivalent definitions of the Tate pairing. Let $P\in E(K)$ and $\xi\in H^1(G_{\bar{K}/K},E(\bar{K}))$. Take any integer $m\geq 1$ which kills ξ and consider the short exact sequence

$$0\to E(K)/mE(K)\xrightarrow{\delta} H^1(G_{\bar{K}/K},E[m]) \to H^1(G_{\bar{K}/K},E(\bar{K}))[m]\to 0.$$

First we push P forward to get an element $\delta P\in H^1(G_{\bar{K}/K},E[m])$. Next we choose an element $\eta\in H^1(G_{\bar{K}/K},E[m])$ which maps to ξ. Then the cup product $\delta P\cup\eta$ is in $H^2(G_{\bar{K}/K},E[m]\otimes E[m])$. Finally we use the Weil pairing $e_m : E[m]\otimes E[m]\to\boldsymbol{\mu}_m$ to get the desired cohomology class

$$e_m(\delta P\cup\eta)\in H^2(G_{\bar{K}/K},\boldsymbol{\mu}_m)\hookrightarrow H^2(G_{\bar{K}/K},\bar{K}^*) = \mathrm{Br}(K)\cong\mathbb{Q}/\mathbb{Z}.$$

Note that the last isomorphism is the identification of the Brauer group of K with $\mathbb{Q}/\mathbb{Z}$ provided by local class field theory.

The global duality theorem is only fully satisfactory when Ш is known to be finite.

Global Duality Theorem. (Cassels) *Let K be a number field and let E/K be an elliptic curve. There is an alternating bilinear pairing*

$$Ш(E/K)\times Ш(E/K)\longrightarrow\mathbb{Q}/\mathbb{Z}$$

whose kernel on either side is the group of divisible elements of Ш(E/K). In particular, if Ш(E/K) is finite, then the pairing is perfect and the order of Ш(E/K) is a perfect square.

The definition of the pairing on Ш is considerably more complicated, so we do not give it here.

§16. Rational Torsion and the Image of Galois

Let E/K be an elliptic curve defined over a number field. The ℓ-adic representation

$$\rho_\ell : G_{\bar{K}/K} \longrightarrow \mathrm{Aut}(T_\ell(E)) \cong \mathrm{GL}_2(\mathbb{Z}_\ell)$$

determines many of the arithmetic properties of E. If E has complex multiplication, ρ_ℓ can be described in terms of class field theory. The following two important results give a further description of ρ_ℓ.

Theorem. (Serre) *Assume the E does not have complex multiplication.*
(a) *The image of ρ_ℓ is of finite index in $\mathrm{GL}_2(\mathbb{Z}_\ell)$ for all primes ℓ.*
(b) *The image of ρ_ℓ is equal to $\mathrm{GL}_2(\mathbb{Z}_\ell)$ for all but finitely many primes ℓ.*

Theorem. (Faltings) *Let E/K and E'/K be elliptic curves. Then the natural map*

$$\mathrm{Hom}_K(E, E') \otimes \mathbb{Z}_\ell \longrightarrow \mathrm{Hom}(T_\ell(E), T_\ell(E'))^{G_{\bar{K}/K}}$$

is an isomorphism.

It is conjectured that the total index of the ρ_ℓ's is bounded independently of the curve E. That is, for a fixed number field K and any non-CM elliptic curve E/K, the quantity

$$\prod_\ell [\mathrm{GL}_2(\mathbb{Z}_\ell) : \rho_\ell(G_{\bar{K}/K})]$$

is bounded by a number depending only on K. In particular, the torsion subgroup $E(K)_{\mathrm{tors}}$ should be bounded independently of E. This last statement has recently been proven.

Theorem. (a) (Mazur) *Let $E/\mathbb{Q}$ be an elliptic curve. Then $E(\mathbb{Q})_{\mathrm{tors}}$ is one of the following 15 groups:*

$$\begin{array}{ll} \mathbb{Z}/n\mathbb{Z} & \text{with } 1 \le n \le 10 \text{ or } n = 12, \text{ or} \\ \mathbb{Z}/2\mathbb{Z} \times \mathbb{Z}/2n\mathbb{Z} & \text{with } 1 \le n \le 4. \end{array}$$

(b) (Kamienny, Mazur, Merel) *Let K be a number field of degree d. Then there is a constant $c(d)$ so that for every elliptic curve E/K, the torsion subgroup of E/K satisfies $\#E(K)_{\mathrm{tors}} \le c(d)$.*

§17. Tate Curves

Let K be a local field which is complete with respect to a non-archimedean absolute value $|\cdot|_v$. The analytic parametrization $\mathbb{C}/L \to E(\mathbb{C})$ of an elliptic curve over $\mathbb{C}$ does not have a direct non-archimedean analogue, because K has no discrete subgroups. However, the situation changes when one considers $\mathbb{C}^*/q^{\mathbb{Z}}$, since any $q \in K^*$ with $|q|_v < 1$ will generate a discrete subgroup. It turns out that suitably normalized q-expansions of $\wp$, $\wp'$ and G_{2k} give a v-adic analytic isomorphism from $K^*/q^{\mathbb{Z}}$ to an elliptic curve E_q defined over K. However, not all elliptic curves over K arise in this fashion, as can be seen by examining the j-invariant

$$j(E_q) = j(q) = q^{-1} + 744 + 196884q + 21493760q^2 + \cdots.$$

It is clear that $|j(q)|_v > 1$, and so E_q must have multiplicative reduction.

Theorem. (Tate) *Let $q \in K^*$ with $|q|_v < 1$. There is an elliptic curve E_q/K and a $G_{\bar{K}/K}$-equivariant v-analytic isomorphism*

$$\phi : \bar{K}^*/q^{\mathbb{Z}} \longrightarrow E_q(\bar{K}).$$

The set of curves $\{E_q : q \in K^*,\ |q|_v < 1\}$ is exactly the set of elliptic curves over K with split multiplicative reduction.

If E/K satisfies $|j(E)|_v > 1$ but does not have split multiplicative reduction, then it is isomorphic over $\bar{K}$ to some E_q. More precisely, there is a unique quadratic extension L/K such that E is isomorphic to E_q over L, and then

$$E(K) \cong \{u \in L^* : \mathrm{N}_{L/K}(u) \in q^{\mathbb{Z}}\} \,/\, q^{\mathbb{Z}}.$$

Further, the extension L/K is unramified if and only if E has non-split multiplicative reduction.

§18. HEIGHTS AND DESCENT

Let K be a number field and let M_K be the set of inequivalent absolute values on K, suitably normalized. The *height on* $\mathbb{P}^n$ is the function

$$h : \mathbb{P}^n(K) \longrightarrow [0,\infty), \qquad h([x_0,\dots,x_n]) = \sum_{v \in M_K} \log \max_{0\le i\le n} |x_i|_v.$$

With the appropriate normalization, the height is independent of the choice of homogeneous coordinates and of the field K. For this reason, h is often called the *absolute logarithmic height.* The height on an elliptic curve E/K given by a Weierstrass equation is

$$h : E(\bar{K}) \to [0,\infty), \qquad h(P) = h([x_P, 1]).$$

Proposition. *The height on an elliptic curve E/K has the following properties:*
(i) $h(mP) = m^2h(P) + O(1)$ *for all* $P \in E(\bar{K})$.
(ii) $h(P+Q) + h(P-Q) = 2h(P) + 2h(Q) + O(1)$ *for all* $P, Q \in E(\bar{K})$.
(iii) *For any H, the set $\{P \in E(K) : h(P) \le H\}$ is finite.*
(The $O(1)$ constants depend on E and, in (i)*, also on m.)*

The *canonical* (or *Néron-Tate*) *height on* E/K is defined by the limit

$$\hat{h} : E(\bar{K}) \to [0,\infty), \qquad \hat{h}(P) = \lim_{n\to\infty} 4^{-n}h(2^nP).$$

Theorem. *The canonical height is a positive semi-definite quadratic form on $E(K)$ with the following properties:*

(i) $\hat{h}(P) = h(P) + O(1)$ *for all* $P \in E(\bar{K})$.

(ii) $\hat{h}(P) = 0$ *if and only if* $P \in E_{\text{tors}}$.

Further, $\hat{h}$ *extends* $\mathbb{R}$*-linearly to give a positive definite quadratic form on the vector space* $E(K) \otimes \mathbb{R}$.

Using the canonical height, it is easy to complete the proof of the Mordell-Weil theorem.

Proof (of the Mordell-Weil theorem). The weak Mordell-Weil theorem says that $E(K)/mE(K)$ is finite, so let $P_1, \ldots, P_n \in E(K)$ be coset representatives. Let $H = \max \hat{h}(P_i)$. I claim that the set

$$S = \{P \in E(K) : \hat{h}(P) \leq H\}$$

is a generating set for $E(K)$. Note this set is finite, since $\hat{h} = h + O(1)$. Suppose that it does not generate. Let $Q \in E(K)$ be a point of minimal canonical height not in the span of S. By assumption, $Q = P_i + mR$ for some i and some $R \in E(K)$. Then R cannot be in the span of S, so

$$\begin{aligned} \hat{h}(Q) \leq \hat{h}(R) = \frac{1}{m^2}\hat{h}(mR) &= \frac{1}{m^2}\hat{h}(Q - P_i) \\ &\leq \frac{2}{m^2}\big(\hat{h}(Q) + \hat{h}(P_i)\big) \leq \frac{2}{m^2}\big(\hat{h}(Q) + H\big). \end{aligned}$$

This implies that

$$\hat{h}(Q) \leq \frac{2}{m^2 - 2} H \leq H,$$

which says that $Q \in S$. This contradiction completes the proof.

The bilinear form associated to the canonical height is denoted

$$\langle P, Q \rangle_E = \frac{1}{2}\big(\hat{h}(P + Q) - \hat{h}(P) - \hat{h}(Q)\big).$$

Using this, the *elliptic regulator of* E/K is defined to be the quantity

$$R(E/K) = \det\big(\langle P_i, P_j \rangle_E\big)_{1 \leq i,j \leq r},$$

where $P_1, \ldots, P_r$ is a basis for $E(K)/E(K)_{\text{tors}}$. The elliptic regulator satisfies $R(E/K) > 0$.

§19. THE CONJECTURE OF BIRCH AND SWINNERTON-DYER

The conjecture of Birch and Swinnerton-Dyer relates the L-series of an elliptic curve to many of its other arithmetic invariants. For simplicity, we will restrict ourselves to $K = \mathbb{Q}$. Let $E/\mathbb{Q}$ be an elliptic curve, and let

$$\Omega_\infty = \int_{E(\mathbb{R})} |\omega|,$$

where ω is the invariant differential on a minimal Weierstrass equation. Further, for each prime p, let

$$\Omega_p = \#E(\mathbb{Q}_p)/E_0(\mathbb{Q}_p).$$

(Thus if E has good reduction at p, then $\Omega_p = 1$. It is possible to express the Ω_p's as the values of p-adic integrals, very much analogous to the archimedean integral defining Ω_∞.)

Conjecture of Birch and Swinnerton-Dyer. *Let $E/\mathbb{Q}$ be an elliptic curve.*

(a) $\operatorname{ord}_{s=1} L(E/\mathbb{Q}, s) = \operatorname{rank} E(\mathbb{Q})$.

(b) *Let $r = \operatorname{rank} E(\mathbb{Q})$. Then*

$$\lim_{s\to 1} \frac{L(E/\mathbb{Q}, s)}{(s-1)^r} = \Omega_\infty \prod_p \Omega_p \cdot \frac{R(E/\mathbb{Q}) \cdot \#\text{Ш}(E/\mathbb{Q})}{(\#E(\mathbb{Q})_{\text{tors}})^2}.$$

§20. COMPLEX MULTIPLICATION

Recall that an elliptic curve E is said to have *complex multiplication* if its endomorphism ring $\operatorname{End}(E)$ is strictly larger than $\mathbb{Z}$. If this happens, then the algebra $K = \operatorname{End}(E) \otimes \mathbb{Q}$ is a quadratic imaginary field and $R = \operatorname{End}(E)$ is an order in K. Fix a Weierstrass equation for E of the form

$$E : y^2 = x^3 + Ax + B \qquad \text{with discriminant } \Delta = -16(4A^3 + 27B^2) \neq 0,$$

and define the *Weber function on E* to be the function

$$\phi_E(P) = \begin{cases} (AB/\Delta)x(P) & \text{if } j(E) \neq 0, 1728, \\ (A^2/\Delta)x(P)^2 & \text{if } j(E) = 1728, \\ (B/\Delta)x(P)^3 & \text{if } j(E) = 0. \end{cases}$$

(One can check that ϕ_E does not depend on the choice of Weierstrass equation.)

Theorem. *With notation as above, suppose that R is the full ring of integers of K.*
(a) *The j-invariant $j(E)$ is an algebraic integer.*
(b) *The field $H = K\big(j(E)\big)$ is the Hilbert class field of K (i.e., H is the maximal abelian unramified extension of K).*
(c) *The field $H\big(\{\phi_E(T) : T \in E_{\mathrm{tors}}\}\big)$ is the maximal abelian extension K^{ab} of K.*

It is possible to describe the action of $G_{K^{\mathrm{ab}}/K}$ on the numbers $\phi_E(T)$ via the Artin map, although this is most efficiently done using an adelic formulation. We will be content to describe the action on $j(E)$. For each prime ideal $\mathfrak{p}$ of K, let $F_{\mathfrak{p}} \in G_{H/K}$ be the Frobenius element corresponding to $\mathfrak{p}$. Further, choose a lattice $L \subset \mathbb{C}$ so that there is a analytic isomorphism $\mathbb{C}/L \cong E(\mathbb{C})$, and define a new elliptic curve $\mathfrak{p} * E$ to be the elliptic curve corresponding to the lattice $\mathfrak{p}^{-1}L$. Then the action of $G_{H/K}$ on $j(E)$ is determined by the relation

$$j(E)^{F_{\mathfrak{p}}} = j(\mathfrak{p} * E).$$

Associated to an elliptic curve E/F with complex multiplication by the full ring of integers of K is a Grössencharacter

$$\psi_{E/F} : \mathbf{A}_F^* \longrightarrow K^*$$

roughly determined by the condition that for each prime $\mathfrak{P}$ of F, the map

$$[\psi_{E/F}(\mathfrak{P})] : E \longrightarrow E$$

reduces modulo $\mathfrak{P}$ to the $\mathfrak{P}$-Frobenius map on the reduced curve $\tilde{E}$. We also recall that to any Grössencharacter $\psi : \mathbf{A}_F^* \to K^*$ is attached the *Hecke L-series*

$$L(s, \psi) = \prod_{\mathfrak{P}} (1 - \psi(\mathfrak{P})\,\mathrm{N}\,\mathfrak{P}^{-s})^{-1},$$

which has an analytic continuation to all of $\mathbb{C}$ and satisfies a functional equation.

Theorem. (Deuring) *Let E/F be an elliptic curve with complex multiplication by the full ring of integers of K.*
(a) *If $K \subset F$, then*

$$L(E/F, s) = L(s, \psi_{E/F})L(s, \overline{\psi_{E/F}}).$$

(b) *If $K \not\subset F$, let $F' = FK$. Then*

$$L(E/F, s) = L(s, \psi_{E/F'}).$$

§21. INTEGRAL POINTS

Let K be a number field, let S be a finite set of places of K including all archimedean places, and let R_S be the ring of S-integers of K. Let

$$E : y^2 + a_1xy + a_3y = x^3 + a_2x^2 + a_4x + a_6$$

be a Weierstrass equation for E/K with integral coefficients

$$a_1, a_2, a_3, a_4, a_6 \in R_S,$$

and consider the set of S-integral points on E,

$$E(R_S) = \{P = (x, y) \in E(K) \, : \, x, y \in R_S\}.$$

More generally, we can look at S-integral points relative to an arbitrary coordinate function on E. A fundamental theorem of Siegel says that such sets are finite.

Theorem. (Siegel) *For any non-constant function $f \in K(E)$, the set of S-integral points of E relative to f,*

$$E_f(R_S) = \{P \in E(K) \, : \, f(P) \in R_S\},$$

is a finite set.

Siegel actually proves a more precise statement. To avoid introducing too much notation, we will only describe it for $K = \mathbb{Q}$ and $R_S = \mathbb{Z}$.

Theorem. (Siegel) *Let $E/\mathbb{Q}$ be an elliptic curve, let $f \in \mathbb{Q}(E)$ be a non-constant function, and for each point $P \in E(\mathbb{Q})$, write*

$$f(P) = a_P/b_P \qquad \text{with } a_P, b_P \in \mathbb{Z} \text{ and } \gcd(a_P, b_P) = 1.$$

(If $f(P) = \infty$, set $a_P = 1$ and $b_P = 0$.) Then

$$\lim_{\substack{P \in E(\mathbb{Q}) \\ h(f(P)) \to \infty}} \frac{\log |a_P|}{\log |b_P|} = 1.$$

Siegel's theorems use methods from the theory of Diophantine approximation and are not effective. Baker used his results on linear forms in logarithms to give effective bounds for the size of integral points on elliptic curves. These bounds have been improved over the years, but are still quite large.

Shafarevich used the finiteness of S-integral points on the curve $y^2 = x^3 + D$ to prove the following finiteness theorem for elliptic curves with prescribed bad reduction.

Theorem. *Fix a finite set S of primes of K. Then there are only finitely many K-isomorphism classes of elliptic curves E/K which have good reduction at all primes not in S.*

Faltings subsequently proved that the same result is true for curves of any fixed genus and for abelian varieties of any fixed dimension.

ACKNOWLEDGMENTS. I would like to thank Rob Gross, Alice Silverberg, John Tate, and Rob Tubbs for pointing out some inaccuracies in the original draft of these notes.

REFERENCES

1. J.W.S. Cassels, *Lectures on Elliptic Curves*, Student Texts 24, London Mathematical Society, Cambridge University Press, 1991.
2. D. Husemöller, *Elliptic Curves*, Springer-Verlag, 1987.
3. A. Knapp, *Elliptic Curves*, Math. Notes 40, Princeton University Press, 1992.
4. N. Koblitz, *Introduction to Elliptic Curves and Modular Forms*, Springer-Verlag, 1984.
5. S. Lang, *Elliptic Curves: Diophantine Analysis*, Springer-Verlag, 1978.
6. ———, *Elliptic Functions*, Graduate Texts in Math., vol. 112, 2nd edition, Springer-Verlag, New York, 1987.
7. A. Silverberg, *Galois representations attached to points on Shimura varieties*, Séminaire de Théorie des Nombres (Sinnou David, ed.), Paris 1990–91, Progress in Math. 108, Birkhäuser, 1993, pp. 221–240.
8. J.H. Silverman, *The Arithmetic of Elliptic Curves*, Graduate Texts in Math., vol. 106, Springer-Verlag, Berlin and New York, 1986.
9. ———, *Advanced Topics in the Arithmetic of Elliptic Curves*, Graduate Texts in Math., vol. 151, Springer-Verlag, Berlin and New York, 1994.
10. J. Tate, *The arithmetic of elliptic curves*, Inventiones Math. **23** (1974), 179–206.

MODULAR CURVES, HECKE CORRESPONDENCES, AND L-FUNCTIONS

DAVID E. ROHRLICH

In memory of my father
George F. Rohrlich
January 6, 1914 – August 21, 1995

These notes on Eichler-Shimura theory are intended for a reader who is familiar with elliptic curves and perhaps slightly acquainted with modular forms. The primary sources are [8], [19], and [20]. I am deeply indebted to Jaap Top for taking my place at the conference on very short notice and to Glenn Stevens for making the necessary arrangements with tact and understanding. I am also grateful to both of them for a careful reading of the text and for several comments which improved the final version.

1. MODULAR CURVES

Throughout, the term "curve" will mean "absolutely irreducible variety of dimension one." If $\mathbb{k}$ is a field, then $\mathbb{k}(t)$ denotes the field of rational functions over $\mathbb{k}$.

1.1. The modular curve $X_0(N)$. Let N be a positive integer. The modular curve $X_0(N)$ may be defined as follows. First choose an elliptic curve E over $\mathbb{Q}(t)$ such that $j(E) = t$. Then choose a point of order N on E and let C be the cyclic group which it generates. The subfield of $\overline{\mathbb{Q}(t)}$ fixed by the group

$$\{\sigma \in \mathrm{Gal}(\overline{\mathbb{Q}(t)}/\mathbb{Q}(t)) : \sigma(C) = C\}$$

is a finite extension K of $\mathbb{Q}(t)$, and it turns out that K contains no proper algebraic extension of $\mathbb{Q}$: in other words, if we think of $\overline{\mathbb{Q}}$ as the algebraic closure of $\mathbb{Q}$ *inside* an algebraic closure of K, then $\overline{\mathbb{Q}} \cap K = \mathbb{Q}$. It follows that K is the function field of a smooth projective curve over $\mathbb{Q}$; this is $X_0(N)$.

The simplest nonvacuous example is the case $N = 2$. Let us choose E to be the curve

$$y^2 = 4x^3 - \frac{27t}{t-1728}x - \frac{27t}{t-1728},$$

Partially supported by NSF grant DMS-9396090

so that K is the extension of $\mathbb{Q}(t)$ generated by a root of the equation

$$x^3 - \frac{27t}{4(t-1728)}x - \frac{27t}{4(t-1728)} = 0.$$

Viewed as a cubic in x, the left-hand side is an Eisenstein polynomial at the place $t = 0$ of $\mathbb{Q}(t)$ with discriminant $2^2 3^{12} t^2/(t-1728)^3 \notin \mathbb{Q}(t)^{\times 2}$. Therefore K is a nonnormal cubic extension of $\mathbb{Q}(t)$. We also see that the place $t = 0$ is totally ramified in K, while the places $t = 1728$ and $t = \infty$ each split into two places, one ramified of degree 2 and the other unramified. A calculation using the Hurwitz genus formula then shows that the genus of $X_0(2)$ is 0. By itself, this says little, because over $\mathbb{Q}$ there are infinitely many mutually nonisomorphic smooth projective curves of genus 0. However, it is easy to see that $X_0(2)$ has a rational point: for example, observe that at either place of K above $t = \infty$, the residue class field is $\mathbb{Q}$. It follows that $X_0(2)$ is isomorphic to $\mathbf{P}^1$ over $\mathbb{Q}$.

Returning to the general case, we must still verify that $\overline{\mathbb{Q}} \cap K = \mathbb{Q}$ and that up to isomorphism K is independent of the choice of E and C. The verification will ultimately lead us to modular functions.

We begin with some notation and conventions. Let $\mathbb{k}$ be a field of characteristic not dividing N. Given a Galois extension $\mathbb{k}'$ of $\mathbb{k}$ containing the group μ_N of N-th roots of unity, we shall write $\kappa : \mathrm{Gal}(\mathbb{k}'/\mathbb{k}) \to (\mathbb{Z}/N\mathbb{Z})^\times$ for the character giving the action of $\mathrm{Gal}(\mathbb{k}'/\mathbb{k})$ on μ_N:

$$\sigma(\zeta) = \zeta^{\kappa(\sigma)} \qquad (\sigma \in \mathrm{Gal}(\mathbb{k}'/\mathbb{k}), \zeta \in \mu_N).$$

Suppose now that E is an elliptic curve over $\mathbb{k}$. Let $E[N] \subset E(\overline{\mathbb{k}})$ denote the subgroup of points of order dividing N, and write $\mathbb{k}(E[N])$ for the finite Galois extension of $\mathbb{k}$ generated by the coordinates relative to some generalized Weierstrass equation for E over $\mathbb{k}$ of the affine points on E of order dividing N. After fixing an ordered basis for $E[N]$ over $\mathbb{Z}/N\mathbb{Z}$, we may identify the natural embedding of $\mathrm{Gal}(\mathbb{k}(E[N])/\mathbb{k})$ in $\mathrm{Aut}(E[N])$ with a faithful representation

$$\rho : \mathrm{Gal}(\mathbb{k}(E[N])/\mathbb{k}) \hookrightarrow \mathrm{GL}(2, \mathbb{Z}/N\mathbb{Z}).$$

The formalism of the Weil pairing shows that $\mathbb{k}(E[N])$ contains μ_N and that the determinant of ρ is κ. In particular, if $\mathbb{k}$ itself contains μ_N, then κ is trivial and ρ is a representation

$$\mathrm{Gal}(\mathbb{k}(E[N])/\mathbb{k}) \hookrightarrow \mathrm{SL}(2, \mathbb{Z}/N\mathbb{Z}).$$

Theorem 1. *If E is an elliptic curve over $\mathbb{C}(t)$ with $j(E) = t$, then the representation on N-division points is an isomorphism*

$$\mathrm{Gal}(\mathbb{C}(t, E[N])/\mathbb{C}(t)) \cong \mathrm{SL}(2, \mathbb{Z}/N\mathbb{Z}).$$

Theorem 1 will be proved later. For now we derive consequences. The first consequence is that if E is an elliptic curve over $\mathbb{Q}(\mu_N)(t)$ with invariant t, then it is still true that the representation on N-division points is an isomorphism

$$\mathrm{Gal}(\mathbb{Q}(t, E[N])/\mathbb{Q}(\mu_N)(t)) \cong \mathrm{SL}(2, \mathbb{Z}/N\mathbb{Z}).$$

For we know that the left-hand side is *embedded* in the right-hand side, but on field-theoretic grounds we also have

$$[\mathbb{Q}(t, E[N]) : \mathbb{Q}(\mu_N)(t)] \geq [\mathbb{C}(t, E[N]) : \mathbb{C}(t)].$$

Hence the conclusion follows from Theorem 1. Next suppose that E is an elliptic curve over $\mathbb{Q}(t)$ with invariant t. Then E can be viewed as an elliptic curve over $\mathbb{Q}(\mu_N)(t)$, and consequently the representation on N-division points

$$\rho : \mathrm{Gal}(\mathbb{Q}(t, E[N])/\mathbb{Q}(t)) \hookrightarrow \mathrm{GL}(2, \mathbb{Z}/N\mathbb{Z})$$

sends $\mathrm{Gal}(\mathbb{Q}(t, E[N])/\mathbb{Q}(\mu_N)(t))$ onto $\mathrm{SL}(2, \mathbb{Z}/N\mathbb{Z})$. Since $\det \rho = \kappa$, the image of ρ also contains a set of coset representatives for $\mathrm{SL}(2, \mathbb{Z}/N\mathbb{Z})$ in $\mathrm{GL}(2, \mathbb{Z}/N\mathbb{Z})$. Therefore the image of ρ is all of $\mathrm{GL}(2, \mathbb{Z}/N\mathbb{Z})$, and we obtain the first part of the following assertion:

Corollary. *If E is an elliptic curve over $\mathbb{Q}(t)$ with invariant t, then the representation on N-division points is an isomorphism*

$$\mathrm{Gal}(\mathbb{Q}(t, E[N])/\mathbb{Q}(t)) \cong \mathrm{GL}(2, \mathbb{Z}/N\mathbb{Z}),$$

and $\overline{\mathbb{Q}} \cap \mathbb{Q}(t, E[N]) = \mathbb{Q}(\mu_N)$.

The equation $\overline{\mathbb{Q}} \cap \mathbb{Q}(t, E[N]) = \mathbb{Q}(\mu_N)$ also follows from the argument just given. Indeed, put $L = \overline{\mathbb{Q}} \cap \mathbb{Q}(t, E[N])$, and suppose that L is strictly larger than $\mathbb{Q}(\mu_N)$. Then $L(t)$ is strictly larger than $\mathbb{Q}(\mu_N)(t)$, and consequently

$$[\mathbb{Q}(t, E[N]) : L(t)] < |\mathrm{SL}(2, \mathbb{Z}/N\mathbb{Z})|.$$

But as before, the left-hand side of this inequality is *a priori* greater than or equal to $[\mathbb{C}(t, E[N]) : \mathbb{C}(t)]$, and we have a contradiction.

We can now clarify some points in the definition of $X_0(N)$. Recall that the recipe for the function field K of $X_0(N)$ was as follows: Choose an elliptic curve E over $\mathbb{Q}(t)$ with $j(E) = t$, and then choose a cyclic subgroup C of $E[N]$; write G for the Galois group of $\mathbb{Q}(t, E[N])$ over $\mathbb{Q}(t)$ and H for the subgroup of G preserving C; then K is the fixed field of H. Now any point of order N on E is the second basis vector in some ordered basis for

$E[N]$ over $\mathbb{Z}/N\mathbb{Z}$, and consequently we may identify G with $\mathrm{GL}(2, \mathbb{Z}/N\mathbb{Z})$ in such a way that H is identified with the lower triangular group

$$\left\{ \begin{pmatrix} a & 0 \\ b & d \end{pmatrix} : a, d \in (\mathbb{Z}/N\mathbb{Z})^\times, b \in \mathbb{Z}/N\mathbb{Z} \right\}.$$

Since the determinant maps the lower triangular group *onto* $(\mathbb{Z}/N\mathbb{Z})^\times$, we have $\mathbb{Q}(\mu_N) \cap K = \mathbb{Q}$. Substituting $\mathbb{Q}(\mu_N) = \overline{\mathbb{Q}} \cap \mathbb{Q}(t, E[N])$ in this equation, we obtain $\overline{\mathbb{Q}} \cap K = \mathbb{Q}$, as claimed earlier. We also see that up to isomorphism over $\mathbb{Q}(t)$, the field K is independent of the choice of C: for if we change the basis of $E[N]$ over $\mathbb{Z}/N\mathbb{Z}$, then we conjugate H inside G, and hence we conjugate (in the field-theoretic sense) K inside $\mathbb{Q}(t, E[N])$. It remains to examine the dependence of K on E. Quite generally, if E is an elliptic curve over a field $\mathbb{k}$ of characteristic not dividing N, let $\mathbb{k}(E[N]/\pm)$ denote the extension of $\mathbb{k}$ generated by the x-coordinates of the affine points on E of order dividing N. Then $\mathbb{k}(E[N]/\pm)$ is the fixed field of

$$\{\sigma \in \mathrm{Gal}(\mathbb{k}(E[N])/\mathbb{k}) : \sigma(P) = \pm P \text{ for every } P \in E[N]\}.$$

Returning to the case at hand, we can say that $\mathbb{Q}(t, E[N]/\pm)$ is the subfield of $\mathbb{Q}(t, E[N])$ corresponding to the subgroup $\{\pm I\}$ of $\mathrm{GL}(2, \mathbb{Z}/N\mathbb{Z})$; thus

$$\mathrm{Gal}(\mathbb{Q}(t, E[N]/\pm)/\mathbb{Q}(t)) \cong \mathrm{GL}(2, \mathbb{Z}/N\mathbb{Z})/\{\pm I\}.$$

Suppose now that E' is another elliptic curve over $\mathbb{Q}(t)$ with $j(E') = t$. Then E and E' differ by a quadratic twist, and consequently so do the associated representations

$$\mathrm{Gal}(\overline{\mathbb{Q}(t)}/\mathbb{Q}(t)) \longrightarrow \mathrm{GL}(2, \mathbb{Z}/N\mathbb{Z})$$

provided that bases for $E[N]$ and $E'[N]$ are chosen compatibly. It follows that the fields $\mathbb{Q}(t, E[N]/\pm)$ and $\mathbb{Q}(t, E'[N]/\pm)$ are equal and that the associated isomorphisms from

$$\mathrm{Gal}(\mathbb{Q}(t, E[N]/\pm)/\mathbb{Q}(t)) \quad \text{and} \quad \mathrm{Gal}(\mathbb{Q}(t, E'[N]/\pm)/\mathbb{Q}(t))$$

to $\mathrm{GL}(2, \mathbb{Z}/N\mathbb{Z})/\{\pm I\}$ are identical. Now the lower triangular subgroup of $\mathrm{GL}(2, \mathbb{Z}/N\mathbb{Z})$ contains $\{\pm I\}$, so that K is a subfield of $\mathbb{Q}(t, E[N]/\pm)$. Hence K can be characterized as the subfield of $\mathbb{Q}(t, E[N]/\pm)$ fixed by the image of the lower triangular subgroup in $\mathrm{GL}(2, \mathbb{Z}/N\mathbb{Z})/\{\pm I\}$, and this characterization is independent of the choice of E.

1.2. Other modular curves. More generally, let H be any subgroup of $\mathrm{GL}(2, \mathbb{Z}/N\mathbb{Z})$ satisfying two conditions:

(i) $-I \in H$.
(ii) The determinant $H \longrightarrow (\mathbb{Z}/N\mathbb{Z})^\times$ is surjective.

If K is the subfield of $\mathbb{Q}(t, E[N])$ fixed by H, then the same argument shows that K is the function field of a smooth projective curve $X(H)$ over $\mathbb{Q}$ which up to isomorphism is independent of the choice of E. As we have just seen, $X_0(N)$ corresponds to the choice

$$H = \left\{ \begin{pmatrix} a & 0 \\ b & d \end{pmatrix} : a, d \in (\mathbb{Z}/N\mathbb{Z})^\times,\ \ b \in \mathbb{Z}/N\mathbb{Z} \right\};$$

another example is the curve $X_1(N)$ corresponding to

$$H = \left\{ \begin{pmatrix} a & 0 \\ b & \pm 1 \end{pmatrix} : a \in (\mathbb{Z}/N\mathbb{Z})^\times,\ \ b \in \mathbb{Z}/N\mathbb{Z} \right\}.$$

These two examples will be the primary focus throughout, and while we shall initially emphasize $X_0(N)$, it will ultimately be $X_1(N)$ which provides the broader context for the application to L-functions. We note in passing that if P is a point of order N on E and C is the subgroup which it generates, then the function field of $X_1(N)$ can be identified with the subfield of $\overline{\mathbb{Q}(t)}$ fixed by $\{\sigma \in \mathrm{Gal}(\overline{\mathbb{Q}(t)}/\mathbb{Q}(t)) : \sigma(P) = \pm P\}$, just as the function field of $X_0(N)$ coincides with the subfield of $\overline{\mathbb{Q}(t)}$ fixed by $\{\sigma \in \mathrm{Gal}(\overline{\mathbb{Q}(t)}/\mathbb{Q}(t)) : \sigma(C) = C\}$.

By construction, a modular curve $X(H)$ with function field K comes equipped with a distinguished morphism to $\mathbf{P}^1$ over $\mathbb{Q}$, namely the morphism corresponding to the inclusion of $\mathbb{Q}(t)$ in K. It is conventional to refer to the finite set of points on $X(H)$ lying over the point $t = \infty$ of $\mathbf{P}^1$ as *cusps*. If we remove the cusps from $X(H)$, then we obtain an affine curve $Y(H)$, which is usually denoted $Y_0(N)$ in the case of $X_0(N)$ and $Y_1(N)$ in the case of $X_1(N)$. To illustrate the notion of a cusp, let us prove that $X_0(N)$ has at least one cusp rational over $\mathbb{Q}$. Choose an elliptic curve E over $\mathbb{Q}(t)$ with invariant t and split multiplicative reduction at the place $t = \infty$ of $\mathbb{Q}(t)$. For example, E could be the curve

$$y^2 + xy = x^3 - \frac{36}{t - 1728}x - \frac{1}{t - 1728},$$

because the covariants c_4 and c_6 of this equation satisfy $-c_4/c_6 = 1$. Denote the place $t = \infty$ of $\mathbb{Q}(t)$ simply by ∞, identify the completion of $\mathbb{Q}(t)$ at ∞ with $\mathbb{Q}((1/t))$, and pick an extension of ∞ to a place of $\overline{\mathbb{Q}(t)}$, so that the corresponding decomposition subgroup $\mathrm{Gal}(\overline{\mathbb{Q}(t)}/\mathbb{Q}(t))_\infty$ of $\mathrm{Gal}(\overline{\mathbb{Q}(t)}/\mathbb{Q}(t))$ is identified with the Galois group of $\overline{\mathbb{Q}((1/t))}$ over $\mathbb{Q}((1/t))$. By the theory of Tate curves, there is an isomorphism of $\mathrm{Gal}(\overline{\mathbb{Q}(t)}/\mathbb{Q}(t))_\infty$-modules

$$E(\overline{\mathbb{Q}((1/t))}) \cong \overline{\mathbb{Q}((1/t))}^\times / q^{\mathbb{Z}},$$

where $q \in \mathbb{Q}((1/t))$ is a uniformizer and $q^{\mathbb{Z}}$ denotes the infinite cyclic group generated by q. It follows in particular that

$$E[N] \cong \mu_N \oplus (q^{1/N})^{\mathbb{Z}}/q^{\mathbb{Z}}$$

as $\mathrm{Gal}(\overline{\mathbb{Q}(t)}/\mathbb{Q}(t))_\infty$-modules. Let $\{P_1, P_2\}$ be the basis for $E[N]$ which corresponds under the preceding isomorphism to the basis $\{q^{1/N} \bmod q^{\mathbb{Z}}, \zeta\}$, where ζ is a generator of μ_N. Relative to this basis, the action of the group $\mathrm{Gal}(\overline{\mathbb{Q}(t)}/\mathbb{Q}(t))_\infty$ on $E[N]$ is represented by matrices of the form

$$\begin{pmatrix} 1 & 0 \\ * & \kappa(g) \end{pmatrix} \qquad (g \in \mathrm{Gal}(\overline{\mathbb{Q}(t)}/\mathbb{Q}(t))_\infty).$$

In particular, the cyclic group C generated by P_2 is preserved by the decomposition group $\mathrm{Gal}(\overline{\mathbb{Q}(t)}/\mathbb{Q}(t))_\infty$, whence the latter is contained in the subgroup of $\mathrm{Gal}(\overline{\mathbb{Q}(t)}/\mathbb{Q}(t))$ fixing the function field of $X_0(N)$. Thus the residue class degree of the restriction of ∞ to this function field is 1, and so the field of rationality of the corresponding cusp is $\mathbb{Q}$.

1.3. Moduli interpretation of $X_0(N)$ and $X_1(N)$. Let $\Bbbk$ be an algebraically closed field, and consider pairs $(\mathcal{E}, \mathcal{C})$ consisting of an elliptic curve $\mathcal{E}$ over $\Bbbk$ and a cyclic subgroup $\mathcal{C} \subset \mathcal{E}[N]$ of order N. An isomorphism from a pair $(\mathcal{E}_1, \mathcal{C}_1)$ to a pair $(\mathcal{E}_2, \mathcal{C}_2)$ is an isomorphism from $\mathcal{E}_1$ to $\mathcal{E}_2$ sending $\mathcal{C}_1$ to $\mathcal{C}_2$. We write $[\mathcal{E}, \mathcal{C}]$ for the isomorphism class containing $(\mathcal{E}, \mathcal{C})$ and $\mathrm{Ell}_0(N)(\Bbbk)$ for the set of all isomorphism classes. Also, if S is any subset of $\mathbf{P}^1(\Bbbk)$, then $\mathrm{Ell}_0(N)(\Bbbk)_S$ denotes the set of all isomorphism classes $[\mathcal{E}, \mathcal{C}] \in \mathrm{Ell}_0(N)(\mathbb{C})$ such that $j(\mathcal{E}) \notin S$.

One can also consider pairs of the form $(\mathcal{E}, \mathcal{P})$, where $\mathcal{P}$ is a point of order N on $\mathcal{E}$. An isomorphism from $(\mathcal{E}_1, \mathcal{P}_1)$ to $(\mathcal{E}_2, \mathcal{P}_2)$ is an isomorphism from $\mathcal{E}_1$ to $\mathcal{E}_2$ sending $\mathcal{P}_1$ to $\mathcal{P}_2$. Note the equality of isomorphism classes $[\mathcal{E}, \mathcal{P}] = [\mathcal{E}, -\mathcal{P}]$. We write $\mathrm{Ell}_1(N)(\Bbbk)$ for the set of isomorphism classes and $\mathrm{Ell}_1(N)(\Bbbk)_S$ for the subset consisting of those $[\mathcal{E}, \mathcal{P}]$ such that $j(\mathcal{E}) \notin S$.

Next observe that if X is a modular curve and E is an elliptic curve over $\mathbb{Q}(t)$ with invariant t, then E can be viewed as an elliptic curve over the function field of X, because the latter field is naturally an extension of $\mathbb{Q}(t)$. In particular, since a point $x \in X(\mathbb{C})$ determines a discrete valuation ring $\mathcal{O}_x$ of the complex function field of X, one can ask whether E has good reduction at the maximal ideal $\mathfrak{m}_x$ of $\mathcal{O}_x$: if so, then reduction modulo $\mathfrak{m}_x$ yields an elliptic curve E_x over $\mathbb{C}$. After extending $\mathcal{O}_x$ to a discrete valuation ring of $\mathbb{C}(t, E[N])$ in some way, we can also consider the reduction map on N-division points $E[N] \longrightarrow E_x[N]$, which is injective. Thus if P is a point of order N on E and C is the cyclic subgroup which it generates, then the reductions P_x and C_x are defined, and are still of order N.

Given a subset S of $\mathbf{P}^1(\mathbb{C})$, let $\mathbf{P}^1(\mathbb{C})_S$ denote the complement of S in $\mathbf{P}^1(\mathbb{C})$ and $X(\mathbb{C})_S$ the inverse image of $\mathbf{P}^1(\mathbb{C})_S$ in $X(\mathbb{C})$. For example, if $S = \{\infty\}$, then $X_0(N)(\mathbb{C})_S = Y_0(N)(\mathbb{C})$.

Proposition 1. *Let E be an elliptic curve over $\mathbb{Q}(t)$ with $j(E) = t$, and let S be a subset of $\mathbf{P}^1(\mathbb{C})$ containing all places where E has bad reduction. Fix an ordered basis for $E[N]$ over $\mathbb{Z}/N\mathbb{Z}$, let P be the second element of this basis, and let C be the cyclic group of order N generated by P. Then the* *map* $x \mapsto [E_x, C_x]$ *defines a bijection of* $X_0(N)(\mathbb{C})_S$ *onto* $\mathrm{Ell}_0(N)(\mathbb{C})_S$. *The same is true if* $X_0(N)$, $\mathrm{Ell}_0(N)$, *and* C *are replaced by* $X_1(N)$, $\mathrm{Ell}_1(N)$, *and* P *respectively.*

Proof. First of all, S necessarily contains 0, 1728, and ∞. Indeed E differs from the curve

$$y^2 + xy = x^3 - \frac{36}{t-1728}x - \frac{1}{t-1728}$$

by a quadratic twist over $\mathbb{Q}(t)$, and therefore the discriminant of any equation for E differs from the discriminant of the above equation by a sixth power in $\mathbb{Q}(t)^\times$. Since the above equation has discriminant $t^2/(t-1728)^3$, it follows that E has bad reduction at 0, 1728, and ∞.

As usual, we identify $\mathrm{Gal}(\mathbb{Q}(t, E[N])/\mathbb{Q}(t))$ with $\mathrm{GL}(2, \mathbb{Z}/N\mathbb{Z})$, and similarly $\mathrm{Gal}(\mathbb{C}(t, E[N])/\mathbb{C}(t))$ with $\mathrm{SL}(2, \mathbb{Z}/N\mathbb{Z})$. Let $K \subset \mathbb{Q}(t, E[N])$ be the subfield corresponding to

$$H = \left\{ \begin{pmatrix} a & 0 \\ b & d \end{pmatrix} : a, d \in (\mathbb{Z}/N\mathbb{Z})^\times, b \in \mathbb{Z}/N\mathbb{Z} \right\},$$

so that $\mathbb{C}K$ is the function field of $X_0(N)$ over $\mathbb{C}$. Also let X be a curve with complex function field $\mathbb{C}(t, E[N])$ and $\pi : X \longrightarrow X_0(N)$ the corresponding morphism. Given a point $t_0 \in \mathbf{P}^1(\mathbb{C})_S$, a point $x_0 \in X_0(N)(\mathbb{C})$ lying over t_0, and a point $\hat{x}_0 \in X(\mathbb{C})$ lying over x_0, we have a bijection

$$(1) \qquad \begin{aligned} \mathrm{SL}(2, \mathbb{Z}/N\mathbb{Z})/(H \cap \mathrm{SL}(2, \mathbb{Z}/N\mathbb{Z})) &\longrightarrow \{\text{fiber of } X_0(N)(\mathbb{C}) \text{ over } t_0\} \\ g(H \cap \mathrm{SL}(2, \mathbb{Z}/N\mathbb{Z})) &\longmapsto \pi(g\hat{x}_0), \end{aligned}$$

because the extension $\mathbb{C}(t, E[N])/\mathbb{C}(t)$ is unramified outside the places of $\mathbb{C}(t)$ where E has bad reduction, and hence in particular outside S. On the other hand, the maps

$$(2) \qquad \begin{aligned} \mathrm{SL}(2, \mathbb{Z}/N\mathbb{Z})/(H \cap \mathrm{SL}(2, \mathbb{Z}/N\mathbb{Z})) &\longrightarrow \mathrm{GL}(2, \mathbb{Z}/N\mathbb{Z})/H \\ g(H \cap \mathrm{SL}(2, \mathbb{Z}/N\mathbb{Z})) &\longmapsto gH \end{aligned}$$

and

$$(3) \qquad \begin{aligned} \mathrm{GL}(2, \mathbb{Z}/N\mathbb{Z})/H &\longrightarrow \{\text{cyclic subgroups of } E \text{ of order } N\} \\ gH &\longmapsto gC \end{aligned}$$

are also bijections, as is the map

$$(4) \qquad \{\text{cyclic subgroups of } E \text{ of order } N\} \longrightarrow \{\text{cyclic subgroups of } E_{x_0} \text{ of order } N\}$$

afforded by reduction modulo $\mathfrak{m}_{x_0}$. Now if $x = \pi(g\hat{x}_0)$, then $E_x = E_{x_0}$ and $(gC)_{x_0} = C_x$ (note that these identifications are meaningful since the residue class field $\mathbb{C}$ is a subring of $\mathcal{O}_{x_0}$ and $\mathcal{O}_x$). Therefore composing the inverse of (1) with (2), (3), and (4), we see that the map

$$\{\text{fiber of } X_0(N)(\mathbb{C}) \text{ over } t_0\} \longrightarrow \{\text{cyclic subgroups of } E_{x_0} \text{ of order } N\}$$
$$x \longmapsto C_x$$

is a bijection. But $j(E_{x_0}) = t_0 \neq 0, 1728$, whence $\mathrm{Aut}(E_{x_0}) = \{\pm 1\}$. Thus an automorphism of E_{x_0} sends each cyclic subgroup of E_{x_0} to itself. Consequently

$$\left\{\begin{matrix}\text{cyclic subgroups of } E_{x_0}\\ \text{of order } N\end{matrix}\right\} \to \{[\mathcal{E}, \mathcal{C}] \in \mathrm{Ell}_0(N)(\mathbb{C})_S : j(\mathcal{E}) = t_0\},$$
$$C \longmapsto [E_{x_0}, C]$$

is also a bijection, and composing this map with the previous one we see that $x \mapsto [E_x, C_x]$ is a bijection from the fiber of $X_0(N)(\mathbb{C})$ over $t_0 \in \mathbf{P}^1(\mathbb{C})_S$ to the subset of $\mathrm{Ell}_0(N)(\mathbb{C})_S$ consisting of isomorphism classes with j-invariant t_0.

The argument for $X_1(N)$ is similar. The only additional wrinkle is that P is not actually defined over the function field of $X_1(N)(\mathbb{C})$, but only over a quadratic extension. Thus P_x depends on a choice of a valuation ring lying over $\mathcal{O}_x$. However, if we make the alternate choice then we simply replace P_x by $-P_x$. Since $[E_x, P_x] = [E_x, -P_x]$, the map $x \mapsto [E_x, P_x]$ is still well defined.

1.4. Proof of Theorem 1: a preliminary reduction. We claim that to prove Theorem 1 it suffices to prove the following:

Proposition 2. *There exists an elliptic curve E over $\mathbb{C}(t)$ with invariant t such that*

$$[\mathbb{C}(t, E[N]/\pm) : \mathbb{C}(t)] = |\mathrm{SL}(2, \mathbb{Z}/N\mathbb{Z})/\{\pm I\}|.$$

Granting Proposition 2, let us see how to deduce Theorem 1. Suppose that E' is any elliptic curve over $\mathbb{Q}(t)$ with invariant t. Then E' differs from E by at most a quadratic twist, so that $\mathbb{C}(t, E[N]/\pm) = \mathbb{C}(t, E'[N]/\pm)$. Consequently

$$[\mathbb{C}(t, E'[N]/\pm) : \mathbb{C}(t)] = [\mathbb{C}(t, E[N]/\pm) : \mathbb{C}(t)] = |\mathrm{SL}(2, \mathbb{Z}/N\mathbb{Z})/\{\pm I\}|,$$

and the embedding $\mathrm{Gal}(\mathbb{C}(t, E'[N]/\pm)/\mathbb{C}(t)) \longrightarrow \mathrm{SL}(2, \mathbb{Z}/N\mathbb{Z})/\{\pm I\}$ is an isomorphism. Hence the image of the representation

$$\mathrm{Gal}(\mathbb{C}(t, E'[N])/\mathbb{C}(t)) \longrightarrow \mathrm{SL}(2, \mathbb{Z}/N\mathbb{Z})$$

contains a set of coset representatives for $\{\pm I\}$ in $\mathrm{SL}(2, \mathbb{Z}/N\mathbb{Z})$ and so in particular contains either the matrix

$$\begin{pmatrix} 0 & -1 \\ 1 & 0 \end{pmatrix}$$

or its negative. Since the square of this matrix is $-I$, we find that the image is all of $\mathrm{SL}(2, \mathbb{Z}/N\mathbb{Z})$, and Theorem 1 follows.

1.5. Modular functions. It remains to prove Proposition 2. The argument will use the theory of modular functions. We begin by recalling the relevant definitions.

Let $\mathfrak{H}$ denote the complex upper half-plane, and let $\mathrm{GL}^+(2, \mathbb{R})$ denote the subgroup of $\mathrm{GL}(2, \mathbb{R})$ consisting of matrices with positive determinant. We consider the usual action of $\mathrm{GL}^+(2, \mathbb{R})$ on $\mathfrak{H}$ by fractional linear transformations: for

$$\gamma = \begin{pmatrix} a & b \\ c & d \end{pmatrix} \in \mathrm{GL}^+(2, \mathbb{R})$$

and $z \in \mathfrak{H}$ put

$$\gamma z = \frac{az + b}{cz + d}.$$

If f is a function on $\mathfrak{H}$ we write $f \circ \gamma$ for the function $z \mapsto f(\gamma z)$. Now suppose that Γ is a subgroup of finite index in $\mathrm{SL}(2, \mathbb{Z})$. A modular function for Γ is a meromorphic function f on $\mathfrak{H}$ satisfying two conditions:

(i) $f \circ \gamma = f$ for $\gamma \in \Gamma$.

(ii) Given $\delta \in \mathrm{SL}(2, \mathbb{Z})$, let M be a positive integer such that

$$(f \circ \delta)(z + M) = (f \circ \delta)(z).$$

(Such an integer exists by (i), because Γ has finite index in $\mathrm{SL}(2, \mathbb{Z})$ and consequently some power of the matrix $\delta \left(\begin{smallmatrix} 1 & 1 \\ 0 & 1 \end{smallmatrix}\right) \delta^{-1}$ belongs to Γ.) Let F be the meromorphic function on the punctured unit disk $D^\circ = \{q \in \mathbb{C} : 0 < |q| < 1\}$ defined by

$$f(\delta z) = F(e^{2\pi i z/M}) \qquad (z \in \mathfrak{H}).$$

Then F extends to a meromorphic function on the full unit disk $D = \{q \in \mathbb{C} : |q| < 1\}$.

Like any meromorphic function on the punctured disk, the function F in (ii) is represented by a Laurent series

$$F(q) = \sum_{n \in \mathbb{Z}} a(n) q^n$$

for q near 0, and the content of (ii) is that F does not have an essential singularity at $q = 0$. Thus (ii) is equivalent to the condition that for every $\delta \in \mathrm{SL}(2,\mathbb{Z})$, the function $f \circ \delta$ has a Fourier series expansion of the form

$$f(\delta z) = \sum_{n \geq n_0} a(n) e^{2\pi i n z/M} \qquad (\mathrm{Im}(z) \gg 0)$$

with $n_0 \in \mathbb{Z}$. Of course if $\Gamma = \mathrm{SL}(2,\mathbb{Z})$ then $f \circ \delta = f$ by (i), and to verify (ii) it suffices to check that f itself has such a Fourier expansion.

The usual operations of addition and multiplication of meromorphic functions make the set of modular functions for Γ into a field $\mathfrak{M}(\Gamma)$. For the proof of Proposition 2 we are interested in the cases where Γ is the group $\mathrm{SL}(2,\mathbb{Z})$ (also denoted $\Gamma(1)$) or the group $\Gamma(N)$, the kernel of the reduction-modulo-N map $\mathrm{SL}(2,\mathbb{Z}) \to \mathrm{SL}(2,\mathbb{Z}/N\mathbb{Z})$. The quotient group

$$\Gamma(1)/\{\pm I\}\Gamma(N) \cong \mathrm{SL}(2,\mathbb{Z}/N\mathbb{Z})/\{\pm I\}$$

acts as a group of automorphisms of $\mathfrak{M}(\Gamma(N))$ with fixed field $\mathfrak{M}(\Gamma(1))$, and therefore $\mathfrak{M}(\Gamma(N))$ is a Galois extension of $\mathfrak{M}(\Gamma(1))$ with Galois group isomorphic to $\mathrm{SL}(2,\mathbb{Z}/N\mathbb{Z})/\{\pm I\}$. For the record, we make an explicit identification

$$\theta : \mathrm{SL}(2,\mathbb{Z}/N\mathbb{Z})/\{\pm I\} \longrightarrow \mathrm{Gal}(\mathfrak{M}(\Gamma(N))/\mathfrak{M}(\Gamma(1)))$$

by declaring that if $\gamma \in \mathrm{SL}(2,\mathbb{Z})$ and $f \in \mathfrak{M}(\Gamma(N))$ then

$$\theta([\gamma])(f) = f \circ \gamma^{\mathrm{t}},$$

where $[\gamma]$ denotes the image of γ in $\mathrm{SL}(2,\mathbb{Z}/N\mathbb{Z})/\{\pm I\}$ and γ^{t} denotes the transpose of γ.

An essential point for the proof of Proposition 2 is that $\mathfrak{M}(\Gamma(1))$ is generated over $\mathbb{C}$ by a single nonconstant function, the j-function. Like the functions g_2 and g_3 appearing in the formula

$$j = 1728 \frac{g_2^3}{g_2^3 - 27g_3^2}$$

which defines it, the j-function should be viewed in the first instance as a function of a lattice variable $\mathcal{L} \subset \mathbb{C}$. We obtain functions of $z \in \mathfrak{H}$ by putting $g_2(z) = g_2(\mathcal{L}_z)$, $g_3(z) = g_3(\mathcal{L}_z)$, and $j(z) = j(\mathcal{L}_z)$, where

$$\mathcal{L}_z = z\mathbb{Z} \oplus \mathbb{Z}.$$

Now as functions of lattices, g_2 and g_3 are defined by the formulas

$$g_2(\mathcal{L}) = 60 \sum_{\substack{\omega \in \mathcal{L} \\ \omega \neq 0}} \omega^{-4} \quad \text{and} \quad g_3(\mathcal{L}) = 140 \sum_{\substack{\omega \in \mathcal{L} \\ \omega \neq 0}} \omega^{-6},$$

which give at once the behavior of g_2 and g_3 under homothety:

$$g_2(\lambda\mathcal{L}) = \lambda^{-4}g_2(\mathcal{L}), \quad \text{and} \quad g_3(\lambda\mathcal{L}) = \lambda^{-6}g_3(\mathcal{L}) \quad \text{for all } \lambda \in \mathbb{C}^\times.$$

It follows that as a function of lattices, j is invariant under homothety. Furthermore, if $z \in \mathfrak{H}$ and $\gamma = \begin{pmatrix} a & b \\ c & d \end{pmatrix} \in \mathrm{SL}(2,\mathbb{Z})$, then

$$\mathcal{L}_z = z\mathbb{Z} \oplus \mathbb{Z} = (az+b)\mathbb{Z} \oplus (cz+d)\mathbb{Z} = \lambda\mathcal{L}_{\gamma z}$$

with $\lambda = cz+d$, whence $j(\gamma z) = j(z)$. This is condition (i) in the definition of modular functions; to verify (ii) we write $z = x+iy$ and observe that

$$\lim_{y\to\infty} g_2(x+iy) = 120\sum_{n\geq 1} n^{-4} \quad \text{and} \quad \lim_{y\to\infty} g_3(x+iy) = 280\sum_{n\geq 1} n^{-6}$$

uniformly in x. Thus the holomorphic functions G_2 and G_3 on D° such that $g_2(z) = G_2(e^{2\pi i z})$ and $g_3(z) = G_3(e^{2\pi i z})$ extend holomorphically to D, and consequently the function $J = G_2^3/(G_2^3 - 27G_3^2)$ extends at least meromorphically to D. Hence j is a modular function for $\mathrm{SL}(2,\mathbb{Z})$. Now the calculation

$$(120\sum_{n\geq 1} n^{-4})^3 - 27(280\sum_{n\geq 1} n^{-6})^2 = (120\pi^4/90)^3 - 27(280\pi^6/945)^2 = 0$$

shows that $J(q)$ actually has a pole at $q = 0$, and a more thorough analysis reveals that the pole is simple with residue 1. Therefore j has an expansion of the form

$$j(z) = \frac{1}{q} + \text{power series in } q$$

for $q = e^{2\pi i z}$ near 0. In fact the Fourier expansion of j holds for all q in the unit disk, i.e., for all $z \in \mathfrak{H}$, because j is holomorphic on $\mathfrak{H}$: indeed the properties of the Weierstrass $\wp$-function show that $g_2^3 - 27g_3^2$ is nowhere vanishing as a function of lattices and hence also as a function of $z \in \mathfrak{H}$. From the fact that j is holomorphic on $\mathfrak{H}$ with only a simple pole as a Laurent series in q, one deduces that for any $f \in \mathfrak{M}(\Gamma(1))$ there exist polynomials $P(t), Q(t) \in \mathbb{C}(t)$ with $P(t) \neq 0$ such that $P(j)f - Q(j)$ is holomorphic on $\mathfrak{H}$ and

$$\lim_{y\to\infty} P(j(z))f(z) - Q(j(z)) = 0$$

uniformly in x. An application of the maximum principle on a suitably truncated fundamental domain for $\mathrm{SL}(2,\mathbb{Z})\backslash\mathfrak{H}$ then gives $P(j)f = Q(j)$, whence $\mathfrak{M}(\Gamma(1)) = \mathbb{C}(j)$ as claimed.

1.6. Elliptic functions. To summarize, $\mathfrak{M}(\Gamma(1)) = \mathbb{C}(j)$, and $\mathfrak{M}(\Gamma(N))$ is a Galois extension of $\mathbb{C}(j)$ with Galois group

$$\mathrm{Gal}\big(\mathfrak{M}(\Gamma(N))/\mathfrak{M}(\Gamma(1))\big) \cong \mathrm{SL}(2, \mathbb{Z}/N\mathbb{Z})/\{\pm I\}.$$

Consider the elliptic curve

$$E : y^2 = 4x^3 - \frac{27j}{j - 1728}x - \frac{27j}{j - 1728}$$

over $\mathbb{C}(j)$. We will show that $\mathbb{C}(j, E[N]/\pm)$ coincides with $\mathfrak{M}(\Gamma(N))$ when both fields are viewed inside a fixed algebraic closure of $\mathbb{C}(j)$. Thsi will prove Proposition 2.

The additional ingredient needed at this point is the Weierstrass parametrization of elliptic curves over $\mathbb{C}$. Let $\mathcal{L}$ be a lattice in $\mathbb{C}$ and consider the elliptic curve

$$\mathcal{E}^{\mathrm{Wst}} : Y^2 = 4X^3 - g_2(\mathcal{L})X - g_3(\mathcal{L}).$$

We recall that the Weierstrass $\wp$-function

$$\wp(u; \mathcal{L}) = \frac{1}{u^2} + \sum_{\substack{\omega \in \mathcal{L} \\ \omega \neq 0}} \frac{1}{(u + \omega)^2} - \frac{1}{\omega^2}$$

affords a complex analytic group isomorphism

$$\begin{aligned} \mathbb{C}/\mathcal{L} &\longrightarrow \mathcal{E}^{\mathrm{Wst}}(\mathbb{C}) \\ u + \mathcal{L} &\longmapsto (\wp(u; \mathcal{L}), \wp'(u; \mathcal{L})), \end{aligned}$$

where $(\wp(u; \mathcal{L}), \wp'(u; \mathcal{L}))$ is to be interpreted as the point at infinity if $u \in \mathcal{L}$. For present purposes we must modify the classical normalization slightly. Assume that $j(\mathcal{L}) \neq 0, 1728$ and consider the elliptic curve

$$\mathcal{E} : y^2 = 4x^3 - \frac{27j(\mathcal{L})}{j(\mathcal{L}) - 1728}x - \frac{27j(\mathcal{L})}{j(\mathcal{L}) - 1728}.$$

Let $(g_2(\mathcal{L})/g_3(\mathcal{L}))^{3/2}$ denote a fixed square root of $(g_2(\mathcal{L})/g_3(\mathcal{L}))^3$. On rewriting the relation $j(\mathcal{L}) = 1728 g_2(\mathcal{L})^3/(g_2(\mathcal{L})^3 - 27g_3(\mathcal{L})^2)$ in the form

$$\frac{g_2(\mathcal{L})^3}{g_3(\mathcal{L})^2} = \frac{27j(\mathcal{L})}{j(\mathcal{L}) - 1728},$$

we see that the change of variables

$$X = (g_3(\mathcal{L})/g_2(\mathcal{L}))x, \qquad Y = (g_3(\mathcal{L})/g_2(\mathcal{L}))^{3/2}y$$

transforms the equation for $\mathcal{E}$ into the equation for $\mathcal{E}^{\mathrm{Wst}}$. Thus we can replace the map $u+\mathcal{L} \mapsto (\wp(u;\mathcal{L}), \wp'(u;\mathcal{L}))$ by the map

$$u+\mathcal{L} \longmapsto \left(\frac{g_2(\mathcal{L})}{g_3(\mathcal{L})} \wp(u;\mathcal{L}), \left(\frac{g_2(\mathcal{L})}{g_3(\mathcal{L})}\right)^{3/2} \wp'(u;\mathcal{L}) \right)$$

to obtain a complex analytic group isomorphism of $\mathbb{C}/\mathcal{L}$ onto $\mathcal{E}(\mathbb{C})$. In particular, if we fix a basis $\{\omega_1, \omega_2\}$ for $\mathcal{L}$, then the numbers

$$x_{r,s}(\mathcal{L}) = \frac{g_2(\mathcal{L})}{g_3(\mathcal{L})} \wp\left(\frac{r\omega_1 + s\omega_2}{N}; \mathcal{L} \right) \qquad (r,s \in \mathbb{Z},\ \ (r,s) \not\equiv (0,0) \bmod N)$$

are the x-coordinates of the affine N-division points on $\mathcal{E}$. Now as a function of u, $\wp(u;\mathcal{L})$ is periodic with respect to $\mathcal{L}$, even, and of degree 2 when viewed as a map $\mathbb{C}/\mathcal{L} \longrightarrow \mathbf{P}^1(\mathbb{C})$. Therefore

$$x_{r,s}(\mathcal{L}) = x_{r',s'}(\mathcal{L}) \Longleftrightarrow (r,s) \equiv \pm(r',s') \bmod N.$$

Letting $\mathcal{R}$ denote the set of orbits of $(\mathbb{Z}/N\mathbb{Z})^2 - \{(0,0)\}$ under the negation map, we see that if (r,s) runs over a set of representatives in $\mathbb{Z}^2$ for the distinct elements of $\mathcal{R}$, then the numbers $x_{r,s}(\mathcal{L})$ are distinct.

Now let $P(w;A,B) \in \mathbb{Z}[w,A,B]$ be the N-th division polynomial, a universal polynomial with the property that $P(w_0;A,B) = 0$ if and only if w_0 is the x-coordinate of an affine N-division point on the elliptic curve $y^2 = 4x^3 + Ax + B$. Applying this property to the elliptic curve $\mathcal{E}$, we find that

$$P\left(x_{r,s}(\mathcal{L}); \frac{27j(\mathcal{L})}{j(\mathcal{L}) - 1728}, \frac{27j(\mathcal{L})}{j(\mathcal{L}) - 1728} \right) = 0$$

whenever $j(\mathcal{L}) \neq 0, 1728$. In particular, let us take $\mathcal{L} = \mathcal{L}_z$, where $j(z) \neq 0, 1728$. Setting

$$f_{r,s}(z) = \frac{g_2(z)}{g_3(z)} \wp\left(\frac{r+sz}{N}; \mathcal{L}_z \right) \qquad (r,s \in \mathbb{Z},\ \ (r,s) \not\equiv (0,0) \bmod N),$$

we have $f_{r,s}(z) = x_{r,s}(\mathcal{L}_z)$ and consequently

$$P\left(f_{r,s}(z); \frac{27j(z)}{j(z) - 1728}, \frac{27j(z)}{j(z) - 1728} \right) = 0.$$

Since this equation holds for all z such that $j(z) \neq 0, 1728$, it holds identically; in other words

$$P\left((f_{r,s}; \frac{27j}{j - 1728}, \frac{27j}{j - 1728} \right) = 0$$

in the field of meromorphic functions on $\mathfrak{H}$. Therefore the functions $f_{r,s}$ are x-coordinates of affine N-division points on the curve E over $\mathbb{C}(j)$ with which we started. In fact the functions $f_{r,s}$ comprise all such x-coordinates:

Proposition 3. *The set of x-coordinates of affine N-division points on the elliptic curve*

$$E: y^2 = 4x^3 - \frac{27j}{j-1728}x - \frac{27j}{j-1728}$$

coincides with the set of functions $f_{r,s}(z) = (g_2(z)/g_3(z))\wp\left(\frac{r+sz}{N}; \mathcal{L}_z\right)$ *in any algebraic closure of* $\mathbb{C}(j)$ *containing these functions. Therefore*

$$\mathbb{C}(j, E[N]/\pm) = \mathbb{C}(j, \{f_{r,s}\}).$$

Proof. As (r, s) runs over a set of representatives for $\mathcal{R}$ the functions $f_{r,s}$ are all distinct, because their values are distinct at any z such that $j(z) \neq 0, 1728$. Since each function $f_{r,s}$ is the x-coordinate of an affine N-division point on E, and since the number of such x-coordinates, like the number of functions $f_{r,s}$, is $|\mathcal{R}|$, we conclude that the functions $f_{r,s}$ are precisely the x-coordinates of the affine N-division points on E.

1.7. Completion of the proof. The proof of Proposition 2 and hence of Theorem 1 is completed by combining Proposition 3 with the following:

Proposition 4. $\mathfrak{M}(\Gamma(N)) = \mathbb{C}(j, \{f_{r,s}\})$.

Proof. Let us use the notations $f_{r,s}$ and $f_{(r,s)}$ interchangeably. The proof rests on two assertions:

(i) $f_{(r,s)} \circ \gamma = f_{(r,s)\gamma}$ for $\gamma \in \mathrm{SL}(2, \mathbb{Z})$.
(ii) There is a meromorphic function on D which extends the meromorphic function $F_{r,s}$ on D° defined by $f_{r,s}(z) = F_{r,s}(e^{2\pi iz/N})$.

Assertion (i) follows after a calculation from the relations

$$g_2(c\mathcal{L}) = c^{-4}g_2(\mathcal{L}), \quad g_3(c\mathcal{L}) = c^{-6}g_3(\mathcal{L}), \quad \text{and} \quad \wp(cu, c\mathcal{L}) = c^{-2}\wp(u, \mathcal{L}).$$

For (ii) one uses the definition of $\wp(u, \mathcal{L})$ as a sum over lattice points to show that $\lim_{y\to\infty} f_{r,s}(x+iy)$ exists uniformly in x. Now (i) implies that

$$f_{r,s} \circ \gamma = f_{r,s} \quad \text{for } \gamma \in \Gamma(N),$$

while (i) and (ii) together imply that if $\delta \in \mathrm{SL}(2, \mathbb{Z})$ then the meromorphic function F on D° defined by

$$f_{r,s}(\delta z) = F(e^{2\pi iz/N})$$

extends to a meromorphic function on D (put $(r', s') = (r, s)\delta$; then $F = F_{r',s'}$). Therefore $f_{r,s}$ belongs to $\mathfrak{M}(\Gamma(N))$. To see that the $f_{r,s}$ actually generate $\mathfrak{M}(\Gamma(N))$, we use (i) again: if the field inclusion $\mathbb{C}(j, \{f_{r,s}\}) \subset \mathfrak{M}(\Gamma(N))$ were proper, then $\mathbb{C}(j, \{f_{r,s}\})$ would be fixed by a nontrivial subgroup of the Galois group

$$\Gamma(1)/\{\pm I\}\Gamma(N) \cong \mathrm{SL}(2, \mathbb{Z}/N\mathbb{Z})/\{\pm I\}.$$

But a subgroup of $\mathrm{SL}(2, \mathbb{Z}/N\mathbb{Z})/\{\pm I\}$ which acts trivially on $\mathcal{R}$ is trivial.

1.8. A normalized basis. The arguments just completed lead to a nearly canonical choice of basis for $E[N]$ and hence to a nearly canonical identification of $\mathrm{Gal}(\mathbb{C}(j, E[N])/\mathbb{C}(j))$ with $\mathrm{SL}(2, \mathbb{Z}/N\mathbb{Z})$ for any elliptic curve E over $\mathbb{C}(j)$ with invariant j. To formulate the result, let us say that E has good reduction at a point $z \in \mathfrak{H}$ if E has good reduction at the place $j = j(z)$ of $\mathbb{C}(j)$. The reduction of E will be denoted E_z. Recall that we have fixed an isomorphism

$$\theta : \mathrm{SL}(2, \mathbb{Z}/N\mathbb{Z})/\{\pm I\} \longrightarrow \mathrm{Gal}(\mathfrak{M}(\Gamma(N))/\mathbb{C}(j))$$

by requiring that for $\gamma \in \mathrm{SL}(2, \mathbb{Z})$ and $f \in \mathfrak{M}(\Gamma(N))$,

$$\theta([\gamma])(f) = f \circ \gamma^{\mathrm{t}},$$

where $[\gamma]$ denotes the image of γ in $\mathrm{SL}(2, \mathbb{Z}/N\mathbb{Z})/\{\pm I\}$.

Proposition 5. *Let E be an elliptic curve over $\mathbb{C}(j)$ with invariant j, and view $\mathbb{C}(j, E[N]/\pm)$ and $\mathfrak{M}(\Gamma(N))$ as subfields of a fixed algebraic closure of $\mathbb{C}(j)$.*

(i) $\mathbb{C}(j, E[N]/\pm) = \mathfrak{M}(\Gamma(N))$. *In particular, for any $z \in \mathfrak{H}$, evaluation at z defines a place of $\mathbb{C}(j, E[N]/\pm)$. Henceforth we fix a place $\hat{z}$ of $\mathbb{C}(j, E[N])$ extending evaluation at z on $\mathbb{C}(j, E[N]/\pm)$, and if E has good reduction at z and $P \in E[N]$, then $P_z \in E_z(\mathbb{C})$ denotes the reduciton of P at $\hat{z}$.*

(ii) *There is a basis $\{P_1, P_2\}$ for $E[N]$, unique up to replacement by $\{-P_1, -P_2\}$, with the following properties:*

(a) *Let $\rho : \mathrm{Gal}(\mathbb{C}(j, E[N])/\mathbb{C}(j)) \longrightarrow \mathrm{SL}(2, \mathbb{Z}/N\mathbb{Z})$ be the isomorphism corresponding to $\{P_1, P_2\}$ and $\rho^{\pm} : \mathrm{Gal}(\mathbb{C}(j, E[N]/\pm)/\mathbb{C}(j)) \longrightarrow \mathrm{SL}(2, \mathbb{Z}/N\mathbb{Z})/\{\pm I\}$ the induced isomorphism. Then $\rho^{\pm} = \theta^{-1}$.*

(b) *If $z \in \mathfrak{H}$ is a point where E has good reduction, then there is a complex analytic group isomorphism of $\mathbb{C}/\mathcal{L}_z$ onto $E_z(\mathbb{C})$ sending $1/N + \mathcal{L}_z$ to $(P_2)_z$.*

Proof. (i) Since $\mathbb{C}(j, E[N]/\pm)$ depends on E only up to quadratic twist, this follows from Propositions 3 and 4.

(ii) Choose an equation for E over $\mathbb{C}(j)$ of the form

$$c(j)y^2 = 4x^3 - \frac{27j}{j - 1728}x - \frac{27j}{j - 1728},$$

where $c(t) \in \mathbb{C}[t]$ is a polynomial with simple zeros. Then E has good reduction at $z \in \mathfrak{H}$ if and only if $j(z) \neq 0, 1728$ and $c(j(z)) \neq 0$. Now we have seen (in the case $c(t) = 1$, and hence in general) that the x-coordinates of the affine points of order N on E are the functions $f_{r,s}$ with $(r, s) \in \mathbb{Z}^2$ and $(r, s) \not\equiv 0 \bmod N$. Thus for each such pair (r, s) there is a point $P_{r,s} \in E[N]$ such that $x(P_{r,s}) = f_{r,s}$. Of course the definition of $P_{r,s}$

represents an arbitrary choice from among two possibilities. We also make an arbitrary choice of square roots $(g_2(z)/g_3(z))^{3/2}$ and $c(j(z))^{1/2}$ at each point $z \in \mathfrak{H}$ where E has good reduction, and we let $\lambda_z : \mathbb{C}/\mathcal{L}_z \to E_z(\mathbb{C})$ denote the complex analytic group isomorphism afforded by the map

$$u \longmapsto \left(\frac{g_2(z)}{g_3(z)} \wp(u; \mathcal{L}_z), \left(\frac{g_2(z)}{g_3(z)} \right)^{3/2} \frac{\wp'(u; \mathcal{L}_z)}{c(j(z))^{1/2}} \right).$$

Now choose any point $z_0 \in \mathfrak{H}$ where E has good reduction, and let P_1 and P_2 be the preimages of $\lambda_{z_0}(z_0/N + \mathcal{L}_{z_0})$ and $\lambda_{z_0}(1/N + \mathcal{L}_{z_0})$ respectively under the isomorphism $P \mapsto P_{z_0}$ of $E[N]$ onto $E_{z_0}[N]$. Then $\{(P_1)_{z_0}, (P_2)_{z_0}\}$ is a basis for $E_{z_0}[N]$ and *a fortiori* $\{P_1, P_2\}$ is a basis for $E[N]$. We claim that

$$rP_1 + sP_2 = \pm P_{r,s}. \tag{1}$$

Since the reduction map is injective on torsion, it suffices to check that

$$(rP_1 + sP_2)_{z_0} = \pm (P_{r,s})_{z_0}.$$

The left-hand side is $\lambda_{z_0}\left(\dfrac{rz_0 + s}{N} + \mathcal{L}_{z_0} \right)$, while the right-hand side has x-coordinate $f_{r,s}(z_0)$. Therefore equality holds.

To verify (a), take $\sigma \in \mathrm{Gal}(\mathbb{C}(j, E[N]/\pm)/\mathbb{C}(j))$, choose an element $\tilde{\sigma} \in \mathrm{Gal}(\mathbb{C}(j, E[N])/\mathbb{C}(j))$ which restricts to σ, and select $\gamma \in \mathrm{SL}(2, \mathbb{Z})$ so that the image of γ in $\mathrm{SL}(2, \mathbb{Z}/N\mathbb{Z})$ is $\rho(\tilde{\sigma})$. Then $\rho^{\pm}(\sigma) = [\gamma]$, and the identity to be proved is $\theta([\gamma]) = \sigma$. Since the $f_{r,s}$ generate $\mathfrak{M}(\Gamma(N))$ over $\mathbb{C}(j)$ (Proposition 4), it suffices to check that $\theta([\gamma])(f_{r,s}) = \sigma(f_{r,s})$. Write

$$\gamma \begin{pmatrix} r \\ s \end{pmatrix} = \begin{pmatrix} r' \\ s' \end{pmatrix}.$$

As we have seen in the proof of Proposition 4,

$$\theta([\gamma])(f_{r,s}) = f_{r,s} \circ \gamma^{\mathrm{t}} = f_{(r',s')}.$$

On the other hand, $r'P_1 + s'P_2 = \pm P_{r',s'}$, whence

$$f_{r',s'} = x(r'P_1 + s'P_2) = x(\rho(\tilde{\sigma})(rP_1 + sP_2)) = \sigma(x(rP_1 + sP_2)).$$

By (1), the last term is $\sigma(f_{r,s})$, and (a) follows.

For (b), suppose that E has good reduction at z. Since the x-coordinate of $(P_2)_z$ is $f_{0,1}(z)$, we have $\lambda_z(1/N + \mathcal{L}_z) = \pm(P_2)_z$. Hence either λ_z or $-\lambda_z$ sends $1/N + \mathcal{L}_z$ to $(P_2)_z$.

Finally, suppose that $\{P_1', P_2'\}$ is another basis for $E[N]$ with properties (a) and (b). Choose a point $z \in \mathfrak{H}$ where E has good reduction, and let

$\lambda'_z : \mathbb{C}/\mathcal{L}_z \longrightarrow E_z(\mathbb{C})$ be a complex analytic group isomorphism sending $1/N + \mathcal{L}_z$ to $(P'_2)_z$. Then $\lambda_z^{-1} \circ \lambda'_z \in \mathrm{Aut}(\mathbb{C}/\mathcal{L}_z)$. Since E has good reduction at z we have $j(z) \neq 0, 1728$ and consequently $\mathrm{Aut}(\mathbb{C}/\mathcal{L}_z) = \{\pm 1\}$. Hence after replacing $\{P_1, P_2\}$ by $\{-P_1, -P_2\}$ if necessary, we may assume that $P'_2 = P_2$. Then the change-of-basis matrix sending $\{P_1, P_2\}$ to $\{P'_1, P'_2\}$ is a lower triangular matrix with 1 in the lower right-hand entry. Furthermore, conjugation by this matrix induces the identity on $\mathrm{SL}(2, \mathbb{Z}/N\mathbb{Z})/\{\pm I\}$, because $\{P'_1, P'_2\}$ also has property (a). It follows that the change-of-basis is the identity, as desired.

1.9. Quotients of the upper half-plane. We will use Proposition 5 to realize the modular curves as compactified quotients of $\mathfrak{H}$. First we must recall how such quotients are given the structure of a Riemann surface.

Put

$$\mathfrak{H}^* = \mathfrak{H} \cup \mathbf{P}^1(\mathbb{Q}).$$

The action of $\mathrm{SL}(2, \mathbb{Z})$ on $\mathfrak{H}$ by fractional linear transformations extends to an action on $\mathfrak{H}^*$ preserving $\mathbf{P}^1(\mathbb{Q})$, and if Γ is any subgroup of finite index in $\mathrm{SL}(2, \mathbb{Z})$, then we denote the respective orbit spaces of $\mathfrak{H}^*$, $\mathfrak{H}$, and $\mathbf{P}^1(\mathbb{Q})$ under Γ by $\Gamma\backslash\mathfrak{H}^*$, $\Gamma\backslash\mathfrak{H}$, and $\Gamma\backslash\mathbf{P}^1(\mathbb{Q})$. Thus

$$\Gamma\backslash\mathfrak{H}^* = (\Gamma\backslash\mathfrak{H}) \cup \left(\Gamma\backslash\mathbf{P}^1(\mathbb{Q})\right).$$

Since Γ has finite index in $\mathrm{SL}(2, \mathbb{Z})$ and $\mathrm{SL}(2, \mathbb{Z})$ acts transitively on $\mathbf{P}^1(\mathbb{Q})$ the set $\Gamma\backslash\mathbf{P}^1(\mathbb{Q})$ is finite.

We would like to put a topology on $\Gamma\backslash\mathfrak{H}^*$. First we put a topology on $\mathfrak{H}^*$ itself. Given $y_0 > 0$ and $c \in \mathbf{P}^1(\mathbb{Q})$, choose a matrix $\delta \in \mathrm{SL}(2, \mathbb{Z})$ such that $c = \delta\infty$, and put

$$U_{y_0} = \{x + iy : x \in \mathbb{R}, y > y_0\} \subset \mathfrak{H},$$

$$U^\circ_{c,y_0} = \delta(U_{y_0}), \quad \text{and} \quad U_{c,y_0} = U^\circ_{c,y_0} \cup \{c\}.$$

The sets U°_{c,y_0} and U_{c,y_0} depend only on c and y_0, not on the choice of δ, because U_{y_0} is preserved by the stabilizer of ∞ in $\mathrm{SL}(2, \mathbb{Z})$, namely $\{\pm \left(\begin{smallmatrix} 1 & n \\ 0 & 1 \end{smallmatrix}\right) : n \in \mathbb{Z}\}$. We make $\mathfrak{H}^*$ into a topological space by choosing as a basis of open sets all sets of the following two types:

(a) open subsets U of $\mathfrak{H}$,
(b) subsets of $\mathfrak{H}^*$ of the form U_{c,y_0}.

Then the quotient toplology on $\Gamma\backslash\mathfrak{H}^*$ corresponding to the natural projection

$$\pi : \mathfrak{H}^* \longrightarrow \Gamma\backslash\mathfrak{H}^*$$

makes $\Gamma\backslash\mathfrak{H}^*$ into a compact Hausdorff space.

The next step is to make $\Gamma\backslash\mathfrak{H}^*$ into a compact Riemann surface. Let $\mathcal{F}$ be the sheaf of continuous complex-valued functions on $\Gamma\backslash\mathfrak{H}^*$, and $\mathcal{F}_x$

the stalk at a point x. We think of $\mathcal{F}_x$ as the set of equivalence classes of pairs (f, V), where V is an open neighborhood of x and f is a continuous complex-valued function on V, two pairs (f, V) and (g, W) being equivalent if f and g coincide on $V \cap W$. To make $\Gamma\backslash\mathfrak{H}^*$ into a Riemann surface, we must define a subsheaf $\mathcal{O}$ of $\mathcal{F}$ to serve as the complex structure sheaf. We define $\mathcal{O}$ by specifying that its stalk $\mathcal{O}_x$ at x is the subring of $\mathcal{F}_x$ consisting of those equivalence classes which contain a pair (f, V) of one of the following two types:

(a) There exists $z \in \mathfrak{H}$ and an open neighborhood U of z in $\mathfrak{H}$ such that $x = \pi(z)$, $V = \pi(U)$, and $f \circ \pi$ is holomorphic on U.
(b) There exists $c \in \mathbf{P}^1(\mathbb{Q})$ and $y_0 > 0$ such that $x = \pi(c)$, $V = \pi(U_{c,y_0})$, and $f \circ \pi$ satisfies the following condition. Choose $\delta \in \mathrm{SL}(2, \mathbb{Z})$ such that $c = \delta\infty$, and let M be a positive integer such that $(f \circ \pi \circ \delta)(z + M) = (f \circ \pi \circ \delta)(z)$ for $z \in U_{y_0}$. (Such an integer exists because Γ has finite index in $\mathrm{SL}(2, \mathbb{Z})$ and π is invariant under Γ.) Put $r = e^{-2\pi y_0/M}$ and let F be the function on the punctured disk $D^\circ(r) = \{q \in \mathbb{C} : 0 < |q| < r\}$ such that

$$(f \circ \pi \circ \delta)(z) = F(e^{2\pi i z/M}).$$

Then F is holomorphic on $D^\circ(r)$ and extends to a holomorphic function on the full disk $D(r) = \{q \in \mathbb{C} : |q| < r\}$.

One can check that with this definition of $\mathcal{O}$, every point x of $\Gamma\backslash\mathfrak{H}^*$ has an open neighborhood V such that the ringed space $(V, \mathcal{O}|_V)$ is isomorphic to the ringed space of an open disk in $\mathbb{C}$. (The verification requires a little care if x is the image of an elliptic fixed point of Γ, i.e., if $x = \pi(z)$ for some $z \in \mathfrak{H}$ which is fixed by an element of Γ different from $\pm I$.) Granting that this is so, we conclude that $\mathcal{O}$ gives $\Gamma\backslash\mathfrak{H}^*$ the structure of a Riemann surface. Furthermore, and this is now the key point, the definitions have been constructed in such a way that the map

$$f \longmapsto (f \circ \pi)|\mathfrak{H}$$

identifies the function field of $\Gamma\backslash\mathfrak{H}^*$ with $\mathfrak{M}(\Gamma)$. Note that both $\Gamma\backslash\mathfrak{H}^*$ and $\mathfrak{M}(\Gamma)$ depend only on $\overline{\Gamma}$, the image of Γ in $\mathrm{SL}(2, \mathbb{Z})/\{\pm I\}$.

1.10. Modular curves as quotients of the upper half-plane. Given a modular curve $X(H)$, we shall now produce a subgroup Γ of $\mathrm{SL}(2, \mathbb{Z})$ such that the Riemann surfaces $X(H)(\mathbb{C})$ and $\Gamma\backslash\mathfrak{H}^*$ are isomorphic. By assumption, H is a subgroup of $\mathrm{GL}(2, \mathbb{Z}/N\mathbb{Z})$ satisfying two conditons: $-I \in H$ and $\det : H \to (\mathbb{Z}/N\mathbb{Z})^\times$ is surjective. We let $\Gamma \subset \mathrm{SL}(2, \mathbb{Z})$ be the transpose of the inverse image of $H \cap SL(2, \mathbb{Z}/N\mathbb{Z})$ under the reduction map $\mathrm{SL}(2, \mathbb{Z}) \to \mathrm{SL}(2, \mathbb{Z}/N\mathbb{Z})$.

Proposition 6. *With H and Γ as above, the Riemann surfaces $X(H)(\mathbb{C})$ and $\Gamma\backslash\mathfrak{H}^*$ are isomorphic.*

Proof. Let E be an elliptic curve over $\mathbb{Q}(j)$ with invariant j, and identify $\mathrm{Gal}(\mathbb{Q}(j,E[N])/\mathbb{Q}(j))$ with $\mathrm{GL}(2,\mathbb{Z}/N\mathbb{Z})$ using a basis for $E[N]$ as in Proposition 5. The function field of $X(H)$ over $\mathbb{Q}$ is the subfield K of $\mathbb{Q}(j,E[N]/\pm)$ fixed by H, whence the function field of the Riemann surface $X(H)(\mathbb{C})$ is $\mathbb{C}K$. Now our identification of $\mathrm{Gal}(\mathbb{Q}(j,E[N])/\mathbb{Q}(j))$ with $\mathrm{GL}(2,\mathbb{Z}/N\mathbb{Z})$ affords an identification

$$\mathrm{Gal}(\mathbb{C}(j,E[N]/\pm)/\mathbb{C}(j)) \cong \mathrm{SL}(2,\mathbb{Z}/N\mathbb{Z})/\{\pm I\},$$

and the hypotheses on H imply that $\mathbb{C}K$ is the subfield of $\mathbb{C}(j,E[N]/\pm)$ fixed by $(H\cap \mathrm{SL}(2,\mathbb{Z}/N\mathbb{Z}))/\{\pm I\}$. Applying parts (i) and (ii)(a) of Proposition 5, we deduce that $\mathbb{C}K = \mathfrak{M}(\Gamma)$, whence the result follows from the fact that a compact Riemann surface is determined up to isomorphism by its function field.

In particular, put

$$\Gamma_0(N) = \left\{\begin{pmatrix} a & b \\ c & d\end{pmatrix} \in \mathrm{SL}(2,\mathbb{Z}) : c \equiv 0 \bmod N\right\}$$

and

$$\Gamma_1(N) = \left\{\begin{pmatrix} a & b \\ c & d\end{pmatrix} \in \mathrm{SL}(2,\mathbb{Z}) : c \equiv 0 \bmod N, a, d \equiv 1 \bmod N\right\}.$$

Then $X_0(N)(\mathbb{C}) \cong \Gamma_0(N)\backslash\mathfrak{H}^*$ and $X_1(N)(\mathbb{C}) \cong \Gamma_1(N)\backslash\mathfrak{H}^*$ (in the latter case we use the fact that $\Gamma\backslash\mathfrak{H}^*$ depends only on $\overline{\Gamma}$). Now consider pairs $(\mathcal{T},\mathcal{C})$ consisting of a one-dimensional complex torus $\mathcal{T}$ and a cyclic subgroup $\mathcal{C}$ of $\mathcal{T}$ of order N. An isomorphism from one pair $(\mathcal{T}_1,\mathcal{C}_1)$ to another $(\mathcal{T}_2,\mathcal{C}_2)$ is a complex analytic group isomorphism from $\mathcal{T}_1$ to $\mathcal{T}_2$ sending $\mathcal{C}_1$ onto $\mathcal{C}_2$. We denote the isomorphism class containing $(\mathcal{T},\mathcal{C})$ by $[\mathcal{T},\mathcal{C}]$ and the set of all isomorphism classes by $\mathrm{Tori}_0(N)$. For a point $\mathcal{P}$ of order N on $\mathcal{T}$ we make the analogous definitions of $(\mathcal{T},\mathcal{P})$, $[\mathcal{T},\mathcal{P}]$, and $\mathrm{Tori}_1(N)$.

Proposition 7. *Let E be an elliptic curve over $\mathbb{Q}(j)$ with invariant j, and let S be a subset of $\mathbf{P}^1(\mathbb{C})$ containing all places where E has bad reduction. Fix an ordered basis for $E[N]$ over $\mathbb{Z}/N\mathbb{Z}$, let P be the second element of this basis, and let C be the cyclic group of order N generated by P. Then there is an isomorphism of Riemann surfaces $X_0(N)(\mathbb{C}) \cong \Gamma_0(N)\backslash\mathfrak{H}^*$ such that the diagram*

$$\begin{array}{ccc} X_0(N)(\mathbb{C})_S & \longrightarrow & \mathrm{Ell}_0(N)(\mathbb{C})_S \\ \big\downarrow & & \big\downarrow \\ \Gamma_0(N)\backslash\mathfrak{H} & \longrightarrow & \mathrm{Tori}_0(N) \end{array}$$

commutes, where:

- *The top horizontal arrow is the bijection* $x \mapsto [E_x, C_x]$ *of Proposition 1.*
- *The bottom horizontal arrow is a bijection and has the form*

$$[z] \longmapsto [\mathbb{C}/\mathcal{L}_z, \langle 1/N + \mathcal{L}_z \rangle] \qquad (z \in \mathfrak{H}),$$

  where $[z]$ *denotes the class of* z *in* $\Gamma_0(N)\backslash\mathfrak{H}$ *and* $\langle 1/N + \mathcal{L}_z \rangle$ *denotes the cyclic subgroup of* $\mathbb{C}/\mathcal{L}_z$ *generated by the coset of* $1/N + \mathcal{L}_z$.
- *The left vertical arrow is the restriction to* $X_0(N)(\mathbb{C})_S$ *of the isomorphism* $X_0(N)(\mathbb{C}) \cong \Gamma_0(N)\backslash\mathfrak{H}^*$.
- *The right vertical arrow is the restriction to* $\mathrm{Ell}_0(N)(\mathbb{C})_S$ *of the bijection from* $\mathrm{Ell}_0(N)(\mathbb{C})$ *to* $\mathrm{Tori}_0(N)$ *given by* $[\mathcal{E}, \mathcal{C}] \mapsto [\mathcal{E}(\mathbb{C}), \mathcal{C}]$.

The same is true if $X_0(N)$, $\mathrm{Ell}_0(N)$, $\Gamma_0(N)$, $\mathrm{Tori}_0(N)$, C, *and* $\langle 1/N + \mathcal{L}_z \rangle$ *are replaced by* $X_1(N)$, $\mathrm{Ell}_1(N)$, $\Gamma_1(N)$, $\mathrm{Tori}_1(N)$, P, *and* $1/N + \mathcal{L}_z$.

Proof. Without loss of generality we may assume that P is the second basis vector in a basis for $E[N]$ chosen as in Proposition 5. Then the only statement requiring proof is the bijectivity of the bottom horizontal arrow. The cases $\Gamma_0(N)$ and $\Gamma_1(N)$ are similar; we deal with the latter. Suppose that

$$[\mathbb{C}/\mathcal{L}_z, 1/N + \mathcal{L}_z] = [\mathbb{C}/\mathcal{L}_{z'}, 1/N + \mathcal{L}_{z'}].$$

Then there exists $\omega \in \mathbb{C}^\times$ so that $\mathcal{L}_{z'} = \omega\mathcal{L}_z$ and $1/N \equiv \omega/N \pmod{\omega\mathcal{L}_z}$. The first condition implies that $\{\omega, \omega z\}$ is a basis for $\mathcal{L}_{z'}$. Hence we can write

$$\begin{cases} z' = \omega(az + b) \\ 1 = \omega(cz + d) \end{cases}$$

with integers a, b, c, d satisfying $ad - bc = \pm 1$. Since $z' = (az + b)/(cz + d)$ and z and z' both have positive imaginary part, it follows that $ad - bc = 1$. Substituting $1 = \omega(cz + d)$ in the congruence $1/N \equiv \omega/N \pmod{\omega\mathcal{L}_z}$, we find that $c \equiv 0 \pmod N$ and $d \equiv 1 \pmod N$, whence $a \equiv 1 \pmod N$ also since $ad - bc = 1$. Thus $z' = \gamma z$ with

$$\gamma = \begin{pmatrix} a & b \\ c & d \end{pmatrix} \in \Gamma_1(N),$$

and consequently $[z] = [z']$. Next suppose that $[\mathcal{T}, \mathcal{P}] \in \mathrm{Tori}_1(N)$. Write $\mathcal{T} = \mathbb{C}/\mathcal{L}$ and $\mathcal{P} = \omega/N + \mathcal{L}$ with $\omega \in \mathcal{L}$. After replacing ω by another element of $\omega + N\mathcal{L}$, we may assume that ω is primitive, so that ω is part of a basis $\{\omega', \omega\}$ for $\mathcal{L}$. Put $z = \pm\omega'/\omega$, where the sign is chosen so that $\mathrm{Im}(z) > 0$. Multiplication by ω^{-1} gives an isomorphism of $(\mathbb{C}/\mathcal{L}, \mathcal{P} + \mathcal{L})$ onto $(\mathbb{C}/\mathcal{L}_z, 1/N + \mathcal{L}_z)$, whence $[\mathcal{T}, \mathcal{P}]$ coincides with $[\mathbb{C}/\mathcal{L}_z, 1/N + \mathcal{L}_z]$.

2. HECKE CORRESPONDENCES

By a *correspondence* on a smooth projective curve X we shall mean a triple $T = (Z, \varphi, \psi)$, where Z is a smooth projective curve and φ and ψ are nonconstant morphisms $Z \to X$. We say that T is defined over a field $\mathbb{k}$ if X, Z, φ, and ψ are all defined over $\mathbb{k}$. We view an automorphism δ of X as a special case of a correspondence by putting $Z = X$, $\varphi = \mathrm{id}_X$, and $\psi = \delta$.

2.1. The Hecke correspondences on $X_0(N)$. Let N be a positive integer, p a prime number, and M the least common multiple of N and p. Choose an elliptic curve E over $\mathbb{Q}(t)$ with invariant t, and fix a basis for $E[M]$ over $\mathbb{Z}/M\mathbb{Z}$, whence an identification of $\mathrm{Gal}(\mathbb{Q}(t, E[M])/\mathbb{Q}(t))$ with $\mathrm{GL}(2, \mathbb{Z}/M\mathbb{Z})$. We consider the subgroup

$$H_p = \left\{ \begin{pmatrix} a & c \\ b & d \end{pmatrix} \in \mathrm{GL}(2, \mathbb{Z}/M\mathbb{Z}) : c \equiv 0 \bmod N,\ \ b \equiv 0 \bmod p \right\}.$$

Since $-I \in H_p$ and $\det(H_p) = (\mathbb{Z}/M\mathbb{Z})^\times$, the fixed field of H_p is the function field of a smooth projective curve over $\mathbb{Q}$, which we shall denote $X_0(N,p)$. The Hecke correspondence T_p on $X_0(N)$ is a correspondence over $\mathbb{Q}$ of the form

$$T_p = (X_0(N,p), \varphi_p, \psi_p),$$

where the morphisms $\varphi_p, \psi_p : X_0(N,p) \to X_0(N)$ must now be defined.

The definition of φ_p is straightforward. Let K_p and K denote the fixed fields of H_p and

$$H = \left\{ \begin{pmatrix} a & c \\ b & d \end{pmatrix} \in \mathrm{GL}(2, \mathbb{Z}/M\mathbb{Z}) : c \equiv 0 \bmod N \right\}$$

respectively. Then $H_p \subset H$, whence $K \subset K_p$. The latter inclusion is an inclusion of function fields and so corresponds to a morphism of curves

$$\varphi_p : X_0(N,p) = X(H_p) \longrightarrow X(H).$$

But $X(H)$ is $X_0(N)$, because the kernel of the reduction map

$$\mathrm{GL}(2, \mathbb{Z}/M\mathbb{Z}) \to \mathrm{GL}(2, \mathbb{Z}/N\mathbb{Z})$$

is a subgroup of H and the image of H in $\mathrm{GL}(2, \mathbb{Z}/N\mathbb{Z})$ is the lower triangular group. Therefore φ_p is a morphism from $X_0(N,p)$ to $X_0(N)$.

The definition of ψ_p is more subtle. It corresponds to an inclusion of function fields $K' \hookrightarrow K_p$, where K' is a subfield of K_p which is isomorphic to, but distinct from, K. To define K', let us recall once again that our identification of $\mathrm{Gal}(\mathbb{Q}(t, E[M])/\mathbb{Q}(t))$ with $\mathrm{GL}(2, \mathbb{Z}/M\mathbb{Z})$ rests on a choice of basis for $E[M]$ over $\mathbb{Z}/M\mathbb{Z}$ and hence in particular on a decomposition of $E[M]$ as a direct sum of cyclic subgroups of order M:

$$E[M] = C_1 \oplus C_2.$$

Let C denote the cyclic subgroup of C_2 of order N, and let Π denote the cyclic subgroup of C_1 of order p. Then C and Π are stable under H_p, hence defined over K_p. In particular, since Π is defined over K_p there is an elliptic curve E/Π defined over K_p together with an isogeny

$$\lambda : E \longrightarrow E/\Pi$$

over K_p with kernel Π. Furthermore, E/Π has a cyclic subgroup of order N defined over K_p, namely the subgroup $\lambda(C)$. Now put

$$t' = j(E/\Pi) \in K_p,$$

and let E' be an elliptic curve over $\mathbb{Q}(t')$ with invariant t'. Then there is an isomorphism $\theta : E/\Pi \to E'$ over $\overline{K_p}$, and θ is unique up to sign because t' is transcendental, hence $\neq 0, 1728$. It follows that the group $C' = \theta(\lambda(C))$ is a cyclic subgroup of E' of order N which is independent of the choice of θ. Furthermore, C' is defined over K_p because $\lambda(C)$ is defined over K_p and $\sigma \circ \theta \circ \sigma^{-1} = \pm\theta$ for $\sigma \in \mathrm{Gal}(\overline{K_p}/K_p)$. Thus K_p contains the field K' fixed by

$$\{\sigma \in \mathrm{Gal}(\overline{\mathbb{Q}(t')}/\mathbb{Q}(t')) : \sigma(C') = C'\}.$$

Since K' is isomorphic to the function field of $X_0(N)$ we obtain the desired morphism ψ_p from $X_0(N,p)$ to $X_0(N)$.

2.2. The Hecke correspondences on $X_1(N)$. *Mutatis mutandis*, the same construction yields a correspondence

$$T_p = (X_1(N,p), \varphi_p, \psi_p)$$

on $X_1(N)$, where $X_1(N,p)$ is the modular curve determined by the subgroup

$$H_p = \left\{ \begin{pmatrix} a & c \\ b & d \end{pmatrix} \in \mathrm{GL}(2, \mathbb{Z}/M\mathbb{Z}) : \begin{matrix} c \equiv 0 \bmod N, \quad b \equiv 0 \bmod p \\ d \equiv \pm 1 \bmod N \end{matrix} \right\}$$

of $\mathrm{GL}(2, \mathbb{Z}/M\mathbb{Z})$. Put

$$H = \left\{ \begin{pmatrix} a & c \\ b & d \end{pmatrix} \in \mathrm{GL}(2, \mathbb{Z}/M\mathbb{Z}) : c \equiv 0 \bmod N, \quad d \equiv \pm 1 \bmod N \right\}$$

and write K_p and K for the subfields of $\mathbb{Q}(t, E[M])$ fixed by H_p and H. Then φ_p is the morphism $X_1(N,p) \to X_1(N)$ corresponding to the inclusion of K in K_p. To define ψ_p, let $\{P_1, P_2\}$ be our chosen basis for $E[M]$ and put $P = (M/N)P_2$. Also let Π be the group of order p generated by $(M/p)P_1$. As before, there is an elliptic curve E/Π over K_p and an isogeny $\lambda : E \to E/\Pi$ over K_p with kernel Π. Since K is contained in K_p and the set $\{\pm P\}$ is stable under $\mathrm{Gal}(\mathbb{Q}(t, E[M])/K)$, it follows that $\{\pm\lambda(P)\}$ is stable under $\mathrm{Gal}(\mathbb{Q}(t, E[M])/K_p)$. Putting $t' = j(E/\Pi)$ as before, we see that if E' is any elliptic curve over $\mathbb{Q}(t')$ with invariant t' and $\theta : E/\Pi \to E'$ is any isomorphism over $\overline{K_p}$, then the point $P' = \theta(\lambda(P))$ has order N and $\{\pm P\}$ is defined over K_p. Hence K_p contains K', the field fixed by

$$\{\sigma \in \mathrm{Gal}(\overline{\mathbb{Q}(t')}/\mathbb{Q}(t')) : \sigma(P') = \pm P'\}.$$

Since K' is isomorphic to the function field of $X_1(N)$ we obtain a morphism ψ_p from $X_1(N,p)$ to $X_1(N)$.

2.3. Moduli interpretation of the Hecke correspondences. We denote the free abelian group on a set W by $\mathrm{Div}(W)$. In particular, if X is a smooth projective curve over an algebraically closed field $\Bbbk$, then $\mathrm{Div}(X(\Bbbk))$ is the usual group of divisors on $X(\Bbbk)$. Given a correspondence $T = (Z, \varphi, \psi)$ on X, we use the same letter T to denote the map

$$\begin{aligned} X(\Bbbk) &\longrightarrow \mathrm{Div}(X(\Bbbk)) \\ x &\longmapsto \sum_{\substack{z \in Z \\ \varphi(z) = x}} (\mathrm{mult}_z \varphi)\psi(z), \end{aligned}$$

where $\mathrm{mult}_z\varphi$ is the ramification index of φ at z. In the case of the Hecke correspondence T_p we shall give a formula for this map which displays its effect on isomorphism classes $[\mathcal{E}, \mathcal{C}]$ and $[\mathcal{E}, \mathcal{P}]$. First a point of notation.

Suppose that $\mathcal{E}$ is an elliptic curve over an algebraically closed field $\Bbbk$ and Λ is a subgroup of $\mathcal{E}$ of order p. In keeping with our usage thus far, we shall write $\mathcal{E}/\Lambda$ for an elliptic curve which is the image of a separable isogeny with domain $\mathcal{E}$ and kernel Λ. Note that $\mathcal{E}/\Lambda$ is unique up to isomorphism. Now if $\lambda : \mathcal{E} \to \mathcal{E}/\Lambda$ is a separable isogeny with kernel Λ and $\mathcal{C}$ is a cyclic subgroup of $\mathcal{E}$ of order N which intersects Λ trivially (a vacuous condition if N is prime to p), then we obtain a well-defined isomorphism class

$$[\mathcal{E}/\Lambda, (\mathcal{C} + \Lambda)/\Lambda] \in \mathrm{Ell}_0(N)(\Bbbk)$$

by putting $[\mathcal{E}/\Lambda, (\mathcal{C} + \Lambda)/\Lambda] = [\lambda(\mathcal{E}), \lambda(\mathcal{C})]$. To see that $[\mathcal{E}/\Lambda, (\mathcal{C} + \Lambda)/\Lambda]$ is independent of the choice of λ, suppose that $\lambda' : \mathcal{E} \to \mathcal{E}/\Lambda$ is another such isogeny. Then there is an automorphism θ of $\mathcal{E}/\Lambda$ such that $\lambda' = \theta \circ \lambda$, whence $[\lambda'(\mathcal{E}), \lambda'(\mathcal{C})] = [\lambda(\mathcal{E}), \lambda(\mathcal{C})]$. Similarly, if $\mathcal{P}$ is a point of order N on $\mathcal{E}$ such that the cyclic subgroup $\langle\mathcal{P}\rangle$ generated by $\mathcal{P}$ intersects Λ trivially, then we define

$$[\mathcal{E}/\Lambda, \mathcal{P} + \Lambda] \in \mathrm{Ell}_1(N)(\Bbbk)$$

by putting $[\mathcal{E}/\Lambda, \mathcal{P} + \Lambda] = [\lambda(\mathcal{E}), \lambda(\mathcal{P})]$.

Proposition 8. *Let E be an elliptic curve over $\mathbb{Q}(t)$ with invariant t. Let S, S', and S'' be subsets of $\mathbf{P}^1(\mathbb{C})$ containing all places where E has bad reduction and such that*

$$\varphi_p^{-1}(X_0(N)(\mathbb{C})_S) \subset X_0(N,p)(\mathbb{C})_{S'}$$

and

$$\psi_p(X_0(N,p)(\mathbb{C})_{S'}) \subset X_0(N)(\mathbb{C})_{S''}.$$

Fix an ordered basis for $E[N]$ over $\mathbb{Z}/N\mathbb{Z}$, let P be the second element of this basis, and let C be the cyclic group generated by P. Then the diagram

$$\begin{array}{ccc} X_0(N)(\mathbb{C})_S & \xrightarrow{T_p} & \mathrm{Div}(X_0(N)(\mathbb{C})_{S''}) \\ \downarrow & & \downarrow \\ \mathrm{Ell}_0(N)(\mathbb{C}) & \longrightarrow & \mathrm{Div}(\mathrm{Ell}_0(N)(\mathbb{C})) \end{array}$$

commutes, where the left vertical arrow is the map $x \mapsto [E_x, C_x]$ of Proposition 1, the right vertical arrow is the corresponding homomorphism between free abelian groups, and the bottom horizontal arrow is the map

$$[\mathcal{E},\mathcal{C}] \longmapsto \sum_{\substack{[\mathcal{E}[p]:\Lambda]=p \\ \mathcal{C}\cap\Lambda=\{0\}}} [\mathcal{E}/\Lambda, (\mathcal{C}+\Lambda)/\Lambda],$$

the sum being taken over subgroups Λ of index p in $\mathcal{E}[p]$ which intersect C trivially. The same is true if $X_0(N)$ is replaced by $X_1(N)$, $\mathrm{Ell}_0(N)$ by $\mathrm{Ell}_1(N)$, the left vertical arrow by $x \mapsto [E_x, P_x]$, and the bottom horizontal arrow by

$$[\mathcal{E},\mathcal{P}] \longmapsto \sum_{\substack{[\mathcal{E}[p]:\Lambda]=p \\ \langle\mathcal{P}\rangle\cap\Lambda=\{0\}}} [\mathcal{E}/\Lambda, \mathcal{P}+\Lambda],$$

where $\langle\mathcal{P}\rangle$ denotes the cyclic subgroup generated by $\mathcal{P}$.

Proof. For $x \in X_0(N)(\mathbb{C})_S$ the formula

$$T_p(x) = \sum_{\substack{z\in Z \\ \varphi_p(z)=x}} (\mathrm{mult}_z\varphi_p)\psi_p(z)$$

can be written simply as

$$T_p(x) = \sum_{z\in\varphi_p^{-1}(x)} \psi_p(z),$$

because the morphism $\varphi_p : X_0(N,p) \to X_0(N)$ is unramified outside S: indeed the corresponding extension of function fields K_p/K is contained inside the extension $\mathbb{Q}(t, E[M])/\mathbb{Q}(t)$ and is therefore unramified outside the places where E has bad reduction. Here M denotes the least common multiple of N and p, as before.

Consider triples $(\mathcal{E}, \mathcal{C}, \Lambda)$, where $\mathcal{E}$ is an elliptic curve over $\mathbb{C}$, $\mathcal{C}$ is a cyclic subgroup of $\mathcal{E}$ of order N, and Λ is a cyclic subgroup of $\mathcal{E}$ of order p which intersects $\mathcal{C}$ trivially. We write $[\mathcal{E}, \mathcal{C}, \Lambda]$ for the isomorphism class of $(\mathcal{E}, \mathcal{C}, \Lambda)$ and $\mathrm{Ell}_0(N,p)(\mathbb{C})$ for the set of isomorphism classes. If we define maps φ and ψ from $\mathrm{Ell}_0(N,p)(\mathbb{C})$ to $\mathrm{Ell}_0(N)(\mathbb{C})$ by

$$\varphi([\mathcal{E}, \mathcal{C}, \Lambda]) = [\mathcal{E}, \mathcal{C}]$$

and

$$\psi([\mathcal{E}, \mathcal{C}, \Lambda]) = [\mathcal{E}/\Lambda, (\mathcal{C} + \Lambda)/\Lambda],$$

then the map

$$[\mathcal{E}, \mathcal{C}] \longmapsto \sum_{\substack{[\mathcal{E}[p]:\Lambda]=p \\ \mathcal{C}\cap\Lambda=\{0\}}} [\mathcal{E}/\Lambda, (\mathcal{C} + \Lambda)/\Lambda],$$

in the statement of the proposition has the form

$$x \longmapsto \sum_{z\in\varphi^{-1}(x)} \psi(z).$$

Therefore it suffices to check that the diagrams

$$\begin{array}{ccc} X_0(N,p)(\mathbb{C})_{S'} & \xrightarrow{\varphi_p} & X_0(N)(\mathbb{C})_S \\ \downarrow & & \downarrow \\ \mathrm{Ell}_0(N,p)(\mathbb{C}) & \xrightarrow[\varphi]{} & \mathrm{Ell}_0(N)(\mathbb{C}) \end{array}$$

and

$$\begin{array}{ccc} X_0(N,p)(\mathbb{C})_{S'} & \xrightarrow{\psi_p} & X_0(N)(\mathbb{C})_{S''} \\ \downarrow & & \downarrow \\ \mathrm{Ell}_0(N,p)(\mathbb{C}) & \xrightarrow[\psi]{} & \mathrm{Ell}_0(N)(\mathbb{C}) \end{array}$$

commute, where the right vertical arrows are the maps $x \mapsto [E_x, C_x]$ of Proposition 1 and the left vertical arrows are given by

$$\begin{aligned} X_0(N,p)(\mathbb{C})_{S'} &\longrightarrow \mathrm{Ell}_0(N,p)(\mathbb{C}) \\ z &\longmapsto [E_z, C_z, \Pi_z]. \end{aligned}$$

As before, Π denotes a subgroup of E of order p which intersects C trivially, and the subscript z indicates reduction modulo the maximal ideal of the discrete valuation ring corresponding to z in the complex function field of $X_0(N,p)$. Now the commutativity of the first diagram amounts to the equation

$$[E_z, C_z] = [E_{\varphi_p(z)}, C_{\varphi_p(z)}],$$

which follows from the compatibility of reduction at a good place with base extension. To verify that the second diagram commutes put $t' = j(E/\Pi)$, choose an elliptic curve E' over $\mathbb{Q}(t')$ with invariant t', the image of $(C+\Pi)/\Pi$ under an isomorphism $E/\Pi \to E'$. We must check that

$$[E_z/\Pi_z, (C_z+\Pi_z)/\Pi_z] = [E'_{\psi_p(z)}, C'_{\psi_p(z)}]. \tag{1}$$

We verify this equation in two steps. First,

$$[E_z/\Pi_z, (C_z+\Pi_z)/\Pi_z] = [(E/\Pi)_z, ((C+\Pi)/\Pi)_z] \tag{2}$$

by the compatibility of reduction with isogenies. Next let θ be an isomorphism from E/Π to E', so that $C' = \theta((C+\Pi)/\Pi)$. Then the reduction of θ gives an isomorphism from $(E/\Pi)_z$ to $(E')_{\psi_p(z)}$ sending $((C+\Pi)/\Pi)_z$ to $(C')_{\psi_p(z)}$, whence

$$[(E/\Pi)_z, ((C+\Pi)/\Pi)_z] = [(E')_{\psi_p(z)}, (C')_{\psi_p(z)}]. \tag{3}$$

Together, (2) and (3) give (1). The argument for $X_1(N)$ is similar.

2.4. The Hecke correspondences on the upper half-plane. Given $d \in (\mathbb{Z}/N\mathbb{Z})^\times$ we write $\langle d\rangle$ to denote any element of $\Gamma_0(N)$ with lower right-hand entry congruent to d modulo N. Also, if $d \in \mathbb{Z}$ is an integer prime to N then we write $\langle d\rangle$ for $\langle d \bmod N\rangle$.

Proposition 9. *There is a commutative diagram*

$$\begin{array}{ccc} X_0(N)(\mathbb{C}) & \xrightarrow{T_p} & \mathrm{Div}(X_0(N)(\mathbb{C})) \\ \downarrow & & \downarrow \\ \Gamma_0(N)\backslash\mathfrak{H}^* & \longrightarrow & \mathrm{Div}(\Gamma_0(N)\backslash\mathfrak{H}^*), \end{array}$$

where the left vertical arrow is the isomorphism of Proposition 7, the right vertical arrow is the corresponding homomorphism of free abelian groups, and the bottom horizontal arrow is the map

$$[z] \longmapsto \begin{cases} \sum_{\nu=0}^{p-1}[(z+\nu)/p] + [pz] & \text{if } p \nmid N \\ \sum_{\nu=0}^{p-1}[(z+\nu)/p] & \text{if } p|N. \end{cases}$$

The same is true if $X_0(N)$ is replaced by $X_1(N)$ and $\Gamma_0(N)$ by $\Gamma_1(N)$ provided that the bottom horizontal arrow is modified as follows: in the case $p \nmid N$, the term $[pz]$ is replaced by $[\langle p\rangle pz]$.

Proof. By a continuity argument it suffices to check that the diagram commutes when $X_0(N)(\mathbb{C})$ is replaced by $X_0(N)(\mathbb{C})_S$ for some finite set S. Propositions 7 and 8 then reduce the problem to the following: Given $[\mathcal{T}, \mathcal{C}] \in \mathrm{Tori}_0(N)$ with $\mathcal{T} = \mathbb{C}/\mathcal{L}_z$ and $\mathcal{C} = \langle 1/N + \mathcal{L}_z\rangle$, show that

$$\sum_{\substack{[\mathcal{T}[p]:\Lambda]=p \\ \mathcal{C}\cap\Lambda=\{0\}}} [\mathcal{T}/\Lambda, (\mathcal{C}+\Lambda)/\Lambda]$$

coincides with

$$\sum_{\nu=0}^{p-1} [\mathbb{C}/\mathcal{L}_{(z+\nu)/p}, \langle 1/N + \mathcal{L}_{(z+\nu)/p}\rangle] + [\mathbb{C}/\mathcal{L}_{pz}, \langle 1/N + \mathcal{L}_{pz}\rangle],$$

the last term being omitted if p divides N. Now the subgroups of order p in $\mathbb{C}/\mathcal{L}_z$ which intersect $\langle 1/N + \mathcal{L}_z\rangle$ trivially are

$$\langle (z+\nu)/p + \mathcal{L}_z\rangle \qquad (0 \le \nu \le p-1)$$

and also

$$\langle 1/p + \mathcal{L}_z\rangle$$

if $p \nmid N$. Furthermore, for $\mathcal{T} = \mathbb{C}/\mathcal{L}_z$ and $\Lambda = \langle (z+\nu)/p + \mathcal{L}_z\rangle$ we have $\mathcal{T}/\Lambda \cong \mathbb{C}/\mathcal{L}_{(z+\nu)/p}$. Hence the only point to check is that if $p \nmid N$ then

$$[\mathbb{C}/(z\mathbb{Z} \oplus p^{-1}\mathbb{Z}), \langle 1/N + (z\mathbb{Z} \oplus p^{-1}\mathbb{Z})\rangle] = [\mathbb{C}/\mathcal{L}_{pz}, \langle 1/N + \mathcal{L}_{pz}\rangle].$$

This holds because multiplication by p maps $\mathbb{C}/(z\mathbb{Z}\oplus p^{-1}\mathbb{Z})$ isomorphically onto $\mathbb{C}/\mathcal{L}_{pz}$ and because the cosets of $1/N$ and p/N generate the same subgroup of $\mathbb{C}/\mathcal{L}_{pz}$ for $p \nmid N$.

The argument for $\Gamma_1(N)$ is much the same, except that in the case $p \nmid N$ one must check that

$$[\mathbb{C}/\mathcal{L}_{pz}, p/N + \mathcal{L}_{pz}] = [\mathbb{C}/\mathcal{L}_{\langle p\rangle pz}, 1/N + \mathcal{L}_{\langle p\rangle pz}].$$

Put $w = pz$ and $\gamma = \langle p\rangle$, and write $\gamma = \begin{pmatrix} a & b \\ c & d\end{pmatrix}$. Since $\mathcal{L}_w = w\mathbb{Z} \oplus \mathbb{Z}$ and $\mathcal{L}_{\gamma w} = \frac{aw+b}{cw+d}\mathbb{Z} \oplus \mathbb{Z}$, multiplication by $cw+d$ defines an isomorphism from $\mathbb{C}/\mathcal{L}_{\gamma w}$ to $\mathbb{C}/\mathcal{L}_w$ sending $1/N + \mathcal{L}_{\gamma w}$ to $(cw+d)/N + \mathcal{L}_w$. The latter coset coincides with $p/N + \mathcal{L}_w$, because

$$(cw+d)/N - p/N = (c/N)w + (d-p)/N \in \mathcal{L}_w.$$

2.5. The diamond automorphisms. The next construction pertains only to $X_1(N)$. Choose an elliptic curve E over $\mathbb{Q}(t)$ with invariant t, and make the usual identification of $\mathrm{Gal}(\mathbb{Q}(t, E[N])/\mathbb{Q}(t))$ with $\mathrm{GL}(2, \mathbb{Z}/N\mathbb{Z})$, and of the function field K of $X_1(N)$ with the fixed field of

$$H = \left\{ \begin{pmatrix} a & 0 \\ b & \pm 1 \end{pmatrix} \in \mathrm{GL}(2, \mathbb{Z}/N\mathbb{Z}) : a \in (\mathbb{Z}/N\mathbb{Z})^\times, b \in (\mathbb{Z}/N\mathbb{Z}) \right\}.$$

Since H is normal in the lower triangular subgroup B, the quotient group B/H acts as a group of automorphisms of K and hence of $X_1(N)$. We shall identify B/H with $(\mathbb{Z}/N\mathbb{Z})^\times/\{\pm 1\}$ via the map sending the coset of $\begin{pmatrix} a & 0 \\ b & d \end{pmatrix}$ modulo H to the coset of d modulo $\{\pm 1\}$. The automorphism of $X_1(N)$ corresponding to the coset of $d \in (\mathbb{Z}/N\mathbb{Z})^\times$ modulo $\{\pm 1\}$ will be denoted $\langle d \rangle$. Of course we have already used the symbol $\langle d \rangle$ to denote an element of $\Gamma_0(N)$. The next two propositions show that the notations are consistant and that $\langle d \rangle$ may also be used for the bijection from $\mathrm{Ell}_1(N)(\mathbb{C})$ to itself given by $[\mathcal{E}, \mathcal{P}] \mapsto [\mathcal{E}, d\mathcal{P}]$. The proofs are left to the reader.

Proposition 10. *Let E be an elliptic curve over $\mathbb{Q}(t)$ with invariant t, and let S be a subset of $\mathbf{P}^1(\mathbb{C})$ containing all places where E has bad reduction. Fix an ordered basis for $E[N]$ over $\mathbb{Z}/N\mathbb{Z}$ and let P be the second element of this basis. For $d \in (\mathbb{Z}/N\mathbb{Z})^\times$ the diagram*

$$\begin{array}{ccc} X_1(N)(\mathbb{C})_S & \xrightarrow{\langle d \rangle} & X_1(N)(\mathbb{C})_S \\ \downarrow & & \downarrow \\ \mathrm{Ell}_1(N)(\mathbb{C}) & \longrightarrow & \mathrm{Ell}_1(N)(\mathbb{C}) \end{array}$$

commutes, where the bottom horizontal arrow is the map $[\mathcal{E}, \mathcal{P}] \mapsto [\mathcal{E}, d\mathcal{P}]$ and the vertical arrows are the map $x \mapsto [E_x, C_x]$ of Proposition 1.

Proposition 11. *There is a commutative diagram*

$$\begin{array}{ccc} X_1(N)(\mathbb{C}) & \xrightarrow{\langle d \rangle} & X_1(N)(\mathbb{C}) \\ \downarrow & & \downarrow \\ \Gamma_1(N)\backslash\mathfrak{H}^* & \longrightarrow & \Gamma_1(N)\backslash\mathfrak{H}^*, \end{array}$$

where the bottom horizontal arrow is the map $[z] \mapsto [\langle d \rangle z]$ and the vertical arrows are the isomorphism of Proposition 7.

For the application to L-functions toward which we are heading it is enough to consider the curve $X_1(N)$, and henceforth $X_0(N)$ will drop out of sight.

2.6. Hecke correspondences and the Frobenius automorphism. Let p be a prime not dividing N, and fix a prime ideal $\mathfrak{p}$ of $\overline{\mathbb{Q}}$ lying over p. Write $\overline{\mathbb{F}}_p$ for the residue class field of $\mathfrak{p}$. Reduction modulo $\mathfrak{p}$ will be denoted by a tilde: for example, if $\mathcal{E}$ is an elliptic curve over $\overline{\mathbb{Q}}$ with good reduction at $\mathfrak{p}$ then $\tilde{\mathcal{E}}$ denotes the elliptic curve over $\overline{\mathbb{F}}_p$ obtained from $\mathcal{E}$ by reduction modulo $\mathfrak{p}$. We define sets $\mathrm{Ell}_1(N)(\overline{\mathbb{Q}})$ and $\mathrm{Ell}_1(N)(\overline{\mathbb{F}}_p)$ by replacing $\mathbb{C}$ by $\overline{\mathbb{Q}}$ or $\overline{\mathbb{F}}_p$ respectively in the definition of $\mathrm{Ell}_1(N)(\mathbb{C})$. Thus $\mathrm{Ell}_1(N)(\overline{\mathbb{Q}})$ can be identified with the subset of $\mathrm{Ell}_1(N)(\mathbb{C})$ consisting of classes $[\mathcal{E},\mathcal{P}]$ such that $j(\mathcal{E}) \in \overline{\mathbb{Q}}$. Also, we let $\mathrm{Ell}_1(N)(\overline{\mathbb{Q}})_{\mathrm{gd}}$ denote the subset of $\mathrm{Ell}_1(N)(\overline{\mathbb{Q}})$ consisting of classes $[\mathcal{E},\mathcal{P}]$ such that $\mathcal{E}$ has good reduction at $\mathfrak{p}$. Under our assumption that p does not divide N we have a well-defined map

$$\begin{aligned} \mathrm{Ell}_1(N)(\overline{\mathbb{Q}})_{\mathrm{gd}} &\longrightarrow \mathrm{Ell}_1(N)(\overline{\mathbb{F}}_p) \\ [\mathcal{E},\mathcal{P}] &\longmapsto [\tilde{\mathcal{E}},\tilde{\mathcal{P}}], \end{aligned}$$

because reduction modulo $\mathfrak{p}$ is injective on N-torsion.

We recall that an elliptic curve $\mathcal{E}$ over $\overline{\mathbb{Q}}$ is said to have *ordinary* good reduction at $\mathfrak{p}$ if $\tilde{\mathcal{E}}[p]$ has order p. If $\mathcal{E}$ has ordinary reduction at $\mathfrak{p}$ then reduction modulo $\mathfrak{p}$ defines a surjective map

$$\mathcal{E}[p] \longrightarrow \tilde{\mathcal{E}}[p]$$

with kernel a subgroup of $\mathcal{E}[p]$ of order (or index) p. In addition $\mathcal{E}[p]$ has exactly p other subgroups of index p.

We use a superscript p to indicate the image of an object under the Frobenius automorphism of $\overline{\mathbb{F}}_p$ and a superscript p^{-1} to indicate the image under the inverse of the Frobenius automorphism.

Proposition 12. *Let $\mathcal{E}$ be an elliptic curve over $\overline{\mathbb{Q}}$ with ordinary reduction at $\mathfrak{p}$, and let Λ_0 be the kernel of the reduction map*

$$\mathcal{E}[p] \longrightarrow \tilde{\mathcal{E}}[p].$$

Let $\mathcal{P}$ be a point of order N on $\mathcal{E}$. If Λ is a subgroup of $\mathcal{E}[p]$ of index p, then

$$[\widetilde{\mathcal{E}/\Lambda},\widetilde{\mathcal{P}+\Lambda}] = \begin{cases} [\tilde{\mathcal{E}}^p,\tilde{\mathcal{P}}^p] & \text{if } \Lambda = \Lambda_0, \\ [\tilde{\mathcal{E}}^{p^{-1}},p\tilde{\mathcal{P}}^{p^{-1}}] & \text{if } \Lambda \neq \Lambda_0. \end{cases}$$

Proof. Let $\lambda_0 : \mathcal{E} \to \mathcal{E}/\Lambda_0$ be an isogeny with kernel Λ_0 and $\mu_0 : \mathcal{E}/\Lambda_0 \to \mathcal{E}$ the dual isogeny. The image of $(\mathcal{E}/\Lambda_0)[p]$ under μ_0 is Λ_0. Indeed, since μ_0 is a p-isogeny, the image of $(\mathcal{E}/\Lambda_0)[p]$ under μ_0 is a group of order p; but if this group were not contained in Λ_0 then the image of $(\mathcal{E}/\Lambda_0)[p]$ under $\lambda_0 \circ \mu_0$ would not be zero, contradicting the fact that $\lambda_0 \circ \mu_0$ is multiplication by p. Now consider the commutative diagram

$$\begin{array}{ccccc} \mathcal{E}[p] & \xrightarrow{\lambda_0} & (\mathcal{E}/\Lambda_0)[p] & \xrightarrow{\mu_0} & \mathcal{E}[p] \\ \downarrow & & \downarrow & & \downarrow \\ \tilde{\mathcal{E}}[p] & \xrightarrow{\tilde{\lambda}_0} & \widetilde{\mathcal{E}/\Lambda_0}[p] & \xrightarrow{\tilde{\mu}_0} & \tilde{\mathcal{E}}[p], \end{array}$$

where the vertical arrows are reduction modulo $\mathfrak{p}$ and hence surjective. Since the image of $(\mathcal{E}/\Lambda_0)[p]$ under μ_0 is Λ_0, the commutativity of the right-hand square shows that $\widetilde{\mu_0}$ is zero on $\widetilde{\mathcal{E}/\Lambda_0[p]}$. Now $\tilde{\mathcal{E}}$ is ordinary by assumption, and since $\widetilde{\mathcal{E}/\Lambda_0}$ is isogenous to $\tilde{\mathcal{E}}$, it too is ordinary. Hence the fact that $\widetilde{\mu_0}$ is zero on $\widetilde{\mathcal{E}/\Lambda_0[p]}$ means that $\widetilde{\mu_0}$ is a *separable* p-isogeny. But multiplication by p is inseparable. Consequently $\widetilde{\lambda_0}$ is an inseparable (and hence purely inseparable) isogeny of degree p, and we can write $\widetilde{\lambda_0} = \theta \circ \beta$, where β is the Frobenius endomorphism of degree p and $\theta : E^p \to \widetilde{\mathcal{E}/\Lambda_0}$ is an isomorphism. Therefore

$$[\widetilde{\lambda_0}(\tilde{\mathcal{E}}), \widetilde{\lambda_0}(\tilde{\mathcal{P}})] = [\beta(\tilde{\mathcal{E}}), \beta(\tilde{\mathcal{P}})].$$

But the left-hand side is $[\widetilde{\mathcal{E}/\Lambda_0}, \widetilde{\mathcal{P}+\Lambda_0}]$, and the right-hand side is $[\tilde{\mathcal{E}}^p, \tilde{\mathcal{P}}^p]$. Hence we get the stated equality.

Now suppose that $\Lambda \neq \Lambda_0$. Choose an isogeny $\lambda : \mathcal{E} \to \mathcal{E}/\Lambda$ with kernel Λ, and consider the curve $\lambda(\mathcal{E})$ $(= \mathcal{E}/\Lambda)$ together with its subgroup $\lambda(\Lambda_0)$ of order p. Since $\lambda(\mathcal{E})$ is isogenous to $\mathcal{E}$, it has ordinary reduction at $\mathfrak{p}$, and consequently the kernel of reduction mod $\mathfrak{p}$ on $\lambda(\mathcal{E})[p]$ is a subgroup of order p. The calculation

$$\widetilde{\lambda(\Lambda_0)} = \tilde{\lambda}(\widetilde{\Lambda_0}) = \tilde{\lambda}(\{0\}) = \{0\}$$

shows that $\lambda(\Lambda_0)$ is contained in this subgroup and hence coincides with it, since both have order p. Therefore we can apply to $\lambda(\mathcal{E})$ and $\lambda(\Lambda_0)$ what we have already proved for $\mathcal{E}$ and Λ_0:

$$[\widetilde{\lambda(\mathcal{E})/\lambda(\Lambda_0)}, \widetilde{\lambda(\mathcal{P})+\lambda(\Lambda_0)}] = [\widetilde{\lambda(\mathcal{E})}^p, \widetilde{\lambda(\mathcal{P})}^p]. \tag{1}$$

But if $\mu : \lambda(\mathcal{E}) \to \lambda(\mathcal{E})/\lambda(\Lambda_0)$ is any isogeny with kernel $\lambda(\Lambda_0)$, then by definition

$$[\lambda(\mathcal{E})/\lambda(\Lambda_0), \lambda(\mathcal{P})+\lambda(\Lambda_0)] = [(\mu\circ\lambda)(\mathcal{E}), (\mu\circ\lambda)(\mathcal{P})].$$

We choose μ to be the isogeny dual to λ, and then $[(\mu\circ\lambda)(\mathcal{E}), (\mu\circ\lambda)(\mathcal{P})]$ becomes $[\mathcal{E}, p\mathcal{P}]$. Thus the left-hand side of (1) is $[\tilde{\mathcal{E}}, p\tilde{\mathcal{P}}]$, and (1) can be rewritten

$$[\tilde{\mathcal{E}}, p\tilde{\mathcal{P}}] = [\widetilde{\mathcal{E}/\Lambda}^p, \widetilde{\mathcal{P}+\Lambda}^p]. \tag{2}$$

Applying the inverse Frobenius automorphism to (2) we obtain the stated formula.

Let $\mathrm{Ell}_1(N)(\overline{\mathbb{Q}})_{\mathrm{ord}}$ be the subset of $\mathrm{Ell}_1(N)(\overline{\mathbb{Q}})_{\mathrm{gd}}$ consisting of classes $[\mathcal{E}, \mathcal{P}]$ such that $\mathcal{E}$ has ordinary reduction at $\mathfrak{p}$. Reduction of isomorphism classes

$$\mathrm{Ell}_1(N)(\overline{\mathbb{Q}})_{\mathrm{ord}} \longrightarrow \mathrm{Ell}_1(N)(\overline{\mathbb{F}}_p)$$
$$[\mathcal{E}, \mathcal{P}] \longmapsto [\tilde{\mathcal{E}}, \tilde{\mathcal{P}}],$$

extends uniquely to a homomorphism

$$\mathrm{red}_{\mathfrak{p}} : \mathrm{Div}(\mathrm{Ell}_1(N)(\overline{\mathbb{Q}})_{\mathrm{ord}}) \longrightarrow \mathrm{Div}(\mathrm{Ell}_1(N)(\overline{\mathbb{F}}_p)).$$

Proposition 13. *Let $\sigma_{\mathfrak{p}} \in \mathrm{Gal}(\overline{\mathbb{Q}}/\mathbb{Q})$ be a Frobenius element at $\mathfrak{p}$. Then*

$$T_p = \sigma_{\mathfrak{p}} + p\langle p\rangle\sigma_{\mathfrak{p}}^{-1}$$

when both sides are regarded as maps

$$\mathrm{Ell}_1(N)(\overline{\mathbb{Q}})_{\mathrm{ord}} \longrightarrow \mathrm{Div}(\mathrm{Ell}_1(N)(\overline{\mathbb{Q}})_{\mathrm{ord}})/\mathrm{Ker}(\mathrm{red}_{\mathfrak{p}}).$$

Proof. Let $\mathcal{E}$ be an elliptic curve over $\overline{\mathbb{Q}}$ with ordinary reduction at $\mathfrak{p}$, and let $\mathcal{P}$ be a point of order N on $\mathcal{E}$. We must show that $T_p([\mathcal{E},\mathcal{P}])$ and $(\sigma_{\mathfrak{p}} + p\langle p\rangle\sigma_{\mathfrak{p}}^{-1})([\mathcal{E},\mathcal{P}])$ have the same image under $\mathrm{red}_{\mathfrak{p}}$. Now Proposition 8 gives

$$T_p([\mathcal{E},\mathcal{P}]) = \sum_{\Lambda}[\mathcal{E}/\Lambda, \mathcal{P}+\Lambda],$$

where the sum runs over subgroups $\Lambda \subset \mathcal{E}[p]$ of index p which have trivial intersection with the cyclic group generated by $\mathcal{P}$. On the other hand,

$$(\sigma_{\mathfrak{p}} + p\langle p\rangle\sigma_{\mathfrak{p}}^{-1})([\mathcal{E},\mathcal{P}]) = [\mathcal{E}^{\sigma_{\mathfrak{p}}}, \mathcal{P}^{\sigma_{\mathfrak{p}}}] + p[\mathcal{E}^{\sigma_{\mathfrak{p}}^{-1}}, p\mathcal{P}^{\sigma_{\mathfrak{p}}^{-1}}].$$

Thus we must check that

$$\sum_{\Lambda}[\widetilde{\mathcal{E}/\Lambda}, \widetilde{\mathcal{P}+\Lambda}] = [\tilde{\mathcal{E}}^{p}, \tilde{\mathcal{P}}^{p}] + p[\tilde{\mathcal{E}}^{p^{-1}}, p\tilde{\mathcal{P}}^{p^{-1}}].$$

This follows from Proposition 12, because the sum on the left-hand side has $p+1$ terms, exactly one of which coincides with the kernel of $\mathcal{E}[p] \to \tilde{\mathcal{E}}[p]$.

Given a smooth projective curve X and a correspondence $T = (Z, \varphi, \psi)$ on X, we use the same letter T to denote the endomorphism $\psi_* \circ \varphi^*$ of its Jacobian variety $\mathrm{Jac}(X)$. If $\Bbbk$ is an algebraically closed field then the map on points $\mathrm{Jac}(X)(\Bbbk) \to \mathrm{Jac}(X)(\Bbbk)$ can be obtained from $T : X(\Bbbk) \to \mathrm{Div}(X(\Bbbk))$ by extending the latter map to $\mathrm{Div}(X(\Bbbk))$, restricting to the subgroup $\mathrm{Div}^0(X(\Bbbk))$ of divisors of degree 0, and then passing to divisor classes.

We are concerned with the case $X = X_1(N)$, $T = T_p$. We denote the Jacobian variety $\mathrm{Jac}(X_1(N))$ simply by $J_1(N)$. Also, the automorphism of $J_1(N)$ induced by the diamond automorphism $\langle d\rangle$ of $X_1(N)$ will be denoted by the same symbol $\langle d\rangle$. If ℓ is a prime and n a positive integer, then the ring of endomorphisms of $J_1(N)$ acts on the abelian group $J_1(N)[\ell^n]$. So does $\mathrm{Gal}(\overline{\mathbb{Q}}/\mathbb{Q})$, because $J_1(N)$ is defined over $\mathbb{Q}$. In sketching a proof of the next statement we shall simply quote what we need from the work of Igusa [10], in particular the fact that $X_1(N)$ has good reduction at primes not dividing N.

Theorem 2. *Let $\sigma_{\mathfrak{p}} \in \mathrm{Gal}(\overline{\mathbb{Q}}/\mathbb{Q})$ be a Frobenius element at $\mathfrak{p}$. Then for $\ell \neq p$ and $n \geq 1$,*

$$T_p = \sigma_{\mathfrak{p}} + p\langle p\rangle \sigma_{\mathfrak{p}}^{-1}$$

as endomorphisms of $J_1(N)[\ell^n]$.

Proof. Let $\widetilde{X_1(N)}$ denote the reduction of $X_1(N)$ modulo p. As a map on points, the reduction map $X_1(N)(\overline{\mathbb{Q}}) \to \widetilde{X_1(N)}(\overline{\mathbb{F}}_p)$ is compatible with the map $[\mathcal{E}, \mathcal{P}] \mapsto [\tilde{\mathcal{E}}, \tilde{\mathcal{P}}]$ in a sense which we shall now describe.

Let $\mathbb{Z}[t]_{(p)}$ denote the localization of $\mathbb{Z}[t]$ at the prime ideal generated by p. We say that an elliptic curve E over $\mathbb{Q}(t)$ has good reduction at p if there is a generalized Weierstrass equation for E over $\mathbb{Z}[t]_{(p)}$ with discriminant a unit of $\mathbb{Z}[t]_{(p)}$. The reduction of this equation modulo $p\mathbb{Z}[t]_{(p)}$ then defines an elliptic curve $\tilde{E}$ over $\mathbb{F}_p(t)$. Now let E be an elliptic curve over $\mathbb{Q}(t)$ with invariant t and good reduction at p. For example we can take E to be the curve defined by the equation

$$y^2 + xy = x^3 - \frac{36}{t-1728}x - \frac{1}{t-1728}$$

of discriminant $t^2/(t-1728)^3$. Let P be a point of order N on E, let $\tilde{P} \in \tilde{E}[N]$ be its reduction modulo $\mathfrak{p}$, and let $\tilde{K}$ be the fixed field of

$$\{\sigma \in \mathrm{Gal}(\overline{\mathbb{F}_p(t)}/\mathbb{F}_p(t)) : \sigma(P) = \pm P\},$$

where $\overline{\mathbb{F}_p(t)}$ denotes a separable algebraic closure of $\mathbb{F}_p(t)$. Then $\widetilde{X_1(N)}$ is characterized up to isomorphism as the smooth projective curve over $\mathbb{F}_p$ with function field $\tilde{K}$. Furthermore, by viewing $\tilde{E}$ as an elliptic curve over $\tilde{K}$ one obtains a reduction map

$$\begin{aligned} \tilde{E}(\overline{\mathbb{F}}_p)_{S'} &\longrightarrow \mathrm{Ell}_1(N)(\overline{\mathbb{F}}_p)_{S'} \\ x &\longmapsto [(\tilde{E})_x, (\tilde{P})_x] \end{aligned}$$

for any subset S' of $\mathbf{P}^1(\overline{\mathbb{F}})$ containing the places where $\tilde{E}$ has bad reduction. Let S be the inverse image of S' under the reduction map $\mathbf{P}^1(\overline{\mathbb{Q}}) \longrightarrow \mathbf{P}^1(\overline{\mathbb{F}})$. Then the diagram of reduction maps

$$\begin{array}{ccc} X_1(N)(\overline{\mathbb{Q}})_S & \longrightarrow & \mathrm{Ell}_1(N)(\overline{\mathbb{Q}}) \\ \downarrow & & \downarrow \\ \widetilde{X_1(N)}(\overline{\mathbb{F}}_p)_{S'} & \longrightarrow & \mathrm{Ell}_1(N)(\overline{\mathbb{F}}_p) \end{array}$$

commutes.

Henceforth we take S' to be the set of places where $\tilde{E}$ has bad or supersingular reduction. Note that S' is a finite set. The commutativity of the above diagram allows us to replace $\mathrm{Ell}_1(N)(\overline{\mathbb{Q}})_{\mathrm{ord}}$ and $\widetilde{\mathrm{Ell}_1(N)(\overline{\mathbb{F}})}$ in the statement of Proposition 13 by $X_1(N)(\overline{\mathbb{Q}})_S$ and $\widetilde{X_1(N)}(\overline{\mathbb{F}}_p)_{S'}$ respectively.

Now let $\widetilde{J_1(N)}$ denote the reduction of $J_1(N)$ modulo p, identifiable with the Jacobian of $\widetilde{X_1(N)}$. There is a commutative diagram

$$\begin{array}{ccc} \mathrm{Div}^0(X_1(N)(\overline{\mathbb{Q}})_S) & \longrightarrow & J_1(N)(\overline{\mathbb{Q}}) \\ \downarrow & & \downarrow \\ \mathrm{Div}^0(\widetilde{X_1(N)}(\overline{\mathbb{F}}_p)_{S'}) & \xrightarrow{\alpha} & \widetilde{J_1(N)}(\overline{\mathbb{F}}_p) \end{array}$$

in which the vertical arrows are reduction modulo $\mathfrak{p}$ and the horizontal arrows send a divisor to the point on the Jacobian representing its divisor class. Since S' is finite, α is surjective. Let $L \in J_1(N)(\overline{\mathbb{Q}})$ be a torsion point of ℓ-power order; we must show that

$$(T_p - \sigma_{\mathfrak{p}} - p\langle p\rangle\sigma_{\mathfrak{p}}^{-1})(L) = 0.$$

In fact it is enough to show that this equation holds after reduction modulo $\mathfrak{p}$, because reduction mod $\mathfrak{p}$ is injective on ℓ-torsion. Write $\tilde{L} = \alpha(\tilde{D})$ with $D \in \mathrm{Div}^0(X_1(N)(\overline{\mathbb{Q}})_S)$. According to Proposition 13, the point $(T_p - \sigma_{\mathfrak{p}} - p\langle p\rangle\sigma_{\mathfrak{p}}^{-1})(D)$ reduces to 0 modulo $\mathfrak{p}$, and consequently so does the divisor $(T_p - \sigma_{\mathfrak{p}} - p\langle p\rangle\sigma_{\mathfrak{p}}^{-1})(L)$.

3. L-FUNCTIONS

Theorem 2 is at best an approximation to the Eichler-Shimura relations, because it refers only to Frobenius *elements* of $\mathrm{Gal}(\overline{\mathbb{Q}}/\mathbb{Q})$, not to the Frobenius *correspondence* in characteristic p (cf. [19], p. 17, formulas (I) and (II)). Nevertheless, it suffices for the application to L-functions, to which we now turn.

3.1. The Hasse-Weil conjecture. Originally conceived of as an assertion about the zeta function of a smooth projective variety over a number field, the conjecture has since evolved into a more general statement about L-functions of motives. Here we shall restrict our attention to motives of a very special kind, namely motives afforded by H^1 of an abelian variety over $\mathbb{Q}$ and more generally products of such motives with Artin motives. To begin with we take the Artin motive to be trivial. Let A be an abelian variety of dimension g defined over $\mathbb{Q}$, and recall that for every prime number ℓ one has an ℓ-adic representation

$$\rho_\ell : \mathrm{Gal}(\overline{\mathbb{Q}}/\mathbb{Q}) \to \mathrm{Aut}(V_\ell(A)) \cong \mathrm{GL}(2g, \mathbb{Q}_\ell),$$

where $V_\ell(A) = \mathbb{Q}_\ell \otimes_{\mathbb{Z}_\ell} T_\ell(A)$ and $T_\ell(A)$ is the Tate module of A:

$$T_\ell(A) = \varprojlim_n A[\ell^n].$$

We let ρ_ℓ^* denote the *contragredient* representation on the *dual space* $V_\ell^*(A)$ of $V_\ell(A)$. Given a prime number p, one defines a polynomial $P_p(A,t) \in \mathbb{Z}[t]$ by the formula

$$P_p(A,t) = \det\left(1 - t\rho_\ell^*(\sigma_\mathfrak{p}^{-1})|V_\ell^*(A)^{I(\mathfrak{p})}\right),$$

where ℓ is any prime number different from p, $I(\mathfrak{p})$ and $\sigma_\mathfrak{p}$ denote respectively the inertia group and a Frobenius element of some prime ideal $\mathfrak{p}$ of $\overline{\mathbb{Q}}$ lying over p, and

$$V_\ell^*(A)^{I(\mathfrak{p})} = \{v \in V_\ell^*(A) : \rho_\ell^*(g)v = v \text{ for all } g \in I(\mathfrak{p})\}.$$

That $P_p(A,t)$ is independent of the choice of $\mathfrak{p}$ and $\sigma_\mathfrak{p}$ follows by a straightforward verification from the conjugacy under $\mathrm{Gal}(\overline{\mathbb{Q}}/\mathbb{Q})$ of the prime ideals lying over a given rational prime. Far deeper is the fact that $P_p(A,t)$ belongs to $\mathbb{Z}[t]$ and is independent of the choice of $\ell \neq p$. Indeed we are able to make this assertion for *all* p, and not just for the p where A has good reduction, precisely because we have confined ourselves to the case of abelian varieties, for which Grothendieck's semistable reduction theorem [9] is available: in the case of an arbitrary smooth projective variety, the analogues of $P_p(A,t)$ – defined using ℓ-adic cohomology groups $H_\ell^i(*)$ rather than the Tate module – are not yet known to be independent of ℓ when p is a prime of bad reduction and $i > 1$ (for $i = 1$ the ℓ-adic cohomology group is dual to the Tate module of the Albanese, so we are back to the case of abelian varieties). Now write

$$P_p(A,t) = \prod_{i=1}^{2g}(1 - \alpha_{i,p}t)$$

with complex numbers $\alpha_{i,p}$. One has

$$|\alpha_{i,p}| \leq \sqrt{p} \quad (1 \leq i \leq 2g)$$

with equality if p is a prime of good reduction, whence the Euler product

$$L(A,s) = \prod_p P_p(A,p^{-s})^{-1}$$

converges in the region $\mathrm{Re}(s) > 3/2$. Another consequence of the semistable reduction theorem is that one can associate to A a well-defined conductor $N(A)$ and sign $W(A) = \pm 1$ (cf. [17]). The definition of the "root number" $W(A)$ requires the theory of local epsilon factors [6].

Conjecture 1. *Put*

$$\Lambda(A,s) = N(A)^{s/2}((2\pi)^{-s}\Gamma(s))^g L(A,s).$$

Then $\Lambda(A,s)$ has an analytic continuation to an entire function of order one satisfying the functional equation

$$\Lambda(A,s) = W(A)\Lambda(A,2-s).$$

It is also useful to have at hand a slightly less precise formulation of the conjecture, evocative of the state of affairs which prevails when $H^1_\ell(A)$ is replaced by the cohomology of an arbitrary smooth projective variety:

Conjecture 1*. *There exist:*

- *a finite set S of prime numbers containing all primes where A has bad reduction,*
- *for each $p \in S$, a polynomial*

$$P_p^*(A,t) = \prod_{i=1}^{2g}(1-\alpha_{i,p}^* t) \in \mathbb{Z}[t]$$

with $|\alpha_{i,p}^*| < p$ for all i,

- *a positive integer $N^*(A)$, and*
- *a sign $W^*(A) \in \{\pm 1\}$,*

such that if

$$L^*(A,s) = \prod_{p \notin S} P_p(A,p^{-s})^{-1} \cdot \prod_{p \in S} P_p^*(A,p^{-s})^{-1}$$

and

$$\Lambda^*(A,s) = N^*(A)^{s/2}((2\pi)^{-s}\Gamma(s))^g L^*(A,s)$$

then $\Lambda^*(A,s)$ has an analytic continuation to an entire function of order one satisfying the functional equation

$$\Lambda^*(A,s) = W^*(A)\Lambda^*(A,2-s).$$

We have included a bound on $\alpha_{i,p}^*$ in the statement of Conjecture 1* to ensure that if Conjecture 1 is true then $N^*(A)$, $W^*(A)$, and $P_p^*(A,t)$ coincide respectively with $N(A)$, $W(A)$, and $P_p(A,t)$. Indeed for all good p (and hence in particular for all $p \notin S$) we already have the stronger information that $|\alpha_{i,p}| = \sqrt{p}$, so that the stated bound on $\alpha_{i,p}^*$ affords a uniform estimate

$$p > \begin{cases} |\alpha_{i,p}| & (p \notin S) \\ |\alpha_{i,p}^*| & (p \in S); \end{cases}$$

but a remark of Deligne-Serre ([7], p. 515, Lemme 4.9) then shows that $N^*(A)$, $W^*(A)$, and the $P_p^*(A,t)$ are uniquely determined by the functional equation, whence these quantities coincide with the corresponding quantities in Conjecture 1 whenever the latter conjecture is satisfied.

3.2. Modular forms. Quite apart from its significance for the arithmetic of abelian varieties, Conjecture 1 asserts the existence of a class of Dirichlet series with Euler products and functional equations. Such Dirichlet series arise naturally in the theory of modular forms.

Let k be a positive integer. Given a holomorphic function f on $\mathfrak{H}$ and a matrix

$$\gamma = \begin{pmatrix} a & b \\ c & d \end{pmatrix} \in \mathrm{GL}^+(2, \mathbb{R}),$$

we put

$$(f|_k\gamma)(z) = \frac{\det(\gamma)}{(cz+d)^k} f(\gamma z).$$

This formula defines a right action of $\mathrm{GL}^+(2,\mathbb{R})$ on the space of holomorphic functions on $\mathfrak{H}$. Now let Γ be a subgroup of finite index in $\mathrm{SL}(2,\mathbb{Z})$. A *modular form of weight k for* Γ is a holomorphic function f on $\mathfrak{H}$ satisfying two conditions:

(i) $f|_k\gamma = f$ for $\gamma \in \Gamma$.
(ii) For every $\delta \in \mathrm{SL}(2,\mathbb{Z})$ the function $f|_k\delta$ has a Fourier expansion of the form

$$(f|_k\delta)(z) = \sum_{n \geq 0} a(n) e^{2\pi i n z/M}.$$

If for every $\delta \in \mathrm{SL}(2,\mathbb{Z})$ the coefficient $a(0)$ in (ii) is 0 then f called a *cusp form.* The vector space of modular forms of weight k for Γ will be denoted $\mathrm{M}_k(\Gamma)$ and the subspace of cusp forms $\mathrm{S}_k(\Gamma)$. These spaces are finite-dimensional. We remark in passing that in condition (ii) the phrase "$\delta \in \mathrm{SL}(2,\mathbb{Z})$" can be replaced by "$\delta \in \mathrm{GL}^+(2,\mathbb{Q})$", where $\mathrm{GL}^+(2,\mathbb{Q}) = \mathrm{GL}(2,\mathbb{Q}) \cap \mathrm{GL}^+(2,\mathbb{R})$. This is simply a matter of writing an element of $\mathrm{GL}^+(2,\mathbb{Q})$ as the product of an element of $\mathrm{SL}(2,\mathbb{Z})$ and an upper triangular matrix. It follows in particular that if Γ is normalized by a matrix $\delta \in \mathrm{GL}^+(2,\mathbb{Q})$ then the spaces $\mathrm{M}_k(\Gamma)$ and $\mathrm{S}_k(\Gamma)$ are stable under the map $f \mapsto f|_k\delta$.

Let us now specialize to the case $\Gamma = \Gamma_1(N)$. In this case we denote the spaces $\mathrm{M}_k(\Gamma)$ and $\mathrm{S}_k(\Gamma)$ simply by $\mathrm{M}_k(N)$ and $\mathrm{S}_k(N)$. Furthermore, given a character χ of $(\mathbb{Z}/N\mathbb{Z})^\times$ we let $\mathrm{M}_k(N,\chi)$ and $\mathrm{S}_k(N,\chi)$ be the subspaces of $\mathrm{M}_k(N)$ and $\mathrm{S}_k(N)$ consisting of f such that

$$f|_k\gamma = \chi(d) f$$

for

$$\gamma = \begin{pmatrix} a & b \\ cN & d \end{pmatrix} \in \Gamma_0(N).$$

(Implicit in the notation $\chi(d)$ is the usual identification of characters of $(\mathbb{Z}/N\mathbb{Z})^\times$ with Dirichlet characters modulo N.) Another way to describe the subspaces $M_k(N,\chi)$ and $S_k(N,\chi)$ is to say that they are the χ-eigenspaces for the "diamond operators" $f \mapsto f|_k\langle d\rangle$. In this approach d denotes an element of $(\mathbb{Z}/N\mathbb{Z})^\times$, and the operator $\langle d\rangle$ is defined by setting

$$f|_k\langle d\rangle = f|_k\gamma$$

for any $\gamma \in \Gamma_0(N)$ which reduces modulo N to a matrix with d as lower right-hand entry. In view of the isomorphism

$$\begin{aligned} \Gamma_0(N)/\Gamma_1(N) &\longrightarrow (\mathbb{Z}/N\mathbb{Z})^\times \\ \text{coset of } \begin{pmatrix} a & b \\ cN & d \end{pmatrix} &\longmapsto d \pmod N \end{aligned}$$

the diamond operators give a well-defined action of $(\mathbb{Z}/N\mathbb{Z})^\times$ on $M_k(N)$ and $S_k(N)$, and consequently we have eigenspace decompositions

$$M_k(N) = \oplus_\chi M_k(N,\chi)$$

and

$$S_k(N) = \oplus_\chi S_k(N,\chi)$$

where χ runs over Dirichlet characters modulo N. Note that if χ is the trivial character then $M_k(N,\chi)$ and $S_k(N,\chi)$ coincide with $M_k(\Gamma_0(N))$ and $S_k(\Gamma_0(N))$ respectively.

Henceforth we restrict our attention to cusp forms. To see why cusp forms give rise to Dirichlet series with functional equations, observe that the matrix

$$W_N = \begin{pmatrix} 0 & -1 \\ N & 0 \end{pmatrix}$$

normalizes $\Gamma_1(N)$, whence $f|_kW_N \in S_k(N)$ if $f \in S_k(N)$. In fact

$$\begin{pmatrix} 0 & -1 \\ N & 0 \end{pmatrix}\begin{pmatrix} a & b \\ cN & d \end{pmatrix}\begin{pmatrix} 0 & -1 \\ N & 0 \end{pmatrix}^{-1} = \begin{pmatrix} d & -c \\ -bN & a \end{pmatrix},$$

so that $f|_kW_N \in S_k(N,\overline{\chi})$ if $f \in S_k(N,\chi)$. Now write

$$f(z) = \sum_{n\geq 1} a(n)e^{2\pi inz}$$

and

$$(f|_kW_N)(z) = \sum_{n\geq 1} b(n)e^{2\pi inz},$$

and put

$$A(s) = N^{s/2}(2\pi)^{-s}\Gamma(s)\sum_{n\geq 1} a(n)n^{-s}$$

and

$$B(s) = N^{s/2}(2\pi)^{-s}\Gamma(s)\sum_{n\geq 1} b(n)n^{-s}.$$

As Hecke observed, these Dirichlet series converge absolutely in some right half-plane and can be analytically continued by a method which goes back to Riemann's paper on the Riemann zeta function: The usual interchange of summation and integration shows that

$$A(s) = \int_0^\infty f(it/\sqrt{N})t^s \frac{dt}{t},$$

whence

$$\begin{aligned} A(s) &= \int_0^1 f(it/\sqrt{N})t^s \frac{dt}{t} + \int_1^\infty f(it/\sqrt{N})t^s \frac{dt}{t} \\ &= \int_1^\infty \left(f(i/(t\sqrt{N}))t^{-s} + f(it/\sqrt{N})t^s\right)\frac{dt}{t} \end{aligned}$$

on making the change of variables $t \mapsto 1/t$ in the integral from 0 to 1. Since

$$f\left(\frac{i}{t\sqrt{N}}\right) = (it)^k (f|_k W_N)(it/\sqrt{N})$$

we obtain

$$A(s) = \int_1^\infty \left(i^k (f|_k W_N)(it/\sqrt{N})t^{k-s} + f(it/\sqrt{N})t^s\right)\frac{dt}{t}.$$

But $(W_N)^2 = -NI$, and consequently $f|_k(W_N)^2 = (-1)^k f$. Hence one can repeat the preceding calculation with $A(s)$ replaced by $B(s)$, f by $f|_k W_N$, and $f|_k W_N$ by $(-1)^k f$, and a comparison of the resulting expressions for $A(s)$ and $B(s)$ yields:

Proposition 14. *The functions $A(s)$ and $B(s)$ have analytic continuations to entire functions of order one satisfying the functional equation $A(s) = i^k B(k-s)$.*

We have avoided calling the Dirichlet series $\sum a(n)n^{-s}$ and $\sum b(n)n^{-s}$ as L-functions, because as yet we have imposed no condition to guarantee the existence of an Euler product. For this we need the Hecke operators.

3.3. Hecke operators. Given a prime number p, let Δ_p denote the set of 2×2 matrices with integer coefficients and determinant p which are congruent modulo N to a matrix of the form

$$\begin{pmatrix} 1 & * \\ 0 & p \end{pmatrix}.$$

It is immediate from the definition that Δ_p is stable under left and right multiplication by $\Gamma_1(N)$, and elementary calculations show that if $\Gamma = \Gamma_1(N)$ and

$$\delta_p = \begin{pmatrix} 1 & 0 \\ 0 & p \end{pmatrix}$$

then Δ_p has the one-term double-coset decomposition

$$\Delta_p = \Gamma \delta_p \Gamma$$

and the following decomposition as a disjoint union of right cosets:

$$\Delta_p = \begin{cases} \bigcup_{\nu=0}^{p-1} \Gamma \begin{pmatrix} 1 & \nu \\ 0 & p \end{pmatrix} \ \cup \ \Gamma \langle p \rangle \begin{pmatrix} p & 0 \\ 0 & 1 \end{pmatrix}, & \text{if } p \nmid N \\ \bigcup_{\nu=0}^{p-1} \Gamma \begin{pmatrix} 1 & \nu \\ 0 & p \end{pmatrix}, & \text{if } p | N \end{cases}$$

(recall that if p does not divide N then $\langle p \rangle$ denotes an arbitrary element of $\Gamma_0(N)$ with lower right-hand entry congruent to p modulo N). Of course if $\gamma \in \Gamma_1(N)$ and $\{\delta\}$ is any set of representatives for the distinct right cosets of $\Gamma_1(N)$ in Δ_p then $\{\delta\gamma\}$ is another such set, because Δ_p is stable under right multiplication by $\Gamma_1(N)$.

The p-th Hecke operator

$$T_p : \mathrm{S}_k(N) \longrightarrow \mathrm{S}_k(N)$$

is defined by the formula

$$f|_k T_p = p^{k/2-1} \sum_{\delta} f|_k \delta,$$

where δ runs over a set of representatives for the distinct right cosets of $\Gamma_1(N)$ in Δ_p. The definition is independent of the choice of coset representatives because $f \in \mathrm{S}_k(N)$. Furthermore, $f|_k T_p$ does belong to $\mathrm{S}_k(N)$, because right multiplication by any $\gamma \in \Gamma_1(N)$ sends one set of right coset representatives to another. For much the same reason, T_p commutes with the diamond operators $\langle d \rangle$, whence each subspace $\mathrm{S}_k(N, \chi)$ is stable under T_p: since $\Gamma_0(N)$ normalizes both Δ_p and $\Gamma_1(N)$, conjugation by an element of $\Gamma_0(N)$ sends one set of right coset representatives for $\Gamma_1(N)$ in Δ_p to another. To exhibit the effect of T_p on Fourier expansions, suppose that $f \in \mathrm{S}_k(N, \chi)$ and write

$$f(z) = \sum_{n \geq 1} a(n) q^n$$

with $q = e^{2\pi i z}$. A straightforward calculation using the right coset representatives listed above gives

$$(f|_k T_p)(z) = \sum_{n \geq 1} a(pn) q^n + \chi(p) p^{k-1} \sum_{n \geq 1} a(n) q^{pn}.$$

Note that if p divides N then $\chi(p)$ is to be interpreted as 0 in keeping with the usual conventions for Dirichlet characters modulo N. In the literature T_p is often denoted U_p in this case and the preceding formula is written

$$(f|_k U_p)(z) = \sum_{n\geq 1} a(pn)q^n \qquad (p|N).$$

The notation U_p has the advantage of forestalling an ambiguity which in principle could arise when $N = pM$, $p \nmid M$, and $f \in \mathrm{S}_k(M)$: in this situation the expression $f|_k T_p$ can have two possible meanings depending on whether we regard f as belonging to $\mathrm{S}_k(M)$ or to $\mathrm{S}_k(N)$. Nevertheless, we shall continue to use the notation T_p for all primes p, leaving the appropriate interpretation to context.

By a *Hecke eigenform* we shall mean a nonzero element of $\mathrm{S}_k(N, \chi)$ which is an eigenvector of the operators T_p for all primes p. If $f = \sum a(n)q^n$ is a Hecke eigenform and λ_p is the eigenvalue of T_p on f, then the above formula for $f|_k T(p)$ gives

$$a(pn) - \lambda_p a(n) + \chi(p)p^{k-1}a(n/p) = 0 \qquad (n \geq 1),$$

where $a(n/p)$ is understood to be 0 if n is not divisible by p. Taking $n = 1$ we see that $a(p) = \lambda_p a(1)$, so that $a(1) = 0$ implies $a(p) = 0$. More generally, using induction on the total number of prime factors of n one finds that if $a(1) = 0$ then $a(n) = 0$ for all $n \geq 1$, whence $f = 0$. Therefore:

Proposition 15. *If $f = \sum_{n\geq 1} a(n)q^n$ is a Hecke eigenform then $a(1) \neq 0$.*

A Hecke eigenform $f = \sum a(n)q^n$ is said to be *normalized* if $a(1) = 1$. The proposition implies that if f is any Hecke eigenform then some scalar multiple of f is normalized. For a normalized eigenform the relation $a(p) = \lambda_p a(1)$ becomes $\lambda_p = a(p)$, whence the recursion formula for $a(n)$ becomes

$$a(pn) - a(p)a(n) + \chi(p)p^{k-1}a(n/p) = 0.$$

Taking $n = p^{\nu-1}$ with $\nu \geq 1$ one sees that

$$a(p^\nu) - a(p)a(p^{\nu-1}) + \chi(p)p^{k-1}a(p^{\nu-2}) = 0,$$

and then taking $n = p^{\nu-1}m$ with m relatively prime to p one deduces by induction on ν that $a(p^\nu m) = a(p^\nu)a(m)$. A further induction on the number of distinct prime factors of some l relatively prime to m shows that $a(lm) = a(l)a(m)$. In other words, the function $n \mapsto a(n)$ is multiplicative; the associated formal Dirichlet series has an Euler product:

$$\sum_{n\geq 1} a(n)n^{-s} = \prod_p \Big(\sum_{\nu\geq 0} a(p^\nu)p^{-\nu s}\Big).$$

On the other hand, the recursion relation for $a(p^\nu)$ amounts to the formal identity

$$\sum_{\nu\geq 0} a(p^\nu)p^{-\nu s} = (1 - a(p)p^{-s} + \chi(p)p^{k-1}p^{-2s})^{-1},$$

and substitution in the preceding equation gives one direction of the following equivalence (the other is obtained by reversing the argument):

Proposition 16. *For an element* $f(z) = \sum_{n\geq 1} a(n)e^{2\pi inz}$ *of* $\mathrm{S}_k(N)$, *the following are equivalent:*

(i) f *is a normalized Hecke eigenform.*

(ii) $\sum_{n\geq 1} a(n)n^{-s} = \prod_p (1 - a(p)p^{-s} + \chi(p)p^{k-1-2s})^{-1}$.

If f is a normalized Hecke eigenform then the Dirichlet series in (ii) is called the L-function of f and denoted $L(f,s)$. ¿From Proposition 14 we know that there is a functional equation relating the L-function of f to a Dirichlet series associated to $f|_k W_N$, but we do not know that the latter Dirichlet series has an Euler product. Thus it remains to find conditions under which both f and $f|_k W_N$ are Hecke eigenforms. Such conditions are provided by the theory of new forms. The starting point is to define a suitable inner product on $\mathrm{S}_k(N)$.

3.4. The Petersson inner product. Put

$$\mathrm{S}_k = \bigcup_{\Gamma} \mathrm{S}_k(\Gamma),$$

where the union is taken over all subgroups of finite index in $\mathrm{SL}(2,\mathbb{Z})$. We define an inner product $(*,*)$ on S_k as follows. Given $f, g \in \mathrm{S}_k$, choose a subgroup Γ of finite index in $\mathrm{SL}(2,\mathbb{Z})$ such that f and g both belong to $\mathrm{S}_k(\Gamma)$, and put

$$(f,g) = [\mathrm{SL}(2,\mathbb{Z}) : \Gamma]^{-1} \int_{\Gamma\backslash \mathrm{GL}^+(2,\mathbb{R})} (f|_k r)(i)\overline{(g|_k r)(i)}dr,$$

where dr denotes the measure on $\Gamma\backslash\mathrm{GL}^+(2,\mathbb{R})$ afforded by a Haar measure on $\mathrm{GL}^+(2,\mathbb{R})$ (recall that $\mathrm{GL}^+(2,\mathbb{R})$ is unimodular – a left Haar measure is a right Haar measure). Using the fact that f and g are cusp forms, one can check that the integral is absolutely convergent. Furthermore, the factor $[\mathrm{SL}(2,\mathbb{Z}) : \Gamma]^{-1}$ in front of the integral guarantees that the value of (f,g) is independent of the choice of Γ. Now if

$$r = \begin{pmatrix} y & x \\ 0 & 1 \end{pmatrix}$$

with $y > 0$ then

$$(f|_k r)(i)\overline{(f|_k r)(i)} = |f(x+iy)|^2 y^k,$$

and consequently $(f,f) > 0$ if $f \neq 0$. Thus $(*,*)$ is in fact an inner product. Since we have not specified a choice of Haar measure on $\mathrm{GL}^+(2,\mathbb{R})$, we have defined $(*,*)$ only up to a scalar multiple; this suffices for our purposes.

Next we observe that if $f, g \in \mathrm{S}_k$ and $\delta \in \mathrm{GL}^+(2,\mathbb{Q})$ then $f|_k\delta$ and $g|_k\delta^{-1}$ both belong to S_k and

$$(f|_k\delta, g) = (f, g|_k\delta^{-1}).$$

Indeed choose Γ of finite index in $\mathrm{SL}(2,\mathbb{Z})$ so that $f, g \in \mathrm{S}_k(\Gamma)$. Then the groups $\Gamma' = \Gamma \cap \delta^{-1}\Gamma\delta$ and $\Gamma'' = \Gamma \cap \delta\Gamma\delta^{-1}$ also have finite index in $\mathrm{SL}(2,\mathbb{Z})$ and satisfy $\delta\Gamma'\delta^{-1} = \Gamma''$. Hence we can express $(f|_k\delta, g)$ and $(f, g|_k\delta^{-1})$ as integrals over $\Gamma'\backslash\mathrm{GL}^+(2,\mathbb{R})$ and $\Gamma''\backslash\mathrm{GL}^+(2,\mathbb{R})$ respectively, and the stated formula follows from the left-invariance of Haar measure on $\mathrm{GL}^+(2,\mathbb{R})$. More generally, taking $\gamma, \gamma' \in \Gamma$ and replacing δ by $\delta\gamma'$, we find that

$$(f|_k\gamma\delta\gamma', g) = (f, g|_k\delta^{-1}),$$

because $f|_k\gamma = f$ and $g|_k(\gamma')^{-1} = g$.

Let us apply the preceding formula with $\Gamma = \Gamma_1(N)$ and $\delta \in \Delta_p$, where p is a prime not dividing N. Since Δ_p is equal to a single double coset of Γ, we have $\Delta_p = \Gamma\delta\Gamma$ and consequently

$$(f|_k\delta', g) = (f, g|_k\delta^{-1})$$

for any $\delta' \in \Delta_p$. It follows that

$$(f|_kT_p, g) = (p+1)p^{k/2-1}(f, g|_k\delta^{-1}), \tag{1}$$

because $f|_kT_p$ is the sum of $p+1$ terms of the form $p^{k/2-1}f|_k\delta'$. Take $\delta = \delta_p$ in (1), and as usual, let $\langle p\rangle$ denote any element of $\Gamma_0(N)$ with lower right-hand entry congruent to p modulo N (and hence with upper left-hand entry congruent to p^{-1} modulo N). Since

$$(pI)\langle p\rangle(\delta_p)^{-1} \in \Delta_p$$

formula (1) becomes

$$(f|_kT_p, g) = (p+1)p^{k/2-1}(f, (g|_k\langle p\rangle^{-1})|_k\delta) \tag{2}$$

with some new element δ of Δ_p. On the other hand, repeating a previous argument we see that $(f, (g|_k\langle p\rangle^{-1})|_k\gamma\delta\gamma')$ is independent of $\gamma, \gamma' \in \Gamma$, and consequently that

$$(p+1)p^{k/2-1}(f, (g|_k\langle p\rangle^{-1})|_k\delta) = (f, (g|_k\langle p\rangle^{-1})|_kT_p). \tag{3}$$

Together, (2) and (3) give

$$\check{T}_p = \langle p\rangle^{-1}T_p,$$

where $\check{T}_p$ denotes the adjoint of T_p on $\mathrm{S}_k(N)$ with respect to $(*,*)$. Since the diamond operators commute with the Hecke operators, we conclude that for p not dividing N the operators T_p are normal. We also obtain:

Proposition 17. *Let $f \in S_k(N,\chi)$ be a Hecke eigenform and p a prime not dividing N. If λ_p is the eigenvalue of T_p on f then $\overline{\lambda_p} = \overline{\chi}(p)\lambda_p$.*

As a commuting family of normal operators, the operators T_p $(p \nmid N)$ are simultaneously diagonalizable on $S_k(N)$. However, simultaneous diagonalization of the T_p for all primes p, including those dividing N, is a more delicate matter and is possible in general only on a subspace of $S_k(N)$, the subspace of new forms.

3.5. New forms. Consider positive integers M and r such that M divides N properly and r divides N/M, and put

$$V_r = \begin{pmatrix} r & 0 \\ 0 & 1 \end{pmatrix}.$$

The calculation

$$\begin{pmatrix} r & 0 \\ 0 & 1 \end{pmatrix} \begin{pmatrix} a & b \\ cN & d \end{pmatrix} \begin{pmatrix} r & 0 \\ 0 & 1 \end{pmatrix}^{-1} = \begin{pmatrix} a & br \\ cN/r & d \end{pmatrix}$$

shows that the map $f \mapsto f|_k V_r$ sends $S_k(M)$ to $S_k(N)$, indeed each subspace $S_k(M,\chi)$ to the corresponding subspace of $S_k(N)$. In fact a glance at Fourier expansions shows that if p does not divide N then $(f|_k T_p)|_k V_r = (f|_k V_r)|_k T_p$, so that $f \mapsto f|_k V_r$ sends eigenvectors of T_p to eigenvectors of T_p. The need for a distinction between "old forms" and "new forms" arises because this last assertion fails for p dividing N.

The *space of old forms of level* N is by definition the subspace $S_k(N)^{\text{old}}$ of $S_k(N)$ spanned by the images of the maps $f \mapsto f|_k V_r$ as M and r vary over all integers satisfying the divisibility conditions stated above. In other words,

$$S_k(N)^{\text{old}} = \text{span}\left(\bigcup_{\substack{M|N \\ M<N}} \bigcup_{r|N/M} \{f|_k V_r : f \in S_k(M)\} \right).$$

A Hecke eigenform belonging to $S_k(N)^{\text{old}}$ is called an *old form of level* N. The *space of new forms of level* N, denoted $S_k(N)^{\text{new}}$, is the orthogonal complement of $S_k(N)^{\text{old}}$ in $S_k(N)$ relative to the Petersson inner product. A Hecke eigenform belonging to $S_k(N)^{\text{new}}$ is called a *new form of level* N, and a normalized new form of level N is called a *primitive form of level* N. Let $\text{Prim}_k(N)$ denote the set of primitive forms of weight k and level N. One of the main theorems of the theory of new forms is that $\text{Prim}_k(N)$ is a basis for $S_k(N)^{\text{new}}$; as a corollary one deduces that the set

$$\bigcup_{M|N} \bigcup_{r|N/M} \{f|_k V_r : f \in \text{Prim}_k(M)\}$$

is a basis for all of $S_k(N)$. Results such as these are important to mention here because they show that the theory of new forms is nonvacuous, but for present purposes the result of primary interest is the following theorem, which will lead us to a functional equation for the L-function of a primitive form:

Theorem 3. *Given $f \in \mathrm{Prim}_k(N)$, $g \in S_k(N)$, and a finite set S of prime numbers such that g is an eigenvector of T_p for $p \notin S$, suppose that the eigenvalues of T_p on f and g coincide for $p \notin S$. Then g is a scalar multiple of f.*

For the proof, the reader is referred to the literature on new forms: Atkin-Lehner [2], Casselman [4], Li [14], and Miyake [15]. The application to L-functions starts from the observation that if $f \in S_k(N, \chi)$ and we set

$$\check{f}(z) = \overline{f(-\overline{z})}$$

then $\check{f} \in S_k(N, \overline{\chi})$. This follows from the identity $-\overline{\gamma z} = \gamma'(-\overline{z})$, where $\gamma \in \mathrm{GL}^+(2, \mathbb{R})$ and γ' is obtained from γ by negating the diagonal entries. One also verifies that the map $f \mapsto \check{f}$ is unitary with respect to $(*, *)$ and preserves $S_k(N)^{\mathrm{old}}$, whence it preserves $S_k(N)^{\mathrm{new}}$ as well. Now at the level of Fourier expansions the map $f \mapsto \check{f}$ has the form

$$\sum_{n \geq 1} a(n) e^{2\pi i n z} \longmapsto \sum_{n \geq 1} \overline{a(n)} e^{2\pi i n z}.$$

Hence on applying complex conjugation to the formal identity in part (ii) of Proposition 16, we see that if f is a normalized Hecke eigenform, then so is $\check{f}$. Since $S_k(N)^{\mathrm{new}}$ is stable under $f \mapsto \check{f}$ we conclude that $\mathrm{Prim}_k(N)$ is stable under this map also.

Suppose now that $f \in S_k(N, \chi)$. We shall compare $\check{f}$ and $f|_k W_N$. For a prime p not dividing N, let Δ'_p denote the set consisting of 2×2 matrices with integer coefficients and determinant p which are congruent modulo N to a matrix of the form

$$\begin{pmatrix} p & * \\ 0 & 1 \end{pmatrix}.$$

A calculation shows that

$$W_N \Delta_p W_N^{-1} = \Delta'_p = \Delta_p \langle p \rangle^{-1}.$$

Since W_N normalizes $\Gamma_1(N)$ we deduce that if $\{\delta\}$ is a set of representatives for the distinct right cosets of $\Gamma_1(N)$ in Δ_p then both $\{W_N \delta W_N^{-1}\}$ and $\{\delta \langle p \rangle^{-1}\}$ are sets of representatives for the distinct right cosets of $\Gamma_1(N)$ in Δ'_p. It follows that

$$f|_k W_N|_k T_p = \overline{\chi(p)} f|_k T_p|_k W_N.$$

Thus f is an eigenvector of T_p for all primes p not dividing N then so is $f|_k W_N$, and if λ_p is the eigenvalue of T_p on f then $\lambda'_p = \overline{\chi(p)\lambda_p}$ is the eigenvalue of T_p on $f|_k W_N$. Referring to Proposition 17, we see that $\lambda'_p = \overline{\lambda_p}$, and then Theorem 3 implies that $f|_k W_N$ is a scalar multiple of $\check{f}$. We shall write the scalar in question as $i^{-k}W(f)$, so that

$$f|_k W_N = i^{-k} W(f) \check{f}.$$

Then Proposition 14 gives:

Proposition 18. *Given* $f \in \mathrm{Prim}_k(N)$, *put*

$$\Lambda(f,s) = N^{s/2}(2\pi)^{-s}\Gamma(s)L(f,s).$$

Then $\Lambda(f,s)$ *has an analytic continuation to an entire function of order one satisfying the functional equation* $\Lambda(f,s) = W(f)\Lambda(\check{f}, k-s)$.

We have reached the limits of what can be done to suggest a possible connection between modular forms and Conjecture 1 on the basis of formal analytic properties alone. The next step is to make a connection between modular forms and modular curves, or at least between cusp forms of weight 2 and regular differentials on modular curves.

3.6. Differentials and cusp forms of weight 2. To begin with let Γ be any subgroup of finite index in $\mathrm{SL}(2,\mathbb{Z})$ and let π denote the restriction to $\mathfrak{H}$ of the natural map $\mathfrak{H}^* \to \Gamma\backslash\mathfrak{H}^*$. If ω is a regular differential on $\Gamma\backslash\mathfrak{H}^*$ then $\pi^*\omega = f(z)dz$ for some function f on $\mathfrak{H}$. We claim that the functions f which arise in this way are characterized by the following conditions:

(o) f is holomorphic.

(i) $f(\gamma z)d(\gamma z) = f(z)dz$ for $\gamma \in \Gamma$.

(ii) Suppose that $\delta \in \mathrm{SL}(2,\mathbb{Z})$, and let M be a positive integer such that

$$(f \circ \delta)(z+M)d(\delta(z+M)) = (f\circ\delta)(z)d(\delta z)$$

(such an integer exists by (i)). Let F be the holomorphic function on the punctured unit disk $D^\circ = \{q \in \mathbb{C} : 0 < |q| < 1\}$ defined by

$$f(\delta z)\frac{d}{dz}\delta z = F(e^{2\pi i z/M}) \qquad (z \in \mathfrak{H}).$$

Then F extends to a holomorphic function on the full unit disk $D = \{q \in \mathbb{C} : 0 < |q| < 1\}$ vanishing at 0.

Indeed (i) says that the differential $f(z)dz$ on $\mathfrak{H}$ descends to a differential on $\Gamma\backslash\mathfrak{H}$, while (o) is the condition for the descended differential to be holomorphic (at an elliptic fixed point of Γ the equivalence between the holomorphy of ω and the holomorphy of f requires a small verification). As for (ii), its content is that the descended differential extends holomorphically from $\Gamma\backslash\mathfrak{H}$ to $\Gamma\backslash\mathfrak{H}^*$. Again there is a small verification: if we assume without loss of generality that M is minimal, then the change of variables $w = \delta z$, $q = e^{2\pi i \delta^{-1} w/M}$ defines a local parameter at $\delta\infty$, and condition (ii) is a consequence of the fact that $dw = \frac{dq}{q} \cdot \frac{d}{dz}\delta z \cdot \frac{M}{2\pi i}$. Thus (o), (i), and (ii) do characterize the functions on $\mathfrak{H}$ obtained by pulling back regular differentials from $\Gamma^*\backslash\mathfrak{H}$. Now if $\gamma = \begin{pmatrix} a & b \\ c & d \end{pmatrix}$ is an element of $\mathrm{GL}^+(2,\mathbb{R})$ then

$$d(\gamma z) = \frac{\det(\gamma)}{(cz+d)^2} dz.$$

Therefore condition (i) can be rewritten

$$f|_2\gamma = f \qquad (\gamma \in \Gamma),$$

while in (ii) the requirement is the existence, for any $\delta \in \mathrm{SL}(2,\mathbb{Z})$, of a Fourier series expansion of the form

$$(f|_2\delta)(z) = \sum_{n\geq 1} a(n) e^{2\pi i n z/M}.$$

Returning to the equation $\pi^*\omega = f(z)dz$, we conclude that as ω runs over the space of regular differentials on $\Gamma\backslash\mathfrak{H}^*$ the function f runs over the space of cusp forms of weight 2 for Γ.

Let us now specialize to the case $\Gamma = \Gamma_1(N)$. We shall write $H^0(\Omega^1_{X_1(N)})$ for the space of regular differentials on $X_1(N)$ defined over $\mathbb{Q}$, and similarly $H^0(\Omega^1_{X_1(N)/\mathbb{C}})$ for the corresponding space over $\mathbb{C}$, so that

$$H^0(\Omega^1_{X_1(N)/\mathbb{C}}) = \mathbb{C} \otimes_{\mathbb{Q}} H^0(\Omega^1_{X_1(N)}).$$

The isomorphism just described gives an identification

$$H^0(\Omega^1_{X_1(N)/\mathbb{C}}) \cong \mathrm{S}_2(\Gamma_1(N)),$$

and on the right-hand side we have an action of the Hecke operators T_p. As we shall now explain, the Hecke correspondences determine operators on the left-hand side (to be denoted T_p also) such that the above isomorphism respects the action of T_p. Quite generally, if $T = (Z, \varphi, \psi)$ is a correspondence on a smooth projective curve X, then T gives rise to the operator

$$\begin{aligned} H^0(\Omega^1_X) &\longrightarrow H^0(\Omega^1_X) \\ \omega &\longmapsto \mathrm{tr}_\varphi(\psi^*\omega), \end{aligned}$$

where tr_φ is the trace on differentials associated to the morphism φ: if $K \subset L$ is the inclusion of function fields afforded by φ then any $\alpha \in H^0(\Omega^1_Z)$ has the form $\alpha = udv$ with $u \in L$ and $v \in K$, and by definition, $\mathrm{tr}_\varphi(\alpha) = \mathrm{tr}_{L/K}(u)dv$. Returning to the case at hand, we see that the Hecke correspondence $T_p = (X_1(N,p), \varphi_p, \psi_p)$ on $X_1(N)$ determines an operator T_p on $H^0(\Omega^1_{X_1(N)})$ and hence by extension of scalars an operator on $H^0(\Omega^1_{X_1(N)/\mathbb{C}})$.

Proposition 19. *The canonical isomorphism*

$$H^0(\Omega^1_{X_1(N)/\mathbb{C}}) \cong \mathrm{S}_2(N)$$

commutes with the action of T_p.

Proof. We begin with a general remark. Suppose that X is a smooth projective curve over a subfield of $\mathbb{C}$ and $T = (Z, \varphi, \psi)$ is a correspondence on X. Let d be the degree of φ. Since the base field is a subfield of $\mathbb{C}$, we can discuss the correspondence T in the language of Riemann surfaces, and in particular we can speak of the local analytic sections $(\varphi^{-1})_{x_0,i}$ $(1 \leq i \leq d)$ of φ in a neighborhood of some unramified point $x_0 \in X(\mathbb{C})$. Then

$$\mathrm{tr}_\varphi(\alpha) = \sum_{1 \leq i \leq d} ((\varphi^{-1})_{x_0,i})^* \alpha$$

locally at x_0. Thus for $\omega \in H^0(\Omega^1_{X/\mathbb{C}})$ we have

$$T(\omega) = \sum_{1 \leq i \leq d} (\varphi_i^{-1})^*_{x_0,i} \psi^* \omega.$$

This formula makes it possible to compute the action of T on differentials directly from a knowledge of the map $T : X(\mathbb{C}) \longrightarrow \mathrm{Div}(X(\mathbb{C}))$. Indeed if the latter map is given locally by

$$x \longmapsto \sum_{1 \leq i \leq d} (\rho_{x_0,i}(x))$$

with analytic functions $\rho_{x_0,i}$, then after a permutation of indices we have $\rho_{x_0,i} = \psi \circ (\varphi_i^{-1})_{x_0,i}$ and consequently

$$T(\omega) = \sum_{1 \leq i \leq d} \rho^*_{x_0,i} \omega$$

locally at x_0.

In the case at hand we can identify $X_1(N)(\mathbb{C})$ with $\Gamma_1(N)\backslash\mathfrak{H}^*$, and in the coordinate z of $\mathfrak{H}$ the map $T_p : X_1(N)(\mathbb{C}) \longrightarrow \mathrm{Div}(X_1(N)(\mathbb{C}))$ is given by

$$[z] \longmapsto \begin{cases} \sum_{\nu=0}^{p-1} [(z+\nu)/p] + [\langle p \rangle pz] & \text{if } p \nmid N \\ \sum_{\nu=0}^{p-1} [(z+\nu)/p] & \text{if } p | N \end{cases}$$

(Proposition 9). This map can be written simply as

$$[z] \longmapsto \sum_{\delta} [\delta z],$$

where δ runs over a set of representatives for the right cosets of $\Gamma_1(N)$ in Δ_p. Now we have already observed that if f is a function on $\mathfrak{H}$ and $\gamma \in \mathrm{GL}^+(2,\mathbb{R})$ then $f(\gamma z)d(\gamma z) = (f|_2\gamma)(z)dz$. Thus for $f \in \mathrm{S}_2(N)$ we can write

$$\sum_{\delta} f(\delta z)d(\delta z) = \sum_{\delta} (f|_2\delta)(z)dz = (f|_2T_p)(z)dz,$$

the factor $p^{k/2-1}$ in the definition of T_p being 1 for $k = 2$. This proves Proposition 19.

Using Proposition 11 one proves the analogous statement for the diamond operators:

Proposition 20. *The canonical isomorphism*

$$H^0(\Omega^1_{X_1(N)/\mathbb{C}}) \cong \mathrm{S}_2(N)$$

commutes with the action of $\langle d \rangle$ *for* $d \in (\mathbb{Z}/N\mathbb{Z})^\times$.

One consequence of Propositions 19 and 20 is that $\mathrm{S}_2(N)$ has a $\mathbb{Q}$-form stable under the operators T_p and $\langle d \rangle$, because $H^0(\Omega^1_{X_1(N)/\mathbb{C}})$ has such a $\mathbb{Q}$-form, namely $H^0(\Omega^1_{X_1(N)})$. It follows that if $\mathrm{S}_2(N,\chi)$ contains a Hecke eigenform on which the operator T_p has eigenvalue λ_p then for any $\sigma \in \mathrm{Aut}(\mathbb{C})$ the space $\mathrm{S}_2(N,\chi^\sigma)$ contains a Hecke eigenform on which the operator T_p has eigenvalue λ_p^σ. Now the Fourier coefficients of a normalized Hecke eigenform are polynomials with integer coefficients in the Hecke eigenvalues and the character values. Hence we can define an action of $\mathrm{Aut}(\mathbb{C})$ on the set of normalized Hecke eigenforms in $\mathrm{S}_2(N)$ by the rule

$$f = \sum_{n\geq 1} a(n)q^n \longmapsto f^\sigma = \sum_{n \geq 1} a(n)^\sigma q^n.$$

We claim that if f is a new form of level N then so is f^σ. Suppose on the contrary that f^σ is an old form. Then there is a proper divisor M of N and an element $g = \sum_{n\geq 1} b(n)q^n$ of $\mathrm{Prim}_2(M)$ such that $a(p)^\sigma = b(p)$ for $p \nmid N$. Then $a(p) = b(p)^{\sigma^{-1}}$ for $p \nmid N$, whence $g^{\sigma^{-1}}$ is a normalized Hecke eigenform in $\mathrm{S}_2(M)$ which has the same Hecke eigenvalues as f for $p \nmid N$. This contradicts Theorem 3, proving the claim. We conclude that the map $f \mapsto f^\sigma$ defines an action of $\mathrm{Aut}(\mathbb{C})$ on $\mathrm{Prim}_2(N)$. Since $\mathrm{Prim}_2(N)$ is finite, it follows that if $f \in \mathrm{Prim}_2(N)$ then the field generated by the Fourier coefficients of f has finite degree over $\mathbb{Q}$. We denote this field $\mathbb{E}_f$.

3.7. The Hecke algebra. Given a smooth projective curve X, we will let $\mathrm{Corr}(X)$ denote the free abelian group on the set of isomorphism classes of correspondences on X. If $T = (Z, \varphi, \psi)$ and $T' = (Z', \varphi', \psi')$ are correspondences on X we define the product of their isomorphism classes $[T'] \cdot [T]$ by the formula

$$[T'] \cdot [T] = \sum_W [\widetilde{W}, \varphi \circ \mathrm{pr}_Z \circ \nu_W, \psi' \circ \mathrm{pr}_{Z'} \circ \nu_W],$$

where W runs over the irreducible components of

$$Z'' = \{(z, z') \in Z \times Z' : \psi(z) = \varphi'(z')\},$$

$\nu_W : \widetilde{W} \to W$ is the normalization map, and $\mathrm{pr}_Z : Z'' \to Z$, $\mathrm{pr}_{Z'} : Z'' \to Z'$ are the projections. Extending this product to $\mathrm{Corr}(X)$ by $\mathbb{Z}$-linearity, we make $\mathrm{Corr}(X)$ into a $\mathbb{Z}$-algebra. We shall view $\mathrm{Aut}(X)$ as a subgroup of the multiplicative group of $\mathrm{Corr}(X)$ by identifying $\psi \in \mathrm{Aut}(X)$ with the isomorphism class of the correspondence (X, id_X, ψ) on X.

In the case of $X_1(N)$ we are interested in the subalgebra of $\mathrm{Corr}(X_1(N))$ generated over $\mathbb{Z}$ by the isomorphism classes of all Hecke correspondences T_p and all diamond automorphisms $\langle d \rangle$. We denote this subalgebra by $\mathbb{T}$, and refer to it as the Hecke algebra (of level N). Furthermore, we use the same symbol $\mathbb{T}$ and the same term "Hecke algebra" for the image of $\mathbb{T}$ under the canonical embedding of $\mathrm{Corr}(X_1(N))$ in $\mathrm{End}(J_1(N))$, and we likewise identify the opposite algebra $\mathbb{T}^{\mathrm{opp}}$ with its image in $\mathrm{End}(H^0(\Omega^1_{X_1(N)}))$. Alternatively, we can view $\mathbb{T}$ itself as acting on the dual space of $H^0(\Omega^1_{X_1(N)})$, or we can consider $\mathbb{T}$ to be acting on $H^0(\Omega^1_{X_1(N)})$ on the right. This last point of view is consistent with our identification of $H^0(\Omega^1_{X_1(N)/\mathbb{C}})$ with $\mathrm{End}(\mathrm{S}_2(N))$ (Propositions 19 and 20), and we may therefore think of $\mathbb{T}$ as the subring of $\mathrm{End}(\mathrm{S}_2(N))$ generated over $\mathbb{Z}$ by the Hecke operators and diamond operators on $\mathrm{S}_2(N)$. It follows in particular that $\mathbb{T}$ is commutative, so that $\mathbb{T}$ and $\mathbb{T}^{\mathrm{opp}}$ are canonically isomorphic and every left $\mathbb{T}$-module is a right $\mathbb{T}$-module.

The next step is to associate a quotient ring $\mathbb{T}_f$ of $\mathbb{T}$ to each $f \in \mathrm{Prim}_2(N)$. Consider the ring homomorphism $\lambda_f : \mathbb{T} \to \mathbb{C}$ such that $f|_2 T = \lambda_f(T) f$ for $T \in \mathbb{T}$, and let $\mathbb{I}_f$ be the kernel of λ_f. We set

$$\mathbb{T}_f = \mathbb{T}/\mathbb{I}_f.$$

Thus $\mathbb{T}_f$ is the quotient of $\mathbb{T}$ by the annihilator ideal of f. Write $f(z) = \sum_{n \geq 1} a(n) q^n$, and recall that $\mathbb{E}_f = \mathbb{Q}(\{a(n) : n \geq 1\})$. If $\mathrm{S}_2(N, \chi)$ is the character space to which f belongs then λ_f induces an isomorphism

$$\mathbb{Q} \otimes_{\mathbb{Z}} \mathbb{T}_f \longrightarrow \mathbb{E}_f$$

sending $T_p + \mathbb{I}_f$ to $a(p)$ and $\langle d \rangle + \mathbb{I}_f$ to $\chi(d)$. Let A_f be the abelian variety over $\mathbb{Q}$ defined by

$$A_f = J_1(N)/\mathbb{I}_f J_1(N).$$

The action of $\mathbb{T}$ on $J_1(N)$ induces an action of $\mathbb{T}_f$ on A_f and hence on each $V_\ell(A_f)$.

Proposition 21. *The image of $\mathbb{T}_f$ in $\mathbb{E}_f$ is an order of $\mathbb{E}_f$, and $V_\ell(A_f)$ is a free module of rank two over $\mathbb{Q}_\ell \otimes_{\mathbb{Z}} \mathbb{T}_f$. In particular, A_f is an abelian variety of dimension $[\mathbb{E}_f : \mathbb{Q}]$, and A_f is an elliptic curve if and only if the Fourier coefficients of f are rational.*

Proof. The second statement is contained in the first because $V_\ell(A_f)$ is a vector space of dimension $2\dim(A_f)$ over $\mathbb{Q}_\ell$, while

$$\dim_{\mathbb{Q}_\ell} \mathbb{Q}_\ell \otimes_{\mathbb{Z}} \mathbb{T}_f = \operatorname{rank}_{\mathbb{Z}} \mathbb{T}_f = [\mathbb{E}_f : \mathbb{Q}].$$

To prove the first statement we start with the observation that as a subring of $\operatorname{End}(J_1(N))$, the Hecke algebra $\mathbb{T}$ acts on $H_1(J_1(N)(\mathbb{C}), \mathbb{Z})$ and consequently also on $H_1(X_1(N)(\mathbb{C}), \mathbb{Z})$, the two homology groups being isomorphic via the map on homology induced by the embedding of $X_1(N)(\mathbb{C})$ in $J_1(N)(\mathbb{C})$. Denoting the complex dual of $H^0(\Omega^1_{X_1(N)/\mathbb{C}})$ by $H^0(\Omega^1_{X_1(N)/\mathbb{C}})^*$, we see that the standard isomorphism of complex tori

$$J_1(N)(\mathbb{C}) \cong \frac{H^0(\Omega^1_{X_1(N)/\mathbb{C}})^*}{H_1(X_1(N)(\mathbb{C}), \mathbb{Z})}$$

is actually an isomorphism of $\mathbb{T}$-modules. Hence so is the isomorphism

$$J_1(N)(\mathbb{C}) \cong \mathrm{S}_2(N)^*/\Lambda, \tag{1}$$

where Λ is the image of $H_1(X_1(N)(\mathbb{C}), \mathbb{Z})$ when we identify $H^0(\Omega^1_{X_1(N)/\mathbb{C}})^*$ with $\mathrm{S}(2, N)^*$. The fact that the lattice Λ in $\mathrm{S}_2(N)^*$ is stable under $\mathbb{T}$ already shows that the eigenvalues of $\mathbb{T}$ on $\mathrm{S}_2(N)$ are algebraic integers, because eigenvalues are preserved under transpose. It follows that the image of $\mathbb{T}_f$ in $\mathbb{E}_f$ is an order.

Next put

$$V_f = \mathrm{S}_2(N)/(\mathrm{S}_2(N)|_2\mathbb{I}_f),$$

where $\mathrm{S}_2(N)|_2\mathbb{I}_f$ denotes the space of all $g|_2T$ with $g \in \mathrm{S}_2(N)$ and $T \in \mathbb{T}$. We identify V_f^* with the quotient of $\mathrm{S}_2(N)^*$ by $\mathbb{I}_f\mathrm{S}_2(N)^*$, and we let Λ_f be the lattice in V_f^* corresponding to $\Lambda/\mathbb{I}_f\Lambda$ under this identification. Then (1) induces an isomorphism of $\mathbb{T}_f$-modules

$$A_f(\mathbb{C}) \cong V_f^*/\Lambda_f. \tag{2}$$

We claim that V_f (hence also V_f^*) is a free module of rank one over $\mathbb{C} \otimes \mathbb{T}_f$. Granting the claim, we deduce that V_f^* is free of rank two over $\mathbb{R} \otimes \mathbb{T}_f$. Since $V_f^* = \mathbb{R} \otimes \Lambda_f$ it follows that there is a sublattice $\Lambda'_f \subset \Lambda_f$ which is free of rank two over $\mathbb{T}_f$. But (2) gives

$$T_\ell(A_f) \cong \varprojlim_n (\ell^{-n}\Lambda_f)/\Lambda_f \cong \mathbb{Z}_\ell \otimes \Lambda_f.$$

Therefore

$$V_\ell(A_f) \cong \mathbb{Q}_\ell \otimes \Lambda_f \cong \mathbb{Q}_\ell \otimes \Lambda'_f,$$

and the proposition follows.

It remains to prove the claim. The semisimple ring $\mathbb{C}\otimes\mathbb{E}_f$ is canonically a product

$$\mathbb{C} \otimes \mathbb{E}_f = \prod_\sigma \mathbb{C},$$

where the factors are indexed by the distinct embeddings of $\mathbb{E}_f$ in $\mathbb{C}$. Projection onto the factor corresponding to σ gives a character $\mathrm{pr}_\sigma : \mathbb{C}\otimes\mathbb{E}_f \to \mathbb{C}$ sending $T_p + \mathbb{I}_f$ to $a(p)^\sigma$ and $\langle d\rangle + \mathbb{I}_f$ to $\chi(d)^\sigma$, and a simple $\mathbb{C}\otimes\mathbb{E}_f$-module is a one-dimensional complex vector space on which $\mathbb{C}\otimes\mathbb{E}_f$ acts through one of the characters pr_σ. Now as a finitely generated $\mathbb{C}\otimes\mathbb{E}_f$-module V_f is a direct sum of simple modules and is therefore spanned over $\mathbb{C}$ by eigenvectors with eigencharacters of the form pr_σ. Suppose that $v \in V_f$ is such an eigenvector. Then v is in particular an eigenvector for the family of operators $\mathcal{T}_f = \{T_p + \mathbb{I}_f : p \nmid N\}$. But the action of $\mathcal{T}_f$ on $V_f = \mathrm{S}_2(N)/(\mathrm{S}_2(N)|_2\mathbb{I}_f)$ is induced by the action of $\mathcal{T} = \{T_p : p \nmid N\}$ on $\mathrm{S}_2(N)$, and as a commuting family of normal operators $\mathcal{T}$ acts semisimply on $\mathrm{S}_2(N)$. It follows that v is the image in V_f of some $\mathcal{T}$-eigenvector $g \in \mathrm{S}_2(N)$. Then Theorem 3 implies that g is a scalar multiple of one of the cusp forms f^σ. It follows that the restriction to $\oplus_\sigma\mathbb{C}f^\sigma$ of the natural map of $S_2(N)$ onto V_f is surjective. But the restriction is also injective, because

$$(\oplus_\sigma\mathbb{C}f^\sigma) \cap (\mathrm{S}_2(N)|_2\mathbb{I}_f) \subset (\oplus_\sigma\mathbb{C}f^\sigma) \cap (\mathrm{S}_2(N)^{\mathrm{new}}|_2\mathbb{I}_f) = \{0\}$$

by the theory of new forms. Therefore V_f is isomorphic to $\oplus_\sigma\mathbb{C}f^\sigma$ as a $\mathbb{C}\otimes\mathbb{E}_f$-module and is consequently free of rank one.

We are now ready to compute the Euler factor of A_f at a prime of good reduction:

Theorem 4. *Let $f \in \mathrm{S}_2(N,\chi)$ be a primitive cusp form of level N, with Fourier expansion*

$$f(z) = \sum_{n\geq 1} a(n)e^{2\pi i n z}.$$

If p is a prime not dividing N then

$$P_p(A_f,t) = \prod_\sigma (1 - a(p)^\sigma t + \chi(p)^\sigma p t^2),$$

where σ runs over the distinct embeddings of $\mathbb{E}_f$ in $\mathbb{C}$.

Proof. Fix a prime ℓ and let ρ_ℓ denote the natural representation

$$\rho_\ell : \mathrm{Gal}(\overline{\mathbb{Q}}/\mathbb{Q}) \longrightarrow V_\ell(A_f).$$

It will suffice to prove that for a prime p not dividing ℓN we have

$$(1)\qquad \det(xI - \rho_\ell(\sigma_\mathfrak{p})) = \prod_\sigma (x^2 - a(p)^\sigma x + \chi(p)^\sigma p),$$

where x is an indeterminate, $\mathfrak{p}$ is a prime ideal of $\overline{\mathbb{Q}}$ lying over p, and $\sigma_\mathfrak{p} \in \mathrm{Gal}(\overline{\mathbb{Q}}/\mathbb{Q})$ is the Frobenius automorphism at $\mathfrak{p}$. Indeed the left-hand side of (1) coincides with the characteristic polynomial of $\rho_\ell^*(\sigma_\mathfrak{p}^{-1})$ on $V_\ell(A_f)$, because a matrix and its transpose have the same characteristic polynomial. Also $V_\ell(A_f) = V_\ell(A_f)^{I(\mathfrak{p})}$ by the criterion of Néron-Ogg-Shafarevich [18]. Hence if x is replaced by $1/t$ and the equation multiplied by $t^{2[\mathbb{E}_f:\mathbb{Q}]}$ then (1) becomes the stated formula for $P_p(A_f, t)$, valid for any prime p not dividing $N\ell$. Since ℓ was arbitrary the stated formula follows for any prime p not dividing N.

To prove (1) we recall a fact from linear algebra. Suppose that B is an $(mn) \times (mn)$ matrix which can be written as an $m \times m$ block matrix $B = (B^{ij})$ with $n \times n$ blocks B^{ij}. Suppose further that the ring generated over $\mathbb{Z}$ by the matrices B^{ij} is commutative. Then

$$\det B = \det(\det_{m\times m}(B)),$$

where $\det_{m\times m}(B)$ denotes the determinant of the $m \times m$ matrix over $\mathbb{Z}[B^{ij}]$ with ij-entry equal to B^{ij}. On replacing B by $xI - B$ we obtain the formula

$$\det(xI - B) = \det(\det_{m\times m}(xI - B))$$

for the characteristic polynomial of B.

To apply this formula, recall that $V_\ell(A_f)$ is a free module of rank two over $\mathbb{Q}_\ell \otimes \mathbb{T}_f$ and observe that $\rho_\ell(\sigma_\mathfrak{p})$ is a $\mathbb{Q}_\ell \otimes \mathbb{T}_f$-linear transformation of $V_\ell(A_f)$. Let $\det_{\mathbb{Q}_\ell\otimes\mathbb{T}_f}(xI - \rho_\ell(\sigma_\mathfrak{p}))$ denote the characteristic polynomial of $\rho_\ell(\sigma_\mathfrak{p})$ as a $\mathbb{Q}_\ell \otimes \mathbb{T}_f$-linear map. Then

$$\det(xI - \rho_\ell(\sigma_\mathfrak{p})) = N_{\mathbb{Q}_\ell\otimes\mathbb{T}_f/\mathbb{Q}_\ell}(\det_{\mathbb{Q}_\ell\otimes\mathbb{T}_f}(xI - \rho_\ell(\sigma_\mathfrak{p})),$$

where $N_{\mathbb{Q}_\ell\otimes\mathbb{T}_f/\mathbb{Q}_\ell}$ is the norm from $\mathbb{Q}_\ell \otimes \mathbb{T}_f[x]$ to $\mathbb{Q}_\ell[x]$ (which coincides on $\mathbb{T}_f[x]$ with the norm from $\mathbb{T}_f[x]$ to $\mathbb{Q}[x]$). To prove (1) it suffices to show that

$$(2)\qquad \det_{\mathbb{Q}_\ell\otimes\mathbb{T}_f}(xI - \rho_\ell(\sigma_\mathfrak{p})) = x^2 - T_p x + \langle p\rangle p,$$

because $N_{\mathbb{T}_f/\mathbb{Q}}(x^2 - T_p x + \langle p\rangle p)$ is the right-hand side of (1).

Write

$$\mathbb{Q}_\ell \otimes \mathbb{T}_f = \prod_{\lambda|\ell} \mathbb{E}_{f,\lambda},$$

where λ runs over the places of $\mathbb{E}_f$ dividing ℓ and $\mathbb{E}_{f,\lambda}$ idenotes the completion of $\mathbb{E}_f$ at λ. Also put $\rho_\lambda = \mathrm{pr}_\lambda \circ \rho_\ell$, where pr_λ is the projection map from $\mathbb{Q}_\ell \otimes \mathbb{T}_f$ to the factor $\mathbb{E}_{f,\lambda}$ on the right-hand side. Then equation (2) is equivalent to a system of equations indexed by the places λ, namely the equations

$$\det(xI - \rho_\lambda(\sigma_\mathfrak{p})) = x^2 - T_{p,\lambda}x + \langle p\rangle_\lambda p \tag{3}$$

with $T_{p,\lambda} = \mathrm{pr}_\lambda(T_p)$ and $\langle p\rangle_\lambda = \mathrm{pr}_\lambda(\langle p\rangle)$. It follows from Theorem 2 that the right-hand side of (3) annihilates $\rho_\lambda(\sigma_\mathfrak{p})$. Furthermore, if a nonscalar 2×2 matrix over a field is annihilated by a monic polynomial of degree 2 then that polynomial is its characteristic poynomial. Therefore (3) holds whenever $\rho_\lambda(\sigma_\mathfrak{p})$ is nonscalar. Now fix a place λ_0 dividing ℓ and let P_0 be the set of primes p not dividing $N\ell$ such that $\rho_{\lambda_0}(\sigma_\mathfrak{p})$ is scalar (note that this condition is independent of the choice of $\mathfrak{p}$). It remains to show that (3) holds for $\lambda = \lambda_0$ and all $p \in P_0$.

Let $e_0 \in \mathbb{Q}_\ell \otimes \mathbb{T}_f$ be the idempotent which generates the kernel of the map

$$\prod_{\lambda \neq \lambda_0} \mathrm{pr}_\lambda : \mathbb{Q}_\ell \otimes \mathbb{T}_f \longrightarrow \prod_{\lambda \neq \lambda_0} \mathbb{E}_{f,\lambda},$$

and choose an integer $\nu \geq 0$ such that the element $d_0 = \ell^\nu e_0$ belongs to $\mathbb{Z}_\ell \otimes \mathbb{T}_f$. Since $V_\ell(A_f)$ is free of rank two over $\mathbb{Q}_\ell \otimes \mathbb{T}_f$ it follows that the $\mathbb{Z}_\ell$-module

$$d_0 T_\ell(A_f) \cong \varprojlim_n d_0 A_f[\ell^n]$$

has a $\mathbb{Z}_\ell$-submodule of finite index which is free of rank $2[\mathbb{E}_{f,\lambda_0} : \mathbb{Q}_\ell]$. In particular, putting

$$A_f[\ell^\infty] = \bigcup_{n \geq 1} A_f[\ell^n],$$

we see that $d_0 A_f[\ell^\infty]$ is infinite.

Put $L = \mathbb{Q}(d_0 A_f[\ell^\infty])$. Then the torsion subgroup of $A_f(L)$ contains $d_0 A_f[\ell^\infty]$ and is consequently infinite. Hence a theorem of Ribet [16] implies that L is not contained in the maximal cyclotomic extension of $\mathbb{Q}$. Therefore the group $G = \mathrm{Gal}(L/\mathbb{Q})$ is nonabelian. Let $\mathrm{Frob}_L(P_0)$ be the set of Frobenius elements of prime ideals of L lying over primes in P_0, and let H be the closure of the subgroup of G generated by $\mathrm{Frob}_L(P_0)$. We claim that H is abelian, whence H is a proper subgroup of G. Indeed ρ_{λ_0} can be viewed as a faithful representation of G on $d_0 V_\ell(A_f)$, and since the restriction of ρ_{λ_0} to H is scalar the claim follows. Now let $\overline{P_0}$ be the complement of P_0 in the set of prime numbers not dividing $N\ell$. Also let $\mathrm{Frob}_L(\overline{P_0}) \subset G$ be the set of Frobenius elements of prime ideals of L lying over primes in $\overline{P_0}$. Then the Chebotarev density theorem implies that the set $G - H$ is contained in the closure of $\mathrm{Frob}_L(\overline{P_0})$. Since any group is generated by the complement of a proper subgroup, it follows that the subgroup generated by $\mathrm{Frob}_L(\overline{P_0})$ is dense in G.

Next we consider two continuous homomorphisms $\mathrm{Gal}(\overline{\mathbb{Q}}/\mathbb{Q}) \longrightarrow \mathbb{E}_{f,\lambda_0}^\times$. The first, to be denoted κ_{λ_0}, is obtained by composing the ℓ-adic cyclotomic character $\mathrm{Gal}(\overline{\mathbb{Q}}/\mathbb{Q}) \longrightarrow \mathbb{Z}_\ell^\times$ with the inclusion of $\mathbb{Z}_\ell^\times$ in $\mathbb{E}_{f,\lambda_0}^\times$. For the second character, we compose the canonical surjection

$$\mathrm{Gal}(\overline{\mathbb{Q}}/\mathbb{Q}) \longrightarrow \mathrm{Gal}(\mathbb{Q}(\mu_N)/\mathbb{Q}) \cong (\mathbb{Z}/N\mathbb{Z})^\times$$

with the map

$$\begin{aligned}(\mathbb{Z}/N\mathbb{Z})^\times &\longrightarrow \mathbb{T}_f^\times\\ d &\longmapsto \langle d\rangle\end{aligned}$$

followed by pr_{λ_0}. This second character will be written $\sigma \mapsto \langle\sigma\rangle_{\lambda_0}$. Note that if p is a prime not dividing $N\ell$ then $\kappa_{\lambda_0}(\sigma_\mathfrak{p}) = p$ and $\langle\sigma_\mathfrak{p}\rangle_{\lambda_0} = \langle p\rangle_{\lambda_0}$. On the other hand, if p happens to belong to $\overline{P_0}$, then $\det\rho_\lambda(\sigma_\mathfrak{p}) = \langle p\rangle_{\lambda_0}p$, because equation (3) holds for $\lambda = \lambda_0$ and $p \in \overline{P_0}$. Therefore, writing $\mathrm{Frob}_{\overline{\mathbb{Q}}}(\overline{P_0})$ for the set of Frobenius elements of prime ideals of $\overline{\mathbb{Q}}$ lying over primes in $\overline{P_0}$, we have

$$\det\rho_{\lambda_0}(\sigma_\mathfrak{p}) = \langle\sigma_\mathfrak{p}\rangle_{\lambda_0}\kappa_{\lambda_0}(\sigma_\mathfrak{p})$$

for $\sigma_\mathfrak{p} \in \mathrm{Frob}_{\overline{\mathbb{Q}}}(\overline{P_0})$. Since both sides of this equation are continuous, equality holds on the closure of the subgroup of $\mathrm{Gal}(\overline{\mathbb{Q}}/\mathbb{Q})$ generated by $\mathrm{Frob}_{\overline{\mathbb{Q}}}(\overline{P_0})$. Let us consider the image of this subgroup under the natural map $\mathrm{Gal}(\overline{\mathbb{Q}}/\mathbb{Q}) \to \mathrm{Gal}(L/\mathbb{Q})$. The image of $\mathrm{Frob}_{\overline{\mathbb{Q}}}(\overline{P_0})$ is $\mathrm{Frob}_L(\overline{P_0})$, and we saw above that the subgroup of $\mathrm{Gal}(L/\mathbb{Q})$ generated by $\mathrm{Frob}_L(\overline{P_0})$ is dense in $\mathrm{Gal}(L/\mathbb{Q})$. Thus the closure of the subgroup of $\mathrm{Gal}(\overline{\mathbb{Q}}/\mathbb{Q})$ generated by $\mathrm{Frob}_{\overline{\mathbb{Q}}}(\overline{P_0})$ maps onto $\mathrm{Gal}(L/\mathbb{Q})$. We conclude that if p is any prime not dividing $N\ell$ then $\det\rho_{\lambda_0}(\sigma_\mathfrak{p}) = \langle p\rangle_{\lambda_0}p$.

We can now prove that equation (3) holds for $\lambda = \lambda_0$ and all $p \in P_0$. Indeed if B is any nonzero 2×2 matrix over a field which is annihilated by a monic polynomial of degree 2, and if the constant term of that polynomial is the determinant of B, then the polynomial is the characteristic polynomial of B. This completes the proof.

3.8. Modular abelian varieties. Let us now complete the train of thought initiated in Theorem 4. Let $f \in \mathrm{S}_2(N,\chi)$ be a primitive cusp form of level N with Fourier expansion

$$f(z) = \sum_{n\geq 1} a(n)e^{2\pi i n z}.$$

For a prime p dividing N we define

$$P_p^*(A_f,t) = \prod_\sigma (1 - a(p)^\sigma t).$$

We also put $g = [\mathbb{E}_f : \mathbb{Q}]$, $N^*(A_f) = N^g$, and $W^*(A_f) = \prod_\sigma W(f^\sigma)$, so that

$$L^*(A_f, s) = \prod_\sigma L(f^\sigma, s)$$

and

$$\Lambda^*(A_f, s) = (N^*(A_f))^{s/2}((2\pi)^{-s}\Gamma(s))^g L^*(A_f, s).$$

Then

$$\Lambda^*(A_f, s) = \prod_\sigma \Lambda(f^\sigma, s).$$

Now according to Proposition 18, each $\Lambda(f^\sigma, s)$ has an analytic continuation to an entire function of order one satisfying the functional equation $\Lambda(f^\sigma, s) = W(f^\sigma)\Lambda(f^{\sigma\rho}, 2-s)$, where $\rho \in \mathrm{Aut}(\mathbb{C})$ denotes complex conjugation. Since composition with ρ merely permutes the distinct embeddings of $\mathbb{E}_f$ in $\mathbb{C}$, we deduce that $\Lambda^*(A_f, s)$ has an analytic continuation to an entire function of order one satisfying the functional equation

$$\Lambda^*(A_f, s) = W^*(A_f)\Lambda^*(A_f, 2-s).$$

Consequently A_f satisfies Conjecture 1*. However, it follows from a theorem of Carayol [3] (completing work of Deligne [5], Ihara [11], and Langlands [12]) that $P_p(A_f, t) = \prod_\sigma(1 - a(p)^\sigma t)$ for p dividing N, and furthermore that $N(A_f) = N^g$ and $W(A_f) = \prod_\sigma W(f^\sigma)$. Thus a stronger assertion holds:

Theorem 5. *For $f \in \mathrm{Prim}_2(N)$ and $g = [\mathbb{E}_f : \mathbb{Q}]$ the invariants*

$$L(A_f, s), \quad N(A_f), \quad \textit{and} \quad W(A_f)$$

coincide with

$$\prod_\sigma L(f^\sigma, s), \quad N^g, \quad \textit{and} \quad \prod_\sigma W(f^\sigma)$$

respectively, where σ runs over the distinct embeddings of $\mathbb{E}_f$ in $\mathbb{C}$. Consequently A_f satisfies Conjecture 1.

Let Prim_2 denote the union of the sets $\mathrm{Prim}_2(N)$ over all positive integers N, and let A be an abelian variety over $\mathbb{Q}$. If A is isogenous over $\mathbb{Q}$ to a product of abelian varieties of the form A_f with $f \in \mathrm{Prim}_2$, then we call A a *modular abelian variety*, or in the case of dimension one, a *modular elliptic curve*. Since the L-function, conductor, and root number of A depend on A only up to isogeny over $\mathbb{Q}$, and since all three of these invariants respect products, we deduce:

Corollary. *If A is a modular abelian variety then A satisfies Conjecture 1.*

In the remaining paragraphs we discuss a partial converse to the corollary in the case of dimension one, the converse being contingent on a suitable strengthening of Conjecture 1.

3.9. Conjecture 1 with twists. As we have already mentioned, Conjecture 1 is a special case of a more general hypothesis about L-functions of motives. We shall now state a slight extension of Conjecture 1 (still far from the general case) in which we allow twists of the motives in Conjecture 1 by Artin motives. For the application we have in mind it would suffice to consider Artin motives corresponding to Dirichlet characters, but specializing the context in this way does not seem to simplify the formulation.

Consider as before an abelian variety A over $\mathbb{Q}$ together with its associated family of ℓ-adic representations $\{\rho_\ell\}$. In addition, let τ be a continuous finite-dimensional complex representation of $\mathrm{Gal}(\overline{\mathbb{Q}}/\mathbb{Q})$, and let $\mathbb{E}_\tau \subset \mathbb{C}$ be a finite extension of $\mathbb{Q}$ such that τ is realizable on an $\mathbb{E}_\tau$-vector space W. If λ is a place of $\mathbb{E}_\tau$ lying over some ℓ and $\mathbb{E}_{\tau,\lambda}$ is the completion of $\mathbb{E}_\tau$ at λ then we obtain a representation $\rho_\ell \otimes \tau$ of $\mathrm{Gal}(\overline{\mathbb{Q}}/\mathbb{Q})$ on the $\mathbb{E}_{\tau,\lambda}$-vector space

$$U_\lambda = (\mathbb{E}_{\tau,\lambda} \otimes_{\mathbb{Q}_\ell} V_\ell(A)) \otimes (\mathbb{E}_{\tau,\lambda} \otimes_{\mathbb{E}_\tau} W).$$

Given a prime p, we choose $\ell \neq p$ and put

$$P_p(A,\tau,t) = \det\left(1 - t(\rho_\ell \otimes \tau)(\sigma_\mathfrak{p})|U_\lambda^{I(\mathfrak{p})}\right).$$

As before, the semistable reduction theorem implies that the coefficients of $P_p(A,\tau,t)$ lie in $\mathbb{E}_\tau$ and are independent of ℓ and λ. Furthermore, the complex numbers $\alpha_{i,p}$ in the factorization

$$P_p(A,\tau,t) = \prod_{i=1}^{2g\dim\tau} (1 - \alpha_{i,p}t).$$

still satisfy

$$|\alpha_{i,p}| \leq \sqrt{p} \quad (1 \leq i \leq 2g\dim\tau),$$

so that the Euler product

$$L(A,\tau,s) = \prod_p P_p(A,\tau,p^{-s})^{-1}$$

converges for $\mathrm{Re}(s) > 3/2$. Also, the conductor $N(A,\tau)$ of the compatible family $\{\rho_\ell \otimes \tau\}_\ell$ is defined, as is the root number $W(A,\tau)$, which is a complex number of absolute value 1 (no longer necessarily equal to ± 1 unless τ is equivalent to its contragredient τ^*). If the conductors $N(A)$ and $N(\tau)$ of A and τ are relatively prime, then

$$N(A,\tau) = N(A)^{\dim\tau} N(\tau)^{2g}$$

and

$$W(A,\tau) = \det\tau((-1)^g N(A)) W(A)^{\dim\tau} W(\tau)^{2g},$$

where in the second equation $W(\tau)$ is the root number of τ and $\det\tau$ is thought of as a Dirichlet character.

Conjecture 2. *Put* $\Lambda(A,\tau,s) = N(A,\tau)^{s/2}((2\pi)^{-s}\Gamma(s))^{g\dim\tau}L(A,\tau,s)$. *Then* $\Lambda(A,\tau,s)$ *has an analytic continuation to an entire function of order one satisfying the functional equation*

$$\Lambda(A,\tau,s) = W(A,\tau)\Lambda(A,\tau^*,2-s).$$

For $A = A_f$ and certain τ with solvable image a statement along these lines follows from the Rankin-Selberg method and the theory of base change (cf. [1], [13], [21]). If $A = A_f$ and τ is one-dimensional then Conjecture 2 is subsumed in the results of Carayol [3].

3.10. Epilogue: the Shimura-Taniyama conjecture. Let us now consider Conjecture 2 in the special case where $\dim A$ and $\dim\tau$ are both one. Thus A is an elliptic curve and τ can be identified with a primitive Dirichlet character χ. We shall further assume that the integers $N = N(A)$ and $r = N(\chi)$ are relatively prime, whence

$$N(A,\chi) = Nr^2$$

and

$$W(A,\chi) = \chi(-N)W(A)W(\chi)^2.$$

In this setting the assertion of Conjecture 2 has a particularly elementary formulation. To begin with, let us put

$$a(p) = \begin{cases} 1 - |\tilde{A}(\mathbb{F}_p)| + p & \text{if } A \text{ has good reduction at } p \\ 1 & \text{if } A \text{ has split multiplicative reduction at } p \\ -1 & \text{if } A \text{ has nonsplit multiplicative reduction at } p \\ 0 & \text{if } A \text{ has additive reduction at } p. \end{cases}$$

Then the Euler factors of A are determined by the elementary rule

$$P_p(A,t) = \begin{cases} 1 - a(p)t + pt^2 & \text{if } A \text{ has good reduction at } p \\ 1 - a(p)t & \text{if } A \text{ has bad reduction at } p. \end{cases}$$

Therefore

$$L(A,s) = \prod_{p\nmid N(A)} (1 - a(p)p^{-s} + p^{1-2s})^{-1} \cdot \prod_{p\mid N(A)} (1 - a(p)p^{-s})^{-1}.$$

Furthermore, since we are assuming that r is relatively prime to N, the L-function $L(A,\chi,s)$ coincides with the naive twist of $L(A,s)$ by χ: if we write $L(A,s)$ as a Dirichlet series

$$L(A,s) = \sum_{n\geq 1} a(n)n^{-s},$$

then

$$L(A,\chi,s) = \sum_{n\geq 1} \chi(n)a(n)n^{-s}.$$

Thus in the case at hand Conjecture 2 asserts that the function

$$\Lambda(A,\chi,s) = (Nr^2)^{s/2}(2\pi)^{-s}\Gamma(s)\sum_{n\geq 1}\chi(n)a(n)n^{-s}$$

is entire of order one and satisfies the functional equation

$$\Lambda(A,\chi,s) = \chi(-N)W(A)W(\chi)^2\Lambda(A,\overline{\chi},2-s).$$

Now compare this assertion to condition (i) of the following result, which is a version of Weil's converse to Hecke theory specialized to the case of weight 2 and trivial character:

Theorem 6. *Let N be a positive integer and $a(1), a(2), a(3), \ldots$ a sequence of complex numbers satisfying the formal identity*

$$\sum_{n\geq 1} a(n)n^{-s} = \prod_{p\nmid N}(1-a(p)p^{-s}+p^{1-2s})^{-1}\cdot\prod_{p\mid N}(1-a(p)p^{-s})^{-1}.$$

Suppose furthermore that

$$|a(p)| < \begin{cases} 2p & \text{if } p\nmid N \\ p & \text{if } p\mid N, \end{cases}$$

so that the Dirichlet series and Euler product actually converge for $\mathrm{Re}(s) > 2$. *Put*

$$f(z) = \sum_{n\geq 1} a(n)e^{2\pi i n z}.$$

Then the following are equivalent:

(i) *There exists a complex number $W(f)$ of absolute value 1 such that for every positive integer r prime to N and every primitive Dirichlet character χ modulo r, the function*

$$\Lambda(f,\chi,s) = (Nr^2)^{s/2}(2\pi)^{-s}\Gamma(s)\sum_{n\geq 1}\chi(n)a(n)n^{-s}$$

has an analytic continuation to an entire function of order one satisfying the functional equation

$$\Lambda(f,\chi,s) = \chi(-N)W(f)W(\chi)^2\Lambda(f,\overline{\chi},2-s).$$

(ii) *f is a primitive cusp form of weight 2 for $\Gamma_0(N)$.*

Theorem 6 can be pieced together from Weil [22], Deligne-Serre ([7], p. 515, Lemme 4.9), and the theory of new forms ([2],[4],[14],[15]). It applies in particular to the situation at hand, because if $a(p)$ is the coefficient of p^{-s} in the L-series $L(A, s)$ of an elliptic curve A over $\mathbb{Q}$, then

$$|a(p)| \leq \begin{cases} 2\sqrt{p} & \text{if } p \nmid N \\ 1 & \text{if } p | N, \end{cases}$$

which is a stronger estimate than that required by the hypothesis of the theorem. Thus conditions (i) and (ii) are equivalent for $L(A, s)$, and if we grant Conjecture 2 then it follows that there is a primitive cusp form f for $\Gamma_0(N)$ of weight 2 such that $L(f, s) = L(A, s)$. Now this equation implies in particular that the Fourier coefficients of f are rational, whence $\mathbb{E}_f = \mathbb{Q}$ and A_f is an elliptic curve. Furthermore, Theorem 5 gives $L(A_f, s) = L(A, s)$, and then the isogeny theorem of Faltings implies that A is isogenous over $\mathbb{Q}$ to A_f. Thus A is a modular elliptic curve. To summarize, if we grant Conjecture 2, then we are forced to believe:

Conjecture 3. *Every elliptic curve over* $\mathbb{Q}$ *is modular.*

REFERENCES

1. J. Arthur and L. Clozel, *Simple Algebras, Base Change, and the Advanced Theory of the Trace Formula*, Annals of Math. Studies 120, Princeton Univ. Press, Princeton, 1989.
2. A. O. L. Atkin and J. Lehner, *Hecke operators on* $\Gamma_0(m)$, Math. Ann. **185** (1970), 134 – 160.
3. H. Carayol, *Sur les représentations ℓ-adiques associées aux formes modulaires de Hilbert*, Ann. Sci. Ec. Norm. Sup. **19** (1986), 409 – 468.
4. W. Casselman, *On some results of Atkin and Lehner*, Math. Ann. **201** (1973), 301 – 314.
5. P. Deligne, *Formes modulaires et représentations ℓ-adiques*, Séminaire Bourbaki, Lect. Notes in Math. 1799, Springer-Verlag, 1971, pp. 139 – 172.
6. P. Deligne, *Les constantes des équations fonctionelles des fonctions L*, Modular Functions of One Variable, II, Lect. Notes in Math. 349, Springer-Verlag, 1973, pp. 501–595.
7. P. Deligne and J-P. Serre, *Formes modulaires de poids* 1, Ann. Sci. Ec. Norm. Sup. **7** (1974), 507 -530.
8. M. Eichler, *Quaternäre quadratische Formen und die Riemannsche Vermutung für die Kongruenzzetafunktion*, Arch. Math. **5** (1954), 355 – 366.
9. A. Grothendieck, *Modèles de Néron et monodromie*, Groupes de Monodromie en Géometrie Algébrique, Lect. Notes in Math. 288, Springer-Verlag, 1971, pp. 313 – 523.
10. J. Igusa, *Kroneckerian model of fields of elliptic modular functions*, Amer. J. Math. **81** (1959).
11. Y. Ihara, *Hecke polynomials as congruence ζ-functions in elliptic modular case*, Ann. Math. **85** (1967).
12. R. P. Langlands, *Modular forms and ℓ-adic representations*, Modular Functions of One Variable, II, Lect. Notes in Math. 349, Springer-Verlag, 1973, pp. 361–500.
13. R. P. Langlands, *Base Change for* GL(2), Annals of Math. Studies 96, Princeton Univ. Press, Princeton, 1980.

14. W. W. Li, *Newforms and functional equations*, Math. Ann. **212** (1975), 285 - 315.
15. T. Miyake, *On automorphic forms on* GL_2 *and Hecke operators*, Ann. Math. **94** (1971), 174 - 189.
16. K. Ribet, *Torsion points of abelian varieties in cyclotomic extensions*, L'Enseignement Math. **27** (1981), 315 - 319.
17. J-P. Serre, *Facteurs locaux des fonctions zêta des variétés algébriques (définitions et conjectures)*, Séminaire Delange-Poitou-Pisot 1969/70 no. 19.
18. J-P. Serre and J. Tate, *Good reduction of abelian varieties*, Ann. Math. **88** (1968), 492 - 517.
19. G. Shimura, *Correspondances modulaires et les fonctions* ζ *de courbes algébriques*, J. Math. Soc. Japan **10** (1958), 1 - 28.
20. G. Shimura, *Introduction to the Arithmetic Theory of Automorphic Functions*, Iwanami Shoten and Princeton University Press, Princeton, 1971.
21. J. Tunnell, *Artin's conjecture for representations of octahedral type*, Bull. AMS **5** (1981), 173 - 175.
22. A. Weil, *Über die Bestimmung Dirichletscher Reihen durch Funktionalgleichungen*, Math. Ann. **168** (1967), 149 - 156.

GALOIS COHOMOLOGY

LAWRENCE C. WASHINGTON

In these lectures, we give a very utilitarian description of the Galois cohomology needed in Wiles' proof. For a more general approach, see any of the references.

First we fix some notation. For a field K, let $\bar{K}$ be a separable closure of K and let $G_K = \mathrm{Gal}(\bar{K}/K)$. For a prime p, let $G_p = G_{\mathbb{Q}_p}$, where $\mathbb{Q}_p$ is the field of p-adic numbers, and let $I_p \subset G_p$ be the inertia group.

Let G be a group, usually either finite or profinite, and let X be an abelian group on which G acts. Such an X will be called a G-module. If there are topologies to consider, we assume the action is continuous, though we shall mostly ignore continuity questions except to say that all maps, actions, etc. are continuous when they should be.

§1. H^0, H^1, AND H^2

We start with

$$H^0(G, X) = X^G = \{x \in X | gx = x \text{ for all } g \in G\}.$$

For example, G_K acts on $\bar{K}^\times$ and

$$H^0(G_K, \bar{K}^\times) = K^\times.$$

For another example, let μ_n denote the group of n-th roots of unity. Then

$$H^0(G_{\mathbb{Q}}, \mu_n) = \begin{cases} \{\pm 1\} & \text{if } 2|n, \\ 1 & \text{if } 2 \nmid n. \end{cases}$$

Occasionally, for a finite group G, we will need the modified Tate cohomology group

$$\hat{H}^0(G, X) = X^G / \mathrm{Norm}(X),$$

where $\mathrm{Norm}(x) = \sum_{g \in G} gx$ (if X is written additively). For example, if X is an abelian group of odd order on which $\mathrm{Gal}(\mathbb{C}/\mathbb{R})$ acts, then $\mathrm{Norm}(X) \supseteq 2(X^G) = X^G$, so $\hat{H}^0(\mathrm{Gal}(\mathbb{C}/\mathbb{R}), X) = 0$.

We now skip $H^1(G, X)$ in order to give a brief description of $H^2(G, X)$. Define

$$H^2(G, X) = \text{ cocycles/coboundaries},$$

where a cocycle is a map (of sets) $f : G \times G \to X$ satisfying

$$\delta f = f(g_1, g_2g_3) - f(g_1g_2, g_3) + g_1 \cdot f(g_2, g_3) - f(g_1, g_2) = 0,$$

and where f is a coboundary if there is a map $h : G \to X$ such that

$$f(g_1, g_2) = g_1 \cdot h(g_2) - h(g_1g_2) + h(g_1) = \delta h.$$

This definition might seem a little strange; we will give a slightly different form of it later after we define $H^1(G, X)$.

Here is an example. Let p be prime and let $G = G_p$. Let $a, b \in \mathbb{Q}_p^\times$ with a not a square. Define

$$f(g_1, g_2) = \begin{cases} b & \text{if } g_1\sqrt{a} = -\sqrt{a} \text{ and } g_2\sqrt{a} = -\sqrt{a}, \\ 1 & \text{otherwise.} \end{cases}$$

It is easy to check that $f : G_p \times G_p \to \mathbb{Q}_p^\times$ satisfies the cocycle condition, hence yields an element of $H^2(G_p, \mathbb{Q}_p^\times)$. Suppose b is a norm from $\mathbb{Q}_p(\sqrt{a})$, so $b = x^2 - ay^2$ for some $x, y \in \mathbb{Q}_p$. Let $h(g) = x + y\sqrt{a}$ if $g\sqrt{a} = -\sqrt{a}$ and $h(g) = 1$ otherwise. Then

$$f(g_1, g_2) = (g_1 h(g_2)) h(g_1)/h(g_1g_2),$$

so the element of H^2 we obtain is trivial. Conversely, it can be shown that if this element is trivial, then b is a norm from $\mathbb{Q}_p(\sqrt{a})$. Recall the Hilbert symbol $(a, b)_p$, which equals 1 if b is a norm from $\mathbb{Q}_p(\sqrt{a})$ and equals -1 otherwise. Thus the above cohomology class we obtain is essentially the same as the Hilbert symbol. We also have $(a, b)_p = 1$ if and only if $x_1^2 - ax_2^2 - bx_3^2 + abx_4^2 = 0$ has a non-zero solution in $\mathbb{Q}_p$. Equivalently, $(a, b)_p = 1$ if and only if the generalized quaternion algebra $\mathbb{Q}_p[i, j, k]$, with $i^2 = a$, $j^2 = b$, $k^2 = -ab$, $ij = k$, etc., is isomorphic to the algebra of two-by-two matrices over $\mathbb{Q}_p$ (rather than being a division algebra). In general, $H^2(G_K, \bar{K}^\times)$ is known as the Brauer group and classifies central simple algebras over the field K. We will need the following result.

Proposition 1. *Let p be a prime number. Then $H^2(G_p, \bar{\mathbb{Q}}_p^\times) \simeq \mathbb{Q}/\mathbb{Z}$.*

This result is an important result in local class field theory. For a proof, see [Se]. In our example, the cohomology class of f is 0 if $(a, b)_p = 1$ and is $\frac{1}{2} \bmod \mathbb{Z}$ if $(a, b)_p = -1$.

We now turn our attention to H^1, which is the most important for us. Define

$$H^1(G, X) = \text{ cocycles/coboundaries},$$

where a cocycle is a map $f : G \to X$ satisfying $f(g_1g_2) = f(g_1) + g_1 f(g_2)$ (a "crossed homomorphism") and where f is a coboundary if there exists $x \in X$ such that $f(g) = gx - x$.

Before continuing, we write the cocycle conditions in a different form that perhaps seems more natural. For a 2-cocycle f, let

$$F(a,b,c) = a \cdot f(a^{-1}b, a^{-1}c),$$

where $a, b, c \in G$. Then $F(ga, gb, gc) = g \cdot F(a,b,c)$ and the cocycle condition becomes

$$F(a,b,c) - F(a,b,d) + F(a,c,d) - F(b,c,d) = 0.$$

For a 1-cocycle f, let $F(a,b) = a \cdot f(a^{-1}b)$. Then $F(ga, gb) = g \cdot F(a,b)$ and the cocycle condition reads

$$F(a,b) - F(a,c) + F(b,c) = 0.$$

We can even describe H^0 in this manner: a 0-cocycle is a map f from the one point set to X, hence simply an element x of X, that satisfies $gx - x = 0$. If we let $F(a) = ax$, then $F(ga) = g \cdot F(a)$ and $F(a) - F(b) = 0$ for all $a, b \in G$. In all three cases, the coboundary condition says that F is the coboundary of a function from the next lower dimension. For example, the function F for a 2-coboundary is of the form $H(a,b) - H(a,c) + H(b,c)$ for a function H satisfying $H(ga, gb) = g \cdot H(a,b)$ (explicitly, $H(a,b) = a \cdot h(a^{-1}b)$ in the above notation). It should now be clear how to define higher cohomology groups $H^n(G,X)$ for $n \geq 3$. With one exception, we will not need these higher groups, and in this one exception, the element we need will be 0; therefore, we may safely ignore them for the present exposition.

A fundamental fact that will be used quite often is the following. Suppose

$$0 \to A \to B \to C \to 0$$

is a short exact sequence of G-modules. Then there is a long exact sequence of cohomology groups (write $H^r(X)$ for $H^r(G,X)$)

$$\begin{aligned} 0 \to H^0(A) \to H^0(B) \to H^0(C) \to H^1(A) \\ \to H^1(B) \to H^1(C) \to H^2(A) \to H^2(B) \to \cdots. \end{aligned}$$

The proof is a standard exercise in homological algebra.

Let's return to $H^1(G,X)$. Suppose the action of G is trivial, so $gx = x$ for all g and x. Then cocycles are simply homomorphisms $G \to X$. A coboundary $f(g) = gx - x$ is the 0-map. Therefore we have proved the useful fact that

$$H^1(G,X) = \mathrm{Hom}(G,X) \quad \text{if the action of } G \text{ is trivial.}$$

Here "Hom" means (continuous) homomorphisms of groups. For example, let K be a field and let $G = G_K$. Then G_K acts trivially on $\mathbb{Z}/2\mathbb{Z}$, so $H^1(G_K, \mathbb{Z}/2\mathbb{Z}) = \text{Hom}(G_K, \mathbb{Z}/2\mathbb{Z})$, which corresponds to the separable quadratic (or trivial) extensions of K; namely, if f is a non-trivial homomorphism, then the fixed field of the kernel of f is a quadratic extension. The trivial homomorphism corresponds to the trivial extension K/K.

Suppose now that G is a finite cyclic group: $G = \langle g \rangle$ with $g^n = 1$. The cocycle relation yields by induction that

$$f(g^i) = (1 + g + g^2 + \cdots + g^{i-1})f(g).$$

Therefore $f(1) = f(g^n) = \text{Norm}(f(g))$. The cocycle condition easily implies that $f(1) = 0$, so $f(g)$ is in the kernel of Norm. Any such choice for $f(g)$ yields a cocycle via the above formula. A coboundary corresponds to $f(g) = (g-1)x$ for some $x \in X$. Therefore

$$H^1(G, X) \simeq \text{(Kernel of Norm)}/(g-1)X \quad \text{for a finite cyclic group } G.$$

As an example, consider a $G_{\mathbb{R}}$-module X of odd order. Let c be complex conjugation. Write $X = \frac{1+c}{2}X \oplus \frac{1-c}{2}X$. Note that $\frac{1-c}{2}X$ is the kernel of Norm $= 1 + c$, and is also equal to $(c-1)X$. Therefore $H^1(G_{\mathbb{R}}, X) = 0$. More generally, it can be shown that if G and X are finite with relatively prime orders, then $H^i(G, X) = 0$ for all $i > 0$, and also for $i = 0$ if we use the modified groups $\hat{H}^0(G, X)$.

When G is infinite cyclic, or is the profinite completion of an infinite cyclic group, and X is finite, then there is a similar description. Let g be a (topological) generator. Let $x \in X$ be arbitrary. There are $k, n > 0$ such that $g^n x = x$ and $kx = 0$. Define a cocycle by $f(g^i) = (1 + g + \cdots + g^{i-1})x$ for $i > 0$. If $i > j$ and $i \equiv j \mod kn$, then $g^j + \cdots g^{i-1}$ is a multiple of $1 + g^n + \cdots + g^{n(k-1)}$, which kills x. Therefore $f(g^i)$ depends only on $i \mod kn$, so f extends to a continuous cocycle on all of G. Since, as above, every cocycle must be of this form, we have

$$H^1(G, X) \simeq X/(g-1)X$$

when G is (the profinite closure of) an infinite cyclic group and X is finite. This result will be applied later to the case where $\mathbb{F}$ is a finite field and $G = \text{Gal}(\bar{\mathbb{F}}/\mathbb{F})$, which is generated by the Frobenius map.

Let L/K be a finite extension of fields with cyclic Galois group G generated by g. Then G acts on $L^\times$. The famous Hilbert Theorem 90 says that if $x \in L^\times$ has Norm 1 then $x = gy/y$ for some $y \in L^\times$. This is precisely the statement that $H^1(G, L^\times) = 0$. More generally, we have

$$H^1(\text{Gal}(L/K),\ L^\times) = 0$$

for any Galois extension of fields L/K ([Se]).

Let $n \geq 1$ be prime to the characteristic of the field K and consider the exact sequence of G_K-modules

$$1 \to \mu_n \to \bar{K}^\times \to \bar{K}^\times \to 1$$

induced by the n-th power map. The long exact sequence of cohomology groups includes the portion

$$H^0(G_K, \bar{K}^\times) \to H^0(G_K, \bar{K}^\times) \to H^1(G_K, \mu_n) \to H^1(G_K, \bar{K}^\times),$$

where the first map is the n-th power map. Since the last group is 0, we find that

$$H^1(G_K, \mu_n) \simeq K^\times/(K^\times)^n.$$

Explicitly, let $a \in K^\times$ and fix an nth root α of a. Then $g \mapsto g\alpha/\alpha$ defines a cocycle and hence an element of $H^1(G_K, \mu_n)$. When $\mu_n \subseteq K$, $H^1(G_K, \mu_n)$ becomes $\mathrm{Hom}(G_K, \mu_n)$, which corresponds (in an obvious many to one fashion) to cyclic extensions of K of degree dividing n, and α is a Kummer generator for this extension (and, correspondingly, there are several Kummer generators mod nth powers for each extension). When $n = 2$, note that $\mathbb{Z}/2\mathbb{Z}$ and μ_2 are isomorphic as G_K-modules, and we find that $H^1(G_K, \mu_2)$ classifies quadratic extensions of K, though in a slightly different manner than $H^1(G_K, \mathbb{Z}/2\mathbb{Z})$.

§2. PRELIMINARY RESULTS

Suppose H is a (closed) normal subgroup of a group G and X is a G-module. Then X^H is a module for G/H in the obvious way. A cocycle for G/H can also be regarded as a cocycle for G ("inflation") by composing with the map $G \to G/H$. A cocycle for G can be regarded as a cocycle for H by restriction. Also, G/H acts on $H^1(H, X)$ by the formula $f^g(h) = g \cdot f(g^{-1}hg)$, where f is a cocycle and g is a representative of a coset in G/H. An easy calculation shows that if g' is another representative of the coset of g then $f^{g'}$ and f^g differ by a coboundary, so the action is well-defined.

Proposition 2 (Inflation-Restriction). *There is an exact sequence*

$$0 \to H^1(G/H, X^H) \to H^1(G, X) \to H^1(H, X)^{G/H} \to H^2(G/H, X^H) \to H^2(G, X).$$

This is the exact sequence of terms of low degree in the Hochschild-Serre spectral sequence, hence is sometimes referred to by that name. For a proof, and the definition of the map from H^1 to H^2, see [Sh].

For example, let p be a prime and let $G = G_p$. Let $H = I_p = \mathrm{Gal}(\bar{\mathbb{Q}}_p/\mathbb{Q}_p^{\mathrm{unr}})$, where $\mathbb{Q}_p^{\mathrm{unr}}$ is the maximal unramified extension of $\mathbb{Q}_p$, so

I_p is the inertia subgroup of G_p, and $G_p/I_p \simeq \mathrm{Gal}(\bar{\mathbb{F}}_p/\mathbb{F}_p)$. The beginning of the above sequence implies that

$$H^1(G_p/I_p, X^{I_p}) \simeq \mathrm{Ker}\Big(H^1(G_p, X) \to H^1(I_p, X)\Big).$$

Thus we can regard $H^1(G_p/I_p, X^{I_p})$ as the subgroup of $H^1(G_p, X)$ consisting of those cohomology classes that become trivial when restricted to the inertia subgroup; hence, we call these the unramified classes. For example, when $X = \mathbb{Z}/2\mathbb{Z}$, the unramified classes are those homomorphisms from G_p to $\mathbb{Z}/2\mathbb{Z}$ that are 0 on I_p, hence that can be identified with homomorphisms from G_p/I_p to $\mathbb{Z}/2\mathbb{Z}$. There are two such homomorphisms, the 0 homomorphism and the one corresponding to the unique unramified quadratic extension of $\mathbb{Q}_p$ (or of $\mathbb{F}_p$). This is well-known, but is also a consequence of the following, which often allows us to calculate the order of the group of unramified classes, since $H^0(G_p, X) = X^{G_p}$.

Lemma 1. *Let X be finite. Then $\#H^1(G_p/I_p, X^{I_p}) = \#H^0(G_p, X)$ (and both are finite).*

Proof. There is an exact sequence

$$0 \to X^{G_p} \to X^{I_p} \xrightarrow{(\mathrm{Frob}-1)} X^{I_p} \to X^{I_p}/(\mathrm{Frob}-1)X^{I_p} \to 0.$$

The exactness at the first X^{I_p} follows from the fact that if $x \in X^{I_p}$ and $(\mathrm{Frob}-1)x = 0$, then x is fixed by both I_p and Frob, which (topologically) generate G_p. The first term gives $H^0(G_p, X)$ and the last term gives $H^1(G_p/I_p, X^{I_p})$. The result follows easily. □

The last preliminary topic that we need is cup products. In general, suppose X_1, X_2, and X_3 are G-modules, and there is a G-module homomorphism $\Phi : X_1 \otimes X_2 \to X_3$. The cup product is a map

$$H^i(G, X_1) \times H^j(G, X_2) \to H^{i+j}(G, X_3).$$

We define the cup product only when $i + j = 2$, since this is the main case we need. Let $f_1 \in H^2(G, X_1)$, so we may regard f_1 as (being represented by) a map $f_1 : G \times G \to X_1$. Let $x_2 \in X_2^G = H^0(G, X_2)$. Then $f_3 = f_1 \cup x_2$ is the 2-cocycle satisfying $f_3(g_1, g_2) = \Phi(f_1(g_1, g_2) \otimes x_2)$. The cup product of H^0 and H^2 is defined similarly. Now let $\phi_k \in H^1(G, X_k)$ for $k = 1, 2$. Define

$$(\phi_1 \cup \phi_2)(g_1, g_2) = \Phi\big(\phi_1(g_1) \otimes g_1\, \phi_2(g_2)\big).$$

It is easy to see that this defines a 2-cocycle, hence an element of $H^2(G, X_3)$.

For example, let $a, b \in \mathbb{Q}_p^\times$. Let $\phi \in H^1(G_p, \mathbb{Z}/2\mathbb{Z})$ be defined by $\phi(g) = 0$ if $g(\sqrt{a}) = \sqrt{a}$ and $\phi(g) = 1$ otherwise. Define $\psi \in H^1(G_p, \mu_2)$ by $\psi(g) = g(\sqrt{b})/\sqrt{b}$. We may regard $\mu_2 \simeq \mathrm{Hom}(\mathbb{Z}/2\mathbb{Z}, \mu_2)$ as the dual of

$\mathbb{Z}/2\mathbb{Z}$; hence there is a map $\mathbb{Z}/2\mathbb{Z} \otimes \mu_2 \to \mu_2 \subset \bar{\mathbb{Q}}_p^\times$. Therefore $\phi \cup \psi \in H^2(\mathbb{Q}_p, \bar{\mathbb{Q}}_p^\times)$. Fix a square root $\sqrt{b}$ and let $h(g) = (g\sqrt{b})^{\phi(g)}$. A calculation shows that $\phi \cup \psi$ multiplied times the coboundary $h(g_1) \cdot g_1 h(g_2)/h(g_1 g_2)$ equals the cocycle f defined earlier, the one corresponding to the Hilbert symbol $(a,b)_p$. In fact, this cup product is one way to define the Hilbert symbol; see [Se]. We now have a pairing

$$H^1(G_p, \mathbb{Z}/2\mathbb{Z}) \times H^1(G_p, \mu_2) \longrightarrow H^2(G_p, \bar{\mathbb{Q}}_p^\times) \simeq \mathbb{Q}/\mathbb{Z}.$$

The non-degeneracy of this pairing is equivalent to the non-degeneracy of the Hilbert symbol.

Now let p be odd and consider the group $H^1(G_p/I_p, \mathbb{Z}/2\mathbb{Z})$ of unramified classes. Assume a is not a square. The element ϕ is in this group if $\sqrt{a}$ generates an unramified extension (in fact, the unique quadratic extension) of $\mathbb{Q}_p$, which means we may assume a is a p-adic unit. We have $(a,b)_p = 1$ $\iff$ b is a norm from $\mathbb{Q}_p(\sqrt{a})$ $\iff$ b is a square times a p-adic unit (this follows from the fact that p is a uniformizer for $\mathbb{Q}_p(\sqrt{a})$) $\iff$ the cocycle ψ is unramified. Therefore, the unramified classes in $H^1(\mathbb{Q}_p, \mu_2)$ form the annihilator of the unramified classes in $H^1(\mathbb{Q}_p, \mathbb{Z}/2\mathbb{Z})$ under the above pairing. All of this will be greatly generalized in the next section.

§3. LOCAL TATE DUALITY

Let p be prime and let X be a G_p-module of finite cardinality n. Let

$$X^* = \mathrm{Hom}_{\mathbb{Z}}(X, \mu_n),$$

where G_p acts on X^* by $(gx^*)(x) = g(x^*(g^{-1}x))$. Note that $X \otimes X^* \simeq \mu_n \subseteq \bar{\mathbb{Q}}_p^\times$ as G_p-modules.

Theorem 1 (Local Tate Duality). (a) *The groups $H^i(G_p, X)$ are finite for all $i \geq 0$, and $= 0$ for $i \geq 3$.*
(b) *For $i = 0, 1, 2$, the cup product gives a non-degenerate pairing*

$$H^i(G_p, X) \times H^{2-i}(G_p, X^*) \to H^2(G_p, \bar{\mathbb{Q}}_p^\times) \simeq \mathbb{Q}/\mathbb{Z}.$$

(c) *If p does not divide the order of X then the unramified classes*

$$H^1(G_p/I_p, X^{I_p}) \quad \text{and} \quad H^1(G_p/I_p, (X^*)^{I_p})$$

are the exact annihilators of each other under the pairing $H^1(G_p, X) \times H^1(G_p, X^*) \to \mathbb{Q}/\mathbb{Z}$.

Proof. For a proof, see [Mi].

For the archimedean prime, the groups $H^i(G_{\mathbb{R}}, X)$ are finite for all i. If we use the modified group $\hat{H}^0$ in place of H^0, then we have $\#\hat{H}^0(G_{\mathbb{R}}, X) = \#H^i(G_{\mathbb{R}}, X)$ for all $i > 0$. There is a non-degenerate pairing

$$H^1(G_{\mathbb{R}}, X) \times H^1(G_{\mathbb{R}}, X^*) \to \mathbb{Q}/\mathbb{Z},$$

and also

$$\hat{H}^0(G_{\mathbb{R}}, X) \times H^2(G_{\mathbb{R}}, X^*) \to \mathbb{Q}/\mathbb{Z}$$

(and with $\hat{H}^0$ and H^2 reversed); note that we use the modified $\hat{H}^0$ here also.

Another result we need evaluates Euler characteristics.

Proposition 3. *Let p be prime and let X be a finite G_p-module. Then*

$$\frac{\#H^1(G_p, X)}{\#H^0(G_p, X) \cdot \#H^2(G_p, X)} = \frac{\#H^1(G_p, X)}{\#H^0(G_p, X) \cdot \#H^0(G_p, X^*)} = p^{v_p(\#X)}.$$

Proof. The first equality follows from Theorem 1. For a proof of the proposition, see [Mi].

By using Theorem 1 and Proposition 3, we can evaluate $\#H^1(G_p, X)$ and $\#H^2(G_p, X)$ in terms of $\#H^0(G_p, X)$ and $\#H^0(G_p, X^*)$. These are much easier to calculate in most cases.

§4. Extensions and deformations

The main reason that Galois cohomology arises in Wiles' work is that certain cohomology groups can be used to classify deformations of Galois representations. In order to explain this, we need a few concepts.

Suppose G is a group acting on an abelian group M, and assume in addition that M is a free module of rank n over a ring R (commutative with 1), and the action of G commutes with the action of R. The action of G is then given by a homomorphism

$$\rho : G \to \mathrm{GL}_n(R).$$

This yields an action of G on $M_n(R)$, the ring of $n \times n$ matrices, via $x \mapsto \rho(g)x\rho(g)^{-1}$. Let $\mathrm{Ad}\,\rho$ denote $M_n(R)$ (or $\mathrm{End}_R(M)$) with this action. We also will need the submodule $\mathrm{Ad}^0\,\rho$ consisting of matrices with trace 0.

An *extension* of M by M will mean a short exact sequence

$$0 \longrightarrow M \xrightarrow{\alpha} E \xrightarrow{\beta} M \longrightarrow 0,$$

where E is an $R[G]$-module and α and β are $R[G]$-homomorphisms. The equivalence of two extensions is given by a commutative diagram

$$\begin{array}{ccccccccc} 0 & \longrightarrow & M & \xrightarrow{\alpha_1} & E_1 & \xrightarrow{\beta_1} & M & \longrightarrow & 0 \\ & & {\scriptstyle =}\downarrow & & {\scriptstyle \gamma}\downarrow & & {\scriptstyle =}\downarrow & & \\ 0 & \longrightarrow & M & \xrightarrow{\alpha_2} & E_2 & \xrightarrow{\beta_2} & M & \longrightarrow & 0, \end{array}$$

where γ is an $R[G]$-isomorphism. The set of equivalence classes of such extensions is denoted $\mathrm{Ext}^1(M,M)$.

Let $R[\epsilon]$ denote the ring $R[T]/(T^2)$ (so $\epsilon^2=0$). An *infinitesimal deformation* of ρ is an extension of ρ to

$$\rho' : G \to \mathrm{GL}_n(R[\epsilon])$$

such that ρ' maps to ρ under the map $\epsilon \mapsto 0$. Two such infinitesimal deformations ρ' and ρ'' are equivalent if there is a matrix $A \equiv I \bmod \epsilon$ such that $A\rho' A^{-1} = \rho''$. The idea behind this is that we want to fit ρ into a family of representations. Suppose, for example, that R is a local ring with maximal ideal $\mathcal{M}$, and that we can extend ρ to $\tilde{\rho} : G \to \mathrm{GL}_n(R[T])$ (or $R[[T]]$ if R is complete). Then we can evaluate T at anything in the maximal ideal $\mathcal{M}$ and get a representation congruent to $\rho \bmod \mathcal{M}$. The infinitesimal deformations are the first steps in the direction of constructing such families.

Proposition 4. *The following sets are in one-one correspondence.*
(a) $H^1(G, \mathrm{Ad}\,\rho)$.
(b) $\mathrm{Ext}^1(M,M)$.
(c) *Equivalence classes of infinitesimal deformations of* ρ.

Proof. Consider an extension $0 \to M \xrightarrow{\alpha} E \xrightarrow{\beta} M \to 0$. Since M is free over R, there is an R-module homomorphism $\phi : M \to E$ such that $\beta \circ \phi = \mathrm{id}_M$. Let $g \in G$ and $m \in M$. Since β is an $R[G]$-homomorphism, $g\phi(g^{-1}m) - \phi(m)$ is in $(\mathrm{Ker}\,\beta)$. Let $T_g : M \to M$ be defined by

$$T_g(m) = \alpha^{-1}\big(g\phi(g^{-1}m) - \phi(m)\big).$$

It is easy to check that $T_{g_1g_2} = T_{g_1} + g_1 T_{g_2}$, where the action of G is the one on $\mathrm{Ad}\,\rho$. Therefore $g \mapsto T_g$ gives an element of $H^1(G, \mathrm{Ad}\,\rho)$. If we have two equivalent extensions and ϕ_1 and ϕ_2 are the corresponding maps, and T_1 and T_2 are the corresponding cocycles, then $(T_2)_g - (T_1)_g = g\psi - \psi$, where $\psi = \alpha^{-1}\gamma^{-1}(\phi_2 - \gamma\phi_1) : M \to M$. Therefore $T_2 - T_1$ is a coboundary for $\mathrm{Ad}\,\rho$, hence T_1 and T_2 represent the same class in $H^1(G, \mathrm{Ad}\,\rho)$. Therefore we have a well-defined map $\mathrm{Ext}^1(M,M) \to H^1(G, \mathrm{Ad}\,\rho)$.

Note that the trivial extension $E = M \oplus M$ (as $R[G]$-modules) yields the trivial cohomology class.

We remark that this method of obtaining cocycles is fairly standard; namely, take an element, such as ϕ, in a bigger set, in this case $\mathrm{Hom}(M,E)$, and form $g\phi - \phi$. Something of this form will automatically satisfy the cocycle condition, but of course we also want $g\phi - \phi$ to be in the original set. When ϕ itself is in the original set, in this case $\mathrm{Ad}\,\rho$, the cocycle is a coboundary.

Now suppose we have two extensions E_1 and E_2 and corresponding cohomology classes T_1 and T_2, and suppose these classes are equal. Then there exists an R-map $\psi : M \to M$ such that $(T_2)_g - (T_1)_g = g\psi - \psi$. Let $e_1 \in E_1$. We can uniquely write $e_1 = \alpha_1(m) + \phi_1(m')$ with $m, m' \in M$. Define $\gamma(e_1) = \alpha_2(m) + \phi_2(m') - \alpha_2(\psi(m'))$. A calculation shows that $\gamma : E_1 \to E_2$ is an $R[G]$-homomorphism that makes the appropriate diagram commute (and is therefore an isomorphism, by the Snake Lemma); hence the extensions are equivalent. We have proved that the map $\mathrm{Ext}^1(M, M) \to H^1(G, \mathrm{Ad}\,\rho)$ is an injection.

Finally, let $g \to C(g) \in \mathrm{Ad}\,\rho$ be a cocycle. Let $E = M \otimes_R R[\epsilon] = \epsilon M \oplus M$. We regard $\rho(g)$ as an element of $\mathrm{GL}_n(R[\epsilon])$ via the natural containment $\mathrm{GL}_n(R) \subseteq \mathrm{GL}_n(R[\epsilon])$. The matrix $I + \epsilon C(g)$ is also in $\mathrm{GL}_n(R[\epsilon])$, so we define

$$\rho'(g) = \big(I + \epsilon C(g)\big)\rho(g).$$

This is easily seen to be a homomorphism, and gives an action of G on E. We have the short exact sequence

$$0 \longrightarrow M \xrightarrow{\ \epsilon\ } E \longrightarrow M \longrightarrow 0.$$

Let $\phi : M \to E = \epsilon M \oplus M$ be the map to the second summand. Then the above recipe gives

$$T_g(m) = \epsilon^{-1}\Big(\big(1 + \epsilon C(g)\big)\rho(g)\,\phi\big(\rho(g)^{-1}m\big) - \phi(m)\Big) = C(g)(m).$$

Therefore this extension yields the cocycle C, so the map $\mathrm{Ext}^1(M, M) \to H^1(G, \mathrm{Ad}\,\rho)$ is surjective.

The above shows that a cocycle yields an infinitesimal deformation. Conversely, if $\rho' : G \to \mathrm{GL}_n(R[\epsilon])$ extends ρ, define $C(g)$ by $I + \epsilon C(g) = \rho'(g)\rho(g)^{-1}$. An easy calculation shows that C is a cocycle. The identity

$$(I + \epsilon A)(I + \epsilon C)\,\rho\,(I - \epsilon A) = \big(I + \epsilon(A - \rho A \rho^{-1} + C)\big)\rho$$

shows that equivalence of deformations corresponds to equivalence of cohomology classes. Note that the trivial cohomology class corresponds to the trivial deformation $\rho' = \rho$. This completes the proof. □

One of the themes in Wiles' work is to consider deformations with various restrictions imposed. By the above, this corresponds to considering cohomology classes lying in certain subsets of $H^1(G, \mathrm{Ad}\,\rho)$. For the moment, we consider two such examples.

Example 1. Suppose we want to consider deformations where the determinant remains unchanged. Note that $\det((I + \epsilon C)\rho) = (1 + \epsilon \mathrm{Tr}(C))\det\rho$. Keeping the determinant unchanged is equivalent to having $C \in \mathrm{Ad}^0\,\rho$. Since $\mathrm{Ad}(\rho) = \mathrm{Ad}^0\,\rho \oplus R$, where R represents the scalar matrices with trivial action of G, we have $H^1(G, \mathrm{Ad}\,\rho) = H^1(G, \mathrm{Ad}^0\,\rho) \oplus H^1(G, R)$. From the above, $H^1(G, \mathrm{Ad}^0\,\rho)$ gives the classes of infinitesimal deformations with fixed determinant.

Example 2. Let p be prime and consider a cohomology class

$$C \in H^1(G_p/I_p, (\operatorname{Ad}\rho)^{I_p}),$$

which is the kernel of the restriction map $H^1(G_p, \operatorname{Ad}\rho) \to H^1(I_p, \operatorname{Ad}\rho)$. Let ρ' be the corresponding deformation. Then ρ' restricted to I_p is (equivalent to) the trivial deformation: $\rho'|_{I_p} = \rho|_{I_p}$. Therefore ρ' is unramified at p if and only if ρ is unramified at p (i.e., $\rho|_{I_p}$ is trivial). Moreover, if ρ is ramified, all the ramification of the deformation ρ' comes from that of ρ. We will often require certain cohomology classes to be unramified in order to control the ramification of the corresponding deformations of ρ.

§5. GENERALIZED SELMER GROUPS

Let X be a $G_{\mathbb{Q}}$-module. Eventually, X will be $\operatorname{Ad}^0 \rho$, but for the moment we do not need to make this restriction. As indicated above, we want to study cohomology classes in $H^1(G_{\mathbb{Q}}, X)$ with various local restrictions. For each place ℓ of $\mathbb{Q}$, including the archimedean one, we may regard the group G_ℓ as a subgroup of $G_{\mathbb{Q}}$. There are many ways to do this, but all the results we obtain will be independent of these choices. We have the restriction maps

$$\operatorname{res}_\ell : H^1(G_{\mathbb{Q}}, X) \to H^1(G_\ell, X).$$

Let $\mathcal{L} = \{L_\ell\}$ be a family of subgroups $L_\ell \subseteq H^1(G_\ell, X)$ as ℓ runs through all places of $\mathbb{Q}$, with $L_\ell = H^1(G_\ell/I_\ell, X^{I_\ell})$ for all but finitely many ℓ. Such a family will be called a collection of local conditions. Define the generalized Selmer group

$$H^1_{\mathcal{L}}(\mathbb{Q}, X) = \{x \in H^1(G_{\mathbb{Q}}, X) \mid \operatorname{res}_\ell(x) \in L_\ell \text{ for all } \ell\}.$$

Let $\mathcal{L}^* = \{L_\ell^\perp\}$, where $L_\ell^\perp$ is the annihilator of L_ℓ under the Tate pairing. By Theorem 1, $L_\ell^\perp = H^1(G_\ell/I_\ell, X^{*I_\ell})$ for all but finitely many ℓ. The following result is crucial in Wiles' proof. It was inspired by work of Ralph Greenberg [Gr].

Theorem 2. *The group $H^1_{\mathcal{L}}(\mathbb{Q}, X)$ is finite, and*

$$\frac{\#H^1_{\mathcal{L}}(\mathbb{Q}, X)}{\#H^1_{\mathcal{L}^*}(\mathbb{Q}, X^*)} = \frac{\#H^0(G_{\mathbb{Q}}, X)}{\#H^0(G_{\mathbb{Q}}, X^*)} \prod_{\ell \leq \infty} \frac{\#L_\ell}{\#H^0(G_\ell, X)}.$$

Note that $\#H^0(G_\ell, X) = \#H^1(G_\ell/I_\ell, X^{I_\ell})$ by Lemma 1, so almost all factors in the product are 1. The formulation of the theorem is that of [DDT], which differs slightly from that of [Wi]. An easy exercise, using Theorem 1 and Proposition 3, shows that the two versions are equivalent.

We sketch the proof of the theorem at the end of the paper.

In the applications, $\mathcal{L}$ is chosen so that $H^1_{\mathcal{L}^*} = 0$. Since the terms on the right are fairly easy to work with, we obtain information about the group

$H^1_{\mathcal{L}}$, which for appropriate X describes deformations of representations with certain local conditions.

To show how the formula may be used, we now give an application in a fairly concrete setting. The techniques are much in the spirit of those used by Wiles. Let $X = \mathbb{Z}/p^n\mathbb{Z}$ (with trivial Galois action), where p is an odd prime. Let S be a finite set of primes containing p and ∞. For $\ell \in S$, let $L_\ell = H^1(G_\ell, \mathbb{Z}/p^n\mathbb{Z})$. For $\ell \notin S$, let $L_\ell = H^1(G_\ell/I_\ell, \mathbb{Z}/p^n\mathbb{Z})$. Then $L_\ell^\perp = 0$ for $\ell \in S$ and $L_\ell^\perp = H^1(G_\ell/I_\ell, \mu_{p^n})$ for $\ell \notin S$. Consider $H^1_{\mathcal{L}^*}(\mathbb{Q}, \mu_{p^n})$.

From above, we know that every element of $H^1(G_\mathbb{Q}, \mu_{p^n})$ is represented by a cocycle of the form $g \mapsto g\alpha/\alpha$, where $\alpha^{p^n} = a \in \mathbb{Q}^\times$. To be in $H^1_{\mathcal{L}^*}$, it must be unramified everywhere. Since

$$H^1(I_\ell, \mu_{p^n}) = H^1(G_{\mathbb{Q}_\ell^{unr}}, \mu_{p^n}) \simeq (\mathbb{Q}_\ell^{unr})^\times/((\mathbb{Q}_\ell^{unr})^\times)^{p^n},$$

where $\mathbb{Q}_\ell^{unr}$ is the maximal unramified extension of $\mathbb{Q}_\ell$, this implies that $v_\ell(\alpha) \equiv 0 \mod p^n$ for all ℓ. Therefore $a = p^n$th power in $\mathbb{Q}$ (we can ignore ± 1 since p is odd) and the cocycle represents the trivial cohomology class. It follows that $H^1_{\mathcal{L}^*}(\mathbb{Q}, \mu_{p^n}) = 0$.

We now evaluate the right side of the formula. First,

$$\#H^0(G_\mathbb{Q}, \mathbb{Z}/p^n\mathbb{Z}) = \#\mathbb{Z}/p^n\mathbb{Z} = p^n.$$

Since we chose p to be odd, $H^0(G_\mathbb{Q}, \mu_{p^n}) = 0$. In the product, the terms for $\ell \notin S$ are all 1. When $\ell \neq \infty$ is in S, the factor is

$$\frac{\#H^1(G_\ell, \mathbb{Z}/p^n\mathbb{Z})}{\#H^0(G_\ell, \mathbb{Z}/p^n\mathbb{Z})} = \#H^0(G_\ell, \mu_{p^n}) \cdot \ell^{v_\ell(p^n)}$$

by Proposition 3. The number of p^nth roots of unity in $\mathbb{Q}_\ell$ is $(\ell - 1, p^n)$, so this is the order of $H^0(G_\ell, \mu_{p^n})$. Since $\#\,\mathrm{Hom}(G_\mathbb{R}, \mathbb{Z}/p^n\mathbb{Z}) = 1$, the factor for $\ell = \infty$ is $1/p^n$. Putting everything together, we find

$$\#H^1_{\mathcal{L}}(\mathbb{Q}, \mathbb{Z}/p^n\mathbb{Z}) = p^n \prod_{\ell \in S\setminus\infty} (\ell - 1, p^n).$$

Note that $H^1(G_\mathbb{Q}, \mathbb{Z}/p^n\mathbb{Z}) = \mathrm{Hom}(G_\mathbb{Q}, \mathbb{Z}/p^n\mathbb{Z})$ classifies cyclic extensions of degree dividing p^n, and $H^1_{\mathcal{L}}(\mathbb{Q}, \mathbb{Z}/p^n\mathbb{Z})$ gives those extensions that are unramified outside S.

We already have a good supply of such extensions coming from subfields of cyclotomic fields. For each finite prime $\ell \in S$, there is a cyclic extension of degree $(\ell - 1, p^n)$ contained in the ℓ-th cyclotomic field. There is also a cyclic extension of degree p^n contained in the p^{n+1}st cyclotomic field. These extensions are disjoint, so we obtain an abelian extension of exponent p^n and degree $p^n \prod_{\ell \in S}(\ell - 1, p^n)$. The Galois group of this extension has this many homomorphisms into $\mathbb{Z}/p^n\mathbb{Z}$, so all homomorphisms of $G_{\mathbb{Q}}$ into $\mathbb{Z}/p^n\mathbb{Z}$ unramified outside S are obtained from subfields of cyclotomic fields. By enlarging S arbitrarily, we find that every cyclic extension of $\mathbb{Q}$ of degree dividing p^n is contained in a cyclotomic field. The same analysis may be done for powers of 2 with the same result. Since every finite abelian group is a product of cyclic groups of prime power order, we obtain the Kronecker-Weber theorem that every abelian extension of $\mathbb{Q}$ is contained in a cyclotomic field. (Of course, this proof is by no means elementary, since the full power of class field theory is used in the proof of Theorem 2.)

As in the proof of the Kronecker-Weber theorem just given, it will sometimes be necessary to enlarge the set of primes at which ramification is allowed. The following estimates how much the Selmer group increases.

Proposition 5. *Let p be prime and suppose $\#X$ is a power of p. Let $\mathcal{L} = \{L_\ell\}$ be a collection of local conditions and let $q \neq p$ be a prime for which $L_q = H^1(G_q/I_q, X^{I_q})$. Define a new collection $\mathcal{L}' = \{L'_\ell\}$ of local conditions by $L'_\ell = L_\ell$ if $\ell \neq q$ and $L'_q = H^1(G_q, X)$. Then*

$$\frac{\#H^1_{\mathcal{L}'}(\mathbb{Q}, X)}{\#H^1_{\mathcal{L}}(\mathbb{Q}, X)} \leq \#H^0(G_q, X^*).$$

Proof. Since $L_q'^{\perp} = 0$, the conditions defining $H^1_{\mathcal{L}'^*}$ are more restrictive than those defining $H^1_{\mathcal{L}^*}$, so $H^1_{\mathcal{L}'^*}$ has order less than or equal to the order of $H^1_{\mathcal{L}^*}$. When $\mathcal{L}$ is changed to $\mathcal{L}'$ in Theorem 2, all factors on the right remain the same except the one for q, which changes from 1 to $\#H^1(G_q, X)/\#H^0(G_p, X)$. By Proposition 3, this equals $\#H^0(G_q, X^*)$, since $q \nmid \#X$. The result follows easily. $\square$

§6. LOCAL CONDITIONS

From now on, fix a finite set Σ of primes (including ∞, though this will not be important). Let p be an odd prime and assume R is a finite ring of cardinality a power of p. We will work with $X = \mathrm{Ad}^0 \rho$, where $\rho : G_{\mathbb{Q}} \to \mathrm{GL}_2(R)$ is a 2-dimensional representation. We also assume ρ is an odd representation. For our present purposes, we take this to mean that if c is (any choice of) complex conjugation, then the matrix $\rho(c)$ is similar to $\begin{pmatrix} 1 & 0 \\ 0 & -1 \end{pmatrix}$.

Define a collection of local conditions as follows:

$$L_\ell = H^1(G_\ell/I_\ell, (\mathrm{Ad}^0 \rho)^{I_\ell}) \quad \text{for } \ell \notin \Sigma,\ \ell \neq p,$$

$$L_\ell = H^1(G_\ell, \mathrm{Ad}^0 \rho) \quad \text{for } \ell \in \Sigma,\ \ell \neq p,$$

L_p will be specified later.

In other words, if we think in terms of infinitesimal deformations, we allow as little ramification as possible at the primes $\neq p$ outside Σ, the ramification at those places being due to ramification in ρ. At the primes $\ell \neq p$ in Σ

we allow arbitrary ramification. At p we want to control what happens a little more carefully, depending on properties of ρ.

In the formula of Theorem 2, we need to evaluate, or at least estimate, the factors $\#L_\ell/\#H^0(G_\ell, \mathrm{Ad}^0\,\rho)$ corresponding to the various primes.

- The factors for the primes $\ell \notin \Sigma$ with $\ell \neq p$ are all 1 by Lemma 1.
- The factor for the infinite prime is easy. Since $G_{\mathbb{R}}$ has order 2 and $\mathrm{Ad}^0\,\rho$ has odd order, $H^1(G_{\mathbb{R}}, \mathrm{Ad}^0\,\rho) = 0$. Therefore L_∞ is a subgroup of the trivial group, hence trivial. We may assume that $\rho(c) = \begin{pmatrix} 1 & 0 \\ 0 & -1 \end{pmatrix}$. Since $\rho(c)A\rho(c)^{-1} = A$ is equivalent to A being diagonal, we see that $H^0(G_{\mathbb{R}}, \mathrm{Ad}^0\,\rho)$ has order $\#R$. Therefore the factor for ∞ is $1/\#R$.
- Let $\ell \in \Sigma$, $\ell \neq p, \infty$. Then, as in the proof of Proposition 5, we have
$$\frac{\#H^1(G_\ell, \mathrm{Ad}^0\,\rho)}{\#H^0(G_\ell, \mathrm{Ad}^0\,\rho)} = \#H^0(G_\ell, (\mathrm{Ad}^0\,\rho)^*).$$

§7. CONDITIONS AT p

Ordinary representations. Suppose $\rho|_{G_p}$ has the form (for some choice of basis) $\begin{pmatrix} \psi_1\epsilon & * \\ 0 & \psi_2 \end{pmatrix}$, where ψ_1 and ψ_2 are unramified characters (with values in $R^\times$), and ϵ is now the cyclotomic character (not the infinitesimal element from above) giving the action of G_p on the p-power roots of unity. Let W^0 be the additive subgroup of $\mathrm{Ad}^0\,\rho$ given by matrices of the form $\begin{pmatrix} 0 & * \\ 0 & 0 \end{pmatrix}$.

Lemma 2. *G_p acts on W^0 by multiplication by $\psi_1\epsilon/\psi_2$.*

Proof.
$$\begin{pmatrix} \psi_1\epsilon & * \\ 0 & \psi_2 \end{pmatrix}\begin{pmatrix} 0 & b \\ 0 & 0 \end{pmatrix}\begin{pmatrix} \psi_1\epsilon & * \\ 0 & \psi_2 \end{pmatrix}^{-1} = \begin{pmatrix} 0 & \psi_1\epsilon b/\psi_2 \\ 0 & 0 \end{pmatrix}.$$

Lemma 3. $\#H^0(G_p, (W^0)^*) = \#R/\left(\frac{\psi_1}{\psi_2}(\mathrm{Frob}_p) - 1\right)R$.

Proof. An element of $(W^0)^*$ is a group homomorphism $\phi : R \to \mu_{p^n}$ (for some sufficiently large n), and ϕ is fixed by G_p if and only if $\phi(gr) = g\phi(r)$ for all $g \in G_p$ and $r \in R$. By Lemma 2, this means $\phi(\frac{\psi_1\epsilon}{\psi_2}r) = \epsilon\phi(r)$. Note that ϵ takes values in the image of $\mathbb{Z}_p$ in R, which is the same as the image of $\mathbb{Z}$ in R. Therefore we can regard ϵ as an integer that is also a unit in R, and consequently obtain $\phi(\frac{\psi_1}{\psi_2}r) = \phi(r)$. Since ψ_1 and ψ_2 are unramified, it suffices to check this for $g = \mathrm{Frob}_p$, so we let $\alpha = \frac{\psi_1}{\psi_2}(\mathrm{Frob}_p)$. We need ϕ to satisfy $\phi((\alpha - 1)r) = 0$ for all r. This says that ϕ is a group homomorphism from $R/(\alpha-1)R$ to $\mu_{p^n}\mathbb{Z}$. The number of such homomorphisms is $\#R/(\alpha-1)R$. $\square$

We now look at two choices for L_p.

Choice 1. $L_p = \mathrm{Ker}\Big(H^1(G_p, \mathrm{Ad}^0\rho) \to H^1(I_p, \mathrm{Ad}^0\rho/W^0)\Big)$
In terms of infinitesimal deformations ρ', this requires $\rho'|_{I_p}$ always to be equivalent to the form $\begin{pmatrix} \epsilon & * \\ 0 & 1 \end{pmatrix}$. This case will be used, for example, in the case of an elliptic curve with good ordinary reduction at p.

Consider the diagram

$$\begin{array}{ccc} & & H^1(G_p, \mathrm{Ad}^0\rho) \\ & & \downarrow u \\ 0 \to H^1(G_p/I_p, (\mathrm{Ad}^0\rho/W^0)^{I_p}) & \to & H^1(G_p, \mathrm{Ad}^0\rho/W^0) \\ & & \xrightarrow{\text{res}} H^1(I_p, \mathrm{Ad}^0\rho/W^0)^{G_p/I_p}. \end{array}$$

Then $L_p = \mathrm{Ker}(\mathrm{res}\circ u)$ and $H^1(G_p, \mathrm{Ad}^0\rho)/L_p \simeq \mathrm{Im}(\mathrm{res}\circ u)$.

From the exact sequence,

$$\begin{aligned} \#\,\mathrm{Im}(\mathrm{res}\circ u) &\geq \#\,\mathrm{Im}\,u/\#H^1(G_p/I_p, (\mathrm{Ad}^0\rho/W^0)^{I_p}) \\ &= \#\,\mathrm{Im}\,u/\#H^0(G_p, \mathrm{Ad}^0\rho/W^0), \end{aligned}$$

the last equality following from Lemma 1. The exact sequence (with $H^i(X) = H^i(G_p, X)$)

$$\begin{aligned} 0 \to H^0(W^0) \to H^0(\mathrm{Ad}^0\rho) \to H^0(\mathrm{Ad}^0\rho/W^0) \\ \to H^1(W^0) \to H^1(\mathrm{Ad}^0\rho) \to \mathrm{Im}\,u \to 0 \end{aligned}$$

yields $\#\,\mathrm{Im}\,u$ as the alternating product of the orders of the other terms, and we obtain

$$\begin{aligned} \frac{\#L_p}{\#H^0(G_p, \mathrm{Ad}^0\rho)} &= \frac{\#H^1(G_p, \mathrm{Ad}^0\rho)}{\#H^0(G_p, \mathrm{Ad}^0\rho)\,\#\,\mathrm{Im}(\mathrm{res}\circ u)} \\ &\leq \frac{\#H^1(G_p, \mathrm{Ad}^0\rho)\#H^0(G_p, \mathrm{Ad}^0\rho/W^0)}{\#H^0(G_p, \mathrm{Ad}^0\rho)\,\#\,\mathrm{Im}\,u} \\ &= \frac{\#H^1(G_p, W^0)}{\#H^0(G_p, W^0)} \\ &= \#R\cdot\#H^0(G_p, (W^0)^*). \end{aligned}$$

The last equality follows from Proposition 3. Combining this with Lemma 3, we obtain

$$\frac{\#L_p}{\#H^0(G_p, \mathrm{Ad}^0\rho)} \leq \#R\cdot\#\left[R/(\frac{\psi_1}{\psi_2}(\mathrm{Frob}_p)-1)R\right].$$

Choice 2. $L_p = \operatorname{Ker}\Big(H^1(G_p, \operatorname{Ad}^0\rho) \to H^1(G_p, \operatorname{Ad}^0\rho/W^0)\Big)$
This is used when working with an elliptic curve that has bad multiplicative reduction at p. It is similar to the previous case, except that it specifies what happens on all of G_p. Actually, in this case ("ordinary but not flat" [DDT], or "strict" [Wi]) we could use the same L_p as before, by a result of Diamond [Wi, Proposition 1.1], but the present choice is more convenient for our calculations. By the calculations just completed, but with the new choice of L_p, we have $H^1(G_p, \operatorname{Ad}^0\rho)/L_p \simeq \operatorname{Im} u$ and

$$\frac{\#L_p}{\#H^0(G_p, \operatorname{Ad}^0\rho)} = \frac{\#R \cdot \#H^0(G_p, (W^0)^*)}{\#H^0(G_p, \operatorname{Ad}^0\rho/W^0)}.$$

In the case where this will be applied, we will have

$$\psi_1 = \psi_2,$$

so $\#H^0(G_p, (W^0)^*) = \#R$ by Lemma 3. Also, we will have a matrix

$$\rho(g) = \begin{pmatrix} \psi_1\epsilon & y \\ 0 & \psi_2 \end{pmatrix} \text{ with } y \in R^\times$$

in the image of $\rho|_{G_p}$. Since

$$\begin{pmatrix} \psi_1\epsilon & y \\ 0 & \psi_2 \end{pmatrix}\begin{pmatrix} a & * \\ c & -a \end{pmatrix}\begin{pmatrix} \psi_1\epsilon & y \\ 0 & \psi_2 \end{pmatrix}^{-1} = \begin{pmatrix} a + \frac{cy}{\psi_1\epsilon} & * \\ \frac{\psi_2 c}{\psi_1\epsilon} & -a - \frac{cy}{\psi_1\epsilon} \end{pmatrix},$$

it follows that an element of $\operatorname{Ad}^0\rho/W^0$ fixed by G_p is represented by a diagonal matrix. Therefore $\#H^0(G_p, \operatorname{Ad}^0\rho/W^0) = \#R$. Putting things together, we obtain

$$\frac{\#L_p}{\#H^0(G_p, \operatorname{Ad}^0\rho)} = \#R.$$

Flat representations. This is a more technical situation that must be used in the case of an elliptic curve with good supersingular reduction. Let $L_p = H^1_f(G_p, \operatorname{Ad}^0\rho)$ be those cohomology classes in $H^1(G_p, \operatorname{Ad}^0\rho)$ representing extensions $0 \to M \to E \to M \to 0$ in the category of $R[G_p]$-modules attached to finite flat group schemes over $\mathbb{Z}_p$. We also assume that $R = \mathcal{O}/\lambda^n$, where $\mathcal{O}$ is the ring of integers in a finite extension of $\mathbb{Q}_p$ and λ generates the maximal ideal. The theory of Fontaine-Lafaille implies that

$$\frac{\#L_p}{\#H^0(G_p, \operatorname{Ad}^0\rho)} = \#R.$$

§8. PROOF OF THEOREM 2

We first address a technical point. Let Σ be a finite set of primes and let $\mathbb{Q}_\Sigma$ be the maximal extension of $\mathbb{Q}$ unramified at the primes not in Σ. Let X be a module for G_Σ =Gal($\mathbb{Q}_\Sigma/\mathbb{Q}$). Then X is also a module for $G_\mathbb{Q}$ that is unramified outside Σ. Some papers, for example [Wi], consider $H^1(G_\Sigma, X)$, while others, for example [DDT], consider the classes of $H^1(G_\mathbb{Q}, X)$ unramified outside Σ. Fortunately, the two groups are isomorphic. In the following, we will find it more convenient to work with $H^1(G_\Sigma, X)$.

Proposition 6. $H^1(G_\Sigma, X) \simeq \mathrm{Ker}\Big(H^1(G_\mathbb{Q}, X) \to \prod_{\ell \notin \Sigma} H^1(I_\ell, X)\Big).$

Proof. The following diagram commutes (the top row is inflation-restriction).

$$\begin{array}{ccccccc} 0 \longrightarrow & H^1(G_\Sigma, X) & \longrightarrow & H^1(G_\mathbb{Q}, X) & \longrightarrow & H^1(\mathrm{Gal}(\bar{\mathbb{Q}}/\mathbb{Q}_\Sigma), X) \\ & & & \downarrow & & \downarrow{=} \\ & & & \prod_{\ell \notin \Sigma} \mathrm{Hom}(I_\ell, X) & \xleftarrow{\phi} & \mathrm{Hom}(\mathrm{Gal}(\bar{\mathbb{Q}}/\mathbb{Q}_\Sigma), X). \end{array}$$

The map ϕ is injective since a homomorphism that is 0 on I_ℓ for all $\ell \notin \Sigma$ must vanish on the smallest normal subgroup generated by all such I_ℓ, which is Gal($\bar{\mathbb{Q}}/\mathbb{Q}_\Sigma$). The result follows easily. □

Proposition 7. *If X is finite then $H^1(G_\Sigma, X)$ is finite.*

Proof. Choose an open normal subgroup H of G_Σ such that H acts trivially on X. Let K be the fixed field of H. The group $H^1(H, X) = \mathrm{Hom}(H, X)$ is finite since it classifies Galois extensions of K, unramified outside Σ, with Galois group isomorphic to a subgroup of X, and there are only finitely many such extensions by a theorem of Hermite-Minkowski. Since G_Σ/H is finite, the group $H^1(G_\Sigma/H, X)$ is finite by its definition. The result now follows from the inflation-restriction sequence. □

Corollary. *$H^1_\mathcal{L}(\mathbb{Q}, X)$ is finite.*

Proof. The group is isomorphic to a subgroup of $H^1(G_\Sigma, X)$. □

Let X be a finite module for $G_\mathbb{Q}$. Fix a set Σ containing ∞, all the prime divisors of $\#X$, and all primes such that I_p does not act trivially on X. There exists an open subgroup that acts trivially on X. This subgroup corresponds to some finite extension $K/\mathbb{Q}$, and the inertia group of any prime not ramifying in K acts trivially on X. Therefore we can take Σ to be finite. Let Σ_f be the set of finite primes in Σ. For an integer $r = 0, 1, 2$, let

$$\alpha_r : H^r(G_\Sigma, X) \longrightarrow \hat{H}^r(G_\mathbb{R}, X) \times \prod_{\ell \in \Sigma_f} H^r(G_\ell, X)$$

be induced by the restriction maps, where $\hat{H}^r(G_{\mathbb{R}}, X)$ is the modified Tate cohomology group (when $r > 0$, let $\hat{H}^r = H^r$). By Theorem 1, $\hat{H}^r(G_{\mathbb{R}}, X) \times \prod H^r(G_\ell, X)$ is the dual of $\hat{H}^{2-r}(G_{\mathbb{R}}, X^*) \times \prod H^{2-r}(G_\ell, X^*)$, so we may dualize the map

$$H^{2-r}(G_\Sigma, X^*) \to \hat{H}^{2-r}(G_{\mathbb{R}}, X^*) \times \prod_{\ell \in \Sigma_f} H^{2-r}(G_\ell, X^*)$$

to obtain

$$\beta_r : \hat{H}^r(G_{\mathbb{R}}, X) \times \prod_{\ell \in \Sigma_f} H^r(G_\ell, X) \longrightarrow H^{2-r}(G_\Sigma, X^*)^\vee,$$

where $A^\vee = \mathrm{Hom}(A, \mathbb{Q}/\mathbb{Z})$ is the dual of an abelian group A. Let

$$\mathrm{Ker}^r(G_\Sigma, X) = \mathrm{Ker}\, \alpha_r.$$

Proposition 8. *There is a non-degenerate canonical pairing*

$$\mathrm{Ker}^2(G_\Sigma, X) \times \mathrm{Ker}^1(G_\Sigma, X^*) \to \mathbb{Q}/\mathbb{Z}.$$

Proof. The pairing can be defined as follows. Let $f \in \mathrm{Ker}^2$ and $g \in \mathrm{Ker}^1$. For $\ell \in \Sigma$, we can write $\mathrm{res}_\ell f = \delta\phi_\ell$ and $\mathrm{res}_\ell g = \delta\psi_\ell$, where $\phi_\ell : G_\ell \to X$, $\psi_\ell \in X^*$, and δ is the coboundary map of the appropriate dimension. It can be shown that the cup product $f \cup g = 0 \in H^3(G_\Sigma, \mathbb{Q}_\Sigma^\times)$, so $f \cup g = \delta h$ for an appropriate h. Then

$$(f \cup \psi_\ell) - h = (\phi_\ell \cup g) - h + \delta(\phi_\ell \cup \psi_\ell),$$

hence $(f \cup \psi_\ell) - h$ and $(\phi_\ell \cup g) - h$ represent the same class

$$x_\ell \in H^2(G_l, \mathbb{Q}_\ell^\times) \simeq \mathbb{Q}/\mathbb{Z},$$

and x_ℓ is independent of the choices involved. Define

$$< f, g >= \sum_{\ell \in \Sigma} x_\ell \in \mathbb{Q}/\mathbb{Z}.$$

The proof of the non-degeneracy is much more difficult. See [Mi]. □

Proposition 9. *α_0 is injective, β_2 is surjective, and for $r = 0, 1, 2$, we have $\mathrm{Im}\, \alpha_r = \mathrm{Ker}\, \beta_r$.*

Proof. For a proof, see [Mi].

This can all be summarized in the following.

Proposition 10 (Poitou-Tate). *The following nine-term sequence is exact:*

$$0 \to H^0(G_\Sigma, X) \xrightarrow{\alpha_0} \hat{H}^0(G_\mathbb{R}, X) \times \prod_{\ell \in \Sigma_f} H^0(G_\ell, X) \xrightarrow{\beta_0} H^2(G_\Sigma, X^*)^\vee$$
$$\to H^1(G_\Sigma, X) \xrightarrow{\alpha_1} \prod_{\ell \in \Sigma} H^1(G_\ell, X) \xrightarrow{\beta_1} H^1(G_\Sigma, X^*)^\vee$$
$$\to H^2(G_\Sigma, X) \xrightarrow{\alpha_2} \prod_{\ell \in \Sigma} H^2(G_\ell, X) \xrightarrow{\beta_2} H^0(G_\Sigma, X^*)^\vee \to 0,$$

where the unlabeled arrows are maps defined by the non-degeneracy of the pairing in Proposition 8.

It is also possible to work with infinite sets Σ, but then some restrictions need to be made on the direct products involved.

We can now prove Theorem 2. The definition of the Selmer group yields the exact sequence

$$0 \to H^1_{\mathcal{L}^*}(\mathbb{Q}, X^*) \to H^1(G_\Sigma, X^*) \to \prod_\Sigma H^1(G_\ell, X^*)/L_\ell^\perp.$$

Dualizing (i.e., $\mathrm{Hom}(-, \mathbb{Q}/\mathbb{Z})$) and using the pairing of Theorem 1 yields

$$0 \leftarrow H^1_{\mathcal{L}^*}(\mathbb{Q}, X^*)^\vee \leftarrow H^1(G_\Sigma, X^*)^\vee \leftarrow \prod L_\ell.$$

Splicing this into the nine-term sequence yields

$$0 \to H^0(G_\Sigma, X) \xrightarrow{\alpha_0} \hat{H}^0(G_\mathbb{R}, X) \times \prod_{\ell \in \Sigma_f} H^0(G_\ell, X) \xrightarrow{\beta_0} H^2(G_\Sigma, X^*)^\vee$$
$$\to H^1_\mathcal{L}(\mathbb{Q}, X) \xrightarrow{\alpha_1} \prod_{\ell \in \Sigma} L_\ell \xrightarrow{\beta_1} H^1(G_\Sigma, X^*)^\vee \to H^1_{\mathcal{L}^*}(\mathbb{Q}, X^*)^\vee \to 0.$$

Therefore

$$\frac{\#H^1_\mathcal{L}(\mathbb{Q}, X)}{\#H^1_{\mathcal{L}^*}(\mathbb{Q}, X^*)} = \frac{\#H^0(G_\Sigma, X)\,\#H^2(G_\Sigma, X^*)^\vee\,\#(1+c)X}{\#H^1(G_\Sigma, X^*)} \prod_{\ell \in \Sigma} \frac{\#L_\ell}{\#H^0(G_\ell, X)},$$

where we have used the fact for $\ell = \infty$ that

$$\hat{H}^0(G_\mathbb{R}, X) = H^0(G_\mathbb{R}, X)/(1+c)X.$$

We now need the following formula for what may be regarded as a global Euler characteristic.

Proposition 11. *Let X be finite. The groups $H^r(G_\Sigma, X)$, $r = 0, 1, 2$, are finite, and*

$$\frac{\#H^0(G_\Sigma, X)\,\#H^2(G_\Sigma, X)}{\#H^1(G_\Sigma, X)} = \frac{\#H^0(G_\mathbb{R}, X)}{\#X}.$$

Proof. For a proof, see [Mi, p. 82].

Since $H^2(G_\Sigma, X^*)$ is finite, it has the same order as its dual. Also, $H^0(G_\Sigma, X) = X^{G_\Sigma} = X^{G_\mathbb{Q}} = H^0(G_\mathbb{Q}, X)$. Therefore the proposition, applied to X^*, reduces the proof to the following.

Lemma 4. $\#(1+c)X \cdot \#H^0(G_\mathbb{R}, X^*) = \#X^*$.

Proof. The (non-degenerate) pairing $X \times X^* \to \mu_n$ satisfies $\langle cx, cx^* \rangle = c\langle x, x^* \rangle = \langle x, x^* \rangle^{-1}$, from which it follows that $\langle (1+c)x, x^* \rangle = \langle x, (1-c)x^* \rangle$. Therefore x^* is fixed by $c \iff (1-c)x^* = 0 \iff \langle x, (1-c)x^* \rangle = 0$ for all $x \iff \langle (1+c)x, x^* \rangle = 0$ for all x. Therefore $H^0(G_\mathbb{R}, X^*)$ is the exact annihilator of $(1+c)X$, hence is dual to $X/(1+c)X$. The result follows easily. □

References

[CF] J.W.S. Cassels and A. Fröhlich, *Algebraic number theory*, Acad. Press, New York, 1967.

[DDT] H. Darmon, F. Diamond, and R. Taylor, *Fermat's Last Theorem* (1995), preprint.

[Gr] R. Greenberg, *Iwasawa theory for p-adic representations*, Algebraic number theory — in honor of K. Iwasawa (J. Coates et al., eds.), Advanced studies in pure mathematics **17**, Academic Press, Boston, 1989.

[Ha] K. Haberland, *Galois cohomology of algebraic number fields*, VEB Deutscher Verlag der Wissenschaften, Berlin, 1978.

[Mi] J.S. Milne, *Arithmetic duality theorems,* Perspectives in Mathematics **1**, Academic Press, Boston, 1986.

[Po] G. Poitou, *Cohomologie galoisienne des modules finis*, Dunod, Paris, 1967.

[Se] J.-P. Serre, *Local fields* (translated by M. Greenberg), Springer-Verlag, New York-Heidelberg-Berlin, 1979.

[Sh] S. Shatz, *Profinite groups, arithmetic, and geometry*, Annals of Mathematics Studies **67**, Princeton Univ. Press, 1972.

[Ta] J. Tate, *Proc. International Cong. Math., Stockholm, 1962*, pp. 234–241.

[Wi] A. Wiles, *Modular elliptic curves and Fermat's Last Theorem*, Annals of Math. **141** (1995), 443–551.

FINITE FLAT GROUP SCHEMES

JOHN TATE

INTRODUCTION

The kernel of an isogeny of degree n of abelian varieties of dimension g is, at a place of good reduction, a finite flat group scheme of order n^{2g} over the local ring of the place. That is perhaps the main reason for studying finite flat group schemes, although they are interesting enough in their own right, and it is in any case the reason a discussion of them appears in this volume. For that reason also, the commutative case is the most important for us, and it is in that case that the theory is most interesting and highly developed by far. Nevertheless we do not assume commutativity at the beginning and develop the basics of the theory without that assumption.

We use the language of schemes, but without much loss of generality we can, and mostly do, restrict to the affine case, because a finite morphism of schemes is affine. Thus only very elementary scheme theory is needed — not much more than the equivalence between the category of affine schemes and the category of rings with arrows reversed. By *ring* or *algebra* in this paper we mean one which is *commutative* with *unity*, unless mention is made to the contrary. If R is a noetherian ring, a finite flat group scheme G over R (that is, over Spec(R)) is of the form $G = \text{Spec}(A)$, where A is a commutative Hopf algebra over R which is locally free of finite rank as R-module. In essence, our topic is the theory of such Hopf algebras. Although we treat the case of a general noetherian base ring as far as possible, the reader will not lose much by restricting to the case in which R is a discrete valuation ring or a field, in which case even the commutative algebra involved is quite elementary.

Beyond the very general properties of group schemes, the only more special results we treat (in §4) are some of Raynaud's, over valuation rings of mixed characteristic. For the more refined theory in characteristic p, we refer the reader to [deJ]

In dealing with group schemes it is extremely convenient to use some basic categorical concepts, in particular, the fact that attaching to an object G in a category C the contravariant set functor represented by G embeds C as a full subcategory of the category $\widehat{C}$ of all such functors. It is often easier to describe the functor represented by a group scheme than to describe the group scheme or Hopf algebra itself.

§1. Group objects in a category

The subject of this section is very clearly explained, with a few more details in [B-L-R, §4.1]. Other sources are, among many, [SGA3, Exp.I] and [SS]. Let $\mathcal{C}$ be a category with finite products and in particular a final object, the empty product, which we will denote by S in anticipation of the case in which $\mathcal{C} = (\mathbf{Sch}/S)$ is the category of schemes over a base scheme S. Let G be an object of $\mathcal{C}$ and $m : G \times G \to G$ a "law of composition" on G. This m induces, for every T in $\mathcal{C}$, a law of composition on the set

$$G(T) := \mathrm{Hom}_{\mathcal{C}}(T, G)$$

in an obvious way, because by definition of the product $G \times G$ we have $(G \times G)(T) = G(T) \times G(T)$. Explicitly, writing the induced law on $G(T)$ multiplicatively, we have $g_1 g_2 = m \circ (g_1, g_2)$, where $(g_1, g_2) : T \to G \times G$ is the unique arrow such that $\mathrm{pr}_i \circ (g_1, g_2) = g_i$ for $i = 1, 2$. (Here pr_1 and pr_2 are the two projections $G \times G \to G$.) A morphism $f : T' \to T$ induces a map $f^* : G(T) \to G(T')$ by $f^*(g) := g \circ f$, and this map preserves the law of composition in the sense that $f^*(g_1 g_2) = f^*(g_1) f^*(g_2)$, because $(g_1, g_2) \circ f = (g_1 \circ f, g_2 \circ f)$. In other words the association $T \mapsto G(T)$ is a contravariant functor from $\mathcal{C}$ to the category of magmas (a magma is a set with a law of composition).

The following four facts are easily checked and are left to the reader.

(1.1) Associativity. The magma $G(T)$ is associative for every T if and only if the equality $(\mathrm{pr}_1 \mathrm{pr}_2) \mathrm{pr}_3 = \mathrm{pr}_1 (\mathrm{pr}_2 \mathrm{pr}_3)$ holds in $G(G \times G \times G)$, i.e., if and only if the following diagram is commutative

$$\text{a)} \quad \begin{array}{ccc} G \times G \times G = G \times (G \times G) & \xrightarrow{\mathrm{id} \times m} & G \times G \\ \| & & \\ (G \times G) \times G & & \Big\downarrow m \\ {\scriptstyle m \times \mathrm{id}} \Big\downarrow & & \\ G \times G & \xrightarrow{\quad m \quad} & G \, . \end{array}$$

(1.2) Unit elements. The magmas $G(T)$ have two-sided unit elements e_T (necessarily unique), and these units are preserved by the morphisms $f^* : G(T) \to G(T')$, if and only if there is a point $\varepsilon \in G(S)$ (recall that S is the final object in $\mathcal{C}$) such that the equality $\pi^*(\varepsilon) \cdot \mathrm{id} = \mathrm{id} = \mathrm{id} \cdot \pi^*(\varepsilon)$ holds in $G(G)$, where $\pi = \pi_G$ is the unique arrow $G \to S$, that is, if and only if each triangle in the following diagram commutes

$$\text{b)} \quad \begin{array}{ccc} G = G \times S & \xrightarrow{\mathrm{id} \times \varepsilon} & G \times G \\ \| & & \\ S \times G & \searrow & \Big\downarrow m \\ {\scriptstyle \varepsilon \times \mathrm{id}} \Big\downarrow & & \\ G \times G & \xrightarrow{\quad m \quad} & G \, . \end{array}$$

When that is the case, $\varepsilon_T := \pi_T^*(\varepsilon)$ is the unit in $G(T)$ for each T.

(1.3) Inverses. Suppose the magmas $G(T)$ have two sided units ε_T preserved by the f^*'s. Then the necessary and sufficient condition that every element $g \in G(T)$ have a left inverse for every T is that the element $\mathrm{id} = \mathrm{id}_G \in G(G)$ have a left inverse in $G(G)$, i.e., that there exist an element $\mathrm{inv} \in G(G)$ such that $\mathrm{inv} \cdot \mathrm{id}_G = \varepsilon_G$, or in other words such that the diagram

$$\text{c)} \qquad \begin{array}{ccc} G \times G & \xrightarrow{\mathrm{inv} \times \mathrm{id}} & G \times G \\ {\scriptstyle \Delta}\uparrow & & \downarrow{\scriptstyle m} \\ G & \xrightarrow{\varepsilon \circ \pi} & G \end{array}$$

is commutative. Then $(\mathrm{inv} \circ g) \cdot g = \varepsilon_T$ for every $g \in G(T)$, any T.

(1.4) Commutativity. The magmas $G(T)$ are commutative if and only if the equality $\mathrm{pr}_1 \mathrm{pr}_2 = \mathrm{pr}_2 \mathrm{pr}_1$ holds in $G(G \times G)$, i.e., if and only if the diagram

$$\text{d)} \qquad \begin{array}{ccc} G \times G & \xrightarrow{\tau = (\mathrm{pr}_2, \mathrm{pr}_1)} & G \times G \\ & {\scriptstyle m} \searrow \quad \swarrow {\scriptstyle m} & \\ & G & \end{array}$$

commutes, where τ is the automorphism interchanging the factors on the product.

(1.5) Definition. A *group object in $\mathcal{C}$*, or a *$\mathcal{C}$-group* is an object G in $\mathcal{C}$ together with a morphism $m : G \times G \to G$ such that the induced law of composition $G(T) \times G(T) \to G(T)$ makes $G(T)$ a group for every T in $\mathcal{C}$. A $\mathcal{C}$-group G is *commutative* if the group $G(T)$ is commutative for every T. A *homomorphism of $\mathcal{C}$-groups* $G \to G'$ is a morphism $G \to G'$ in the category $\mathcal{C}$ such that, for every object T in $\mathcal{C}$, the induced map $G(T) \to G'(T)$ given by $g \mapsto \varphi \circ g$ is a homomorphism of groups.

From the above discussion it is clear that a pair (G, m) is a group object if and only if the diagram a) is commutative and there exist morphisms $\varepsilon : S \to G$ and $\mathrm{inv} : G \to G$ such that diagrams b) and c) commute. Of course ε and inv are unique if they exist. And (G, m) is a commutative group object if and only if in addition diagram d) commutes.

Suppose (G, m) and (G', m') are two group objects in $\mathcal{C}$. In order that a morphism $\varphi : G \to G'$ be a $\mathcal{C}$-group homomorphism it is necessary and sufficient that the equality $\varphi_*(\mathrm{pr}_1 \mathrm{pr}_2) = \varphi_*(\mathrm{pr}_1)\varphi_*(\mathrm{pr}_2)$ hold in $G'(G \times G)$, i.e., that the diagram

$$\begin{array}{ccc} G \times G & \xrightarrow{\varphi \times \varphi} & G' \times G' \\ {\scriptstyle m}\downarrow & & \downarrow{\scriptstyle m'} \\ G & \xrightarrow{\varphi} & G' \end{array}$$

be commutative.

(1.6) Group object = Group functor. Suppose we are given an object G in $\mathcal{C}$, and, instead of a "morphic law of combination" $m : G \times G \to G$, we are given for each T in $\mathcal{C}$ a group structure on $G(T)$ such that for each $f : T' \to T$ the induced map $f^* : G(T) \to G(T')$ is a group homomorphism. Then there is a unique $m : G \times G \to G$ which induces the given group structure on $G(T)$ for each T. The unicity of m follows from the fact that an $m : G \times G \to G$ can be recovered from the law of composition it induces on the set $G(G \times G)$, as the product for that law of the two projections; $m = \mathrm{pr}_1\,\mathrm{pr}_2$. On the other hand, it is easy to check that that choice of m does induces the given law of combination in $G(T)$ for each T. The point of this paragraph is that a group object in $\mathcal{C}$ is the same thing as a contravariant functor from $\mathcal{C}$ to the category (**Gr**) of groups such that the underlying functor from $\mathcal{C}$ to (**Sets**) is *representable*, i.e., isomorphic to a functor of the form $T \to G(T)$ for some object G of $\mathcal{C}$.

Similarly, if G and G' are $\mathcal{C}$-groups, then to give a homomorphism of $\mathcal{C}$-groups $\varphi : G \to G'$ is the "same" as to give a homomorphism of the functors they represent, that is, to give for each T in $\mathcal{C}$ a group homomorphism

$$\varphi_T : G(T) \to G'(T)$$

such that $f^* \circ \varphi_T = \varphi_{T'} \circ f^*$ for every morphism $f : T' \to T$ of objects in $\mathcal{C}$. One recovers $\varphi \in \mathrm{Hom}_{\mathcal{C}}(G, G') = G'(G)$ as the image of the identity in $\mathrm{Hom}_{\mathcal{C}}(G, G) = G(G)$ under the map $\varphi_G : G(G) \to G'(G)$.

(1.7) Kernels. A simple example of the use of **(1.6)** is the construction of kernels. Let $\varphi : G \to G'$ be a homomorphism of group objects in $\mathcal{C}$. Let us define a *kernel of* φ to be a homomorphism of group objects $\alpha : H \to G$ such that, for every T in $\mathcal{C}$, the sequence

$$0 \longrightarrow H(T) \xrightarrow{\alpha_*} G(T) \xrightarrow{\varphi_*} G'(T)$$

is exact. Such an H exists if the fiber product indicated by the following diagram exists in $\mathcal{C}$:

$$\begin{array}{ccc} H = G \times_{G'} S & \xrightarrow{\mathrm{pr}_2} & S \\ {\scriptstyle \alpha=\mathrm{pr}_1}\downarrow & & \downarrow{\scriptstyle \varepsilon'} \\ G & \xrightarrow{\varphi} & G'. \end{array}$$

Then the lefthand vertical arrow $\alpha = \mathrm{pr}_1$ identifies the set $H(T)$ with $\mathrm{Ker}(G(T) \to G'(T))$, because $S(T) = \{\pi_T\}$ is a singleton for each T, and $\varepsilon' \circ \pi_T = \varepsilon_T$ is the unit in $G'(T)$. This identification makes $H(T)$ a group in a functorial way so that $\mathrm{pr}_1 : H \to G$ is a kernel for φ.

Thus if the category $\mathcal{C}$ has fiber products, then $\mathrm{Ker}\,\varphi$ exists for every φ. We leave to the reader to check that it is unique up to a unique isomorphism and that, in the notation above, if H' is any group object in $\mathcal{C}$, then to give a homomorphism $H' \to H$ is the "same" as to give a homomorphism $H' \to G$ whose composition with φ is the trivial homomorphism

$$H' \to S \xrightarrow{\varepsilon'} G',$$

i.e., the sequence

$$0 \longrightarrow \mathrm{Hom}(H', H) \longrightarrow \mathrm{Hom}(H', G) \longrightarrow \mathrm{Hom}(H', G')$$

is exact.

(1.8) Cokernels. The question of coset spaces and cokernels cannot be treated in the same simple-minded way. Even if we assume that $\varphi : G \to G'$ is an injective homomorphism of commutative group objects, the functor $T \mapsto \mathrm{Coker}(\varphi_T) = G'(T)/\varphi G(T)$ is rarely representable. The situation is analogous to the case of sheaves of abelian groups, in which the naive cokernel is only a presheaf in general, not a sheaf. In the commutative case, one can characterize the desired cokernel as a $\mathcal{C}$-group H with a homomorphism $G' \to H$ such that, for every $\mathcal{C}$-group H', the sequence

$$0 \longrightarrow \mathrm{Hom}(H, H') \longrightarrow \mathrm{Hom}(G', H') \longrightarrow \mathrm{Hom}(G, H')$$

is exact. But to show the existence of such an H and to prove it has other desirable properties is often a serious problem. In case $\mathcal{C} = (\mathbf{Sch}/S)$, the category of schemes over a base scheme S, and G is a finite flat closed subgroup scheme of G', then the problem was solved by Grothendieck; we discuss the matter in §3.

§2. GROUP SCHEMES. EXAMPLES

We now specialize to the case of the category $(\mathbf{Sch}/S)$ of schemes over a base scheme S.

(2.1) Definition. An *S-group scheme*, or simply *S-group*, is a group object in $(\mathbf{Sch}/S)$.

We will denote the category of S-group schemes by $(\mathbf{Gr}/S)$.

(2.2) Hopf Algebras. For us, S will usually be affine, say $S = \mathrm{Spec}(R)$, and we will often replace S by R in the notation and terminology, writing $(\mathbf{Sch}/R)$ and R-group scheme, etc. Let $G = \mathrm{Spec}\, A$ be an affine R-scheme. In view of the arrow-reversing equivalence between the category of commutative R-algebras and the category of affine R-schemes, to make G into an R-group scheme is to give R-algebra homomorphisms

$$\tilde{m} : A \longrightarrow A \otimes_R A \qquad \tilde{\varepsilon} : A \longrightarrow R \qquad \widetilde{\mathrm{inv}} : A \longrightarrow A\ ,$$

corresponding to the morphisms $m, \varepsilon, \mathrm{inv}$ discussed in §1, which make commutative the diagrams, let's call them ã), $\tilde{\text{b}}$), c̃), obtained from diagrams a), b), c) by reversing arrows, replacing S by R, G by A, $\times$ by $\otimes_R$, and putting $\sim$ on the labels of the arrows, with $\tilde{\Delta} : A \otimes_R A \to A$ induced by the multiplication in the ring A. One calls $\tilde{m}$ the *comultiplication*, $\tilde{\varepsilon}$ the *augmentation*, or *counit*, and $\widetilde{\mathrm{inv}}$ the *antipode*. A commutative R-algebra A with unit which is furnished with homomorphisms $\tilde{m}$, $\tilde{\varepsilon}$, $\widetilde{\mathrm{inv}}$ satisfying the stated commutative diagram conditions is called a *commutative Hopf algebra*. Thus the category of affine R-group schemes is antiequivalent to the category of commutative Hopf algebras over R, with the obvious definition of homomorphism of Hopf algebras. Commutative Hopf algebras, especially over fields, have been extensively studied for a long time in connection with the theory of affine algebraic groups. Cocommutative Hopf algebras have been around a long time also — examples are group algebras, enveloping algebras of Lie algebras and the one originally studied by Hopf — the homology of a manifold M with a product operation $M \times M \to M$. Some general references for these types of Hopf algebras are [A], [C-S], [M-M], [Sw] and [W]. But it's only in recent times that important Hopf algebras which are neither commutative nor cocommutative have been discovered, usually as deformations of commutative ones, and are being studied seriously ([Dr], [SS-SS]), under the name "quantum groups." But in this paper all the Hopf algebras we encounter will be either commutative or cocommutative, mostly the former.

(2.3) The Augmentation Ideal. Let $G = \mathrm{Spec}(A)$ be an affine R-group scheme. The kernel of the augmentation map $\tilde{\varepsilon}$ is an ideal $I = I_G$ in A called the *augmentation ideal*. As R-module we have $A = R \cdot 1 \oplus I$, direct sum, because the canonical map $R \to A$ splits the exact sequence $0 \to I \to A \to R \to 0$. Thus $A \otimes A = R \oplus (I \otimes 1) \oplus (1 \otimes I) \oplus (I \otimes I)$. An important fact about the comultiplication is that

$$\tilde{m}(f) - f \otimes 1 - 1 \otimes f \in I \otimes I \ , \quad \text{for } f \in I \ ,$$

as one sees by applying the maps $\tilde{\varepsilon} \otimes \mathrm{id}$ and $\mathrm{id} \otimes \tilde{\varepsilon}$ whose kernels $I \otimes A$ and $A \otimes I$ have intersection $I \otimes I$.

(2.4) First examples; $\mathbb{G}_A$ and $\mathbb{G}_m$. Let $G = \mathrm{Spec}(A)$ be an affine scheme over a ring R. To give G an R-group structure it suffices, as explained in §1, to give a group structure on a functor

$$T \to G(T) = \mathrm{Hom}_{(\mathbf{Sch}/R)}(T, G)$$

from R-schemes T to sets. *One does not have to construct m and show the existence of ε and* inv *such that diagrams* a), b), c) *commute*; one recovers $m : G \times G \to G$ as the composition of the two projections pr_1 and pr_2 in the group $G(G \times G)$ and similarly ε and inv. Since G is affine, we can restrict

T to be affine if we wish, and will usually do so. If $T = \operatorname{Spec} B$, then we write $G(B) := G(T) = \operatorname{Hom}_{R\text{-alg}}(A, B)$. Thus, to make G an R-group is to make $B \mapsto \operatorname{Hom}_{R\text{-alg}}(A, B)$ a functor from R-algebras to groups. Then the comultiplication $\tilde{m} : A \to A \otimes_R A$ is obtained as the composition in the group $\operatorname{Hom}_{R\text{-alg}}(A, A \otimes_R A)$ of the two maps

$$\widetilde{\mathrm{pr}_1} : a \mapsto a \otimes 1 \quad \text{and} \quad \widetilde{\mathrm{pr}_2} : a \to 1 \otimes a \; .$$

Here are some standard examples.

The additive group $\mathbb{G}_a$. Let $\mathbb{G}_a = \operatorname{Spec}(R[u])$, u an indeterminate. For each commutative R-algebra B, the map $f \mapsto f(u)$ identifies

$$\operatorname{Hom}_{R\text{-alg}}(R[u], B)$$

with B itself. The additive group structure on B for varying B makes $\mathbb{G}_a$ an R-group, with comultiplication $\tilde{m}$ determined by $\tilde{m}(u) = u \otimes 1 + 1 \otimes u$. Not surprisingly, one finds $\tilde{\varepsilon}(u) = 0$, and $\widetilde{\mathrm{inv}}(u) = -u$. More generally, if M is any R-module, and $A = \operatorname{Sym}_R(M)$ its symmetric algebra over R, then $\operatorname{Hom}_{R\text{-alg}}(A, B) = \operatorname{Hom}_{R-\mathrm{mod}}(M, B)$ is a commutative group under addition for each R-algebra B. Thus $\operatorname{Spec}(A)$ is a commutative R-group. Taking for M the free R-module Ru on one generator u, we recover $\mathbb{G}_a$.

The multiplicative group $\mathbb{G}_m$. Let $\mathbb{G}_m = \operatorname{Spec}(R[u, u^{-1}])$. For each R-algebra B, the map $f \mapsto f(u)$ identifies $\operatorname{Hom}_{R\text{-alg}}(R[u, u^{-1}], B)$ with the multiplicative group B^* of invertible elements of B. Thus $\mathbb{G}_m$ is an R-group, with

$$\tilde{m}(u) = (u \otimes 1)(1 \otimes u) = u \otimes u, \quad \tilde{\varepsilon}(u) = 1, \quad \text{and} \quad \widetilde{\mathrm{inv}}(u) = u^{-1}.$$

This example has at least two important generalizations which we discuss in the next paragraphs.

(2.5) The general linear group GL_n. Let n be an integer ≥ 0, and let $U = (u_{ij})$ and $V = (v_{ij})$ be two $n \times n$ matrices with independent indeterminate entries. In the polynomial ring of $2n^2$ variables

$$R[u, v] = R[u_{11}, u_{12}, \dots , v_{nn}],$$

let J be the ideal generated by the n^2 entries of the matrix $UV - I$, and let $A = R[u, v]/J$. Then $f \mapsto f(U)$ gives a bijection between $\operatorname{Hom}_{R\text{-alg}}(A, B)$ and the group $\mathrm{GL}_n(B) := (\mathbb{M}_n(B))^*$ of invertible $n \times n$ matrices with entries in B, because a right inverse of a square matrix is unique if it exists, and is a left inverse as well. Thus $\operatorname{Spec}(A)$ is an R-group scheme, denoted by GL_n. For $n = 1$, we recover $\mathbb{G}_m = \mathrm{GL}_1$.

A *linear representation of degree n* of an R-group scheme G is a homomorphism of R-group schemes $G \to \mathrm{GL}_n$. To give such a homomorphism is the same as to give an invertible $n \times n$ matrix (a_{ij}) of sections of O_G such that $\tilde{m}(a_{ik}) = \sum_{j=1}^{n} a_{ij} \otimes a_{jk}$.

Exercise. Generalize GL_n in the following way. Instead of $\mathbb{M}_n(R)$, take D to be any (not necessarily commutative) R-algebra which is free of finite rank as R-module. Show that there is an affine R-group scheme, call it $\mathbf{D}^*$, such that $\mathbf{D}^*(B) = (D \otimes_R B)^*$ for every commutative R-algebra B.

(2.6) Diagonalizable group schemes. If X is an ordinary commutative group, we denote by $R[X] = \oplus_{x \in X} Rx$ the group algebra of X over R. The association $X \mapsto D(X) := \mathrm{Spec}(R[X])$ is a contravariant functor from the category $(\mathbf{Ab})$ of abelian groups to the category of R-schemes. In fact, it is naturally a functor to commutative R-group schemes because the identifications

$$(D(X))(B) = \mathrm{Hom}_{R\text{-alg}}(R[X], B) = \mathrm{Hom}_{(\mathbf{Ab})}(X, B^*)$$

gives us a commutative group structure on the functor $B \mapsto D(X)(B)$ for each X. Hence $D(X)$ is a commutative R-group scheme. On the basis elements $x \in X$ of $R[X]$ we have

$$\tilde{m}(x) = x \otimes x, \quad \tilde{\varepsilon}(x) = 1, \quad \text{and} \quad \widetilde{\mathrm{inv}}(x) = x^{-1}$$

as in easily checked. A special case is $\mathbb{G}_m = \mathrm{Spec}(R[u, u^{-1}]) = D(\mathbb{Z})$, because $R[u, u^{-1}]$ is the group algebra of the infinite cyclic group generated by u.

If X is a finite abelian group of order n, then $R[X]$ is a free R-module of rank n, and $D(X)$ is finite and flat over R and is therefore an example of the type of R-group scheme which is our main concern in this paper.

Suppose

$$X \longrightarrow Y \longrightarrow Z \longrightarrow 0$$

is an exact sequence of abelian groups. Then

$$0 \longrightarrow \mathrm{Hom}(Z, B^*) \longrightarrow \mathrm{Hom}(Y, B^*) \longrightarrow \mathrm{Hom}(X, B^*)$$

is exact for every B, and consequently the corresponding sequence of group schemes

$$0 \longrightarrow D(Z) \longrightarrow D(Y) \longrightarrow D(X)$$

is exact (meaning that $D(Z) \to D(Y)$ is a kernel of $D(Y) \to D(X)$ in the sense of §1).

(2.7) The group schemes $\boldsymbol{\mu}$. Let n be an integer ≥ 1. The R-group scheme $D(\mathbb{Z}/n\mathbb{Z})$ is denoted by $\boldsymbol{\mu}_n$ and is called the scheme of n-th roots of unity over R. The dual of the exact sequence

$$\mathbb{Z} \xrightarrow{n} \mathbb{Z} \longrightarrow \mathbb{Z}/n\mathbb{Z} \longrightarrow 0$$

is

$$0 \longrightarrow \boldsymbol{\mu}_n \longrightarrow \mathbb{G}_m \xrightarrow{n} \mathbb{G}_m \, .$$

Thus μ_n is the kernel of raising to the n-th power in $\mathbb{G}_m$. For each R-algebra B we have $\mu_n(B) = \{b \in B \mid b^n = 1\}$. The arrow $\mu_n \to \mathbb{G}_m$ corresponds to the algebra map $R[u, u^{-1}] \to R[u, u^{-1}]/(u^n - 1)$ and identifies μ_n with a closed finite flat subgroup scheme of $\mathbb{G}_m$ of order n, in the sense of §3.

Suppose the abelian group X is finitely generated. Then X is isomorphic to a finite product of cyclic groups, hence $D(X)$ to a finite product of copies of $\mathbb{G}_m$ and μ_n's, for various n, and therefore to a closed subgroup of a product $\mathbb{G}_m^r$ of copies of $\mathbb{G}_m$. Viewing $\mathbb{G}_m^r$ as the closed subgroup of GL_r consisting of diagonal matrices we obtain a faithful linear representation of $D(X)$ which identifies $D(X)$ with a diagonal closed subgroup scheme of GL_r. That is the reason the group schemes $D(X)$ are called "diagonalizable."

(2.8) Base change. Let U be an S-scheme. If T is an S-scheme we sometimes write $U_T := U \times_S T$ for the "base change from S to T of U." Every T-scheme V is an S-scheme in a natural way, and $U_T(V) = U_S(V)$. Thus, if G is an S-group scheme, then the functor $V \mapsto G_T(V)$ is a group functor on $(\mathbf{Sch}/T)$, and hence G_T is a T-group scheme. Every scheme S is uniquely a $(\operatorname{Spec}\mathbb{Z})$-scheme, and all our examples so far are the canonical base changes from $\mathbb{Z}$ to R of group schemes over $\mathbb{Z}$. That is why the groups $\mathbb{G}_a(B)$, $\mathbb{G}_m(B)$, etc., depend only on B as a ring; i.e., as a $\mathbb{Z}$-algebra, and not on B as an R-algebra. From now on we will let $\mathbb{G}_a$, $\mathbb{G}_m$, GL_n, $D(X)$ stand for the versions over $\mathbb{Z}$ and will write $(\mathbb{G}_a)_S := \mathbb{G}_a \times S$, etc. for their base change to a scheme S.

For an S-scheme T, if we denote by $B_T = \Gamma(T, O_T)$ the ring of sections of the structure sheaf of T, we have

$$
\begin{aligned}
(\mathbb{G}_a)_S(T) &= \mathbb{G}_a(T) = B_T \qquad \text{(additive group)}\\
(\mathbb{G}_m)_S(T) &= \mathbb{G}_m(T) = B_T^* \qquad \text{(multiplicative group)}\\
(\mathrm{GL}_n)_S(T) &= \mathrm{GL}_n(T) = \mathrm{GL}_n(B_T) = \mathbb{M}_n(B_T)^*\\
D(X)_S(T) &= D(X)(T) = \mathrm{Hom}_{(\mathbf{Ab})}(X, B_T^*) = \mathrm{Hom}_{(\mathbf{Ab})}(X, \mathbb{G}_m(T))\,.
\end{aligned}
$$

(2.9) Characters and group-like elements. Let G be an S-group scheme. A *character of* G is a homomorphism of S-group schemes

$$\chi : G \to (\mathbb{G}_m)_S$$

or, what is the same, a non-vanishing section of the structure sheaf O_G of G for which the equality $m^*\chi = (\mathrm{pr}_1^*\,\chi)(\mathrm{pr}_2^*\,\chi)$ holds on $G \times_S G$. These characters form a subgroup

$$\mathrm{Hom}_{(\mathbf{Gr}/S)}(G, (\mathbb{G}_m)_S) \quad \text{of} \quad \mathrm{Hom}_{(\mathbf{Sch}/S)}(G, (\mathbb{G}_m)_S) = \mathbb{G}_m(G).$$

If $S = \operatorname{Spec}(R)$ and $G = \operatorname{Spec}(A)$ are affine, then a character χ of G is an invertible element of A such that $\tilde{m}\chi = (\chi \otimes 1)(1 \otimes \chi) = \chi \otimes \chi$. Such an element of a Hopf algebra A is called *group-like*. The group-like elements of A form a subgroup of A^*, the group of characters of $\mathrm{Spec}(A)$ defined over R.

The functor

$$T \longmapsto \mathrm{Hom}_{(\mathbf{Gr}/T)}(G_T, (\mathbb{G}_m)_T)$$

is a contravariant functor from $(\mathbf{Sch}/S)$ to $(\mathbf{Ab})$. If it is representable, the representing commutative S-group scheme is called the character group scheme of G.

If G' is the character group scheme of G, then for each T in $(\mathbf{Sch}/S)$ we have a pairing

$$(*) \qquad G'(T) \times G(T) \longrightarrow \mathbb{G}_m(T)$$

given by the map

$$G'(T) = \mathrm{Hom}_{(\mathbf{Gr}/T)}(G_T, (\mathbb{G}_m)_T) \longrightarrow \mathrm{Hom}_{(\mathbf{Gr})}(G(T), \mathbb{G}_m(T)) .$$

The pairings $(*)$ are compatible with base change $T' \to T$. Conversely, given S-group schemes G and G', a collection of pairings $(*)$ compatible with base change determines a homomorphism of $G'(T)$ into the group of homomorphisms of the *functor* G_T into the *functor* $(\mathbb{G}_m)_T$, hence a homomorphism of $G'(T)$ into $\mathrm{Hom}_{(\mathbf{Gr}/T)}(G_T, (\mathbb{G}_m)_T)$ for each T. If these homomorphisms are isomorphisms, then the pairings $(*)$ identify G' with the character group scheme of G.

(2.10) The duality between X_S and $D(X)_S$. Let X be a set, S a scheme. The *constant S-scheme* X_S attached to X is by definition the disjoint union $X_S = \coprod_{x \in X} S_x$ of copies S_x of S indexed by X. Then for an S-scheme T, an element $f \in X_S(T)$, that is, a morphism of S-schemes $f : T \to X_S$, is determined by the collection of subsets $U_x = f^{-1}(S_x)$ of T. These subsets are open, disjoint, and cover T. The restriction of f to U_x is the unique morphism $U_x \to S_x \approx S$. Such a covering determines and is determined by the locally constant X-valued function φ on T taking the value x on U_x for each $x \in X$. In this way, $X_S(T)$ is identified with the set of locally constant functions $\varphi : T \to X$. If T is non-empty and connected, then $X_S(T) = X$.

Since $X_S = \coprod_{x \in X} S_x$ we have

$$\Gamma(X_S, O_{X_S}) = \prod_{x \in X} \Gamma(S_x, O_{S_x})$$

and since $S_x = S$ for all x, this is simply the ring of functions on X with values in $\Gamma(S, O_S)$. The scheme X_S is affine if and only if $S = \mathrm{Spec}(R)$ is affine and X is finite (or $R = (0)$), in which case $X = \mathrm{Spec}(A)$, where $A = \mathrm{Map}(X, R)$ is the ring of R-valued functions on X.

Suppose now X is a group. Then $X(T)$ is a group under value-wise composition of the locally constant functions $\varphi : T \to X$, so X_S is a group scheme, the *constant* S-group scheme determined by X. It is easy to check that a section χ of O_{X_S} is a character of X_S if and only if, when viewed as above as a function on X with values in $\Gamma(S, O_S)$, χ is a group homomorphism $X \to \Gamma(S, O_S)^*$.

Suppose X is a commutative group. Then such a homomorphism χ is a point of $D(X)$ with values in S, so the group of characters of X_S is $D(X)(S)$. The same is true after base change $T \to S$. Hence $D(X)_S$ is the character group scheme of X_S (hence the notation: $D(X)$ = dual of X).

Slightly less tautological is the fact that X_S is the character group scheme of $D(X)_S$. The pairing

$$D(X)(T) \times X(T) \longrightarrow \Gamma(T, O_T)^* = \mathbb{G}_m(T)$$

takes $\chi \times \varphi$ into "$\chi \circ \varphi$," by which we mean the section of O_T which coincides with the section $\chi(x)$ on the set $\varphi^{-1}(x) = U_x$ for each $x \in X$. This pairing gives a homomorphism

$$X(T) \longrightarrow \mathrm{Hom}_{(\mathbf{Gr}/T)}(D(X)_T, (\mathbb{G}_m)_T)$$

for each T. To show it is an isomorphism for all T it is enough to show it is for T affine, because each side, as functor of T, is a sheaf in the Zariski topology. Suppose $T = \mathrm{Spec}(B)$, so $D(X)_T = \mathrm{Spec}(B[X])$. The character of $D(X)_T$ corresponding to $\varphi \in X(T)$ is the section of $O_{D(X)_T}$ which is x on the open set $\varphi^{-1}(x)$, that is, is the group-like element $\sum_x e_{\varphi,x} x \in B[X]$, where $e_{\varphi,x}$ is the idempotent in B which is the "characteristic function" of $U_x = \varphi^{-1}(x)$. On the other hand, it is easy to check that every group-like element of the Hopf-algebra $B[X]$ is of the form $\sum_x e_x x$, where $\{e_x, x \in X\}$ is a family of orthogonal idempotents in B, indexed by X, whose sum is 1. For more details, see Grothendieck's discussion in the first sections of [SGA3, II].

(2.11) Derivations. Suppose $G = \mathrm{Spec}(A)$ is an affine R-group scheme. Let I be the augmentation ideal and $\tilde{m} : A \to A \otimes A$ the comultiplication, as usual. Let $\pi : A = R1 \oplus I \to I/I^2$ be the R-linear map killing $R1$ and projecting I.

Proposition. *Let M be an A-module and $\psi : M \otimes A \to M$ the map giving the action of A on M. The map $\lambda \mapsto \psi \circ ((\lambda \circ \pi) \otimes \mathrm{id}) \circ \tilde{m}$ is an isomorphism from* $\mathrm{Hom}_{(R\text{-mod})}(I/I^2, M)$ *to* $\mathrm{Der}_R(A, M)$, *the module of R-linear derivations $A \to M$.*

Corollary. *The map $(\pi \otimes \mathrm{id}) \circ \tilde{m} : A \to (I/I^2) \otimes_R A = \Omega^1_{A/R}$ is a universal R-linear derivation for A.*

We sketch a proof. For more details see for example [W], 11.3. The corollary follows from the proposition because the map $\lambda \mapsto \psi \circ (\lambda \otimes \mathrm{id})$ is a bijection from $\mathrm{Hom}_{(R\text{-mod})}(I/I^2, M)$ to $\mathrm{Hom}_{(A\text{-mod})}((I/I^2) \otimes_R A, M)$.

Let B be an R-algebra and N a B-module. Make $B \oplus N$ a B-algebra with $N^2 = (0)$ and let $j : B \oplus N \to B$ be the projection killing N. The induced group homomorphism $j_* : G(B \oplus N) \to G(B)$ is a projection to the subgroup $G(B)$. Hence $G(B \oplus N) = H \rtimes G(B)$, semidirect product, where $H = \mathrm{Ker}(j_*)$. For $x \in G(B)$, let N_x denote N viewed as A-module via $x : A \to B$. The coset Hx is the set of all homomorphisms $A \to B \oplus N$ lifting $x : A \to B$. A standard computation shows that these are the maps of the form $x \oplus \delta$ where $\delta : A \to N_x$ is an R-linear derivation.

Let $\varepsilon_B : A \to R \to B$ be the identity in $G(B)$. For $\delta \in \mathrm{Der}_R(A, N_{\varepsilon_B})$ and $x \in G(B)$ define δ_x by $(\varepsilon_B \oplus \delta)x = (x \oplus \delta_x)$. Consideration of the group $G(B \oplus N)$ shows that the map $\delta \mapsto \delta_x$ is a bijection from $\mathrm{Der}_R(A, N_{\varepsilon_B})$ to $\mathrm{Der}_R(A, N_x)$, and working out the group law explicitly one finds the formula $\delta_x = \psi \circ (\delta \otimes x) \circ \tilde{m}$, where $\psi : N \otimes B \to N$ is the map giving the action of B on N. (Exercise: Show that the map $\delta \mapsto \varepsilon_B \oplus \delta$ is a group isomorphism from $\mathrm{Der}_R(A, N_{\varepsilon_B})$ to H.) On the other hand, from the definitions one checks that the map $\lambda \mapsto \lambda \circ \pi$ is a bijection from $\mathrm{Hom}_{(R\text{-mod})}(I/I^2, N)$ to $\mathrm{Der}_R(A, N_{\varepsilon_B})$. Taking $B = A$, $N = M$, $x = \mathrm{id}$ and putting things together gives the proposition.

Proposition. *Let $D \in \mathrm{Der}_R(A, A)$ be a derivation of the R-algebra A, and let $\lambda : I/I^2 \to A$ be the R-linear map corresponding to D as in the proposition just proved. Then D is right invariant if and only if $\lambda(I/I^2) \subset R1$, in which case $\varepsilon_A \circ D = \lambda \circ \pi$, and D is the unique invariant derivation of A such that $D(f) \equiv \lambda\pi(f) \pmod I$ for all $f \in I$.*

Proof. For each point $x \in G(B)$, any B, $D(x) := x \circ D$ is in $\mathrm{Der}_R(A, B_x)$. We say D is *right invariant* if $(x \oplus D(x))y = (xy \oplus D(xy))$ for all $x, y \in G(B)$, any B, or, equivalently, if $D(x) = D(\varepsilon_B)_x$ for all $x \in G(B)$ in the notation δ_x of the previous paragraph. As usual, this condition will hold for all B, x if it holds for $B = A$, $x = \mathrm{id}$, that is, if $D = \psi \circ ((\varepsilon_A \circ D) \otimes \mathrm{id}) \circ \tilde{m}$. Hence D is invariant if and only if $\varepsilon_A \circ D = \lambda \circ \pi$. For arbitrary D a computation shows that $\varepsilon_A \circ D = \varepsilon_A \circ \lambda \circ \pi$, and the proposition follows.

§3. Finite flat group schemes; passage to quotient

Throughout this section, S is a locally noetherian base scheme. An S-scheme X is finite and flat over S if and only if O_X is locally free of finite rank as O_S-module, that is, if and only if there is a covering of S by affine open subsets U such that the morphisms $X \mid U \to U$ are of the form $\mathrm{Spec}(A) \to \mathrm{Spec}(R)$ with A free of finite rank as R-module. This rank is a locally constant function n on S with integer values ≥ 0 which we call the *order* of X over S.

Notation. We denote the order of X over S by $[X : S]$, and sometimes write simply "$[X : S] = n$" to indicate that X is finite and flat over S and that n is its order.

3.1 Proposition. (i) *Suppose* $X \to Y \to S$ *are morphisms of schemes and suppose* $[X : Y] = m$ *and* m *is a constant* > 0. *Then* X *is finite flat over* S *if and only if* Y *is, in which case* $[X : S] = [X : Y][Y : S]$, *as functions on* Y.

(ii) *If* $[X_i : S] = n_i$, $i = 1, 2$, *then* $[X_1 \times_S X_2 : S] = n_1 n_2$.

(iii) *If* $[X : S] = n$, *then* $[X \times_S T : T] = n$ *for every* S*-scheme* T.

Proof. (i) Since $m > 0$, $X \to Y$ is faithfully flat, hence $X \to S$ flat implies $Y \to S$ flat. Since S is noetherian, $X \to S$ finite implies $Y \to S$ finite.

The rest of (i), and (ii) and (iii), are left to the reader.

Finite flat S-group schemes are our main concern. So far our only examples are the constant group schemes X_S attached to a finite group X and their duals $D(X)_S$ for X abelian. Both X_S and $D(X)_S$ have the same (constant) order as the group X. In particular, for each integer $n \geq 1$, both $(\mathbb{Z}/n\mathbb{Z})_S$ and $(\mu_n)_S$ have order n.

If $S = \mathrm{Spec}(R)$ is affine then finite flat S-schemes are affine. We are ultimately interested in the case R is the ring of integers in a local field, in particular the case $R = \mathbb{Z}_p$. Therefore we will limit the discussion to the affine case and will often assume $G = \mathrm{Spec}(A)$ with A free over R, not only locally free, which is automatic if R is a local ring.

Note that if $[G : S] = [A : R] = n$, then the augmentation ideal I (cf. 2.3) is locally free of rank $n - 1$ as R-module. This makes the case $n = 2$ very easy to analyze.

(3.2) Example – exercise; G of order 2. Suppose R is a ring and $G = \mathrm{Spec}(A)$ is an affine R-scheme such that $[G : R] = 2$, with an associative law of combination

$$m : G \times_R G \to G$$

for which there is a 2-sided unit $\varepsilon : S = \mathrm{Spec}(R) \to G$, but not necessarily an inverse. Let

$$I = \mathrm{Ker}(\tilde{\varepsilon} : A \to R).$$

Then I is an invertible (= locally free of rank 1) R-module. Assume I is free with basis element x, so $A = R + Rx$ is a free R-module of rank 2 with basis $\{1, x\}$. The ring structure of A is determined by the element $a \in R$ such that $x^2 = ax$. As discussed in **(2.3)**, the comultiplication $\tilde{m}$ must be of the form

$$\tilde{m}(1) = 1 \otimes 1 = 1 \quad \text{and} \quad \tilde{m}(x) = x \otimes 1 + 1 \otimes x + b(x \otimes x)$$

for some $b \in R$. Check that for $\tilde{m} : A \to A \otimes_R A$ to be a homomorphism of R-algebras, it is necessary and sufficient that $(ab + 1)(ab + 2) = 0$ in R. Assuming that is the case, G is a commutative and associative R-magma scheme with two sided unit, representing the functor

$$G(B) = \{y \in B \mid y^2 = ay\}$$

for R-algebras B, with law of composition $*$ in $G(B)$ defined by

$$y * z = y + z + byz,$$

with unit element $y = 0$. The elements $e_1 = ab + 2$ and $e_2 = -ab - 1$ are orthogonal idempotents in R whose sum is 1, so $S = S_1 \coprod S_2$ is a disjoint union of open affine subschemes such that $ab = -2$ on S_1 and $ab = -1$ on S_2. Hence we can without loss of generality treat those cases separately. Check that if $ab = -2$, then G is an R-group scheme with $y * y = 0$ for all $y \in G(B)$, all B, but if $ab = -1$, then G is *not* an R-group scheme, but is a monoid with $y * y = y$, all $y \in G(B)$, all B.

For each pair of elements a, b in R such that $ab = -2$, let $G_{a,b}$ denote the R-group scheme just introduced. For example, $G_{-2,1} = (\boldsymbol{\mu}_2)_R$ because $G_{-2,1}(B) = \{y \in B \mid (1+y)^2 = 1\}$ and $1 + y * z = (1+y)(1+z)$. On the other hand, $G_{1,-2} \approx (\mathbb{Z}/2\mathbb{Z})_R$ as is easily checked.

Check that

$$G_{a,b} \approx G_{\alpha,\beta} \iff \exists\, u \in R^* \text{ such that } \alpha = ua \text{ and } \beta = u^{-1}b.$$

Thus, if 2 is invertible in R, then all $G_{a,b}$'s are isomorphic to the constant group scheme $(\mathbb{Z}/2\mathbb{Z})_R$. If $R = \mathbb{Z}$ or $\mathbb{Z}_2$, then $G_{a,b} \approx (\mathbb{Z}/2\mathbb{Z})_R$ or $(\boldsymbol{\mu}_2)_R$. If $R = \mathbb{Z}_2[2^{1/17}]$, then there are exactly 18 types of finite flat R-group schemes of order 2, up to isomorphism. If R is an integral domain of characteristic 2, then the types of $G_{a,b}$'s are: one $G_{0,0}$, and one $G_{a,0}$ and one $G_{0,a}$ for each non-zero principal ideal (a) in R; in particular, if R is a field of characteristic 2, there are three types of R-groups of order 2, $(\mathbb{Z}/2\mathbb{Z})_R$, $(\boldsymbol{\mu}_2)_R$, and $(\boldsymbol{\alpha}_2)_R := G_{0,0}$.

If $u, v, w \in R$ and $uvw = -2$, then there are pairings on the functors

$$G_{u,vw}(B) \times G_{v,uw}(B) \longrightarrow G_{uv,w}(B)$$

given by

$$(y_u, y_v) \longmapsto y_u y_v\ .$$

If $w = 1$, check that the pairings

$$G_{u,v} \times G_{v,u} \longrightarrow (\boldsymbol{\mu}_2)_R \subset (\mathbb{G}_m)_R$$

identify $G_{u,v}$ with the character group scheme of $G_{v,u}$ (cf. 2.9).

(3.3) Passage to quotient by a group scheme of finite order. Let H be an S-group scheme and X a scheme over S. A *right action of H on X* is a morphism $a : X \times_S H \to X$ such that, for every S-scheme T the induced map $X(T) \times H(T) \to X(T)$ is a right action of the group $H(T)$ on the set $X(T)$, i.e., satisfies the rules $x(h_1h_2) = (xh_1)h_2$ and $x \cdot 1 = x$. We will say such an action is *strictly free* if the morphism

$$(\mathrm{id}, a) : X \times_S H \to X \times_S X,$$

i.e., the morphism inducing $(x, h) \mapsto (x, xh)$ on the functors, is not only injective on the functors, but is a closed immersion. Given a right action of H on X we will say that a morphism $f : X \to Y$ is *constant on orbits* if

$$f \circ a = f \circ \mathrm{pr}_1 : X \times_S H \to Y,$$

that is, if $f(xh) = f(x)$, all $x \in X(T)$, $h \in H(T)$, all T.

3.4 Theorem (Grothendieck). *Suppose H finite flat over S locally noetherian acts strictly freely on X of finite type over S in such a way that every orbit is contained in an affine open set. Then the category of morphisms $X \to Z$ which are constant on orbits has an initial object; in other words there exists an S-scheme Y and a morphism $u : X \to Y$ constant on orbits such that for every morphism $v : X \to Z$ which is constant on orbits there is a unique morphism $f : Y \to Z$ such that $v = f \circ u$. (Of course the morphism $u : X \to Y$ is then unique up to a unique isomorphism; we denote it by $u : X \to X/H$ and call it the canonical morphism from X to the* orbit scheme *or the* quotient of X by H.*) The morphism $u : X \to Y = X/H$ has the following further properties:*

(i) *X is finite flat over X/H and $[X : (X/H)] = [H : S]$.*
(ii) *For every S-scheme T the map $X(T)/H(T) \to (X/H)(T)$ is injective.*
(iii) *If $S = \mathrm{Spec}(R)$, $H = \mathrm{Spec}(B)$ and $X = \mathrm{Spec}(A)$ are affine, then $X/H = \mathrm{Spec}(A_0)$, where A_0 is the subring of A where the two homomorphisms $\widetilde{\mathrm{pr}_1}, \tilde{a} : A \to A \otimes_R B$ coincide.*

Remarks. This theorem is a special case of results of Grothendieck ([Gro], [SGA3, I, Exp.V]). As he realized, the theorem has really nothing to do with group schemes; one can replace the morphism $(\mathrm{pr}_1, a) : X \times_S H \hookrightarrow X \times_S X$ by any closed subscheme $\mathcal{R}$ of the product $X \times_S X$ which is the graph of an equivalence relation such that $\mathrm{pr}_1 : \mathcal{R} \to X$ is finite and flat. As suggested by (iii) it is easy to construct the scheme which will be X/H. The difficulty is to show that it has the characterizing property and satisfies (i) and (ii) as well. For a concise proof we refer the reader to Raynaud's short article [R1], which we recommend also as an excellent introduction to the general problem of passage to quotients. For more general results we refer the reader to [K-M].

(3.5) Quotient group scheme, coset spaces. For us, the main application of 3.4 is the case in which $X = G$ is an S-group scheme, $H \subset G$ is a finite flat closed subgroup scheme, and the action $a : G \times_S H \to G$ is the restriction of the group law $m : G \times_S G \to G$. We call G/H the *scheme of left cosets* of H in G. If G/H is finite and flat over S we call its order, $[(G/H) : S]$, the *index of H in G* and denote it by $[G : H]$. Suppose $[H : S] = m$ is constant. Then $m > 0$ because H has a unit section. By part (i) of **(3.4)** we conclude that $[G : (G/H)] = m$ and then by **(3.2)** that

G is finite flat over S if and only if G/H is, in which case,

$$[G:S] = [G:(G/H)][(G/H):S] = [H:S][G:H] ,$$

that is, "order of group = order of subgroup × index of subgroup."

The definition of a *left* action of a group scheme G on a scheme X is clear, and it is easy to see that G acts naturally on the left of the scheme G/H in a unique way such that the diagram

$$\begin{array}{ccc} G \times G & \xrightarrow{\mathrm{id} \times u} & G \times (G/H) \\ {\scriptstyle m}\big\downarrow & & \big\downarrow{\scriptstyle \text{left action of } G} \\ G & \xrightarrow{u} & G/H \end{array}$$

commutes. If H is *normal* in G, i.e., if H acts trivially on G/H, then we get a morphism $G/H \times G/H \to G/H$ which makes G/H an S-group scheme and $u : G \to G/H$ an S-group homomorphism. The sequence $0 \to H \to G \xrightarrow{u} G/H \to 0$ is exact in the sense of **(1.7)** and **(1.8)** and also in the sense that u is faithfully flat and $H = \operatorname{Ker} u$. Perhaps a simpler approach to these matters, one we have been avoiding, perhaps wrongly, is that advocated by Raynaud [R1] of identifying a group scheme G with the sheaf for the fppf (faithfully flat finite presentation) topology which it represents, and using Grothendieck's theory of faithfully flat descent [Gro]. Then the quotient group G/H represents the quotient sheaf, and the exact sequence in question is simply an exact sequence of sheaves of groups.

(3.6) The fundamental group $\pi_1(S, \alpha)$ and finite étale S-group schemes. A morphism $Y \to S$ is *finite étale* if it is finite flat and *unramified* in the sense that for each point $s \in S$ the fiber $Y_s := Y \times_S \{s\}$ is the spectrum of a separable algebra over the residue field $\kappa(s)$ of s, that is, Y_s is reduced, and for each point $y \in Y_s$ the corresponding residue field extension $\kappa(y)/\kappa(s)$ is separable; in other words, the inequalities in the following display are equalities:

$$(*) \qquad \begin{aligned} [Y:S](s) := [Y_s : \{s\}] &\geq [(Y_s)_{\mathrm{red}} : \{s\}] \\ &= \sum_{y \in Y_s} [\kappa(y) : \kappa(s)] \\ &\geq \sum_{y \in Y_s} [\kappa(y) : \kappa(s)]_{\mathrm{sep}}. \end{aligned}$$

Let $\alpha : \operatorname{Spec}(\Omega) \to S$ be a geometric point of S centered at s, that is, an embedding $\tilde{\alpha} : \kappa(s) \hookrightarrow \Omega$ of $\kappa(s)$ into an algebraically closed field Ω. The set $Y(\alpha)$ of geometric points of Y mapping to α has cardinality

$$\sum_{y \in S} [\kappa(y) : \kappa(s)]_{\mathrm{sep}}.$$

From $(*)$ we conclude that for a finite flat S-scheme Y the inequality $[Y:S](s) \geq \#Y(\alpha)$ holds for a geometric point α of S centered at s, and that Y is étale over S if and only if that inequality is an equality for all geometric points α of S.

Let $(\mathbf{FEt}/S)$ denote the category of finite étale S-schemes. Here is a quick review of the description of $(\mathbf{FEt}/S)$ in terms of the fundamental group of S. A convenient reference is [M]; see also [SGA1] and [Mu]. For simplicity we assume S *non-empty* and *connected*. Let α be a geometric point of S. The *fundamental group* $\pi = \pi_1(S,\alpha)$ *of* S *at the geometric point* α can be defined as the group of automorphisms of the functor $Y \mapsto Y(\alpha)$ from $(\mathbf{FEt}/S)$ to $(\mathbf{Sets})$. An element $\sigma \in \pi$ is a collection of permutations σ_Y of the sets $Y(\alpha)$, one for each $Y \in (\mathbf{FEt}/S)$, such that for every $(\mathbf{FEt}/S)$-morphism $Y \to Y'$, the induced map $Y(\alpha) \to Y'(\alpha)$ commutes with the σ_Y's. Then π is a profinite group, that is, a compact Hausdorff topological group in which the open subgroups (those which contain $\mathrm{Ker}(\pi \to \mathrm{Perm}(Y(\alpha)))$ for some Y) form a fundamental system of neighborhoods of 1.

Let $(\mathbf{F}\pi\text{-}\mathbf{sets})$ denote the category of finite sets X with a continuous action of π on them. By construction, each $Y(\alpha)$ is an object in $(\mathbf{F}\pi\text{-}\mathbf{sets})$. Grothendieck's theorem is that *the functor* $Y \to Y(\alpha)$ *from* $(\mathbf{FEt}/S)$ *to* $(\mathbf{F}\pi\text{-}\mathbf{sets})$ *is an equivalence of categories.* This functor commutes with cartesian products and disjoint sums; in particular, expressing Y as disjoint union of its connected components corresponds to expressing $Y(\alpha)$ as a union of orbits for the action of π.

The fundamental group $\pi_1(S,\alpha)$ is a functor of geometrically pointed connected noetherian schemes (S,α). A morphism $f: T \to S$ induces a homomorphism $f_*: \pi_1(T,\beta) \to \pi_1(S,f(\beta))$ in a natural way so that the base change functor $Y \mapsto Y_T = Y \times_S T$ from $(\mathbf{FEt}/S)$ to $(\mathbf{FEt}/T)$ corresponds under the equivalence of categories to the process of viewing a $\pi_1(S,f(\beta))$-set as a $\pi_1(T,\beta)$-set via the homomorphism f_*.

The fundamental group $\pi_1(S,\alpha)$ is determined up to an inner automorphism by the scheme S. If α' is another geometric point of S, the functors $Y \mapsto Y(\alpha)$ and $Y \mapsto Y(\alpha')$ are isomorphic; an isomorphism between them is the analog of a homotopy class of paths from α to α', and induces an isomorphism $\pi_1(S,\alpha) \xrightarrow{\sim} \pi_1(S,\alpha')$.

If $k = \kappa(s)$, and α is given by the embedding $\tilde{\alpha}: k \hookrightarrow \Omega$, the group $\mathrm{Aut}_k(\Omega)$ acts on the left of Ω, so on the right of $\mathrm{Spec}(\Omega)$, so, for each Y in $(\mathbf{FEt}/S)$, on the left of $Y(\alpha) = \mathrm{Hom}_{(\mathbf{Sch}/S)}(\mathrm{Spec}\,\Omega, Y)$. This action gives a homomorphism of $\mathrm{Aut}_k(\Omega)$ into $\pi_1(S,\alpha)$ which factors through the quotient $\mathrm{Gal}(k_s/k)$ of $\mathrm{Aut}_k(\Omega)$, where k_s is the separable algebraic closure of k in Ω, and thereby induces a natural homomorphism $\mathrm{Gal}(k_s/k) \to \pi_1(S,\alpha)$. It is a nice exercise in Galois theory to show that if $S = \{s\} = \mathrm{Spec}(k)$, this homomorphism is an isomorphism, and the equivalence of categories above does hold. The reverse equivalence in this case is given by $X \mapsto \mathrm{Spec}(\mathrm{Map}_\pi(X,k_s))$, where for a finite set X with a continuous action of

$\pi = \text{Gal}(k_s/k)$ on it we denote by $\text{Map}_\pi(X, k_s)$ the k-algebra of maps $X \to k_s$ commuting with π.

Getting back to our business of group schemes, the upshot of all this is that the category of finite étale group schemes over a noetherian base scheme S has a simple description. If α is a geometric point of S, the functor $G \mapsto G(\alpha)$ is an equivalence of that category with the category of finite groups with a continuous operation of $\pi_1(S, \alpha)$.

Let G be a finite flat S-group scheme. Then G/S is étale if and only if the sheaf of relative differentials $\Omega^1_{G/S}$ is zero (cf. e.g., [M], Ch.1, Prop.3.5). Hence, by **(2.11)**, G/S is étale if and only if $\mathcal{I} = \mathcal{I}^2$, where $\mathcal{I} \subset \mathcal{O}_G$ is the augmentation ideal sheaf. Equivalently, G/S is étale if and only if the unit section $\varepsilon(S) = \text{Spec}(\mathcal{O}_G/\mathcal{I})$ is open (and closed) in G. This is true, because if $\mathcal{I} = \mathcal{I}^2$, then $I_x = (0)$ for $x \in \text{Spec}(\mathcal{O}/\mathcal{I})$, by Nakayama's Lemma, hence the complement of $\text{Spec}(\mathcal{O}/\mathcal{I})$ is the support of $\mathcal{I}$ and is closed.

We will soon see that every finite flat S-group G whose order $[G : S]$ is invertible on S is étale.

(3.7) The connected-étale exact sequence over a Henselian local ring. In this section we assume $S = \text{Spec}(R)$ is the spectrum of a henselian local ring R, for example, a field or a complete discrete valuation ring. For some basic properties of hensel rings which we use here see for example [M,I,§4]. Let M be the maximal ideal of R, $k = R/M$ the residue field, and $s = \text{Spec}(k)$ the closed point of S.

Our aim in this section is to prove the following four things about a finite flat S-group scheme G.

(I). Let G^0 be the connected component of the identity in G. Then G^0 is the spectrum of a henselian local R-algebra with the same residue field as R and is a flat closed normal subgroup scheme of G such that the quotient in the sense of **(3.4)**, $G^{et} := G/G^0$, is étale. We call the exact sequence

$$0 \longrightarrow G^0 \longrightarrow G \longrightarrow G^{et} \longrightarrow 0$$

the *connected-étale sequence for* G. It can be characterized by the fact that every homomorphism from G to an étale S-group scheme factors through $G \to G^{et}$, and G^0 is the kernel of that homomorphism.

(II). If the residue characteristic of R is 0, then $G^0 = S$ and $G = G^{et}$. If it is $p > 0$, then the order $[G^0 : S]$ of G^0 is a power of p. (It follows immediately from this that if $[G : S]$ is invertible in S, then $[G^0 : S] = 1$ and $G = G^{et}$ is étale over S. The same is true over an arbitrary base scheme, by passage to the henselizations (or localizations) of its local rings.)

(III). If $R = k$ is a field, and $n = [G : S]$, then G is killed by n, that is, $x^n = 1$ for $x \in G(B)$, for every k-algebra B. (In the next section, we will give Deligne's proof that a *commutative* finite locally free group scheme over *any* base is killed by its order.)

(IV). If R is a perfect field, the homomorphism $G \to G^{et}$ has a section and G is a semidirect product, $G = G^0 \rtimes G^{et}$.

Suppose T is a scheme finite over S. Then $T = \mathrm{Spec}(A)$ with A a finite R-algebra. Since R is henselian we have $A = \prod_{i=1}^r A_i$ with each A_i a local hensel ring ([M], loc.cit.). Accordingly,

$$T = \coprod_{i=1}^{r} T_i$$

is a finite disjoint union of open subschemes $T_i = \mathrm{Spec}(A_i)$, each of which is the spectrum of a local hensel R-algebra. In particular, the T_i are connected; they are the *connected components* of T. For each i, let t_i be the closed point of T_i and $k_i = \kappa(t_i)$ its residue field.

Let α be the geometric point of S corresponding to an algebraic closure $\bar{k}$ of $k = \kappa(s)$. Let $\pi = \mathrm{Aut}_k(\bar{k}) = \mathrm{Gal}(k_s/k)$. Then π acts on

$$T(\alpha) = \mathrm{Hom}_{R\text{-alg}}(A, \bar{k}) = \coprod_{i=1}^{r} \mathrm{Hom}_k(k_i, \bar{k})$$

through its action on $\bar{k}$. The functor $T \mapsto T(\alpha)$ from finite S-schemes to finite π-sets commutes with products and disjoint unions. From this several things are obvious:

1) The $T_i(\alpha)$ are the orbits for the action of π on $T(\alpha)$; in particular, T is connected if and only if π acts transitively on $T(\alpha)$.
2) $T_i \times_S T_j$ is connected $\Leftrightarrow$ either $T_i(\alpha)$ or $T_j(\alpha)$ is a singleton $\Leftrightarrow$ either k_i or k_j is pure inseparable over k.
3) The connected components of the closed fiber $T_s = T \times_S \{s\}$ are the closed fibers $(T_i)_s$ of the connected components of T.

Suppose now G is a finite S-group scheme. Let G^0 be the connected component of G which contains the image of the identity section $\varepsilon : S \to G$. Then S is a closed subscheme of the local scheme G^0 so they have the same residue field, k. From 2) above it follows that for each connected component G_i of G the product $G_i \times_S G^0$ is connected. Its image G_iG^0 under the law of composition $m : G \times_S G \to G$ is connected and contains $G_iS = G_i$ so is equal to G_i. In particular $G^0G^0 = G^0$. Also, the inverse morphism inv preserves G^0 because it is an automorphism of the scheme S preserving ε. Hence G^0 is an open and closed subgroup scheme of G. To show it is normal in G it suffices to show that the map

$$G \times_S G^0 \to G, \qquad (g, g^0) \mapsto gg^0g^{-1},$$

has image in G^0. This is true because $G \times_S G^0 = \coprod_i G_i \times_S G^0$, and for each i the image of $G_i \times_S G^0$ is connected and contains the unit section.

Suppose now G is flat as well as finite over S. Then each connected component of G is flat, so G^0 is a flat normal subgroup scheme of G and we can form the quotient S-group scheme $G^{et} := G/G^0$ as in **(3.4)**, **(3.5)**. As remarked in 3.5, the fact that G is flat implies that G^{et} is flat and $[G : S] = [G^{et} : S][G^0 : S]$. Since G^0 is open in G, the unit section $G^0/G^0 = S$ is open in $G/G^0 = G^{et}$, and this implies G^{et} is étale, as remarked at the end of **(3.6)**.

To finish the proof of (I), note that there is no non-trivial homomorphism of a connected S-group scheme to an étale one, because such a homomorphism would factor through the identity component of the étale one, which is the unit section S. Thus a homomorphism of G into an étale S-group H has G^0 in its kernel, so factors through G/G^0.

Over a hensel local base the functor $Y \mapsto Y_s$ is an equivalence between $(\mathbf{FEt}/S)$ and $(\mathbf{FEt}/\{s\})$; equivalently, the homomorphism

$$\pi = \pi_1(\{s\}, \alpha) \to \pi_1(S, \alpha)$$

induced by the inclusion $\{s\} \hookrightarrow S$ is an isomorphism. Therefore a finite étale S-group scheme H is determined by the π-group $H(\alpha)$ which can be an arbitrary finite group on which π acts continuously.

The π-group corresponding to G^{et} is $G(\alpha)$. Indeed, the homomorphism $G(\alpha) \to G^{et}(\alpha)$ is surjective because G is finite over G^{et}, and is injective because its kernel $G^0(\alpha)$ has only one element.

Segment II. To prove **(II)** we can assume $G = G^\circ$ is connected and $R = k$, a field. Then $G = \mathrm{Spec}(A)$, with A a finite dimensional local k-algebra. The maximal ideal of A is the augmentation ideal I and is nilpotent. Let $\{x_i\}$, $1 \le i \le r = \dim_k(I/I^2)$, be a family of elements of I whose residues $\tilde{x}_i$ form a basis for the k-vector space I/I^2. By **(2.11)**, there exist right invariant derivations $D_i : A \to A$, $i \le i \le r$, such that $D_i x_j \equiv \delta_{ij} \pmod{I}$. By the product rule we have $D_i I^\nu \subset I^{\nu-1}$, and the D_i's induce derivations $\tilde{D}_i$ of degree -1 on the graded ring $\mathbf{Gr}_I(A) = \oplus_{\nu=0}^{\infty} I^\nu/I^{\nu+1} = k[\tilde{x}_1, \ldots, \tilde{x}_r]$. Let X_i, $1 \le i \le r$, be independent variables, $k[X] = k[X_1, \ldots, X_r]$ and $\varphi : k[X] \to \mathbf{Gr}_I(A)$ the k-algebra homomorphism given by $\varphi(X_i) = \tilde{x}_i$.

Lemma 3.7.1. *If* char $k = 0$, *then* φ *is an isomorphism; if* char $k = p > 0$, *then* φ *induces an isomorphism* $k[X]/(X_1^p, \ldots, X_r^p) \tilde{\to} \mathbf{Gr}_I(A)/(\tilde{x}_1^p, \ldots, \tilde{x}_r^p)$.

Proof. We have $D_i\varphi = \varphi\frac{\partial}{\partial X_i}$, because these two k-linear derivations coincide on the generators X_i of $k[X]$. Let $J = \mathrm{Ker}\,\varphi$ if char $k = 0$ and $J = \varphi^{-1}(\tilde{x}_1^p, \ldots, \tilde{x}_r^p)$ if char $k = p > 0$. Then J is a homogeneous ideal in $k[X]$, stable by $\frac{\partial}{\partial X_i}$ for each i, not equal to $k[X]$, and containing $(X_1^p, \ldots, X_r^p)$ if char $k = p > 0$. Let $P = \sum c_{\nu_1, \ldots, \nu_r} X_1^{\nu_1} \cdots X_r^{\nu_r} \in J$. Then

$$\nu_1!\nu_2! \cdots \nu_r! c_{\nu_1, \ldots, \nu_r} = \left(\left(\frac{\partial}{\partial X_1}\right)^{\nu_1} \cdots \left(\frac{\partial}{\partial X_r}\right)^{\nu_r} P \right)(0, 0, \ldots, 0)$$

is in J because J is homogeneous. Since $1 \notin J$ it follows that $c_{\nu_1,\dots,\nu_r} = 0$ for all $(\nu_1, \dots, \nu_r)$ if char $k = 0$, and for $(\nu_1, \dots, \nu_r)$ such that $\nu_i < p$ for each i, if char $k = p > 0$. Thus $J = (0)$ or $J = (X_1^p, \dots, X_r^p)$ in the two cases, as claimed.

Lemma 3.7.1 proves **(II)** if char $k = 0$ because $k[X]$ is finite dimensional only if $r = 0$, $k[X] = k$. If char $k = p > 0$ we use induction on the order of G and

Lemma 3.7.2. *Suppose* char $k = p > 0$. *Let* $B = A/(x_1^p, \dots, x_r^p)$. *Then* $H = \operatorname{Spec}(B)$ *is a finite flat normal subgroup scheme of* G *of order* p^r.

Proof. The closed subscheme $H \subset G$ is flat because k is a field. It is a normal subgroup scheme because it is the kernel of the Frobenius homomorphism $F : G \to G^{(p)}$. Recall that for a scheme X over k, $X^{(p)}$ denotes the base change of X from k to k corresponding to the homomorphism $x \mapsto x^p$ of k into itself. For a k-algebra B we have $X^{(p)}(B) = X(B')$, where B' denotes the ring B, viewed as k-algebra with elements $c \in k$ acting on B' via the p-th power of their action on B. The map $F : B \to B'$ defined by $F(b) = b^p$ is then a k-algebra homomorphism, and the corresponding homomorphism of functors

$$F_* : X(B) \longrightarrow X(B') = X^{(p)}(B)$$

induces a morphism of k-schemes which is given by raising the coordinates of a point to the p-th power and which we denote by $F : G \to G^{(p)}$.

If $X = \operatorname{Spec}(k[x_1, \dots, x_r])$ is an affine k-group scheme with augmentation ideal I generated by the coordinate functions x_i, as is the case with our G, then F is a group homomorphism (because F_* above is), and $\operatorname{Ker} F$ is represented by the closed finite flat subscheme $\operatorname{Spec}(k[x_1, \dots, x_r]/(x_1^p, \dots, x_r^p))$, which is therefore a normal subgroup scheme. For more on Frobenius maps see [SGA3, Exp.VII$_{\text{A}}$,4].

Statement **(II)** in case char $k = p > 0$ now follows. If $r = 0$, i.e., $[G : k] = 1$ there is nothing to prove. Otherwise, in the notation of Lemma 3.7.2, $[H : k] = p^r > 1$ and $[G : k] = p^r[(G/H) : k]$. By induction, $[(G/H) : k]$ is a power of p, so the same holds for G.

Notation. If $G = \operatorname{Spec}(A)$ is a group scheme and m an integer, we let $[m] : A \to A$ denote the homomorphism corresponding to raising to the $m-th$ power in G. Thus, $([m](f))(x) = f(x^m)$.

Lemma 3.7.3. *Suppose the ground ring R satisfies $pR = 0$ for some prime p. Let $G = \operatorname{Spec}(A)$ be a finite free R-group scheme, or more generally, a closed R-subgroup scheme of $(GL_n)_R$ for some n, with augmentation ideal I. Then $[p]I \subset I^p$.*

Remark. I learned this lemma and its very simple proof from a preprint of F. Andreatta and R. Schoof in which they use it to prove that a finite flat group scheme over the ring of dual numbers $k[\varepsilon]$ (k a field, $\varepsilon^2 = 0$) is killed by its order. They tell me that they learned it from Bas Edixhoven.

Proof. Let $U = (u_{ij})$ be an $n \times n$ matrix with independent indeterminate entries u_{ij}. Then $(GL_n)_R = \mathrm{Spec}(B)$ where $B = R[u_{ij}, 1/\det(U)]$. If $G = \mathrm{Spec}(A)$ is a finite flat R-group scheme such that A is a free R-module of rank n, then the action of G on itself by translations gives an imbedding of G as a closed subgroup scheme of $(GL_n)_R$ — the "regular representation" of G. If (f_i), $1 \le i \le n$ is a basis for A over R, such an imbedding is given by the homomorphism of R-algebras $\phi : B \to A$ such that $\phi(u_{ij}) = a_{ij}$, where the a_{ij} are defined by $\tilde{m}(f_j) = \sum_{i=1}^{n} f_i \otimes a_{ij}$. (This is a representation using *right* translations; if $y \in G(R')$, R' an R-algebra, then the automorphism τ_y of $A \otimes_R R' = \sum f_i \otimes R'$ discussed in **(2.11)** is given by $\tau_y(f_j) = \sum f_i a_{ij}(y)$.) The homomorphism ϕ is surjective because $f_j = \sum \tilde{\varepsilon}(f_i) a_{ij}$ for each j.

Suppose more generally that $G = \mathrm{Spec}(A)$ is any closed subgroup scheme of $(GL_n)_R$ and $\phi : B \to A$ the corresponding homomorphism. Let J be the augmentation ideal in B, generated by the entries of the matrix $U - I_n = (u_{ij} - \delta_{ij}) = (v_{ij})$, say. We have $U^p = ([p](u_{ij}))$. Therefore $(([p](v_{ij})) = ([p](u_{ij})) - (\delta_{ij}) = U^p - I_n = (U - I_n)^p = (v_{ij})^p$, which shows that $[p]J \subset J^p$. Since J is the inverse image of I under the surjective map $\phi B \to A$, I is the image of J in A and the lemma follows. •

We can now prove **(III)**. If $H = G/N$ and H, G and N are finite flat R-group schemes, and H and N are killed by their orders, then so is G. The equivalence of categories discussed at the end of **(3.6)** shows that a finite étale group scheme is killed by its order. Hence to show that G is killed by its order it suffices by **(I)** to show that its connected component G^0 is killed by its order. Suppose therefore G is connected and $R = k$, a field. By **(II)** we can suppose the characteristic of k is $p > 0$, and the order of G is $q = p^m$ for some m. Then Lemma 3.7.3 applies, and in the notation of that lemma we have $[p](I) \subset I^p$. Iterating m times gives $[q](I) \subset I^q$. But in an Artin local ring of length q with maximal ideal I one has $I^q = (0)$. Hence $[q](I) = (0)$. This means that $[q](f) = f(1) = [0](f)$ as claimed in **(III)**.

To prove **(IV)** we assume k is a field of characteristic $p > 0$ and $G = \mathrm{Spec}(A)$ is a finite k-group scheme. Let N be the nilradical of A, so $G_{\mathrm{red}} = \mathrm{Spec}(A/N)$. Suppose G_{red} is étale over k, which is automatic if k is perfect. Then $G_{\mathrm{red}} \times G_{\mathrm{red}}$ is reduced so that the map

$$G_{\mathrm{red}} \times G_{\mathrm{red}} \hookrightarrow G \times G \longrightarrow G$$

factors through G_{red} and induces a k-group scheme structure on G_{red}. Let $\alpha = \mathrm{Spec}(\bar{k})$ as in the beginning of 3.7. The isomorphisms $G_{\mathrm{red}}(\alpha) = G(\alpha) = G^{et}(\alpha)$ show that the restriction to G_{red} of the map $G \to G^{et}$ is an isomorphism. Hence $G \simeq G^0 \rtimes G_{\mathrm{red}}$ is a semidirect product, as claimed.

(3.8) The dual Hopf algebra and Cartier duality. We denote the dual of an R-module M by $M' := \mathrm{Hom}_{R\text{-mod}}(M, R)$, and sometimes write $\langle m', m\rangle = m'(m)$ for $m \in M$ and $m' \in M'$. If M and N are locally free of finite rank, then the natural homomorphisms $M \to (M')'$ and $M \otimes N' \to (M \otimes N)'$ are isomorphisms. Thus $M \to M'$ is an anti-equivalence of the category of all such modules with itself, which commutes with tensor products. It follows easily that if $G = \mathrm{Spec}(A)$ is a finite flat R-group scheme, then the dual A' of the commutative Hopf algebra A is a cocommutative Hopf algebra in which the multiplication $A' \otimes A' \to A'$ is dual to the comultiplication $\tilde{m} : A \to A \otimes A$ and vice versa, the unit element in A' is the counit $\tilde{\varepsilon} : A \to R$ in A and vice versa, and the antipodes are dual to each other. We leave the details to the reader with the following remarks. That the dual of an algebra is a coalgebra and vice versa is easily checked. The requirement that the comultiplication and counit should be unitary algebra homomorphisms can be expressed as the commutativity of the following diagram.

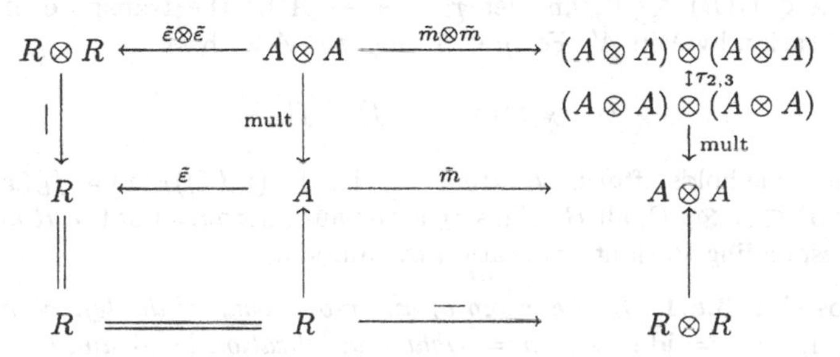

(Here $\tau_{2,3}$ is defined by $\tau_{2,3}(x_1 \otimes x_2 \otimes x_3 \otimes x_4) = x_1 \otimes x_3 \otimes x_2 \otimes x_4$.) The symmetry of the diagram upon reversing arrows shows that the condition holds in A' since it holds in A.

If $G = H_R$ is the constant R-group scheme associated with a finite group H, then A is the ring of R-valued functions on H, and $A' = R[H]$ is the group algebra of H over R. The pairing is the obvious one:

$$\Big\langle \sum_{x \in H} r_x x, f \Big\rangle = \sum_{x \in H} r_x f(x)$$

In the general case it is useful to think of A and A' as the analogs, respectively, of the ring of functions on G and the group algebra of G. In fact, an element $f \in A$ does give a function f_B on $G(B)$ with values in B for each R-algebra f_B, if we put $f_B(x) := x(f)$ for each point $x : A \to B$ in $G(B)$, and f is determined by these functions, in fact by f_A, since $f = f_A(\mathrm{id})$, where $\mathrm{id} \in G(A) = G(G)$ is the identity map.

Although A' is not the group algebra of $G(R)$, nevertheless the inclusion

$$G(R) = \mathrm{Hom}_{(R\text{-alg})}(A, R) \subset \mathrm{Hom}_{(R\text{-mod})}(A, R) = A'$$

identifies $G(R)$ with the multiplicative group of *group-like* elements of A', that is, the group of invertible elements λ in A' such that λ is mapped to $\lambda \otimes \lambda$ by the comultiplication in A' . This is routine to verify. The group law in $G(R)$ is given by multiplication in A' because it is dual to the comultiplication $\tilde{m}$ in A. Again by duality, an element $\lambda \in A'$ is group-like if and only if it is invertible in A' and the map $\lambda : A \to R$ is multiplicative. (Assuming lambda is multiplicative, one checks that it is invertible if and only if $\lambda(1) = 1$.)

The formation of the dual Hopf algebra A' commutes with base extension; for each R-algebra B we can identify $A' \otimes_R B$ with$(A \otimes_R B)'$, where the second prime $(')$ is relative to the base B. Thus $G(B)$ is the group of group-like elements in the Hopf algebra $A'_B := A' \otimes_R B$ over B, for each R-algebra B. We denote the B-linear pairing $A'_B \times A_B \to B$ by $\langle\, , \rangle_B$. Then for $f \in A$ or A_B and $x \in G(B) \subset A'_B$ the value of f at x is $f_B(x) = \langle x, f\rangle_B \in B$. In particular, for $f \in A$ and $\mathrm{id} \in G(A) \subset A' \otimes A$ we have $f = f_A(\mathrm{id}) = \langle \mathrm{id}, f \otimes 1\rangle_A$.

Let $\lambda \in G(R) \subset A'$, and let $\tau_\lambda : A \to A$ be the transpose of right multiplication by λ in A'. For $\mu \in A'$ and $f \in A$ we have

$$\langle \mu, \tau_\lambda(f)\rangle = \langle \mu\lambda, f\rangle = f(\mu\lambda)$$

and the same holds after base extension, that is, $(\tau_\lambda(f))_B(x) = f_B(x\lambda)$ for $x \in G(B) \subset A' \otimes_R B$, all B. Thus τ_λ is the automorphism of the R-algebra A corresponding to right translation by lambda.

Proposition 3.8.1. *In the group of automorphisms of the left A'-module $A' \otimes_R A$, let $\tau := \mathrm{id}_{A'} \otimes \tau_\lambda$, $\rho :=$ right multiplication by* id *and $\ell :=$ right multiplication by $\lambda \otimes 1$. Then $\tau\rho\tau^{-1}\rho^{-1} = \ell$.*

Proof. Taking $\phi = \tau_\lambda$ in the lemma below we find that $\tau_{(\mathrm{id})} = \ell(\mathrm{id}) = \mathrm{id} \cdot (\lambda \otimes 1)$. Hence, for $X \in A' \otimes_R A$ we have (since τ_λ, hence also τ, is a ring automorphism) $\tau\rho(X) = \tau(X \cdot \mathrm{id}) = \tau(X) \cdot \tau(\mathrm{id}) = \tau(X) \cdot \mathrm{id} \cdot (\lambda \otimes 1) = \ell\rho\tau(X)$.

Lemma 3.8.2. *Let $\phi : A \to A$ be an R-linear map and let $\phi' : A' \to A'$ be its transpose. Then* $(\mathrm{id}_{A'} \otimes \phi)(\mathrm{id}) = (\mathrm{id}_A \otimes \phi')(\mathrm{id})$.

Proof. We leave this bit of linear algebra to the reader. In fact, each side of the stated equality is equal to the element of $A' \otimes A$ which corresponds to ϕ and to ϕ' under the canonical isomorphisms $A' \otimes A \approx \mathrm{End}_R(A) \approx \mathrm{End}_R(A')$.

The left A'-module $A' \otimes A$ is free of rank $n := [G : R]$, the order of G and **(3.8.1)** shows that the "constant" matrix λI_n is a commutator in the group $\mathrm{GL}_n(A')$. If A' is commutative, that is, G is commutative, then we can use the determinant homomorphism $\mathrm{GL}_n(A') \to (A')^*$ to conclude that $\lambda^n = 1$. The same holds for $\lambda \in G(B) \subset A'_B$ for an arbitrary base ring extension $R \to B$. Thus a *commutative* finite flat group scheme is killed by its order. The above is Deligne's proof of that fact, presented perhaps in a less comprehensible way than in [O-T].

I do not know whether a non-commutative finite flat group scheme is killed by its order. This is true if $R = k$ is a field (cf. **(3.7)III**), and hence if R has no nilpotent elements. As remarked (loc. cit.), Andreatta and Schoof have proved it for the ring of dual numbers $R = k[\varepsilon]$, $\varepsilon^2 = 0$.

Suppose now G is commutative. Then A' is commutative, so that $G' :=$ Spec(A') makes sense and is a finite flat commutative R-group scheme of the same order as G. The functor $G \mapsto G'$ is an anti-equivalence of the category of finite flat commutative R-group schemes with itself, such that $(G')'$ is canonically and functorially isomorphic to G. This *Cartier duality* is a vast generalization of the classical duality of finite abelian groups. As explained in **(2.9)**, the group-like elements of A'_B are the characters of G' defined over B, and it follows from the above that G is the character group scheme of G' in the sense of **(2.9)**, that is, represents the functor

$$B \longmapsto \mathrm{Hom}_{(\mathbf{Gr}/B)}(G'_B, (\mathbb{G}_m)_B)$$

By symmetry, G' is the character group scheme of G. For each R-algebra B, the pairing

$$G(B) \times G'(B) \to \mathbb{G}_m(B) = B^*$$

is given by the symbol $\langle\, , \rangle_B$, if we imbed $G(B)$ and $G'(B)$ in A'_B and A_B respectively as above. On the other hand, if we view these pairings as a bimultiplicative invertible function on $G \times_R G' = \mathrm{Spec}(A \otimes_R A')$, the function is the element id $\in A \otimes A' = \mathrm{End}_R(A)$ corresponding to the identity map of A, because $\langle \lambda \otimes f, \mathrm{id} \rangle = \langle \lambda, f \rangle$, as one easily checks, and the same holds after base extension.

An application: Why non-abelian simple groups are étale. Serre and Raynaud have explained to me why finite flat group schemes which are very non-commutative tend to be étale. The point is that if G/S is not étale at a point $s \in S$, then $p = \mathrm{char}(\kappa(s))$ is not 0, and over the henselization R^h of the local ring $R = O_{S,s}$ of s, the connected component G^0 of G is a *normal* subgroup scheme of p-power order, the normality of which works against non-abelianness. For example, suppose S is a normal scheme with field of fractions K and the general fiber G_K of G is étale (which is automatic if $\mathrm{char}(K) = 0$). Let $\bar{K}$ be an algebraic closure of K. Then in the situation just discussed, we will have $R \subset R^h \subset \bar{K}$, and if G is not étale at s, then, for $p = \mathrm{char}(\kappa(s))$, the finite group $G(\bar{K})$ will have a non-trivial subgroup $G^0(\bar{K})$ which is of p-power order by **(II)** above, and is normal (and also stable under the action of the decomposition group $\pi_s = \mathrm{Aut}(\bar{K}/R^h)$). If $G(\bar{K})$ has no such subgroup, then G is étale at s. Thus if $G(\bar{K})$ has no normal p-subgroup for every prime p, for example, if $G(\bar{K})$ is a non-abelian simple group, then G/S is étale.

§4. Raynaud's results on commutative p-group schemes

This part is taken entirely from Raynaud's great paper [R2]

(4.1) Prolongations. In this section we assume for simplicity that our ground ring R is a discrete valuation ring of mixed characteristic. Let K be its field of fractions, π a prime element, $k = R/\pi R$ the residue class field, p the residue characteristic, v the normalized valuation ($v(\pi) = 1$), and $e = v(p)$ the absolute ramification index.

Let $G_0 = \mathrm{Spec}(A_0)$ be a finite commutative K-group scheme. By a *prolongation* of G_0 (to $\mathrm{Spec}\,R$) we mean a finite flat R-group scheme G whose generic fiber is G_0. The isomorphism classes of prolongations of G_0 are represented by the R-group schemes G of the form $G = \mathrm{Spec}(A)$, for A a finite R-sub-algebra of A_0, containing R and spanning A_0, such that $c(A) \subset A \otimes_R A$, where $c : A_0 \to A_0 \otimes A_0$ is the comultiplication in A_0. (Exercise: Why is the existence of an inverse automatic in this situation?)

Let $G_0^D = \mathrm{Spec}(A_0^D)$ be the Cartier dual of G_0 and let

$$\langle \cdot, \cdot \rangle : A_0 \times A_0^D \longrightarrow K$$

be the canonical bilinear map. The Cartier dual of a prolongation $G = \mathrm{Spec}(A)$ is $G^D = \mathrm{Spec}(A^D)$, where $A^D \subset A_0^D$ is the complementary module to A, that is

$$A^D := \{\lambda \in A_0^D : \langle \lambda, f \rangle \in R \text{ for all } f \in A\} \approx \mathrm{Hom}_{R\text{-mod}}(A, R).$$

The multiplication $A_0^D \otimes_K A_0^D \to A_0^D$ is the transpose of the comultiplication c. Hence the condition $c(A) \subset A \otimes_R A$ is equivalent to $A^D \supset A^D A^D$; a finite R-submodule A of A^0 which contains R and spans A^0 is the ring of a prolongation if and only if both A and its complementary module A^D are closed under multiplication.

The prolongations of G_0 are partially ordered. If $G = \mathrm{Spec}(A)$ and $G' = \mathrm{Spec}(A')$ are two prolongations, we write $G \geq G'$ if $A \supset A'$, that is, if there is a morphism $G \to G'$ (necessarily unique) inducing the identity on G_0.

Proposition 4.1.1. *Two prolongations of G_0 have a* sup *and an* inf.

Proof. If $G' = \mathrm{Spec}(A')$ and $G'' = \mathrm{Spec}(A'')$ are two prolongations with $A', A'' \subset A_0$, let $A = A'A''$ be the R-algebra generated by A' and A''. Then

$$c(A) = c(A')c(A'') \subset (A' \otimes A')(A'' \otimes A'') = A'A'' \otimes A'A'' = A \otimes A.$$

Hence $G = \mathrm{Spec}(A)$ is a prolongation of G_0 and it is obviously a least upper bound for G' and G'' in the partially ordered set of all such prolongations. Cartier duality reverses order, so $\inf(G', G'') = \left(\sup(G'^D, G''^D)\right)^D$ is a greatest lower bound.

Corollary. *If G_0 has a prolongation, then it has a maximal one G^+ and a minimal one G^-.*

Proof. G^+ exists because the rings of prolongations are R-orders in the separable K-algebra A_0, so are all contained in the maximal order, the integral closure of R in A_0. By duality, there is also a minimal prolongation.

(4.2) Dévissage. Let

$$0 \longrightarrow G_0' \longrightarrow G_0 \longrightarrow G_0'' \longrightarrow 0$$

be a short exact sequence of finite K-group schemes, and let

$$A_0' \longleftarrow A_0 \hookleftarrow A_0''$$

be the corresponding K-algebra picture. Suppose $G = \mathrm{Spec}(A)$ is a prolongation of G_0. Let A' be the image of A in A_0' and put $G' = \mathrm{Spec}(A')$, the "scheme-theoretic closure of G_0' in G." Obviously G' is a closed subgroup scheme of G prolonging G_0', and is the unique one such that the inclusion $G' \subset G$ extends the given $G' \hookrightarrow G_0$. The quotient $G'' := G/G'$ is a prolongation of $G_0'' = G/G'$. By induction on the order of G, this proves part (a) of:

Proposition 4.2.1. *Suppose G is a prolongation of G_0 and $\left(G_0^{(j)}\right)_{0\le j\le n}$ is a composition series for G_0.*
(a) *There exist unique prolongations $G^{(j)}$ of the $G_0^{(j)}$ such that the inclusions $G_0^{(j)} \subset G_0^{(j+1)}$ induce closed immersions $G^{(j)} \subset G^{(j+1)}$. The quotient $G^{(j+1)}/G^{(j)}$ is a prolongation of $G_0^{(j+1)}/G_0^{(j)}$ for each j.*
(b) *Suppose H is another prolongation of G_0, such that $G \ge H$. Then $G^{(j+1)}/G^{(j)} \ge H^{(j+1)}/H^{(j)}$ for each j, and if equality holds for the quotients for each j, then $G = H$.*

Proof. (of (b)) By induction on the length n of the series, it suffices to consider the case $n = 2$, in which case we have a diagram

$$\begin{array}{ccccccccc} 0 & \longrightarrow & G^{(1)} & \longrightarrow & G & \longrightarrow & G/G^{(1)} & \longrightarrow & 0 \\ & & \downarrow & & \downarrow & & \downarrow & & \\ 0 & \longrightarrow & H^{(1)} & \longrightarrow & H & \longrightarrow & H/H^{(1)} & \longrightarrow & 0 \end{array}$$

with exact rows, and must conclude that the central vertical arrow is an isomorphism if the outer two are. This is true in an abelian category, hence also for our group schemes which form a full subcategory of a category of sheaves. In our special case, another way to see this is to note that the discriminant ideal of G is determined by the discriminant ideals of the quotients $G^{(j)}/G^{(j+1)}$, cf. [O-T].

(4.3) F-vector space schemes. Let $\bar{K}$ be an algebraic closure of K and $\mathcal{G} = \mathrm{Gal}(\bar{K}/K)$. Let G be a finite flat commutative R-group scheme. Since $\mathrm{char}(K) = 0$, the generic fiber $G_0 = G_K$ is determined by the $\mathcal{G}$-module $G_0(\bar{K}) = G(\bar{K})$. Let us call G *simple* if G_0 is simple, or what is the same, if $G(\bar{K})$ is a simple $\mathcal{G}$-module.

Lemma 4.3.1. *Suppose R is strictly henselian and G is simple of p-power order. Then the pro-p Sylow subgroup $\mathcal{P}$ of $\mathcal{G}$ is normal, with procyclic quotient $\mathcal{G}_{\mathrm{tame}}$, and $\mathcal{G}$ acts on $G(\bar{K})$ through $\mathcal{G}_{\mathrm{tame}}$.*

Proof. This is well-known; $\mathcal{G}$ is an inertia group and $\mathcal{P}$ the ramification group fixing the maximal tame extension of K. The subgroup of $G(\bar{K})$ of points fixed by $\mathcal{P}$ is stable by $\mathcal{G}$ since $\mathcal{P}$ is normal, and is of order divisible by p by the usual counting argument, hence is all of $G(\bar{K})$ by simplicity.

Suppose G is simple and $\mathcal{G}$ acts on $G(\bar{K})$ through an abelian quotient group, as in the lemma. Then $G(\bar{K})$, as simple module over the commutative ring $\mathbb{Z}[\mathcal{G}]$, is a 1-dimensional vector space over a residue field F of $\mathbb{Z}[\mathcal{G}]$. This field F is necessarily finite, having the same number of elements as $G(\bar{K})$, i.e., as the order of G. Scalar multiplication by any element of F^* commutes with the action of $\mathcal{G}$ on $G(\bar{K})$ and therefore induces an automorphism of the K-scheme G_0. These automorphisms may not extend to G, but they certainly extend to the maximal and minimal prolongations G^+ and G^- of G_0, by "transport of structure." Thus G^+ and G^- are F-module schemes over R in the sense of the following paragraph.

Let F be an associative ring with unity. By an F-module scheme over a base scheme S one means a commutative S-group scheme G together with a unitary ring homomorphism $F \to \mathrm{End}(G)$. This makes $G(T)$ an F-module for every T over S, functorially in T, and in this way an F-module scheme is the same thing as a representable functor from $(\mathbf{Sch}/S)$ to the category of F-modules. As a matter of notation, if $G = \mathrm{Spec}(A)$ is an affine F-module scheme, we denote by $[t] : A \to A$ the Hopf algebra endomorphism corresponding to the action of t on G. Thus

$$([t]f)(x) = f(tx) \quad \text{for } f \in A \text{ and } x \in G(T), \text{ any } T.$$

If F is a finite field, we will call an F-module scheme which is finite flat of the same order as F a *Raynaud F-module scheme.* The discussion above proves:

Proposition 4.3.2. *Suppose G_0 is a simple commutative K-group scheme of p-power order which has a prolongation G. Suppose R is strictly henselian or, more generally, that $\mathcal{G}$ acts on $G(\bar{K})$ through an abelian quotient group. Then*

$$F := \mathrm{End}(G_0) = \mathrm{End}(G^+) = \mathrm{End}(G^-)$$

is a finite field and G_0, G^+, and G^- are Raynaud F-module schemes.

(4.4) The classification theorem.. Let F be a finite field of characteristic p and denote by $q = p^r$ the number of elements of F. Let $\mu = \mu_{q-1}(\bar{K})$ be the group of $(q-1)^{\text{st}}$ roots of 1. Let $M = \operatorname{Hom}(F^*, \mu)$ be the group of "multiplicative characters" of F, and extend each $\chi \in M$ to all of F by putting $\chi(0) = 0$. If $\mu \subset R$, then there are r special elements $\chi \in M$, called *fundamental characters*, such that the map

$$F \xrightarrow{\chi} R \longrightarrow k = R/\pi R$$

is a homomorphism of fields. If χ is one such character, the others are χ^{p^i}, $0 \le i < r$, and $\chi^{p^r} = \chi$. Hence the fundamental characters form a set $(\chi_i)_{i \in \mathcal{I}}$ indexed by a principal homogeneous space $\mathcal{I}$ over $\mathbb{Z}/r\mathbb{Z}$, in such a way that $\chi_{i+1} = \chi_i^p$.

Theorem 4.4.1. *Let f, μ and $(\chi_i)_{i \in \mathcal{I}}$ be as above and suppose $\mu \subset R$.*
(a) *Let $(\delta_i)_{i \in \mathcal{I}}$ be elements of R such that $0 \le v(\delta_i) \le e = v(p)$ for each i. Let A be the R-algebra presented by generators X_i, $i \in \mathcal{I}$, and relations $X_i^p = \delta_i X_{i+1}$, $i \in \mathcal{I}$. Then there is a unique F-module structure on the R-scheme $G = \operatorname{Spec} A$ such that $[s]X_i = \chi_i(s)X_i$ for each $i \in \mathcal{I}$ and $s \in F$, and with that structure G is a Raynaud F-module scheme.*
(b) *Every Raynaud F-module scheme over R is of the type described in (a).*
(c) *Suppose G and G' are two such schemes defined, respectively, by the equations $X_i^p = \delta_i X_{i+1}$ and the equations $X_i'^p = \delta_i' X_{i+1}'$. The F-module scheme homomorphisms $G' \to G$ correspond to the R-algebra homomorphisms of the form $X_i \mapsto a_i X_i'$, where $(a_i)_{i \in \mathcal{I}}$ is a family of elements of R such that $a_{i+1}\delta_i = a_i^p \delta_i'$ for each $i \in \mathcal{I}$.*

Proof. We prove (b) first. Let $G = \operatorname{Spec} A$ be a Raynaud F-module scheme over R. The geometric generic fiber $G_{\bar{K}} = \operatorname{Spec}(A \otimes_R \bar{K})$ is the constant scheme $G(\bar{K})_{\bar{K}}$ of the 1-dimensional F-vector space $G(\bar{K})$ and is therefore F-isomorphic to the constant scheme $F_{\bar{K}}$. Choosing an isomorphism, we can view $A \otimes_R \bar{K}$ as the ring of $\bar{K}$-valued functions on F, which has a $\bar{K}$-basis consisting of the constant function 1 and the multiplicative characters $\chi \in M$, or, what is the same, consisting of the monomials

$$\prod_{i \in \mathcal{I}} \chi_i^{\nu_i}, \quad 0 \le \nu_i \le p-1,$$

if we interpret $\prod_i \chi_i^0$ as 1. Since R contains $(q-1)^{-1}$ and the $(q-1)^{\text{st}}$ roots of 1, the action of F^* on the augmentation ideal I of A gives a direct sum decomposition

$$I = \bigoplus_{\chi \in M} I_\chi, \quad \text{where} \quad I_\chi = \{f \in I : [s]f = \chi(s)f \text{ for all } s \in F^*\}.$$

Each of the R-modules I_χ is of rank 1 because $I_\chi \otimes \bar{K} = \bar{K}_\chi$. For each fundamental character χ_i, choose a basis element X_i for I_{χ_i}, and write

$X_i = c_i\chi_i$, $c_i \in \bar{K}^*$. Since obviously $I_\chi^p \subset I_{\chi^p}$ for each χ, we have $X_i^p = \delta_i X_{i+1}$ with $\delta_i = c_i^p/c_{i+1} \in R$. Hence the R-subalgebra of A generated by the X_i has a basis consisting of the monomials

$$\prod_{i\in\mathcal{I}} X_i^{\nu_i} = \prod_{i\in\mathcal{I}} c_i^{\nu_i} \cdot \prod_{i\in\mathcal{I}} \chi_i^{\nu_i}.$$

To show this algebra is all of A, we consider the Cartier dual $G^D = \operatorname{Spec}(A^D)$. Identify $A^D \otimes \bar{K}$ with the group algebra $\bar{K}[F]$. Denoting by λ^s the basis element of $\bar{K}[F]$ corresponding to $s \in F$, so that $\lambda^{s+t} = \lambda^s\lambda^t$, the linear duality is given by $\langle \lambda^s, f\rangle = f(s)$, and it is easy to check that the basis of the augmentation ideal in $K[F]$ which is dual to the basis $(\chi)_{\chi\in M}$ of $I \otimes \bar{K}$ is then $(e_\chi)_{\chi\in M}$, where

$$e_\chi = \frac{1}{q-1}\sum_{s\in F}\chi^{-1}(s)(\lambda^s - 1).$$

Put $Y_i = c_i^{-1}e_{\chi_i}$. Then $I_{\chi_i}^D = RY_i$, and there are constants $\gamma_i \in R$ such that $Y_i^p = \gamma_i Y_{i+1}$. For each $\chi = \prod \chi_i^{\nu_i}$, $0 \le \nu_i \le p-1$, not all $\nu_i = 0$, we have

$$\langle \prod Y_i^{\nu_i}, \prod X_i^{\nu_i}\rangle = \langle \prod e_{\chi_i}^{\nu_i}, \prod \chi_i^{\nu_i}\rangle = \langle \prod e_{\chi_i}^{\nu_i}, \chi\rangle = w_\chi,$$

say, and

$$\gamma_i\delta_i = \langle Y_i^p, X_i^p\rangle = \langle e_{\chi_i}^p, \chi_{i+1}\rangle = w_i,$$

say, with w_χ and w_i in R depending only on F, i.e., on $\mathcal{G}$, not on G. These w's are unchanged by automorphism of F, hence $w_\chi = w_{\chi^p}$ and $w_i = w$ is independent of i. By considering $(F_R)^D \bmod p$, Raynaud proves

$$w_\chi \equiv \prod_{i\in\mathcal{I}} \nu_i! \pmod{pR} \qquad \text{and} \qquad w \equiv -p \pmod{p^2R}.$$

Since the w_χ are units in R, it follows that $R\big[(X_i)_{i\in\mathcal{I}}\big]$ and $R\big[(Y_i)_{i\in\mathcal{I}}\big]$ are complementary modules contained, respectively, in A and A^D and therefore equal, respectively, to A and A^D, since A and A^D are complementary. This proves (b)

Let E be the field of $(q-1)^{\text{st}}$ roots of 1, and let $\mathcal{O}$ be the local ring of a place of E above p. Obviously the congruences for the w_χ and w above, which express properties of the constant scheme $F_\mathcal{O}$, are the key to the proof just given — the rest is formal. Another approach to these congruences, which gives explicit expressions for the w's, uses the self-duality of F given by an additive character $\psi(s) = \zeta^{\operatorname{Tr}(s)}$, where ζ is a primitive p^{th} root of 1. This duality gives an isomorphism $F_{\mathbb{Q}(\zeta)} \xrightarrow{\sim} F_{\mathbb{Q}(\zeta)}^D$ which, over $E(\zeta)$, is given on the Hopf algebras by $e_\chi \mapsto \tau(\chi)\chi$, where

$$\tau(\chi) = \frac{1}{q-1}\sum_{s\in F}\chi^{-1}(s)(\psi(s) - 1) = \psi(e_\chi).$$

From this it follows that $w = \tau^{p-1}$, where $\tau = \tau(\chi_i)$, any i, and that, for $\chi = \prod_i \chi_i^{\nu_i}$ as usual,

$$w_\chi = \frac{\tau^{\sum \nu_i}}{\tau(\chi)}.$$

Up to powers of $q-1$, the $\tau(\chi)$'s are Gauss sums and the w_χ's are Jacobi sums. From this point of view, the congruences above are equivalent to classical results of Stickelberger on the leading term of the $(\zeta-1)$-adic expansion of Gauss sums in the ring $\mathcal{O}[\zeta]$.

We now turn to the proof of 4.4.1(a). Let δ_i, $A = R[(X_i)]$, and $G = \operatorname{Spec}(A)$ be as in (a). Fix $j \in \mathcal{I}$. Eliminating X_i for $i \neq j$ in the system of equations $X_i^p = \delta_i X_{i+1}$, one finds that the map $P \mapsto X_j(P)$ is a bijection between $G(\bar{K})$ and the set of solutions $x \in \bar{K}$ of the equation $x^q = \Delta_j x$, where $\Delta_j = \prod_{k=0}^{r-1} \delta_{i-1-k}^{p^k}$. From this it is easy to see that the set $G(\bar{K})$ has a unique structure of (1-dimensional) F-vector space such that $X_i(sP) = \chi_i(s)X_i(P)$ holds for $s \in F$ and $P \in G(\bar{K})$.

Choose a point $P \neq 0$ in $G(\bar{K})$ and put $c_i = X_i(P)$. Note that the action of $\mathcal{G} = \operatorname{Gal}(\bar{K}/K)$ on $G(\bar{K})$ is F-linear via the homomorphism $\phi : \mathcal{G} \to F^*$ for which $\chi(\phi(\sigma)) = c_i^{\sigma-1}$. Viewing A as a ring of functions on F via the isomorphism $s \mapsto sP$ from F to $G(\bar{K})$, so that X_i becomes the function $c_i\chi_i$, one checks as in the proof of (b), using congruence properties of the w's, that A is stable under the comultiplication in $F_{\bar{K}}$, because its complementary module $A \otimes_R K$ is the ring $R[(Y_i)]$ with $Y_i = c_i^{-1} e_{\chi_i}$.

Part (c) is left to the reader.

In case $q = 2$, $F = \mathbb{Z}/2\mathbb{Z}$, the results of §3.2 give a generalization of Theorem 4.4.1. The corresponding generalization is proved in [R] for arbitrary q. Let R be any local ring of residue characteristic p which is an algebra over the ring $\mathcal{O}$ above. Then Raynaud F-module schemes G over R are described by families δ_i, γ_i, $i \in \mathcal{I}$, of elements of R such that $\gamma_i\delta_i = w$ for each i, as follows. We have $G = \operatorname{Spec}(A)$ and $G^D = \operatorname{Spec}(A^D)$, where

$$A = R[(X_i)];\ X_i^p = \delta_i X_{i+1} \qquad \text{and} \qquad A^D = R[(Y_i)];\ Y_i^p = \gamma_i Y_{i+1}.$$

For $\chi \in M$, put $\chi = \prod \chi_i^{\nu_i}$, $0 \le \nu_i \le p-1$, not all $\nu_i = 0$, and put $X_\chi = \prod X_i^{\nu_i}$ and $Y_\chi = \prod Y_i^{\nu_i}$. The pairing $A \times A^D \to R$ is the one for which $(X_\chi)_{\chi \in M}$ and $(w_\chi^{-1} Y_\chi)_{\chi \in M}$ are dual bases of I and I^D. The comultiplication in A is most easily described as the transpose of the multiplication in A^D. The explicit formula for it, given in [R], is a bit messy, involving not only the w_χ^{-1}'s, but also some products of the γ_i's which come from using the relations $Y_i^p = \gamma_i Y_{i+1}$ as necessary to express a product $Y_{\chi'}Y_{\chi''}$ of two basic monomials as a constant times $Y_{\chi'\chi''}$.

The proof of the generalization is not hard. For existence, for example, one simply redoes the proof of 4.4.1(a) in the following generic situation. Let $K = E((\delta_i))$ be the field of rational functions in r variables δ_i over E and

$R = \mathcal{O}[\delta_i, \gamma_i]$, where $\gamma_i = w/\delta_i$. Then exactly as in the proof of 4.4.1(a), one shows that

$$\operatorname{Spec}\big(R[X_i];\ X_i^p = \delta_i X_{i+1}\big)$$

is a Raynaud F-module whose extensions give the ones which were to be constructed.

(4.5) Applications. In this section we prove two fundamental results of Raynaud which are used in [C].

Theorem 4.5.1. *Suppose $e < p-1$. (There are no other assumptions on R other than that it be a discrete valuation ring of mixed characteristic $(0,p)$.) Let G be a commutative finite flat R-group scheme of p-power order. Then G is, up to isomorphism, the unique prolongation of its generic fiber G_K.*

Proof. If G is not unique, then $G^+ > G^-$. Such a strict inequality is preserved by passage to an extension of R, so we can assume without loss of generality that R is strictly henselian. Also, by dévissage (Proposition 4.2.1(b)), we can assume that G is simple. By Proposition 4.3.2, then, G^+ and G^- are Raynaud F-module schemes for the finite field $F = \operatorname{End}(G_K)$. Since R is strictly henselian, $\mu \subset R$. Therefore G^+ and G^- are Raynaud F-module schemes of the type described in Theorem 4.4.1. Taking $G' = G^+$ and $G = G^-$ in Theorem 4.4.1(c), we find that there are constants $a_i \in R$, $a_i \neq 0$, such that $\delta_i = a_i^p a_{i+1}^{-1} \delta_i'$ for $i \in \mathcal{I}$. Choosing an i such that $v(a_i)$ is maximal, we conclude $e \geq (p-1)v(a_i)$, which, if $p-1 > e$, implies $v(a_i) = 0$. Hence all a_i are units in R and $G = G'$ as was to be shown.

Corollary. *Write $\mathcal{C}$ for the category of commutative finite flat R-group schemes of p-power order. Let G and H be objects in $\mathcal{C}$. If $p-1 > e$, then the natural map*

$$\operatorname{Hom}_{\mathcal{C}}(G,H) \longrightarrow \operatorname{Hom}_{\mathcal{G}}\big(G(\bar{K}), H(\bar{K})\big)$$

is bijective and the natural map

$$\operatorname{Ext}_{\mathcal{C}}(G,H) \longrightarrow \operatorname{Ext}_{\mathcal{G}}\big(G(\bar{K}), H(\bar{K})\big)$$

is injective

Proof. The injectivity of the map on Hom's is obvious and doesn't require $e < p-1$. For surjectivity, let G_0 and H_0 be the generic fibers and $u_0 : G_0 \to H_0$ a homomorphism. We must show u_0 has a prolongation $u : G \to H$, assuming $e < p-1$. There are homomorphisms

$$G \longrightarrow G/(\overline{\operatorname{Ker}(u_0)} \text{ in } G) \xrightarrow{\sim} (\overline{\operatorname{Image}(u_0)} \text{ in } G) \hookrightarrow H,$$

where $(\overline{X} \text{ in } Y)$ is a temporary notation for the scheme-theoretic closure of X in Y (see the discussion before Proposition 4.2.1). The key point is the isomorphism in the middle. Both groups it connects prolong the same general fiber Image(u_0) and they are therefore isomorphic, by the theorem.

Also by unicity of prolongations, an exact sequence

$$0 \to G' \to G \to G'' \to 0$$

of prolongations is determined up to isomorphism by the sequence of its generic fibers, so the Ext map is injective

Which K-schemes G_0 have a prolongation? In case G_0 is a Raynaud F-module scheme and $\mu \subset R$, the classification theorem gives the following answer.

Theorem 4.5.2. *Let F be a finite field with $q = p^r$ elements. Suppose $\mu := \mu_{q-1}(\bar{K}) \subset K$ and let $\chi_i : F^* \to \mu$ be a fundamental character. Let G_0 be a Raynaud F-module scheme over K and $\phi : \mathcal{G} \to F^*$ the character giving the action of $\mathcal{G} = \mathrm{Gal}(\bar{K}/K)$ on $G_0(\bar{K})$. Then G_0 has a prolongation (or, as one says, the representation $\mathcal{G} \to \mathrm{GL}_1(F)$ is "flat") if and only if there is an element $\Delta \in K^*$ such that*

$$\chi_i(\phi(\sigma)) = (\Delta^{1/(q-1)})^{\sigma-1} \quad \text{and} \quad v(\Delta) = \sum_{k=0}^{r} n_k p^k,$$

with integers n_k in the range $0 \le n_k \le e$.

Proof. Suppose G_0 has a prolongation G. Replacing G by G^+ if necessary, we may assume G is a Raynaud F-module scheme over R, hence is of the type described in Theorem 4.4.1(a) via equations $X_i^p = \delta_i X_{i+1}$. Let $x_i = X_i(P)$ for some point $P \in G(\bar{K})$. The relations $x_i^p = \delta_i x_{i+1}$ imply $x_i^q = \Delta_i x_i$, where

$$\Delta_i = \delta_{i-1}\delta_{i-2}^p \cdots \delta_{i-r}^{p^{r-1}}.$$

Choosing $P \neq 0$, we have $x_i \neq 0$, hence $x_i^{q-1} = \Delta_i$. On the other hand,

$$x_i^\sigma = \chi_i(\phi(\sigma))x_i \quad \text{for } \sigma \in \mathcal{G},$$

so the condition of the theorem is satisfied with $\delta = \delta_i$.

Conversely, given Δ with $v(\Delta)$ as in the theorem, we can construct δ's giving a prolongation by putting $\delta_{i-1-k} = \pi^{n_k}$ for $2 \le k \le r-1$ and defining δ_{i-1} by the equation $\Delta_i = \delta_{i-1}\delta_{i-2}^p \cdots \delta_{i-r}^{p^{r-1}}$.

The theorem just proved applies in particular in the case R is strictly henselian and G_0 simple of p-power order, and gives immediately in that case a result conjectured by Serre which was a main motivation for Raynaud's work, and which, in case $R = \mathbb{Z}_p$, is Theorem 1.7 of [C].

REFERENCES

[A] Abe, E., *Hopf Algebra*, Cambridge Tracts in Math., Cambridge Press Syndicate, 1980.

[B-L-R] Bosch, S., Lutkebohmert, W., Raynaud, M., *Néron Models*, Springer-Verlag, 1990.

[C-S] Chase, S.U., Sweedler, M.E., *Hopf Algebras and Galois Theory*, Lecture Notes in Math., vol. 97, Springer, 1969.

[C] Conrad, B., *The Flat Deformation Functor*, this volume.

[deJ] de Jong, A.J., *Finite locally free group schemes in characteristic p and Dieudonne modules*, Invent. Math. **114** (1993), 89–137.

[D] Demazure, M., *Lectures on p-divisible groups*, Lecture Notes in Math., vol. 302, Springer.

[Dr] Drinfeld, V.G., *Quantum Groups*, Proc. Int. Cong. Math. (Berkeley 1986), AMS, Providence, 1987, pp. 798–820.

[F] Fontaine, J-M., *Il n'y a pas de variété abélienne sur* $\mathbb{Z}$, Invent. Math. **81** (1985), 515–538.

[Gre] Greither, C., *Extensions of finite group schemes and Hopf Galois theory over a complete discrete valuation ring*, Math. Zeit. **210** (1992), 37–67.

[Gro] Grothendieck, A., *Technique de descente et théorèmes d'existence en géometrie algebrique I–IV*, Exposés in Seminaire Bourbaki, 1959–1961.

[SGA1] Grothendieck et al., *Revêtements étales et groupe fondamental*, Lecture Notes in Math., vol. 224, Springer, 1971.

[SGA2] ———, *Schémas en groupes I, II, III*, Lectures Notes in Math., vol. 151, 152, 153, Springer, 1970.

[K-M] Keel, S., Mori, S., *Quotients by groupoids.*

[L] Lang, S., *Algebra*, Third edition, Addison-Wesley, Reading, MA, 1993.

[M] Milne, J.S., *Etale Cohomology*, Princeton U. Press, 1980.

[M-M] Milnor, J.W., Moore, J.C., *On the Structure of Hopf Algebras*, Ann. of Math., Series 2, **81** (1965), 211–264.

[M-O] Mumford, D., Oort, F., *Deformations and Liftings of Finite, Commutative Group Schemes*, Invent. Math. **5** (1968), 317–334.

[Mu] Murre, J., *Lectures on an introduction to Grothendieck's theory of the fundamental group*, Lecture Notes, Tata Inst. of Fundamental Research, Bombay, 1967.

[O] Oort, F., *Commutative Group Schemes*, Lecture Notes in Math., vol. 15, Springer, 1966.

[O-T] Oort, F., Tate, J., *Group schemes of prime order*, Ann. Scient. Ec. Norm. Sup., 4e serie, t.3 (1970), 1–21.

[R1] Raynaud, M., *Passage au quotient par une relation d'équivalence plate*, Proceedings of a Conference on Local Fields, Springer-Verlag, 1967, pp. 78–85.

[R2] ———, *Schémas en groupes de type* $(p, \dots, p)$, Bull. Soc. Math. France **102** (1974), 241–280.

[SS] Shatz, S.S., *Group Schemes, Formal Groups, and p-Divisible Groups*, Arithmetic Geometry (Cornell, G. and Silverman, J.H., eds.), Chapter III, Springer-Verlag, 1986, pp. 29–78.

[SS-SS] Shnider, S., Sternberg, S., *Quantum Groups – from coalgebras to Drinfeld algebras*, International Press, Inc., 1993.

[Sw] Sweedler, M.E., *Hopf Algebras*, W.A. Benjamin, Inc., New York, 1969.

[T1] Tate, J., *p-Divisible Groups*, Proceedings of a Conference on Local Fields, Springer-Verlag, 1967, pp. 158–183.

[T2] ———, *A review of non-archimedean elliptic functions*, Elliptic Curves, Modular Forms, and Fermat's Last Theorem (Coates and Yau, eds.), International Press, 1995, pp. 162–184.

[W] Waterhouse, W.C., *Introduction to Affine Group Schemes*, Springer-Verlag, 1979.

THREE LECTURES ON THE MODULARITY OF $\bar{\rho}_{E,3}$ AND THE LANGLANDS RECIPROCITY CONJECTURE

STEPHEN GELBART

WILES' work on Fermat's Last Theorem is based on methods due to FALTINGS, FREY, LANGLANDS, MAZUR, RIBET, SERRE, TAYLOR, and others. My purpose in these Lectures is to explain how the (automorphic representation theoretic methods and) results of LANGLANDS come into the proof, and how these results themselves are proved. An Introduction to each of the Lectures describes more of the topics discussed; but the titles already speak for themselves:

Lecture I: "The Modularity of $\bar{\rho}_{E,3}$ and Automorphic Representations of Weight One"

Lecture II: "The Langlands Program: Some Results and Methods"

Lecture III: "Proof of the Langlands-Tunnell Theorem"

Acknowledgements

I am grateful to V. Berkovich, E. Lapid, K. Ribet, D. Rohrlich, K. Rubin, Z. Rudnick, and especially J. Tunnell and R. Livne, for helpful comments and suggestions. Special thanks are due to the attentive abilities of Miriam Abraham for the typing of this manuscript. The work on this paper was partly done through the support of the Minerva Foundation, Germany.

Lecture I
The Modularity of $\bar{\rho}_{E,3}$ and Automorphic Representations of Weight One

Abstract
The following result plays a small but key step in Wiles' proof of the Shimura-Taniyama-Weil Conjecture:

Proposition 1.4. *For an elliptic curve E over $\mathbb{Q}$, let*

$$\bar{\rho}_{E,p} : G_{\mathbb{Q}} \longrightarrow \mathrm{GL}_2(\mathbb{F}_p)$$

denote the natural representation of $G_{\mathbb{Q}} = \mathrm{Gal}(\bar{\mathbb{Q}}/\mathbb{Q})$ on the points of $E(\bar{\mathbb{Q}})$ of order p. Then if $p = 3$, and $\bar{\rho}_{E,3}$ is irreducible, it must also follow that $\bar{\rho}_{E,3}$ is modular, i.e., there exists a normalized eigen-cuspform

$$f(z) = \sum_{n=1}^{\infty} a_n e^{2\pi i n z}$$

of weight two, and a prime λ of $\bar{\mathbb{Q}}$ containing 3, such that

$$a_q \equiv \mathrm{trace}(\bar{\rho}_{E,3}(\mathrm{Fr}_q)) \pmod{\lambda}$$

for almost all primes q. (Fr_q *is explained below.)*

Our main purpose in this Lecture is to explain how this result follows from the following special case of Langlands' Reciprocity Conjecture for Artin L-functions:

Theorem 1.3. (cf. [La1] and [Tu]). *Suppose that the continuous representation*

$$\sigma : G_{\mathbb{Q}} \longrightarrow \mathrm{GL}_2(\mathbb{C})$$

is "odd," irreducible, and has solvable image in $\mathrm{PGL}_2(\mathbb{C})$. *(Here odd means that if τ denotes complex conjugation in $G_{\mathbb{Q}}$, then $\det(\sigma(\tau)) = -1$.) Then there exists a normalized eigen-cuspform*

$$g(z) = \sum_{n=1}^{\infty} b_n e^{2\pi i n z}$$

of weight one such that

$$b_q = \mathrm{trace}(\sigma(\mathrm{Fr}_q))$$

for all but finitely many primes q.

As we shall see, the *proof* of this result requires working not only over an arbitrary number field, but also with automorphic cuspidal *representations* (in place of classical cusp forms). Thus the second half of this Lecture will be devoted to recalling the basic representation theory required to reformulate Theorem 1.3 as follows:

Theorem 2.6. *For each irreducible representation*

$$\sigma : G_{\mathbb{Q}} \longrightarrow \mathrm{GL}_2(\mathbb{C})$$

which is odd and solvable, there is an automorphic "weight one" cuspidal representation of $\mathrm{GL}_2(\mathbb{A}_{\mathbb{Q}})$*, call it* $\pi(\sigma)$*, with the property that*

$$\mathrm{trace}(t_{\pi_q}) = \mathrm{trace}(\sigma(\mathrm{Fr}_q))$$

for almost every q*. (Here* t_{π_q} *denotes the Langlands class in* $\mathrm{GL}_2(\mathbb{C})$ *associated to the unramified local component* π_q *of* $\pi(\sigma) = \otimes\pi_p$*, and "weight one" means that* π_∞ *is the principal series representation of* $\mathrm{GL}_2(\mathbb{R})$ *induced from the characters 1 and* sgn.*)*

§1. The Modularity of $\bar{\rho}_{E,3}$

1.1. Galois Representations mod p

Let E denote a fixed elliptic curve defined over $\mathbb{Q}$. For a chosen prime p, let $E[p]$ denote the subgroup of $E(\bar{\mathbb{Q}})$ consisting of points of order p. Then $E[p] \cong \mathbb{F}_p^2$, regarded as a two-dimensional vector space over $\mathbb{F}_p$. The natural action of the Galois group

$$G_{\mathbb{Q}} = \mathrm{Gal}(\bar{\mathbb{Q}}/\mathbb{Q})$$

on $E[p]$ consequently gives rise to a continuous representation

$$\bar{\rho}_{E,p} : G_{\mathbb{Q}} \longrightarrow \mathrm{GL}_2(\mathbb{F}_p) \approx \mathrm{Aut}(E[p]),$$

which is uniquely defined up to its isomorphism class. That $\bar{\rho}_{E,p}$ encodes much of the arithmetic of E is clear from the two following crucial properties of $\bar{\rho}_{E,p}$:

(a) Write

$$\omega_p : G_{\mathbb{Q}} \longrightarrow \mathbb{F}_p^\times$$

for the character giving the action of $G_{\mathbb{Q}}$ on the p-th roots of unity μ_p. Then

$$\det(\bar{\rho}_{E,p}) = \omega_p; \tag{1.1.1}$$

this results from the existence of a "Weil pairing" $E[p] \times E[p] \longrightarrow \mu_p$, compatible with the action of $G_{\mathbb{Q}}$ and such that $\bigwedge^2_{\mathbb{F}_p}(E[p]) \approx \mu_p$ (cf. §V.2 of [Silv]).

(b) If q is any prime number, and Q is a prime of $\bar{\mathbb{Q}}$ dividing q, let Fr_q denote the canonical *Frobenius* conjugacy class in D_Q/I_Q (the quotient of the decomposition group at Q by the inertia group at Q). Then

$$\mathrm{trace}\ \bar{\rho}_{E,p}(\mathrm{Fr}_q) \equiv q + 1 - \#(E(\mathbb{F}_q)) \pmod p \tag{1.1.2}$$

for almost all primes q, namely those where $\bar{\rho}_{E,p}$ is trivial on (any) I_Q (i.e. those q where $\bar{\rho}_{E,p}$ is *unramified*).

N.B. (i) The invariants trace $\bar{\rho}_{E,p}(\mathrm{Fr}_q)$ (and also $\det \bar{\rho}_{E,p}(\mathrm{Fr}_q)$) are well-defined elements of $\mathbb{F}_p$ precisely when $\bar{\rho}_{E,p}$ is unramified at q.

(ii) The identity (1.1.2) essentially amounts to the Riemann hypothesis for elliptic curves over finite fields (proved by Hasse; cf. §V.2 of [Silv]).

(iii) Alternatively, the primes q for which (1.1.2) holds can be characterized as those which are different from p and such that E has "good reduction mod$q$." Equivalently, let K be the kernel of $\bar{\rho}_{E,p}$, and $\bar{\mathbb{Q}}^K := \mathbb{Q}(E[p])$ the corresponding finite Galois extension of $\mathbb{Q}$; then

$$\mathrm{Gal}(\mathbb{Q}(E[p])/\mathbb{Q}) \approx \mathrm{Im}\, \bar{\rho}_{E,p},$$

and (1.1.2) holds exactly for those q which are unramified in $\mathbb{Q}(E[p])$ (equivalently, those q such that $\bar{\rho}_{E,p}$ is trivial on I_q).

1.2. The Modularity of $\bar{\rho}_{E,p}$

Let $S_k(\Gamma_0(N), \varepsilon)$ denote the vector space of modular cusp forms $f(z)$ of weight $N \geq 1$ and character $\varepsilon : (\mathbb{Z}/N\mathbb{Z})^\times \longrightarrow \mathbb{C}^\times$.

Definition. We call $\bar{\rho}_{E,p}$ *modular* if there exists some (normalized) eigenform

$$f(z) = \sum_{n=1}^{\infty} a_n e^{2\pi i n z} \in S_2(\Gamma_0(N), \varepsilon)$$

(for some N and ε), and a prime λ of $\bar{\mathbb{Q}}$ containing p, such that

$$a_q \equiv q + 1 - \#(E(\mathbb{F}_q)) \pmod{\lambda}$$

for almost all primes q.

Recall that Wiles' goal was to prove that E itself is *modular*, i.e., for some weight two $f(z)$ as above, the *identity* (as opposed to congruence)

$$a_q = q + 1 - \#(E(\mathbb{F}_q))$$

holds for almost all q. As discussed elsewhere, what Wiles actually proves is Mazur's "Modular Lifting Conjecture":

If p is a prime such that
(i) $\bar{\rho}_{E,p}$ is irreducible, and
(ii) $\bar{\rho}_{E,p}$ is modular,
THEN E ITSELF IS MODULAR.

More precisely, Wiles proves that (a) the Modular Lifting Conjecture is true for $p = 3$ and 5 when E is a semistable elliptic curve, and (b) the Modular Lifting Conjecture for $p = 3$ and 5 already implies the Taniyama-Shimura-Weil Conjecture (that E is modular).

Our modest goal is to explain how the theory of automorphic forms is used to prove that for $p = 3$, the second hypothesis of the Modular Lifting Conjecture automatically follows from the first, i.e., if $\bar{\rho}_{E,3}$ is irreducible, then it is modular.

1.3. The Theorem of Langlands-Tunnell

The crucial ingredient in proving the modularity of $\bar{\rho}_{E,3}$ is the following:

Theorem 1.3. (cf. [La1] and [Tu]) *Suppose*

$$\sigma : G_{\mathbb{Q}} \longrightarrow \mathrm{GL}_2(\mathbb{C})$$

is a continuous, irreducible two dimensional representation whose image in $\mathrm{PGL}_2(\mathbb{C})$ *is a solvable group. Suppose moreover that* σ *is "odd" in the sense that*

$$\det(\sigma(\tau)) = -1.$$

(τ is an automorphism in $G_{\mathbb{Q}}$ defined by complex conjugation.) Then there exists a (normalized)

$$g(z) = \sum_{n=1}^{\infty} b_n e^{2\pi i n z} \in S_1(\Gamma_0(N), \psi)$$

(for some N and ψ), such that f is an eigenform for all the Hecke operators, and

$$b_q = \mathrm{trace}(\sigma(\mathrm{Fr}_q)) \tag{1.3.1}$$

for almost all primes q.

Remarks. (1) Because *any* continuous representation

$$\sigma : G_{\mathbb{Q}} \longrightarrow \mathrm{GL}_2(\mathbb{C})$$

factors through some finite Galois group $\mathrm{Gal}(K/\mathbb{Q})$, its image in $\mathrm{GL}_2(\mathbb{C})$ is finite, and its image *in* $\mathrm{PGL}_2(\mathbb{C})$ is just one of the symmetry groups of a regular polyhedron in $\mathbb{R}^3$ (cf. section 13 of [Shaf]). From this it is deduced that the image of any *irreducible* σ in $\mathrm{PGL}_2(\mathbb{C})$ is either A_5 (the *icosahedral* case), S_4 (the *octahedral* case), A_4 (the *tetrahedral* case), or D_{2n} (the *dihedral* case). As we shall recall in §5.3, in the *dihedral* case the existence of the required weight one from $g(z)$ above is essentially due to much earlier work of Hecke and Maass. Hence in dealing with "solvable" σ, the theorem of Langlands and Tunnell is ultimately concerned with "just" the tetrahedral and octahedral possibilities.

(2) The relevant theorems of [La] and [Tu] do not actually produce the required modular form $g(z)$, but rather a certain automorphic representation $\pi(\sigma)$. Using the fact that $\det(\sigma(\tau)) = -1$, we shall explain in §4.2 of Lecture II how this automorphic representation produces $g(z)$ itself. In the meantime, we take the above theorem as given, and use it to prove the modularity of $\bar{\rho}_{E,3}$.

1.4. Proof of the Modularity of $\bar{\rho}_{E,3}$

More precisely, we need to prove:

Proposition 1.4. *If $\bar{\rho}_{E,3}$ is irreducible, then it is modular, i.e., there exists a normalized eigenform*

$$f(z) = \sum_{n=1}^{\infty} a_n e^{2\pi i n z}$$

of weight two, and a prime λ of $\bar{\mathbb{Q}}$ containing 3, such that

$$a_q \equiv q + 1 - \#E(\mathbb{F}_q) \pmod{\lambda}$$

for almost all primes q.

The *strategy of proof* is simple. First one "lifts" $\bar{\rho}_{E,3}$ to a complex representation $\sigma : G_{\mathbb{Q}} \longrightarrow \mathrm{GL}_2(\mathbb{C})$ to which the Theorem of Langlands-Tunnell is applicable; this produces a modular form $g(z)$ of weight *one* whose Fourier coefficients b_q are almost everywhere equal to trace(Fr_q). Then one multiplies g by an Eisenstein series of weight one, whose (non-trivial) Fourier coefficients are all congruent to 0 mod 3; this essentially produces the required form of weight *two* whose Fourier coefficients are *congruent* to trace(Fr_q) modulo some divisor of 3 (and hence also congruent to $q + 1 - \#E(\mathbb{F}_q)$, by virtue of (1.1.2)).

Because of the importance of Proposition 1.4, we shall go through its proof carefully (expanding on the single paragraph allotted it in Chapter V of [W1]). We note that the idea of applying "Langlands-Tunnell" in this context goes back to Serre (cf. [Se], §5.3, page 220).

Step 1. Extend $\bar{\rho}_{E,3} : G_{\mathbb{Q}} \longrightarrow \mathrm{GL}_2(\mathbb{F}_3)$ to a complex representation

$$\sigma : G_{\mathbb{Q}} \longrightarrow \mathrm{GL}_2(\mathbb{C})$$

by composing $\bar{\rho}_{E,3}$ with a specific (injective) homomorphism

$$\Psi : \mathrm{GL}_2(\mathbb{F}_3) \hookrightarrow \mathrm{GL}_2(\mathbb{Z}(\sqrt{-2})) \subset \mathrm{GL}_2(\mathbb{C})$$

described below.

Following [RuSi], we introduce Ψ directly through the formulas

$$\Psi\begin{pmatrix} -1 & 1 \\ -1 & 0 \end{pmatrix} = \begin{pmatrix} -1 & 1 \\ -1 & 0 \end{pmatrix}$$

and

$$\Psi\begin{pmatrix} 1 & -1 \\ 1 & 1 \end{pmatrix} = \begin{pmatrix} 1 & -1 \\ -\sqrt{-2} & -1+\sqrt{-2} \end{pmatrix}.$$

Here $\alpha = \begin{pmatrix} -1 & 1 \\ -1 & 0 \end{pmatrix}$ and $\beta = \begin{pmatrix} 1 & -1 \\ 1 & 1 \end{pmatrix}$ are two convenient generators of $\mathrm{GL}_2(\mathbb{F}_3)$. Once it is checked that the above formulas indeed preserve the required relations, it is immediately seen that the resulting homomorphism

$$\Psi : \mathrm{GL}_2(\mathbb{F}_3) \longrightarrow \mathrm{GL}_2(\mathbb{Z}(\sqrt{-2})) \subset \mathrm{GL}_2(\mathbb{C})$$

is the identity upon reduction mod$(1+\sqrt{-2})$. In particular,

$$\text{trace}(\Psi(g)) \equiv \text{trace}(g) \pmod{1+\sqrt{-2}} \tag{1.4.1}$$

and

$$\det(\Psi(g)) \equiv \det(g) \; \big(\text{mod } 3 = (1+\sqrt{-2})(1-\sqrt{-2})\big). \tag{1.4.2}$$

N.B. This representation

$$\Psi : \mathrm{GL}_2(\mathbb{F}_3) \longrightarrow \mathrm{GL}_2(\mathbb{C})$$

is really just one of the (three) so-called *cuspidal* representations of the group $\mathrm{GL}_2(\mathbb{F}_3)$; compare, for example, [PS1] §10.

Step 2. Check that

$$\sigma = \Psi \circ \bar\rho_{E,3} : G_{\mathbb{Q}} \longrightarrow \mathrm{GL}_2(\mathbb{C})$$

is "odd," irreducible and solvable.

Let us first check that $\bar\rho_{E,3}$ itself has odd determinant. On the one hand, (1.1.1) implies

$$\det(\rho_{E,3}(\tau)) = \omega_3(\tau),$$

and it is clear that $\omega_3(\tau) = -1$. On the other hand, $\det \sigma(\tau)$ is *a priori* ± 1, since $\tau^2 = 1$, and (1.4.2) implies $\det(\sigma(\tau)) \equiv \det \bar\rho_{E,3}(\tau) \pmod 3$. So since $-1 \not\equiv 1 \pmod 3$, we must have $\det(\sigma(\tau)) = -1$, as required. As for the "solvable" assertion, just recall that

$$\mathrm{PGL}_2(\mathbb{F}_3) \approx S_4;$$

this says that the image of $\sigma = \Psi \circ \rho_{E,3}$ in $\mathrm{PGL}_2(\mathbb{C})$ is a subgroup of S_4, hence itself solvable.

Now what about irreducibility? From the fact that $\det \bar\rho_{E,3}(\tau) = -1$, it follows that $\bar\rho_{E,3}$ has distinct eigenvalues in $\mathbb{F}_3$ (namely 1 and -1). We claim this implies $\bar\rho_{E,3}$ is *absolutely* irreducible, i.e., irreducible over $\overline{\mathbb{F}}_3$ as well as $\mathbb{F}_3$. Indeed, the only matrices in $M_2(\overline{\mathbb{F}}_3)$ which can commute with $\bar\rho_{E,3}(G_{\mathbb{Q}})$ (in particular $\begin{pmatrix} 1 & 0 \\ 0 & -1 \end{pmatrix}$ *and* some non-diagonal matrix $\bar\rho_{E,3}(g)$) are the scalar matrices λI themselves. Hence by Schur's Lemma, $\bar\rho_{E,3}$ is absolutely irreducible.

Now suppose that the *complex* representation $\sigma = \Psi \circ \bar\rho_{E,3}$ is *not* irreducible. We claim this implies its image in $\mathrm{GL}_2(\mathbb{C})$ must be abelian. Indeed, any complex representation of a finite (or compact) group is completely reducible. In the case of σ, this means σ is the sum of two characters, and this clearly implies that its image in $\mathrm{GL}_2(\mathbb{C})$ is abelian.

On the other hand, as $\bar\rho_{E,3}$ is absolutely irreducible, the only matrices commuting with its image in $\mathrm{GL}_2(\overline{\mathbb{F}}_3)$ must be scalar ones (again by

Schur's Lemma). So pulling back through the embedding Ψ, we conclude $\bar{\rho}_{E,3}$ has both an abelian and irreducible image in $\mathrm{GL}_2(\overline{\mathbb{F}}_3)$, an obvious contradiction. Thus $\sigma = \Psi \circ \bar{\rho}_{E,3}$ must after all be irreducible.

Step 3. Apply Theorem 1.3 to get a normalized eigenform

$$g(z) = \sum_{n=1}^{\infty} b_n e^{2\pi i n z} \quad \text{in some} \quad S_1(\Gamma_0(N_1), \varepsilon_1)$$

with

$$b_q = \mathrm{trace}(\sigma(\mathrm{Fr}_q)) \quad \text{for almost all primes } q. \tag{1.4.3}$$

Remark. Recall that for any normalized new form of weight $k \geq 1$ (and character ψ), the Fourier coefficients a_n (together with the values $\psi(n)$) lie in the ring of integers O_K of some number field K (of finite degree over $\mathbb{Q}$). In particular, for our form $g(z)$ above, it makes sense to discuss congruence conditions on the coefficients b_n modulo a prime ideal $\mathfrak{p}$ of (the appropriate) O_K.

Now recall that (1.4.1) implies

$$\begin{aligned}\mathrm{trace}(\sigma(\mathrm{Fr}_q)) &= \mathrm{trace}(\Psi \cdot \sigma(\mathrm{Fr}_q)) \\ &\equiv \mathrm{trace}(\bar{\rho}_{E,3}(\mathrm{Fr}_q)) \pmod{1+\sqrt{-2}}\end{aligned}$$

So by (1.1.2), (1.4.3), and the above Remark, we have

$$b_q \equiv q + 1 - \#E(\mathbb{F}_q) \pmod{\mathfrak{p}} \tag{1.4.4}$$

for almost every q, and $\mathfrak{p}$ some prime of $\bar{\mathbb{Q}}$ containing $(1+\sqrt{-2})$ (and hence 3). In other words, we've proven that $\bar{\rho}_{E,3}$ is modular, but with the *hitch* that $\sum_{n=1}^{\infty} b_n e^{2\pi i n z}$ is of weight *one* instead of *two*! To remedy this, we follow two ideas, going back respectively to Shimura, and Deligne-Serre.

Step 4. Pick a modular (non-cuspidal) form E of weight 1, such that $E \equiv 1$ (3); the product

$$g(z)E(z) = \sum_{n=1}^{\infty} c_n e^{2\pi i n z}$$

is of weight 2 (for some level N and character ψ), and

$$c_n \equiv b_n \pmod{\mathfrak{p}}$$

(for $\mathfrak{p}$ a prime of $\bar{\mathbb{Q}}$ lying above 3...).

More explicitly, take

$$E(z) = E_{1,\chi}(z) = 1 + 6 \sum_{n=1}^{\infty} \sum_{d|n} \chi(d) e^{2\pi i n z}$$

where

$$\chi(d) = \begin{cases} 0 & \text{if } d \equiv 0 \pmod 3 \\ 1 & \text{if } d \equiv 1 \pmod 3 \\ -1 & \text{if } d \equiv -1 \pmod 3 \end{cases}$$

is an *odd* Dirichlet character mod 3 (i.e., $\chi(-1) = -1$). Then $E_{1,\chi} \in M_1(\Gamma_0(3), \chi)$, and $gE_{1,\chi}$ belongs to $S_2(\Gamma_0(3N_1), \varepsilon_1\chi)$. Furthermore, each "non-constant" Fourier coefficient of $E_{1,\chi}$ is divisible by 3 (in fact 6), so it easily follows that

$$c_n \equiv b_n \pmod{\mathfrak{p}}.$$

Note that $g(z)E(z)$ is the *product* of a normalized eigenform with an Eisenstein series, but not itself such an eigenform. If it were, we would (by (1.4.4)) have completed our task of proving $\bar{\rho}_{E,3}$ modular. To finish the job, we need the following result of Deligne and Serre:

Lemma. (cf. §6.10 of [DS]) *Suppose*

$$f_1(z) = \sum_{n=1}^{\infty} c_n e^{2\pi i n z}$$

is a normalized element of $S_k(\Gamma_0(N), \psi)$, and K is a finite extension of $\mathbb{Q}$ whose ring of integers contains the coefficients c_n and $\psi(n)$. Suppose $\mathfrak{p}$ is a prime ideal of O_K containing 3, and that f_1 is a mod $\mathfrak{p}$ eigenform, i.e., there exists b_n such that $T_n f_1 - b_n f_1 \equiv 0 \pmod{\mathfrak{p}}$ for all n. Then there exists an f in $S_k(\Gamma_0(N), \psi)$, and d_n, such that for all n,

$$T_n f = d_n f,$$

and

$$d_n \equiv c_n \pmod{\mathfrak{p}'}$$

for some prime $\mathfrak{p}'$ dividing $\mathfrak{p}$.

We want to apply this Lemma to our modular form $g(z)E_1(z)$ of weight 2. Since the constant term in the Fourier expansion of $E_1(z)$ is 1, this gE is indeed normalized, i.e., $c_1 = 1$. Let K be its corresponding number field with prime ideal $\mathfrak{p}$ (dividing 3). Since $E \equiv 1 \pmod 3$, we have

$$T_n f_1 \equiv b_n f_1 \pmod{\mathfrak{p}}$$

(since $T_n g = b_n g$) for all n. Thus the Lemma is indeed applicable (taking $f_1 = g(z)E_1(z)$), and it produces a normalized form $f \in S_2(\Gamma_0(N), \psi)$ such that $T_p f = a_p f$ for all p, and $a_p \equiv c_p \equiv b_p \pmod{p}$. In particular, for almost all q,

$$a_q \equiv q + 1 - \#E(\mathbb{F}_q) \pmod{\mathfrak{p}'},$$

as required.

§2. Automorphic Representations of Weight One

2.1. For σ an irreducible, "odd," solvable two-dimensional representation of $G_{\mathbb{Q}}$, the relevant results of Langlands and Tunnell do not directly produce the required modular form $g(z)$, but rather — as already suggested — a certain automorphic cuspidal representation $\pi(\sigma)$, which is shown to *correspond* to such a form $g(z)$. An honest explanation of how this works requires (at least) the exposition of representation theory given in the pages below. Roughly speaking, a classical eigen-cusp form g, of weight one, and fixed level and character, generates a *collection* of irreducible representations

$$\{\pi_p\}_{p\leq\infty}$$

of the "local" groups $\mathrm{GL}_2(\mathbb{Q}_p)$ — each of which is uniquely determined by the data attached to $g(z)$; moreover this collection comprises what is called an *automorphic cuspidal representation* π of the adele group $\mathrm{GL}_2(\mathbb{A}_{\mathbb{Q}})$.

Once the notion of "automorphic form" is liberated from its classical (upper half-plane) setting, it seems completely natural to take the further step of replacing $\mathbb{Q}$ by an arbitrary number field F, and GL_2 by any GL_n, $n \geq 1$. Thus one ultimately views "Dirichlet characters mod n," Hecke characters of a number field, classical modular forms, and even "Maass cusp forms" (cf. Remark 2.5.5) as manifestations of one and the same kind of global object, namely, an *automorphic representation of* GL_n *over a number field.* It is this language which Langlands used to formulate the following general *Langlands Reciprocity Conjecture* (LRC): for any n-dimensional representation σ of $\mathrm{Gal}(\bar{F}/F)$ (or more generally the *Weil group* W_F), there is a corresponding automorphic representation

$$\pi(\sigma) = \otimes_v \pi_v$$

of $\mathrm{GL}_n(\mathbb{A}_F)$ such that for almost all the primes v of F,

$$\operatorname{trace}\sigma(\mathrm{Fr}_v) = \operatorname{trace}(t_{\pi_v})$$

(with t_{π_v} the Langlands class of π_v in $\mathrm{GL}_n(\mathbb{C})$); moreover, if σ is irreducible, then $\pi(\sigma)$ is *cuspidal.*

Our main goal in Lectures II and III will be to describe the ideas behind the statement and the *proof* of this "Strong Artin Conjecture" in case $n = 2$ and σ is solvable, namely:

Theorem 2.1. *To each irreducible*

$$\sigma : W_F \longrightarrow \mathrm{GL}_2(\mathbb{C})$$

with solvable image in $\mathrm{PGL}_2(\mathbb{C})$, *there corresponds an automorphic cuspidal representation* $\pi(\sigma) = \otimes\pi_v$ *of* $\mathrm{GL}_2(\mathbb{A}_F)$ *with the property that (its central character equals* $\det\sigma$ *and)*

$$\operatorname{trace}(t_{\pi_v}) = \operatorname{trace}\sigma(\mathrm{Fr}_v) \quad \textit{for almost every prime } v \textit{ of } F.$$

Our more modest goal in the rest of this Lecture is to explain how Theorem 2.1 gives the required classical result (Theorem 1.3) *when* $F = \mathbb{Q}$, σ *factors through the Galois group* $G_{\mathbb{Q}}$, *and* $\det(\sigma)$ *is odd.* One step is to show that for such σ, the corresponding $\pi(\sigma)$'s are "automorphic representations of $GL_2(\mathbb{A}_{\mathbb{Q}})$ of weight one." This step will actually be postponed until we discuss the correspondence $\sigma \longrightarrow \pi(\sigma)$ in earnest in Lecture II; cf. Proposition 4.2. A second step is to show there is a one-one correspondence between these "automorphic representations of weight one" and the classical new forms of weight one required in Theorem 1.3 (cf. Proposition 2.5).

2.2. Archimedean Representation Theory (GL_2)

Let $G = G_\infty$ denote $GL_2(\mathbb{R})$, and $\mathfrak{g}$ its complexified Lie algebra. Let $K = K_\infty = O(2, \mathbb{R})$ denote the real 2×2 orthogonal group, a maximal compact subgroup of G.

Definition. An *admissible representation* of G on a Hilbert space H is a homomorphism

$$\pi : G \longrightarrow GL(H)$$

such that (i) the map $(g, v) \longrightarrow \pi(g)v$ from $G \times H \longrightarrow H$ is continuous, and (ii) ("admissibility") the restriction of π to K contains each irreducible unitary representation of K with *finite* multiplicity (recall that each such representation is automatically *finite* dimensional).

Remark. For an admissible representation $\pi : G \longrightarrow \mathrm{Aut}(H)$, let $V = V_K \subset H$ denote the subspace of *K-finite vectors*, i.e., those v in H whose translates under $\pi(K)$ span a finite-dimensional space. Such vectors are *not* preserved by the action of $\pi(G)$, but they are by the corresponding (differentiated) action of the *Lie algebra* $\mathfrak{g}$ *of* G. In fact, as a representation space jointly for the action of $\mathfrak{g}$ and K, V enjoys certain "compatibility" properties which ensure that it is (what's called) a $(\mathfrak{g}, K)$-module. The advantage of these modules is that they are "algebraic" linear objects as opposed to "analytic" Lie group theoretic objects, and yet they accurately reflect the nature and properties of π. For example, π is *irreducible* in the usual sense (that H has no *Hilbert space* subspaces invariant under $\pi(G)$) if and only if V is irreducible in the algebraic sense that it has no *vector space* subspaces invariant under both $\pi(K)$ and $d\pi(\mathfrak{g})$). This leads to an equivalence between the natural categories of irreducible admissible representations of G and irreducible $(\mathfrak{g}, K)$ modules. In particular, whenever convenient, we allow ourselves to confuse π with the $(\mathfrak{g}, K)$ module V.

Example 2.1. Let μ_1 and μ_2 denote two characters of $\mathbb{R}^\times$ (i.e., *not* necessarily "unitary" homomorphisms from $\mathbb{R}^\times$ to $\mathbb{C}^\times$). Let $H = H(\mu_1, \mu_2)$ denote the Hilbert space of functions $f : G \longrightarrow \mathbb{C}$ such that

$$f\left(\begin{pmatrix} a_1 & x \\ 0 & a_2 \end{pmatrix} g\right) = \left| \frac{a_1}{a_2} \right|^{1/2} \mu_1(a_1)\mu_2(a_2)f(g)$$

for all $\begin{pmatrix} a_1 & x \\ 0 & a_2 \end{pmatrix}$ in the Borel subgroup B consisting of upper triangular matrices of G, and such that

$$\|f\|^2 = \int_K |f(k)|^2 dk < \infty .$$

(Strictly speaking, one first looks at *continuous* such f, and then defines $H(\mu_1, \mu_2)$ to be the closure of such functions with respect to the norm $\|f\|$....) Then the right regular action of G on functions f in $H(\mu_1, \mu_2)$ defines an *admissible* representation of G which is denoted $\pi(\mu_1, \mu_2)$ and called the representation of G *induced from the character* $\mu_1\mu_2$ (of B). The corresponding $(\mathfrak{g}, K)$ module V_K consists of finite linear combinations of smooth functions ϕ_k defined by

$$\phi_k(g) = \mu_1(a_1)\mu_2(a_2)e^{-ik\theta}$$

if g has "Iwasawa decomposition"

$$g = \begin{pmatrix} a_1 & x \\ 0 & a_2 \end{pmatrix} r(\theta).$$

(Here $r(\theta)$ denotes the rotation element $\begin{pmatrix} \cos\theta & -\sin\theta \\ \sin\theta & \cos\theta \end{pmatrix}$ in K, and there is an obvious compatibility condition coming from the elements $\pm I$ in $B \cap K$, namely $\mu_1\mu_2(-1) = e^{-ik\pi}$.) From this picture it is a simple matter to prove that $\pi(\mu_1, \mu_2)$ is *irreducible* if and only if $\mu_1\mu_2^{-1}(x) \neq x^p \operatorname{sgn}(x)$ with p a non-zero integer. Slightly less transparent is the proof of the following:

Fact. *Every* irreducible admissible representation π of G is (equivalent to) a $\pi(\mu_1, \mu_2)$, or an irreducible subrepresentation thereof. For example, suppose $\mu_1\mu_2^{-1}(x) = x^p \operatorname{sgn}(x)$ with p a positive integer. Then $H(\mu_1, \mu_2)$ contains exactly one *invariant* subspace, namely

$$\{\ldots, \phi_{-p-3}, \phi_{-p-1}, \phi_{p+1}, \phi_{p+3}, \ldots\},$$

and the restriction of $\pi(\mu_1, \mu_2)$ to this subspace realizes an *irreducible discrete series representation* of "lowest weight $p+1$."

Concluding Remarks. (a) All the above notions, and most of the results, hold for an arbitrary semisimple or reductive Lie group; in particular, every irreducible admissible representation of such a group is still realizable as a subrepresentation of the analogous induced representation. (The theory for $\mathrm{GL}_2(\mathbb{R})$ is essentially Bargmann's, and the general theory essentially due to Harish-Chandra; for many more details, see the recent survey paper of Knapp [Kn].)

(b) For any irreducible admissible representation π, an analogue of Schur's Lemma implies that there is a character of $\mathbb{R}^\times$, denoted ω_π and called *the central character of* π, such that

$$\pi\begin{pmatrix} r & 0 \\ 0 & r \end{pmatrix} = \omega_\pi(r)I \quad \text{for all } r \in \mathbb{R}^\times.$$

2.3. *P*-adic Representation Theory (GL_2)

In this Section, F is a p-adic field with a ring of integers O_F, and $G = GL_2(F)$. As in the real case, *most* of the notions and facts we review below extend not only to GL_n, but to an arbitrary reductive p-adic group; however, we recall here only those facts (even for GL_2) which are really needed in the sequel.

Definition 2.3.1. An *admissible representation* of G on a *vector space* V is a homomorphism

$$\pi : G \longrightarrow GL(V)$$

such that (i) the stabilizer in G of any v in V is open, and (ii) ("admissibility") for any compact open subgroup $K^0 \subset G$, the space

$$\{v \in V : \pi(k)v = v \text{ for all } k \in K^0\}$$

is *finite*-dimensional.

Remark. Suppose π is an *irreducible unitary* representation of G in some Hilbert space H (same definition as in the case of Lie groups), and V_K is its subspace of K-finite vectors (for any compact open K, say $K = GL_2(O_F)$). Then (by a Theorem of J. Bernstein) the restriction of $\pi(G)$ to $V = V_K$ produces an *admissible* representation of G in the above sense. Thus the p-adic notion of an admissible representation (on a vector space V) is a natural analogue of the archimedean notion of a $(\mathfrak{g}, K)$ module. What's special in the p-adic case is that G itself acts on V_K (and there is no need to go to the Lie algebra...).

Example (of $\pi(\mu_1, \mu_2)$). For each pair of characters $\mu_i : F^\times \longrightarrow \mathbb{C}^\times$, the induced representations $\pi(\mu_1, \mu_2)$ in $H(\mu_1, \mu_2)$ are defined just as in the archimedean case. But now reducibility is possible *only* if $\mu_1\mu_2^{-1}(x) = |x|^{\pm 1}$, i.e., there is no room for many "discrete series" representations to appear as subrepresentations of $\pi(\mu_1, \mu_2)$. This reflects the fact that there are now representations which are absolutely *cuspidal*, i.e., they cannot be constructed in this simple way.

Fortunately, we shall not need to discuss the cuspidal representations in these Lectures, but rather only those representations which are as far from being cuspidal as possible!

Definition 2.3.1. An irreducible admissible representation π of G is *unramified* (or *of conductor zero*) if its space of $K = GL_2(O_F)$ fixed vectors is non-empty.

In this case, it is known that the space of K-fixed vectors is *one-dimensional*, and that π is either one-dimensional (of the form $\chi \circ \det$, for some *unramified* character χ of $F^\times$), or an irreducible $\pi(\mu_1, \mu_2)$ with μ_1 and μ_2 unramified characters of $F^\times$. As we shall recall in the next paragraph, it is these latter representations which typically play a crucial role in the adelic theory of automorphic representations.

Concluding Remark. Even if some π does not have "conductor zero," there will still exist a smallest (positive) integer N (called the *conductor of* π) such that the space

$$\{v \in V : \pi \begin{pmatrix} a & b \\ c & d \end{pmatrix} v = \omega_\pi(a)v \text{ for all } k \in K^N\}$$

is non-empty. (Here K^N denotes the Hecke congruence subgroup of K consisting of $\begin{pmatrix} a & b \\ c & d \end{pmatrix}$ with $c \equiv 0 \pmod{\mathfrak{p}^N}$.) This fact was first proved in [Cas], and then generalized to GL_n in [J-PS-S1]; if $N = \text{conductor}(\pi)$, then it is also known (as in the case $N = 0$) that the space V^{K^N} is automatically one-dimensional.

2.4. Adelic Representations (GL_2)

This is mostly a matter of putting together the local representation theory.

Suppose we are given a collection of irreducible admissible representations $\{\pi_p\}$ of the local groups $G_v = \mathrm{GL}_2(\mathbb{Q}_p)$, such that π_p is unramified for all but finitely many p. Then the *restricted* tensor product

$$\bigotimes_{p \leq \infty} \pi_p$$

makes sense as a representation of the restricted product $\mathrm{GL}_2(\mathbb{A}_\mathbb{Q}) = \prod'_{p \leq \infty} \mathrm{GL}_2(\mathbb{Q}_p)$, and defines an irreducible "admissible" representation π in a sense that can be made precise; cf. §4.c of [Gel] or §9 of [JL] for more details. Conversely, any irreducible "admissible" (and in particular any irreducible *unitary*) representation π of $\mathrm{GL}_2(\mathbb{A}_\mathbb{Q})$ is uniquely factorizable as

$$\pi = \otimes \pi_p,$$

with almost every π_p unramified. Moreover, the obvious analogues of these statements hold for an arbitrary number field F.

Example $G = \mathrm{GL}_1$. In this case, we consider a collection of *characters* χ_v of $F_v^\times$, one for each prime of F, such that almost all the χ_v's are unramified, i.e., trivial on $O_v^\times$. Then

$$\chi = \prod_v \chi_v$$

defines a nice (continuous) character of the ideles $\mathbb{A}_F^\times = \mathrm{GL}_1(\mathbb{A}_F)$, and every such character thus arises (cf. Proposition 7-1-12 of [Gold]). Of course, in number theory one is primarily concerned with characters χ of $\mathbb{A}_F^\times$ which are trivial on $F^\times$, i.e., with "grossencharacters," or characters of the *idele class group*

$$\mathrm{GL}_1(F)\backslash \mathrm{GL}_1(\mathbb{A}_F^\times) = F^\times\backslash \mathbb{A}^\times.$$

These are the "automorphic representations" of GL(1) over F.

Special Example. Since $\mathbb{Q}$ has class number one,

$$(*) \qquad \mathbb{A}_\mathbb{Q}^\times = \mathbb{Q}^\times \cdot \mathbb{R}^+ \cdot \prod_{p<\infty} \mathbb{Z}_p^\times .$$

Thus a *Dirichlet character* $\psi : (\mathbb{Z}/N\mathbb{Z})^\times \longrightarrow \mathbb{C}^\times$ determines a *grossencharacter* χ_ψ as follows: for any $p < \infty, \psi$ can be pulled back to a character χ_p of $\mathbb{Z}_p^\times$ through the canonical homomorphism from $\mathbb{Z}_p^\times$ to $(\mathbb{Z}/N\mathbb{Z})^\times$; the product $\prod_{p<\infty} \chi_p$ then defines a character of $\prod_{p<\infty} \mathbb{Z}_p^\times$, and hence by (*) a character χ_ψ of $\mathbb{A}_\mathbb{Q}^\times$ *trivial on* $\mathbb{R}^+$ *as well as* $\mathbb{Q}^\times$. In this way one obtains a grossencharacter of $\mathbb{A}_\mathbb{Q}^\times$ of *finite order*, and all such grossencharacters arise in this way for suitably large N. (Note that χ_p is unramified for the primes p not dividing N.)

Following this lead for GL_1, it is clear that one should define *automorphic representations* for $G = \mathrm{GL}_n$ in terms of the quotient space

$$\mathrm{GL}_n(F) \backslash \mathrm{GL}_n(\mathbb{A}).$$

For simplicity, we give a precise definition only for $G = \mathrm{GL}_2$ and *cuspidal* automorphic representations.

Definition. Fix a grossencharacter ω of F, and let $L_0^2(G(F) \backslash G(\mathbb{A}), \omega)$ denote the (closure of the) space of all continuous $\varphi : G(\mathbb{A}) \longrightarrow \mathbb{C}$ such that

(i) $\varphi(\gamma g z) = \omega(z)\varphi(g)$

for all $\gamma \in G(F)$ and $z \in Z(\mathbb{A}) = \left\{\begin{pmatrix} z & 0 \\ 0 & z\end{pmatrix}\right\}$ (the center...);

(ii) $$\int_{Z(\mathbb{A})G(F)\backslash G(\mathbb{A})} |\varphi(g)|^2 dg < \infty \,; \qquad \text{and}$$

(iii) φ is *cuspidal*, i.e., for any g in $G(\mathbb{A})$,

$$\int_{F\backslash \mathbb{A}} \varphi\left(\begin{pmatrix} 1 & x \\ 0 & 1\end{pmatrix} g\right) dx = 0.$$

Then an irreducible admissible (necessarily unitary) representation $\pi = \otimes\pi_v$ of $G(\mathbb{A})$ is called *automorphic cuspidal* if there is some ω such that π is equivalent to an irreducible summand of the right regular representation of $G(\mathbb{A})$ in $L_0^2(G(F) \backslash G(\mathbb{A}), \omega)$.

Remarks. (i) In a completely similar way, the notion of an automorphic cuspidal representation π can be defined for $G = \mathrm{GL}_n$, and more generally, for an arbitrary reductive algebraic group G. In this generality, it is a Theorem of Gelfand and Piatetski-Shapiro (cf. [GGPS]) that the right regular representation R_0 of $G(\mathbb{A})$ in $L_0^2(G(F) \setminus G(\mathbb{A}), \omega)$ decomposes *discretely*, and with finite multiplicities, i.e.,

$$R_0 = \bigoplus m_\pi \pi, \quad \text{with} \quad m_\pi < \infty.$$

For $G = \mathrm{GL}_n$, it is actually known that this multiplicity is *one* (cf. [JL], [GK], and [Shal]), but for other groups this need no longer hold. For general G, it is at least still true that any automorphic cuspidal representation has a factorization of the type

$$\pi = \otimes \pi_v$$

discussed above, with π_v almost everywhere an unramified representation of $G(F_v)$.

(ii) There is also the notion of an *automorphic* (*not* necessarily cuspidal) representation of $G(\mathbb{A})$, but as we shall not focus on these in the sequel, we refrain from giving a precise definition. Suffice it to say that such π are the irreducible admissible representations of $G(\mathbb{A})$ which are built out of cuspidal representations of the Levi components of G by way of *induction* (and taking of quotients); for example, for GL_2, the automorphic (non-cuspidal) representations are the quotients of the induced representations $\pi(\mu_1, \mu_2)$ with μ_1 and μ_2 grossencharacters of F (viewed as "cuspidal representations" of the diagonal subgroup of GL_2). For a thorough discussion, see [BoJa] and [La2].

2.5. A Dictionary (Between the Classical and Modern Theories for GL$_2$)

In the last paragraph, we recalled how classical Dirichlet characters correspond to certain automorphic representations of GL(1), namely those grossencharacters which are of finite order. We now describe an analogous result for GL(2).

Proposition. *There is a 1-1 correspondence between normalized new forms*

$$f(z) = \sum_{n=1}^{\infty} a_n e^{2\pi i n z}$$

in $S_k(\Gamma_0(N), \psi)$, and irreducible automorphic cuspidal representations

$$\pi = \otimes_p \pi_p$$

of $G(\mathbb{A})$ such that:

(a) the central character of π is χ_ψ,

(b) π_p is unramified for all $p \nmid N$, and

(c) π_∞ has "lowest weight k."
Moreover, if $N = \prod p_i^{\alpha_i}$, and $p|N$, then π_p has conductor α_i; but if $p\nmid N$, then $\pi_p = \pi(\mu_1, \mu_2)$, with μ_1, μ_2 unramified characters of $\mathbb{Q}_p^\times$ such that

$$p^{\frac{k-1}{2}}(\mu_1(p) + \mu_2(p)) = a_p. \tag{2.5.1}$$

Remark 2.5.2. In case $k > 1$, then π_∞ is a *discrete series* representation of "lowest weight k," sitting inside some $\pi_\infty(\mu_1, \mu_2)$ (cf. Fact 2.2); this discrete series representation has central character *trivial* on $\mathbb{R}^+$, and is denoted D_k. However, if $k = 1$, then π_∞ is *not* a *sub*representation of any $\pi_\infty(\mu_1, \mu_2)$, but rather equals the full induced ("principal series") representation π_∞ (1, sgn) (see the proof below). In this case, the corresponding representation $\pi = \otimes\pi_p$ (with $\pi_\infty = \pi_\infty$ (1, sgn)) is called an *automorphic cuspidal representation of weight one.*

Corollary of the Proposition. *A normalized new form in $S_1(\Gamma_0(N), \psi)$, with eigenvalues $\{a_p\}$, is one and the same thing as an automorphic cuspidal representation $\otimes\pi_p$ of weight one such that (cf. (2.5.1))*

$$a_p = \mu_1(p) + \mu_2(p), \quad \textit{for all } p\nmid N.$$

N.B. Here μ_1 and μ_2 are the two unramified characters of $\mathbb{Q}_p^\times$ inducing the unramified representation $\pi_p = \pi_p(\mu_1, \mu_2)$ of $\mathrm{GL}_2(\mathbb{Q}_p)$. From now on, we rewrite the above relation in the more suggestive form

$$a_p = \text{trace }(t_{\pi_p}) \tag{2.5.3}$$

with

$$t_{\pi_p} = \begin{pmatrix} \mu_1(p) & 0 \\ 0 & \mu_2(p) \end{pmatrix} \quad \text{in} \quad \mathrm{GL}_2(\mathbb{C})$$

the so-called *Langlands class of π_p.*
Sketch of the Proof (of the Proposition). The *first step* is at the level of *functions.* Using the decomposition

$$G(\mathbb{A}_\mathbb{Q}) = \mathrm{GL}_2(\mathbb{Q})\,\mathrm{GL}_2^*(\mathbb{R}) \prod_{p<\infty} K_p^N,$$

(analogous to (*) for GL_1), one defines for any f in $S_k(\Gamma_0(N), \psi)$ a function φ_f on $G(\mathbb{A}_\mathbb{Q})$ by

$$\varphi_f(\gamma g_\infty k_0) = f(g_\infty(i))j(g_\infty, i)^{-k}\chi_\psi(k_0); \tag{2.5.4}$$

here

$$K_p^N = \left\{ \begin{pmatrix} a & b \\ c & a \end{pmatrix} \in K_p = \mathrm{GL}_2(\mathbb{Z}_p) : c \equiv 0(N) \right\}$$

and

$$j(g_\infty, z) = (cz+d)(\det\, g_\infty)^{-\frac{1}{2}} \quad \text{if} \quad g_\infty = \begin{pmatrix} a & b \\ c & a \end{pmatrix}.$$

It is now a standard matter (cf. [Cas] and [De]) to check that this map $f \longrightarrow \varphi_f$ is an isomorphism from $S_k(\Gamma_0(N), \psi)$ to the space of smooth functions $\{\varphi\}$ on $\mathrm{GL}_2(\mathbb{A}_\mathbb{Q})$ such that
(i) $\varphi(\gamma g) = \varphi(g)$ for all $g \in \mathrm{GL}_2(\mathbb{Q})$;
(ii) $\varphi(gk) = (\chi_\psi)(k)\varphi$ for all $k \in K_p^N$;
(iii) $\varphi(gr(\theta)) = e^{-ik\theta}\varphi(g)$ for all $r(\theta) = \begin{pmatrix} \cos\theta & -\sin\theta \\ \sin\theta & \cos\theta \end{pmatrix} \in G_\infty$;
(iv) $\varphi(zg) = \chi_\psi(z)\varphi(g)$ for all $z \in Z(\mathbb{A})$;
(v) If X denotes the "pushing down" differential operator corresponding to the element $\begin{pmatrix} i & 1 \\ 1 & -i \end{pmatrix}$ of $\mathfrak{g}$, then

$$X \cdot \varphi = 0$$

(this is the condition reflecting the holomorphy of f);
(vi) φ is of "moderate growth" on $G(\mathbb{A})$ (relecting the holomorphy of f "at the cusps"); and
(vii) φ is cuspidal, i.e.,

$$\int_{\mathbb{F}\backslash\mathbb{A}} \varphi\left(\begin{pmatrix} 1 & x \\ 0 & 1 \end{pmatrix}\right) dx \equiv 0$$

(reflecting the cuspidality of f...).

Conditions (i)–(vii) imply that φ_f belongs to $L_0^2(G(\mathbb{Q})\backslash G(\mathbb{A}), \chi_\psi)$. The *second step* of the proof is then to show that φ_f generates (under right translation by $G(\mathbb{A})$) an *irreducibly* invariant subspace π_f (of $L_0^2(\chi_\psi)$) whose corresponding representation $\pi_f = \otimes\pi_p$ is as claimed; one must also show that all such representations are thus obtained. These arguments are explained in §5 of [Ge1], but only really for the case $k > 1$, where π_∞ is the discrete series representation D_k. In case $k = 1$, one must argue as follows (in order to identify π_∞).

Suppose $\mu_i(x) = |x|^{s_i}\,\mathrm{sgn}(x)^{\varepsilon_i}$, and $\pi_\infty = \pi(\mu_1, \mu_2)$. Then $H(\mu_1, \mu_2)$ consists only of functions ϕ_k with k of the same parity as $\varepsilon_1 + \varepsilon_2$. (In particular, $\pi(1, \mathrm{sgn})$ consists only of "odd" functions....) A straightforward computation (á la Bargmann...) also shows that

$$X \cdot \phi_k = \left(\frac{s_1 - s_2 + 1}{2} - \frac{k}{2}\right)\phi_k$$

and

$$\phi_k\left(g\begin{pmatrix} r & 0 \\ 0 & r \end{pmatrix}\right) = r^{s_1+s_2}\phi_k(g), \quad \text{for } r > 0.$$

This means that if ω_{π_∞} is to be trivial on $\mathbb{R}^+$, we must have $s_1 = -s_2 = s$, and if $X \cdot \phi_1$ is to be 0, we must have $s_1 = s_2 = s = 0$, i.e., $\pi_\infty = \pi(1, \mathrm{sgn})$ as claimed.

Concerning the converse direction, we recall the following. Suppose $\pi = \otimes\pi_p$ is an irreducible subrepresentation of $L_0^2(G(\mathbb{Q} \setminus G(\mathbb{A}_\mathbb{Q}), \chi_\psi)$ (for some grossencharacter χ_ψ of finite order), and π_∞ is "of weight k" ($k \geq 1$). If the *conductor* $c(\pi) = \prod_{p<\infty} c(\pi_p)$ of π is N, let φ_π denote the function in the space H_π of π which is right K_p^N invariant for all p. (This function is uniquely determined up to a scalar; cf. the Concluding Remark of §2.3.) Then via the correspondence $\varphi_\pi \longrightarrow f_\pi$ (inverse to (2.5.4)), we obtain from π the required *new form* of weight k.

Remark 2.5.5. The fact that the principal series $\pi(1, \mathrm{sgn})$ leads to a *holomorphic* form $f(z)$ (of weight one...) relies on the critical confluence of conditions

$$X\phi = 0$$

and

$$\phi(gr(\theta)) = e^{-i\theta}\phi(g).$$

Indeed, these two conditions together imply that the function f defined by

$$f(z) = \phi(g_\infty(i))j(g_\infty, i)$$

will be holomorphic on the upper half-plane $\mathfrak{h}$. The delicacy of this point can be appreciated by examining what would happen if we took π_∞ to be the principal representation $\pi(1,1)$ (or $\pi(\mathrm{sgn}, \mathrm{sgn})$) in place of $\pi(1, \mathrm{sgn})$. In this case, only ϕ_k's of even parity occur in $H(\mu_1, \mu_2)$, $X \cdot \phi_k = \left(\frac{1-k}{2}\right)\phi_k$ is *never* zero, and no *holomorphic* f's arise in $\mathfrak{h}$. The crucial point now is that ϕ_0 will be K_∞*-invariant*, and hence directly define a function $f(z)$ on $\mathfrak{h}$, with the property that

$$\Delta f = -y^2\left(\frac{\partial^2}{\partial x^2} + \frac{\partial^2}{\partial y^2}\right) f = 1/4f.$$

Indeed, let D denote the standard *Casimir operator* in the center of the universal enveloping algebra of $\mathfrak{g}$, which for K_∞-invariant functions corresponds exactly to the *Laplace-Beltrami* operator Δ above. Then the action of D in $H(\mu_1, \mu_2)$ (with $\mu_2(x) = |x|^{s_i}\,\mathrm{sgn}(x)^{\varepsilon_i}$) is given by the formula

$$D \cdot \phi_k = \frac{(s_1 - s_2)^2 - 1}{4}\,\phi_k$$

(cf. Lemma 5.6 of [JL], page 166, keeping in mind that our Casimir is 1/2 theirs). Thus the same reasoning as used above (to show that an automorphic cuspidal representation $\otimes\pi_p$ with $\pi_\infty = \pi(1, \mathrm{sgn})$ corresponds to a classical cusp form of weight 1) shows also that *a cuspidal representation* $\otimes\pi_p$ *of* $\mathrm{GL}_2(\mathbb{A}_\mathbb{Q})$ *with* $\pi_\infty = \pi(1,1)$ *(or* $\pi(\mathrm{sgn}, \mathrm{sgn})$*) corresponds to a Maass cusp form of "eigenvalue 1/4."*

2.6. Reformulation of Theorem 1.3

Suppose

$$\sigma : G_{\mathbb{Q}} \longrightarrow \mathrm{GL}_2(\mathbb{C})$$

is a continuous, irreducible, two dimensional "odd" representation whose image in $\mathrm{PGL}_2(\mathbb{C})$ is solvable. Then there exists an automorphic cuspidal representation $\pi(\sigma) = \otimes_p \pi_p$ of $\mathrm{GL}_2(\mathbb{A}_{\mathbb{Q}})$ which is of weight one, central character $\det \sigma$, and such that for almost all p, $\pi_p = \pi(\mu_1, \mu_2)$ is unramified with

$$\text{(2.6.1)} \qquad \operatorname{trace} \sigma(\mathrm{Fr}_p) = \operatorname{trace}(t_{\pi_p}) = \mu_1(p) + \mu_2(p).$$

Remarks. **(2.6.2).** Recall that the matrix

$$\begin{pmatrix} \mu_1(p) & 0 \\ 0 & \mu_2(p) \end{pmatrix} = t_{\pi_p}$$

is the *Langlands class* in $\mathrm{GL}_2(\mathbb{C})$ attached to the unramified representation π_p.

(2.6.3) According to Corollary 2.5, the existence of an automorphic cuspidal $\pi = \otimes \pi_p$ as above implies the existence of a *new* form $f = \sum a_n e^{2\pi i n z}$ in some $S_1(\Gamma_0(N), \psi)$ with

$$a_p = \text{trace } \sigma(\mathrm{Fr}_p)$$

for almost all p. Thus this representation theoretic reformulation of Theorem 1.3 indeed implies Theorem 1.3.

(2.6.4) In the next few lectures we shall explain how the more general Theorem 2.1 is proved; this will imply Theorem 2.6, in case $F = \mathbb{Q}$, σ factors through $G_{\mathbb{Q}}$, and $\det \sigma$ is odd, for it is a simple matter to see that $\pi(\sigma)$ is then actually of weight one, i.e., $\pi_\infty = \pi(1, \mathrm{sgn})$ (cf. Proposition 4.2 of Lecture II).

(2.6.5) *"Strong Multiplicity One"* for GL(2) asserts that two automorphic cuspidal representations π and π' are equivalent as soon as they are equivalent almost everywhere, i.e.,

$$\pi_p \cong \pi'_p \quad \text{for almost all } p.$$

This fact is explicitly used, together with *multiplicity one* for GL(2), in the proof that a new form f generates an *irreducible* subspace π_f of L_0^2. In the statement of Theorem 2.6 (or 2.1), it also implies that $\pi(\sigma)$ (if it exists at all) is unique. Indeed, once the central character is fixed, condition (2.6.1), which holds almost everywhere, uniquely determines π_p. Similarly, in the classical version (Theorem 1.3) of Langlands-Tunnell, the new form $g(z)$ is uniquely determined by the condition (1.3.1) which fixes its eigenvalues almost everywhere; this reflects the fact that the theory of *new* forms is one and the same thing as the strong multiplicity one result coupled with the notion of conductors! (See [Cas] or [Gel] for a further explanation of this point.)

Lecture II
The Langlands Program: Some Results and Methods

Abstract

We start this lecture by describing the *Local Langlands Conjecture (LLC) for* GL(n) over a local field F. In case n is a prime, this is a *Theorem* over any field F, known as the "Local Langlands Correspondence." Thus we can (and will) describe the resulting correspondence in some detail for $n = 2$, and apply it to *refine* (and generalize) the two-dimensional Langlands Reciprocity Conjecture (LRC) as follows:

To each continuous irreducible representation

$$\sigma : W_F \longrightarrow \mathrm{GL}_2(\mathbb{C})$$

of the *Weil* group of a number field F, there is associated an automorphic cuspidal representation $\pi(\sigma) = \otimes\pi_v$ of $\mathrm{GL}_2(\mathbb{A}_F)$ with the property that

$$\pi_v \longleftrightarrow \sigma_v$$

for every place v. (Here σ_v, a two-dimensional representation of the local Weil group W_{F_v}, is the "Langlands parameter" of the corresponding representation π_v of $\mathrm{GL}_2(F_v)$; for almost every v, σ_v and π_v are unramified, and the relation $\sigma_v \longleftrightarrow \pi_v$ reduces to the more familiar relation

$$\sigma_v(\mathrm{Fr}_v) \sim t_{\pi_v}$$

in $\mathrm{GL}_2(\mathbb{C})$.)

As we shall see, the "classical version" of the Langlands-Tunnell Theorem (Theorem 1.3 of 2.6) follows immediately from the proof of the general Reciprocity Conjecture in the solvable case; indeed, when $F = \mathbb{Q}$, and σ factors through $G_{\mathbb{Q}}$ and is "odd," $\pi(\sigma)$ must be automorphic of weight one, i.e., $\pi(\sigma_\infty) = \pi_\infty(1, \mathrm{sgn})$ (cf. Proposition 4.2).

In the second half of this lecture, we also begin to collect the automorphic results required for proving the global LRC in the two-dimensional solvable case. As we shall see, all these results, as well as the LRC itself, are but special realizations of a "Principle of Functoriality with respect to the L-group," namely:

Langlands' Functoriality Conjecture

Given two reductive F-groups G and G' (with G' quasi-split), and a morphism

$$\rho : {}^LG \longrightarrow {}^LG'$$

between their L-groups, there is a corresponding mapping of automorphic representations

$$\pi \longrightarrow \pi(\rho) = \otimes\pi'_v$$

from G to G', such that for almost every v, ρ takes the Langlands class t_{π_v} in ${}^LG$ to the Langlands class $t_{\pi'_v} = t_{\pi_v(\rho)}$ in ${}^LG'$.

§3. The Local Langlands Correspondence for $GL(2)$

3.1. The Archimedean Case

We assume $F = \mathbb{R}$. (For the simpler case of $F = \mathbb{C}$, which we do not need, see Remark 3.1.2 below). In this case, the *Weil Group* W_F is an extension of $\mathbb{C}^\times$ by $\mathbb{Z}/2\mathbb{Z}$ given by

$$W_{\mathbb{R}} = \mathbb{C}^\times \cup j\mathbb{C}^\times,$$

where $j^2 = -1$, and $jcj^{-1} = \bar{c}$, and the natural surjection

$$\varphi : W_{\mathbb{R}} \longrightarrow \mathrm{Gal}(\mathbb{C}/\mathbb{R})$$

is given by $\varphi(\mathbb{C}^\times) = 1$ and $\varphi(j\mathbb{C}^\times) = \tau$ (complex conjugation). We are interested in the set of equivalence classes $\Phi(\mathrm{GL}_n/\mathbb{R})$ of n-dimensional complex representations σ of $W_{\mathbb{R}}$ whose images consist of semisimple elements in $\mathrm{GL}_n(\mathbb{C})$.

Example 3.1.1. The *one-dimensional* representations of $W_{\mathbb{R}}$ are of the form $\mu \sim (t, \varepsilon)$, taking z in $\mathbb{C}^\times$ to $|z|_{\mathbb{C}}^t$, $t \in \mathbb{C}$, and j to $\varepsilon = \pm 1$. (Indeed, if $\mu(j) = w$, then on $\mathbb{C}^\times, \mu(\bar{z}) = \mu(jzj^{-1}) = w\mu(z)w^{-1} = \mu(z) = z^t\bar{z}^s$ with $t = s$, i.e., $\mu(z) = r^{2t} = |z|^t$; also $\mu(-1) = 1 = \mu(j^2) = w^2 \Longrightarrow w = \pm 1$.) On the other hand, the *two-dimensional irreducible* representations of $W_{\mathbb{R}}$ are all induced from some character

$$z \longrightarrow |z|_{\mathbb{C}}^t \left(\frac{z}{|z|}\right)^m$$

of $\mathbb{C}^\times$, with t arbitrary in $\mathbb{C}$, *and* $m \geq 1$ *an integer.* Clearly these representations are "semisimple." It is also easy to show that every n-dimensional semisimple representation σ of $W_{\mathbb{R}}$ is a direct sum of these one and two-dimensional irreducible representations.

Theorem. The Local Langlands Correspondence for $\mathrm{GL}_n(\mathbb{R})$. *There is a well defined bijection*

$$\sigma \longleftrightarrow \pi(\sigma)$$

between $\Phi(\mathrm{GL}_n/\mathbb{R})$, *the set of classes of n-dimensional semisimple complex representations σ of $W_{\mathbb{R}}$, and* $\Pi(\mathrm{GL}_n/\mathbb{R})$, *the set of classes of irreducible admissible representations π of* $\mathrm{GL}_n(\mathbb{R})$; *moreover, the L and ε factors assigned to σ and π are preserved by this correspondence.*

Remarks. The existence of this correspondence, formulated and proved more generally by Langlands for an arbitrary reductive Lie group, is the subject matter of [La3]; the fact that L and ε factors may be defined for σ

and π *in the context of* GL_n, and then preserved by this correspondence, is discussed in [Ja1].

Example of $\mathrm{GL}_2(\mathbb{R})$. Suppose first that σ is the sum of two one-dimensional representations (i.e., characters) $\mu_i \sim (t_1, \varepsilon_1)$ as in Example (3.1.1). Then $\pi(\sigma)$ is taken to be the unique irreducible *quotient* of the (induced representation) $\pi(\mu_1, \mu_2)$, where $\mu_i(x) = |x|_{\mathbb{R}}^{t_1}(\mathrm{sgn}(x))^{\varepsilon_1}$, and the order of t_1, t_2 is arranged so that $\mathrm{Re}(t_1) \geq \mathrm{Re}(t_2)$. For example, if $\sigma = (\frac{1}{2}, 0) \oplus (-\frac{1}{2}, 0)$, then $\pi(\sigma)$ is the trivial representation, whereas if $\sigma = (0,0) \oplus (0,1)$ (resp. $(0,0) \oplus (0,0)$) then $\pi(\sigma)$ is the irreducible principal series representation $\pi(1, \mathrm{sgn})$ (of "lowest weight 1") (resp. the class 1 principal series representation $\pi(1,1)$, with Casimir eigenvalue $\lambda = -\frac{1}{4}$). On the other hand, suppose now that σ is the irreducible two-dimensional representation of Ex. 3.1.1, with parameters t and $m \geq 1$. Then $\pi(\sigma)$ is taken to be the *discrete series representation* $D_{m+1} \otimes |\det(\ )|_{\mathbb{R}}^{t}$, with D_{m+1} of lowest weight $m+1$ and trivial central character.

Remark 3.1.2.. In case $F = \mathbb{C}$, the Weil group is just $\mathbb{C}^\times$ and each n dimensional semisimple representation σ is just a sum of characters μ_i of the form $\left(\frac{z}{\bar{z}}\right)^{m_i} |z|_{\mathbb{C}}^{t_i}$ with $m_i \in \mathbb{Z}$. In this case, there are *no* discrete series, and to each σ as above, the corresponding $\pi(\sigma)$ is just the unique irreducible quotient of $\mathrm{Ind}\, \mu_1 \mu_2 \cdots \mu_n$, with the μ_i's arranged so that

$$\mathrm{Re}(t_1) \geq \cdots \geq \mathrm{Re}(t_n).$$

3.2. The p-adic case

In case F is a p-adic field, its *Weil group* W_F is a dense *subgroup* of $\mathrm{Gal}(\bar{F}/F)$, equipped with an isomorphism

$$F^\times \xleftarrow{\sim} W_F^{ab}.$$

In particular, the *one*-dimensional (complex) representations of W_F are again identified with the irreducible admissible representations (i.e. characters) of $F^\times = \mathrm{GL}(1, F)$, just as in the archimedean case. However, unlike in the archimedean case, there are now irreducible representations of W_F of *arbitrary* dimension (reflecting the existence of extensions of F of arbitrary degree...). This fact considerably complicates the representation theory — and concommitent local Langlands correspondence — for $\mathrm{GL}(n)$. Fortunately, for our purposes, we don't need to describe the full Langlands correspondence; instead, we need only the following:

Theorem 3.2. *For each two-dimensional "semisimple" representation σ of W_F, there is exactly one irreducible admissible representation $\pi = \pi(\sigma)$ of $\mathrm{GL}_2(F)$ with*

$$\omega_\pi(\alpha) := \pi \begin{pmatrix} \alpha & 0 \\ 0 & \alpha \end{pmatrix} = \det \sigma(\alpha) I, \tag{3.2.1}$$

and such that for all characters χ of $F^\times$, and ψ_F of F,

$$L(s, \pi \otimes \chi) = L(s, \sigma \otimes \chi),$$
$$L(s, \tilde{\pi} \otimes \chi^{-1}) = L(s, \tilde{\sigma} \otimes \chi^{-1}),$$
$$\varepsilon(s, \pi \otimes \chi, \psi_F) = \varepsilon(s, \sigma \otimes \chi, \psi_F).$$

Moreover, all irreducible admissible π thus arise. (Here the L and ε factors on the left-hand side are those of Jacquet-Langlands, and those on the right-hand side the local factors of [La4]; "$\sim$" denotes the contragredient representation.)

Remark 3.2.2. The *existence* parts and exhaustion of the Theorem are easy, except for the case of irreducible σ (which is due — for arbitrary F — to Kutzko [Kut]). The *uniqueness* part is Corollary 2.19 of [JL], and the resulting bijection

$$\sigma \longleftrightarrow \pi(\sigma)$$

amounts to *the Langlands correspondence for* GL(2).

Caution. Missing in the image of the map

$$\sigma \longrightarrow \pi(\sigma)$$

just described are the "special representations" of $GL_2(F)$. Although they can be obtained by considering representations of the Weil-Deligne group W'_F in place of W_F (see, for example, [Ta] or [Kud]), we prefer to ignore these representations as they play no crucial role in the sequel. In fact, for the global applications we have in mind to the Reciprocity Conjecture, it is crucial to make explicit only the following *unramified* part of the correspondence.

Example 3.2.3. Recall that if k denotes the residue field of F, then W_F consists of those elements of $\mathrm{Gal}(\bar{F}/F)$ whose image in $\mathrm{Gal}(\bar{k}/k)$ is an integer power of the Frobenius automorphism generator of $\mathrm{Gal}(\bar{k}/k)$. Thus the inertia subgroup I of $\mathrm{Gal}(\bar{F}/F)$ is contained in W_F, and a representation $\sigma : W_F \longrightarrow GL_2(\mathbb{C})$ is called *unramified* if it is trivial on I. In this case, since $I \setminus W_F \cong \mathbb{Z}$ (integral powers of the generator of $\mathrm{Gal}(\bar{F}/F)$), σ is completely determined by where it takes the (class of a) Frobenius element Fr of W_F. So suppose (after conjugation, if necessary) that

$$\sigma(\mathrm{Fr}) = \begin{pmatrix} p^{-s_i} & \\ & p^{-s_2} \end{pmatrix} \quad \text{in } GL_2(\mathbb{C}),$$

with $s_1, s_2 \in \mathbb{C}$. Then the corresponding representation $\pi(\sigma)$ of $GL_2(F)$ will be the unramified induced representation $\pi(\mu_1, \mu_2)$, with $\mu_i(x) = | x |^{s_i}$, i.e., the Langlands class

$$t_{\pi(\sigma)} = \begin{pmatrix} \mu_1(p) & 0 \\ 0 & \mu_2(p) \end{pmatrix}$$

will be conjugate to $\sigma(\mathrm{Fr})$. (More precisely if $\pi(\mu_1, \mu_2)$ is itself reducible, then $\pi(\sigma)$ will be the unique irreducible *unramified quotient* of $\pi(\mu_1, \mu_2)$, *perforce* one-dimensional....)

§4. The Langlands Reciprocity Conjecture (LRC)

4.1. Reformulation of Theorem 2.6 ("Langlands-Tunnell")

For F a global field, the *Weil group* W_F maps surjectively onto the Galois group $\mathrm{Gal}(\bar{F}/F)$, and there is a canonical isomorphism

$$W_F^{ab} \cong {}_{F^\times}\backslash^{\mathbb{A}_F^\times}.$$

For each place v of F, there is also an injection $W_{F_v} \longrightarrow W_F$, defining a map

$$\sigma \longrightarrow \sigma_v$$

from the two dimensional semisimple representations of W_F to those of W_{F_v} (cf. [Ta] for background). For a given σ, almost all the resulting σ_v's will be *unramified*, and these unramified σ_v's uniquely determine σ. (This is "strong multiplicity one" on the "Galois side.")

Using the *local Langlands correspondence for* GL(2), we can now attach to any nice $\sigma : W_F \longrightarrow \mathrm{GL}_2(\mathbb{C})$ a global representation $\pi(\sigma)$ of $\mathrm{GL}_2(\mathbb{A}_F)$, namely

$$\pi(\sigma) = \otimes \pi(\sigma_v).$$

The thrust of the Conjecture below is that this $\pi(\sigma)$ must be automorphic.

Conjecture (LRC). *Suppose*

$$\sigma : W_F \longrightarrow \mathrm{GL}_2(\mathbb{C})$$

is irreducible. Then there exists an automorphic cuspidal representation $\pi = \otimes \pi_v$ *of* $\mathrm{GL}_2(\mathbb{A}_F)$ *such that*

$$\pi_v = \pi(\sigma_v) \quad \textit{for all } v.$$

(In particular, the Hecke-Jacquet-Langlands L-function

$$L(s, \pi) = \prod_v L(s, \pi_v)$$

attached to π *— which is known by [JL] to be entire — will equal the Artin L-function* $L(s, \sigma) = \prod_v L(s, \sigma_v)$*.)*

Remarks. (1) This conjecture is actually *equivalent* (via the "converse theorem" for L-functions on GL(2)) to Artin's conjecture for two dimensional irreducible σ; cf. 5.3.1 below. Thus this LRC is sometimes called the "Strong Artin Conjecture."

(2) If σ is *reducible* and the sum of two grossencharacters μ_1 and μ_2, then there is easily seen to be an automorphic (non-cuspidal) representation $\pi = \otimes\pi_v$ of $\mathrm{GL}(2, \mathbb{A}_F)$ with $\pi_v = \pi(\sigma_v)$ for all v, namely the induced "Eisensteinian" representation $\pi(\mu_1, \mu_2)$ (or appropriate irreducible quotient thereof).
(3) When $F = \mathbb{Q}$, and $\sigma : W_{\mathbb{Q}} \longrightarrow \mathrm{GL}_2(\mathbb{C})$ factors through $G_{\mathbb{Q}}$, the above form of the LRC clearly implies the "almost everywhere" version which we stated in §2.6. According to the Proposition below, these two forms of the LRC are actually equivalent!

Proposition 4.1. *Suppose σ is a two-dimensional representation of W_F, and π is a cuspidal representation of $\mathrm{GL}_2(\mathbb{A}_F)$. Then*

$$\pi_v = \pi(\sigma_v) \quad \textit{for all } v$$

if and only if

$$\mathrm{trace}(t_{\pi_v}) = \mathrm{trace}(\sigma_v(\mathrm{Fr}_v))$$

for almost all v (where both π_v and σ_v are unramified).

Note that this last condition really says that, for the unramified places, $\pi_v = \pi_v(\sigma_v)$ (cf. Example 3.2.3 above). Thus this proposition essentially amounts to "strong multiplicity one" for GL(2); for further discussion, see pages 23–24 of [La].

4.2. Relations with Classical Forms

Proposition. *Suppose $\sigma : G_{\mathbb{Q}} \longrightarrow \mathrm{GL}_2(\mathbb{C})$ is irreducible, and "odd," and let $\pi(\sigma)$ denote the corresponding automorphic cuspidal representation of $\mathrm{GL}_2(\mathbb{A}_{\mathbb{Q}})$ (assuming it exists!). Then $\pi(\sigma)$ corresponds (via the correspondence $f \longleftrightarrow \pi_f$ already described) to a normalized new form*

$$f(z) = \sum_{n=1}^{\infty} a_n e^{2\pi i n z} \in S_1(\Gamma_0(N), \psi)$$

with $N =$ conductor(σ), ψ determined by the central character of $\pi(\sigma)$, and

$$a_p = \mathrm{trace}(\sigma(\mathrm{Fr}_p))$$

for almost every p.

Proof By Proposition 2.5, it suffices to check that $\pi_\infty = \pi_\infty(\sigma_\infty)$ is of the form $\pi(\mu_1, \mu_2)$, with $\mu_1 \equiv 1$ and $\mu_2 = \mathrm{sgn}(\ )$. Equivalently, we must check that σ_∞ is a sum of these two characters. But when viewed as a representation of $W_{\mathbb{R}}$, σ_∞ is clearly trivial on $\mathbb{C}^\times$. This means σ_∞ cannot be induced from a *non-trivial* character of $\mathbb{C}^\times$. Thus σ_∞ must be *reducible* (cf. Example 3.1.1), say the sum of two characters μ_i, with $\mu_i \sim (t_i, \varepsilon_i)$.

Since σ_∞ is trivial on $\mathbb{C}^\times$, it follows that each $t_i = 0$. On the other hand, the assumption $\det\sigma(\tau) = -1$ implies that $\sigma(\tau)$ *is not a scalar*; hence

$$\sigma(\tau) \sim \begin{pmatrix} -1 & 0 \\ 0 & 1 \end{pmatrix},$$

which means $\pi_\infty = \operatorname{Ind} 1 \cdot \operatorname{sgn}$, as claimed.

Concluding Remarks. (1) If $\det\sigma$ is *even*, then by the same reasoning as above, σ_∞ is the sum of two characters, but now either *both* trivial or both the sgn character. Thus one concludes $\pi_\infty = \pi(1,1)$ or $\pi(\operatorname{sgn},\operatorname{sgn})$, and from Remark 2.5.5 it follows that π_σ corresponds to a cuspidal Maass eigenform of eigenvalue $1/4$ for Δ.

(2) In [DS], Deligne and Serre associated to each normalized new form $f(z) = \sum_{n=1}^{\infty} a_n e^{2\pi i n z}$ in $S_1(\Gamma_0(N), \psi)$ an irreducible two-dimensional representation σ of $G_{\mathbb{Q}}$, of conductor N and (odd) determinant ψ, such that

$$\operatorname{trace}\ \sigma(\mathrm{Fr}_p) = a_p$$

for almost all primes p. Taken together with the Langlands reciprocity conjecture for $F = \mathbb{Q}$ and "odd" σ (or equivalently, Artin's conjecture for such σ), their result says that *new forms of weight one are one and the same thing as irreducible, odd two-dimensional representations of $G_{\mathbb{Q}}$ (satisfying Artin's conjecture...).*

(3) One expects an analogue of Deligne-Serre to hold for cuspidal Maass-eigenforms (of eigenvalue $1/4$), i.e., that to each such form there should correspond an irreducible two-dimensional, even representation of $G_{\mathbb{Q}}$ (satisfying Artin's conjecture), with $L(\sigma, s) = L(f, s)$. But this remains an open problem; cf. 4.3 below.

4.3. Representations of W_F vs. "Arithmetic" Automorphic Representations of GL(2)

For further reference, it will be convenient to repeat in a more precise form the classification of two-dimensional "semisimple" representations

$$\sigma : W_F \longrightarrow \mathrm{GL}_2(\mathbb{C})$$

over a number field F.

Proposition. *Each σ as above is classified according to its image in* $\mathrm{PGL}_2(\mathbb{C})$, *called the "type" of the representation:*

(i) Cyclic type: $\mu \oplus \nu$: *σ is the direct sum of the two one-dimensional representations defined by Hecke characters μ and ν.*

(ii) Dihedral type: *σ is irreducible of the form $\operatorname{Ind}_{W_E}^{W_F} \theta$, with θ a character of $E^\times \setminus \mathbb{A}_E^\times$, E a quadratic extension of F, and $\theta \neq \theta^\tau$ for $\tau \neq 1$ in* $\mathrm{Gal}(E/F)$. *(Such representations are also called monomial.)*

(iii) Exceptional type: *The image of σ in $\mathrm{PGL}_2(\mathbb{C})$ is A_4, S_4 or A_5.*

Now let's assume that the LRC holds for all irreducible

$$\sigma : W_F \longrightarrow \mathrm{GL}_2(\mathbb{C}),$$

and ask which automorphic cuspidal representation of $\mathrm{GL}_2(\mathbb{A}_F)$ are of the form $\pi(\sigma)$ for some σ? A *necessary* condition is clearly the following.

Definition. Given an irreducible admissible representation $\pi = \otimes\pi_v$, and a real place v, let $\sigma_v : W_{\mathbb{C}/\mathbb{R}} \longrightarrow \mathrm{GL}_2(\mathbb{C})$ denote the Langlands parameter of π_v. Then π is called of *type* A_{00} (resp. A_0) if the restriction of σ_v to $\mathbb{C}^\times$ is trivial (resp. the sum of characters of the form $z \to z^a\bar{z}^b$ with $a, b, \in \mathbb{Z}$). Alternatively, π of type A_{00} (resp. A_0) is called *of Galois type* (resp. *arithmetic*).

Exmaple 4.3.1. If $\sigma : W_{\mathbb{Q}} \longrightarrow \mathrm{GL}_2(\mathbb{C})$ actually factors through $\mathrm{Gal}(\bar{\mathbb{Q}}/\mathbb{Q})$, then the corresponding cuspidal $\pi(\sigma)$ (if it exists) will be of *type* A_{00} (cf. Proposition 4.2 and the Remarks immediately following it). Conjecturally, one expects that all cuspidal π of $\mathrm{GL}_2(\mathbb{A}_{\mathbb{Q}})$ of type A_{00} are "motivic," i.e., arise in this way; in case $\det(\sigma)$ is odd, this is the result of Deligne-Serre.

On the other hand, cuspidal $\pi(\sigma)$ of *type* A_0 are related to ℓ-adic representations of $G_{\mathbb{Q}}$ or $W_{\mathbb{Q}}$ (or the L-series attached to ℓ-adic cohomology spaces of varieties over $\mathbb{Q}$). This is the subject matter of [Ant] (really a representation theoretic reformulation and strengthening of "Eichler-Shimura" theory). For example, if σ_∞ is induced from the character $z \to z^{-n}\bar{z}^{-m}$ of $\mathbb{C}$, with $n > m \geq 0$, let D_k denote the discrete series representation of $\mathrm{GL}_2(\mathbb{R})$ of lowest weight $k = n - m + 1$ (and appropriate central character). Then Langlands in [La5] associates to $\pi = \otimes\pi_p$ of type A_0 (with $\pi_\infty = \pi(\sigma_\infty) = D_k$) a two-dimensional ℓ-adic representation σ of $G_{\bar{\mathbb{Q}}}$ whose local L and ε factors are (eventually) shown to agree with those of π_p *for all* p; cf. [Car].

§5. The Langlands Functoriality Principle Theory and Results

All the automorphic results used to prove the Reciprocity Conjecture in the two-dimensional solvable case, as well as the LRC itself, are but special realizations of what Langlands calls "functoriality of automorphic forms with respect to the L-group." Hence it seems worthwhile to review some of the necessary background on "functoriality" in this Section.

5.1. L groups and L-factors

Recall that for GL_2, an unramified representation $\pi_p = \pi_p(\mu_1, \mu_2)$ is parametrized by a semisimple conjugacy class in $\mathrm{GL}_2(\mathbb{C})$, namely the Langlands class

$$t_{\pi_p} = \begin{pmatrix} \mu_1(p) & 0 \\ 0 & \mu_2(p) \end{pmatrix}.$$

More generally, an arbitrary irreducible admissible representation π_p is parametrized by a *Langlands parameter*

$$\sigma_p : W_{\mathbb{Q}_p} \longrightarrow \mathrm{GL}_2(\mathbb{C}),$$

and the "local Langlands Conjecture" says that the same should hold for GL_n, over any local field F_v. Namely, each nice representation π of $GL_n(F_v)$ should be attached to a parameter $\sigma_v : W_{F_v} \longrightarrow GL_n(\mathbb{C})$.

For an arbitrary reductive group G, over a local field F_v, Langlands introduced the notion of the *L-group* ${}^LG$ to take the place of $GL_n(\mathbb{C})$ in parametrizing the irreducible admissible representations of $G(F_v)$. Roughly speaking, each nice representation of $G(F_v)$ should be attached to a "semi-simple" homomorphism

$$\varphi : W_{F_v} \longrightarrow {}^LG,$$

and in the case of unramified representations, this should amount to fixing a certain semisimple conjugacy class in ${}^LG$ (again called the *Langlands class* t_{π_v} attached to π_v).

In general, if G is defined over a local or global field F, then ${}^LG$ is a group of the form

$$\hat{G} \rtimes \mathrm{Gal}(\bar{F}/F).$$

Here $\hat{G}$ is the complex Lie group "dual" to the root datum of $G(\mathbb{C})$, and the action of $\mathrm{Gal}(\bar{F}/F)$ on $\hat{G}$ is trivial if and only if G is split over F; cf. §2 of [Bo] for details. It is sometimes convenient to replace $\mathrm{Gal}(\bar{F}/F)$ by $\mathrm{Gal}(E/F)$, where E is any Galois extension of F over which G splits. Indeed, since $\mathrm{Gal}(\bar{F}/E)$ acts trivially on $\hat{G}$, we can take ${}^LG$ to be

$$\hat{G} \rtimes \mathrm{Gal}(E/F)$$

(now a complex reductive Lie group). For example, for $G = \mathrm{GL}(n)$, and F local or global, we can take

$${}^LG = GL_n(\mathbb{C}) \quad \text{or} \quad GL_n(\mathbb{C}) \times \mathrm{Gal}(E/F)$$

for any E. It is also convenient to define a semi-direct product

$${}^LG = \hat{G} \rtimes \Sigma$$

for Σ any group endowed with a homomorphism into $\mathrm{Gal}(\bar{F}/F)$, for example, the Weil group W_F. Henceforth, we deal almost exclusively with this "*Weil*" form" of ${}^LG$.

For the moment, let us also assume that G is *unramified over* F, i.e, quasi-split over the local field F, and split over an unramified extension E. For such a group (like $GL_n(F)$), an irreducible admissible representation is called *unramified* if its restriction to a very special maximal compact subgroup (like $GL_n(O_F)$) contains the identity representation (and then just once). If F_r denotes a Frobenius generator for $\mathrm{Gal}(E/F)$, then the *unramified* representations π of $G(F)$ are in one-to-one correspondence with the semisimple ${}^LG^0$-conjugacy classes t_π in ${}^LG^0 \rtimes \mathrm{Fr} \subset {}^LG$. The resulting bijection

$$\pi \longleftrightarrow t_\pi$$

attaches to each unramified π its *Langlands class* $t_\pi = g \rtimes \mathrm{Fr}$ in ${}^LG$, and it is in terms of these classes, and their matrix realizations, that the general Langlands L-factors are defined.

Definition. By a *representation* of ${}^LG$ is meant a continuous homomorphism $r : {}^LG \longrightarrow \mathrm{GL}_m(\mathbb{C})$ whose restriction to $\hat{G}$ is a morphism of complex Lie groups. Given such an r, and an unramified representation π, one sets

$$L(s, \pi, r) = \det(1 - r(t_\pi)q^{-s})^{-1}, \tag{5.1.1}$$

where q is the order of the residue class field of F.

Example. If $F = \mathbb{Q}_p$, $\pi = \pi_p$ is the local component of a cuspidal representation of $\mathrm{GL}(2, \mathbb{A}_\mathbb{Q})$ associated to a new form in $S_k(SL_2(\mathbb{Z}))$, and $r : \mathrm{GL}_2(\mathbb{C}) \longrightarrow \mathrm{GL}_2(\mathbb{C})$ is the "standard" representation taking g to g, then

$$\begin{aligned} L(s, \pi, r) &= [(1 - \mu_1(p)p^{-s})(1 - \mu_2(p)p^{-s})]^{-1} \\ &= (1 - a_p p^{-s-\frac{k-1}{2}} + p^{-2s})^{-1}, \end{aligned}$$

with $a_p = p^{(k-1)/2}(\mu_1(p) + \mu_2(p))$.

The question remains: what L-factors can be assigned to irreducible admissible π which are *not* unramified? In the case of GL_n (and some other classical groups as well now), the local representation theory of G may be used to directly define L (and ε-factors) $L(s, \pi, r)$ (and $\varepsilon(s, \pi, r)$), at least for r sufficiently close to the standard embedding of ${}^LG$ in some $\mathrm{GL}_d(\mathbb{C})$. For example, for GL_n, and $r : \mathrm{GL}_n(\mathbb{C}) \longrightarrow \mathrm{GL}_n(\mathbb{C})$ the identity, [GoJa] constructs such L and ε factors, and in [Ja] they are related (modulo the local Langlands conjecture for $\mathrm{GL}(n)$) to the L and ε factors of their corresponding Langlands parameter.

In general, a typical *Langlands parameter* φ *for* G is a continuous homomorphism

$$\varphi : W_F \longrightarrow {}^LG \; (= \hat{G} \times W_F)$$

such that $\varphi(w)$ is "semisimple" for each w in W_F, and such that the composition with projection onto W_F induces the identity map. Then to each representation $r : {}^LG \longrightarrow \mathrm{GL}_d(\mathbb{C})$, and class of parameters φ (modulo conjugation by $\hat{G}$), one can attach the Langlands factor

$$L(s, \varphi, r) = \det(I - q^{-s} r(\varphi(\mathrm{Fr})|_{V^I}))^{-1}. \tag{5.1.2}$$

Here Fr is a Frobenius element in W_F, and V^I is the subspace of $\mathbb{C}^d$ invariant under the action of the inertia subgroup. Note that when φ is *unramified*, i.e., trivial on I, then φ is determined by the semisimple element $\varphi(\mathrm{Fr}) = g \rtimes \mathrm{Fr}$ in ${}^LG$ and $L(s, \varphi, r)$ *reduces to the Langlands L-function* $L(s, \pi, r)$, with π the unramified representation of $G(F)$ such that t_π is conjugate to $\varphi(\mathrm{Fr})$. In general, a form of the "local Langlands conjecture" *for* G asserts that any irreducible admissible π is associated to some parameter $\varphi : W_F \longrightarrow {}^LG$, and then $L(s, \pi, r)$ should be $L(s, \varphi, r)$, for any r.

N.B. If G is not quasi-split, its Langlands parameters must satisfy certain additional "rationality" conditions; cf. 8.2(ii) of [Bo].

5.2. Statement of the Functoriality Principle

A homomorphism between L-groups

$$\rho : {}^LG \longrightarrow {}^LG'$$

is called an $L-$*morphism* if it commutes with the natural projections onto W_F. Such a morphism clearly gives rise to a map

$$\varphi \longrightarrow \rho \circ \varphi$$

between the Langlands parameters of G and G', hence (conjecturally) also between representations of G and G' (by the local Langlands conjecture).

Now suppose F is global, and $\pi = \otimes \pi_v$ is an automorphic cuspidal representation of $G(\mathbb{A}_F)$. For almost every place v of F, G_v and π_v will be *unramified*, and the corresponding Langlands class t_π in ${}^LG_v$ defined. Thus the (partial Langlands) L-function $L^S(s, \pi, r)$ can be defined for any representation $r : {}^LG \to \mathrm{GL}_d(\mathbb{C})$ through the formula

$$L^S(s, \pi, r) = \prod_{\substack{v \\ \text{"unramified"}}} L(s, \pi_v, r_v).$$

(Here each r_v arises through composition of the natural embedding

$${}^LG_v = \hat{G} \rtimes W_{F_v} \hookrightarrow {}^LG = \hat{G} \rtimes W_F.)$$

Langlands has shown that this *Euler product* converges in some half-plane, and conjectured that it admits a meromorphic continuation to $\mathbb{C}$, with only finitely many poles in $\mathrm{Re}(s) \geq 0$.

N.B. (1) If one accepts the local Langlands Conjecture, one can also introduce $L(s, \pi_v, r_v)$ at the remaining places (as in (5.1.2)), and define the "completed" functions $L(s, \pi, r)$ (and $\varepsilon(s, \pi, r)$).

(2) If $G = \mathrm{GL}_n$, and $r : {}^LG \to \mathrm{GL}_n(\mathbb{C})$ is the standard representation, then $L^S(s, \pi, r)$ is simply denoted $L^S(s, \pi)$.

The Functoriality Principle. *Suppose that G' is quasi-split, and that $\rho : {}^LG \to {}^LG'$ is a morphism of L-groups. For each v, consider the corresponding commutative diagram*

$$\begin{array}{ccc} {}^LG_v & \xrightarrow{\rho_v} & {}^LG'_v \\ \downarrow & & \downarrow \\ {}^LG & \xrightarrow{\rho} & {}^LG'. \end{array}$$

Then to each automorphic cuspidal representation $\pi = \otimes \pi_v$ of $G(\mathbb{A}_F)$ there corresponds an automorphic representation $\pi' = \otimes \pi'_v$ of $G'(\mathbb{A}_F)$ such that for almost all v (where both π_v and π'_v are unramified)

$$t_{\pi'_v} = \rho_v(t_{\pi_v}).$$

In particular, for any representation $r' : {}^L G' \to \mathrm{GL}_d(\mathbb{C})$,

$$L^S(s, \pi', r') = L^S(s, \pi, r' \circ \rho).$$

Moreover, if one accepts the local Langlands Conjecture for G, *then the Langlands parameter of* π'_v *should be the image (under* ρ_v*) of the Langlands parameter of* π_v *for* every v.

Example. Take $G = \{1\}$, and $G' = \mathrm{GL}(n)$. In this case, a morphism

$$\rho : {}^L G \longrightarrow {}^L G' = \mathrm{GL}(n, \mathbb{C}) \times W_F$$

must be of the form $\rho(1, w) = \sigma(w) \times w$, with a continuous representation

$$\sigma : W_F \longrightarrow \mathrm{GL}_n(\mathbb{C}),$$

(and conversely, any Artin representation $\sigma : W_F \longrightarrow \mathrm{GL}_n(\mathbb{C})$ determines a morphism ρ_σ through this formula). Since $G = \{1\}$, its only automorphic representation π is the trivial one, with Langlands class $1 \rtimes \mathrm{Fr}_v$ for every (finite) v. Thus the Functoriality Conjecture in this case asserts that (for any given σ) there is an automorphic representation $\pi(\sigma) = \otimes \pi_v$ of $\mathrm{GL}_n(\mathbb{A}_F)$ such that

$$t_{\pi_v} = \sigma(\mathrm{Fr}_v)$$

for almost every v.

This example shows that the general Reciprocity Conjecture is but a special instance of the Functoriality Principle. Hence it is clear that this Principle is more a guiding light than a problem to be solved in the near future!

5.3. Established Examples of Functoriality

We collect here some instances of "functoriality" which are required for the proof of Langlands-Tunnell.

(A) Automorphic Induction

This is a generalization of the classical construction of Hecke and Maass, whereby an automorphic cuspidal representation of $\mathrm{GL}_2(\mathbb{Q})$ (a modular form, or Maass form, in their language) is attached to each Hecke character of a *quadratic* extension of $\mathbb{Q}$ (which is purely imaginary or real, respectively).

For a general formulation, fix a number field F, and K a *cyclic* Galois extension of F of degree n. Let G denote the group $\mathrm{Res}_{K/F}\,\mathrm{GL}_1$ (defined by "restriction of scalars" from K to F) and let G' denote the group GL_n. Then G is isomorphic to a maximal F-torus of G', and

$${}^L G = (\mathrm{GL}_1(\mathbb{C}) \times \cdots \times \mathrm{GL}_1(\mathbb{C})) \rtimes W_F,$$

with W_F acting on $\hat{G}$ through its projection onto $\mathrm{Gal}(K/F)$, and (the generator of) $\mathrm{Gal}(K/F)$ acting through cyclic permutation of the n $\mathrm{GL}_1(\mathbb{C})$-factors of $\hat{G}$. Let ρ_1 be the natural homomorphism of ${}^LG$ into the normalizer of a maximal torus of $\hat{G'} = \mathrm{GL}(n, \mathbb{C})$, and define a morphism of L-groups

$$\rho : {}^LG \longrightarrow {}^LG'$$

by $\rho(g) = (\rho_1(g), \rho_2(g))$, where $\rho_2 : {}^LG \to W_F$ is the canonical projection. Note that an automorphic form on G is the same thing as a grossencharacter χ of K, since $G(F) = K^\times$; and when v splits (completely) in K, the representation $\pi_v = \pi_v(\chi)$ (induced from the χ_v's above v) satisfies $t_{\pi_v} = \rho(t_{\chi_v})$ if χ is unramified at v. Thus the principle of functoriality *suggests* the following:

Theorem 5.3.1. *For each grossencharacter χ of K there is an automorphic representation $\pi(\chi)$ of $\mathrm{GL}_n(\mathbb{A}_F)$ whose L-function $L^S(s, \pi)$ equals the Hecke-L-function $L^S(s, \chi)$; moreover, $L^S(s, \pi(\chi))$ is entire (and hence $\pi(\chi)$ is* cuspidal *automorphic) if χ does not factor through the norm map $N_{K/F}$ (equivalently χ is not fixed by the natural action of the Galois group* $\mathrm{Gal}(K/F)$*).*

For $n = 2$ and $F = \mathbb{Q}$, this Theorem follows essentially from the classical work of Hecke and Maass. For $n = 2$ and F arbitrary, it is proved in [JL] (using L-functions), [LL] (using the "stable trace formula"), and [ST] (using theta-series); it also follows from Jacquet's "relative trace formula" (cf. §VIII 4 of [Ge2]). For $n = 3$ it is proved in [J-PS-S2] (using L-functions) and for arbitrary n in [AC] (using the trace formula). The only cases needed in the sequel are $n = 2$ or 3, and here it is simplest to (follow [JL] and [J-PS-S2] and) appeal to the so-called "*Converse Theorem to Hecke Theory.*"

For this, suppose that $\pi = \otimes\pi_v$ is an irreducible admissible representation of $\mathrm{GL}_n(\mathbb{A}_F)$ whose central character is invariant under $F^\times$. If π is actually automorphic, then it is known from "Hecke theory" (cf. [GoJa]) that π *is "nice" relative to any idele class character* ω *of* F, i.e., $L(s, \pi \otimes \omega)$ and $L(s, \tilde{\pi} \otimes \omega^{-1})$ are absolutely convergent in some half-place, admit analytic continuations to the whole s-plane which are bounded in vertical strips, and have a functorial equation relating s to $1 - s$; moreover, if π is cuspidal, then these analytic continuations are also *entire.* For $n = 2$ or 3, the *Converse Theorem* (cf. [JL] and [J-PS-S2]) simply says that the converse to each of these statements is also true.

Remarks 5.3.1. (a) In real life situations, like the application to proving $\pi(\chi)$ automorphic in Theorem 5.3.1, the situation is complicated by the fact that the representation we are trying to prove automorphic may not be easily defined at *every* place, but rather only at almost all places; thus, in fact, a more complicated "almost everywhere" version of the converse theorem is needed; cf. §§13–14 of [J-PS-S2].

(b) In the paper [Co-PS], Cogdell and Piatetski-Shapiro conjecture that the *Converse Theorem* should also hold for any n, with the additional

caveat that for $n \geq 4$, π need only be almost everywhere equivalent to some automorphic representation of $\mathrm{GL}_n(\mathbb{A}_F)$; cf. [He].
(c) If χ (in Theorem 5.3.1) is not fixed by any non-trivial element of $\mathrm{Gal}(K/F)$, then $\mathrm{Ind}_{W_K}^{W_F} \chi = \sigma$ is an *irreducible* n-dimensional representation of W_F with $L(s, \sigma) = L(s, \chi)$ (a Hecke L-series with grossencharacter χ *over* K). Hence Theorem 5.3.1 may be viewed as an affirmation of *the Langlands Reciprocity Conjecture for monomial representations.*
(d) Note that "on the Galois side," *induction* brings a Langlands parameter for GL_1 (over K), namely $\chi : W_K \longrightarrow \mathbb{C}^\times$, to a Langlands parameter for GL_2 over F, namely $\sigma = \mathrm{Ind}\,\chi : W_F \longrightarrow \mathrm{GL}_2(\mathbb{C})$. "On the automorphic side," this map is reflected by the correspondence $\chi \longrightarrow \pi(\chi) = \pi(\sigma)$ (hence the aptness of the terminology "automorphic induction").
(e) Finally, we note that in case $n = 2$ or 3, the "converse theorem approach" to Theorem 5.3.1 does *not* depend on K being a normal (Galois) extension of F. This will be crucial in the application to the Reciprocity Conjecture in the Octahedral case; cf. §7.2.

(B) **The Symmetric Square Lifting**

Let A denote the three dimensional representation of $\mathrm{PGL}_2(\mathbb{C})$ determined by the adjoint action of $\mathrm{PGL}_2(\mathbb{C})$ on the Lie algebra of $SL(2, \mathbb{C})$, and denote the resulting (three-dimensional) representation

$$\begin{array}{ccc} \mathrm{GL}_2(\mathbb{C}) & \xrightarrow{\mathrm{Ad}} & \mathrm{GL}_3(\mathbb{C}) \\ & \searrow \quad \quad \nearrow A & \\ & \mathrm{PGL}_2(\mathbb{C}) & \end{array}$$

of $\mathrm{GL}(2, \mathbb{C})$ by Ad. This representation Ad may be viewed as a natural morphism between the L-groups of GL(2) and GL(3).

Theorem 5.3.2. (The "Symmetric Square" Lift from GL(2) to GL(3); cf. [GeJa]).
(i) *To each cuspidal automorphic representation* π *of* $\mathrm{GL}_2(\mathbb{A}_F)$, *there exists an automorphic representation* Π *of* $\mathrm{GL}_3(\mathbb{A}_F)$ *such that for almost all* v,

$$\Pi_v = \Pi_v(\mathrm{Ad}(\sigma_v))$$

whenever $\pi_v = \pi_v(\sigma_v)$; *equivalently,*

$$t_{\Pi_v} = \mathrm{Ad}(t_{\pi_v}).$$

(ii) *This lift of* π *to* GL(3) *is* cuspidal *automorphic unless* π *is monomial, i.e., of the form* $\pi(\sigma)$, *with* σ *induced from a Hecke character of some* quadratic *extension* K.

Method of Proof The "converse theorem for GL(3)" says that $L^S(s, \pi, \mathrm{Ad})$ will be the L-function of an automorphic representation Π of GL(3) (with

$t_{\Pi_v} = \mathrm{Ad}(t_{\pi_v})\cdots)$ as soon as $L^S(s,\pi,\mathrm{Ad})$ is shown to have the expected analytic properties; moreover, this Π will be cuspidal if and only if all $L^S(s,\Pi\otimes\omega)$'s are *entire*. To establish the required analytic properties, it is shown (following [Sh]) that

$$L^S(s,\pi,\mathrm{Ad}) = A_S(s) \int_{SL_2(F)\backslash SL_2(\mathbb{A})} \varphi_\pi(g)\Theta(g)E(g,s)dg.$$

Hence φ_π belongs to the space of π, $\Theta(g)$ is a theta-function on Weil's metaplectic group, $E(g,s)$ is an Eisenstein series of half-integral weight which is real analytic in g and meromorphic in s, and $A_S(s)$ is a meromorphic function which at the possible poles of $E(g,s)$ can be chosen non-zero.

N.B. The idea of using the integral of an automorphic form to derive analytic properties of its L-function of course goes back to Hecke, and even Riemann. But the idea of mixing automorphic forms in the integral with *Eisenstein series* was first systematically developed by Rankin and Selberg, and is now a flourishing industry; cf. below.

(C) **Rankin-Selberg Products (Especially** $\mathrm{GL}(3)\times\mathrm{GL}(3)$**)**

Underlying this work is the following instance of Langlands functoriality. Viewing $\mathrm{GL}(k,\mathbb{C})$ as the L-group of $\mathrm{GL}(k)$, and $\mathrm{GL}(n,\mathbb{C})\times\mathrm{GL}(m,\mathbb{C})$ as that of $\mathrm{GL}(n)\times\mathrm{GL}(m)$, consider the natural L-group morphism

$$\rho: \mathrm{GL}_n(\mathbb{C})\times\mathrm{GL}_m(\mathbb{C}) \xrightarrow{\otimes} \mathrm{GL}_{nm}(\mathbb{C})$$

given by the tensor product map.

So far, there seems no hope of establishing Langlands functoriality in this case, i.e., of proving the existence of an automorphic Π on GL_{nm} such that $t_{\Pi_v} = t_{\pi_v}\otimes t_{\pi'_v}$ for two given cuspidal representations π' on GL_m and π on GL_n. Indeed, this is an important open problem, whose solution would play a crucial role in finding "the" group whose irreducible representations are expected to parametrize *all* the automorphic cuspidal representations of GL_n (not just those "arithmetic" ones coming from representations of W_F); cf. [Ram] for further discussion along these lines. A big first step, however, was taken by Jacquet and Shalika:

Theorem 5.3.3. (cf. [JaSh1,2] and [Mo-Wald]) *Given cuspidal representations π on GL_n and π' on GL_m, let $L^S(s,\pi\times\pi')$ denote the partial L-function*

$$\prod_{v\notin S}[\det(I-(t_{\pi_v}\otimes t_{\pi'_v})q^{-s}]^{-1}.$$

(i) *$L^S(s,\pi\times\pi')$, originally defined only in some right half-plane, extends to a meromorphic function in all of $\mathbb{C}$, with functional equation relating the value at s to the value at $1-s$.*

(ii) $L^S(s, \pi \times \pi')$ *may be "completed" to an Euler product*

$$L(s, \pi \times \pi') = \prod_{\text{all } v} L(s, \pi_v \times \pi'_v)$$

which is holomorphic on $\operatorname{Re}(s) \geq 1$ *if* $m \neq n$, *and otherwise has a pole at* s *with* $\operatorname{Re}(s) = 1$ *if and only if* $|\det(\ )|^{s-1} \otimes \pi \cong \tilde{\pi}$ *(the* contragredient *of* π'*).*

As already suggested, the proof of this result constitutes a non-trivial representation-theoretic generalization of the classical integral representations of Rankin and Selberg; see [Ja] for the case of GL(2) × GL(2). In the sequel, we need only the case $n = m = 3$.

Concluding Remarks. (1) There is one more example of functoriality needed for the proof of Langlands-Tunnell, namely the theory of *base-change* of Saito, Shintani and Langlands. However, since that theory is so intimately tied up with Artin's conjecture, and its proof relies on the trace formula rather than L-functions, it seems convenient to postpone discussion of it until the last lecture.

(2) There are of course large aspects of the Langlands Program which we have not seriously broached here because they have no *immediate* bearing on Wiles' work. Perhaps the most obvious such topic is the (conjectured) relation between Hasse-Weil zeta-functions of algebraic varieties ("motive" L-functions) and automorphic L-functions of type $L(s, \pi, r)$. For example, in [La6] the zeta-functions of certain Shimura varieties are related to automorphic L-functions of degree 2^n. This "program" represents the beginnings of a higher dimensional analogue of the theory of Eichler-Shimura and has greatly influenced much of the work during the last twenty years in representation theory and the theory of automorphic forms. Among other things, it pushed to the forefront the need to refine and generalize the "Selberg trace formula"; more about this in the next lecture. It also brought into representation theory such crucial but different concepts as "L-indistinguishability," "endoscopy," "L-packets," etc., and encouraged the use of new algebro-geometric methods for counting points on these varieties.

(3) Finally, one should say a *few* words about the relation between the Langlands Program and the Shimura-Taniyama-Weil Conjecture. Personally, I do not think that it is so significant that the Langlands Program actually includes the S-T-W conjecture as a special "example" (and that's why I haven't bothered to broach the topic here). After all, Taniyama obviously made his Conjecture — and Shimura and Weil understood its importance — *before* the Langlands Program was conceived. Also, from the other point of view, it is equally clear that including the S-T-W Conjecture inside the Langlands Program is more incidental than crucial to the Program. Rather the crux of the Program is two pronged: its overall *vision* relating motives of all kinds to automorphic representations, and its

methods which push representation theory to the forefront, and infuse the subject with a seemingly endless string of challenging problems. It is these aspects of the Langlands Program which (albeit indirectly) play a role in the proof of Fermat's Last Theorem.

Lecture III
Proof of the Langlands-Tunnell Theorem

Abstract
Our task is to describe the proof of the following:

Theorem. *Suppose F is a number field and the irreducible representation*

$$\sigma : W_F \longrightarrow \mathrm{GL}_2(\mathbb{C})$$

has a solvable image in $\mathrm{PGL}_2(\mathbb{C})$. *Then there exists a (unique) irreducible automorphic cuspidal representation* $\pi(\sigma) = \otimes\pi_v$ *of* $\mathrm{GL}_2(\mathbb{A}_F)$ *such that*

$$\mathrm{trace}(\sigma(\mathrm{Fr}_v)) = \mathrm{trace}(t_{\pi_v})$$

for almost every v.

The crucial instance of the Functoriality Conjecture required in the proof of this Theorem is the theory of "Base Change" as developed in [La1]. This we describe in §6, along with its proof, which relies heavily on *trace* formula methods. The application of base change to the Langlands Reciprocity theorem is explained in §7, the proof of the actual theorem proceeding in two steps: first the base change (trace formula) methods are exploited to produce the best possible candidate for $\pi(\sigma)$ (which is called $\pi_{ps}(\sigma)$); then the results from the theory of L-functions (recalled in §5) are used to prove that $\pi_{ps}(\sigma)$ actually equals $\pi(\sigma)$.

§6. Base Change Theory

(6.1). Fix E a cyclic extension of the number field F, of prime degree ℓ. Roughly speaking, the theory of "base change" describes the correspondence between automorphic representations of the groups $\mathrm{GL}_n(\mathbb{A}_F)$ and $\mathrm{GL}_n(\mathbb{A}_E)$ which reflects the operation of *restriction* of Galois representations of W_F to W_E. The first results on base change for automorphic forms (or representations) used the theory of *L-functions*, and were restricted to the case of *quadratic* E and GL_2. The introduction of the *trace formula* is due to H. Saito, who dealt with GL_2 and arbitrary cyclic E using the classical language of automorphic *forms*; cf. [Sai]. Immediately after that, Shintani reformulated Saito's results using group representations, and gave the correct *local* definition of base change lifting; cf. [Shin]. Finally, Langlands saw the connection with Artin's conjecture, and reshaped the trace formula proof for GL_2 in a form suitable for the later generalization to GL_n developed by Arthur and Clozel; see [La1] and [AC] for a more detailed history. Since only the case $n = 2$ is required here, we restrict ourselves henceforth to this case.

Definition. Suppose $\pi = \otimes\pi_v$ is an automorphic cuspidal representation of $\mathrm{GL}_2(\mathbb{A}_F)$, and $\Pi = \otimes_w \Pi_w$ is an automorphic representation of $\mathrm{GL}_2(\mathbb{A}_E)$. Then Π is a *base change lift of* π, denoted $\mathrm{BC}_{E/F}(\pi)$, if for each place v of F, and $w|v$, the Langlands parameter attached to Π_w equals the restriction to W_{E_w} of the Langlands parameter $\sigma_v : W_{F_v} \longrightarrow \mathrm{GL}_2(\mathbb{C})$ of π_v.

Remarks. (i) The above (essentially local) definition of base-change lifting is at the level of Langlands parameters rather than representations. The key idea of [Shin] is to define the lift of π_v on $\mathrm{GL}_2(F_v)$ directly in terms of a character identity between π_v and the extension of Π_w to the group $\mathrm{GL}_2(E_w)\rtimes \mathrm{Gal}(E_w/F_v)$. Implicit here is the fact that Π_w is $\mathrm{Gal}(E_w/F_v)$ invariant and hence this extension, call it $\tilde{\Pi}_w$, exists. If τ is a generator of $\mathrm{Gal}(E_w/F_v)$ the character of this identity reads

$$\chi_{\tilde{\Pi}_w}(g\rtimes\tau) = \chi_{\pi_v}(x)$$

whenever $N_{E/F,\sigma}(g) = g^{\tau^{\ell-1}}\cdots g^{\tau}g$ is conjugate in $\mathrm{GL}_2(E_w)$ to a regular semisimple element x of $G(F_v)$.
(ii) *Functoriality.* Let $G = \mathrm{GL}_2$ and set $G' = \mathrm{Res}_{E/F}(G)$. As recalled in §5.3, $\hat{G}'$ is then a product of ℓ copies of $\mathrm{GL}_2(\mathbb{C})$ indexed and permuted by $\mathrm{Gal}(E/F)$. So let

$$\rho : {}^LG = \mathrm{GL}_2(\mathbb{C}) \times W_F \longrightarrow {}^LG' = \hat{G}'\rtimes W_F$$

be the natural morphism which takes $g \times w$ in ${}^LG$ to $(g,\cdots,g)\rtimes w$ in ${}^LG'$. The Functoriality Principle suggests the existence of a map taking automorphic cuspidal representations π of $\mathrm{GL}_2(\mathbb{A}_F)$ to automorphic representations Π of $\mathrm{GL}_2(\mathbb{A}_E) \approx G'(\mathbb{A}_F)$ such that for π_v and Π_w unramified,

$$(*) \qquad t_{\Pi_w} = \rho_v(t_{\pi_v}).$$

Using either definition of lifting given above, it is easy to check that if $\pi_v = \pi(\mu_1,\mu_2)$ (with μ_i an unramified character of F_v) then

$$\mathrm{BC}_{E/F}(\pi_v) = \Pi(\nu_1,\nu_2) \quad \text{with } \nu_i = \mu_i \circ N_{E_w/F_v}.$$

From this it follows that $(*)$ holds, i.e., Base Change is functorial.

N.B.. In verifying that $(*)$ holds, one must keep in mind that $\pi(\nu_1,\nu_2)$ (viewed as a representation of $\mathrm{GL}_2(E_w)$) corresponds first to the Langlands class $g \times \sigma$ in $\mathrm{GL}_2(\mathbb{C}) \times \sigma$, with

$$g = \begin{pmatrix} \mu_1 \circ N(\tilde{\omega}_w) & 0 \\ 0 & \mu_2 \circ N(\tilde{\omega}_w) \end{pmatrix}$$

$$= \begin{pmatrix} \mu_1(\tilde{\omega}_v)^2 & 0 \\ 0 & \mu_2(\tilde{\omega}_v)^2 \end{pmatrix};$$

but the corresponding class in

$${}^LG' = \mathrm{GL}_2(\mathbb{C}) \times \mathrm{GL}_2(\mathbb{C})\rtimes\sigma$$

(viewing $\pi(\nu_1, \nu_2)$ as a representation of $G'(F_v) \approx \mathrm{GL}_2(E_w)$) is

$$\left(\begin{pmatrix} \mu_1(\tilde{\omega}_v) & 0 \\ 0 & \mu_2(\tilde{\omega}_v) \end{pmatrix}, \begin{pmatrix} \mu_1(\tilde{\omega}_v) & 0 \\ 0 & \mu_2(\tilde{\omega}_v) \end{pmatrix}, \sigma\right),$$

i.e., just $\rho(t_{\pi_v})$. Indeed, the Hecke algebras of $\mathrm{GL}_2(E_w)$ and $G'(F_v)$ are the same, and if f_w and f'_w represent the same element in this algebra $\mathcal{H}_w$, then $(f'_w)^v(g_1, \cdots, g_\ell \times \sigma) = f^v_w(g_e \cdots g_2 g_1)$; see 6.3 below for definitions of $\mathcal{H}_w$ and the Satake isomorphism $(f')^v$.
(iii) Because we are assuming E over F cyclic of *prime* degree, each v of F either remains *inert* or splits completely. In the later case, it is clear that $E_w \approx F_v$ for any $w|v$, and the base change lift of π_v is just $\Pi_w \approx \pi_v$. This case being trivial, we usually assume (as above) that we are dealing with the inert local case.

Theorem. (cf. [La1])
(a) *Every cuspidal representation π of* $\mathrm{GL}_2(\mathbb{A}_F)$ *has a unique base change lift to* $\mathrm{GL}_2(\mathbb{A}_E)$*; the lift is itself* cuspidal *(as opposed to "just" automorphic) unless E is* quadratic *over F, and π is monomial (or dihedral) of the form $\pi(\sigma)$, with $\sigma = \mathrm{Ind}_{W_E}^{W_F} \theta$.*
(b) *If two cuspidal representations π and π' have the same base change lift to E, then $\pi' \approx \pi \otimes \omega$ for some character ω of $F^\times N_{E/F}(\mathbb{A}_E^\times) \setminus \mathbb{A}_F^\times$.*
(c) *A cuspidal representation Π of* $\mathrm{GL}_2(\mathbb{A}_E)$ *equals* $\mathrm{BC}_{E/F}(\pi)$ *for some cuspidal π on* $\mathrm{GL}_2(\mathbb{A}_F)$ *if and only if Π is invariant under the natural action of* $\mathrm{Gal}(E/F)$.

In some ways, the *proof* of Base Change is as interesting as the result itself. Since it involves a form of the trace formula which should (and does) generalize, and apply to other instances of functoriality, we devote some time to it below.

(6.2). The Trace Formula of Arthur-Selberg

Recall that the right regular representation R_0 of $G(\mathbb{A}_F)$ in the space of cusp forms $L_0^2(G(F) \setminus G(\mathbb{A}_F), \omega)$ decomposes discretely as

$$R_0 = \sum m_\pi \pi,$$

and it is the cuspidal constituents π which are the building blocks of the theory of automorphic forms on G. What the "trace" in "the trace formula" refers to is the distributional trace of R_0. More precisely, suppose $f(g)$ is any nice compactly supported "test function" on $G(\mathbb{A})$, and define the operator $R_0(f)$ on $L_0^2(G(F) \setminus G(\mathbb{A}), \omega)$ through the formula

$$R_0(f) = \int_{Z(\mathbb{A}) \setminus G(\mathbb{A})} f(g) R_0(g) dg.$$

(For simplicity, assume that the central character ω of R_0 is trivial.) Then clearly

$$\operatorname{trace} R_0(f) = \sum m_\pi \operatorname{trace}(\pi(f));$$

but as we know next to nothing about the π's which occur in R_0, we also know next to nothing about trace$(R_0(f))$. The original idea of the trace formula was to give an *alternative* formula for trace $R_0(f)$, which ultimately gives some of the sought after information about R_0 and its constituents π.

The original trace formula was introduced by Selberg, in the context of a semisimple Lie group G and discrete subgroup Γ (in place of our $G(\mathbb{A})$ and $G(F)$). In his famous 1956 paper [Sel], Selberg first of all described a general formula for the case of *compact* $\Gamma \setminus G$ (equivalently $G(F) \setminus G(\mathbb{A})$); it took the form

$$\operatorname{trace} R_0(f) = \sum_{\pi} m_\pi \operatorname{trace} \pi(f) = \sum_{\{\gamma\}} m_\gamma \Phi_f(\gamma) \tag{6.2.1}$$

with $\{\gamma\}$ running over the conjugacy classes in $G(F)$, and each $\Phi_f(\gamma)$ an "orbital integral"

$$\int_{G_\gamma(\mathbb{A})\setminus G(\mathbb{A})} f(g^{-1}\gamma g)\, dg.$$

Secondly, Selberg treated in detail certain *non*-compact quotient cases such as $SL_2(\mathbb{Z}) \setminus SL_2(\mathbb{R})$, which already required the analytic continuation of Eisenstein series to handle the *continuous* spectrum of L^2 outside L_0^2.

Subsequently, in the 1960's and 70's, Langlands developed a general theory of Eisenstein series valid for any reductive group G, and Arthur used it to develop a general trace formula in the context of not necessarily compact quotients $G(F) \setminus G(\mathbb{A})$. The resulting *trace formula of Arthur* takes the form

$$\sum_{\mathfrak{o}} J_{\mathfrak{o}}(f) = \sum_{\chi} J_\chi(f). \tag{6.2.2}$$

Here the left (or *geometric*) side of Arthur's formula is a sum of special types of equivalence classes in $G(F)$ (generalizing the ordinary notion of equivalence in the case of compact quotient), and the sum on the right (or *spectral*) side is over certain classes of automorphic cuspidal representations of "Levi subgroups" of G (as opposed to just the cuspidal representations of $G(\mathbb{A})$ itself in the case of compact quotient). Although it *looks* like the "trace" has been lost in Arthur's trace formula, this is not really so; certain of the spectral terms $J_\chi(f)$ add up to exactly trace $R_0(f)$, and so making (6.2.2) explicit still (ultimately) gives us the information we seek about trace$(R_0(f))$.

Now instead of focusing efforts on finding an explicit, new formula for trace($R_0(f)$), it was the idea of Langlands to *compare* the trace formulas for *two different* groups G and G', in order to find *relations* (presumably "functorial") between their automorphic representations. For example, this is exactly the strategy exploited in §16 of [JL] to establish the correspondence between automorphic cuspidal representations of $G = D^\times$ and $G' = \mathrm{GL}_2$, D a division quaternion algebra over F. In the case of $G = \mathrm{GL}_2$ and $G' = \mathrm{Res}_{E/F}\, G$ this strategy brings us back to the proof of Theorem 6.1 which we now explain.

(6.3) The Proof of Base Change.

If τ denotes a generator of Gal(E/F), then τ acts naturally on $G'(\mathbb{A}) \approx \mathrm{GL}_2(\mathbb{A}_E)$, and hence on $L^2(G'(F) \setminus G'(\mathbb{A}_F))$ through the rule

$$(\tau \cdot \varphi)(g) = \varphi(g^\tau).$$

We can also define a *twisted* regular representation R^τ through the composition of R with τ. Then for any nice f' on $G'(\mathbb{A}_F)$, there is a "twisted" version of the trace formula of the form

$$\sum_{\mathfrak{o}'} J^\tau_{\mathfrak{o}'}(f') = \sum_{\chi'} J^\tau_{\chi'}(f'), \tag{6.3.1}$$

with the "twisted" trace($R_0^\tau(f')$) hidden inside the right side of (6.3.1), and "twisted" orbital integrals $\Phi^\tau_{f'}(\gamma') = \int f'(g^{-\tau}\gamma g)dg$ on the left. The significance of working with a *twisted* formula for G' is that then only Galois invariant cuspidal representations Π will contribute (to trace($R_0^\tau(f')$))[†]; hence we might indeed establish the desired base change map $\pi \longrightarrow \Pi$ between G and G' by relating (6.3.1) to (6.2.2), and ultimately trace($R_0(f)$) to trace($R_0^\tau(f')$).

The first step is to prove that the left-hand (i.e., *geometric*) sides of the trace formulas for G and G' coincide, at least for certain "matching" f and f'. This matching is a non-trivial *local* step, which first of all requires that the orbital integrals $\Phi_{f_v}(N\gamma')$ on G_v match the *twisted* orbital integrals $\Phi^\tau_{f'_w}(\gamma')$ on G'_w. Moreover, it must be shown that this matching $f' \longrightarrow f$ is compatible with "base change at the unramified places," in the following sense:

If $\mathcal{H}_v$ denotes the Hecke algebra of bi-K_v-invariant (compactly supported smooth) functions on G_v, each f_v in $\mathcal{H}_v$ may be viewed as a function on ${}^LG_v$ through the formula

$$f_v^V(t) = \mathrm{trace}\, \pi_v(f_v)$$

whenever $t_{\pi_v} = t$. (This is the Satake *isomorphism* $f_v \longrightarrow f_v^V$, defined analogously for the Hecke algebra $\mathcal{H}'_w$ of G'_w.) Then the base change map

---

[†] This is because τ permutes the constituents Π of R_0, and a permutation matrix without fixed points has zero trace...

of Hecke algebras, dual to the base change morphism $\rho : {}^LG_v \longrightarrow {}^LG'_w$, is defined by

$$\rho^V : (f'_w)^V \longrightarrow f'_v(g) = (f'_w)^V(\rho(g)),$$

and the compatibility condition mentioned above is that f'_w will match its image $\rho^V(f'_w) = f'_v$ in the above sense, for any f'_w in $\mathcal{H}'_w$. (This is what is known as the *fundamental lemma*, in the context of "base change.")

The next step, which takes a great deal more work on the *spectral* sides of (6.2.2) and (6.3.1), is to conclude from the equality of the geometric trace formulas that

$$\operatorname{trace} R_0(f) \qquad \text{essentially equals} \qquad \operatorname{trace} R_0^\tau(f')$$

for such matching f and f'. Equivalently,

$$\sum_{\substack{\pi \\ \text{cuspidal}}} \operatorname{trace} \pi(f) = \sum_{\substack{\pi' \text{ cuspidal} \\ (\pi')^\tau \approx \pi}} \operatorname{trace}(\tau \circ \pi')(f') \tag{6.3.2}$$

with the sum on the right only over Galois fixed π'.

Now for π_v, π'_w, f_v, f'_w all "unramified," we will have π'_w equal to the base change lift of π_v if and only if

$$\operatorname{trace}(\tau \circ \pi'_w)(f'_w) = \operatorname{trace} \pi_v(f_v)$$

for any f_v the base change image of f'_w as above. (Indeed, in terms of Satake transforms, this last identity reads $(f'_w)^V(t_{\Pi_w}) = f_v^V(t_{\pi_v}) = (f'_w)^V(\rho(t_{\pi_v}))$, i.e., $t_{\Pi_w} = \rho(t_{\pi_v})$, as required.)

In this way, with a "linear independence of characters" argument, (6.3.2) ultimately implies that for a given π occurring on the left-hand side, there must be a π' on the right-hand side which is Galois invariant, and almost everywhere the base change lift of π (and, conversely, all such Galois invariant π' thus arise). Thus the required correspondence is established.

Remark. Whenever the trace formula can be used to establish an instance of functoriality (like base change above), it offers the additional bonus of *characterizing the image* of the automorphic representations in question. This is *not* so for the method of L-functions (witness the example of the lifting from GL(2) to GL(3), where the image is left uncharacterized, or Proposition 7.2 below giving base change for non-Galois cubic E).

§7. Application to Artin's Conjecture

The idea of applying "base change" to attack Artin's Conjecture arises from the following observation. Suppose that for any $\sigma : W_F \longrightarrow \mathrm{GL}_2(\mathbb{C})$ there really is a corresponding cuspidal representation $\pi(\sigma)$ of $\mathrm{GL}_2(\mathbb{A}_F)$. Then it follows from the original definition of base change lifting that

$$\mathrm{BC}_{E/F}(\pi(\sigma)) = \pi(\operatorname{Res} \sigma|_{W_E})$$

for any cyclic extension E of F. This means that if we *start* with σ, and want to find *candidates* for $\pi(\sigma)$, then the thing to do is to pick an E such that $\pi(\operatorname{Res}\sigma|_{W_E})$ is already known to exist, and look among the cuspidal π's such that $\mathrm{BC}_{E/F}(\pi) = \pi(\operatorname{Res}\sigma|_{W_E})$. In this way the following "obvious" strategy unfolds: Among the possible candidates for $\pi(\sigma)$, pick a "best possible" one, call it $\pi_{\mathrm{ps}}(\sigma)$ (for $\pi_{\mathrm{pseudo}}(\sigma)$), and then prove that $\pi_{\mathrm{ps}}(\sigma)$ must equal $\pi(\sigma)$. Roughly speaking, the first step uses the trace formula (via base change), while the second uses L-functions.
Convention. Henceforth, if we are given $\sigma : W_F \longrightarrow \mathrm{GL}_2(\mathbb{C})$, and any field E over F, then by σ_E we denote the restriction of σ to W_E.

(7.1). The Tetrahedral Case

(a) *Choosing* $\pi_{\mathrm{ps}}(\sigma)$

We are given an irreducible representation

$$\sigma : W_F \longrightarrow \mathrm{GL}_2(\mathbb{C})$$

whose image in $\mathrm{PGL}_2(\mathbb{C})$ is isomorphic to A_4. This group is solvable, with composition series

$$A_4 \rhd D_2 \rhd \{e\}.$$

(In general, D_n will denote the dihedral group of $2n$ elements; in this case, D_2 is the Klein 4-group). Since $A_4/D_2 \cong A_3 \cong \mathbb{Z}_3$, the inverse image of D_2 in W_F under the map

$$W_F \longrightarrow A_4 \subset \mathrm{PGL}_2(\mathbb{C})$$

is a (normal) subgroup of index 3, hence the Weil group of a *cubic* extension of F, call it E. Pictorially:

$$\begin{array}{ccccccccc} 1 & \longrightarrow & W_E & \longrightarrow & W_F & \longrightarrow & \mathrm{Gal}(E/F) & \longrightarrow & 1 \\ & & \downarrow & & \downarrow & & \downarrow \wr & & \\ 1 & \longrightarrow & D_2 & \longrightarrow & A_4 & \longrightarrow & \mathbb{Z}_3 & \longrightarrow & 1 \end{array}$$

Thus the resulting representation $\sigma_E : W_E \longrightarrow \mathrm{GL}_2(\mathbb{C})$ is "monomial" in the sense of Proposition 4.3.

Let $\pi(\sigma_E)$ denote the automorphic cuspidal representation of $\mathrm{GL}_2(\mathbb{A}_E)$ attached to this monomial representation by Theorem 5.3.1. This representation of $\mathrm{GL}_2(\mathbb{A}_E)$ is clearly invariant under the action of $\mathrm{Gal}(E/F)$; indeed, $\pi(\sigma_E)^\tau = \pi(\sigma_E^\tau) = \pi(\sigma_E)$. So by (the Base Change) Theorem 6.1, $\pi(\sigma_E)$ will be the base change lift of exactly *three* classes of irreducible cuspidal representations π_i of $\mathrm{GL}_2(\mathbb{A}_F)$, each one related to the other by a twist $\omega \circ \det$ for some character ω of $F^\times N_{E/F}(\mathbb{A}_E^\times) \setminus \mathbb{A}_F^\times$, i.e.,

$$\pi_i = \pi_j \otimes \omega \circ \det .$$

These π_i's are our natural candidates for $\pi(\sigma)$.

Recall that the *central character* of $\pi(\sigma)$ is to be $\det\sigma$. On the other hand, the central character ω_i of each π_i above "base change lifts" to the central character of $\pi(\sigma_E)$, which is $\det\sigma_E = (\det\sigma)\circ N_{E/F}$. Since each $\omega_i = \omega_j\omega^2$ if $\pi_i = \pi_j \otimes \omega\circ\det$, it is clear that *exactly one of these π_i's has central character* $\det\sigma$, and this is the one we choose to be $\pi_{\text{ps}}(\sigma)$.

(b) *Proving* $\pi_{\text{ps}}(\sigma) = \pi(\sigma)$

Write $\pi_{\text{ps}}(\sigma) = \otimes\pi_v$. Then for each v, $\pi_v = \pi_v(\sigma'_v)$ for some

$$\sigma'_v : W_{F_v} \longrightarrow \mathrm{GL}_2(\mathbb{C}),$$

and what we must prove is that

$$\sigma'_v = \sigma_v \tag{7.1.1}$$

for almost every v.

Note that the *restriction* of σ'_v to W_{E_w} (for $w|v$) is by construction the same as the *restriction* of σ_v to W_{E_w}. Thus there is nothing to prove in case v splits (completely) in E, and we henceforth assume E_w *cubic and unramified* over F_v.

If Fr_v denotes a Frobenius element of $\mathrm{Gal}(E_w/F_v)$ we can suppose

$$\sigma_v(\mathrm{Fr}_v) = \begin{pmatrix} a_v & 0 \\ 0 & b_v \end{pmatrix} \qquad \text{and} \qquad \sigma'_v(\mathrm{Fr}_v) = \begin{pmatrix} c_v & 0 \\ 0 & d_v \end{pmatrix}$$

for some a_v, b_v, c_v, d_v in $\mathbb{C}^\times$. Then to prove (7.1.1) it will suffice to prove that $\begin{pmatrix} a_v & 0 \\ 0 & b_v \end{pmatrix}$ is conjugate to $\begin{pmatrix} c_v & 0 \\ 0 & d_v \end{pmatrix}$. But the fact that σ_v and σ'_v have the same restrictions to W_{E_w} means that $\sigma_v(\mathrm{Fr}_v)^3$ is conjugate to $\sigma'_v(\mathrm{Fr}_v)^3$ (since Fr_v^3 belongs to W_{E_w}). Thus

$$\begin{pmatrix} a_v^3 & 0 \\ 0 & b_v^3 \end{pmatrix} \quad \text{is conjugate to} \quad \begin{pmatrix} c_v^3 & 0 \\ 0 & d_v^3 \end{pmatrix}$$

In particular, for some pair of cube roots of 1, say ξ and ξ', either

$$c_v = \xi a_v \quad \text{and} \quad d_v = \xi' b_v,$$

or else

$$c_v = \xi b_v \quad \text{and} \quad d_v = \xi' a_v.$$

We claim now that $\xi' = \xi^2$. Indeed $\pi_{\text{ps}}(\sigma)$ was chosen so that

$$\omega_{\pi_{\text{ps}}(\sigma)} = \det(\sigma).$$

Since this implies $\det \sigma'_v = \det \sigma_v$, we must have $\xi\xi' = 1$, i.e., $\xi' = \xi^2$. So to prove (7.1.1) it will suffice to prove

$$(7.1.2) \qquad \xi = 1.$$

To continue, let us assume (for the moment) that

$$(7.1.3) \qquad \mathrm{Ad} \circ \sigma'_v = \mathrm{Ad} \circ \sigma_v.$$

Since the kernel of $\mathrm{Ad} : \mathrm{GL}_2(\mathbb{C}) \longrightarrow \mathrm{GL}_3(\mathbb{C})$ is precisely the group of scalar matrices $\left\{\begin{pmatrix} \lambda & 0 \\ 0 & \lambda \end{pmatrix}\right\}$, it follows from (7.1.3) that $\sigma_v(\mathrm{Fr}_v)$ and $\sigma'_v(\mathrm{Fr}_v)$ must differ by some scalar $\lambda \neq 0$. Thus

$$\begin{pmatrix} \xi a_v & 0 \\ 0 & \xi^2 b_v \end{pmatrix} \quad \text{is conjugate to} \quad \begin{pmatrix} \lambda a_v & 0 \\ 0 & \lambda b_v \end{pmatrix},$$

and it suffices to prove

$$\lambda = 1.$$

If $\lambda a_v = \xi a_v$ and $\lambda b_v = \xi^2 b_v$ then $\lambda = \xi = \xi^2 = 1$ for the trivial reason that ξ is a cube root of 1. On the other hand, if $\lambda a_v = \xi^2 b_v$ and $\lambda b_v = \xi a_v$, then $\lambda^2 = 1$ (since $a_v = \xi^2/\lambda b_v = (\lambda/\xi)\ b_v$). If $\lambda = -1$, this means that the image of

$$\sigma_v(\mathrm{Fr}_v) = \begin{pmatrix} a_v & 0 \\ 0 & b_v \end{pmatrix} = \begin{pmatrix} a_v & 0 \\ 0 & a_v \end{pmatrix} \begin{pmatrix} 1 & 0 \\ 0 & \xi\lambda \end{pmatrix}$$

in $\mathrm{PGL}_2(\mathbb{C})$ is *of order 6* (since $\xi\lambda$ will then have order 6). But as A_4 has no elements of order 6, this means we are done.

It remains to prove (7.1.3). For this, we note (following Serre) that $\mathrm{Ad} \circ \sigma : W_F \longrightarrow \mathrm{GL}_3(\mathbb{C})$ is a *monomial* representation. In particular, there is a character θ of W_E (not invariant by $\mathrm{Gal}(E/F)$) such that

$$\mathrm{Ad} \circ \sigma = \mathrm{Ind}_{W_E}^{W_F} \theta.$$

This means (again by Theorem 5.3.1, this time with $n = 3$) that there is associated to this irreducible representation $\mathrm{Ad} \circ \sigma$ a *cuspidal* automorphic representation of $\mathrm{GL}_3(\mathbb{A}_F)$, call it Π_1. On the other hand, by the "symmetric square lift" (Theorem 5.3.2) $\pi_{\mathrm{ps}}(\sigma)$ has a lift to $\mathrm{GL}_3(\mathbb{A}_F)$, call it Π_1^*, which is almost everywhere associated to the Langlands parameter $\mathrm{Ad} \circ \sigma'_v$. Thus to prove (7.1.3), it clearly suffices to prove that

$$(7.1.4) \qquad \Pi_1 \approx \Pi_1^*.$$

N.B. The automorphic representation Π_1^* will be *cuspidal* automorphic (by Theorem 5.3.2) if and only if $\pi_{\mathrm{ps}}(\sigma)$ is not monomial. But if $\pi_{\mathrm{ps}}(\sigma)$ were equal to $\pi(\sigma')$ for any irreducible two dimensional (let alone monomial) representation of W_F, we would have to conclude that $\sigma' = \sigma$ (which is impossible, since σ is tetrahedral, not monomial). Therefore Π_1^* is also cuspidal, and the proof of 7.1.4 reduces to the following:

Lemma. *The Rankin-Selberg L-function* $L(s, \Pi_1^* \times \tilde{\Pi}_1)$ *on* $\mathrm{GL}(3) \times \mathrm{GL}(3)$ *has a pole at* $s = 1$ *(and so, by Theorem 5.3.3,* Π_1^* *is indeed isomorphic to* Π_1*).*

Proof By definition

$$L(s, \Pi_1^* \times \tilde{\Pi}_1) = \prod_v L(s, (\Pi_1^*)_v \times (\tilde{\Pi}_1)_v),$$

where for almost every v (namely the "unramified" v),

$$L(s, (\Pi_1^*)_v \times (\tilde{\Pi}_1)_v) = L(s, (\mathrm{Ad} \circ \sigma'_v) \otimes (\mathrm{Ad} \circ \tilde{\sigma}_v)).$$

Keeping in mind that $\mathrm{Ad} \circ \sigma$ is monomial, it is possible to check that we also have

$$L(s, (\Pi_1^*)_v \times (\tilde{\Pi}_1)_v) = L(s, (\Pi_1)_v \times (\tilde{\Pi}_1)_v) \tag{7.1.5}$$

(again for almost every v). Indeed, since $\mathrm{Ad} \circ \sigma$ is induced from θ on E, we have

$$\mathrm{Ad} \circ (\tilde{\sigma}_v) = \bigoplus_{w|v} \mathrm{Ind}_{W_{E_w}}^{W_{F_v}} \theta_w^{-1}.$$

Hence

$$\mathrm{Ad}(\sigma'_v) \otimes \mathrm{Ad}(\tilde{\sigma}_v) = \bigoplus_{w|v} \mathrm{Ind}_{W_{E_w}}^{W_{F_v}} (\theta_w^{-1} \otimes \Sigma'_w)$$

if Σ_w (resp. Σ'_w) denotes the restriction of $\mathrm{Ad}(\sigma_v)$ (resp. $\mathrm{Ad}(\sigma'_v)$) to W_{E_w}. (Here we are using the fact that for σ (resp. Σ) a representation of some group G (resp. a subgroup H),

$$\sigma \otimes \mathrm{Ind}_H^G \Sigma \cong \mathrm{Ind}_H^G (\Sigma \otimes \mathrm{Res}\ \sigma|_H).)$$

Similarly we have

$$\mathrm{Ad}(\sigma_v) \otimes \mathrm{Ad}(\tilde{\sigma}_v) = \bigoplus_{w|v} \mathrm{Ind}_{W_{E_w}}^{W_{F_v}} (\theta_w^{-1} \otimes \Sigma_w).$$

So since $\Sigma_w \cong \Sigma'_w$ almost everywhere (by construction), we indeed have

$$\begin{aligned} L(s, (\Pi_1^*)_v \times (\tilde{\Pi}_1)_v) &= L(s, \mathrm{Ad}(\sigma'_v) \otimes \mathrm{Ad}(\tilde{\sigma}_v)) \\ &= L(s, \mathrm{Ad}(\sigma_v) \otimes \mathrm{Ad}(\tilde{\sigma}_v)) \\ &= L(s, (\Pi_1)_v \times (\tilde{\Pi}_1)_v) \end{aligned}$$

for almost every v.

Using (7.1.5), it remains to show that $\Pi_1^* = \Pi_1$. So suppose (7.1.5) holds for all v outside the finite set S. Then

$$L(s, \Pi_1^* \times \tilde{\Pi}_1) = \left(\prod_{v \in S} \frac{L(s, (\Pi_1^*)_v \times (\tilde{\Pi}_1)_v)}{L(s, (\Pi_1)_v \times (\tilde{\Pi}_1)_v)} \right) \cdot L(s, \Pi_1 \times \tilde{\Pi}_1).$$

But by Theorem 5.3.3, $L(s, \Pi_1 \times \tilde{\Pi}_1)$ has a pole at $s = 1$; moreover, the quotient expression in parentheses above is non-zero at $s = 1$. Therefore $L(s, \Pi_1^* \times \tilde{\Pi}_1)$ also has a pole at $s = 1$, as asserted, and this in turn implies (by the same Theorem 5.3.3) that $\Pi_1^* \cong \Pi_1$.

(7.2). The Octahedral Case

(a) *Choosing* $\pi_{\mathrm{ps}}(\sigma)$

In this case, the image of $\sigma(W_F)$ in $\mathrm{PGL}_2(\mathbb{C})$ is S_4, and the pull-back of the normal subgroup $A_4 \subset S_4$ is the Weil group W_E of a quadratic extension E of F.

Pictorially:

$$\begin{array}{ccccccccc} 1 & \longrightarrow & W_E & \longrightarrow & W_F & \longrightarrow & \mathrm{Gal}(E/F) & \longrightarrow & 1 \\ & & \downarrow & & \downarrow & & \downarrow \wr & & \\ 1 & \longrightarrow & A_4 & \longrightarrow & S_4 & \longrightarrow & \mathbb{Z}_2 & \longrightarrow & 1 \end{array}$$

Since $\sigma_E = \mathrm{Res}\, \sigma \mid_{W_E}$ is now of *tetrahedral* type, we know $\pi(\sigma_E)$ exists as an irreducible cuspidal representation of $\mathrm{GL}_2(\mathbb{A}_E)$ (by the results of the last paragraph). Moreover, we again have $\pi(\sigma_E)$ invariant under the action of $\mathrm{Gal}(E/F)$. So again by Theorem 6.1, we conclude that $\pi(\sigma_E)$ must equal $\mathrm{BC}_{E/F}(\pi_i)$ for (this time) *two* irreducible cuspidal representations π_i of $\mathrm{GL}_2(\mathbb{A}_F)$. The problem now is that we can no longer distinguish these π_i's by their central characters. Indeed, π_1 now equals $\pi_2 \otimes \omega$ for a *quadratic* character of $F^\times \setminus \mathbb{A}^\times$; hence $\omega_{\pi_1} = \omega_{\pi_2}\omega^2 = \omega_{\pi_2}$!

Tunnell's contribution to the "Langlands-Tunnell Theorem" was to get around this problem by appealing to a new kind of base-change which appeared only after the publication of [La1], namely the following result:

Proposition. (cf. [J-PS-S3]) *If L is a cubic* not necessarily *Galois extension of F, then each automorphic cuspidal representation π of* $\mathrm{GL}_2(\mathbb{A}_F)$ *has a base change lift* Π *on* $\mathrm{GL}_2(\mathbb{A}_L)$, *i.e.,* $\Pi = \mathrm{BC}_{L/F}(\pi)$ *is automorphic, and for almost every place v of F, and place w of L dividing v, $\pi_v = \pi_v(\sigma_v)$ implies $\Pi_w = \pi(Res_{L_w/F_v}(\sigma_v))$.*

The proof of [J-PS-S3] uses the theory of L-functions for the groups GL(3) and GL(2) $\times$ GL(3) (and is entirely analogous to Jacquet's original proof of base change for GL_2 over a *quadratic* extension in [Ja]). The idea is to introduce the representation Π on $\mathrm{GL}_2(\mathbb{A}_L)$ through the formula

$$L(s, \Pi \times \chi) = L(s, \pi \times \pi(\chi));$$

here χ is any Hecke character of L, $\pi(\chi)$ is the corresponding automorphic representation of $\mathrm{GL}_3(\mathbb{A}_F)$ (whose existence is assured by Theorem 5.3.1 in the *non-Galois case* — recall Remark 5.3.1 (e)), and $L(s, \pi \times \pi(\chi))$ is the Rankin-Selberg L-function on $\mathrm{GL}(2) \times \mathrm{GL}(3)$. Then one shows that $L(s, \Pi \times \chi)$ has the analytic properties required by the Converse Theorem to ensure that Π is automorphic. (The fact that each Π_w is the base change lift of π_v is relatively easy to check, from the definitions.)

N.B. The trace formula methods of [La1] fail in this context precisely because there may not be any Galois group attached to L over F (hence no way to define the twisted trace $R_0^\tau \cdots$). On the other hand, because L-function methods are used, there is no way to *characterize* the image of this base change map; fortunately, as we shall now see, there is also no need for this in the application Tunnell found for this result.

What Tunnell did in [Tu] is introduce L/F as the cubic (*non-normal*) subextension of K/F fixed by a 2-Sylow subgroup (of order 8) of S_4. (More precisely, L is the cubic subextension fixed by all elements of $\mathrm{Gal}(K/F)$ mapping to this chosen Sylow subgroup.) Then if M is the composition in K of L and E (the quadratic Galois extension chosen above), we have the diagram shown in Figure 1,

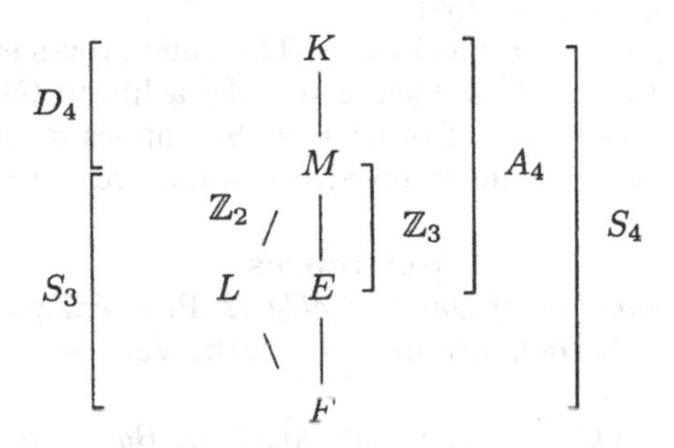

Figure 1

and the crucial:

Lemma. (cf. [Tu], page 174) *There is a unique $i = 1, 2$ such that*

$$\mathrm{BC}_{L/F}(\pi_i) = \pi(\sigma_L)$$

(and this is the π to be designated as $\pi_{\mathrm{ps}}(\sigma)$).

Proof Note first that $\pi(\sigma_L)$ actually exists, since the 2-Sylow subgroup used to define L is just D_4, and therefore σ_L is *monomial*; similarly, $\mathrm{BC}_{L/F}(\pi_i)$ exists for $i = 1, 2$ by the Base Change Theorem quoted above. To prove the Lemma, one appeals to the identity

$$\mathrm{BC}_{M/L}(\mathrm{BC}_{L/F}(\pi_i)) = \pi(\sigma_M) \quad \text{for } i = 1, 2.$$

(This is "transitivity of base change"; it follows immediately from the definition of base change.) Since $\mathrm{BC}_{L/F}(\pi_2)$ and $\mathrm{BC}_{L/F}(\pi_1)$ have the same (quadratic) base change to M, it follows that

$$\mathrm{BC}_{L/F}(\pi_2) \approx \mathrm{BC}_{L/F}(\pi_1) \otimes \omega_{M/L}.$$

Now we claim that the representations $\mathrm{BC}_{L/F}(\pi_i)$ are distinct for $i = 1, 2$. Indeed, if they were not, we would have

$$\mathrm{BC}_{L/F}(\pi_1) \approx \mathrm{BC}_{L/F}(\pi_1) \otimes \omega_{M/L},$$

which by Lemma 11.7 of [La1] implies π_1 is "monomial." By part (b) of Theorem 6.1, this would then imply $\mathrm{BC}_{M/L}(\mathrm{BC}_{L/F}(\pi_1)) = \pi(\sigma_M)$ is *not* cuspidal. But the image of σ_M in $\mathrm{PGL}_2(\mathbb{C})$ is $S_3 \approx D_3$, which means that σ_M itself is monomial and irreducible, i.e., $\pi(\sigma_M)$ *is* cuspidal. This contradiction establishes that $\mathrm{BC}_{L/F}(\pi_1)$ and $\mathrm{BC}_{L/F}(\pi_2)$ are *the* two (distinct) cuspidal representations of $\mathrm{GL}_2(\mathbb{A}_L)$ yielding $\pi(\sigma_M)$ upon base change to M. Since we also have $B_{M/L}(\sigma_L) = \pi(\sigma_M)$, it must be that $\pi(\sigma_L) = \mathrm{BC}_{L/F}(\pi_i)$ for (exactly) one i, as required.

(b) *Proving* $\pi_{\mathrm{ps}}(\sigma) = \pi(\sigma)$.

Write $\pi_{\mathrm{ps}}(\sigma) = \otimes \pi_v(\sigma'_v)$ as before. Then one proves exactly as in the tetrahedral case (but *without* having to take a lift to $\mathrm{GL}_3(\mathbb{C})$) that the non-existence of an element of order 6 in S_4 implies $\sigma_v \cong \sigma'_v$ for almost all v. Since no new ideas are involved, we simply refer the reader to [Tu] for details.

References

[Ant] *Modular Functions of One Variable II*, Proceedings of the Antwerp 1972 Summer School, Lectures in Math. Vol. 349, Springer-Verlag, 1973.

[AC] Arthur, J., and Clozel, L., *Simple Algebras, Base Change, and the Advanced Theory of the Trace Formula*, Annals of Math. Studies, No.120, Princeton University Press, 1989.

[BlRo] Blasius, D., and Rogawski, J., "Zeta functions of Shimura varieties," in *Proc. Symp. Pure Math.*, Vol. 55, Part 2, A.M.S., Providence, 1994, 525–571.

[Bo] Borel, A., "Automorphic L-functions," in *Proc. Symp. Pure Math.*, Vol. 33, Part 2, A.M.S., Providence, 1979, 27–61.

[BoJa] Borel, A., and Jacquet, H., "Automorphic forms and automorphic representations," in *Proc. Symp. Pure Math.* Vol. 33, Part 1, 189–202.

[Car] Carayol, H., "Sur les représentation ℓ-adiques associecs aux formes modulaires de Hilbert," *Ann. Sc. E.N.S.* 19 (1986), 409–468.

[Cas] Casselman, W., "On some results of Atkin and Lehner," *Math. Ann.* 201 (1973), 301–314.

[CoPS] Cogdell, J., and Piatetski-Shapiro, I., "Converse theorems for GL_n, *Pub. Math. I.H.E.S.*, No.79 (1994), 157–214.

[De] Deligne, D., "Formes modulaires et représentations de GL(2)," in *Modular Functions of One Variable, II*, Lecture Notes in Math., Vol. 349, Springer-Verlag, 1973.

[DS] Deligne, P., and Serre, J.-P., "Formes modulaires de poids 1," *Ann. Scient. Ec. Norm. Sup.*, 4e série 7 (1974), 507–530.

[Ge1] Gelbart, S., *Automorphic Forms on Adele Groups*, Annals of Math. Studies, Vol. 83, Princeton University Press, Princeton, 1975.

[Ge2] Gelbart, S., *Lectures on the Arthur-Selberg Trace Formula*, MSRI Preprint No. 041–95, May 1995; Univ. Lecture Ser., Vol. 9, AMS, 1996.

[GeJa] Gelbart, S., and Jacquet, H., "A relation between automorphic forms on GL_2 and GL_3" *Ann. Sci. Ecole Norm. Sup.*, Vol. 11, (1978), 471–541.

[GeLa] Gerardin, P., and Labesse, J.-P., "The solution of a base change problem for GL(2) (following Langlands, Saito, Shintani)," in *Proc. Symp. Pure Math.*, Vol. 33, Part 2, A.M.S., Providence, 1979, 115–133.

[GGPS] Gelfand, I., Graev, M., and Piatetski-Shapiro, I., *Representation Theory and Automorphic Functions*, W.B. Saunders Co., Phila., 1969.

[GK] Gelfand, I., and Kazhdan, D., "Representations of the group $GL(n,k)$," in Proceedings of the Summer School of the Bolyai Janos Math. Soc. on Group Representations, Adam Hilger, London, 1975.

[GoJa] Godement, R., and Jacquet, H., "*Zeta Functions of Simple Algebras*," Lecture Notes in Math., Vol. 260, Springer-Verlag, 1972.

[Gold] Goldstein, L., *Analytic Number Theory*, Prentice-Hall, Inc., Englewood Cliffs, N.J., 1971.

[He] Henniart, G., "Quelques remarques sur les théorèmes réciproque, *Israel Math. Conf. Proceedings*, Vol. 2, The Weizmann Science Press of Israel, 1990, 77–92.

[Ja1] Jacquet, H., "Principal L-functions of the linear group," in *Proc. Symp. Pure Math.*, Vol. 33, Part I, AMS, Providence, 1979, 63–86.

[Ja2] Jacquet, H., *Automorphic Forms on* GL(2)*: II*, Lecture Notes in Mathematics, Vol. 278, Springer-Verlag, New York, 1972

[JL] Jacquet, H., and Langlands, R.P., *Automorphic Forms on* GL(2), Lecture Notes in Math. Vol. 114, Springer-Verlag, 1970.

[J-PS-S1] Jacquet, H., Piatetski-Shapiro, I, and Shalika, J., "Conducteur des représentations du groupe linéaire," *Math. Ann.* 256 (1981), 199–214.

[J-PS-S2] Jacquet, H., Piatetski-Shapiro, I., and Shalika, J., "Automorphic forms on GL(3), I and II," *Annals of Math.* 109 (1979), 169–258.

[J-PS-S3] Jacquet, H., Piatetski-Shapiro, I., and Shalika, J., "Relèvement cubiqne non normal," C.R. Acad. Sci. Paris 292 (1981), 567–579.

[J-Sh1,2] Jacquet, H., and Shalika, J., "On Euler Products and the Classification of Automorphic Representations, I and II," *Amer. J. Math.*, Vol. 103, No.3 (1981), 499–558 and 777–815.

[Kn] Knapp, A.W., "Local Langlands Correspondence: The Archimedean Case," in *Proc. Symp. Pure Math.*, Vol. 55 (1994), Part 2, 393–410.

[Kud] Kudla, S., "Local Langlands correspondence: The non-Archimedean Case," in *Proc. Symp. Pure Math.*, Vol. 55, Part 2, AMS, Providence, 1994, 365–391.

[Kut] Kutzko, P., "The Local Langlands conjecture for GL(2) of a finite field," *Annals of Math.* 112 (1980), 381–412.

[La1] Langlands, R.P., *Base Change for* GL(2), Annals of Math. Studies, Vol. 96, Princeton University Press, Princeton, NJ, 1980.

[La2] Langlands, R.P., "On the notion of an automorphic representation," in *Proc. Symp. Pure Math.*, Vol. 33, Part 2, 203–207.

[La3] Langlands, R.P., "On the classification of irreducible representations of real algebraic groups," in *Representation Theory and Harmonic Analysis on Semi-simple Groups*, (P. Sally and D. Vogan, editors), Math. Surveys and Monographs, Vol. 31, AMS, Providence, 1989, 101–170.

[La4] Langlands, R.P., "On the functional equations of Artin L-functions," mimeographed notes, Yale University; cf. *Rice University Studies*, Vol. 56, No.2, 1970, 23–28.

[La5] Langlands, R.P., "Modular forms and ℓ-adic representations, in [Ant], pp.361–500.

[La6] Langlands, R.P., "Automorphic representations, Shimura varieties and motives," in *Pure Symp. Pure Math.*, Vol. 33, Part 2, A.M.S., Providence, 1979, 205–246.

[Mo-Wald] Moeglin, C. and Waldspurger, J.-L., "Le spectre residuel de GL(n)," *Ann. Sci. École Norm Sup.* (4) 22, (1989), 605–674.

[PS] Piatetski-Shapiro, I., *Complex Representations of* GL$(2, K)$ *for finite Fields* K*, Contemporary Mathematics* Vol. 16, AMS, Providence, 1983.

[Ram] Ramakrishnan, D., "Pure Motives and Automorphic forms," in *Proc. Symp. Pure Math.*, Vol. 55, Part 2, A.M.S., Providence, 1994, 411–446.

[RuSi] Rubin K., and Silverberg, A., "A report on Wiles' Cambridge lectures," *Bull. AMS* (new series) 31, 1994, 15–38.

[Sai] Saito, H., *Automorphic Forms and Extensions of Number Fields*, Lectures in Math., No.8, Kinokuniya Book Store Co. Ltd., Tokyo, Japan, 1975.

[Se] Serre, J.-P., "Sur les représentations modulaires de degré 2 de Gal($\bar{\mathbb{Q}}/\mathbb{Q}$)," *Duke Math. J.* 54 (1987), 179–230.

[Sel] Selberg, A., "Harmonic analysis and discontinuous groups in weakly symmetric Riemannian spaces with applications to Dirichlet series," *J. Ind. Math. Soc.* 20 (1956), 47–87.

[Shaf] Shafarevich, I., *Algebra I*, Encyclopaedia of Mathematical Sciences, Vol. 11 (A. Kostrikin and I. Shafarevich, Editors), Springer-Verlag, 1990.

[Shal] Shalika, J., "The multiplicity one theorem for GL$_n$," *Annals of Math.*, 100 (1974), 171–193.

[Sh] Shimura, G., "On the holomorphy of certain Dirichlet series", Proc. London Math. Soc. 3 (1975), 79–98.

[Shin] Shintani, T., "On liftings of holomorphic cusp forms," in *Proc. Symp. Pure Math.*, Vol. 33, Part 2, A.M.S., Providence, 1979, 97–110.

[Silv] Silverman, J., *The Arithmetic of Elliptic Curves*, Grad. Texts in Math. Vol. 106, Springer-Verlag, 1986.

[ST] Shalika, J., and Tanaka S., "On an explicit construction of a certain class of automorphic forms," *Amer. J. Math.*, Vol. 91 (1969), 1049–1076.

[Ta] Tate, J., "Number theoretic background," in *Proc. Symp. Pure Math.*, Vol. 33, Part 2, A.M.S., Providence, 1979, 3–26.

[Tu] Tunnell, J., "Artin's Conjecture for representations of octahedral type," *Bull. AMS* (new series) 5, 1981, 173–175.

[W1] Wiles, A., "Modular elliptic curves and Fermat's Last Theorem," *Annals of Math.* 142 (1995), 443–551.

SERRE'S CONJECTURE

BAS EDIXHOVEN

The aim of the first section is to state Serre's conjecture and to tell what is presently known about it, without proof. We start by recalling what modular forms are. Then we recall the result, due to Deligne, that to a mod p modular form one can associate a mod p Galois representation. After that we state Serre's conjecture and what we know about it. In Section 2 we will see which cases of it are actually needed in order to prove, following Wiles, that all semi-stable elliptic curves over $\mathbb{Q}$ are modular. In the last two sections we will sketch the proofs in those cases. These notes follow, to some extent, the lectures given by Dick Gross during the conference.

1 Serre's Conjecture: Statement and Results

Let $N \geq 1$ and k be integers and let R be a $\mathbb{Z}[1/N]$-algebra. We will first recall Katz's definition [22] of modular forms of level N and weight k over R. This definition may seem more complicated than necessary, but it is very convenient in order to deal with modular forms over fields of positive characteristic p. For example, one has the Hasse invariant (see [24, §12.4]), which cannot be lifted to characteristic zero as a form of level 1 and weight $p-1$ for p equal to 2 or 3, and one has the derivation $q\,d/dq$ sending modular forms to modular forms, increasing the weight by $p+1$ (p being the characteristic of the finite field). A good reference for more details concerning modular forms in various settings is [14]. That book also contains an account of Serre's conjecture by Darmon.

Let $[\Gamma_1(N)]_R$ denote the category whose objects are pairs $(E/S/R, \alpha)$, with S an R-scheme, E/S a generalized elliptic curve in the sense of [10] and $\alpha\colon (\mathbb{Z}/N\mathbb{Z})_S \to E[N]$ an embedding of group schemes such that the image of α meets all irreducible components of all geometric fibres of E/S. The morphisms from $(E'/S'/R, \alpha')$ to $(E/S/R, \alpha)$ in $[\Gamma_1(N)]_R$ are Cartesian diagrams:

$$\begin{array}{ccc} E' & \rightarrow & E \\ \downarrow & & \downarrow \\ S' & \rightarrow & S \end{array} \tag{1.1}$$

which are compatible with α and α'. For a generalized elliptic curve E over a scheme S we have the invertible $\mathcal{O}_S$-module $\underline{\omega}_{E/S} := 0^*\Omega^1_{E/S}$, obtained by pulling back the sheaf of Kähler differentials of E over S by the zero section 0 in $E(S)$. A modular form f of level N and weight k over R is then a rule, that assigns to every object $(E/S/R, \alpha)$ of $[\Gamma_1(N)]_R$ an element $f(E/S/R, \alpha)$ of $\underline{\omega}^{\otimes k}_{E/S}(S)$, compatible with morphisms in $[\Gamma_1(N)]_R$. The R-module of such modular forms will be denoted by $M(N,k)_R$. Let us give one example: for p a prime, the Hasse invariant is an element of $M(1, p-1)_{\mathbb{F}_p}$.

There are more down to earth ways to describe $M(N,k)_R$. For $N \geq 5$ the category $[\Gamma_1(N)]_R$ has a final object, called $(E_{\text{univ}}/X_1(N)_R, \alpha_{\text{univ}})$, and in that case one simply has $M(N,k)_R = \mathrm{H}^0(X_1(N)_R, \underline{\omega}^{\otimes k})$. Recall that $X_1(N)_R$ is a smooth projective curve over R whose fibres are geometrically irreducible. For $N \geq 1$ and $n \geq 3$ invertible in R one has the description:

$$(1.2) \qquad M(N,k)_R = \mathrm{H}^0(\overline{\mathcal{M}}([\Gamma_1(N), \Gamma(n)]_R), \underline{\omega}^{\otimes k})^G,$$

where $\overline{\mathcal{M}}([\Gamma_1(N), \Gamma(n)]_R)$ denotes the moduli scheme parametrizing triples $(E/S/R, \alpha, \beta)$ of generalized elliptic curves with $\alpha\colon (\mathbb{Z}/N\mathbb{Z})_S \to E[N]$ and $\beta\colon (\mathbb{Z}/n\mathbb{Z})^2_S \to E[n]$ embeddings of group schemes such that the image of $\alpha + \beta$ meets all irreducible components of all geometric fibres, and where G is the group $\mathrm{GL}_2(\mathbb{Z}/n\mathbb{Z})$. This $\overline{\mathcal{M}}([\Gamma_1(N), \Gamma(n)]_R)$ is a smooth projective curve over R, but its fibres are not geometrically irreducible.

In the special case $R = \mathbb{C}$ one easily shows that $M(N,k)_{\mathbb{C}}$ is naturally isomorphic to the space of modular forms defined as certain holomorphic functions on the upper half plane $\mathbb{H}$ (see Chapter III, Section 1.5).

One can show, for example by using the modular form Δ in $M(1,12)_R$, that $M(N,k)_R = 0$ for all $k < 0$.

Let $R \to R'$ be a morphism of $\mathbb{Z}[1/N]$-algebras. Then we have a morphism of R'-modules $M(N,k)_R \otimes_R R' \to M(N,k)_{R'}$. Such a morphism is not always an isomorphism (consider for example $\mathbb{Z} \to \mathbb{F}_2$, $N = 1$ and $k = 1$). However, it is an isomorphism when $R \to R'$ is flat (use (1.2)).

Over $\mathbb{Z}[[q]]$ one has the Tate curve $\mathrm{Tate}(q)$, which is equipped with a basis dt/t of $\underline{\omega}_{\mathrm{Tate}(q)/\mathbb{Z}[[q]]}$. One has an isomorphism of groups:

$$(1.3) \qquad (\mathbb{Z}/N\mathbb{Z})^2 \longrightarrow \mathrm{Tate}(q)[N](\overline{\mathbb{Q}((q))}), \quad (a,b) \mapsto \zeta_N^a (q^{1/N})^b,$$

where $\zeta_N \in \overline{\mathbb{Q}}^*$ is a fixed root of unity of order N. For R a $\mathbb{Z}[1/N, \zeta_N]$-algebra one has the standard q-expansion map $M(N,k)_R \to R[[q]]$ sending f to the series $\sum_{n \geq 0} a_n(f) q^n$ defined by:

$$(1.4) \qquad f(\mathrm{Tate}(q), 1 \mapsto \zeta_N) = \left(\sum_{n \geq 0} a_n(f) q^n \right) (dt/t)^k.$$

This map is injective. The R-modules $M(N,k)_R$ can also be defined as follows. One replaces "generalized elliptic curve" in the definition we gave by "elliptic curve" and one demands that the q-expansions obtained from all points of order N of the Tate curve over $\mathbb{Z}[\zeta_N]((q^{1/N}))$ are power series in $q^{1/N}$ (a priori they are Laurent series). The R-module of cusp forms of level N and weight k is defined to be the submodule $M^0(N,k)_R$ of $M(N,k)_R$ of those f all of whose q-expansions have zero as constant coefficient. Equivalently, they are the forms that vanish on degenerate elliptic curves.

Suppose that $k \geq 1$. The R-modules $M(N,k)_R$ and $M^0(N,k)_R$ are equipped with certain endomorphisms. For $n \geq 1$ one has the Hecke operator T_n, defined in terms of isogenies of degree n. For a in $(\mathbb{Z}/N\mathbb{Z})^*$ one has the diamond operator $\langle a \rangle$, induced by the automorphism of $[\Gamma_1(N)_R]$ that sends $(E/S/R, \alpha)$ to $(E/S/R, a\alpha)$. The action of these operators is given by the usual formulas in terms of q-expansions (see Chapter III, Sections 2.4 and 2.5).

In particular, one has $a_1(T_n(f)) = a_n(f)$. The construction of the T_n in this generality is a bit complicated, especially for $k=1$. See for example [17, §4] for a construction of T_p on $M(N,1)_{\mathbb{F}_p}$. In general, one can construct the T_n as follows. It suffices to construct the T_p with p prime. If p divides N then p is invertible in R and the construction is easy. So suppose that p does not divide N. In this case one proceeds as in [17, §4]. Take f in $M(N,k)_R$, view it as a G-invariant section of $\underline{\omega}^{\otimes k}$ on $\overline{\mathcal{M}}([\Gamma_1(N),\Gamma(n)]_R)$, as in (1.2). Then restrict it to $\mathcal{M}([\Gamma_1(N),\Gamma(n)]_R)$, the complement of the cusps. This is affine, hence this restriction of f is a linear combination, with coefficients in R, of sections of $\underline{\omega}^{\otimes k}$ over $\mathcal{M}([\Gamma_1(N),\Gamma(n)]_{\mathbb{Z}[1/Nn]})$ and one knows what T_p does with those. One verifies that the $T_p(f)$ obtained in this way is again G-invariant and regular at the cusps.

The endomorphisms T_n and $\langle a \rangle$ of $M(N,k)_R$ all commute with each other. For ε a character of $(\mathbb{Z}/N\mathbb{Z})^*$ with values in R^* let $M(N,k,\varepsilon)_R$ denote the R-submodule of $M(N,k)_R$ of elements f such that $\langle a \rangle(f) = \varepsilon(a)f$ for all a; such f will be called forms of type (N,k,ε). If R is a field and f a non-zero element of $M(N,k)_R$ which is an eigenform for all T_n, then the formula $a_1(T_n(f)) = a_n(f)$ implies that the corresponding eigenspace for the T_n has dimension one, and that there is a unique character ε such that f is of type (N,k,ε).

Theorem 1.5 (Deligne) *Let p be a prime, N an integer prime to p, and $k \geq 1$. Let f be a non-zero eigenform in $M(N,k,\varepsilon)_{\overline{\mathbb{F}}_p}$, and let a_n be its eigenvalue for T_n. Then there is a unique semi-simple continuous representation $\rho_f \colon G_{\mathbb{Q}} := \mathrm{Gal}(\overline{\mathbb{Q}}/\mathbb{Q}) \to \mathrm{GL}_2(\overline{\mathbb{F}}_p)$, unramified outside pN, such that for all primes l not dividing pN the (arithmetic) Frobenius element $\rho_f(\mathrm{Frob}_l)$ has trace a_l and determinant $\varepsilon(l)l^{k-1}$.*

(The topology on $\mathrm{GL}_2(\overline{\mathbb{F}}_p)$ in this statement is the discrete one.) In fact, Deligne [8] showed that for an eigenform f of weight $k \geq 2$ and with coefficients in $\overline{\mathbb{Q}}_p$ one even has a p-adic representation ρ_f of $G_{\mathbb{Q}}$ with values in $\mathrm{GL}_2(\overline{\mathbb{Q}}_p)$. The case $k = 2$ had already been treated by Shimura in [43]. A detailed and reasonably elementary proof of Theorem 1.5 can be found in [17].

Let f be as in Theorem 1.5. The fact that every elliptic curve has the automorphism -1 implies that $\varepsilon(-1)f = \langle -1 \rangle(f) = (-1)^k f$, hence that $\varepsilon(-1) = (-1)^k$. It follows that, for σ in $G_{\mathbb{Q}}$ any complex conjugation, $\det(\rho_f(\sigma)) = \varepsilon(-1)(-1)^{k-1} = -1$. A representation of $G_{\mathbb{Q}}$ with this property will be called odd. Serre conjectured [37, (3.2.3)$_?$] that any continuous odd representation $\rho: G_{\mathbb{Q}} \to \mathrm{GL}_2(\overline{\mathbb{F}}_p)$ is isomorphic to some ρ_f. Then of course the question arises how to see from ρ what the possible levels, weights and characters for such f are. So Serre also conjectured [37, (3.2.4)$_?$] that for irreducible ρ, such an f exists of a certain "minimal" type $(N(\rho), k(\rho), \varepsilon(\rho))$. Section 5 of [37] gives a number of examples where this conjecture can be at least partially verified. These examples concern ρ with values in $\mathrm{GL}_2(\mathbb{F}_q)$ with q equal to 2, 3, 4, 7 and 9. In the first two cases, every ρ satisfies [37, (3.2.3)$_?$]. Recently, Shepherd-Barron and Taylor [42] proved similar results for q equal to 4 and 5. In general, very little is known about [37, (3.2.3)$_?$], but there has been a lot of progress on the question of whether [37, (3.2.3)$_?$] implies [37, (3.2.4)$_?$]. Let us mention that [37, §4] gives some spectacular consequences of [37, (3.2.4)$_?$], including Fermat's Last Theorem and variants of it, and Shimura-Taniyama. It seems that no work has been done on the problem posed in Remark 4 on page 197 of [37]. Before we can state Serre's conjecture, we need some terminology.

Let p be a prime and $\rho: G_{\mathbb{Q}} \to \mathrm{GL}_2(\overline{\mathbb{F}}_p)$ a continuous representation. Then the image G of ρ is finite, let's say equal to $\mathrm{Gal}(K/\mathbb{Q})$ with K a finite Galois extension of $\mathbb{Q}$ in $\overline{\mathbb{Q}}$. It follows that ρ is unramified at all but finitely many primes. The number $N(\rho)$ is by definition the Artin conductor of ρ, except that one doesn't take the prime p into account. More precisely, $N(\rho)$ will be a positive integer prime to p. Let l be a prime number different from p. Let I_l denote the inertia subgroup at l of G corresponding to a place of K above l, and let $I_l = I_{l,0} \supset I_{l,1} \supset \cdots$ be the higher ramification subgroups: $I_{l,i}$ is the subgroup of the decomposition group at the chosen place whose elements act trivially on the ring of integers modulo the $(i+1)$th power of the maximal ideal. The valuation of $N(\rho)$ at l, i.e., the number of factors of l in it, is then given by the formula:

$$v_l(N(\rho)) = \sum_{i \geq 0} \frac{1}{[I_{l,0} : I_{l,i}]} \dim(V/V^{I_{l,i}}), \tag{1.6}$$

where V is a two-dimensional $\overline{\mathbb{F}}_p$-vector space with a G-action giving ρ, and where $V^{I_{l,i}}$ denotes its subspace of invariants under $I_{l,i}$. For a motivation for formula (1.6), see [38], Section 19. Let us note, by the way, that replacing $V/V^{I_{l,i}}$ by the kernel of the map to the coinvariants $V \to V_{I_{l,i}}$ would give the same result. For $i > 0$ this is clear since then $I_{l,i}$ is an l-group, for $i = 0$ one uses that $V_{I_{l,0}}$ is obtained by first taking the coinvariants for the l-group $I_{l,1}$ and then the coinvariants for the cyclic group $I_{l,0}/I_{l,1}$.

We are now able to define the character $\varepsilon(\rho)$ and the image of $k(\rho)$ in $\mathbb{Z}/(p-1)\mathbb{Z}$. For each positive integer n let $\chi_n: G_{\mathbb{Q}} \to (\mathbb{Z}/n\mathbb{Z})^*$ denote the cyclotomic character of the nth roots of unity, i.e., the character such that for all σ in $G_{\mathbb{Q}}$ and all x in $\overline{\mathbb{Q}}$ with $x^n = 1$ we have $\sigma(x) = x^{\chi(\sigma)}$. We will often view a character of the group $(\mathbb{Z}/n\mathbb{Z})^*$ as a character of $G_{\mathbb{Q}}$, by composing it with χ_n. With these conventions, Theorem 1.5 implies that for f an eigenform in $M(N,k,\varepsilon)_{\overline{\mathbb{F}}_p}$ one has $\det \circ \rho_f = \varepsilon \chi_p^{k-1}$. This equality shows that $\det \circ \rho_f$ is a character of the group $(\mathbb{Z}/pN\mathbb{Z})^*$. This group is canonically isomorphic to $(\mathbb{Z}/N\mathbb{Z})^* \times \mathbb{F}_p^*$ via the corresponding isomorphism of rings. Under this decomposition, the restriction of $\det \circ \rho_f$ to $(\mathbb{Z}/N\mathbb{Z})^*$ is ε, and the restriction to $\mathbb{F}_p^*$ is χ_p^{k-1}. Let us now go back to the representation ρ. By the definition of $N(\rho)$, it is unramified away from $pN(\rho)$. Since the maximal abelian extension of $\mathbb{Q}$ is the cyclotomic extension, $\det \circ \rho$ can be considered as a character of $(\mathbb{Z}/p^n\mathbb{Z})^* \times (\mathbb{Z}/N(\rho)^m)^*$, for some $n, m \gg 0$. Since the character has values in $\overline{\mathbb{F}}_p^*$, one can take $n = 1$. Comparing formula (1.6) for ρ and $\det \circ \rho$, and using some class field theory, one sees that one can take $m = 1$. Then one defines $\varepsilon(\rho)$ to be the restriction of $\det \circ \rho$ to the factor $(\mathbb{Z}/N(\rho)\mathbb{Z})^*$, and the image of $k(\rho)$ in $\mathbb{F}_p^*$ such that the restriction of $\det \circ \rho$ to $\mathbb{F}_p^*$ is $\chi_p^{k(\rho)-1}$.

Now we get to the exact definition of $k(\rho)$. The first thing to note is that $k(\rho)$ depends only on the restriction of ρ to some inertia subgroup $I_p \subset G_{\mathbb{Q}}$ at p. Hence the level $N(\rho)$ reflects the ramification away from p and the weight $k(\rho)$ the ramification at p. In order to state the definition of $k(\rho)$ we first have to classify the two-dimensional representations over $\overline{\mathbb{F}}_p$ of a decomposition subgroup $G_p \subset G_{\mathbb{Q}}$ at p (this subgroup G_p depends on the choice of a maximal ideal containing p, but since these are permuted transitively by $G_{\mathbb{Q}}$ the choice will not matter). For this classification, we need some terminology.

Let $\overline{\mathbb{Z}}$ be the integral closure of $\mathbb{Z}$ in $\overline{\mathbb{Q}}$. Let us choose a morphism of rings $\overline{\mathbb{Z}} \to \overline{\mathbb{F}}_p$. This gives us a decomposition group G_p (the stabilizer of the kernel). The action of G_p on $\overline{\mathbb{F}}_p$ gives a surjection $G_p \to G_{\mathbb{F}_p}$; its kernel I_p is the inertia group. We can identify G_p with $\mathrm{Gal}(\overline{\mathbb{Q}}_p/\mathbb{Q}_p)$ and I_p with $\mathrm{Gal}(\overline{\mathbb{Q}}_p/\mathbb{Q}_p^{\mathrm{unr}})$, where $\mathbb{Q}_p^{\mathrm{unr}}$ is the maximal unramified extension of $\mathbb{Q}_p$. Let $\mathbb{Q}_p^{\mathrm{tr}}$ be the maximal tamely ramified extension of $\mathbb{Q}_p$. Then the quotient $I_{p,t} := \mathrm{Gal}(\mathbb{Q}_p^{\mathrm{tr}}/\mathbb{Q}_p^{\mathrm{unr}})$ of I_p is called the tame ramification group, and the subgroup $I_{p,\mathrm{w}} := \mathrm{Gal}(\overline{\mathbb{Q}}_p/\mathbb{Q}_p^{\mathrm{tr}})$ of I_p the wild ramification group. The tame extensions of $\mathbb{Q}_p^{\mathrm{unr}}$ are obtained by taking nth roots of elements p, with n prime to p. It follows that $I_{p,\mathrm{t}}$ can be identified with $\varprojlim \mathbb{F}_{p^n}^*$, where the limit is taken over the $n \geq 1$ and where the transition morphisms are norm maps. A character $\phi: I_{p,\mathrm{t}} \to \overline{\mathbb{F}}_p^*$ is called of level n, if n is the smallest integer such that ϕ factors through $\mathbb{F}_{p^n}^*$. The n characters

$$I_{p,\mathrm{t}} \to \mathbb{F}_{p^n}^* \to \overline{\mathbb{F}}_p^*$$

that are induced by embeddings of fields $\mathbb{F}_{p^n} \to \overline{\mathbb{F}}_p$ are called the fundamental characters of level n.

Suppose now that $\rho_p: G_p \to \mathrm{GL}(V)$ is a continuous 2-dimensional representation on a $\overline{\mathbb{F}}_p$-vector space. Let V^{ss} be the semi-simplification of V for the action of G_p. we claim that $I_{p,\mathrm{w}}$ acts trivially on V^{ss}. To see that, let W be an irreducible factor of V^{ss}. Then the subspace $W^{I_{p,\mathrm{w}}}$ of W is stable under G_p because $I_{p,\mathrm{w}}$ is a normal subgroup of it, so either $W^{I_{p,\mathrm{w}}}$ is trivial or it is W. Since the image of G_p in $\mathrm{GL}(W)$ is finite, the representation of G_p on W can be realized over a finite extension of $\mathbb{F}_p$. For such a realization W', the number of fixed points for $I_{p,\mathrm{w}}$ is a multiple of p since the image of $I_{p,\mathrm{w}}$ is a p-group, hence W' has non-trivial $I_{p,\mathrm{w}}$-invariants. This proves that the action of I_p on V^{ss} is given by two characters $\phi, \phi': I_{p,\mathrm{t}} \to \overline{\mathbb{F}}_p^*$. Since $\mathrm{Gal}(\overline{\mathbb{F}}_p/\mathbb{F}_p)$ acts by conjugation on $I_{p,\mathrm{t}}$, it follows that $\{\phi^p, \phi'^p\} = \{\phi, \phi'\}$. This means that there are two cases: either ϕ and ϕ' are both of level one, or ϕ and ϕ' are both of level two and $\phi^p = \phi'$, $\phi'^p = \phi$. In the first case ρ_p is reducible, whereas in the second case ρ_p is absolutely irreducible.

We need one more ingredient before we can define $k(\rho_p)$: the notion of "finiteness at p". Let $\mathbb{F}$ be a finite extension of $\mathbb{F}_p$ such that ρ_p can be realized over $\mathbb{F}$. An $\mathbb{F}$-vector space scheme V over a scheme S is then an S-scheme V, together with a structure of $\mathbb{F}$-vector space on all the sets of points $V(T)$ for S-schemes T, functorially in T. Equivalently, an $\mathbb{F}$-vector space scheme over S is a commutative group scheme over S given with an action of $\mathbb{F}$ on it by endomorphisms. An $\mathbb{F}$-vector space scheme V over $\mathbb{Q}_p$ gives rise to the representation of G_p on the $\mathbb{F}$-vector space $V(\overline{\mathbb{Q}}_p)$. This construction induces an equivalence between the category of $\mathbb{F}$-vector space schemes finite over $\mathbb{Q}_p$ and the category of representations of G_p on finite dimensional $\mathbb{F}$-vector spaces. Concretely, the $\mathbb{F}$-vector space scheme $V_{\mathbb{Q}_p}$ corresponding to ρ_p can be obtained as the quotient of the scheme $\mathbb{F}^2_{\mathbb{Q}_p} \times_{\mathrm{Spec}(\mathbb{Q}_p)} \mathrm{Spec}(\overline{\mathbb{Q}}_p)$ by the group G_p, with σ in G_p acting as $(\rho_p(\sigma)^{-1}, \mathrm{Spec}(\sigma))$. One says that ρ_p is finite at p if $V_{\mathbb{Q}_p}$ can be extended to a finite flat $\mathbb{F}$-vector space scheme over $\mathbb{Z}_p$. It is equivalent to demand that $V_{\mathbb{Q}_p}$ can be extended to a finite flat $\mathbb{F}$-vector space scheme over the ring of integers $\mathbb{Z}_p^{\mathrm{unr}}$ in $\mathbb{Q}_p^{\mathrm{unr}}$, or to demand that it can be extended over $\mathbb{Z}_p$ or $\mathbb{Z}_p^{\rm unr}$ as a finite flat group scheme. Finally, one can formulate the condition that ρ_p be finite at p in Galois theoretic terms, without using group schemes (see [15, Proposition 8.1], and its proof, and [37, (2.4.7)]).

We will now give the definition of $k(\rho_p)$. Let ψ and ψ' be the two fundamental characters of level 2.

Definition 1.7. Let $\rho_p\colon G_p \to \mathrm{GL}(V)$, $V^{\rm ss}$, ϕ and ϕ' be as above. We associate an integer $k(\rho_p)$ to ρ_p as follows.

1. Suppose that ϕ and ϕ' are of level 2. We have:

$$\rho_p|_{I_p} \cong \begin{pmatrix} \phi & 0 \\ 0 & \phi' \end{pmatrix}.$$

   After interchanging ϕ and ϕ' if necessary, we have (uniquely) $\phi = \psi^{a+pb} = \psi^a\psi'^b$ and $\phi' = \psi'^a\psi^b$ with $0 \le a < b \le p-1$. We set $k(\rho_p) = 1 + pa + b$.

2. Suppose that ϕ and ϕ' are of level 1.

   (a) If $\rho_p|_{I_{p,\rm w}}$ is trivial, then we have:

$$\rho_p|_{I_p} \cong \begin{pmatrix} \chi^a & 0 \\ 0 & \chi^b \end{pmatrix}$$

   with $0 \le a \le b \le p-2$. We set $k(\rho_p) = 1 + pa + b$.

   (b) Suppose that $\rho_p|_{I_{p,\rm w}}$ is not trivial. We have:

$$\rho_p|_{I_p} \cong \begin{pmatrix} \chi^\beta & * \\ 0 & \chi^\alpha \end{pmatrix}$$

   for unique α and β with $0 \le \alpha \le p-2$ and $1 \le \beta \le p-1$. We set $a = \min(\alpha,\beta)$, $b = \max(\alpha,\beta)$. If $\chi^{\beta-\alpha} = \chi$ and $\rho_p \otimes \chi^{-\alpha}$ is not finite at p then we set $k(\rho_p) = 1 + pa + b + p - 1$; otherwise we set $k(\rho_p) = 1 + pa + b$.

□

For $\rho\colon G_{\mathbb{Q}} \to \mathrm{GL}_2(\overline{\mathbb{F}}_p)$ continuous, we define $k(\rho)$ to be $k(\rho_p)$, where ρ_p is the restriction of ρ to a decomposition group at p.

Conjecture 1.8 (Serre) *Let p be a prime number and $\rho\colon G_{\mathbb{Q}} \to \mathrm{GL}_2(\overline{\mathbb{F}}_p)$ a continuous irreducible odd representation. Then there exists an eigenform f in $M^0(N(\rho), k(\rho), \varepsilon(\rho))_{\overline{\mathbb{F}}_p}$ such that ρ is isomorphic to ρ_f.*

Several remarks should be made at this point. First of all, it is easy to see that for f an eigenform which is not cuspidal, the representation ρ_f is reducible (restrict f to the cusps, and study the action of the Hecke operators on the cusps). This explains that in the conjecture, one can demand that f be a cuspform.

The second remark is more important. The conjecture, as we state it here (i.e., as suggested in [15, 4.3]), is not equivalent to the one stated by Serre in [37, (3.2.4)?]. The difference comes from the fact that the modular forms over $\overline{\mathbb{F}}_p$ considered in [37] are defined to be those obtained by reduction mod p of forms over $\overline{\mathbb{Q}}$. More precisely, in [37], a cuspidal modular form over $\overline{\mathbb{F}}_p$ of some type (N, k, ε), with N prime to p and $k \geq 2$, is an element in the image of the map $M^0(N, k, \tilde{\varepsilon})_{\overline{\mathbb{Z}}_p} \to M^0(N, k, \varepsilon)_{\overline{\mathbb{F}}_p}$, where $\overline{\mathbb{Z}}_p$ is the integral closure of $\mathbb{Z}_p$ in $\overline{\mathbb{Q}}_p$ and $\tilde{\varepsilon}\colon (\mathbb{Z}/N\mathbb{Z})^* \to \overline{\mathbb{Z}}_p^*$ is the Teichmüller lift of ε, i.e., $\tilde{\varepsilon}$ induces ε and they have the same order. The problem is that these reduction maps are not all surjective. Before we discuss what is known today about Serre's conjecture, we will discuss the differences between Conjecture 1.8 and [37, (3.2.4)?]. It was suggested by Serre in [40] to replace the mod p modular forms in [37] by those defined by Katz, i.e., the ones we are using here. See also [41].

Let us first consider the problem of lifting modular forms from $\overline{\mathbb{F}}_p$ to $\overline{\mathbb{Z}}_p$, without paying attention to the character. Then we have the following result.

Lemma 1.9 *Let p be a prime, $N \geq 1$ prime to p.*

1. *Suppose that $k \geq 2$. Then the map $M^0(N, k)_{\overline{\mathbb{Z}}_p} \to M^0(N, k)_{\overline{\mathbb{F}}_p}$ is surjective if $N \neq 1$ or if $p > 3$.*

2. *The map $M^0(1, k)_{\overline{\mathbb{Z}}_2} \to M^0(1, k)_{\overline{\mathbb{F}}_2}$ is not surjective if and only if $k \geq 12$ and ($k \equiv 1$ mod 2 or $k \equiv 2$ mod 12).*

3. *The map $M^0(1, k)_{\overline{\mathbb{Z}}_3} \to M^0(1, k)_{\overline{\mathbb{F}}_3}$ is not surjective if and only if $k \geq 12$ and $k \equiv 2$ mod 12.*

Proof. Let us prove the first statement; the other two can be proved using the explicit descriptions of the rings of modular forms of level one over $\mathbb{Z}$, $\mathbb{F}_2$ and $\mathbb{F}_3$ found in [9, Proposition 6.2]. Suppose first that $N \geq 5$. Because $\overline{\mathbb{Z}}_p$ is flat over $\mathbb{Z}_p$ and $\overline{\mathbb{F}}_p$ is flat over $\mathbb{F}_p$, it suffices to prove that the reduction map induced by $\mathbb{Z}_p \to \mathbb{F}_p$ is surjective. Consider the long exact cohomology sequence arising from the short exact sequence of sheaves on $X_1(N)_{\mathbb{Z}_p}$:

$$0 \to \underline{\omega}^{\otimes k}(-\text{cusps}) \to \underline{\omega}^{\otimes k}(-\text{cusps}) \to i_*\underline{\omega}^{\otimes k}(-\text{cusps}) \to 0, \tag{1.9.1}$$

where the map $\underline{\omega}^{\otimes k}(-\text{cusps}) \to \underline{\omega}^{\otimes k}(-\text{cusps})$ is multiplication by p and $i\colon X_1(N)_{\mathbb{F}_p} \to X_1(N)_{\mathbb{Z}_p}$ denotes the closed immersion. To get the surjectivity we want, it is sufficient to show that $\mathrm{H}^1(X_1(N)_{\mathbb{F}_p}, \underline{\omega}^{\otimes k}(-\text{cusps})) = 0$, since by Nakayama's lemma and the long exact sequence this implies that $\mathrm{H}^1(X_1(N)_{\mathbb{Z}_p}, \underline{\omega}^{\otimes k}(-\text{cusps})) = 0$. The Kodaira-Spencer isomorphism (see [22, A1.4]) and Serre duality give isomorphisms:

$$\begin{aligned}\mathrm{H}^1(X_1(N)_{\mathbb{F}_p}, \underline{\omega}^{\otimes k}(-\text{cusps})) &= \mathrm{H}^1(X_1(N)_{\mathbb{F}_p}, \Omega \otimes \underline{\omega}^{\otimes k-2}) \\ &= \mathrm{H}^0(X_1(N)_{\mathbb{F}_p}, \underline{\omega}^{\otimes 2-k})^\vee.\end{aligned}$$

So if $k > 2$, this shows what we want, since the degree of $\underline{\omega}$ is positive. The case $k = 2$ is in fact easy, since the Kodaira-Spencer isomorphism identifies weight 2 cuspforms with differential forms, and those can be lifted. Another way to phrase the argument is to say that the dimensions of $M^0(N,k)_{\mathbb{F}_p}$ and $M^0(N,k)_{\mathbb{Q}_p}$ are given by the Riemann–Roch formula since the H^1-term vanishes, and that hence the reduction map is surjective.

Suppose now that $p > 3$. Then

$$M^0(N,k)_{\mathbb{F}_p} = \mathrm{H}^0(\overline{\mathcal{M}}([\Gamma_1(N), \Gamma(3)]_{\mathbb{F}_p}), \underline{\omega}^{\otimes k}(-\text{cusps}))^G,$$

with $G = \mathrm{GL}_2(\mathbb{F}_3)$. Since p does not divide the order of G, the functor $M \mapsto M^G$ from $\mathbb{Z}_p[G]$-modules to $\mathbb{Z}_p$-modules is exact. Combining this with the long exact sequence arising from the short exact sequence

$$0 \to \underline{\omega}^{\otimes k}(-\text{cusps}) \to \underline{\omega}^{\otimes k}(-\text{cusps}) \to i_*\underline{\omega}^{\otimes k}(-\text{cusps}) \to 0$$

on $\overline{\mathcal{M}}([\Gamma_1(N), \Gamma(3)])_{\mathbb{Z}_p}$ gives the result.

Suppose now that $N = 2$ or $N = 4$. Then $p \neq 2$. In these cases, the category $[\Gamma_1(N)]_{\mathbb{Z}_p}$ is the quotient, in the sense of algebraic stacks, for the action of a subgroup G of $\mathrm{GL}_2(\mathbb{Z}/4\mathbb{Z})$ acting on $[\Gamma(4)]_{\mathbb{Z}_p}$. This gives a formula analogous to (1.2). The group G is a 2-group, hence of order prime to p. One can then apply an argument which is similar to the one used in the case $p > 3$ above.

Suppose now that $N = 3$. Then $p \neq 3$. Let $\mathbb{Z}_p[\zeta_3]$ be the subring of $\overline{\mathbb{Z}}_p$ generated by $\mathbb{Z}_p$ and a third root of unity ζ_3. Let $[\Gamma(3)^{\zeta_3-\text{can}}]_{\mathbb{Z}_p[\zeta_3]}$ be the category of generalized elliptic curves over schemes over $\mathbb{Z}_p[\zeta_3]$ with an embedding α of the constant group scheme $(\mathbb{Z}/3\mathbb{Z})^2$, such that the Weil pairing of $\alpha(1,0)$ and $\alpha(0,1)$ equals ζ_3. Then $[\Gamma_1(N)]_{\mathbb{Z}_p[\zeta_3]}$ is the quotient of $[\Gamma(3)^{\zeta_3-\text{can}}]_{\mathbb{Z}_p[\zeta_3]}$ for the action of a group of order 3. This means that one can again use the same argument. □

The proof of this lemma indicates that the case $k = 1$ is very different, since the degree of $\underline{\omega}$ is, as one sees from the Kodaira-Spencer isomorphism, too small to make the H^1-term in the Riemann–Roch theorem vanish. Mestre has indeed found examples with $p > 3$ where the map $M^0(N,1)_{\mathbb{Z}_p} \to M^0(N,1)_{\mathbb{F}_p}$ is not surjective. In these examples one has an eigenform f in $M^0(N,1)_{\overline{\mathbb{F}}_p}$ such that the image of the representation ρ_f is too big to be embeddable in $GL_2(\mathbb{C})$; if ρ could be lifted to characteristic zero this would contradict the theorem of Deligne–Serre (Theorem 4.1 of [11]). Let us note that for a representation ρ as in Conjecture 1.8 it can very well happen that $k(\rho) = 1$; one can check that this is equivalent to ρ being unramified at p. This explains that the weight k_ρ for f that one finds in [37, (3.2.4)$_?$], is not in all cases the same as $k(\rho)$ defined above. The difference between k_ρ and $k(\rho)$ can be summarized as follows, in the notation of Definition 1.7. There are only two cases where they are different; in both cases the characters ϕ and ϕ' are of level 1. In the first case, the restriction of ρ to the wild inertia group $I_{p,\mathrm{w}}$ is trivial and $a = 0 = b$; then $k(\rho) = 1$ and $k_\rho = p$. In the second case $p = 2$, ρ is wildly ramified at 2, $\alpha = 0$, $\beta = 1$ and ρ is not finite at 2; then $k(\rho) = 3$ and $k_\rho = 4$.

Other problems arise if we take the character into account. In his course at the Collège de France, 1987–1988, Serre gave some counter examples against his conjecture [37, (3.2.4)$_?$]. These examples are found by considering the genus two curve $X_1(13)$. On this curve there are two eigenforms of weight 2 over $\mathbb{Z}[\zeta_3]$, and the two corresponding characters are of order 6. The reductions mod 2 and 3 of these eigenforms have characters of order 3 and 2, respectively. One verifies that the Galois representations corresponding to these mod 2 and mod 3 forms are irreducible, and that the weights associated to them equal two (Definition 1.7 and [37, §2] coincide in these cases). In fact, these representations are dihedral, induced from $G_{\mathbb{Q}(\sqrt{-1})}$ and $G_{\mathbb{Q}(\sqrt{-3})}$, respectively. According to [37, (3.2.4)$_?$], the mod 2 and mod 3 reductions of the two eigenforms should have lifts to weight two eigenforms in characteristic zero on $X_1(13)$ with a character of the same order as the reduction. But since the genus is two, there are no such forms.

In the same course at the Collège de France, Serre showed that the only mod p eigenforms f of weight at least 2 that cannot be lifted to an eigenform with the same level, weight and order of character are among those in characteristic 2 or 3, whose representation ρ_f is induced from $G_{\mathbb{Q}(\sqrt{-1})}$ or $G_{\mathbb{Q}(\sqrt{-3})}$, respectively. This result was obtained independently by Carayol, see [4, §4.4], and is usually called Carayol's Lemma. Serre's proof uses a result of Nakajima implying that for $n \geq 3$ prime to p and $k \geq 2$ the $\mathbb{Z}_p[\mathrm{GL}_2(\mathbb{Z}/n\mathbb{Z})]$-module $\mathrm{H}^0(X(n)_{\mathbb{Z}_p}, \underline{\omega}^{\otimes k})$ is projective, if all stabilizers are of order prime to p.

Carayol's proof uses the realization of the Galois representation associated to modular forms in the first cohomology group of certain p-adic sheaves on modular curves over $\overline{\mathbb{Q}}$. His arguments can be adapted to the sheaves $\underline{\omega}^{\otimes k}$. This gives the following result, that we state without proof, and which can also be found in Serre's notes. Yet another version of it can be found in [12, §2].

Proposition 1.10 (Carayol's Lemma) *Let $N \geq 1$, p a prime not dividing N, and $k \geq 2$. Let $\varepsilon\colon (\mathbb{Z}/N\mathbb{Z})^* \to \overline{\mathbb{Z}}_p^*$ be a character with $\varepsilon(-1) = (-1)^k$, and let $\bar{\varepsilon}\colon (\mathbb{Z}/N\mathbb{Z})^* \to \overline{\mathbb{F}}_p^*$ be its reduction. Consider the map $\phi\colon M^0(N,k,\varepsilon)_{\overline{\mathbb{Z}}_p} \to M^0(N,k,\bar{\varepsilon})_{\overline{\mathbb{F}}_p}$. If $p \geq 5$, then ϕ is surjective. If $p = 3$ (resp., $p = 2$) and $f \in M^0(N,k,\bar{\varepsilon})_{\overline{\mathbb{F}}_p}$ is an eigenform with ρ_f irreducible and f not in the image of ϕ, then ρ_f is induced from $\mathbb{Q}(\sqrt{-3})$ (resp., $\mathbb{Q}(\sqrt{-1})$).*

In both proofs of this result it is quite clear where the $\mathbb{Q}(\sqrt{-3})$ and $\mathbb{Q}(\sqrt{-1})$ come from. Suppose for simplicity that $N \geq 5$. In Carayol's proof, it comes from the fact that the points of $X_0(N)_{\overline{\mathbb{Q}}}$ with an automorphism (i.e., an automorphism of the pair $(E/\overline{\mathbb{Q}}, G)$ corresponding to it) of order 3 (resp., order 4) are defined over abelian extensions of $\mathbb{Q}(\sqrt{-3})$ (resp., $\mathbb{Q}(\sqrt{-1})$). In Serre's proof, it comes from the fact that for primes $l \equiv -1 \bmod 3$ (resp., mod 4) there is no elliptic curve with an automorphism of order 6 (resp., 4) fixing a subgroup of order l, implying that if an eigenform is not liftable (in the sense of Proposition 1.10) then it is annihilated by T_l for such l; this implies that the character of ρ_f vanishes on Frob_l for such l; then it follows that ρ_f is induced as stated.

The statement of Carayol's Lemma in [4] is actually different from Proposition 1.10. It says that if an irreducible representation $\rho\colon G_{\mathbb{Q}} \to \mathrm{GL}_2(\overline{\mathbb{F}}_p)$ arises from some eigenform f in $M(N,k,\varepsilon)_{\overline{\mathbb{Q}}_p}$ with $k \geq 2$, then for every character $\varepsilon'\colon (\mathbb{Z}/N\mathbb{Z})^* \to \overline{\mathbb{Q}}_p^*$ inducing the same character $(\mathbb{Z}/N\mathbb{Z})^* \to \overline{\mathbb{F}}_p^*$ as ε, there exists an eigenform f' inducing ρ. In this statement, which does not speak of modular forms mod p, one does not suppose that N is prime to p.

Let us now discuss what is known about Conjecture 1.8 and about its relation to [37, $(3.2.4)_?$].

Proposition 1.11 *Let p be a prime, and let $\rho\colon G_{\mathbb{Q}} \to \mathrm{GL}_2(\overline{\mathbb{F}}_p)$ be continuous, irreducible and odd.*

1. *If $p = 2$ (resp., $p = 3$), suppose that ρ is not induced from $\mathbb{Q}(\sqrt{-1})$ (resp., $\mathbb{Q}(\sqrt{-3})$). Then if ρ satisfies Conjecture 1.8, it satisfies [37, $(3.2.4)_?$].*

2. *If $p = 2$, suppose that the restriction of ρ to a decomposition group at 2 is not an extension of some character by itself. Then if ρ satisfies [37, $(3.2.4)_?$], it satisfies Conjecture 1.8.*

Theorem 1.12 *Let $p > 2$ be a prime, and let $\rho: G_{\mathbb{Q}} \to \mathrm{GL}_2(\overline{\mathbb{F}}_p)$ be continuous, irreducible and odd. Suppose that ρ comes from a modular form of some type. Then ρ satisfies Conjecture 1.8. Moreover, if ρ comes from a mod p modular form of some type (N, k, ε) with N prime to p, then N is a multiple of $N(\rho)$, $k \geq k(\rho)$ and ε is obtained from $\varepsilon(\rho)$ via composition with $\mathbb{Z}/N\mathbb{Z} \to \mathbb{Z}/N(\rho)\mathbb{Z}$.*

The proof of these results is quite long and many people have contributed to it. A complete proof can be found by reading Diamond's article [12], and the references therein. A very good overview of the strategy of the whole proof is given in Ribet's report [32]. In the next section we will see which parts of these results are used in the proof of the conjecture of Shimura–Taniyama for semi-stable elliptic curves over $\mathbb{Q}$ and the proof of Fermat's Last Theorem. In Sections 3 and 4, we will then describe the proofs of those cases. To finish this section, we will briefly recall the history of the proofs of Proposition 1.11 and Theorem 1.12.

For the rest of this section, let p be prime and $\rho: G_{\mathbb{Q}} \to \mathrm{GL}_2(\overline{\mathbb{F}}_p)$ be continuous, irreducible and odd. We will say that ρ is modular of type $(N, k, \varepsilon)_{\overline{\mathbb{F}}_p}$ (resp., $(N, k, \varepsilon)_{\overline{\mathbb{Q}}_p}$) if there is an eigenform f in $M^0(N, k, \varepsilon)_{\overline{\mathbb{F}}_p}$ (resp., $M^0(N, k, \varepsilon)_{\overline{\mathbb{Q}}_p}$) such that f gives the representation ρ. Serre formulated, in a letter to Mestre dated August 13, 1985, a part of his conjectures that, together with the Shimura–Taniyama conjecture, implies Fermat's Last Theorem. Mazur proved, in a letter to Mestre dated August 16, 1985, the following result: suppose moreover that $p > 2$, that ρ is modular of some type $(N, 2, 1)_{\bar{\mathbb{Q}}_p}$, that l is a prime not congruent to 1 mod p, that l divides N but l^2 does not, that ρ is unramified at l if $l \neq p$ and that ρ is finite at p if $l = p$; then ρ is modular of type $(N/l, 2, 1)_{\bar{\mathbb{Q}}_p}$. In 1987, Ribet removed the condition "$l \not\equiv 1 \bmod p$" from Mazur's result, under the assumption that p does not divide N (see [33]). These two results together imply already that Fermat's Last Theorem is a consequence of the Shimura–Taniyama conjecture. Together with Mazur [29], Ribet extended his result to the case where p divides N, but where p^2 does not.

Langlands, Deligne and Carayol have proved [5] that for f a newform the conductor of the system of l-adic representations associated to f is equal to the level of f. From this it follows easily that for f an eigenform in some $M^0(N, k, \varepsilon)_{\overline{\mathbb{Q}}_p}$, and ρ_f the mod p Galois representation that it gives, $N(\rho_f)$ divides N. Carayol [4] and Livné [26] classified, independently, in terms of the admissible irreducible representation of $\mathrm{GL}_2(\mathbb{Q}_l)$ associated to a newform f in some $M^0(N, k, \varepsilon)_{\overline{\mathbb{Q}}_p}$, the cases where the l-adic valuations of $N(\rho_f)$ and N are different (here ρ_f denotes the mod p representation associated to f, ρ_f is supposed to be irreducible and l is a prime different from p). Carayol [4] showed that if ρ is modular of type $(N, k, \varepsilon)_{\overline{\mathbb{Q}}_p}$ and not induced from $\mathbb{Q}(\sqrt{-1})$ (resp., $\mathbb{Q}(\sqrt{-3})$) if $p = 2$ (resp., $p = 3$), then it is modular of type $(N, k, \varepsilon')_{\overline{\mathbb{Q}}_p}$ for all ε' whose mod p reduction equals that of ε and such that $\varepsilon'(-1) = (-1)^k$ (this last condition is implied by the first if $p \neq 2$). This result shows that in order to prove Proposition 1.11 and Theorem 1.12 one need not pay attention anymore to the character, so we will drop it from the notation in what follows.

Suppose now that ρ is modular of some type $(N, k)_{\overline{\mathbb{Q}}_p}$. Then one wants to prove that ρ is modular of type $(N(\rho), k)$. In [4] Carayol reduces the proof of this, for $p \geq 5$ and $k \geq 2$, to the following two statements:

(A) There exists a prime number q not dividing Nl and a newform of type $(N'q, k, \varepsilon')_{\overline{\mathbb{Q}}_p}$ with N' dividing N and ε' trivial mod q, whose associated mod p Galois representation is isomorphic to ρ.

(B) If $l \neq p$ divides N, l^2 does not divide N and l does not divide $N(\rho)$ (i.e., ρ is unramified at l), then ρ is modular of type $(N/l, k)_{\overline{\mathbb{Q}}_p}$.

The first of these two statements is used to switch in certain cases from modular curves to Shimura curves associated to indefinite quaternion algebras over $\mathbb{Q}$, via the Jacquet–Langlands correspondence. The main part of Ribet's article [32] is about establishing some geometric integral version of this correspondence in the case of weight two and trivial character. Statement (A) for weight two and trivial character was proved first by Ribet in [34] and more generally by Diamond for $2 \leq k \leq p+1$ and arbitrary character in [13]. Note that statement (B) for weight two, trivial character and p^2 not dividing N, is the result of Mazur and Ribet above. A crucial point in their proof is that ρ has multiplicity one in the p-torsion of the jacobian $J_0(N)$, in some sense (see Section 3.3). A semi-simplicity result in [2] made it possible for Ribet to prove statement (B) for weight two and trivial character, but without the condition that p^2 does not divide N (see [35] and [32]). In [15] it was shown, using work of Gross [17] and of Coleman and Voloch [7], that one can always adapt the weight, in the following sense: if $p \neq 2$ and ρ is modular of some type $(N, k, \varepsilon)_{\overline{\mathbb{F}}_p}$ with N prime to p, then ρ is modular of type $(N, k(\rho), \varepsilon)_{\overline{\mathbb{F}}_p}$. The definition of $k(\rho)$ makes it clear that the mechanism behind the proof of this result was known to Serre; in Section 4 we will discuss a part of it. This mechanism includes the fact that if ρ is modular of some type $(N, k, \varepsilon)_{\overline{\mathbb{F}}_p}$, with N prime to p, then for some integer a, $\rho \otimes \chi_p^a$ is modular of type $(N, k', \varepsilon)_{\overline{\mathbb{F}}_p}$ with $2 \leq k' \leq p+1$, and ρ is modular of type $(Np^2, 2)_{\overline{\mathbb{Q}}_p}$. It follows from this that in order to prove statement (B), one may assume that the weight is two. This is used in [32] to show statement (B) for $p \geq 3$ and ρ with $\det(\rho) = \chi_p$; Ribet also remarks that he expects his proof to extend without difficulty to $\det(\rho)$

arbitrary. Finally, Diamond [12] proved statement (B) for $p \geq 3$, following [32]. Another proof of statement (B), not using the reduction to weight two, but extending the arguments of [33] to weights k between 2 and $p+1$, was suggested by Jordan and Livné in [21]. The multiplicity one result needed for that is proved in [16].

2 The Cases We Need

Special cases of Theorem 1.12 are used at three different places in the proof of the Shimura–Taniyama conjecture and of Fermat's Last Theorem. First of all, Ribet's proof that the Shimura–Taniyama conjecture implies Fermat's Last Theorem is a special case of Theorem 1.12. We briefly recall the situation. One supposes that Fermat's Last Theorem is not true. Then there exist a prime $p > 3$ and non-zero integers a, b and c that are pairwise relatively prime and satisfy $a^p + b^p + c^p = 0$. This leads, via a construction of Hellegouarch (see [19] and [20]), to a semi-stable elliptic curve E over $\mathbb{Q}$ that is usually called the Frey curve associated to (a^p, b^p, c^p). (This construction is already implicit in work of Klein and Fricke, see [48, p. 196].) Let ρ_p be the representation $G_\mathbb{Q} \to \mathrm{GL}_2(\overline{\mathbb{F}}_p)$ given by the p-torsion of E. It follows from Mazur's work on isogenies between elliptic curves over $\mathbb{Q}$ (see [27] and [28]) that ρ_p is irreducible. Moreover, E has the miraculous property that ρ_p is unramified away from $2p$ and that its ramification at 2 and p is very well-behaved: one has $N(\rho_p) = 2$, $k(\rho_p) = 2$ and $\varepsilon(\rho_p) = 1$; see [37, §4]. The conductor N of E is the product of all primes dividing abc; note that it is square free. If E is modular, i.e., if the Shimura–Taniyama conjecture is true for E, then ρ_p is modular of type $(N, 2, 1)_{\mathbb{Q}_p}$. So in this case it suffices to have Theorem 1.12 for ρ_p that are modular of some type $(N, 2, 1)_{\overline{\mathbb{Q}}_p}$ with N square free and with $p > 3$.

Let us now look where Theorem 1.12 intervenes in Wiles's proof of the Shimura–Taniyama conjecture for semi-stable elliptic curves. One has the following proposition, that was given by Oesterlé in the seminar held in Paris on the work of Wiles.

Proposition 2.1 *Let E be a semi-stable elliptic curve over $\mathbb{Q}$, p be a prime number and $\rho_p: G_\mathbb{Q} \to \mathrm{GL}_2(\mathbb{F}_p)$ the representation obtained by choosing some basis of the p-torsion of E. Then either ρ_p is surjective, or it is reducible and its semi-simplification is isomorphic to the direct sum $1 \oplus \chi_p$ of the trivial character with the cyclotomic character.*

Proof. If $p > 5$ or if the image G of ρ_p has order divisible by p, this is part of Proposition 21 of [39]. Suppose that we are not in this case. We use the same arguments as Serre does. If ρ_p is reducible Serre's argument shows that its semi-simplification is $1 \oplus \chi_p$. So we suppose that ρ_p is irreducible and we have to derive a contradiction. The morphism $\det: G \to \mathbb{F}_p^*$ is surjective because $\det \circ \rho_p = \chi_p$. We have $p = 2$, 3 or 5 and ρ_p is unramified outside p. The image under ρ_p of inertia at p is a non-split Cartan subgroup (i.e., cyclic of order $p^2 - 1$) or "half" of a split Cartan (i.e., a conjugate of the subgroup of elements of the form $\left(\begin{smallmatrix} * & 0 \\ 0 & 1 \end{smallmatrix}\right)$). The first case arises if and only if E has good, supersingular reduction at p.

Suppose that $p = 2$. Then G has order 3 since ρ_p is irreducible. It follows that $\mathbb{Q}$ has a Galois extension of degree 3 which is ramified only at 2; there is no such extension. This finishes the proof for $p = 2$.

Suppose that $p = 3$. Then G has order dividing 16. The absolute irreducibility of ρ_p implies that G has order 8 or 16. But then G has a quotient of order 4, leading to a Galois extension of $\mathbb{Q}$ of degree 4 which is unramified outside 3; again, such an extension does not exist.

Suppose that $p = 5$. By §2.6 of [39], G is contained in the normalizer of a Cartan subgroup or the image $\overline{G}$ of G in $\mathrm{GL}_2(\mathbb{F}_p)/\mathbb{F}_p^*$ is isomorphic to A_4, S_4 or A_5. Since G is of order prime to 5, we cannot have $\overline{G}$ isomorphic to A_5. If $\overline{G}$ is isomorphic to A_4 or S_4 we get a Galois extension of degree 3 of $\mathbb{Q}$ or of $\mathbb{Q}(\sqrt{5})$ which is unramified outside 5, and such extensions do not exist. Hence G is contained in the normalizer N of a Cartan subgroup C. The absolute irreducibility of ρ_p implies that G is not contained in C. This gives us a degree 2 extension K of $\mathbb{Q}$. We claim that K is unramified, which gives us the desired contradiction. The extension K is clearly unramified outside p, so we have to show that $\rho_p(I_p)$ is contained in C. If $\rho_p(I_p)$ is a non-split Cartan subgroup then this Cartan subgroup has to be C (look at the action on the projective line), hence we get what we need. If $\rho_p(I_p)$ is "half" of a split Cartan subgroup, then again $\rho_p(I_p)$ has to be contained in C because of its action on the projective line. □

Let E be a semi-stable elliptic curve over $\mathbb{Q}$. If for a prime number p the representation ρ_p is not surjective, then E admits a $\mathbb{Q}$-rational isogeny of degree p. Hence, if ρ_3 and ρ_5 are both not surjective, E admits a $\mathbb{Q}$-rational isogeny of degree 15, hence defines a $\mathbb{Q}$-rational non-cuspidal point of $X_0(15)$. According to the tables in [1], $X_0(15)$ is an elliptic curve with exactly eight rational points, four of which are cusps. The four non-cuspidal points correspond (up to twist) to elliptic curves of conductor 50 of which one can easily see that no twist of them is semi-stable. So we conclude that at least one among ρ_3 and ρ_5 is surjective.

Suppose that ρ_3 is surjective. Results of Langlands [25] and Tunnell [46] show that ρ_3 is modular of type $(3^m N(\rho_3)^2, 1)_{\mathbb{Q}_3}$, for some $m \geq 0$. This will be discussed in Section 4. In that same section, we will show that ρ_3 is then modular of type $(N(\rho_3), 2, 1)_{\overline{\mathbb{Q}}_3}$ if ρ_3 is finite at 3, and of type $(3N(\rho_3), 2, 1)_{\overline{\mathbb{Q}}_3}$ if ρ_3 is not finite at 3. We will not need the results on companion forms, and we can use an older version of Carayol's Lemma dating back to Mazur's [28]. Carayol's reductions in [4] imply that ρ_3 is modular of type $(N(\rho_3), 2, 1)_{\overline{\mathbb{Q}}_3}$ if ρ_3 is finite at 3, and of type $(3N(\rho_3), 2, 1)_{\overline{\mathbb{Q}}_3}$ if ρ_3 is not finite at 3. Starting at this point, Wiles and Taylor prove (see [47] and [45]) that all deformations of ρ_3 that are ramified at only finitely many primes, that are "semi-stable" at 3 and whose determinant is the cyclotomic character are modular. Hence E is modular. (Let us note that the restriction of ρ_3 to $\mathrm{Gal}(\overline{\mathbb{Q}}/\mathbb{Q}(\sqrt{-3}))$ is absolutely irreducible since its image is $\mathrm{SL}_2(\mathbb{F}_3)$.) So in this case we don't need Ribet's part of the proof of Theorem 1.12.

Suppose that ρ_3 is not surjective. Then ρ_5 is surjective. In this case, Wiles [47] shows that there exists an elliptic curve E' over $\mathbb{Q}$ such that its representation ρ'_3 is irreducible and with ρ'_5 isomorphic to ρ_5 (this follows from the fact that $X(5)_{\overline{\mathbb{Q}}}$ is a disjoint union of four projective lines). The curve E' is semi-stable since this can be read off from ρ'_5, even at the prime 5 itself. The previous argument shows that E' is modular, hence ρ_5 is modular of type $(N', 2, 1)_{\mathbb{Q}_5}$, where N' is the conductor of E'. Note that N' is square free. Ribet's part of the proof of Theorem 1.12 shows that ρ_5 is modular of type $(N(\rho_5), 2, 1)_{\overline{\mathbb{Q}}_5}$ if ρ_5 is finite at 5, and of type $(5N(\rho_5), 2, 1)_{\overline{\mathbb{Q}}_5}$ if ρ_5 is not finite at 5. Wiles and Taylor ([47] and [45]) show that all deformations of ρ_5 that are ramified at only finitely many primes, that are "semi-stable" at 5 and whose determinant is the cyclotomic character are modular. Hence E is modular.

The final conclusion of this section is the following. We need Ribet's part of the proof of Theorem 1.12 in the case where $p > 3$ and ρ is modular of some type $(N, 2, 1)_{\overline{\mathbb{Q}}_p}$ with N square free. We also need Carayol's reductions for $p = 3$ and ρ modular of type $(N^2, 2, 1)_{\overline{\mathbb{Q}}_3}$ with N square free. It should be noted that Carayol's reductions use Mazur's result, so in the next section we will explain that result for $p \geq 3$.

3 Weight Two, Trivial Character and Square Free Level

The aim of this section is to describe a proof of the following theorem.

Theorem 3.1 *Let $p \geq 3$ be prime. Let $\rho: G_{\mathbb{Q}} \to \mathrm{GL}_2(\overline{\mathbb{F}}_p)$ be irreducible and modular of some type $(N, 2, 1)_{\overline{\mathbb{Q}}_p}$ with N square free. If ρ is finite at p, it is modular of type $(N(\rho), 2, 1)_{\overline{\mathbb{Q}}_p}$. If ρ is not finite at p, it is modular of type $(pN(\rho), 2, 1)_{\overline{\mathbb{Q}}_p}$.*

The proof of this theorem has two parts. The first part, due to Mazur, deals with the primes l dividing $N/N(\rho)$ that are not congruent to 1 modulo p. The second part, due to Ribet, deals with an arbitrary $l \neq p$ at the cost of introducing a prime q in the level that one can get rid of by the first part. These two results will be described in the following two sections. For a detailed proof of in fact more general results we refer to [32, §6–§8]; our aim is just to give a good idea of what happens in [32].

3.2 Mazur's Result

Theorem 3.2.1 *Let $p \geq 3$ be prime. Let $\rho: G_{\mathbb{Q}} \to \mathrm{GL}_2(\overline{\mathbb{F}}_p)$ be irreducible and modular of some type $(N,2,1)_{\overline{\mathbb{Q}}_p}$. Suppose that l is a prime not congruent to 1 mod p, that l divides N but l^2 does not, that ρ is unramified at l if $l \neq p$ and that ρ is finite at p if $l = p$. Then ρ is modular of type $(N/l,2,1)_{\overline{\mathbb{Q}}_p}$.*

The proof of this result is by contradiction: let $M := N/l$ and suppose that ρ is not modular of type $(M,2,1)_{\overline{\mathbb{Q}}_p}$. Let $X_0(N)$ denote the model over the localization $\mathbb{Z}_{(l)}$ of $\mathbb{Z}$ at l which is constructed in [10] as a coarse moduli space (see also [24]). Let $J_0(N)_{\mathbb{Q}}$ be the jacobian of $X_0(N)_{\mathbb{Q}}$, and let $J_0(N)$ be its Néron model over $\mathbb{Z}_{(l)}$. Then $J_0(N)$ represents in fact the degree zero part of the relative Picard functor of $X_0(N)$ over $\mathbb{Z}_{(l)}$ by [30]; this implies that one can describe $J_0(N)$ in terms of $X_0(N)$. By [10, V–VI], the fibre $X_0(N)_{\mathbb{F}_l}$ of $X_0(N)$ over $\mathbb{F}_l$ is the union of two copies of the smooth curve $X_0(M)_{\mathbb{F}_l}$, which intersect transversally at the supersingular points. This gives the following "devissage" of $J_0(N)_{\mathbb{F}_l}$. One has the connected component of the identity element $J_0(N)^0_{\mathbb{F}_l}$ of $J_0(N)_{\mathbb{F}_l}$. The quotient $J_0(N)_{\mathbb{F}_l}/J_0(N)^0_{\mathbb{F}_l}$ is a finite etale group scheme over $\mathbb{F}_l$, which is in fact a constant group scheme that we will denote by $\Phi_0(N)_{\mathbb{F}_l}$. By construction we have an exact sequence:

$$0 \to J_0(N)^0_{\mathbb{F}_l} \to J_0(N)_{\mathbb{F}_l} \to \Phi_0(N)_{\mathbb{F}_l} \to 0. \tag{3.2.2}$$

The normalization map $X_0(M)_{\mathbb{F}_l} \coprod X_0(M)_{\mathbb{F}_l} \to X_0(N)_{\mathbb{F}_l}$ induces another short exact sequence:

$$0 \to T_0(N)_{\mathbb{F}_l} \to J_0(N)^0_{\mathbb{F}_l} \to J_0(M)^2_{\mathbb{F}_l} \to 0, \tag{3.2.3}$$

where $T_0(N)_{\mathbb{F}_l}$ is a torus whose character group $\mathrm{Hom}_{\overline{\mathbb{F}}_l}(T_0(N)_{\overline{\mathbb{F}}_l}, \mathrm{G}_{m\overline{\mathbb{F}}_l})$ can be identified with the group of degree zero divisors on $X_0(M)_{\overline{\mathbb{F}}_l}$ with support in the supersingular points. Let $\mathbb{T}_0(N)$ be the Hecke algebra of level N: it is the subring of $\mathrm{End}(J_0(N)_{\mathbb{Q}})$ generated by the Hecke operators T_n, $n \geq 1$. By the Néron property of $J_0(N)$, $\mathbb{T}_0(N)$ acts on it. This action can be described explicitly by correspondences over $\mathbb{Z}_{(l)}$. The short exact sequences (3.2.2) and (3.2.3) are compatible with the action of $\mathbb{T}_0(N)$.

Let us now look what all this has to do with our representation ρ. The fact that ρ is modular of type $(N,2,1)_{\overline{\mathbb{Q}}_p}$ means that there exist a maximal ideal m of $\mathbb{T}_0(N)$, an embedding of $k := \mathbb{T}_0(N)/m$ into $\overline{\mathbb{F}}_p$, an integer $d \geq 1$ and a 2-dimensional k-vector space V with an action by $G_{\mathbb{Q}}$ with $\overline{\mathbb{F}}_p \otimes_k V$ giving ρ, such that $J_0(N)(\overline{\mathbb{Q}})[m]$ is isomorphic to the direct sum V^d of d copies of V as $k[G_{\mathbb{Q}}]$-modules. (Here $J_0(N)(\overline{\mathbb{Q}})[m]$ denotes the kernel of m, i.e., the elements x such that $tx = 0$ for all t in m.) The fact that $J_0(N)(\overline{\mathbb{Q}})[m]$ is semi-simple is proved in [2].

We will now first treat the case where $l \neq p$, which is technically simpler than the case $l = p$. So we suppose that $l \neq p$. Then ρ is unramified at l by hypothesis, and there is a unique finite etale k-vector space scheme W over $\mathbb{Z}_{(l)}$ such that $V \cong W(\overline{\mathbb{Q}})$ as $k[G_{\mathbb{Q}}]$-modules. We choose an injection of V into $J_0(N)(\overline{\mathbb{Q}})[m]$. This gives us an injection of $W_{\mathbb{Q}}$ into $J_0(N)_{\mathbb{Q}}$. The Néron property of $J_0(N)$ implies that this injection extends to a morphism $W \to J_0(N)$, which must be injective since $J_0(N)[p]$ is etale. Consider the image of $W_{\mathbb{F}_l}$ in $\Phi_0(N)_{\mathbb{F}_l}$ under (3.2.2). It was proved in [33] that the action of $\mathbb{T}_0(N)$ on $\Phi_0(N)_{\mathbb{F}_l}$ is "Eisenstein," in the sense that for q prime to N the operator T_q acts as multiplication by $q+1$. Since ρ is irreducible, it follows that the image of $W_{\mathbb{F}_l}$ in $\Phi_0(N)_{\mathbb{F}_l}$ is zero. Hence $W_{\mathbb{F}_l}$ lands in $J_0(N)^0_{\mathbb{F}_l}$. Since we suppose that ρ is not modular of type $(M, 2, 1)_{\overline{\mathbb{Q}}_p}$, the image of $W_{\mathbb{F}_l}$ in $J_0(M)^2_{\mathbb{F}_l}$ is zero. So $W_{\mathbb{F}_l}$ lands in the torus $T_0(N)_{\mathbb{F}_l}$. This has strong consequences for the Frobenius element $\rho(\mathrm{Frob}_l)$. Namely, Frob_l acts on $T_0(N)_{\mathbb{F}_l}(\overline{\mathbb{F}}_l)$ simultaneously as lT_l and as $-lw_l$, with w_l the Atkin–Lehner involution of level l. This implies that $\rho(\mathrm{Frob}_l)$ is in k^* and it follows that $l = \det(\rho(\mathrm{Frob}_l)) = \rho(\mathrm{Frob}_l^2) = l^2$. This contradicts the assumption that l is not congruent to 1 modulo p.

Let us now assume that $l = p$. We have the information that ρ is finite at p. Let $W_{\mathbb{Q}}$ be finite k-vector space scheme over $\mathbb{Q}$ such that $W_{\mathbb{Q}}(\overline{\mathbb{Q}})$ gives V. Then W can be extended to a finite flat k-vector space scheme over $\mathbb{Z}_{(p)}$. Such an extension is unique by results of Raynaud in [31] (here we use that $p \neq 2$). We choose an injection of V into $J_0(N)(\overline{\mathbb{Q}})[m]$. This gives an injection of $W_{\mathbb{Q}}$ into $J_0(N)_{\mathbb{Q}}$. Ribet showed in [33, Lemma 6.2] that this injection extends to an injection of W into $J_0(N)$. Note that this is not just a consequence of the Néron property of $J_0(N)$, since W is not smooth as $\det(\rho) = \chi_p$. Once one has this result the rest of the proof of Theorem 3.2.1 proceeds as in the case $l \neq p$. So it remains to explain the proof of [33, Lemma 6.2]. We may replace $\mathbb{Z}_{(p)}$ by its completion $\mathbb{Z}_p$. Then W sits in a short exact sequence:

$$0 \to W^0 \to W \to W^{\mathrm{et}} \to 0, \tag{3.2.4}$$

where W^0 denotes the connected component of the zero section of W, and where W^{et} is the largest finite etale quotient of W. Grothendieck proved, see [18, 7, IX, §11.6], that the action of $G_{\mathbb{Q}_p}$ on $J_0(N)(\overline{\mathbb{Q}}_p)/J_0(N)^0(\overline{\mathbb{Z}}_p)$ is unramified. This implies that the injection of $W^0_{\mathbb{Q}_p}$ into $J_0(N)_{\mathbb{Q}_p}$ extends to a morphism of W^0 into $J_0(N)^0$, which is a closed immersion by [31]. Then one can finish by applying the Néron property to the amalgamated sum $(W \oplus J_0(N)^0)/W^0$ (this trick is due to Grothendieck, see [18, 7, IX, 5.9.2]).

3.3 Ribet's Result

Theorem 3.3.1 *Let $p \geq 3$ be prime. Let $\rho: G_{\mathbb{Q}} \to \mathrm{GL}_2(\overline{\mathbb{F}}_p)$ be irreducible and modular of some type $(N, 2, 1)_{\overline{\mathbb{Q}}_p}$. Suppose that $l \neq p$ is a prime, that l divides N but l^2 does not, and that ρ is unramified at l. Then there exists a prime number q, not dividing N and congruent to -1 mod p, such that ρ is modular of type $(qN/l, 2, 1)_{\overline{\mathbb{Q}}_p}$.*

We define $M := N/l$. Let q be a prime number not dividing Np with the property that $\rho(\mathrm{Frob}_q) = \rho(c)$, with c a complex conjugation. This implies that q is congruent to -1 modulo p. In what follows we will consider the modular curves $X_0(Mlq)_{\mathbb{Q}}$, $X_0(Ml)_{\mathbb{Q}}$, $X_0(Mq)_{\mathbb{Q}}$, and a certain Shimura curve $C_{\mathbb{Q}}$ that we will now define (see [33, §4] for more details). Let $B_{\mathbb{Q}}$ be a quaternion algebra over $\mathbb{Q}$ of discriminant lq and let B be a maximal order in it. Then $C_{\mathbb{Q}}$ is the Shimura curve associated to $B_{\mathbb{Q}}$ of level $\Gamma_0(M)$. More precisely, this means that $C_{\mathbb{Q}}$ is the coarse moduli scheme for objects $(A/S, \alpha, G)$, with S a $\mathbb{Q}$-scheme, A/S an abelian scheme of relative dimension two, with $\alpha: B \to \mathrm{End}_S(A)$ a morphism of rings and $G \subset A$ a finite flat closed subgroup scheme of rank M^2, which is killed by M and which is stable for the action by B. We define $J_{\mathbb{Q}}$ to be the jacobian of $C_{\mathbb{Q}}$. From the moduli interpretation of $C_{\mathbb{Q}}$ it is clear that one can define Hecke correspondences on $C_{\mathbb{Q}}$, inducing endomorphisms T_n, $n \geq 1$, of $J_{\mathbb{Q}}$. Let $\mathbb{T}$ be the subring of $\mathrm{End}_{\mathbb{Q}}(J_{\mathbb{Q}})$ generated by these T_n.

We have the usual pairs of degeneracy morphisms from $X_0(Mlq)_{\mathbb{Q}}$ to $X_0(Ml)_{\mathbb{Q}}$ and to $X_0(Mq)_{\mathbb{Q}}$, inducing a morphism

$$J_0(Mlq)_{\mathbb{Q}} \to J_0(Ml)^2_{\mathbb{Q}} \times J_0(Mq)^2_{\mathbb{Q}}.$$

The lq-new subvariety $J_0(Mlq)^{l,q\text{-new}}_{\mathbb{Q}}$ of $J_0(Mlq)_{\mathbb{Q}}$ is defined as the connected component of the identity element of the kernel of this last morphism. One knows that $J_{\mathbb{Q}}$ is isogeneous to $J_0(Mlq)^{l,q\text{-new}}_{\mathbb{Q}}$ (this results from trace formula calculations by Eichler, Shimizu, Jacquet–Langlands and Faltings's isogeny theorem). Ribet has given a more precise version of this in terms of the character groups of the torus parts of the reductions mod l and q of the jacobians of the curves under consideration. For G a commutative algebraic group over a field k, let $X(G) := \mathrm{Hom}_{\overline{k}}(T_{\overline{k}}, \mathbf{G}_{m\overline{k}})$ be the character group scheme of the maximal torus $T_{\overline{k}}$ of $G_{\overline{k}}$. Then Ribet constructed a short exact sequence:

$$0 \to X(J_{\mathbb{F}_q}) \to X(J_0(Mlq)_{\mathbb{F}_l}) \to X(J_0(Ml)^2_{\mathbb{F}_l}) \to 0 \tag{3.3.2}$$

which is Hecke-equivariant in the sense that for each $n \geq 1$, the element T_n in $\mathbb{T}_0(Mlq)$ induces the element T_n of $\mathbb{T}$ on $X(J_{\mathbb{F}_q})$. The induced action of T_n on $X(J_0(Ml)^2_{\mathbb{F}_l})$ can be described in terms of a two by two matrix with coefficients in $\mathbb{T}_0(Ml)$. Since l and q play symmetric roles, we also have the following exact sequence:

$$(3.3.3) \qquad 0 \to X(J_{\mathbb{F}_l}) \to X(J_0(Mlq)_{\mathbb{F}_q}) \to X(J_0(Mq)^2_{\mathbb{F}_q}) \to 0.$$

To construct these sequences, Ribet relies heavily on work of Cerednik, Drinfeld and Jordan–Livné concerning the q-adic uniformization of $C_{\mathbb{Q}_q}$. A detailed account of this uniformization can be found in [3]. An amazing feature of these sequences is that they compare character groups of tori over fields of distinct characteristics. Since $J^0_{\mathbb{F}_q}$ is its maximal torus, it follows from (3.3.2) that $\mathbb{T}_0(Mlq)$ acts on $J_{\mathbb{Q}}$ via a (necessarily unique) morphism of rings $\mathbb{T}_0(Mlq) \to \mathbb{T}$ that sends T_n to T_n.

Let η_q be the element $T_q^2 - 1$ of $\mathbb{T}_0(Mlq)$, and let $\Phi_{\mathbb{F}_q}$ be the group of connected components of $J_{\mathbb{F}_q}$. For M a finite abelian group, let $M^* := \mathrm{Hom}_{\mathbb{Z}}(M, \mathbb{Q}/\mathbb{Z})$ be its Pontrjagin dual. Theorem 4.3 of [33] asserts that there is a Hecke equivariant exact sequence:

$$(3.3.4) \qquad 0 \to K_l \to X(J_0(Ml)^2_{\mathbb{F}_l})/\eta_q X(J_0(Ml)^2_{\mathbb{F}_l}) \to \Phi^*_{\mathbb{F}_q} \to C_q \to 0,$$

with K_l and C_q "Eisenstein" in the sense we saw in the previous section. Likewise, one has an exact sequence:

$$(3.3.5) \qquad 0 \to K_q \to X(J_0(Mq)^2_{\mathbb{F}_q})/\eta_l X(J_0(Mq)^2_{\mathbb{F}_q}) \to \Phi^*_{\mathbb{F}_l} \to C_l \to 0$$

with K_q and C_l "Eisenstein". Since ρ is modular of type $(Ml, 2, 1)_{\overline{\mathbb{Q}}_p}$, ρ arises from a maximal ideal of $\mathbb{T}_0(Ml)$, in the way we have seen in Section 3.2. It is not hard to see that then ρ also arises from a maximal ideal m of $\mathbb{T}_0(Mlq)$ (note that we do not claim that ρ arises from a newform whose level is divisible by q). More precisely, we have a maximal ideal m of $\mathbb{T}_0(Mlq)$, an embedding of $k := \mathbb{T}_0(Mlq)/m$ into $\overline{\mathbb{F}}_p$, a two-dimensional k-vector space V with an action by $G_{\mathbb{Q}}$ with $\overline{\mathbb{F}}_p \otimes_k V$ giving ρ, such that $J_0(Mlq)(\overline{\mathbb{Q}})[m]$ is isomorphic to V^λ for some positive integer λ (this λ is called the multiplicity at m of ρ in $J_0(Mlq)$). Let μ be the multiplicity at m of ρ in $J_{\mathbb{Q}}$: $J_{\mathbb{Q}}(\overline{\mathbb{Q}})[m] \cong V^\mu$ (it follows again from [2] that $J_{\mathbb{Q}}(\overline{\mathbb{Q}})[m]$ is semi-simple). It is clear that $\mu \geq 0$.

From now on we suppose that ρ is not modular of type $(Mq, 2, 1)_{\overline{\mathbb{Q}}_p}$. From this assumption and the exact sequences above Ribet then derives that $2\mu \leq \lambda$ and that $2\lambda \leq 2\mu$, which gives a contradiction because we know that $\lambda > 0$. So let us describe the arguments of Ribet. One starts by localizing the exact sequence (3.3.5) at m; this shows that $\Phi_{\mathbb{F}_l,m}$ is zero, because m is not in the support of $X(J_0(Mq)^2_{\mathbb{F}_q})$. This implies that $J^0_{\mathbb{F}_l}(\overline{\mathbb{F}}_l)[m]$ is isomorphic to V^μ (as k-vector spaces), so that:

$$\dim_k(X(J_{\mathbb{F}_l}) \otimes_{\mathbb{T}_0(Mlq)} k) = 2\mu \tag{3.3.6}$$

The exact sequence (3.3.3) shows that:

$$\dim_k(X(J_0(Mlq)_{\mathbb{F}_q}) \otimes_{\mathbb{T}_0(Mlq)} k) = \dim_k(X(J_{\mathbb{F}_l}) \otimes_{\mathbb{T}_0(Mlq)} k). \tag{3.3.7}$$

Next we have the following exact sequence, obtained by replacing N by Mlq in (3.2.2):

$$0 \to J_0(Mlq)^0_{\mathbb{F}_l} \to J_0(Mlq)_{\mathbb{F}_l} \to \Phi_0(Mlq)_{\mathbb{F}_l} \to 0. \tag{3.3.8}$$

Since $\Phi_0(Mlq)_{\mathbb{F}_l}$ is "Eisenstein", it follows that $J_0(Mlq)(\overline{\mathbb{Q}})[m]$ specializes into $J_0(Mlq)^0_{\mathbb{F}_l}(\overline{\mathbb{F}}_l)$. As in (3.2.3), the normalization of $X_0(Mlq)_{\mathbb{F}_l}$ induces a short exact sequence:

$$0 \to T_0(Mlq)_{\mathbb{F}_l} \to J_0(Mlq)^0_{\mathbb{F}_l} \to J_0(Mq)^2_{\mathbb{F}_l} \to 0. \tag{3.3.9}$$

Since ρ is not modular of type $(Mq, 2, 1)_{\overline{\mathbb{Q}}_p}$, it follows that:

$$\dim_k(X(J_0(Mlq)_{\mathbb{F}_l}) \otimes_{\mathbb{T}_0(Mlq)} k) = 2\lambda. \tag{3.3.10}$$

Ribet shows that the Frobenius endomorphism of $J^0_{\mathbb{F}_q}$ is equal to qT_q. It follows from this that $\rho(\mathrm{Frob}_q)$ acts as a scalar (i.e., an element of k) on $J^0_{\mathbb{F}_q}(\overline{\mathbb{F}}_q)[m]$. But by the choice of q, $\rho(\mathrm{Frob}_q)$ is in the conjugacy class of $\left(\begin{smallmatrix} 1 & 0 \\ 0 & -1 \end{smallmatrix}\right)$. It follows that:

$$\dim_k(X(J_{\mathbb{F}_q}) \otimes_{\mathbb{T}_0(Mlq)} k) \le \mu. \tag{3.3.11}$$

The same argument applied to the maximal torus $T_0(Mlq)_{\mathbb{F}_q}$ in $J_0(Mlq)_{\mathbb{F}_q}$ gives:

$$\dim_k(X(J_0(Mlq)_{\mathbb{F}_q}) \otimes_{\mathbb{T}_0(Mlq)} k) \le \lambda. \tag{3.3.12}$$

Lemma 3.3.13 *We have* $\dim_k(X(J_0(Ml)^2_{\mathbb{F}_l}) \otimes_{\mathbb{T}_0(Mlq)} k) \le \mu$.

Proof. First note that η_q is in m. The exact sequence (3.3.4) shows that:

$$\dim_k(X(J_0(Ml)^2_{\mathbb{F}_l}) \otimes_{\mathbb{T}_0(Mlq)} k) = \dim_k(\Phi^*_{\mathbb{F}_q}) \otimes_{\mathbb{T}_0(Mlq)} k = \dim_k(\Phi_{\mathbb{F}_q}[m]).$$

Grothendieck's description [18, 7, IX, §11] of $\Phi_{\mathbb{F}_q}$ gives an exact sequence:

$$0 \to X(J'_{\mathbb{F}_q}) \to X(J_{\mathbb{F}_q})^\vee \to \Phi_{\mathbb{F}_q} \to 0, \tag{3.3.14}$$

where for M a $\mathbb{Z}$-module $M^\vee$ denotes its $\mathbb{Z}$-dual, and where $J'_{\mathbb{F}_q}$ is just $J_{\mathbb{F}_q}$, but with the dual $\mathbb{T}_0(Mlq)$-action: t in $\mathbb{T}_0(Mlq)$ acts as t^*, the dual of the endomorphism given by t (this uses the natural autoduality of jacobians). This makes the sequence (3.3.14) Hecke equivariant. Consider multiplication by p on the exact sequence (3.3.14). Applying the snake Lemma and then taking kernels for m gives an injection of $\Phi_{\mathbb{F}_q}[m]$ into $(X(J'_{\mathbb{F}_q})\otimes_{\mathbb{Z}}\mathbb{F}_p)[m]$. Ribet shows, using more results from [18, 7, IX, §11] and a description of the action of Frob_q on $\mathrm{Hom}(X(J_{\mathbb{F}_q}),\mu_p)$ and on $X(J'_{\mathbb{F}_q})$, that one has an exact sequence:

$$(3.3.15)\quad 0\to \mathrm{Hom}(X(J_{\mathbb{F}_q}),\mu_p)[m]\to J(\overline{\mathbb{Q}})[m]\to (X(J'_{\mathbb{F}_q})\otimes\mathbb{F}_p)[m]\to 0.$$

The middle term in this sequence has dimension 2μ by the definition of μ. The fact that $\rho(\mathrm{Frob}_q)$ has the two distinct eigenvalues 1 and -1 then implies that the other two terms both have dimension μ. □

We can now finish our description of the proof of Theorem 3.3.1. The sequence (3.3.2) gives:

(3.3.16)

$$\begin{aligned}\dim_k(X(J_0(Mlq)_{\mathbb{F}_l})\otimes_{\mathbb{T}_0(Mlq)}k) &\le \dim_k(X(J_{\mathbb{F}_q})\otimes_{\mathbb{T}_0(Mlq)}k)\\ &\quad+\dim_k(X(J_0(Ml)^2_{\mathbb{F}_l})\otimes_{\mathbb{T}_0(Mlq)}k).\end{aligned}$$

Combining (3.3.6), (3.3.7) and (3.3.12) gives $2\mu\le\lambda$. On the other hand, combining (3.3.10), (3.3.16), (3.3.11) and Lemma 3.3.13 gives $2\lambda\le 2\mu$.

4 Dealing with the Langlands–Tunnell Form

Let E be a semi-stable elliptic curve over $\mathbb{Q}$, with ρ_3 irreducible. It follows, Proposition 2.1, that $\rho_3\colon G_{\mathbb{Q}}\to \mathrm{GL}_2(\mathbb{F}_3)$ is surjective. The aim of this section is to show that ρ_3 is modular of type $(N(\rho_3),2,1)_{\overline{\mathbb{Q}}_3}$ if ρ_3 is finite at 3 and of type $(3N(\rho_3),2,1)_{\overline{\mathbb{Q}}_3}$ if ρ_3 is not finite at 3.

4.1 The Theorem of Langland and Tunnell

The first step is to show that ρ_3 is modular, by lifting ρ_3 to a continuous representation of $G_{\mathbb{Q}}$ with values in $\mathrm{GL}_2(\mathbb{C})$, i.e., a continuous representation $\rho\colon G_{\mathbb{Q}}\to\mathrm{GL}_2(\mathbb{C})$ (for the discrete topology on $\mathbb{C}$) such that ρ_3 is a reduction modulo 3 of ρ, and applying a theorem of Langlands and Tunnell to ρ. To do this, we compose ρ_3 with a two-dimensional complex representation of $\mathrm{GL}_2(\mathbb{F}_3)$. The irreducible representations of the groups $\mathrm{GL}_2(k)$, where k is a finite field, are well-known, see for example [6]. If q denotes the cardinality of k, then the irreducible representations have dimensions $1, q-1, q$ or $q+1$. Those of dimension one factor via the determinant. Hence for $q > 3$ there are no faithful two-dimensional representations.

In the case of interest for us, i.e., $q = 3$, there are three irreducible two-dimensional representations. One of them can be realized in $\mathrm{GL}_2(\mathbb{Z})$, but it is not faithful (it factors through the quotient S_3). The other two are Galois conjugates and can be realized in $\mathrm{GL}_2(\mathbb{Z}[\sqrt{-2}])$. Let $\pi\colon \mathrm{GL}_2(\mathbb{F}_3) \to \mathrm{GL}_2(\mathbb{Z}[\sqrt{-2}])$ be one of these, and let $\rho := \pi \circ \rho_3\colon G_{\mathbb{Q}} \to \mathrm{GL}_2(\mathbb{Z}[\sqrt{-2}])$. From the character table given in [6], one sees immediately that π is faithful (our π is one of the $\pi(\Lambda)$ with Λ of order eight), and that the restriction of π to the subgroup $\left(\begin{smallmatrix} * & 0 \\ 0 & 1 \end{smallmatrix}\right)$ is the regular representation. It follows that ρ is odd and that $\det \circ \rho$ is the composition of $\chi_3\colon G_{\mathbb{Q}} \to \mathbb{F}_3^*$ with the embedding of $\mathbb{F}_3^*$ into $\mathbb{C}^*$. Looking again at the character table shows that reduction of ρ modulo one of the two maximal ideals containing 3 gives ρ_3. According to [37, §5.3] the conductor $N(\rho)$ of ρ is $3^m N(\rho_3)^2$, for some $m \geq 0$.

This is explained as follows. Let $l \neq 3$ be a prime. If ρ_3 is unramified at l then ρ is unramified too. Suppose that ρ_3 is ramified at l. Since E is semi-stable, $\rho_3(I_l)$ is a conjugate of the subgroup $\left(\begin{smallmatrix} 1 & * \\ 0 & 1 \end{smallmatrix}\right)$. From the character table one sees that $\pi\left(\begin{smallmatrix} 1 & 1 \\ 0 & 1 \end{smallmatrix}\right)$ has trace -1, hence its eigenvalues are the two roots of unity of order 3. This means that the dimension of the space of I_l-invariants for ρ_3 is one, and that it is zero for ρ. Hence the exponent of l in $N(\rho)$ is twice that of l in $N(\rho_3)$. It will be of use for us to determine the exponent m of 3 in $N(\rho)$ in the case where $\rho_3|_{I_3}$ is isomorphic to the direct sum $\chi_3 \oplus 1$. In this case, in which the ramification is tame, (1.6) shows that $m = 1$. One can see easily that in all other cases $m > 1$.

A deep result of Langlands and Tunnell ([25] and [46]) says that ρ is modular of type $(N(\rho), 1, \det(\rho))_{\mathbb{Z}[\sqrt{-2}]}$, i.e., there exists a cuspidal eigenform f of level $N(\rho)$, of weight 1 and with character $\det(\rho)$ such that $\rho \cong \rho_f$, where $\rho_f\colon G_{\mathbb{Q}} \to \mathrm{GL}_2(\mathbb{C})$ is the representation associated to f by Deligne–Serre in [11] (see Chapter VI, Theorem 1.3). At this point we know that ρ_3 arises from the modular form f.

4.2 Some Purely Mod 3 Arguments

The second step is to show that ρ_3 is modular of type $(N(\rho_3)^2, k(\rho_3), 1)_{\overline{\mathbb{F}}_3}$. First we want to see that ρ is modular of type $(N(\rho_3)^2, k)_{\overline{\mathbb{F}}_3}$, for some $k \geq 1$. This is folklore. Ribet has given a proof of this in [32, §2] for $p > 2$. We remark that there are (unpublished) conceptual short proofs, valid for all p, using either the results of Katz and Mazur in [24] on the reduction mod p of modular curves (the idea is just to restrict the form to a suitable irreducible component of this reduction), or the representation theory of $\mathrm{GL}_2(\mathbb{F}_p)$ on $\mathbb{F}_p$-vector spaces and some group cohomology or etale cohomology.

We can now assume that ρ_3 is modular of type $(N(\rho_3)^2, k)_{\overline{\mathbb{F}}_3}$, for some $k \geq 1$. We want to show that it is modular of type $(N(\rho_3)^2, k(\rho_3))_{\overline{\mathbb{F}}_3}$, i.e., we want to adjust the weight. Theorem 4.5 of [15] says that this adjustment of the weight can always be done; in [15] one can find a detailed proof. We will now describe only the main features of that proof, and indicate a simplification, due to Taylor, in the case of our ρ_3.

The proof of Theorem 4.5 of [15] runs as follows. Let p be a prime number and $\rho: G_{\mathbb{Q}} \to \mathrm{GL}_2(\overline{\mathbb{F}}_p)$ be irreducible, continuous and modular of some level N which is prime to p. Then one knows that there exist a in $\mathbb{Z}/(p-1)\mathbb{Z}$ and k in $\{2, \ldots, p+1\}$ such that

$$\rho \cong \rho_f \otimes \chi_p^a$$

for some eigenform f of type $(N, k)_{\overline{\mathbb{F}}_p}$. Since f has low weight, i.e., $k \leq p+1$, one knows that $\rho_f|_{I_p}$ is an extension of the trivial character by χ_p^{k-1} if f is ordinary (i.e., $T_p f \neq 0$) and that it is the direct sum of ψ^{k-1} and ψ'^{k-1} if f is supersingular (i.e., $T_p f = 0$); here ψ and ψ' are the two fundamental characters of level two that occur in Definition 1.7. The first part of this result is due to Deligne (see [17] for a proof). The second part was first proved by Fontaine (see [15] for a proof). It follows that a and k are completely determined by $\rho|_{I_p}$, except if $\rho|_{I_p}$ is a direct sum of two distinct powers of χ_p or if it is a non-split extension of χ_p^a by χ_p^{a+1}.

In the first case there are two candidates for a; in the second case one cannot decide between the values 2 and $p+1$ for k. In the second case Serre conjectured, and Mazur showed under the assumption that $p > 2$, that one can take $k = 2$ if and only if $\rho \otimes \chi_p^{-a}$ is finite at p. Recall that Mazur's argument was described in §3.2. In the first case, Serre conjectured that both candidates for a should work and called the two forms corresponding to the two values of a companions of each other. The fact that such companion forms exist was proved in [17], [7] and [16], except for some cases when $p = 2$. With these results, it remains to construct a form f' of type $(N, k')_{\overline{\mathbb{F}}_p}$, with k' as small as possible, such that $\rho \cong \rho_{f'}$.

This last equality is equivalent to: $a_l' = l^a a_l$ for all primes l not dividing Np, where the a_l' and a_l are the eigenvalues of f' and f for T_l. There is an operation θ on modular forms over $\overline{\mathbb{F}}_p$ that has the effect of the derivation qd/dq on q-expansions. This derivation θ is described for forms of level one by Swinnerton-Dyer and Serre in [44], and in general, in terms of the Gauss–Manin connection, by Katz in [23]. The properties proved about θ in [23] imply that one knows exactly how to construct f': one should take $f' := A^{-n}\theta^a(f)$ with n as large as possible (here A denotes the Hasse invariant), or the same expression with f replaced by its companion if it exists. The results in [23] imply that one knows the value of n.

Let us now look at our ρ_3. We want to show that we do not need the results on companion forms to show that ρ_3 is modular of type $(N(\rho_3)^2, k(\rho_3))$ (this is an observation of Taylor). We know that ρ_3 or $\rho_3 \otimes \chi_3$ arises from a form of weight k with $2 \leq k \leq 4$. If E has bad or good and ordinary reduction at 3, then $\rho_3|_{I_3}$ is an extension of the trivial character 1 by χ_3; $k(\rho_3) = 2$ if ρ_3 is finite at 3 (this is the case if and only if the exponent of 3 in the discriminant of E is a multiple of 3), otherwise $k(\rho_3) = 4$. If E has good and supersingular reduction at 3, then $\rho_3|_{I_3}$ is the direct sum of ψ and ψ', the two fundamental characters of level two.

Suppose first that E is good and supersingular at 3. Then $\rho_3 \otimes \chi_3$ cannot arise from a form of weight between 2 and 4, because its restriction to I_3 is not of the right form. Hence ρ_3 itself arises from an eigenform f of weight k with $2 \leq k \leq 4$. Because $\det(\rho_3) = \chi_3$, we must have $k = 2$ or $k = 4$. It remains to see that we can take $k = 2$. So suppose that f has weight 4. We claim that then $f = Ag$, where A is the Hasse invariant and g an eigenform of weight two. To prove this, it suffices to show that f vanishes at the supersingular points of the modular curve on which it lives. Now if f would not vanish at all the supersingular points, then the regular differential form $\omega(f)$ on $X_0(3N(\rho_3)^2)_{\overline{\mathbb{F}}_3}$ that will be constructed in the next section has a non-zero residue at some supersingular point. A study of the action of T_3 then shows that $T_3 f = \pm f$, but we have $T_3 f = 0$. The eigenform g has the same eigenvalues as f, hence $\rho_3 \cong \rho_g$.

Suppose now that E has bad or good and ordinary reduction. Then $\rho_3|_{I_3}$ is an extension of 1 by χ_3. Let us first deal with the case where this extension is not split. Then $\rho_3 \otimes \chi_3$ cannot arise from a form of weight between 2 and 4, hence there is a form f of weight 2 or 4 such that $\rho_3 \cong \rho_f$. If $k(\rho_3) = 4$ we have $k = 4$ and there is nothing to prove. So suppose that $k(\rho_3) = 2$. If $k = 2$ there is nothing to prove, so suppose moreover that $k = 4$. Let $\omega(f)$ be the differential form on $X_0(3N(\rho_3)^2)_{\overline{\mathbb{F}}_3}$ as constructed in the next section. Since $\omega(f)$ can be lifted this shows that ρ_3 is modular of type $(3N(\rho_3)^2, 2, 1)_{\overline{\mathbb{Q}}_3}$. Since ρ_3 is finite at 3, Mazur's result (Theorem 3.2.1) shows that ρ_3 is modular of type $(N(\rho_3)^2, 2, 1)_{\overline{\mathbb{Q}}_3}$. It remains to deal with the split case. So suppose that $\rho_3|_{I_3} \cong \chi_3 \oplus 1$. Then we have seen that the weight one form f given by Langlands and Tunnell has level $3N(\rho_3)^2$. Then $fE_{1,\chi}$, with $\chi\colon \mathbb{F}_3^* \to \mathbb{C}^*$ non-trivial, shows that ρ_3 is modular of type $(3N(\rho_3)^2, 2, 1)_{\overline{\mathbb{Q}}_3}$. Theorem 3.2.1 shows that ρ_3 is modular of type $(N(\rho_3)^2, 2, 1)_{\overline{\mathbb{Q}}_3}$.

4.3 Differential Forms and a Version of Carayol's Lemma

In the last section we saw that ρ_3 is modular of type $(N(\rho_3)^2, k(\rho_3), 1)_{\overline{\mathbb{F}}_3}$. Now we want to deduce that ρ_3 is modular of type $(N(\rho_3)^2, 2, 1)_{\overline{\mathbb{Q}}_3}$ if ρ_3 is finite at 3 and of type $(3N(\rho_3)^2, 2, 1)_{\overline{\mathbb{Q}}_3}$ otherwise. It is well known that cuspidal modular forms of a given type $(N, 2, 1)_K$, where K is a field of characteristic zero, correspond via the Kodaira–Spencer isomorphism to differential forms on the modular curve $X_0(N)_K$, i.e, to sections of the sheaf Ω^1 on $X_0(N)_K$. This correspondence is compatible with the action of Hecke operators. We define $N := N(\rho_3)^2$ if ρ_3 is finite at 3 and $N := 3N(\rho_3)^2$ otherwise. We will show in fact that there exists a non-zero differential form ω on $X_0(N)_{\overline{\mathbb{Q}}_3}$ which is an eigenform for the Hecke algebra and whose associated Galois representation $\rho_\omega: G_{\mathbb{Q}} \to \mathrm{GL}_2(\overline{\mathbb{F}}_3)$ is isomorphic to ρ_3. We will construct this ω by first constructing it over $\overline{\mathbb{F}}_3$, which is in fact the hardest part (it involves a version of Carayol's lemma); a standard argument from commutative algebra shows that it can be lifted to an eigenform (see for example [17, §9]).

Let us first do this construction in the case where ρ_3 is finite at 3. So then N is prime to 3 and we have a cuspidal eigenform f of type $(N, 2, 1)_{\overline{\mathbb{F}}_3}$. We want to construct a differential form ω on $X_0(N)_{\overline{\mathbb{F}}_3}$ which is an eigenform for all Hecke operators, with the same eigenvalues as f. The problem is that in general we cannot view f as a section of a sheaf $\underline{\omega}^{\otimes 2}$ on $X_0(N)_{\overline{\mathbb{F}}_3}$ and apply a Kodaira–Spencer isomorphism, because $X_0(N)_{\overline{\mathbb{F}}_3}$ does not carry a universal family of generalized elliptic curves with a level structure of type $\Gamma_0(N)$. We will see that this problem is not just artificial, in the sense that for certain eigenforms of type $(N, 2, 1)_{\overline{\mathbb{F}}_3}$ there does not exist a differential form as we want; it will exist for our f because ρ_3 is not induced from $G_{\mathbb{Q}(\sqrt{-3})}$.

Let $n \geq 3$ be an integer prime to 3. According to (1.2) we can view f as a section of the sheaf $\underline{\omega}^{\otimes 2}$ on the smooth projective curve $X := \overline{\mathcal{M}}([\Gamma_1(N), \Gamma(n)]_{\overline{\mathbb{F}}_3})$ over $\overline{\mathbb{F}}_3$ that carries a universal family of generalized elliptic curves with a point of order N and a trivialization of the n-torsion. Let Y denote the curve $X_0(N)_{\overline{\mathbb{F}}_3}$. Then Y is the quotient of X by the action of the group $G := (\mathbb{Z}/N\mathbb{Z})^* \times \mathrm{GL}_2(\mathbb{Z}/n\mathbb{Z})$ acting on X, and f is a G-invariant section of $\underline{\omega}^{\otimes 2}$. Let $\omega(f)$ be the global differential form on X obtained by applying the Kodaira–Spencer isomorphism to f. The functorial properties of this isomorphism imply that $\omega(f)$ is G-invariant. We would like to conclude that $\omega(f)$ is the pullback of a unique differential form, that we will also call $\omega(f)$, on Y. The uniqueness is guaranteed by the fact that the morphism $\pi: X \to Y$ is separable, which implies that the pullback morphism π^* on differential forms is injective. The problem is the existence. The rest of this section is motivated by [28, II, Lemma 4.4], and its proof.

Let V be the biggest open part of Y over which π is etale. Note that G acts on X via its quotient $\overline{G}$ by the subgroup generated by $(-1,-1)$. The group $\overline{G}$ acts faithfully, hence V is the complement in Y of the image under π of those points of X with non-trivial stabilizer in $\overline{G}$. Since $\pi\colon \pi^{-1}V \to V$ is etale, the restriction of $\omega(f)$ to $\pi^{-1}V$ is the pullback of a unique differential form $\omega(f)$ on V. So we have to show that this $\omega(f)$ has no poles in the complement of V. The information we have is that the pullback of $\omega(f)$ to X has no poles.

Let y be a point of Y over which π is ramified, and let x be in $\pi^{-1}y$. Let s and t be uniformizers at x and y, respectively. Let e be the ramification index at x, i.e., $t = s^e u$ with u in $\mathcal{O}^*_{X,x}$. Let r be the valuation $v_x(dt)$ at x of dt. Since $dt = s^{e-1}(e + u's)ds$, where $u' = du/ds$, one has $r \geq e-1$ with equality if and only if π is tamely ramified at x. We have

$$v_x(\omega(f)) = ev_s(\omega(f)) + r.$$

It follows that $\omega(f)$ is regular at y if π is tamely ramified over y, so it remains to look at those y over which π is wildly ramified. Such points y all correspond to the elliptic curve E of j-invariant zero over $\overline{\mathbb{F}}_3$. The automorphism group of E is isomorphic to $\mathbb{F}_3^* \times \mathrm{GL}_2(\mathbb{F}_2)$; the projection to $\mathbb{F}_3^*$ comes from the action of $\mathrm{Aut}(E)$ on the tangent space at zero of E, the other projection comes from the action on the 2-torsion of E. The invariants e and r are the same at all x at which π is wildly ramified; note that we have $e = 3$ or $e = 6$. A global calculation as in [28, II, §2], using the Hurwitz formula for π, shows that we have $r = e$, which means that the wild ramification is of the mildest form. (In fact, this can also be derived from Table 1 of [28, II, §2].) It follows that $\omega(f)$ has a pole of order at most one at y.

Now the number of y over which π is wildly ramified can be easily computed: it is 1 if $N = 1$, it is 0 if N is divisible by a prime number congruent to -1 modulo 3, and otherwise it is $2^{\nu-1}$ where ν is the number primes dividing N. In the situation of [27, II, Lemma 4.4] this number is zero or one, and it follows that in fact $\omega(f)$ is regular because the sum of its residues must be zero. But if all primes dividing N are 1 modulo 3 and $\nu > 1$, there really exist eigenforms f such that $\omega(f)$ has poles, so we have to show why our $\omega(f)$ is regular.

Suppose that $\omega(f)$ is not regular. Let D be the set of y at which π is wildly ramified, viewed as an effective divisor on Y, and let V be the $\overline{\mathbb{F}}_3$-vector space of $\overline{\mathbb{F}}_3$-valued functions on D. Consider the map

$$R\colon \mathrm{H}^0(Y, \Omega^1(D)) \to V$$

which sends a form to its residues at the y in D. Then the image of $\omega(f)$ is not zero. There is a natural action of the Hecke operators on V which is compatible with R. For $n \geq 1$ the Hecke operator T_n acts on V via isogenies of degree n between elements y of D, hence by endomorphisms of degree n of E. Let σ denote one of the two automorphisms of order 3 of E. An endomorphism ϕ of E of degree n that does not commute with σ contributes zero to the action of T_n on V, since the correspondence inducing T_n is wildly ramified at ϕ. Let l be prime and congruent to -1 mod 3. Then E has no endomorphism of degree l commuting with σ, hence T_l acts as zero on V. It follows that $T_l f = 0$ for all such l, which contradicts that ρ_3 is not induced from $G_{\mathbb{Q}(\sqrt{-3})}$.

Let us now consider the case where ρ_3 is not finite at 3. Then $N = 3N'$, with $N' = N(\rho_3)^2$ prime to 3, and $k(\rho_3) = 4$. In this case $X_0(N)_{\overline{\mathbb{F}}_3}$ has two irreducible components, both isomorphic to $X_0(N')_{\overline{\mathbb{F}}_3}$, which intersect transversally at the supersingular points. The sheaf of Kähler differentials on $X_0(N)_{\mathbb{Z}_3}$ is not locally free of rank one at the double points over $\overline{\mathbb{F}}_3$, and it is better to work with the dualizing sheaf Ω on it. This sheaf Ω can be obtained as follows: let $X_0(N)^{\mathrm{sm}}_{\mathbb{Z}_3}$ be the smooth locus of $X_0(N)_{\mathbb{Z}_3}$ (i.e., the complement of the double points), let $j\colon X_0(N)^{\mathrm{sm}}_{\mathbb{Z}_3} \to X_0(N)_{\mathbb{Z}_3}$ be the inclusion and let Ω^1 be the sheaf of Kähler differentials on $X_0(N)^{\mathrm{sm}}_{\mathbb{Z}_3}$; then $\Omega = j_*\Omega^1$. The dualizing sheaf Ω is locally free of rank one and it is dualizing in the sense of Serre duality. For a more detailed description of Ω in the context of modular curves see [29, §§6–7], [17, §§8–9] and references therein.

Let $Y := X_0(N)_{\overline{\mathbb{F}}_3}$, and let Y_0 (resp., Y_∞) be the irreducible component of Y containing the cusp 0 (resp., ∞). The restrictions of Ω to Y_0 and Y_∞ are the sheaves of Kähler differentials with poles of order at most one at the supersingular points. Let $n \geq 3$ be prime to 3, let $X := \overline{\mathcal{M}}([\Gamma_1(N'), \Gamma(n)]_{\overline{\mathbb{F}}_3})$ and let $G := (\mathbb{Z}/N'\mathbb{Z})^* \times \mathrm{GL}_2(\mathbb{Z}/n\mathbb{Z})$. Then f is a G-invariant section of $\underline{\omega}^4$ on X. It follows that f/A (recall that A is the Hasse invariant) is a rational section of $\underline{\omega}^{\otimes 2}$ with poles of at most order one at the supersingular points. Applying the Kodaira–Spencer isomorphism to f/A gives a G-invariant rational differential form $\omega(f)$ on X which has poles of order at most one at the supersingular points. We identify the quotient of X by G with Y_∞. Since $\omega(f)$ is G-invariant, we can view $\omega(f)$ as a rational differential form on Y_∞, with poles only at the supersingular points and at the points where $X \to Y_\infty$ is ramified. A calculation as above shows that there are only poles of order at most one at the supersingular points. The proof of [17, Prop. 9.3] shows that there exists a section of Ω on $X_0(N)_{\overline{\mathbb{Z}}_3}$ which is an eigenform for the Hecke algebra and whose restriction to Y_∞ is $\omega(f)$.

We end this section with some remarks. First of all, one can show that the Hecke action on the $\overline{\mathbb{F}}_3$-vector space V that occurs in the arguments for

ρ_3 finite at 3, is "Eisenstein": for l prime and not dividing $3N$, T_l acts as $l+1$ on V. Hence it would have been sufficient to use that ρ_3 is irreducible. For more general quotients of X by subgroups of the form $H \times \mathrm{GL}_2(\mathbb{Z}/n\mathbb{Z})$ of G, the vector space V is not necessarily "Eisenstein." This is related to results on groups of connected components of Néron models in [36].

Secondly, instead of studying in detail the wild ramification in the morphism $X \to Y$, we could have used Theorem 3.2.1 as follows. Let q be any prime number that is congruent to -1 mod 3 and that does not divide N. Then, replacing N by qN, one gets $X \to Y$ tamely ramified, hence a differential form on $X_0(Nq)_{\overline{\mathbb{F}}_3}$. This shows that ρ_3 is modular of type $(Nq, 2, 1)_{\overline{\mathbb{Q}}_3}$. Then Mazur's result shows that ρ_3 is modular of type $(N, 2, 1)_{\overline{\mathbb{Q}}_3}$. One reason to give the argument above is to illustrate the problems one gets when interpreting modular forms as differential forms.

4.4 Carayol's Reductions

At this point we know that ρ_3 is modular of type $(N(\rho_3)^2, 2, 1)_{\overline{\mathbb{Q}}_3}$ if ρ_3 is finite at 3 and of type $(3N(\rho_3)^2, 2, 1)_{\overline{\mathbb{Q}}_3}$ otherwise. We want to show that ρ_3 is modular of type $(N(\rho_3), 2, 1)_{\overline{\mathbb{Q}}_3}$ if ρ_3 is finite at 3 and of type $(3N(\rho_3), 2, 1)_{\overline{\mathbb{Q}}_3}$ otherwise. Before explaining how this is done, it is good to recall some results of Langlands, Deligne and Carayol (see [5]). Let p be a prime number and let f be a newform of some type $(N, k, \varepsilon)_{\overline{\mathbb{Q}}_p}$ with $k \geq 2$. Then this gives us a representation $\rho_f \colon G_{\mathbb{Q}} \to \mathrm{GL}_2(\overline{\mathbb{Q}}_p)$, determined by the property that it is unramified outside Np and that for l not dividing Np the Frobenius element $\rho_f(\mathrm{Frob}_l)$ has trace $a_l(f)$. On the other hand, there is also a representation $\pi_f \colon \mathrm{GL}_2(\hat{\mathbb{Z}} \otimes \mathbb{Q}) \to \mathrm{GL}(V)$, with V an infinite dimensional $\overline{\mathbb{Q}}_p$-vector space, associated to f in the following way. Let W be the direct limit, taken over all multiples $n \geq 1$ of N, of the $\overline{\mathbb{Q}}_p$-vector spaces $\mathrm{H}^0(\overline{\mathcal{M}}([\Gamma(n)]_{\overline{\mathbb{Q}}_p}), \underline{\omega}^{\otimes k})$. It is clear that $\mathrm{GL}_2(\hat{\mathbb{Z}})$ acts on W, and it is not hard to see that this action extends to one of $\mathrm{GL}_2(\hat{\mathbb{Z}} \otimes \mathbb{Q})$. Then V is the subspace of W that is generated by the $g(f)$, for g in $\mathrm{GL}_2(\hat{\mathbb{Z}} \otimes \mathbb{Q})$. One knows that V is an irreducible representation of $\mathrm{GL}_2(\hat{\mathbb{Z}} \otimes \mathbb{Q})$ and that $V = \otimes_l' V_l$ is the restricted tensor product, over all primes l, of irreducible admissible representations $\pi_{f,l} \colon \mathrm{GL}_2(\mathbb{Q}_l) \to \mathrm{GL}(V_l)$. The result alluded to describes, for all $l \neq p$, the restriction $\rho_{f,l}$ of ρ_f to some decomposition group at l in terms of $\pi_{f,l}$. It follows from this result that the conductor $N(\overline{\rho}_f)$ of the reduction $\overline{\rho}_f \colon G_{\mathbb{Q}} \to \mathrm{GL}_2(\overline{\mathbb{F}}_p)$ divides N (here we suppose that $\overline{\rho}_f$ is irreducible, since otherwise it is not well defined). Carayol [4] and Livné [26] have classified, in terms of the $\pi_{f,l}$, the $l \neq p$ dividing $N/N(\overline{\rho}_f)$.

The strategy for proving that ρ_3 is modular of type $(N(\rho_3), 2, 1)_{\overline{\mathbb{Q}}_3}$ if ρ_3 is finite at 3 and of type $(3N(\rho_3), 2, 1)_{\overline{\mathbb{Q}}_3}$ otherwise will be the following.

Suppose that we know that ρ_3 is modular of type $(N,2,1)_{\overline{\mathbb{Q}}_3}$ for some N dividing $N(\rho_3)^2$. Let $l \neq 3$ be a prime number and suppose that l^2 divides N. Then we want to show that ρ_3 is modular of type $(N',2,1)_{\overline{\mathbb{Q}}_3}$ for some N' dividing N with l^2 not dividing N'. Of course, this suffices.

So suppose that f is a cuspidal eigenform of type $(N,2,1)_{\overline{\mathbb{Q}}_3}$ with N dividing $N(\rho_3)^2$ and that $l \neq 3$ is a prime number such that l^2 divides N. The newform associated to f has level dividing N, so we may in fact assume that f is a newform. The classification of Carayol and Livné then says that $\rho_{f,l}$ is of one of the following two types:

1. $\rho_{f,l}$ is a direct sum of two ramified characters $\alpha, \beta \colon G_{\mathbb{Q}_l} \to \overline{\mathbb{Q}}_3^*$ whose reductions $\overline{\alpha}, \overline{\beta} \colon G_{\mathbb{Q}_l} \to \overline{\mathbb{F}}_3^*$ are unramified,

2. $\rho_{f,l} = \mathrm{Ind}_{G_K}^{G_{\mathbb{Q}_l}} \psi$ with K the unique unramified quadratic extension of $\mathbb{Q}_l$ and $\psi \colon G_K \to \overline{\mathbb{Q}}_3^*$ a ramified character with unramified reduction $\overline{\psi} \colon G_K \to \overline{\mathbb{F}}_3^*$.

Let us deal with the first case first. Let $\tilde{\chi}_3 \colon G_{\mathbb{Q}} \to \mathbb{Z}_3^*$ be the character giving the action on all roots of unity of 3-power order. Recall that since f has trivial character and weight two we have $\det(\rho_f) = \tilde{\chi}_3$. This implies that $\alpha\beta$ is unramified. There is a unique character ϵ of $\mathbb{F}_l^* = \mathrm{Gal}(\mathbb{Q}(\zeta_l)/\mathbb{Q})$ with values in the kernel of $\overline{\mathbb{Z}}_3^* \to \overline{\mathbb{F}}_3^*$ such that $\alpha\varepsilon$ is unramified. Let f' be the newform corresponding to the twist $f \otimes \varepsilon$ of f by ε. One way to express this is to say that $a_n(f') = a_n(f)\varepsilon(n)$ for all n prime to l. Another way is to say that $\rho_{f'} = \rho_f \otimes \varepsilon$. Anyway, f' is a newform of type $(N/l, 2, \varepsilon^2)_{\overline{\mathbb{Q}}_3}$ giving ρ_3. Since $\overline{\varepsilon} = 1$ and ρ_3 is not induced from $\mathbb{Q}(\sqrt{-3})$ Carayol's Lemma implies that ρ_3 is modular of type $(N/l, 2, 1)_{\overline{\mathbb{Q}}_3}$.

Let us now say something about case 2. The analog of this case with 3 replaced by a prime $p \geq 5$ is treated in [4, §5], and uses the Jacquet–Langlands correspondence to switch to a certain Shimura curve. The generalization to the case $p = 3$, using that ρ_3 is not induced from $\mathbb{Q}(\sqrt{-3})$, is explained in [12, §5]. One might also say that this generalization is done in [4], if one admits that the remarks in [4, §4.4] concerning modular curves also hold for the Shimura curves used in [4, §5]. We will now sketch the argument.

Let q be a prime number not dividing $3N$, such that $\rho_3(\mathrm{Frob}_q)$ is conjugated to $\rho_3(c)$, with c a complex conjugation. Then a result of Ribet (see [34]) says that there exists a newform f' of type $(N'q, 2, 1)_{\overline{\mathbb{Q}}_3}$, with N' dividing N, such that $\overline{\rho}_{f'} \cong \rho_3$ and with $\rho_{f',q}$ special, i.e., $\rho_{f',q}$ is a non-split extension of an unramified character α with $\alpha^2 = 1$ by $\alpha\tilde{\chi}_3$ (see also [47, II, Lemma 2.3]). If $\rho_{f',l}$ is not in case 2, then one applies the method to deal with case 1 to get rid of the l^2 in the level and then one applies Mazur's Theorem 3.2.1 to get rid of q. So we may assume that $\rho_{f',l}$ is in case 2. Let B be the quaternion algebra over $\mathbb{Q}$ with discriminant pq. By the Jacquet–Langlands correspondence and the results of [5], $\rho_{f'}$ can be constructed from the 3-adic Tate module of the jacobian of a Shimura curve of a certain level associated to B. On this Shimura curve one has an action by the group $\mathbb{F}_{l^2}^* \times \mathbb{F}_{q^2}^*$ which is analogous to the action of the diamond operators on modular curves. A version of Carayol's Lemma then shows that ρ_3 actually arises from the quotient of this Shimura curve by that group. Switching back to modular curves by the Jacquet–Langlands correspondence then shows that ρ_3 is modular of type $(N''q, 2, 1)_{\overline{\mathbb{Q}}_3}$, with N'' dividing N/l. Then one finishes by applying Theorem 3.2.1.

References

[1] B.J. Birch and W. Kuyk (editors). *Modular functions of one variable IV.* Springer Lecture Notes in Mathematics 476 (1975).

[2] N. Boston, H.W. Lenstra and K.A. Ribet. *Quotients of group rings arising from two-dimensional representations.* C.R. Acad. Sci. Paris, t. 312, Série I, p. 323–328 (1991).

[3] J.-F. Boutot and H. Carayol. *Uniformisation p-adique des courbes de Shimura: les théorèmes de Cerednik et Drinfeld.* Astérisque 196–197, 45–158 (1991).

[4] H. Carayol. *Sur les représentations galoisiennes mod l attachées aux formes modulaires.* Duke Math. Journal **59** (1989), No. 3, 785–801.

[5] H. Carayol. *Sur les représentations l-adiques associées aux formes modulaires de Hilbert.* Ann. Sci. Ecole Norm. Sup. (4) 19, 409–468 (1986).

[6] P. Cartier. *Détermination des caractères des groupes finis simples: travaux de Lusztig.* Séminaire Bourbaki, Exposé 658 (1986).

[7] R.F. Coleman and J.F. Voloch. *Companion forms and Kodaira-Spencer theory.* Invent. Math. **110** (1992), 263–281.

[8] P. Deligne. *Formes modulaires et représentations l-adiques.* Séminaire Bourbaki, exposé 355, Springer Lecture Notes in Mathematics 179, 139–172, (1971).

[9] P. Deligne. *Courbes elliptiques: formulaire d'après John Tate.* Modular functions of one variable IV, pages 53–73. Springer Lecture Notes in Mathematics 476 (1975).

[10] P. Deligne and M. Rapoport. *Les schémas de modules des courbes elliptiques.* In Modular Functions of One Variable II, pages 143–316. Springer Lecture Notes in Mathematics 349 (1973).

[11] P. Deligne and J-P. Serre. *Formes modulaires de poids 1.* Ann. Sci. Ecole Norm. Sup. (4) 7, 507–530 (1974).

[12] F. Diamond. *The refined conjecture of Serre.* In: Elliptic Curves, Modular Forms and Fermat's Last Theorem, J. Coates, S.T. Yau, eds., International Press, Cambridge, pages 22–37 (1995).

[13] F. Diamond. *Congruence primes for cuspforms of weight $k \geq 2$.* Astérisque 196–197, 205–213 (1991).

[14] F. Diamond and J. Im. *Modular forms and modular curves.* CMS Conf. Proc., AMS Publ., Providence, "Elliptic curves, Galois representations and modular forms", edited by V.K. Murty.

[15] S.J. Edixhoven. *The weight in Serre's conjectures on modular forms.* Invent. Math. **109** (1992), 563–594.

[16] G. Faltings and B. Jordan. *Crystalline cohomology and* $\mathrm{GL}_2(\mathbb{Q})$. Israel Journal of Mathematics **90**, 1–66 (1995).

[17] B.H. Gross. *A tameness criterion for Galois representations associated to modular forms (mod p).* Duke Math. Journal **61** (1990). No. 2, 445–517.

[18] A. Grothendieck. *Séminaire de géométrie algébrique.* Springer Lecture Notes in Mathematics 151, 152, 153, 224, 225, 269, 270, 288, 305, 340, 589.

[19] Y. Hellegouarch. *Courbes elliptiques et équation de Fermat.* Thesis, Besançon, 1972.

[20] Y. Hellegouarch. *Points d'ordre $2p^h$ sur les courbes elliptiques.* Acta Arithmetica XXVI, 253–263 (1975).

[21] B. Jordan and R. Livné. *Conjecture "epsilon" for weight $k > 2$.* Bull. AMS 21, 51–56 (1989).

[22] N.M. Katz. *p-adic properties of modular schemes and modular forms.* In Modular Functions of One Variable III. Springer Lecture Notes in Mathematics 350 (1973).

[23] N.M. Katz. *A result on modular forms in characteristic p.* Springer Lecture Notes in Mathematics 601, 53–61 (1976).

[24] N. Katz and B. Mazur. *Arithmetic moduli of elliptic curves.* Annals of Mathematics studies, study 108. Princeton University Press, 1985.

[25] R.P. Langlands. *Base change for* GL_2. Princeton University Press (1980).

[26] R. Livné. *On the conductors of mod l Galois representations coming from modular forms.* Journal of Number Theory 31, 133–141 (1989).

[27] B. Mazur. *Rational isogenies of prime degree.* Invent. Math. 44, 129–162 (1978).

[28] B. Mazur. *Modular curves and the Eisenstein ideal.* Publications Mathématiques de l'I.H.E.S. 47 (1977).

[29] B. Mazur and K.A. Ribet. *Two-dimensional representations in the arithmetic of modular curves.* Astérisque 196–197, 215–255 (1991).

[30] M. Raynaud. *Spécialisation du foncteur de Picard.* Publications Mathématiques de l'I.H.E.S. 38 (1970).

[31] M. Raynaud. *Schémas en groupes de type* $(p, \ldots, p)$. Bull. Soc. Math. France 102, 241-280 (1974).

[32] K.A. Ribet. *Report on mod l representations of* $\mathrm{Gal}(\overline{\mathbb{Q}}/\mathbb{Q})$. Proceedings of Symposia in Pure Mathematics, **55** (1994), Part 2, 639–676.

[33] K.A. Ribet. *On modular representations of* $\mathrm{Gal}(\overline{\mathbb{Q}}/\mathbb{Q})$ *arising from modular forms.* Invent. Math. 100, 431–476 (1990).

[34] K.A. Ribet. *Congruence relations between modular forms.* Proc. Int. Congr. of Math., Warsaw, 503–514 (1983).

[35] K.A. Ribet. *Lowering the levels of modular representations without multiplicity one.* Internat. Math. Res. Notices, 15–19 (1991).

[36] K.A. Ribet. *Irreducible Galois representations arising from component groups of jacobians.* In: Elliptic curves, modular forms and Fermat's Last Theorem (eds: Coates, Yau), International Press, 1995, pages 131–147.

[37] J-P. Serre. *Sur les représentations de degré 2 de* $\mathrm{Gal}(\overline{\mathbb{Q}}/\mathbb{Q})$. Duke Mathematical Journal 54, No. 1, (1987), 179–230.

[38] J-P. Serre. *Représentations linéaires des groupes finis.* Collection Méthodes, Hermann, Paris (1978).

[39] J-P. Serre. *Propriétés galoisiennes des points d'ordre fini des courbes elliptiques.* Invent. Math. **15** (1972), 259–331.

[40] J-P. Serre. Résumé des cours en 1978–1988, Annuaire du Collège de France (1988), 79–82.

[41] J-P. Serre. *Two letters on quaternions and modular forms (mod p).* Israel Journal of Mathematics **95** (1996), 281–299.

[42] N.I. Shepherd-Barron and R. Taylor. *Mod 2 and mod 5 icosahedral representations.* To appear in the Journal of the A.M.S.

[43] G. Shimura. *Introduction to the theory of automorphic functions.* Iwanami Shoten and Princeton University Press, Princeton, 1971.

[44] H.P.F. Swinnerton-Dyer. *On l-adic representations and congruences for coefficients of modular forms.* Springer Lecture Notes in Mathematics 350, 1–55 (1973).

[45] R. Taylor and A. Wiles. *Ring theoretic properties of certain Hecke algebras.* Annals of Math. **142**, 553–572 (1995).

[46] J. Tunnell. *Artin's conjecture for representations of octahedral type.* Bull. A.M.S. **5**, 173–175 (1981).

[47] A. Wiles. *Modular elliptic curves and Fermat's Last Theorem.* Annals of Math. **142**, 443–551 (1995).

[48] D.S. Kubert and S. Lang. *Modular Units.* Grundlehren der Math. Wiss., **244** (1981), Springer-Verlag.

AN INTRODUCTION TO THE DEFORMATION THEORY OF GALOIS REPRESENTATIONS

BARRY MAZUR

CONTENTS

Part One
Chapter I. Galois representations
§1. The Galois group of a number field, and a way of studying it
§2. What coefficient-rings should we allow for our Galois representations?
§3. Galois representations arise naturally
Chapter II. Group representations
§4. Group representations versus algebra representations
§5. Representations and their characters
§6. "Descent" of group representations. Schur-type Theorems
§7. Characterizing character-functions (a result of Rouquier)
§8. Deformations of a group representation
Chapter III. The deformation theory of Galois representations
§9. Why study "Galois" deformation theory?
§10. The universal "Galois" deformation ring
§11. An alternative description of the deformation problem for group representations (in a slightly more general context)
§12. Representations with coefficient-rings which are Λ-algebras
§13. Is there a relationship between the (formal, say) deformation space of a variety V defined over a number field K, and the deformation space of the various Galois representations occurring in the étale cohomology of V?
Part Two
Chapter IV. Functors and Representability
§14. Fiber products and representability
§15. A functor's-eye view of the Zariski tangent space (the "absolute" case)
§16. The Zariski tangent A-module (the "relative" case)
§17. Continuous Kähler differentials and the (relative) Zariski Tangent Space

Notes for lectures given at the Boston University Conference on Fermat's Last Theorem — August 1995

§18. Schlessinger's representability theorem
§19. Relatively representable subfunctors
§20. Representability results regarding deformation problems attached to group representations
Chapter V. Zariski tangent spaces and deformation problems subject to "conditions"
§21. A cohomological interpretation of Zariski tangent A-modules
§22. The Zariski tangent space of the universal deformation ring as "extensions"
§23. "Deformation conditions"
§24. Determinant conditions
§25. Categorical conditions
Chapter VI. Back to Galois representations
§26. Galois deformation conditions
§27. Passage to the limit
§28. The special case when $\Lambda = A$
§29. A bestiary of local Galois deformation conditions
§30. The condition of being "ordinary"
§31. The condition of being "finite flat"
A brief glossary of terminology and notation
References

Before this conference I had never been to any mathematics gathering where so many people worked as hard or with such high spirits, trying to understand a single piece of mathematics.

The talents of the organizers helped shape this collective activity which concentrated on the work of Wiles [W] and that of Taylor-Wiles [T-W]. Among their ideas for doing this, there is one which I feel should be recorded in these Proceedings (because of its novelty, and its possible usefulness in future conferences): since the lecture hall of the conference held roughly 400 participants, and attendance was usually that high, it was impractical for the audience to be given the opportunity to ask questions either during a lecture or immediately afterwards. The solution to this difficulty was to set up a special "question room," open every evening, in which a team of "experts" would be available to either answer, or explore, any question!

The organizers asked me to lecture about deformation theory; they also gave me a specific assignment meant to fit in with their over-all program: I was asked to include in my lectures an explanation of a particular theorem which describes the Zariski tangent space of the universal deformation rings attached to specific Galois deformation problems.

I hope the introductory material about representations which comprises Part One of these notes, will be helpful to people who want to get a glimpse of this subject, and of its deformations.

As for the general concept of "deformation," it is everywhere in mathematics and to get a quick sense of its various manifestations in the literature, you can consult the annotated bibliography of this subject, compiled by C. Doran (eventually to be included in [D-W]). This can be downloaded by anonymous ftp at ⟨`abel.harvard.edu`⟩ in the directory `mazur_handout`.

Hendrik Lenstra, in his lecture in the conference, recounted that twenty years ago he was firm in his conviction that he DID want to solve Diophantine equations, and that he DID NOT wish to represent functors — and now he is amused to discover himself representing functors in order to solve Diophantine equations! Part Two of these notes is all about functors: here I concentrate on the specific task assigned to me; namely, to show that "first-order infinitesimal" information concerning the universal deformation ring attached to a representation ρ can be expressed in terms of group cohomology (of the adjoint representation of ρ). This is quite a general phenomenon, does not even depend upon the representability of the deformation problem, and has an appropriate variant for deformation problems subject to conditions. To get the easiest-to-state and most-flexible result, it made sense to me to do a bit of housekeeping in the theory: firstly, to formulate a general functorial notion (called **nearly representable** in §18 below) which is general enough so that *every* representation is "nearly representable" and which is stringent enough so that any "nearly representable" functor has well-working Zariski tangent modules; secondly, to give appropriate functorial axioms which cover the bestiary of different specific "conditions" that will eventually be imposed on the various deformation theories, so as to be able to deal with all of them somewhat more systematically. This leads to the formal definition of **deformation condition** given in §23 below. I want to thank B. Conrad, F. Gouvêa, J.-P. Serre, and J. Tate for their valuable suggestions and for their corrections to a preliminary version of this article. I am also grateful to Carol Oliveira for the excellent job she did in typing my manuscript.

Part One

CHAPTER I. GALOIS REPRESENTATIONS

§1. The Galois group of a number field, and a way of studying it. If K is a field and $\overline{K}$ a choice of separable algebraic closure of K, let $G_K = \mathrm{Gal}(\overline{K}/K)$ denote "the" Galois group of K. The group G_K is a profinite topological group with its natural Krull topology, where a base of open subgroups is given by the "fixers" of finite extension fields of K in $\overline{K}$.

Now let K be a number field; i.e., a finite extension of $\mathbf{Q}$. If S is a finite set of non-archimedean places of K, or equivalently of non-zero prime ideals in the ring of integers of K, then $G_{K,S}$ will denote the Galois group of

$$\overline{K}_S := \text{the maximal algebraic extension of } K \text{ in } \overline{K} \text{ unramified outside } S$$

(with no condition imposed at the archimedean primes of K). An equivalent definition of $\overline{K}_S$ is that it is the union of all finite extensions of K in $\overline{K}$ whose relative discriminant is not divisible by any prime outside S. Fixing a finite set of primes S and a number d, the classical theorem of Hermite-Minkowski assures us that there are only a finite number of field extensions of K inside $\overline{K}$ of degree $\leq d$ unramified outside S. But $\overline{K}_S/K$ is often of infinite degree; equivalently, the group $G_{K,S}$ is often infinite (and certainly is so if, for example, S contains all primes of K lying above some prime of $\mathbf{Q}$). The profinite group $G_{K,S}$ (given its Krull topology) is naturally a quotient of G_K. The kernel of the projection homomorphism $G_k \to G_{K,S}$ is the closed normal subgroup generated by all inertia subgroups of G_K attached to places v in S. The Galois group G_K is (countably) infinitely generated as a topological group.

What is the "structure" of $G_{K,S}$ — whatever that means? It is not even known whether or not $G_{K,S}$ is finitely generated as a topological group (although this has been conjectured to be the case by Shafarevich about thirty years ago). Here is a property, weaker than the property of being "topological finitely generated," which is known to hold for the groups $G_{K,S}$ and which will serve us well in our theory below.

Definition. Let p be a prime number, and Π a profinite group. Let us say that Π satisfies the ***p*-finiteness condition** if for all open subgroups $\Pi_0 \subset \Pi$ of finite index, there are only a finite number of continuous homomorphisms from Π_0 to $\mathbf{Z}/p\mathbf{Z}$.

For a discussion of this property and its various equivalent formulations, see [M 1].

The groups $\Pi = G_{K,S}$ satisfy the p-finiteness condition for all prime numbers p. The reason for this is that any open subgroup $\Pi_0 \subset \Pi = G_{K,S}$

of finite index is again of the form G_{K_0,S_0} for some finite field extension K_0/K and the set of continuous homomorphisms,

$$\mathrm{Hom}_{\mathrm{cont}}(G_{K_0,S_0}, \mathbf{Z}/p\mathbf{Z}) = \mathrm{Hom}_{\mathrm{cont}}(G^{\mathrm{ab}}_{K_0,S_0}, \mathbf{Z}/p\mathbf{Z})$$

is finite, as can be proved as an exercise using (you choose!) either some Kummer Theory or a bit of Class Field Theory.

Here, as below, the superscript "ab" means the maximal (profinite) topological quotient group which is abelian; i.e., the quotient by the closure of the commutator subgroup.

Nowadays it is generally understood that the salient "structure" needed to be studied in connection with arithmetic problems is not merely the topological group $G_{K,S}$. Consider this rather more elaborate structure. For each place v of K, an imbedding of $\overline{K}$ in an algebraic closure, $\overline{K}_v$, of the completion of K at v gives us a continuous homomorphism

$$i_v : G_{K_v} \to G_{K,S};$$

a change of imbedding $\overline{K} \subset \overline{K}_v$ changes the homomorphism i_v by conjugation. If v is nonarchimedean and not in S, then the homomorphism i_v factors through the quotient of G_{K_v} by the inertia subgroup I_{K_v} giving us a homomorphism

$$i_v : G_{K_v}/I_{K_v} \to G_{K,S}.$$

Since G_{K_v}/I_{K_v} is canonically isomorphic to G_{k_v}, where k_v is the residue field at v and $\overline{k}_v$ is the residue field of the valuation ring of $\overline{K}_v$, and since G_v has a canonical topological generator φ_v (called the "Frobenius" element: φ_v is the automorphism of $\overline{k}_v$ which sends any element of $\overline{k}_v$ to its $|k_v|$-th power), the mapping i_v is determined by simply giving the image of φ_v under i_v. There is usually no confusion caused by the practice of referring to the image of φ_v under i_v as "the Frobenius element," Frob_v, in $G_{K,S}$ attached to v, with the understanding that such a "Frobenius" element is only unique up to conjugation. If v is real, then G_{K_v} is cyclic of order two, and the "Frobenius element" at such a v will simply mean the image of the nontrivial element of G_{K_v}. We want to study the isomorphism class of the *entire* "package"

— $G_{K,S}$,
— the conjugacy classes of the homomorphisms $i_v : G_{K_v} \to G_{K,S}$ for all places v of K.

Equivalently, we want to understand the package

— $G_{K,S}$,
— the conjugacy classes of the Frobenius elements $\varphi_v \in G_{K,S}$ for all places v of K which are not in S,
— the homomorphisms $i_v : G_{K_v} \to G_{K,S}$ for the finite set of places $v \in S$.

In contrast to our lack of knowledge concerning the topological finite generation of $G_{K,S}$ we know that the local Galois groups G_{K_v} are topologically finitely generated, and we have a fairly developed understanding of some systems of generators and relations for them, thanks to the efforts of Neukirch, Koch, and others.

We also have a reasonably satisfactory understanding of the abelianization $(G_{K,S})^{ab}$, of $G_{K,S}$, as well as of the abelianization of the entire "package" above; this is the principal achievement of Class Field Theory. The special case of this when $K = \mathbf{Q}$ was known earlier (by the turn of the century). Explicitly, if S is a finite set of prime numbers, let μ_S stand for the set of all N-th roots of unity in $\overline{\mathbf{Q}}$ where N ranges through all numbers whose set of prime divisors is contained in S. Then the maximal abelian extension of $\mathbf{Q}$ unramified outside a finite set of primes S is the subfield of $\overline{\mathbf{Q}}$ generated by μ_S (a theorem of Kronecker and Weber). Moreover, we have canonical isomorphisms

$$G^{\mathrm{ab}}_{\mathbf{Q},S} \cong \mathrm{Gal}(\mathbf{Q}(\mu_S)/\mathbf{Q}) \cong \mathrm{Aut}(\mu_S) \cong \prod_{p \in S} \mathbf{Z}_p^*,$$

the second isomorphism above being essentially the content of the result of Gauss which established the "irreducibility of the cyclotomic polynomials." The Frobenius element at any prime number ℓ not in S corresponds, under the above isomorphisms, to that element in $\prod \mathbf{Z}_p^*$ whose p-th coordinate is given by the integer ℓ in $\mathbf{Z}_p^*$, for each $p \in S$.

But can we extend our study of $G_{K,S}$ beyond describing its abelianization? One unavoidable point to contend with, if you want to go further, is that the group G_K is nonabelian — is defined only in reference to a choice of algebraic closure of K — and therefore is difficult to be pinned down more intrinsically than "up to conjugation." A standard tactic (which might be called the "Tannakian approach") suitable for such situations is to try to study *representations* of G_K (up to isomorphism) because the study of representations is insensitive to the fact that we know G_K only up to inner automorphism. From this perspective, one achievement of Class Field Theory has been to provide an adequate theory of *one*-dimensional representations of $G_{K,S}$: i.e., representations into $\mathrm{GL}_1(\mathbf{C})$, the multiplicative group of $\mathbf{C}$ (or, more flexibly but with no more generality, into the multiplicative group of any commutative ring). To go further in our study, we are led, then, to think about **Galois representations**, i.e., continuous homomorphisms,

$$\rho : G_{K,S} \to \mathrm{GL}_N(A) \tag{1}$$

for A some topological ring, and any $N = 1, 2, \ldots$. To understand the "package" above we must understand such representations as *well* as their restrictions to the groups G_{K_v} for all places v of K, i.e., their "local behavior." In particular, if v is a place of K *not in* S, the restriction of ρ to the group G_{K_v} is given by simply giving the conjugacy class of the image, $\rho(\varphi_v)$ of a Frobenius element φ_v under ρ. The trace, $a_v := \mathrm{Trace}_A(\rho(\varphi_v))$, is independent of the choice of Frobenius as it is independent of the representation ρ up to conjugation. It is therefore a well-defined invariant of the equivalence class of the representation ρ, for each choice of v not in S. As we shall see, in many instances, this data

$$v \mapsto a_v \in A \quad \text{for } v \text{ not in } S$$

will be enough to reconstruct ρ up to equivalence. Thanks to the Theorem of Chebotarev, even less data is often sufficient: e.g., one need only give the above data for v ranging through a set of places of density 1 (outside S).

§2. What coefficient-rings should we allow for our Galois representations? Since $G_{K,S}$ is a profinite topological group and since we are requiring the homomorphisms (1) to be continuous, the tightest fit, so to speak, would be if the receiving topological group $\mathrm{GL}_N(A)$ were a profinite topological group as well. I hope this is enough to motivate the following choice:

From now on in this article, a **coefficient-ring** will mean a complete noetherian local ring A with finite residue field k. The choice of k is usually fixed in our discussions. We will consecrate the letter p for the characteristic of k. Such a coefficient-ring A carries its natural profinite topology, a base of open ideals being given by the powers of its maximal ideal m_A:

$$A = \underset{\nu \to \infty}{\text{proj. lim.}}\, A/m_A^\nu.$$

By a **coefficient-ring homomorphism** let us mean a continuous homomorphism of coefficient rings

$$A' \to A$$

such that the inverse image of the maximal ideal m_A is the maximal ideal $m_{A'} \subset A'$ and the induced homomorphism on residue fields is an isomorphism.

If A is a coefficient-ring and p the characteristic of its residue field k, p is topologically nilpotent in A, and so there is a natural ring-homomorphism $\mathbf{Z}_p \to A$. This ring-homomorphism would be a "coefficient-ring homomorphism" if the residue field k were the prime field $\mathbf{F}_p$. In general, let $W(k)$ be the "ring of Witt vectors of k," that is, $W(k)$ is the canonical discrete valuation ring extension of $\mathbf{Z}_p$ which is absolutely unramified and which has residue field equal to k. Any coefficient-ring A with residue field k is naturally endowed with a continuous ("coefficient-ring") homomorphism $W(k) \to A$, which induces the identity on residue fields. (For the construction and basic properties of the ring of Witt vectors, see [Se 1] or [Mat].) Our coefficient-rings are then naturally topological $W(k)$-algebras.

The group $\mathrm{GL}_N(A)$ carries the corresponding profinite topology,

$$\mathrm{GL}(A) = \operatorname*{proj.lim.}_{\nu \to \infty} \mathrm{GL}(A/m_A^\nu),$$

a base of open normal subgroups being the multiplicative group of $N \times N$ matrices with coefficients in A which, when reduced modulo a fixed power of m_A become the identity $N \times N$ matrix.

A continuous homomorphism (1) will be referred to as a **Galois representation** with coefficient-ring A. The integer N is called the **degree** of the representation.

§3. Galois representations arise naturally. Given an elliptic curve E defined over a number field K, and an integer n, by the **group of n-division points of** E, denoted $E[n]$, we mean the group of points of E rational over $\overline{K}$, which lie in the kernel of the homomorphism

$$\begin{array}{ccc} E & \to & E \\ x & \mapsto & n \cdot x \end{array}$$

given by multiplication by n. The group G_K acts naturally as a group of automorphisms of the group $E(\overline{K})$ of $\overline{K}$-rational points of the elliptic curve E, and induces an action of G_K on $E[n]$. Since $E[n]$ is abstractly a product of two cyclic groups of order n, this natural action gives a continuous homomorphism,

$$G_K \to \mathrm{Aut}(E[n]) \cong \mathrm{GL}_2(\mathbf{Z}/n\mathbf{Z})$$

which factors through $G_{K,S}$ where S comprises all prime divisors of n, and primes of bad reduction for E. The induced homomorphism

$$\rho_{E,n} : G_{K,S} \to \mathrm{GL}_2(\mathbf{Z}/n\mathbf{Z})$$

we might call the **n-division point representation attached** to E. Passing to the projective limit of these n-division point representations as n ranges over the multiplicative system of natural numbers, or as n ranges over the direct system of all powers of a fixed prime number p, give representations

$$\rho_E : G_{K,S} \to \mathrm{GL}_2(\hat{\mathbf{Z}}), \quad \text{and}$$
$$\rho_{E,p^\infty} : G_{K,S} \to \mathrm{GL}_2(\mathbf{Z}_p),$$

respectively, where $\hat{\mathbf{Z}}$ is the profinite completion of $\mathbf{Z}$, and $\mathbf{Z}_p$ is the ring of p-adic integers.

Example. The only $n > 1$ for which the n-division point representation attached to an elliptic curve E is "dead easy" to describe directly in terms of the defining Weierstrass form of the equation, $y^2 = g(x)$, for E, is $n = 2$. Here $g(x)$ is a cubic polynomial with distinct roots. The 2-division point representation.

$$\rho_{E,2} : G_{K,S} \to \mathrm{GL}_2(\mathbf{Z}/2\mathbf{Z})$$

factors through the Galois group of the splitting field over K of the polynomial $g(x)$, and $\rho_{E,2}$ factors through the natural representation of the Galois group of that splitting field to the symmetric group S_3 using the isomorphism,

$$S_3 \cong \mathrm{GL}_2(\mathbf{Z}/2\mathbf{Z})$$

which is unique up to conjugation.

More generally: Going back to consideration of general n-division point representations associated to elliptic curves, a construction similar to the one involving elliptic curves, but beginning with an abelian variety of dimension g over a number field K provide Galois representations of degree $2g$ with coefficient rings $\mathbf{Z}/n\mathbf{Z}$, $\hat{\mathbf{Z}}$, and $\mathbf{Z}_p$ as well. If we start with an abelian variety whose ring of endomorphisms rational over K contains a commutative ring $\mathcal{A}$ larger than the ring $\mathbf{Z}$, we may get Galois representations with other coefficient-rings, as well (specifically, quotients and completions of $\mathcal{A}$).

We can construct Galois representations with coefficient-rings $\mathbf{Z}/n\mathbf{Z}$, $\hat{\mathbf{Z}}$, and $\mathbf{Z}_p$ by considering the natural action of G_K on the étale cohomology groups of algebraic varieties defined over K. Related to this, there is the classical theory due to Shimura, Deligne, and Deligne-Serre, which attach to arbitrary classical modular eigenforms (of integral weights ≥ 1) Galois representations of degree 2 with coefficient-rings equal to various completions and quotients of the ring generated by the action of Hecke operators on the space of modular forms of given level and weight.

CHAPTER II. GROUP REPRESENTATIONS

§4. Group representations versus algebra representations.

Given a positive integer N, a coefficient-ring A with residue field k, and a profinite group Π, the set of continuous group-homomorphisms

$$\rho : \Pi \to \mathrm{GL}_N(A)$$

is in one-one correspondence with the set of continuous homomorphisms of A-algebras

$$r : A[[\Pi]] \to M_N(A),$$

where $A[[\Pi]]$ is the completed group-ring of Π with coefficients in A,

$$A[[\Pi]] = \operatorname*{proj.lim.}_{\Pi_0 \subset \Pi} A[\Pi/\Pi_0],$$

where Π_0 runs through all open normal subgroups of finite index in Π, and $A[\Pi/\Pi_0]$ is the usual group-ring of the finite group Π/Π_0 with coefficients in A. Here $M_N(A)$ is the A-algebra of $N \times N$ matrices with entries from A. The correspondence $r \mapsto \rho$ comes by restriction, noting that Π may be identified with a subgroup of the group of multiplicative units $A[[\Pi]]^*$ in the ring $A[[\Pi]]$ and the algebra-homomorphism r restricts to a continuous homomorphism of groups of units,

$$r^* : A[[\Pi]]^* \to M_N(A)^* = GL_N(A).$$

By the "**underlying residual representation**" to ρ, and to r,

$$\overline{\rho} : \Pi \to GL_N(k) \qquad \text{and} \qquad \overline{r} : A[[\Pi]] \to M_N(k),$$

we mean the composition of ρ and of r with the natural projections

$$GL_N(A) \to GL_N(k) \qquad \text{and} \qquad M_N(k) \to M_N(k)$$

respectively.

Proposition. *The residual representation*

$$\overline{\rho} : \Pi \to \mathrm{GL}_N(k)$$

associated to ρ is absolutely irreducible if and only if the homomorphism r is surjective.

Proof. This is well known if A is a field, i.e., if $A = k$: cf. [Bourb 1, Ch. VIII §13 n° 4]. It follows for general coefficient-rings A from Nakayama's Lemma applied to the following diagram of A-modules:

$$\begin{array}{ccc} \mathrm{Image}(r) & \subset & M_N(A) \\ & \searrow & \downarrow \\ & & M_N(k). \end{array}$$

Corollary. (Schur's Lemma) *Let $\rho : \Pi \to \mathrm{GL}_N(A)$ be a continuous representation with coefficient-ring A. If the associated residual representation $\overline{\rho}$ is absolutely irreducible, any matrix in $M_N(A)$ which commutes with all the elements in the image of ρ is a scalar.*

Proof. Since the completion of the A-algebra generated by the image of ρ is equal to the image of r (i.e., is all of $M_N(A)$ by the above proposition) any matrix commuting with all the elements in the image of ρ lies in the center of $M_N(A)$. The fact that such elements are scalar matrices is valid for A any commutative ring with unit; it can be seen by directly checking what it means for a matrix to commute with the basic $N \times N$ matrices E_{ij} (which have a 1 as their entry in the i-th row and j-th column and 0 elsewhere).

§5. Representations and their characters. Keeping the notational conventions of the previous paragraph, let $\rho : \Pi \to \mathrm{GL}_N(A)$ be a representation where A is a coefficient-ring with residue field k of characteristic p. We assume that the underlying residual representation $\overline{\rho} : \Pi \to \mathrm{GL}_N(k)$ is absolutely irreducible (or equivalently, by the proposition in §4, that $r : A[[\Pi]] \to M_N(A)$ is surjective).

Proposition. *Let $\rho' : \Pi \to \mathrm{GL}_N(A)$ be a representation with the same character as ρ, i.e., such that $\mathrm{Trace}_A\, \rho(g) = \mathrm{Trace}_A\, \rho'(g)$ for all $g \in \Pi$. Then ρ' and ρ are equivalent representations.*

See [Ca] and [Se 2]. The following proof is taken from [Se 2].

Proof. Let $r, r' : A[[\Pi]] \to M_N(A)$ be the A-algebra homomorphisms corresponding to ρ and ρ'. By hypothesis, the residual representation $\overline{\rho}$ is absolutely irreducible. We shall first prove that $\overline{\rho}'$, the residual representation associated to ρ', is equivalent to $\overline{\rho}$ and hence is also absolutely irreducible. Let $\overline{\rho}'_{ss}$ denote the semi-simplification of $\overline{\rho}'$. Then $\overline{\rho}$ and $\overline{\rho}'_{ss}$ are semi-simple representations with the same character. It follows (cf. the proof of Th. 30.16 in [C-R])) that the multiplicity of any absolutely irreducible representation ψ occurring in $\overline{\rho}'_{ss}$ is congruent modulo p to the multiplicity of ψ in $\overline{\rho}$. In particular, since $\overline{\rho}$ is absolutely irreducible, the multiplicity of $\overline{\rho}$ in $\overline{\rho}'_{ss}$ is $1 + p \cdot \mu$ for some integer $\mu \geq 0$. But $\overline{\rho}$ and $\overline{\rho}'_{ss}$ are both of the same degree, and therefore $\mu = 0$ and $\overline{\rho}$ is equivalent to $\overline{\rho}'_{ss}$. So $\overline{\rho}$ and $\overline{\rho}'$ are both absolutely irreducible. By the proposition of §4, r and $\overline{r}$ are both surjective. We will be using this latter fact, along with the hypothesis that the character functions of r and of r' are equal; i.e. $\mathrm{Trace}_A(r(\alpha)) = \mathrm{Trace}_A(r'(\alpha))$ for all $\alpha \in A[[\Pi]]$.

Define the A-module homomorphism

$$\varphi : A[[\Pi]] \to M_N(A) \times M_N(A)$$

by the rule $\varphi(x) = (r(x), r'(x))$ and let

$$\Phi \subset M_N(A) \times M_N(A)$$

denote the image of φ. Since r and r' are surjective, the VERY general principle that Serre calls "Goursat's Lemma" applies (cf. [Se 2])[1] which gives a precise description of the image of Φ. The A-submodule Φ is given as follows:

(*) There are two-sided ideals $\mathcal{I} \subset M_N(A)$ and $\mathcal{I}' \subset M_N(A)$ and an isomorphism of A-algebras $f : M_N(A)/\mathcal{I} \to M_N(A)/\mathcal{I}'$ such that Φ is the "graph" of f in the sense that

$$\Phi = \{(\alpha, \alpha') \in M_N(A) \times M_N(A) \mid f(\alpha \cdot \mathcal{I}) = \alpha' \cdot \mathcal{I}'\}.$$

But the only two-sided ideals $\mathcal{J}$ in $M_N(A)$ are of the form $\mathcal{J} = J \cdot M_N(A)$ where $J \subset A$ is an ideal in A. (The proof of this is an exercise: the ideal $J \subset A$ may be taken to be the ideal in A generated by all entries of all the matrices in $\mathcal{J}$.) Therefore the ideals $\mathcal{J}, \mathcal{J}'$ occurring in (*) are of the form $I \cdot M_N(A)$ and $I' \cdot M_N(A)$ for ideals $I, I' \subset A$. Since the annihilators of the (isomorphic) A-modules $M_N(A)/\mathcal{I}$ and $M_N(A)/\mathcal{I}'$ are I and I',

[1] Goursat's Lemma works with $M_N(A)$ replaced by e.g., any A-algebra.

respectively, we have $I = I'$. Now take any element $x \in I$ and consider the $N \times N$ matrix X which has x as entry in its first column and first row, and zeroes elsewhere. Let X' denote the $N \times N$ matrix all of whose entries are zero. The couple of $N \times N$ matrices $(X, X') \in M_N(A) \times M_N(A)$ are in Φ (by the description $(*)$).

Therefore

$$x = \mathrm{Trace}(X) = \mathrm{Trace}(X') = 0$$

i.e., the ideal I vanishes, and Φ is in fact the graph of an actual isomorphism of A-algebras $f : M_N(A) \to M_N(A)$. By [Bourb 2, Ch II §5, ex. 2] any such isomorphism is inner; by construction, we have $f \circ r = r'$. It follows that ρ is equivalent to ρ'.

Corollary. *Let $\rho, \rho' : G_{K,S} \to \mathrm{GL}_N(A)$ be continuous representations. Suppose that one of these representations is residually absolutely irreducible. Suppose further that*

$$\mathrm{Trace}_A\, \rho(\mathrm{Frob}_\ell) = \mathrm{Trace}_A\, \rho'(\mathrm{Frob}_\ell)$$

for ℓ running through a set of prime numbers (outside S) which is of Dirichlet density 1. Then ρ is equivalent to ρ'.

Proof. This comes from combining the above proposition with the Chebotarev Density Theorem.

§6. "Descent" of group representations. Schur-type Theorems.

Suppose we are given a continuous representation

$$\rho : \Pi \to \mathrm{GL}_N(A)$$

such that the associated character function

$$\chi_\rho(x) = \mathrm{Trace}_A(\rho(x)) \in A$$

has the property that it takes its values in a sub-ring $A_0 \subset A$. Is there a "descent" of ρ to A_0, i.e., is there a continuous representation

$$\rho_0 : \Pi \to \mathrm{GL}_N(A_0)$$

such that when one extends scalars from A_0 to A, ρ_0 becomes equivalent to ρ? The answer is YES for our coefficient-rings A, but descent is not necessarily valid for more general rings. For results along these lines, see [M 1] and [G]. The most general (and the most perspicuous!) such result, to date, is to be found in [Ca] and [Se 2]. Here is a brief account of it.

Let R be a general commutative local ring, with maximal ideal m_R and residue field $k = R/m_R$. Let

$$r : R[[\Pi]] \to M_N(R)$$

be the R-algebra homomorphism associated to a continuous representation $\rho : \Pi \to \mathrm{GL}_N(R)$. We assume that r is surjective. Let R_0 be a local sub-ring of R with maximal ideal $m_{R_0} = m_R \cap R_0$ and residue field $k_0 \subset k$. Suppose that the traces, $\mathrm{Trace}_R(r(\alpha))$, of all elements $\alpha \in R[[\Pi]]$ lie in R_0. The "descent question" posed above would have a positive answer if, for example, the image $r(R_0[[\Pi]]) \subset M_N(R)$ were isomorphic, as R_0-algebra, to the matrix algebra $M_N(R_0)$. Now this is not the case in general but it is "almost" the case. The sense in which it is "almost" the case is best explained in terms of Azumaya algebras (cf.[K-O]).

Definition. An **Azumaya Algebra** Σ over R is a finite flat R-algebra such that $\Sigma/m_R\Sigma$ is a central simple algebra over k.

By the **rank** of an Azumaya Algebra over R, one means its rank as R-module. For example, the matrix algebra $M_N(R)$ is an Azumaya Algebra over R of rank N^2.

The fundamental general result is the following

Proposition. (Carayol, Serre) *Let $r : R[[\Pi]] \to M_N(R)$ be a surjective R-algebra homomorphism and let R_0 be a local sub-ring of R as above. Suppose that the traces,* $\mathrm{Trace}_R(r(\alpha))$*, of all elements $\alpha \in R[[\Pi]]$ lie in R_0. Then the image*

$$\mathcal{R}_0 := r(R_0[[\Pi]]) \subset M_N(R)$$

is an Azumaya Algebra over R_0 of rank N^2 such that the natural A-algebra homomorphism $\mathcal{R}_0 \otimes_{R_0} R \to M_N(R)$ is an isomorphism.

Brief proof (and remarks). Carayol and Serre phrase their proposition in slightly greater generality (or at least greater flexibility) in that the domain may be taken to be an algebra (and not necessarily a completed group algebra) and the range may be taken to be a general Azumaya Algebra $\mathcal{R}$ over R (and not necessarily $M_N(R)$). Let, then, $\mathcal{R}$ be a general Azumaya Algebra over R of rank N^2 and $\mathcal{R}_0$ a subring of $\mathcal{R}$ which is an R_0-algebra such that the traces of elements of $\mathcal{R}_0$ lie in $A \cdot R_0$ and such that $R \cdot \mathcal{R}_0 = \mathcal{R}$. From the last condition we see that we can find a set of N^2 elements e_j ($j = 1, \ldots, N^2$) in $\mathcal{R}_0$ which form an R-basis for $\mathcal{R}$. For any element $\alpha \in \mathcal{R}_0$ write

$$\alpha = \Sigma_j \lambda_j \cdot e_j$$

with λ_j in R. We will show that the λ_j's lie in R_0. From the displayed formula, we have

$$(*) \quad \mathrm{Trace}(\alpha \cdot e_k) = \sum_j \lambda_j \cdot \mathrm{Trace}(e_j \cdot e_k) \qquad (\text{for} \quad k = 1, \ldots, N^2)$$

Now since the matrix $(\mathrm{Trace}(e_j \cdot e_k))$ has a determinant which is not in m_R (i.e., is nonzero after reduction to $k = R/m_R$) and since the system of linear equations $(*)$ in the "variables" λ_j has coefficients in R_0, the unique solution $(\lambda_1, \dots, \lambda_{N^2})$ lies in R_0. It follows that the elements e_j $(j = 1, \dots, N^2)$ form a free R_0-basis for $\mathcal{R}_0$.

Consequently,

$$\mathcal{R}_0 \otimes_{R_0} R \to M_N(R)$$

is an R-algebra isomorphism, and $\mathcal{R}_0$ is an R_0-Azumaya Algebra of rank N^2 (because

$$(\mathcal{R}_0/m_{R_0} \cdot \mathcal{R}_0) \otimes_{k_0} k \cong \mathcal{R}/m_R \cdot \mathcal{R}$$

is a central simple algebra over k, and therefore $\mathcal{R}_0/m_{R_0} \cdot \mathcal{R}_0$ is a central simple algebra over k_0).

To apply the above proposition we must know something about Azumaya Algebras over R. A theorem of Azumaya [Az], [K-O] gives us that the Brauer group of a Henselian local ring is isomorphic to that of its residue field. This applies to our situation, for all our coefficient-rings A are Henselian and their residue fields are finite (and therefore they have trivial Brauer group). Thus, our coefficient rings admit no nontrivial Azumaya algebras. We get:

Corollary. *Let $\rho : \Pi \to \mathrm{GL}_N(A)$ be absolutely irreducible, and let $A_0 \subset A$ be a local subring of the coefficient-ring A such that the traces $\mathrm{Trace}_A(\rho(x))$ for all elements $x \in \Pi$ lie in $A_0 \subset A$. Then there is a representation $\rho_0 : \Pi \to \mathrm{GL}_N(A_0)$ which, after extension of scalars from A_0 to A becomes equivalent to ρ.*

But note that the ring A_0 given in the Corollary may have a smaller residue field than that of A; i.e., the injection $A_0 \to A$ may not be a coefficient-ring homomorphism.

Remark. I had given a proof of the above result (cf.[M 1] 1.8 Prop. 4 and Corollaries 1, 2) under a further hypothesis (that the one-dimensional cohomology of the image of $\overline{\rho}$ in $\mathrm{GL}_N(k)$ with coefficients in the adjoint representation $\mathrm{Ad}(\overline{\rho})^\circ$ vanishes). That proof has the disadvantage that it is under this extra hypothesis and that it *uses* the construction of the universal deformation ring of $\overline{\rho}$. In contrast, the above result of Carayol and Serre can itself be used to aid in the construction of the universal deformation ring as in Lenstra and de Smit's construction; or in that of Rouquier, or Nyssen (see §7 below). Compare this also with the construction of universal varieties of representations of algebras given by Procesi in the early 70's ([P 1], [P 2]).

An idle question. From the vantage point of this section, an absolutely irreducible $G_{K,S}$ representation with coefficient-ring A is given by an Azumaya Algebra (equivalently: total matrix algebra of finite rank) over A occurring as a quotient A-algebra of $A[[G_{K,S}]]$. Are there interesting classes of A-algebras (of *infinite* rank over A — analogues of "factor" occurring in the classical theory of Muuray and von Neumann —) which occur as quotients of $A[[G_{K,S}]]$ and which deserve study?

§7. Characterizing character-functions (results of Rouquier, Nyssen). In this section, let $\mathcal{K}$ be any commutative ring, and Π a profinite group. By a **central function** $f : \mathcal{K}[[\Pi]] \to \mathcal{K}$ we mean a $\mathcal{K}$-linear function such that $f(x \cdot y) = f(y \cdot x)$ for all $x, y \in \mathcal{K}[[\Pi]]$. Given a central function f, and a positive integer m, define the function $f_m(x_1, x_2, \ldots, x_m)$ to be the signed-symmetrization of f evaluated on the products of the x_j in all permuted ways. Explicitly,

$$f_m(x_1, x_2, \ldots, x_m) = \sum_{\sigma \in S_m} \operatorname{sign}(\sigma) \cdot f(x_{\sigma(1)}, x_{\sigma(2)}, \ldots, x_{\sigma(m)})$$

where S_m is the symmetric group on m letters. Clearly, f_m is an anti-symmetric $\mathcal{K}$-linear function on $\mathcal{K}[[\Pi]]^m$ with values in $\mathcal{K}$. A central function f is called a **pseudo-character of degree** $N \geq 1$ (see [Rouq]) if, equivalently,

(1) f_N does not vanish identically, but f_{N+1} does vanish identically.
(2) f_m does not vanish identically for all $m \leq N$ and does vanish identically for all $m > N$.

The characters of irreducible representations of finite groups Π yield "pseudo-characters" in the above sense, as was proved by Frobenius [Fr]. The definition of pseudo-character given by Rouquier is a mild modification of the notion of *pseudo-representation* due to Taylor [T], which generalized a prior notion due to Wiles. One says that a pseudo-character f is **irreducible** if f cannot be expressed as the sum of two pseudo-characters whose degrees add up to the degree of f. For a full discussion of this theory, see loc. cit.; see also the preprint of Louise Nyssen [Ny]. See Th. 4.2 of [Rouq] for a proof of the fact that if $\mathcal{K}$ is an algebraically closed field, irreducible pseudo-characters of degree N are precisely the characters of irreducible representations of Π with values in $\mathcal{K}$. Closely related to this result is a characterization of the characters of representations of Π into $\mathrm{GL}_N(\mathcal{K})$ for $\mathcal{K}$ any commutative ring, and in particular, any coefficient-ring (cf. §5, §6 of [Rouq]), leading to a construction of the universal deformation ring by "constructing the universal pseudo-character."

§8. Deformations of a group representation. Let Π be a profinite group. Suppose we are given a coefficient-ring homomorphism

$$h : A_1 \to A_0$$

of two coefficient-rings. Let N be a positive integer and denote by the same letter

$$h : \mathrm{GL}_N(A_1) \to \mathrm{GL}_N(A_0)$$

the induced homomorphism of groups of invertible $N \times N$ matrices. If

$$\rho_0 : \Pi \to \mathrm{GL}_N(A_0)$$

is a continuous homomorphism, a **deformation of ρ_0 to the coefficient-ring A_1** is a *strict equivalence class* of liftings

$$\begin{array}{ccc} \Pi & \xrightarrow{\rho_1} & \mathrm{GL}_N(A_1) \\ & {\scriptstyle\rho_0}\searrow & \downarrow h \\ & & \mathrm{GL}_N(A_0), \end{array}$$

where two liftings ρ_1 and ρ_1' are called **strictly equivalent** if they can be brought one into another by conjugation by elements of $GL_N(A_1)$ in the kernel of h.

Any representation ρ is, of course, a deformation of its underlying residual representation $\overline{\rho}$ to A.

Convention. It is not uncommon in the literature to use the phrase "representation ρ" to mean, at times, a specific homomorphism ρ and at other times an equivalence class of homomorphisms of which ρ is a member. It is probably best not to be too pedantic about this point, if, in every instance where this occurs, the context makes it clear which sense is meant, or else makes it clear that it doesn't matter which sense is meant. We will try to make things clear in what follows, but mention here that whenever we use the phrase "**residual representation**" $\overline{\rho}$", we mean a specific homomorphism, and whenever we are interested in making a specific choice of a *homomorphism* ρ whose underlying residual representation is $\overline{\rho}$, we shall refer to it explicitly as a *lifting* of $\overline{\rho}$; if we want its strict equivalence class we will refer to it as a *deformation* of $\overline{\rho}$.

For a coefficient-ring A, consider the category $\hat{\mathcal{C}}(A)$ whose objects are coefficient-rings A_1 together with a coefficient-ring homomorphism $A_1 \to A$ (which will be sometimes referred to as an **A-augmentation**) and where morphisms are commutative diagrams of coefficient-ring homomorphisms,

$$\begin{array}{ccc} A_1 & \to & A_2 \\ \downarrow & & \downarrow \\ A & = & A. \end{array}$$

(The reason for the ^ in the notation is that we will later be also considering the full sub-category $\mathcal{C}(A)$ whose objects are artinian coefficient-rings A_1 with homomorphism to A). Note that by our hypothesis that the A-augmentation is a coefficient-ring homomorphism, for all objects $A_1 \to A$ in $\hat{\mathcal{C}}(A)$ the residue field of A_1 is equal to k.

Given a coefficient-ring A, a profinite group Π, and a continuous homomorphism

$$\rho : \Pi \to \mathrm{GL}_N(A),$$

define the functor $D_\rho : \hat{\mathcal{C}}(A) \to$ *Sets* by the rule which assigns to any object $A_1 \to A$ of $\hat{\mathcal{C}}(A)$ the set of strict equivalence classes of deformations of ρ to A_1. The phrase "**the deformation problem for** ρ" will refer to the study of this functor. Much of the time we will be interested in the case when $A = k$, the residue field, and $\rho = \overline{\rho}$ is a residual representation (the "absolute" case) but from time to time we will be dealing with the "relative" case, i.e., with a specific lifting of a residual representation $\overline{\rho}$ to a *homomorphism* $\rho : \Pi \to \mathrm{GL}_N(A)$ (and not just a strict equivalence class of liftings of $\overline{\rho}$ to A) and we will be interested in deformations of ρ to coefficient-rings A_1 endowed with a homomorphism to A.

CHAPTER III. THE DEFORMATION THEORY OF GALOIS REPRESENTATIONS

§9. Why study "Galois" deformation theory? We will be principally interested in the case where $\Pi = G_{K,S}$ for some algebraic number field K and finite set of primes S in K. Here are three possible reasons for studying the deformation theory of representations of $G_{K,S}$.

1) First consider **residual representations**, i.e., Galois representations $\overline{\rho} : G_{K,S} \to \mathrm{GL}_N(k)$ where k is a finite field. It takes only a finite amount of data to give a residual representation, and moreover, there are only a finite number of such residual representations (for fixed K, S, N, and k). Attached to a residual representation $\overline{\rho}$ one can consider the whole panoply of Galois representations which are deformations of $\overline{\rho}$. If $\overline{\rho}$ is absolutely irreducible, any member of this panoply comes from a single neat package, namely from a "**universal deformation**" (see below) and in particular, from a single representation into GL_N with coefficients in a single complete noetherian local ring with residue field k. This coefficient ring $R(\overline{\rho})$, uniquely defined up to unique isomorphism by the universal property, is called the **universal deformation ring**, an explicit description of which (and of the universal deformation of $\overline{\rho}$ to it) is tantamount to a systematic "classification" of all Galois representations which are liftings of $\overline{\rho}$. The spectrum, $\mathrm{Spec}\, R(\overline{\rho})$, will be called the **universal deformation space** of $\overline{\rho}$. For some "explicit" easy examples, see [B 1] and [B-M]; for other expository accounts of the deformation theory of Galois representations giving a number of examples, see [B 2], [M 1-3].

2) Given the universal deformation ring $R(\overline{\rho})$ of a residual representations, one can then ask which quotient rings correspond to Galois representations with *particularly desirable properties.* Equivalently, we are asking for the closed subschemes of $\mathrm{Spec}(R(\overline{\rho}))$, "the universal deformation space of $\overline{\rho}$," corresponding to those properties. For example: Which points of the universal deformation space are "modular"[2]? Which come as irreducible representations on the étale cohomology of algebraic varieties?

The recipe for cutting down the "universal deformation" to these more specifically desirable Galois representations is (surprisingly enough!) at

[2]That is, which such points classify representations that are "attached to modular forms?"

least conjecturally nothing more than the "imposition" of local conditions at the ramified primes, and sometimes with the additional prescription of the appropriate global determinant. For example,

i) There is a conjecture I made with Fontaine [F-M] which says that, up to $\overline{\mathbf{Q}}_p$-equivalence, the irreducible Galois representations (with coefficient-ring $A = \overline{\mathbf{Z}}_p$) which come as irreducible constituents of the natural Galois representations on the p-adic étale cohomology of algebraic varieties (allowing integral twists) are *precisely* those whose restriction to the decomposition groups at primes dividing p are *potentially semi-stable.*

ii) There is a somewhat older conjecture for $N = 2$, $K = \mathbf{Q}$, relating Galois representations which are "ordinary at p" to classical modular forms of slope 0 (see [M 2], [M-T] and [G]).

iii) There is the generalization of the conjectures referred to in **ii)** as formulated in [W] (still for $N = 2$ and $K = \mathbf{Q}$).

A good part of this generalized conjecture **iii)** and **ii)** has recently proved by the monumental work in [W] and [T-W], which more than amply answers the question posed by the title of this section.

3) Galois representations are often systematically presented to us "in certain families," these families being continuous, and they are usually even analytic in a p-adic sense. Hida, for example, has an extensive theory which shows that all Galois representations attached to classical modular eigenforms of slope 0 come to us in such families (cf. Hida's book [H] and the bibliography there for the extensive literature about this). Based on Hida's work, and on some numerical investigation, Fernando Gouvêa and I had conjectured that all modular (finite slope) Galois representations come in specific families of this type [G-M]. This conjecture (or at least a qualitative form of it) has very recently been established by Coleman [C]. To "visualize" these families of modular Galois representations and specifically how these families intersect with each other and with the various loci describing various local conditions, it is good (perhaps even essential!) to be working in something like the universal space. Certain families of Galois representations are tightly controlled simply by understanding how they sit in the universal deformation space (cf. [M 3]).

One often has some understanding of the universal deformation space. We shall end this section by citing two examples:

Example 1. (An "unobstructed" case) When $\overline{\rho}$ is an absolutely irreducible representation of degree two, and of *odd* determinant (meaning that if c is a complex conjugation involution in $G_{\mathbf{Q},S}$ then the determinant of $\overline{\rho}(c)$ is -1) and when "the deformation theory for $\overline{\rho}$ is unobstructed"[3] then the universal deformation ring $R(\overline{\rho})$ is isomorphic to a power series ring in three variables over $W(k)$; cf. [M 1]. Here is a specific instance of this. Let $K/\mathbf{Q}$ be the splitting field of the cubic polynomial $X^3 - X + 1$.

[3] for a definition of the notion of "unobstructed deformation theory," cf. [M 1]

The Galois group of this equation is the symmetric group on three letters, and K is unramified over $\mathbf{Q}$ at all primes other than $p = 23$ (and ∞). Let $p = 23$. Since the group S_3 has a faithful representation in $GL_2(\mathbf{F}_p)$ we obtain from this equation an absolutely irreducible residual representation

$$\overline{\rho} : G_{\mathbf{Q},\{23,\infty\}} \to \mathrm{GL}_2(\mathbf{F}_{23}).$$

It has been shown (cf. [M 1]) that this is an "unobstructed deformation problem" and (therefore) that the universal deformation ring of $\overline{\rho}$ is isomorphic to a power series ring $\mathbf{Z}_{23}[[t_1, t_2, t_3]]$ in three variables. For a detailed study of this deformation problem and a general class of unobstructed problems, see [M 1], [B 1], [B-M],

Example 2. (An "obstructed" case) N. Boston and S.V. Ullom [B-U] have studied the interesting deformation theory of the residual representation

$$\overline{\rho} : G_{\mathbf{Q},\{3,7,\infty\}} \to \mathrm{GL}_2(\mathbf{F}_3)$$

coming from the Galois representation on the 3-division points of the elliptic curve $X_0(49)$. Here the universal deformation ring is isomorphic to

$$\mathbf{Z}_3[[t_1, t_2, t_3, t_4]]/((1 + t_4)^3 - 1)$$

whose deformation space then is geometrically reducible, and (after the adjunction of a primitive cube root of unity) splits into three irreducible components (given by specializing $1 + t_4$ to the three cube roots of 1).

§10. The universal "Galois" deformation ring. We mentioned that for absolutely irreducible representations $\overline{\rho}$, there is a universal solution to the problem of classifying deformations of $\overline{\rho}$. Explicitly,

Proposition. *If N is a positive integer and*

$$\overline{\rho} : G_{K,S} \to \mathrm{GL}_N(k)$$

is absolutely irreducible, there is a "universal coefficient-ring" $R = R(\overline{\rho})$ with residue field k, and a "universal" deformation,

$$\rho^{\mathrm{univ}} : G_{K,S} \to \mathrm{GL}_N(R),$$

of $\overline{\rho}$ to R; it is universal in the sense that given any coefficient-ring A with residue field k, and deformation

$$\rho : G_{K,S} \to \mathrm{GL}_N(A)$$

of $\overline{\rho}$ to A, there is one and only one homomorphism $h : R \to A$ inducing the identity isomorphism on residue fields for which the composition of the *universal deformation* ρ^{univ} *with the homomorphism* $GL_N(R) \to \mathrm{GL}_N(A)$ *coming from* h *is equal to the deformation* ρ. *In other terms, the functor*

$$D_{\overline{\rho}} : \begin{pmatrix} \textit{Coefficient-rings} \\ \textit{with residue field } k \end{pmatrix} \longrightarrow \textit{Sets}$$

is representable by R, *i.e.*,

$$D_{\overline{\rho}}(A) \cong \mathrm{Hom}_{W(k)\text{-alg}}(R, A),$$

where $W(k)$ *is the ring of Witt vectors of* k.

Easy but important exercise. If you have never worked with these concepts before, it is very helpful to give a direct proof of this proposition for $N = 1$ and to give an explicit description of the ring $R(\overline{\rho})$ and the universal representation

$$\rho^{\text{univ}} : G_{K,S} \to R(\overline{\rho})^*$$

in the case when $\overline{\rho}$ is of degree 1 (using Class Field Theory). But the word "explicit" in the previous sentence should be taken with a grain of salt, because (if S contains all places of characteristic p) the determination of the Krull dimension of $R(\overline{\rho})$ is equivalent to the determination of the truth or falsity of the Leopoldt Conjecture for p and the number field K. For all this spelled out, see [M 1].

For the proof of this proposition for all N, the reader may consult [M 1], [G], or [D-D-T]. Also, a very detailed discussion of all this is forthcoming in [D-W]. Prior to the work we have just cited, there had already been numerous studies of the local deformation theory, and also of the global variations of representations of finitely generated groups and algebras: see Procesi's [P 1] Chapter IV, Lemma 1.7, and his follow-up article [P 2]; see also the memoir of Lubotzky and Magid [L-M] and the other works cited in the bibliography by Doran (available by anonymous ftp) referred to in the introduction to this article.

Let us simply list some approaches to the proof of this proposition:

1. Via Schlessinger's Criteria: Schlessinger, in [Sch], gives necessary and sufficient criteria for any covariant functor

$$D : \begin{pmatrix} \text{Coefficient-rings} \\ \text{with residue field } k \end{pmatrix} \longrightarrow \text{Sets}$$

to be representable, i.e., for there to exist a coefficient ring $R = R_D$ (not necessarily artinian) and a "universal element" $\xi = \xi_E$ in $D(R)$ satisfying the "universal property" that — for any coefficient ring A and element $\alpha \in D(A)$ there is one and only one ring homomorphism $R \to A$ which is the identity on residue fields and which brings the "universal element" ξ to α.

See §18 below for a "review" of Schlessinger's Criteria. See [M 1] for a proof that Schlessinger's criteria are met within the context of the proposition above. The main "nonformal" ingredients needed to check this are, firstly, Schur's lemma (which is available to us because $\overline{\rho}$ is absolutely irreducible) and secondly (to insure noetherian-ness of R) that the set of deformations of $\overline{\rho}$ to the coefficient-ring $k[\xi]$ (where ξ is nontrivial and has square zero) is finite. This finiteness condition holds in our situation as given by the Corollary in §21.

2. A construction due to Faltings. If you wish to see a description of the universal ring in terms of generators and relations (a description which uses a "far-from-minimal" number of generators and relations, but which has the virtue of being explicitly given in terms of the data) there is a construction of R, and hence also a proof of representability of $D\overline{\rho}$, due to Faltings which does exactly that. For an account of this construction, see for example pp. 56,57 of [D-D-T]; also, the forthcoming [D-W].

3. A construction due to Lenstra and de Smit. For this, see their article [L-de-S] in this volume.

4. Via Universal Characters. Another attitude towards the statement of the proposition above is that it guarantees the existence of a "universal character function" (together, of course, with a "universal ring R" acting as value ring for this character function). Conversely, Rouquier and Nyssen approach the construction of universal deformation rings by dealing directly with pseudo-characters using the results of [Rouq], [Ny] described in §7 above. One shows that the properties of being a character function has a universal solution, thereby giving another construction of the universal deformation ring.

§11. An alternative description of the deformation problem for group representations (in a slightly more general context). Let Π be a profinite group which satisfies the p-finiteness condition of §1. Let k be a finite field of characteristic p. Let $\overline{V}$ be a finite-dimensional k-representation space for Π (and we assume that the action of Π on $\overline{V}$ is continuous). If B is a coefficient-ring with residue field k, by a **deformation** V of $\overline{V}$ to B let us mean a couple (V, α) where V is a free B-module (of finite rank) with continuous G-action and $\alpha : V \otimes_B k \cong \overline{V}$ is an isomorphism as Π-representation spaces. By $D_{\overline{V}}(B)$ let us mean the set of isomorphism classes of deformations of $\overline{V}$ to B; view $D_{\overline{V}}$ as covariant functor from the category of coefficient-rings with residue field k to the category of sets.

$$D_{\overline{V}} : \begin{pmatrix} \text{Coefficient-rings} \\ \text{with residue field } k \end{pmatrix} \longrightarrow \text{Sets}$$

By fixing a k-basis of $\overline{V}$ one may identify the automorphism group $\mathrm{Aut}_k(\overline{V})$ with $\mathrm{GL}_N(k)$ where $N = \dim_k(\overline{V})$ and the Π-action on $\overline{V}$ then gives us a specific (continuous) residual representation $\overline{\rho} : \Pi \to \mathrm{GL}_N(k)$. One then sees directly from the definitions that there is an isomorphism of functors, $D_{\overline{V}} \cong D_{\overline{\rho}}$. The relative problem can also be phrased this way: If A is a coefficient-ring with residue field k, and $\overline{V}$ is a fixed free A-module of rank N with A-linear continuous Π-action, and if $\rho : \Pi \to \mathrm{GL}_N(A)$ is the continuous homomorphism obtained from $\overline{V}$ by choosing an A-basis, then letting

$$\begin{pmatrix} A\text{-augmented coefficient-rings} \\ \text{with residue field } k \end{pmatrix} \longrightarrow \text{Sets}$$

denote the functor which associates to an A-augmented coefficient-ring B with residue field k the set of isomorphism classes of pairs (V, α) where V is a free B-module of rank N endowed with a B-linear continuous Π-action, and $\alpha : V \otimes_B A \cong \overline{V}$ is an isomorphism of $A[[T]]$-modules, we have a natural isomorphism of functors $D_\rho \cong D_{\overline{V}}$.

Now let us return to the *absolute* deformation problem. Let $\overline{V}$ be a finite-dimensional k-representation space for Π (the action of Π being assume continuous) such that the natural mapping

$$k \to \mathrm{End}_{k[\Pi]}(\overline{V})$$

is an isomorphism. This would be the case, by Schur's Lemma, if $\overline{\rho}$ were absolutely irreducible; cf. the Corollary of §4. But there are other important examples of representations $\overline{\rho}$ which satisfy the above condition without being absolutely irreducible. Specifically, let

$$\overline{\rho} : \Pi \longrightarrow \mathrm{GL}_2(k)$$

be a representation equivalent to a representation of the form

$$\overline{\rho}(g) = \begin{bmatrix} \chi(g) & u(g) \\ 0 & \eta(g) \end{bmatrix}$$

which is *not* semisimple (equivalently: such that the image of $\overline{\rho}$ is of order divisible by p), and such that one of the two characters χ or η is nontrivial. Examples of such representations may be found among the residual representations attached to elliptic curves with ordinary reduction over p-adic number fields. Then the Π-representation space $\overline{V}$ attached to $\overline{\rho}$ is not absolutely irreducible, and yet does satisfy the condition displayed above.

The representability proposition of the previous section is valid in this context, that is to say,

Proposition. *$D_{\overline{V}}$ is representable; i.e., there is a coefficient-ring R with residue field k, and a finite free R-module V_R endowed with a continuous Π-action which is a deformation of $\overline{V}$ to R which is universal in the sense that any deformation V of $\overline{V}$ to any coefficient-ring A with residue field k comes from V_R by tensor-product via a unique homomorphism $R \to A$ (which induces the identity on residue fields):*

$$V \cong V_R \otimes_R A.$$

§12. Representations with coefficient-rings which are Λ-algebras. Fix a coefficient-ring Λ with residue field k. For a given profinite group Π and finite-dimensional k-vector space $\overline{V}$ with continuous Π-action, with Π and $\overline{V}$ satisfying the conditions formulated in §11 above, let us ask for deformations of the representation $\overline{V}$ to coefficient-rings A which are Λ-algebras (where the structural algebra homomorphism $\Lambda \to A$ is a coefficient-ring homomorphism). Let $D_{\overline{V},\Lambda}$ denote the "restriction" of the functor $D_{\overline{V}}$ to the category of such coefficient-ring Λ-algebras, i.e., the functor

$$D_{\overline{V},\Lambda} : \begin{pmatrix} \text{Coefficient-ring} \\ \Lambda\text{-algebras} \\ \text{with residue field } k \end{pmatrix} \longrightarrow \text{Sets}$$

associates to the Λ-algebra A, the set $D_{\overline{V}}(A)$ of isomorphism classes of deformations of $\overline{V}$ to A.

Letting R denote the universal deformation ring of the Π-representation $\overline{V}$ (whose existence is guaranteed in the proposition of §11) then:

Proposition. *The functor $D_{\overline{V},\Lambda}$ is representable by $R\hat{\otimes}_{W(k)}\Lambda$, where $\hat{\otimes}$ means "completed tensor product".*

Proof. Before we engage in the proof proper, let us take a minute to review the notion of "completed tensor product." The reason for its involvement in the above proposition is because the (standard) tensor product $R_1 \otimes_{W(k)} R_2$ of two coefficient-rings, R_1 and R_2 (over $W(k)$) is not necessarily a coefficient-ring: it need not be complete. The simple remedy is to complete $R_1 \otimes_{W(k)} R_2$ with respect to the ideal

$$m := \ker(R_1 \otimes_{W(k)} R_2 \to k);$$

one sees easily that $m = m_1 \otimes_{W(k)} R_2 + R_1 \otimes_{W(k)} m_2$ where $m_i \subset R_i$ ($i = 1, 2$) are the maximal ideals. The completion $R_1\hat{\otimes}_{W(k)}R_2$ has the following two descriptions

$$\begin{aligned} R_1\hat{\otimes}_{W(k)}R_2 &= \operatorname*{proj.\,lim.}_{\nu\to\infty}(R_1 \otimes_{W(k)} R_2)/m^\nu \\ &= \operatorname*{proj.\,lim.}_{\nu\to\infty}(R_1/m_1^\nu) \otimes_{W(k)} (R_2/m_2^\nu), \end{aligned}$$

and if $\hat{m} \subset R_1\hat{\otimes}_{W(k)}R_2$ denotes the closure of m, one sees that $R_1\hat{\otimes}_{W(k)}R_2$ is again a complete noetherian local ring with maximal ideal $\hat{m}$ and with residue field k. In particular, the category of coefficient rings (with residue field k) is closed under completed tensor product.

Concretely, if R_1 and R_2 are the quotients of the power series rings $W(k)[[x_1, \dots, x_s]]$ and $W(k)[[y_1, \dots, y_t]]$ by the closed ideals generated by the power series

$$f_1, \dots, f_\nu \in W(k)[[x_1, \dots, x_s]] \quad \text{and} \quad g_1, \dots, g_\mu \in W(k)[[y_1, \dots, y_t]]$$

respectively, then $R_1 \hat{\otimes}_{W(k)} R_2$ is isomorphic to the quotient of the power series ring $W(k)[[x_1, \dots, x_s, y_1, \dots, y_t]]$ by the closed ideal generated by the $\nu + \mu$ power series $f_1, f_2, \dots, f_\nu, g_1, g_2, \dots, g_\mu$.

The proof of the proposition comes from reviewing the definitions involved: the ring $R \hat{\otimes}_{W(k)} \Lambda$ *is* a "coefficient-ring and a Λ-algebra" (a "coefficient-Λ-algebra" for short) and carries a deformation of $\overline{V}$ induced from the universal deformation of $\overline{V}$ to R. Moreover, any deformation of $\overline{V}$ to a coefficient-ring Λ-algebra A is induced from the universal deformation to R via a unique homomorphism $R \to A$ which extends to a unique Λ-algebra homomorphism $R \hat{\otimes}_{W(k)} \Lambda \to A$, establishing the required universal property for $R \hat{\otimes}_{W(k)} \Lambda$.

From now on in these notes, we shall be fixing a coefficient-ring Λ with residue field k of characteristic p, and we will work with Λ as base ring. That is, we deal with the category whose objects are coefficient-Λ-algebras A and morphisms are homomorphisms of coefficient-Λ-algebras: we will study representations with these Λ-algebras as coefficient-rings. The "default" base ring Λ is, of course, just $W(k)$, as discussed in §2.

§13. Is there a relationship between the (formal, say deformation space of a variety V defined over a number field K, and the deformation space of the various Galois representations occurring in the étale cohomology of V?

No. These seem to be quite different animals.

Part Two

CHAPTER IV. FUNCTORS AND REPRESENTABILITY

§14. **Fiber products and representability.** Fix Λ a coefficient-ring with residue field k of characteristic p. Denote by $\hat{\mathcal{C}}_\Lambda(A)$ the category whose objects are coefficient-Λ-algebras which are endowed with a coefficient-Λ-algebra homomorphism to A. Let $\mathcal{C}_\Lambda(A)$ denote the full subcategory of $\hat{\mathcal{C}}_\Lambda(A)$ whose objects are *artinian* coefficient-Λ-algebras (again endowed with an **A-augmentation.** i.e., a coefficient-Λ-algebra homomorphism to A). If A is the residue field k, let us drop it from the notation, i.e., $\hat{\mathcal{C}}_\Lambda(k)$ and $\mathcal{C}_\Lambda(k)$ will be denoted $\hat{\mathcal{C}}_\Lambda$ and $\mathcal{C}_\Lambda$, respectively. The reason for the ^ notation is that any coefficient-ring A may be written as the projective limit of artinian ones:

$$A = \operatorname{proj.lim.}_{n\to\infty} A/m_A^n.$$

If we are out to prove that a given functor, call it D (say on the larger category $\hat{\mathcal{C}}_\Lambda$) is representable (as we shall be!), the representing coefficient-Λ-algebra, call it R, is completely determined by the restriction of the functor to the smaller category $\mathcal{C}_\Lambda$. This is true because

$$\operatorname{Hom}(R, A) = \operatorname{proj.lim.}_{n\to\infty} \operatorname{Hom}(R, A/m_A^n)$$

as sets. It is convenient to do most of our work directly with the smaller category $\mathcal{C}_\Lambda$ if our functors D satisfy the property that

$$D(A) = \operatorname{proj.lim.}_{n\to\infty} D(A/m_A^n) \tag{1}$$

for all coefficient Λ-algebras A. Call such a functor **continuous**. A continuous functor on $\hat{\mathcal{C}}_\Lambda$ is determined by its restriction to $\mathcal{C}_\Lambda$.

Schlessinger calls functors on $\mathcal{C}_\Lambda$ which are represented by objects of the larger category $\hat{\mathcal{C}}_\Lambda$ *pro-representable* (as is only fitting, since they are represented by projective limits of objects on the category on which they are defined) but we will often drop the prefix *"pro-"*.

Given a diagram of sets,

$$\begin{array}{ccc} A & & B \\ \alpha \searrow & & \swarrow \beta \\ & C & \end{array} \tag{2}$$

the "fiber-product" $A \times_C B$ is the subset of the product $A \times B$ consisting of all couples (a, b) such that $\alpha(a) = \beta(b)$. The fiber-product $A \times_C B$ "comes along" with projections to A and to B, and fits into a diamond

$$\begin{array}{ccccc} & & A\times_C B & & \\ & \swarrow & & \searrow & \\ (3) \qquad A & & & & B \\ \alpha & \searrow & & \swarrow & \beta \\ & & C & & \end{array}$$

It is useful to have the accompanying notion of *Cartesian diagram* (of which (3) is the prototype). One says that a diagram of sets

$$\begin{array}{ccccc} & & E & & \\ & \swarrow & & \searrow & \\ (4) \qquad A & & & & B \\ \alpha & \searrow & & \swarrow & \beta \\ & & C & & \end{array}$$

is **cartesian** if the pair of mappings $E \to A$ and $E \to B$ identify the set E with the fiber-product $A \times_C B$; i.e., if the diagrams (3) and (4) are isomorphic (the isomorphism being the identity on similarly labeled sets and mappings).

The notions of fiber-product and cartesian diagram are "categorical" in the sense that if, instead of starting with the diagram of sets (1), we start with a diagram (5) of set-valued covariant functors on any category $\mathcal{C}$,

$$\begin{array}{ccccc} (5) \qquad \underline{A} & & & & \underline{B} \\ \alpha & \searrow & & \swarrow & \beta \\ & & \underline{C} & & \end{array}$$

then the same definitions allow us to talk of the fiber-product $\underline{A}\times_{\underline{C}}\underline{B}$ whose value on any object X of $\mathcal{C}$ is given by the fiber-product of the values of $\underline{A}$ and $\underline{B}$ on X, i.e.,

$$(6) \qquad (\underline{A} \times_{\underline{C}} \underline{B})(X) = \underline{A}(X) \times_{\underline{C}(X)} \underline{B}(X),$$

giving us a diagram of functors

$$\begin{array}{ccccc} & & \underline{A}\times_{\underline{C}} \underline{B} & & \\ & \swarrow & & \searrow & \\ (7) \qquad \underline{A} & & & & \underline{B} \\ \alpha & \searrow & & \swarrow & \beta \\ & & \underline{C} & & \end{array}$$

and allowing us to say, in analogy with our discussion for sets, what it means for a diagram of functors

$$
\begin{array}{ccccc}
 & & \underline{E} & & \\
 & \swarrow & & \searrow & \\
(8) \qquad & \underline{A} & & \underline{B} & \\
\alpha & & \searrow \; \swarrow & & \beta \\
 & & \underline{C} & &
\end{array}
$$

to be **cartesian.**

Even if $\underline{A}, \underline{B}, \underline{C}$ are representable covariant functors on $\mathcal{C}$, (representing objects A, B, C) the fiber-product functor $\underline{A} \times_{\underline{C}} \underline{B}$ may or may not be representable in $\mathcal{C}$; but if it is representable, its representing object, called $A \times_C B$, and coming along with a pair of morphisms $A \times_C B \to A$, $A \times_C B \to B$, is well-defined up to unique isomorphism in $\mathcal{C}$. If this is the case, colloquially one says that the fiber product $A \times_C B$ "exists" in $\mathcal{C}$, and we get a (cartesian) diagram in $\mathcal{C}$:

$$
\begin{array}{ccccc}
 & & A \times_C B & & \\
 & \swarrow & & \searrow & \\
(9) \qquad & A & & B & \\
\alpha & & \searrow \; \swarrow & & \beta \\
 & & C & &
\end{array}
$$

The prototypical example. If you are not familiar with the notion of fiber-product, it might be helpful to note that fiber-products (as defined for any category above) do indeed exist in the category of sets, and these fiber-products are given by the construction given in diagram (3) above. Fiber-products also exist in the category of commutative rings and are given by the analogous construction.

When fiber-products "exist," we may use the bijection (6), turning it around a bit, to provide for us a powerful *necessary condition* for representability. Specifically, suppose that we have a covariant set-valued functor F on our category $\mathcal{C}$. Applying F to diagram (9) gives a diagram of sets.

$$
\begin{array}{ccccc}
 & & F(A \times_C B) & & \\
 & \swarrow & & \searrow & \\
(10) \qquad & F(A) & & F(B) & \\
\alpha & & \searrow \; \swarrow & & \beta \\
 & & F(C) & &
\end{array}
$$

which, if F were representable (say by an element X of $\mathcal{C}$) would be cartesian by (6), i.e., the mapping

$$
(11) \qquad h : F(A \times_C B) \longrightarrow F(A) \times_{F(C)} F(B)
$$

would be a bijection.

The *"earmark"* of representability, then, for a functor F is the property (which I shall refer to as the **Mayer-Vietoris property**) that the morphism h of (11) above is a bijection for *all* cartesian diagrams (9) of the category $\mathcal{C}$. This is germane to our situation for we have the easy

Lemma. *Let $\mathcal{A}$ be a coefficient-Λ-algebra. Fiber products "exist" in the categories $\mathcal{C}_\Lambda(\mathcal{A})$.*

Specifically, if

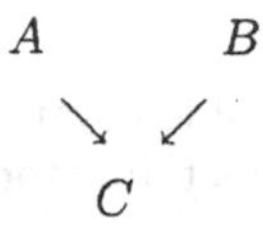

is a diagram of artinian Λ-algebra coefficient-rings with $\mathcal{A}$-augmentation, then the subring

$$A \times_C B \subset A \times B$$

consisting of elements (a, b) such that $\alpha(a) = \beta(b)$ is again a coefficient-Λ-algebra which is artinian. It inherits an $\mathcal{A}$-augmentation, and is the categorical fiber-product.

I am thankful to Brian Conrad for explaining to me that the larger category $\hat{\mathcal{C}}_\Lambda(\mathcal{A})$ is not closed under fiber products, the problem being that the fiber product of elements in $\hat{\mathcal{C}}_\Lambda(\mathcal{A})$ need not be *noetherian*. He suggested the following example. Let k be a field, $A = k[[X, Y]]$, $B = k$, and $C = k[[X]]$, i.e., A and C are the power series rings in the indicated variables over k. Mapping the k-algebra $k[[X, Y]]$ to $k[[X]]$ by sending Y to 0, and mapping the k-algebra k to $k[[X]]$ in the unique manner, we get a diagram

$$\begin{array}{ccc} A = k[[X,Y]] & & k = B \\ & \searrow \quad \swarrow & \\ & C = \; k[[X]] & \end{array}$$

and the fiber-product $A \times_C B$ is given by the sub-ring $k \oplus Y \cdot k[[X, Y]]$ in $k[[X, Y]]$. The maximal ideal of $A \times_C B$ is $Y \cdot k[[X, Y]]$, and the Zariski tangent space of $A \times_C B$ may be identified with the k-vector space $k[[X]]$, which is infinite dimensional; i.e., $A \times_C B$ is not noetherian. In the special case where *both* $A \to C$ and $B \to C$ are surjective morphisms in the category $\hat{\mathcal{C}}_\Lambda$ then the ring $A \times_C B$ is noetherian (see ex. 3.2 of [Mat]) and is again in $\hat{\mathcal{C}}_\Lambda$.

§15. A functor's-eye view of the Zariski tangent space (the "absolute" case). Fix Λ a coefficient ring and R a coefficient ring Λ-algebra. Denoting their maximal ideals $m_\Lambda \subset \Lambda$ and $m_R \subset R$, let us recall the definition of $t^*_R = t^*_{R/\Lambda}$, the "Zariski cotangent space" of the Λ-algebra R,

$$t^*_R := m_R/(m^2_R + (\text{image of } m_\Lambda) \cdot R).$$

The intuition behind this definition is that if one thinks of R as being "functions on some base-pointed space," then m_R may be thought of as those functions vanishing at the base point, and t_R^* is the quotient of m_R by the appropriate ideal (of "higher order terms" of these functions) so as to isolate the "linear parts" of these functions. "Linear" is a key word here, for t_R^* is naturally endowed with the structure of $R/m_R = \Lambda/m_\Lambda$ module, i.e., t_R^* is a vector space over k. As is only fitting one defines the **Zariski tangent (k-vector) space to R** to be the dual k-vector space,

$$t_R = \mathrm{Hom}_k(m_R/(m_R^2 + m_\Lambda \cdot R), k).$$

Since R is noetherian, t_R^* is a finite-dimensional k-vector space and so t_R is naturally the k-dual of t_R^* thereby justifying the notation.

It will be important for us to give a definition of the k-vector space t_R using only the covariant functor, call it F_R, which is represented by R, i.e., the functor

$$B \longmapsto D_R(B) := \mathrm{Hom}_{\hat{\mathcal{C}}_\Lambda}(R, B)$$

for B in $\mathcal{C}_\Lambda$. The key idea is to invoke the Λ-algebra $k[\epsilon]$ defined by the relation $\epsilon^2 = 0$. The algebra $k[\epsilon]$ is a vector space of dimension two over k,

$$k[\epsilon] = k \oplus \epsilon \cdot k, \tag{1}$$

the first subspace in the above direct sum decomposition being generated by the unit element of the algebra $k[\epsilon]$ and the second subspace being the maximal ideal (which has, of course, square zero).

Proposition. *There is a natural isomorphism of k-vector spaces*

$$\mathrm{Hom}_{k\text{-v.sp}}(m_R/(m_R^2 + m_\Lambda \cdot R), k) \cong \mathrm{Hom}_{\Lambda\text{-alg}}(R, k[\epsilon]). \tag{2}$$

(If you have never seen this before, it is more instructive to try to do this as an a exercise, rather than to read the proof below.)

Proof. Since the maximal ideal of $k[\epsilon]$ has square zero, the natural mapping

$$\mathrm{Hom}_{k\text{-alg}}(R/(m_R^2 + m_\Lambda \cdot R), k[\epsilon]) \to \mathrm{Hom}_{\Lambda\text{-alg}}(R, k[\epsilon]) \tag{3}$$

is a bijection. Now the k-algebra $R/(m_R^2 + m_\Lambda \cdot R)$ has a natural direct sum decomposition

$$R/(m_R^2 + m_\Lambda \cdot R) = k \oplus m_R/(m_R^2 + m_\Lambda \cdot R) \tag{4}$$

the first subspace in the above direct sum decomposition being generated by the unit element and the second subspace being the maximal ideal.

Clearly then, any Λ-algebra homomorphism from $R/(m_R^2 + m_\Lambda \cdot R)$ to $k[\epsilon]$ must respect the direct sum decompositions (1) and (4) and (since the homomorphism is constrained to be the identity on the first summand, but may be any k-vector space homomorphism on the second) we have

$$\mathrm{Hom}_{\Lambda\text{-alg}}(R, k[\epsilon]) \cong \mathrm{Hom}_{k\text{-v.sp}}(m_R/(m_R^2 + m_\Lambda \cdot R), \epsilon \cdot k).$$

Identifying the k-vector space $\epsilon \cdot k$ with k, and combining the above isomorphism with (3) yields (2).

This lemma does give us a functorial interpretation of the relative Zariski tangent space, namely:

$$t_R \cong D_R(k[\epsilon]), \tag{5}$$

and allows us, *jumping the gun a bit!*, to make the following definition.

Definition. Let $D : \mathcal{C}_\Lambda \to \mathrm{Sets}$ be any covariant functor such that $D(k)$ consists of a single element. Then the "**Zariski tangent (k-vector) space**" of D, denoted t_D, is the *set* $D(k[\epsilon])$.

In this generality, we cannot yet guarantee a natural k-vector space structure on the set t_D (and this is what I meant by saying that we have "jumped the gun"). Nevertheless we can already see the structure of "scalar multiplication by k" on t_D. Namely, let us notice that the multiplicative monoid $\mathrm{End}_{\Lambda\text{-alg}}(k[\epsilon])$ acts on the set $t_D = D(k[\epsilon])$ by functoriality of D and we have a natural ring homomorphism

$$\begin{aligned} &k \cong \mathrm{End}_{\Lambda\text{-alg}}(k[\epsilon]) \\ &a \mapsto \alpha_a \qquad \text{where} \quad \alpha_a(x \oplus y \cdot \epsilon) = x \oplus a \cdot y \cdot \epsilon. \end{aligned} \tag{6}$$

This multiplicative action of k will be the scalar multiplication in the eventual k-vector space structure of t_D in the special case where this vector space structure can be defined. Let us also point to the structure which will give rise to the law of vector-addition. Namely, define the k-algebra homomorphism which we will simply label " $+$ ":

$$\begin{gathered} k[\epsilon] \times_k k[\epsilon] \xrightarrow{+} k[\epsilon] \\ (x \oplus y_1 \cdot \epsilon, x \oplus y_2 \cdot \epsilon) \mapsto x \oplus (y_1 + y_2) \cdot \epsilon \end{gathered} \tag{7}$$

We need a further hypothesis concerning our functor D. We call it $(\mathbf{T}_k)$ — for "Tangent Space Hypothesis."

> $(\mathbf{T}_k)$ The mapping $h : D(k[\epsilon] \times_k k[\epsilon]) \to D(k[\epsilon]) \times D(k[\epsilon])$ is a bijection.

If D satisfies $(\mathbf{T}_k)$ we define vector addition on the tangent space of D by the composition

$$\begin{array}{ccccc} D(k[\epsilon]) \times D(k[\epsilon]) & \overset{h^{-1}}{\cong} & D(k[\epsilon] \times_k k[\epsilon]) & \xrightarrow{D(+)} & D(k[\epsilon]) \\ \downarrow{=} & & & & \downarrow{=} \\ t_D \times t_D & & \longrightarrow & & t_D. \end{array} \tag{8}$$

§16. The Zariski tangent A-module (the "relative" case). Suppose, now, that we are in the relative case. That is, we have fixed a coefficient Λ-algebra A, and a covariant functor

$$D : \mathcal{C}_\Lambda(A) \to Sets$$

such that the value of the functor D on the A-augmented coefficient-Λ-algebra A itself, $D(A)$, is a single point. Let $A[\epsilon]$ denote the coefficient-Λ-algebra $A[T]/(T^2)$ where $\epsilon = T \bmod (T^2)$. Then $A[\epsilon]$ is free of rank 2 as an A-module, with A-basis $\{1, \epsilon\}$:

$$A[\epsilon] := A \oplus \epsilon \cdot A \tag{9}$$

We view $A[\epsilon]$ as an A-augmented coefficient-Λ-algebra, where the augmentation mapping is passage to the quotient by (ϵ); i.e., it is the projection to the first summand in (9). Then, as in §15, the object $A[\epsilon]$ of $\mathcal{C}_A$ is "an A-module-object" in $\mathcal{C}_A$, in the sense that $A[\epsilon]$ admits an "addition law"

$$\begin{array}{ccc} A[\epsilon] \times_A A[\epsilon] & \xrightarrow{+} & A[\epsilon] \\ (x \oplus y_1 \cdot \epsilon, x \oplus y_2 \cdot \epsilon) & \longmapsto & x \oplus (y_1 + y_2) \cdot \epsilon \end{array} \tag{10}$$

and it may be endowed with scalar multiplication by elements α of A

$$\begin{array}{ccc} A[\epsilon] & \xrightarrow{\cdot\alpha} & A[\epsilon] \\ (x \oplus y \cdot \epsilon) & \longmapsto & (x \oplus \alpha \cdot y \cdot \epsilon) \end{array} \tag{11}$$

where these operations satisfy, formally, all the properties of "A-module operations." Note also that $A[\epsilon] \times_A A[\epsilon]$ is again a coefficient Λ-algebra (the point being that it is still noetherian; see the discussion at the end of §14).

Now suppose that D satisfies the "Tangent A-module Hypothesis":

($\mathbf{T}_A$) The mapping $h : D(A[\epsilon] \times_A A[\epsilon]) \to D(A[\epsilon]) \times D(A[\epsilon])$ is a bijection.

Then we make the analogous

Definition. The **Zariski tangent** A**-module,** denoted $t_{D,A}$, or t_D if A is understood, is given, as a set, by

$$t_{D,A} := D(A[\epsilon])$$

and it inherits an A-module structure as follows:

Letting $D(+)$ denote the mapping induced from (10) above, the addition law in t_D is given by the composition

$$D(A[\epsilon]) \times D(A[\epsilon]) \xrightarrow{h^{-1}} D(A[\epsilon] \times_A A[\epsilon]) \xrightarrow{D(+)} D(A[\epsilon]),$$

while scalar multiplication by $\alpha \in A$ is the mapping induced from (11) above.

Since any coefficient-Λ-algebra has a natural "k-augmentation," we may think of the "absolute case," treated in the previous section as a particular example of the "relative case" under discussion here, just by taking $A = k$.

§17. Continuous Kähler differentials and the (relative) Zariski Tangent Space. Let R be a coefficient-Λ-algebra. Let $\Omega^{\text{cont}}_{R/\Lambda}$ denote the R-module of relative (continuous) Kähler differentials of the Λ-algebra R. We will refer to this module simply as $\Omega_{R/\Lambda}$. For a discussion of the notion of Kähler differentials, and for a list of its basic properties see [Mat] and [Hal], but note that we are working here in a slightly different category than is dealt with in those references. Specifically, we will be dealing exclusively with (topologically profinite) complete noetherian rings, and we will demand that our Kähler differentials respect the appropriate topologies. For a reference that does things in such a topological context, see [Gr] (Chapter 0, §20). Intuitively, $\Omega_{R/\Lambda}$ is the R-module packaging the *"maximum amount of first-order infinitesimal information about the Λ-algebra R."* Somewhat more formally, the R-module $\Omega_{R/\Lambda}$ comes along with a (continuous) derivation

$$d : R \to \Omega_{R/\Lambda} \tag{11}$$

relative to Λ (i.e., such that $d\Lambda = 0$) and it is *universal* for this structure, i.e., it is the universal (topologically profinite) R-module equipped with (continuous) derivation from R, relative to Λ. That is, for any topological R-module M. we have a canonical isomorphism

$$\text{Hom}_{R\text{-mod}}(\Omega_{R/\Lambda}, M) \cong \text{Der}_{\Lambda-\text{alg}}(R, M) \tag{12}$$

where $\text{Der}_\Lambda\text{alg}(R, A)$ is the R-module of continuous derivations from R to the R-module M.

Somewhat more concretely, it can be constructed as follows: Let $R \hat{\otimes}_\Lambda R$ be the completed tensor product of the coefficient-Λ-algebra R with itself, over Λ. Then $R\hat{\otimes}_\Lambda R$ is again a coefficient-Λ-algebra, and the multiplication homomorphism

$$\mu : R \hat{\otimes}_\Lambda R \to R$$
$$x \hat{\otimes} y \mapsto x \cdot y$$

is a surjective homomorphism of coefficient-Λ-algebras. Let $I \subset R \hat{\otimes}_\Lambda R$ denote the kernel of μ. Since μ is continuous, I is a closed ideal in $R \hat{\otimes}_\Lambda R$. We have a continuous homomorphism of Λ-modules $\delta : R \to I$ defined by the equation $\delta(r) = r \hat{\otimes} 1 - 1 \hat{\otimes} r$ for $r \in R$. Since $R \hat{\otimes}_\Lambda R$ is noetherian, I^2 is a closed ideal. The (topologically profinite) $R \hat{\otimes}_\Lambda R$-module structure on I/I^2 "factors through" a canonical R-module structure on I/I^2 (via the topological identification $R \hat{\otimes}_\Lambda R/I = R$) and the mapping

$$d : R \to I/I^2 \tag{13}$$

obtained from δ by projection $I \to I/I^2$ is easily seen to be a derivation relative to Λ. One checks that the continuous derivation (13) is indeed "universal" for continuous derivations of R relative to Λ. Therefore we may take $\Omega_{R/\Lambda} = I/I^2$, and (13) provides us with a construction of the universal derivation (11).

Exercises. **1)** Let P be the power series ring over Λ,

$$P = \Lambda[[X_1, \dots, X_n]].$$

Show that $\Omega_{P/\Lambda}$ is canonically isomorphic to the free P-module of rank n generated by elements $dX_1, \dots, dX_n$ and show that the universal derivation $d : P \to \Omega_{P/\Lambda}$ is the standard differential

$$d : P \to P \cdot dX_1 \oplus P \cdot dX_2 \oplus \cdots \oplus P \cdot dX_N$$

on power series (with $d\lambda = 0$ for all $\lambda \in \Lambda$).
2) Let R be given as the quotient ring of the power series ring P in **1)** by the ideal $I \subset P$ generated by m elements,

$$R = \Lambda[[X_1, \dots, X_n]]/(f_1, \dots, f_m).$$

Show that Ω_R/Λ may be presented as the quotient of the free R-module on n generators $dX_1, \dots, dX_n$ by the sub-R-module generated by the images of $df_1, \dots, df_m$ under the projection

$$P \cdot dX_1 \oplus P \cdot dX_2 \oplus \dots P \cdot dX_n \longrightarrow R \cdot dX_1 \oplus R \cdot dX_2 \oplus \dots R \cdot dX_n.$$

In particular, $\Omega_{R/\Lambda}$ is an R-module of finite type.
3) For R a coefficient-Λ-algebra, show that there is a natural identification between the k-vector space $\Omega_{R/\Lambda} \otimes_R k \cong \Omega_{R/\Lambda} \hat{\otimes}_R k$ and the *cotangent space* of R relative to Λ, i.e., with

$$t^*_{R/\Lambda} := m_R/(m_R^2 + m_\Lambda \cdot R).$$

Give the derivation from R to $t^*_{R/\Lambda}$ which corresponds, under this identification, to the projection of the mapping $d : R \to \Omega_{R/\Lambda}$ to $\Omega_{R/\Lambda} \otimes_R k$.

In the case where the functor D of the last section is prorepresentable by an A-augmented coefficient-Λ-algebra R, then D does satisfy hypothesis $(\mathbf{T}_A)$. To record the dependence of the relative Zariski tangent space

$$t_{D,A} = D(A[\epsilon])$$

on the representation $\rho : R \to A$, we find it useful sometimes to adopt the alternate notation $t_{D,A} = t_{D,\rho}$. We have the following description of $t_{D,\rho}$ which follows directly from the definition.

(14) $t_{D,\rho} \cong$ the subset of $\mathrm{Hom}_{\Lambda\text{-alg}}(R, A[\epsilon])$ consisting of those Λ-algebra homomorphisms whose composition with the projection $A[\epsilon] \to A$ is equal to ρ.

The A-module $t_{D,\rho}$ may also be obtained directly from the R-module of (continuous) Kähler differentials $\Omega_{R/\Lambda}$. Specifically,

Proposition. *Let D be represented by the coefficient-Λ-algebra R (so, in our usual terminology $D = D_R$) and let $\Omega_{R/\Lambda}$ denote the R-module of relative continuous Kähler differentials of the topological Λ-algebra R. Let (A, ρ) be a pair where A is a coefficient-Λ-algebra in $\hat{C}_\Lambda$ and $\rho : R \to A$ is a coefficient-Λ-algebra homomorphism (i.e., ρ an element in $D(A) :=$ $\mathrm{Hom}_{\Lambda\text{-alg}}(R, A)$). We view A in this manner as R-algebra.*

Then we have a natural isomorphism of A-modules:

$$\mathrm{Hom}_{A\text{-mod}}(\Omega_{R/\Lambda} \hat{\otimes}_R A, A) \cong t_{D,\rho}.$$

Note. In particular, if $A = k$ and $\rho = \overline{\rho}$, the original residual representation, so that $t_{D,\overline{\rho}}$ is the Zariski tangent (k-vector) space attached to D, we have:

$$\mathrm{Hom}_k(\Omega_{R/\Lambda} \hat{\otimes}_R k, k) = t_{R/\Lambda} \cong t_{D,\overline{\rho}}.$$

Proof of the Proposition. We have these canonical isomorphisms:

$$\mathrm{Hom}_{A\text{-mod}}(\Omega_{R/\Lambda} \hat{\otimes}_R A, A) \cong \mathrm{Hom}_{R\text{-mod}}(\Omega_{R/\Lambda}, A) \tag{15}$$

$$\mathrm{Hom}_{R\text{-mod}}(\Omega_{R/\Lambda}, A) \cong \mathrm{Der}_\Lambda(R, A) \qquad \text{(by (12))}. \tag{16}$$

Moreover there is a natural injection

$$\imath : \mathrm{Der}_\Lambda(R, A) \to \mathrm{Hom}_{\Lambda\text{-alg}}(R, A[\epsilon])$$

which sends a derivation $\delta \in \mathrm{Der}_\Lambda(R, A)$ to the Λ-algebra homomorphism $\rho_\delta : R \to A[\epsilon]$ given by

$$\rho_\delta(r) = \rho(r) \oplus \epsilon \cdot \delta(r) \in A \oplus \epsilon \cdot A.$$

The injection $\imath$ identifies $\mathrm{Der}_\Lambda(R, A)$ with the subset of $\mathrm{Hom}_{\Lambda\text{-alg}}(R, A[\epsilon])$ consisting of the Λ-algebra homomorphisms from R to $A[\epsilon]$ such that composition with the projection $A[\epsilon] \to A$ yields $\rho : R \to A$. By (14) we then have a bijection of sets

$$\mathrm{Der}_\Lambda(R, A) \cong t_{D,\rho}, \tag{17}$$

and checking back on the definition of the R-module structure of $t_{D,\rho}$ one immediately sees that (17) is an isomorphism of R-modules. Putting (15)–(17) together yields the proposition.

§18. Schlessinger's representability theorem. To get us in the mood, let us begin with a result which is easy to prove (it is a good exercise) but which is sometimes difficult to use because its hypothesis is hard to check:

Grothendieck's Theorem. *Let $D : \mathcal{C}_\Lambda \to$ Sets be a covariant functor such that $D(k)$ consists of a single element. Then D is pro-representable, i.e., $D = D_R$ for some coefficient ring R in $\hat{\mathcal{C}}_\Lambda$, if and only if D satisfies the "Mayer-Vietoris Property," i.e., the mapping*

$$h : D(A \times_C B) \to D(A)\times)_{D(C)} D(B)$$

is an isomorphism for all diagrams

$$(8) \qquad \begin{array}{ccc} A & & B \\ \alpha & \searrow \quad \swarrow & \beta \\ & C & \end{array}$$

in $\mathcal{C}_\Lambda$ and the Zariski tangent space $t_{D,k}$ is finite dimensional over k.

In contrast to this Theorem, which requires the Mayer-Vietoris Property for all diagrams, the theorem of Schlessinger formulated below artfully cuts down the number of diagrams for which one must check the Mayer-Vietoris Property. To prepare for this:—

Definition. A mapping $p : A \to C$ in $\mathcal{C}_\Lambda$ is **small** of if its kernel is a principal ideal annihilated by m_A.

Schlessinger's Theorem. *Let $D : \mathcal{C}_\Lambda \to$ Sets be a covariant functor such that $D(k)$ consists of a single element. Then D is (pro)-representable if and only if these four conditions holds:*

($\mathbf{H}_1$) *h is surjective if $A \to C$ is small (or equivalently: h is surjective if $A \to C$ is surjective).*

($\mathbf{H}_2$) *h is bijective if $A \to C$ is $k[\epsilon] \to k$. (* **Note**: *($\mathbf{H}_2$) implies hypothesis ($\mathbf{T}_k$) of §13 and therefore it implies that the "Zariski tangent space" $t_{D,k}$ is naturally endowed with the structure of k-vector space).*

($\mathbf{H}_3$) *Hypothesis ($\mathbf{T}_k$) holds and $\dim_k(t_{D,k})$ is finite.*

($\mathbf{H}_4$) *h is bijective if $A \to C$ and $B \to C$ are equal, and small.*

For a proof, see [Sch].

In view of our assignment in this conference, *it is almost less important to us* that our functors be representable than that they satisfy hypothesis ($\mathbf{T}_A$). This motivates us to make the following definition (which we state in the relative case):

Definition. Fix a coefficient-Λ-algebra A. A contravariant functor

$$D : \mathcal{C}_\Lambda(A) \to \text{Sets}$$

such that $D(A)$ is a single element will be called **nearly representable** if it satisfies hypothesis ($\mathbf{T}_A$) of §16, together with the following "finiteness" hypothesis ($\mathbf{F}$).

($\mathbf{F}$) The relative Zariski tangent A-module $t_{D,A}$ is of finite type.

Although many of the functors of interest to us in this article will *not* be representable, they will all turn out to be nearly representable, and also they will all satisfy (**H 1**), (**H 2**), and (**H 3**). It is condition (**H 4**) that will, at times, not be satisfied. Functors satisfying (**H 1,2,3**), have an important property which at times, is a reasonable consolation even when they are not prorepresentable. Such functors are, in any case, very nearly pro-representable: a functor D satisfying (**H 1,2,3**) has, in Schlessinger's terms, a **pro-representable hull** (Def. 2.7 of [Sch]). To describe this notion we must define what it means for a morphism of functors $\xi : D' \to D$ on $\mathcal{C}_\Lambda$ (such that $D'(k)$ and $D(k)$ are singletons) to be **smooth.** The morphism ξ is **smooth** if it satisfies the following "lifting property": given any surjection $B \to A$ in $\mathcal{C}_\Lambda$, any element $\alpha' \in D'(A)$ and any lifting of $\alpha = \xi(\alpha') \in D(A)$ to an element $\beta \in D(B)$, there exists an element $\beta' \in D'(B)$ which is a lifting of α' such that $\xi(\beta') = \beta$. Equivalently, we may phrase this lifting property as the condition that the natural mapping.

$$D'(B) \to D'(A) \times_{D(A)} D(B)$$

be surjective for all surjections $B \to A$ in $\mathcal{C}_\Lambda$. It is also equivalent to request the same surjectivity property for surjections $B \to A$ in $\hat{\mathcal{C}}_\Lambda$. For example, if $D' \to D$ is smooth, then it follows that $D'(B) \to D(B)$ is surjective for every B in $\hat{\mathcal{C}}_\Lambda$ (proof: use the lifting property with $A = k$). For a list of the basic properties of smooth morphisms of functors, see Prop. 2.5 of [Sch].

Definition. A pro-representable hull for D is a pair (R, ϵ) where R is a coefficient-Λ-algebra, and $\epsilon : D_R \to D$ is a morphism of functors (where D_R is the functor pro-represented by R) satisfying two properties:

(i) the morphism of functors $\epsilon : D_R \to D$ is *smooth.*

(ii) the induced mapping of Zariski tangent spaces

$$t_R \to t_D$$

is an isomorphism of k-vector spaces.

Schlessinger proves that any (covariant, Set-valued) functor D on $\mathcal{C}_\Lambda$ such that $D(k)$ is a singleton, and which satisfies (**H 1,2,3**) possesses a pro-representable hull (R, ϵ) and, moreover, any two pro-representable hulls of D are *isomorphic* (but they are, in general, only "noncanonically" isomorphic).

§19. Relatively representable subfunctors. Let us be given two covariant functors $\mathcal{D} \subset D$ from the category $\mathcal{C}_\Lambda$ to sets with $\mathcal{D}$ a *subfunctor* of D (and, in particular, $\mathcal{D}(A)$ is a subset of $D(A)$ for all objects A of $\mathcal{C}_A$) such that $\mathcal{D}(k) = D(k)$ is a single element. Let us say that $\mathcal{D} \subset D$ is **relatively representable** if for all diagrams in $\mathcal{C}_\Lambda$,

(8)
$$\begin{array}{ccc} A & & B \\ & \alpha \searrow \quad \swarrow \beta & \\ & C & \end{array}$$

the square

$$\begin{array}{ccc} \mathcal{D}(A \times_C B) & \xrightarrow{h} & \mathcal{D}(A) \times_{\mathcal{D}(C)} \mathcal{D}(B) \\ \cap\big\downarrow & & \big\downarrow\cap \\ D(A \times_C B) & \xrightarrow{h} & D(A) \times_{D(C)} D(B) \end{array}$$

is cartesian.

The terminology "*relatively representable*" is justified by the

Exercise. In the above context, if $\mathcal{D} \subset D$ is relatively representable, then $\mathcal{D}$ satisfies $(\mathbf{H}_i)$ if D does (this is true for each of the i's ($= 1, 2, 3, 4$) separately). Also, $\mathcal{D}$ satisfies $\mathbf{T}_k$ if D does, and $\mathcal{D}$ is nearly representable if D is. If D is representable by a coefficient Λ-algebra R_D then $\mathcal{D}$ is representable by a quotient-Λ-algebra $R_{\mathcal{D}}$ of R_D.

A hint for this statement is given by the following fact:

Lemma. *Let R be a coefficient-Λ-algebra, and $D_R : \mathcal{C}_\Lambda \to$ Sets the functor represented by R; i.e., $D_R(A) := \mathrm{Hom}_\Lambda(R, A)$. Let $\varphi : R_1 \to R_2$ be a homomorphism of coefficient-Λ-algebras and denote by φ again the natural transformation of functors on $\mathcal{C}_\Lambda$ which is induced by φ, $\varphi : D_{R_2} \to D_{R_1}$. Then these two properties are equivalent.*

(i) *The ring-homomorphism $\varphi : R_1 \to R_2$ is surjective.*
(ii) *The natural transformation $\varphi : D_{R_2} \to D_{R_1}$ is injective; i.e., we may identify D_{R_2} with a subfunctor of D_{R_1}.*

Proof of Lemma. Clearly (i) implies (ii). To see that (ii) implies (i) note that since φ is a homomorphism of coefficient-Λ-algebras it induces the identity on residue fields. Since R_i are complete noetherian local rings it then suffices to show that $\varphi : R_1 \to R_2$ is surjective on Zariski tangent spaces, or, since these k-vector spaces are finite-dimensional, we must show, dually, that the mapping induced by φ,

$$D_{R_2}(k[\epsilon]) \to D_{R_1}(k[\epsilon]),$$

is injective, which it is by (ii).

§20. Representability results regarding deformation problems attached to group representations. Let Π be a profinite group satisfying the p-finiteness condition, k a finite field of characteristic p, Λ a coefficient-ring with residue field k, and

$$\overline{\rho} : \Pi \to \mathrm{GL}_N(k)$$

a continuous representation. Recall the "*absolute*" Λ-*deformation problem for $\overline{\rho}$*. This is given by the functor

$$D_{\overline{\rho}} : \hat{\mathcal{C}}_\Lambda \to \text{Sets} \tag{1}$$

which associates to each coefficient-Λ-algebra B the set $D_{\overline{\rho}}(B)$ of deformation of $\overline{\rho}$ to B.

Also, for a given choice of *lifting* $\rho : \Pi \to \mathrm{GL}_N(A)$ of $\overline{\rho}$ to a coefficient-Λ-algebra A (i.e., ρ is an actual homomorphism, not just a strict equivalence class) we have the *"relative" Λ-deformation problem* (relative to this lifting ρ), given by the functor

$$D_\rho : \hat{\mathcal{C}}_\Lambda(A) \to \text{Sets} \tag{2}$$

which associates to each A-augmented coefficient-Λ-algebra B the set

$$D_\rho(B) := \text{the set of deformations of } \rho \text{ to } B.$$

If A is a coefficient-Λ-algebra and n a positive integer, let A_n be the artinian quotient coefficient-Λ-algebra, $A_n := A/m_A^n$.

Proposition 1 (Continuity). *Let $\rho : \Pi \to \mathrm{GL}_N(A)$ be a lifting of $\overline{\rho}$ to a coefficient-Λ-algebra A, and $\rho_n : \Pi \to \mathrm{GL}_N(A_n)$ the induced continuous homomorphism for $n \geq 1$. The functor $D_\rho : \hat{\mathcal{C}}(A) \to$ Sets is continuous in the sense that it satisfies (1) of §14; i.e.,*

$$D_\rho(B) = \operatorname*{proj.\,lim.}_{n\to\infty} D_{\rho_n}(B_n)$$

for all A-augmented coefficient-Λ-algebras B. The functor $D_{\overline{\rho}} : \hat{\mathcal{C}}_\Lambda \to$ Sets is continuous.

Proof. Let $i : D_\rho(B) \to \text{proj.\,lim.}\, D_{\rho_n}(B_n)$ denote the natural mapping, which we will show to be bijective. We adopt the interpretation of the functor D_ρ (and of the functors D_{ρ_n}) given in §11. That is, letting W be the underlying A-module of rank N endowed with the Π-representation ρ, then for any A-augmented coefficient-Λ-algebra B, the set $D_\rho(B)$ is the set of isomorphism classes of pairs (V, α) where V is a free B-module of rank N, endowed with continuous (B-linear) Π-action, and $\alpha : V \otimes_B A \to W$ is an isomorphism of $A[[\Pi]]$-modules. We use the same notation with the subscript "n" to describe the sets $D_{\rho_n}(B_n)$. Let $\{(V_n, \alpha_n)\}_{n\geq 1}$ denote a cofinal system in the projective system $\{D_{\rho_n}(B_n)\}_{n\geq 1}$. So, for each $n \geq 1$, V_n is a $B_n[[\Pi]]$-module, free of rank N over B_n and

$$\alpha_n : V_n \otimes_{B_n} A_n \to W \otimes_A A_n$$

is an isomorphism of $A[[\Pi]]$-modules. Cofinality is expressed by the *existence* of an isomorphism β_n of $B_n[[\Pi]]$-modules,

$$\beta_n : V_{n+1} \otimes_{B_{n+1}} B_n \xrightarrow{\cong} V_n$$

which "fits" into a commutative diagram

$$\begin{array}{ccc} V_{n+1} \otimes_{B_{n+1}} A_{n+1} & \xrightarrow{\alpha_{n+1}} & W \otimes_A A_{n+1} \\ {\scriptstyle \beta_n \otimes \pi_n} \downarrow & & \downarrow {\scriptstyle 1 \otimes \pi_n} \\ V_n \otimes_{B_n} A_n & \xrightarrow{l\alpha_n} & W \otimes_A A_n \end{array}$$

where π_n is the natural projection. Compiling the $B[[\Pi]]$-modules V_n via the compositions of the natural projections $V_{n+1} \to V_{n+1} \otimes_{B_n+1} B_n$ and with (*a choice of*) β_n for each $n \geq 1$, we see by Nakayama's Lemma that the $B[[\Pi]]$-module obtained in the (projective) limit,

$$V := \text{proj. lim.}\, V_n,$$

is a free B-module of rank N. The limit of the $\alpha_n' s$ gives us an isomorphism of $A[[\Pi]]$-modules $\alpha : V \otimes_B A \to W$.

This proves surjectivity of the mapping i. As for injectivity, let (V, α) and (V', α') be two representatives of elements in $D_\rho(B)$ such that we have isomorphisms $\gamma_n : V_n \to V_n'$ of $B_n[[\Pi]]$-modules for each $n \geq 1$, such that the γ_n's "fit" into commutative diagrams.

$$\begin{array}{ccc} V_1 \otimes_{B_n} A_n & \xrightarrow{\gamma_n \otimes 1} & V_n' \otimes_{B_n} A_n \\ {\scriptstyle \alpha_n} \downarrow & & \downarrow {\scriptstyle \alpha_n'} \\ W_n & = & W_n \end{array}$$

Then the projective limit of the γ_n's yield an isomorphism between the couples (V, α) and (V', α').

In view of Proposition 1, the functor D_ρ (resp. $D_{\overline{\rho}}$) is representable if and only if its restriction to the subcategory $\mathcal{C}_\Lambda(A)$ of artinian objects (resp. $\mathcal{C}_\Lambda$) is "pro-representable." Regarding the absolute deformation problem, we have:

Proposition 2. *Let $\overline{\rho} : \Pi \to \mathrm{GL}_N(k)$ be a continuous residual representation, with k a finite residue field of characteristic p, and Π a profinite group satisfying the p-finiteness condition. Fix Λ a coefficient-ring with residue field k.*
(i) *The functor $D_{\overline{\rho}}$ (restricted to $\mathcal{C}_\Lambda$) satisfies* $(\mathbf{H}_1)$, $(\mathbf{H}_2)$, $(\mathbf{H}_3)$.
(ii) *If $\overline{\rho}$ is absolutely irreducible, then the functor $D_{\overline{\rho}}$ is representable.*

Our relative deformation problems are all "nearly representable" and we shall formally state this fact in two "strengths":

Proposition 3a ("weak near representability"). *For every artinian coefficient Λ-algebra A, and every lifting $\rho : \Pi \to \mathrm{GL}_N(A)$ of $\overline{\rho}$ to A, the relative functor D_ρ is nearly representable in the sense of* §*18.*

and

Proposition 3b ("strong near representability"). *For every coefficient Λ-algebra A, and every lifting $\rho : \Pi \to \mathrm{GL}_N(A)$ of $\overline{\rho}$ to A, the relative functor D_ρ is nearly representable in the sense of* §*18.*

We have separated the two statements above because the strong statement requires a somewhat more elaborate proof than the weak one does, and it is only the weak statement that we shall actually use in this article.

We should include, in the above list of representability results the relationship between the "representability" of the absolute deformation problem and the representability of the corresponding collection of "relative" problems. Namely, if $\overline{\rho}$ is absolutely irreducible and R is the ("universal") coefficient-Λ-algebra representing the functor $D_{\overline{\rho}}$ (such an R existing, by part **(ii)** of Proposition 2), then fixing a lifting ρ of $\overline{\rho}$ to a coefficient-Λ-algebra A, we may view the coefficient-Λ-algebra R as "A-augmented" via the homomorphism $R \to A$ which classifies ρ. We have

Proposition 4. *If $\overline{\rho}$ is absolutely irreducible and R is the ("universal") coefficient-Λ-algebra representing the functor $D_{\overline{\rho}}$, then for every lifting $\rho : \Pi \to \mathrm{GL}_N(A)$ of $\overline{\rho}$ to a coefficient-Λ-algebra A, which satisfies the ("minimality") property that the coefficient-Λ-algebra A is generated by the traces of ρ, the functor $D_\rho : \mathcal{C}_\Lambda(A) \to$* Sets *is prorepresentable by the A-augmented coefficient-Λ-algebra R. That is, for each A-augmented coefficient-Λ-algebra B, there is a natural one-to-one correspondence between the set of deformations of ρ to B and the set of A-augmented coefficient-Λ-algebra homomorphisms from R to B.*

Remarks. We will not give the proof of Proposition 2, which has been written up in various places (e.g., [M 1]); item (ii) in its assertion requires Schlessinger's Theorem. More germane to our purposes in these notes, really, are Propositions 3 and 4 whose proofs we will give, "independent of Proposition 2," in full detail. Proposition 2 (i) implies that the absolute deformation problem for any residual representation $\overline{\rho}$ (with Π satisfying the p-finiteness hypothesis) possesses a pro-representable hull in the sense of Schlessinger [Sch]; see the discussion about this given in §21 below.

Proofs of Propositions 3a and 3b. Let A be any coefficient-Λ-algebra, and $\rho : \Pi \to \mathrm{GL}_N(A)$ a homomorphism. The first step is to show that our functor D_ρ satisfies hypothesis $(\mathbf{T}_A)$. That is, we must show that

$$D_\rho(A[\epsilon] \times_A A[\epsilon]) \xrightarrow{h} D_\rho(A[\epsilon]) \times_{D_\rho(A)} D_\rho(A[\epsilon]) = D_\rho(A[\epsilon]) \times D_\rho(A[\epsilon])$$

is a bijection. That h is surjective is straightforward. We must show injectivity of h, which is in fact also straightforward, but here it is. Let

$$\gamma, \delta : \Pi \to \mathrm{GL}_N(A[\epsilon] \times_A A[\epsilon])$$

be homomorphisms representing two elements of $D_\rho(A[\epsilon] \times_A A[\epsilon])$ which map to the same element under h. Let $\gamma_1, \delta_1 : \Pi \to GL_N(A[\epsilon])$ be the homomorphisms obtained from γ, δ by composing them with the homomorphism

$$\mathrm{GL}_N(A[\epsilon] \times_A A[\epsilon]) \to \mathrm{GL}_N(A[\epsilon])$$

induced by projection to the first factor $A[\epsilon] \times_A A[\epsilon] \to A[\epsilon]$, and similarly let $\gamma_2, \delta_2 : \Pi \to \mathrm{GL}_N(A[\epsilon])$ be the ones induced by projection to the second factor.

The corresponding homomorphism $\Pi \to \mathrm{GL}_N(A)$ obtained by projecting all the way to $\mathrm{GL}_N(A)$ are both equal to ρ.

Since $h(\gamma) = h(\delta)$ there are elements $\alpha_i \in GL_N(A[\epsilon])$ which "intertwine" γ_i with δ_i for $i = 1, 2$, and which project to the identity in $\mathrm{GL}_N(A)$. It follows that there is an element $\alpha \in \mathrm{GL}_N(A[\epsilon] \times_A A[\epsilon])$ projecting to α_1 and α_2 under the first and second projections respectively, and this α "intertwines" γ with δ, showing that γ is strictly equivalent to δ, i.e., that h is injective.

To conclude the proof of Propositions 3a and 3b, we must show that the Zariski tangent A-module t_{D_ρ} is of finite type for an artinian coefficient-Λ-algebra A, and for *any* coefficient-Λ-algebra A, respectively. This will be done in Propositions 2a and 2b of the next section.

Proof of Proposition 4. We shall show that if $\overline{\rho}$ is absolutely irreducible, then for every A-augmented coefficient-Λ-algebra B, the natural mapping

$$(3) \qquad D_\rho(B) \to \{x \in D_{\overline{\rho}}(B) \mid x \mapsto \text{class of } \rho \text{ in } D_{\overline{\rho}}(A)\}$$

is a one-to-one correspondence, this statement being equivalent to the statement of our proposition. First, let us show injectivity of (3), not using the "minimality" assumption on A. For this, let $\rho_1, \rho_2 : \Pi \to \mathrm{GL}_N(B)$ be two homomorphisms projecting to ρ after composition with the map $\mathrm{GL}_N(B) \to \mathrm{GL}_N(A)$, and which are assumed to be strictly equivalent relative to $\overline{\rho}$. We must show them to be strictly equivalent relative to ρ. Let $\beta \in \mathrm{GL}_N(B)$ be the element that intertwines ρ_1 to ρ_2, and let $\alpha \in \mathrm{GL}_N(A)$ be the projection of β. Since α commutes with the image of ρ, by Schur's Lemma, α is a scalar matrix in A, i.e., $\alpha = a \cdot I_N$ where $a \in A^*$ and I_N is the $N \times N$ identity matrix. Since α comes by projection from *some* matrix in $\mathrm{GL}_N(B)$ it follows that the element $\alpha \in A^*$ is in the image of B, and let $b \in B$ be some element which projects to $a \in A^*$. Since A is a local ring, and $B \to A$ is a mapping of local rings, it follows that $b \in B^*$, and we may form $\beta' = b^{-1} \cdot \beta \in \mathrm{GL}_N(B)$ which projects to the identity in $\mathrm{GL}_N(A)$ and intertwines ρ_1 to ρ_2, showing that ρ_1 is indeed strictly equivalent to ρ_2 relative to ρ.

As for surjectivity, we must show that if we are given a homomorphism $\rho_1 : \Pi \to GL_N(B)$ which after composition with $\mathrm{GL}_N(B) \to \mathrm{GL}_N(A)$ is in the same strict equivalence class (relative to $\overline{\rho}$) of ρ in $D_{\overline{\rho}}(A)$, then we can find a homomorphism $\rho_2 : \Pi \to \mathrm{GL}_N(B)$ which is strictly equivalent to ρ_1 relative to $\overline{\rho}$ and which, under composition with $\mathrm{GL}_N(B) \to \mathrm{GL}_N(A)$, projects to the homomorphism ρ itself. For this, just note that by the minimality property on A, if $D_{\overline{\rho}}(B)$ is non-empty (which we may assume) the homomorphism $B \to A$ is surjective, and therefore an element α in $\ker\{GL_N(A) \to GL_N(k)\}$ intertwining the image of ρ_1 with ρ may be lifted to an element $\beta \in \ker\{GL_N(B) \to \mathrm{GL}_N(k)\}$. Conjugating ρ_1 with the inverse of β, one gets the desired ρ_2.

CHAPTER V. ZARISKI TANGENT SPACES AND DEFORMATION PROBLEMS SUBJECT TO "CONDITIONS"

§21. A cohomological interpretation of Zariski tangent A-modules.

One of the basic tools of deformation theory is the cohomological interpretation of the Zariski tangent spaces that occur. This allows us at times to "control" the somewhat abstract universal deformation rings that occur by means of concrete cohomological calculations.

To describe this, fix Π a profinite group satisfying the p-finiteness condition,

$$\overline{\rho} : \Pi \to \mathrm{GL}_N(k)$$

a continuous residual representation with k a finite field of characteristic p, and Λ a coefficient-ring with residue field k. For the discussion below, we fix a specific *homomorphism*

$$\rho : \Pi \to \mathrm{GL}_N(A),$$

where A is a coefficient-Λ-algebra and we consider the deformation problem relative to ρ, i.e., the functor

$$D_\rho : \hat{\mathcal{C}}_\Lambda(A) \to \text{Sets}.$$

More specifically, we consider its Zariski tangent A-module, $t_{D_\rho,A}$ which we shall more simply denote t_ρ.

The cohomological interpretation we are referring to is as follows. Let V be the free A-module of rank N, $V = A^N$, endowed with A-linear Π-action given via composition of $\rho : \Pi \to \mathrm{GL}_N(A)$ with the natural action of $\mathrm{GL}_N(A)$ on V. Let $\mathrm{End}_A(V)$ denote the free A-module (of rank N^2) consisting of A-linear endomorphisms of V. The action of Π on V induces an action (the "adjoint action") of Π on $\mathrm{End}_A(V)$: the formula being

$$(g \cdot e)(v) = \rho(g)(e(\rho(g)^{-1}(v))$$

where $g \in \Pi$, $e \in \mathrm{End}_A(V)$, and $v \in V$.

Proposition 1. *There is a natural isomorphism of A-modules*

$$t_\rho \cong H^1(\Pi, \mathrm{End}_A(V)).$$

Here, and in what follows, we use 'continuous' cohomology.

Sketch of Proof. Let $\Gamma := \ker\{\mathrm{GL}_N(A[\epsilon]) \to \mathrm{GL}_N(A)\}$ so that we have a short exact sequence of groups

$$(4) \qquad 1 \to \Gamma \to \mathrm{GL}_N(A[\epsilon]) \to \mathrm{GL}_N(A) \to 1$$

with a natural splitting, coming from the injection

$$\mathrm{GL}_N(A) \subset \mathrm{GL}_N(A[\epsilon]),$$

so that we may view $\mathrm{GL}_N(A[\epsilon])$ as a semi-direct product $\mathrm{GL}_N(A) \ltimes \Gamma$. Moreover, letting $M_N(A)$ denote the underlying additive group of the A-algebra of $N \times N$ matrices with entries in A, there is a natural isomorphism of (commutative) groups

$$\begin{array}{ccccccc} \Gamma & = & 1+\epsilon \cdot M_N(A) & \cong & M_N(A) & \cong & \mathrm{End}_A(V). \\ & & (1+\epsilon \cdot m & \mapsto & m) & & \end{array}$$

Using these isomorphisms, one may rewrite $\mathrm{GL}_N(A[\epsilon])$ as the semi-direct product

$$\mathrm{GL}_N(A[\epsilon]) \cong \mathrm{GL}_N(A) \ltimes M_N(A),$$

where the action of $\mathrm{GL}_N(A)$ on $M_N(A)$ is by conjugation, i.e., is the standard "adjoint action."

Now the set of deformations, $D_\rho(A[\epsilon])$, lifting ρ to $A[\epsilon]$ is the set of strict equivalence classes (relative to ρ) of homomorphisms ρ' fitting into the diagram

$$\Pi \xrightarrow{\rho'} \mathrm{GL}_N(A[\epsilon]) \cong \mathrm{GL}_N(A) \ltimes \mathrm{End}_A(V) \to \mathrm{GL}_N(A)$$

where the composition above is ρ. "Strict equivalence" means, of course, that we will be considering the "orbit" of the set of such homomorphisms ρ', under the action of the subgroup $\Gamma \subset \mathrm{GL}_N(A[\epsilon])$ via conjugation.

But let us first ignore the equivalence relation and study the set of homomorphisms ρ' which lift ρ as above. There is a chosen such homomorphism, call it ρ_0: namely the composition of ρ with the natural imbedding $\mathrm{GL}_N(A) \subset \mathrm{GL}_N(A[\epsilon])$. For any other ρ', define the **difference cocycle**

$$c_{\rho'} : \Pi \to \Gamma \cong \mathrm{End}_A(V)$$

by $c_{\rho'}(g) = \rho'(g) \cdot \rho_0(g)^{-1} \in \Gamma$ for $g \in \Pi$. Check first that this construction $\rho' \mapsto c_{\rho'}$ provides a bijection between the set of liftings ρ' of ρ, and the set $Z^1(\Pi, \mathrm{End}_A(V))$ of 1-cocycles on Π with values in the Π-module $\mathrm{End}_A(V) \cong M_N(A)$, where the action of Π on $\mathrm{End}_A(V)$ is the "adjoint action" as described above. Then check that under this bijection, liftings ρ' and ρ'' of $\overline{\rho}$ are strictly equivalent if and only if their associated cocycles $c_{\rho'}$ and $c_{\rho''}$ are cohomologous. Finally, one must check that the resulting bijection $t_\rho \cong H^1(\Pi, \mathrm{End}_A(V))$ is A-linear.

We are now in a position to prove the following result, thereby concluding the proof of Proposition 3a of §20.

Proposition 2a ("weak finiteness"). *Let A be an artinian coefficient-Λ-algebra. Then the Zariski tangent A-module t_ρ is finite.*

Proof of Proposition 2a. Let A be artinian. It suffices to show that the A-module $H^1(\Pi, \mathrm{End}_A(V))$ is finite. Let $\Pi_\circ \subset \Pi$ be the kernel of ρ. Since A is artinian, $\Pi_\circ$ is an open subgroup of finite index in Π. The A-module $H^1(\Pi, \mathrm{End}_A(V))$ fits into an exact sequence

$$H^1(\Pi/\Pi_\circ, \mathrm{End}_A(V)) \to H^1(\Pi, \mathrm{End}_A(V)) \to \mathrm{Hom}(\Pi_\circ, \mathrm{End}_A(V)),$$

the left-most A-module being finite since both $\Pi/\Pi_\circ$ and $\mathrm{End}_A(V)$ are finite, and the right-most A-module being finite since Π satisfies the p-finiteness condition.

Remark. In the special case where $\overline{\rho}$ is absolutely irreducible, with universal deformation ring $R = R(\overline{\rho})$, the isomorphisms of the Proposition of §17 and of Proposition 1 above give:

$$\mathrm{Hom}_A(\Omega_{R/\Lambda} \hat{\otimes}_R A, A) \cong t_\rho \cong H^1(\Pi, M_N(A)).$$

In particular, taking $A = k$ and $\rho = \overline{\rho}$, the Zariski tangent k-vector space of the local ring R has the following cohomological interpretation:

$$\mathrm{Hom}_k(m_R/(p, {m_R}^2), k) \cong H^1(\Pi, M_N(k)).$$

We now can sketch the proof of the following "strong finiteness" result, thereby concluding the proof of Proposition 3b of §20.

Proposition 2b ("strong finiteness"). *Let A be any coefficient-Λ-algebra. Then the Zariski tangent A-module t_ρ is finite.*

Proof. The functor $D_{\overline{\rho}}$ satisfies Schlessinger's conditions (**H 1,2,3**) as given in Proposition 2 of §20 above and as proved in [M 1]. By Schlessinger's Theorem (cf. the discussion in §18 above) $D_{\overline{\rho}}$ possesses a *pro-representable hull* (R, ξ). We have then the smooth morphism of functors

$$\xi : D_R \to D_{\overline{\rho}}$$

(which induces an isomorphism on Zariski tangent k-vector spaces). Since

$$\xi(A) : D_R(A) \to D_{\overline{\rho}}(A)$$

is surjective, we may (and do) choose some element $\tilde{\rho} \in D_R(A)$ whose image under $\xi(A)$ is $\rho \in D_{\overline{\rho}}(A)$. Now, referring back to the terminology of §17, we have the *relative Zariski tangent A-modules* for our two functors, D_R and $D_{\overline{\rho}}$ (relative to $\overline{\rho}$ and ρ, respectively) the definitions of which we now review:

$t_{D_R,\tilde{\rho}} \cong$ the subset of $\mathrm{Hom}_{\Lambda\text{-alg}}(R, A[\epsilon])$ consisting of those Λ-algebra homomorphisms whose composition with the projection $A[\epsilon] \to A$ is equal to $\tilde{\rho}$.

$t_\rho = t_{D_{\bar\rho},\rho} \cong$ the subset of $D_{\bar\rho}(A[\epsilon])$ consisting of those Λ-algebra homomorphisms whose composition with the projection $A[\epsilon] \to A$ is equal to ρ.

Since $A[\epsilon] \to A$ is surjective, and since the morphism ξ is smooth, ξ induces a surjection h of A-modules

$$h : t_{D_R,\bar\rho} \to t_\rho.$$

By the proposition of §17, we have a natural isomorphism of A-modules:

$$t_{D_R,\bar\rho} \cong \mathrm{Hom}_{A\text{-mod}}(\Omega_{R/\Lambda} \hat{\otimes}_R A, A)$$

and, in particular, since $\Omega_{R/\Lambda}$ is of finite type as R-module, we see that $t_{D_R,\bar\rho}$ is of finite type as A-module, and therefore, by noetherianness of A and the surjectivity of h, it follows that t_ρ is of finite type over A.

For amusement, consider the corollary of Proposition 2b below. I apologize in advance for (a) the clearly too round-about method of the proof of this corollary and (b) for not pausing long enough in these notes to prove anything stronger than:

Corollary. (finiteness result for cohomology A-modules of profinite groups satisfying the p-finiteness condition): *Let A be a coefficient ring with (finite) residue field of characteristic p. Let Π be a profinite group satisfying the p-finiteness condition. Let M be a free A-module of finite rank with a continuous A-linear action of Π, and let $H^1(\Pi, M)$ denote the continuous (1-dimensional) cohomology group of Π with coefficients in the $A[\Pi]$-module M. We give $H^1(\Pi, M)$ its natural A-module structure. Then $H^1(\Pi, M)$ is of finite type as A-module.*

Proof. Let V be the free A-module $M \oplus A$ endowed with Π action which is the direct sum of the given action on M and the trivial action on A. Let $N-1$ denote the rank of the free A-module M, so that N is the rank of V. Let

$$\rho : \Pi \to \mathrm{Aut}_A(V) \approx \mathrm{GL}_N(A)$$

be the associated representation, and let $\overline{\rho}$ be its associated residual representation. We view $\mathrm{End}_A(V)$ as an $A[\Pi]$-module where the action of Π is the adjoint action, and we let $H^1(\Pi, \mathrm{End}_A(V))$ denote the continuous (1-dimensional) cohomology of Π with coefficients in the $A[\Pi]$-module $\mathrm{End}_A(V)$. Since $H^1(\Pi, M)$ is a direct summand of $H^1(\Pi, \mathrm{End}_A(V))$ and since A is noetherian, it suffices to show that $H^1(\Pi, \mathrm{End}_A(V))$ is of finite type as A-module. But by Proposition 1, we have an A-module isomorphism $H^1(\Pi, \mathrm{End}_A(V)) \cong t_\rho$ and by Proposition 2, t_ρ is of finite type over A.

§22. The Zariski tangent space of the universal deformation ring as "extensions". Another way of stating the proposition of the previous section which is, perhaps, more revealing than the way we have stated it and which is particularly useful in some contexts is the following:

Proposition. *Let A be a coefficient-Λ-algebra. There are natural isomorphisms of A-modules*

$$t_\rho \cong H^1(\Pi, \mathrm{End}_A(V)) \cong \mathrm{Ext}_{A[\Pi]}(V, V).$$

Here, $\mathrm{Ext}_{A[\Pi]}(V, V)$ mean Ext^1 in the category of profinite $A[\Pi]$-modules with continuous Π-action and we will abbreviate it to the shorter notation: $\mathrm{Ext}_\Pi(V, V)$. We will directly compare the two end-modules,

$$t_\rho = D_\rho(A[\epsilon]) \overset{s}{\cong} \mathrm{Ext}_\Pi(V, V). \tag{5}$$

Given a deformation V_1 of V to $A[\epsilon]$, by restricting the ring of scalars of V_1 from $A[\epsilon]$ to A (via the injection $\imath : A \to A[\epsilon]$) we may view V_1 as free A-module of rank $2N$, with an A-linear continuous action of the group Π. Identifying the $A[\Pi]$-modules $\epsilon \cdot V_1$ and $V_1/\epsilon \cdot V_1$ with V (in the natural manner) we then see V_1 as an extension of V by V in the category of profinite $A[\Pi]$-modules with continuous Π-action:

$$\mathcal{E} : 0 \to V \xrightarrow{\alpha} V_1 \xrightarrow{\beta} V \to 0$$

Sending the element of $t_\rho = D_\rho(A[\epsilon])$ which corresponds to the isomorphism class of V_1 to the element of $\mathrm{Ext}_\Pi(V, V)$ corresponding to $\mathcal{E}$ gives a well-defined mapping

$$s : D_\rho(A[\epsilon]) \to \mathrm{Ext}_\Pi(V, V),$$

which is easily seen to be an A-module homomorphism. Going the other way is equally direct: given an extension $\mathcal{E}$, one imposes an $A[\epsilon]$-module structure on V_1 in the evident manner (multiplication by ϵ is given by $\beta\alpha$) enabling us to view V_1 as deformation of V to $A[\epsilon]$. This gives the isomorphism (5).

Remark. The second isomorphism displayed in the proposition above can also be seen as coming from the "natural" isomorphism

$$H^1(\Pi, \mathrm{End}_A(V)) \cong \mathrm{Ext}_\Pi(V, V) \tag{6}$$

which arises from the degeneration of the Spectral Sequence

$$H^p(\Pi, \underline{\mathrm{Ext}}^q(V, V)) \Longrightarrow \mathrm{Ext}_\Pi^{p+q}(V, V) \tag{7}$$

where $\underline{\mathrm{Ext}}^q$ denotes "sheaf-Ext"; that is, $\underline{\mathrm{Ext}}^q = \mathrm{Ext}_A^q$, meaning "$\mathrm{Ext}^q$ in the category of profinite topological A-modules." Since V is a free A-module, $\underline{\mathrm{Ext}}^q(V, V) = 0$ for $q > 0$, and since the Π-module $\underline{\mathrm{Ext}}^0(V, V)$ is just $\mathrm{End}(V)$ with the adjoint Π-action, the Spectral Sequence (7) degenerates, yielding an isomorphism (6). But this is somewhat "learned" and one is left with the chore of making precise the identifications involved, if one wishes to check that the isomorphisms of the proposition in §21, and (5) and (6) are compatible.

§23. "Deformation conditions". It is often the case that one wishes to restrict the *deformation problem* to deformations of $\bar{\rho}$ which satisfy certain conditions. This will occupy us much in what is to come. In order to treat issues that arise here in a uniform, rather than ad hoc, manner it is convenient to pause and ask what general form these "conditions" will take. Here is a suggestion that covers all conditions that we will encounter in this article.

Fix N, Π, and Λ. Let us consider the category $F_N = F_N(\Lambda;\Pi)$ whose objects are pairs (A, V) consisting of an artinian coefficient-Λ-algebra A, and a free A-module V of rank N endowed with an A-linear continuous Π-action. A morphism in F_N from an object (A, V) to an object (A_1, V_1) consists of a pair of morphisms $A \to A_1$ (of artinian coefficient-Λ-algebras) and $V \to V_1$ (of A-modules) inducing an isomorphism of A-modules $V \otimes_A A_1 \cong V_1$ which is compatible with Π-action in the evident sense. Consider, now, the following three conditions on a full sub-category $\mathcal{D}F_N \subset F_N$:

(1) For any morphism $(A, V) \to (A_1, V_1)$ in F_N, if (A, V) is an object of $\mathcal{D}F_N$ then so is (A_1, V_1).

(2) Let A, B, C be artinian coefficient-Λ-algebras fitting into a diagram

$$\begin{array}{ccc} A & & B \\ \alpha \searrow & & \swarrow \beta \\ & C & . \end{array}$$

Consider an object $(A \times_C B, V)$ in F_N and let V_A, V_B denote the tensor products of V with respect to the natural projections $1 \times_C \beta$ and $\alpha \times_C 1$ from $A \times_C B$ to A and B, respectively. Then $(A \times_C B, V)$ is an object in $\mathcal{D}F_N$ if and only if both (A, V_A) and (B, V_B) are objects of $\mathcal{D}F_N$.

(3) For any morphism $(A, V) \to (A_1, V_1)$ in F_N, if (A_1, V_1) is an object of $\mathcal{D}F_N$ and $A \to A_1$ is injective, then (A, V) is an object of $\mathcal{D}F_N$.

Definition. Fix Λ, Π, and a continuous residual representation

$$\bar{\rho} : \Pi \to \mathrm{GL}_N(k).$$

We denote by $\bar{V}$ the N-dimensional k-vector space k^N with k-linear Π-action given by $\bar{\rho}$. By a **deformation condition** $\mathcal{D}$ for $\bar{\rho}$, we mean a full subcategory $\mathcal{D}F_N \subset F_N$ satisfying (1)–(3) and containing $(k, \bar{V})$ as object.

Suppose that we are given a deformation condition $\mathcal{D}F_N \subset F_N$. For a coefficient-$\Lambda$-algebra A, and homomorphism

$$\rho : \Pi \to \mathrm{GL}_N(A)$$

lifting $\bar{\rho}$, let $V(\rho)$ denote the free A-module A^N with Π-action given by ρ. If $(A, V(\rho))$ is in $\mathcal{D}F_N$ then we will say that ρ is of type $\mathcal{D}$, and its strict equivalence class is a $\mathcal{D}$-deformation of type $\bar{\rho}$. Define a subfunctor

$$\mathcal{D}_\rho \subset D_\rho : \mathcal{C}_\Lambda(A) \longrightarrow Sets$$

by the following rule:

For any artinian A-augmented coefficient-Λ-algebra B, and element $\xi \in D_\rho(B)$ representing a strict equivalence class of liftings ρ_1 relative to ρ to B, then ξ is in the subset $\mathcal{D}_\rho(B) \subset D_\rho(B)$ if and only if ρ_1 is of type $\mathcal{D}$. Extend the functor $\mathcal{D}_\rho$ to $\hat{\mathcal{C}}_\Lambda(A)$ by continuity; i.e., for B any A-augmented coefficient-Λ-algebra, let $\mathcal{D}_\rho(B) \subset D_\rho$ denote the subset,

$$\operatorname*{proj.\,lim.}_{n\to\infty} \mathcal{D}_\rho(B_n) \subset \operatorname*{proj.\,lim.}_{n\to\infty} D_\rho(B_n).$$

Proposition. *If $\mathcal{D}$ is a deformation condition for $\overline{\rho}$, and $\rho : \Pi \to \mathrm{GL}_N(A)$ is a lifting of $\overline{\rho}$, then the subfunctor $\mathcal{D}_\rho \subset D_\rho$ is relatively representable.*

Proof. This is immediate from conditions (1) and (2) alone.

Corollary. *If $\mathcal{D}$ is a deformation condition for $\overline{\rho}$, then the functor $\mathcal{D}_{\overline{\rho}}$ (on $\mathcal{C}_\Lambda$) satisfies $(\mathbf{H}_1)$, $(\mathbf{H}_2)$, $(\mathbf{H}_3)$. The functor $\mathcal{D}_{\overline{\rho}}$ has a "pro-representable hull" in the sense of Schlessinger (cf. the discussion of §18 above). For ρ any lifting of $\overline{\rho}$ the functor $\mathcal{D}_\rho$ is nearly representable. If $\overline{\rho}$ is absolutely irreducible, then $\mathcal{D}_{\overline{\rho}}$ is pro-representable (by a quotient ring of the ring pro-representing $D_{\overline{\rho}}$).*

Proof. This is straightforward from the relevant definitions coupled with the discussion of §18, the Exercise and Lemma in §19, and Proposition 2 of §20.

Fix a lifting ρ of $\overline{\rho}$ to a coefficient-Λ-algebra A which is of type $\mathcal{D}$. Since $\mathcal{D}_\rho$ is nearly representable, we may speak of the tangent A-module to $\mathcal{D}_\rho$ which is a sub-A-module of t_ρ (necessarily of finite type) and we denote it

$$t_{\mathcal{D},\rho} \subset t_\rho.$$

If $\overline{\rho}$ is absolutely irreducible and $R = R(\overline{\rho})$ is the universal deformation ring for deformations of $\overline{\rho}$ to coefficient Λ-algebras, and if $R_{\mathcal{D}}$ is the quotient-ring of R which represents the subfunctor $\mathcal{D}$ then (by §17) we have the following commutative diagram of A-modules in which vertical maps are isomorphisms

$$\begin{array}{ccc} t_{\mathcal{D},\rho} & \subset & t_\rho \\ \cong\downarrow & & \cong\downarrow \\ \mathrm{Hom}_A(\Omega_{R_{\mathcal{D}}/\Lambda} \hat{\otimes}_{R_{\mathcal{D}}} A, A) & \subset & \mathrm{Hom}_A(\Omega_{R/\Lambda} \hat{\otimes}_R A, A). \end{array}$$

We may also quote the Proposition of §21 which identifies the A-modules t_ρ and $H^1(\Pi, \mathrm{End}_A(V))$ for $V = V(\rho)$. Therefore the sub-A-module $t_{\mathcal{D},\rho}$ of t_ρ is identified with some sub-A-module of $H^1(\Pi, \mathrm{End}_A(V))$. Let us call this sub-module $H_{\mathcal{D}}{}^1(\Pi, \mathrm{End}_A(V))$ so that we have a commutative square,

$$\begin{array}{ccc} t_{\mathcal{D},\rho} & \subset & t_\rho \\ \cong\downarrow & & \cong\downarrow \\ \mathrm{Hom}^1_{\mathcal{D}}(\Pi, \mathrm{End}_A(V)) & \subset & H^1(\Pi, \mathrm{End}_A(V)). \end{array}$$

But of course this notation, $H_{\mathcal{D}}{}^1(\Pi, \mathrm{End}_A(V))$, is nothing more than a "promissory note" that our theory be eventually required to deal with: in any concrete instance of a deformation condition $\mathcal{D}$, it will be our chore to describe, in group-cohomologial terms if possible, the sub-A-module

$$H^1_{\mathcal{D}}(\Pi, \mathrm{End}_A(V)) \subset H^1(\Pi, \mathrm{End}_A(V)).$$

In the next two sections we will consider examples of this.

§24. Determinant conditions. Keep to our standard notation, so A is a coefficient-Λ-algebra and $V(\rho) = V = A^N$ the free A-module of rank N with Π-action induced by ρ.

Let $\delta : \Pi \to \Lambda^*$ be a continuous homomorphism. Say that V has "**determinant** δ" if the action of Π on the N-fold wedge product $\bigwedge^N(V)$ of the A-module V is given via the continuous homomorphism

$$\delta_A : \Pi \xrightarrow{\delta} \Lambda^* \to A^*$$

where the homomorphism $\Lambda^* \to A^*$ comes from the Λ-algebra structure of A. Explicitly, $g(w) = \delta_A(g) \cdot w$ for $g \in \Pi$, and $w \in \bigwedge^N(V)$. Let us keep the quotation-marks around the phrase "determinant δ" to remind us that the actual determinant of our representation takes values in a different ring than δ does. If ρ' is a deformation of ρ and ρ' has "determinant δ" then so does ρ.

Let $\overline{\rho}$ be the underlying residual representation attached to ρ, so $\overline{\rho}$ also has "determinant δ." Consider the property $\mathcal{D}$ which is defined by imposing the condition on a homomorphism $\rho : \Pi \to \mathrm{GL}_N(A)$ that it be of "determinant δ."

Proposition. *The condition $\mathcal{D}$ (of being of "determinant δ") is a "deformation condition." If $\rho : \Pi \to \mathrm{GL}_N(A)$ is a continuous homomorphism of "determinant δ" and $V = V(\rho)$, we have*

$$H^1_{\mathcal{D}}(\Pi, \mathrm{End}_A(V)) = H^1(\Pi, \mathrm{End}^\circ_A(V)) \subset H^1(\Pi, \mathrm{End}_A(V))$$

where $\mathrm{End}^\circ_A(V) \subset \mathrm{End}_A(V)$ is the sub A-module of "traceless" endomorphisms (endomorphisms whose trace is zero). The sub-A-module $\mathrm{End}^\circ_A(V)$ is stable under the action of Π, the cohomology group $H^1(\Pi, \mathrm{End}^\circ_A(V))$ being computed with respect to this action.

Proof. The first sentence in our proposition is straightforward to show. The second requires going back to the proof of the Proposition of §21. Consider the commutative diagram

$$\begin{array}{ccc} 1 + \epsilon \cdot M_N(A) & \xrightarrow{j} & M_N(A) \\ {\scriptstyle \det}\downarrow & & \downarrow{\scriptstyle \mathrm{Trace}} \\ 1 + \epsilon \cdot A & \xrightarrow{j} & A \end{array}$$

where the mappings labeled $\jmath$ are the natural isomorphisms of groups given by $\jmath(1+\epsilon\cdot x)=x$.

Let $\rho_0 : \Pi \to \mathrm{GL}_N(A[\epsilon])$ be the composition of ρ with the natural injection induced by $\imath : A \to A[\epsilon]$.

For any lifting ρ' of ρ to $A[\epsilon]$, and any $g \in \Pi$, note that $\rho'(g)\cdot\rho_0(g)^{-1}$ lies in the subgroup $1+\epsilon\cdot M_N(A) \subset \mathrm{GL}_N(A[\epsilon])$. Now recall the "difference cocycle" constructed in the proof of the proposition of §21, i.e.,

$$c_{\rho'} : \Pi \to M_N(A)$$

given in our present notation by $c_{\rho'}(g) = \jmath(\rho'(g)\cdot\rho_0(g)^{-1})$. By the commutativity of diagram (6) we have

$$\mathrm{Trace}(c_{\rho'}(g)) = \jmath(\det(\rho'(g))\cdot\det(\rho(g))^{-1})$$

and, in particular, the image of $c_{\rho'}$ lies in $M_N(A)^\circ$ if and only if ρ' has "determinant δ."

In the special case of an absolutely irreducible residual representation $\overline{\rho}$ of "determinant δ," if $R = R(\overline{\rho})$ is the universal Λ-algebra deformation ring for $\overline{\rho}$, and

$$\rho^{\mathrm{univ}} : \Pi \to \mathrm{GL}_N(R)$$

the universal deformation, one can describe explicitly $R_{\mathcal{D}}$, the quotient Λ-algebra of R universal for deformations of $\overline{\rho}$ which are of "determinant δ," in the following elementary way: Let

$$\delta^{\mathrm{univ}} : \Pi \to R^*$$

denote the determinant of ρ^{univ}, and let δ_R denote the composition of our character δ with the natural homomorphism $u : \Lambda^* \to R^*$ which comes from the Λ-algebra structure on R:

$$\delta_R : \Pi \xrightarrow{\delta} \Lambda^* \xrightarrow{u} R^*.$$

Then $R_{\mathcal{D}} = R/I$ where I is the ideal in R generated by the elements

$$\delta_R(g) - \delta^{\mathrm{univ}}(g) \in R$$

for all $g \in \Pi$ (equivalently, for a system of topological generators g of Π).

§25. Categorical conditions (Ramakrishna's Theory cf. [Ram]). In this section we fix Λ a coefficient-ring, and Π a profinite group. Let $\mathrm{Rep}_\Lambda(\Pi)$ denote the category of Λ-modules of finite length which are endowed with continuous, Λ-linear action of Π. Let $\mathcal{P}$ be a full subcategory of $\mathrm{Rep}_\Lambda(\Pi)$ which is closed under passage to subobjects, quotients, and direct sums. The following proposition shows that any such full subcategory determines "deformation conditions":

Proposition 1. *Let $\mathcal{P}$ be a full subcategory of* $\mathrm{Rep}_\Lambda(\Pi)$ *closed under subobjects, quotients, and direct sums. For N any positive integer let $\mathcal{P}F_N$ denote the full subcategory of the category F_N (as in §23) comprising those objects (A, V) such that V viewed as Λ-linear representation of Π, lies in $\mathcal{P}$. Then $\mathcal{P}F_N$ is a "deformation condition" as defined in §23 above.*

Proof. We must show that $\mathcal{P}F_N$ satisfies the two conditions

(1) For any morphism $(A, V) \to (A_1, V_1)$ in F_N, if V is an object of $\mathcal{P} \subset \mathrm{Rep}_\Lambda(\Pi)$ then so is V_1.
(2) If A, B, and C are in $\mathcal{C}_\Lambda$, and $\alpha : A \to C$ and $\beta : B \to C$ are morphisms in $\mathcal{C}_\Lambda$, with $A \times_C B$ the associated fiber product, then for any object $(A \times_C B, V)$ in F_N, let V_A, V_B denote the tensor products of V with respect to the natural projections from $A \times_C B$ to A and B, respectively. Then V is in $\mathcal{P}$ if and only if V_A and V_B are both in $\mathcal{P}$.

But (1) holds since $\mathcal{P}$ is closed under passage to direct sums and quotients and (2) holds because $\mathcal{P}$ is closed under passage to subobjects, direct sums, and quotients.

Remark. In practice, the example of particular interest to us (and the principal application in [Ram]) is in the case where $\Pi = G_{\mathbf{Q},S}$ with $p \in S$ and the condition $\mathcal{P}$ on our representations is the condition of being **finite flat at** p. See the brief discussion of this condition for $\Pi = G_{\mathbf{Q}_\ell}$, with $\ell \neq 2$, in §31 below. Explicitly, let $\mathcal{P} = \mathcal{P}_{\mathrm{ffl}}$ be the full subcategory of $\mathrm{Rep}(G_{\mathbf{Q}_\ell})$ whose objects are the $G_{\mathbf{Q}_\ell}$-representations whose representation spaces are isomorphic to the $G_{\mathbf{Q}_\ell}$-Galois modules which can be given as the generic fibers of finite flat group schemes over $\mathrm{Spec}(\mathbf{Z}_\ell)$. Then $\mathcal{P}_{\mathrm{ffl}}$ is closed under passage to subobjects, quotients, and direct sums, and consequently gives rise to a "deformation condition."

Fixing a full subcategory $\mathcal{P}$ (closed under sub-objects, quotient-objects, and direct sums) of $\mathrm{Rep}_\Lambda(\Pi)$, by the above proposition, we have that $\mathcal{P}$ determines deformation conditions for each positive integer N. Therefore, whenever we are given a coefficient-Λ-algebra A and a continuous homomorphism $\rho : \Pi \to \mathrm{GL}_N(A)$ such that $V = V(\rho)$ is in $\mathrm{Rep}_\Lambda(\Pi)$, the deformation problem $\mathcal{D}_\rho$ (obtained by restricting to deformations whose finite-length quotients lie in $\mathcal{P}$) is nearly representable, and has a tangent A-module

$$t_{\mathcal{D},\rho} \cong H^1_{\mathcal{D}}(\Pi, \mathrm{End}(V)) \subset H^1(\Pi, \mathrm{End}(V)).$$

In terms of the description of $H^1(\Pi, \mathrm{End}(V))$ as an "Ext-group" given in §22, we may give a direct categorical description of the sub-A-module $H^1_{\mathcal{D}}(\Pi, \mathrm{End}(V))$, as follows.

If V, V' are (continuous) $A[\Pi]$-modules of finite length (as A-modules) which, as $\Lambda[\Pi]$-modules, are elements in $\mathcal{P}$, the subset

$$\mathrm{Ext}_{A[\Pi],\mathcal{P}}(V, V') \subset \mathrm{Ext}_{A[\Pi]}(V, V')$$

of extensions

$$\epsilon : 0 \to V' \to \mathcal{E} \to V \to 0$$

where $\mathcal{E}$ is an $A[\Pi]$-module which, as $\Lambda[\Pi]$-modules, is a member of $\mathcal{P}$ is easily seen to be a sub-A-module of $\mathrm{Ext}_{A[\Pi]}(V, V')$: the sum of two elements,

$$\epsilon_1 : 0 \to V' \to \mathcal{E}_1 \to V \to 0$$

$$\epsilon_2 : 0 \to V' \to \mathcal{E}_2 \to V \to 0$$

in $\mathrm{Ext}_A[\Pi](V, V')$ is obtained by restricting the direct sum $\mathcal{E}_1 \oplus \mathcal{E}_2$ to the diagonal in $V \oplus V$ and then passing to the quotient by the diagonal in $V' \oplus V'$. Since $\mathcal{P}$ is closed under direct sum, sub-object, and quotient, if $\mathcal{E}_1$, $\mathcal{E}_2$ are in $\mathcal{P}$, then the sum of the extensions ϵ_1, ϵ_2 remains in $\mathrm{Ext}_{A[\Pi],\mathcal{P}}(V', V)$, which is also closed under scalar multiplication by A.

Proposition 2. *Let V, V' be (continuous) $A[\Pi]$-modules of finite length (as A-modules). We have the commutative diagram where the vertical maps are isomorphisms*

$$\begin{array}{ccc} t_{\mathcal{D},\rho} & \subset & t_\rho \\ \cong\downarrow & & \cong\downarrow \\ \mathrm{Ext}_{A[\Pi],\mathcal{P}}(V, V') & \subset & \mathrm{Ext}_{A[\Pi]}(V, V'). \end{array}$$

Proof. After the Proposition of §22 and the above discussion, this is evident.

CHAPTER VI. BACK TO GALOIS REPRESENTATIONS

§26. Galois deformation conditions. Fix a coefficient-ring Λ of residual characteristic p, and positive integer N. Let K be a number field and S a finite set of (non-archimedean) primes of K which contain all primes of residual characteristic p. By a **Global Galois deformation problem** (for Λ, N, K, S as above) let us mean that for each prime $\lambda \in S$ we specify a "local deformation problem," i.e., a deformation condition in the sense of §23 for the decomposition group $\Pi = G_{K_\lambda}$.

More explicitly, we must specify (for each $\lambda \in S$) a full subcategory $\mathcal{D}_\lambda F_N(\Lambda; G_{K_\lambda})$ in the category $F_N(\Lambda; G_{K_\lambda})$ which satisfies properties (1) and (2) of §23.

Attached to such data, we may define a full subcategory $\mathcal{D}F_N(\Lambda; G_{K,S})$ of $F_N(\Lambda; G_{K,S})$ by restricting to objects (A, V) of $F_N(\Lambda; G_{K,S})$ such that when, for each $\lambda \in S$, they are considered objects of $F_N(\Lambda; G_{K_\lambda})$ they lie in the subcategory $\mathcal{D}_\lambda F_N(\Lambda; G_{K_\lambda})$.

By a **Global Galois deformation problem with fixed determinant** we mean a slight variant of the above. Namely, for a character

$$\delta : G_{K,S} \to \Lambda^*,$$

we let $\mathcal{D}F_N(\Lambda; G_{K,S}; \delta)$ denote the full subcategory of $\mathcal{D}F_N(\Lambda; G_{K,S})$ comprised of objects (A, V) of "determinant δ."

Proposition 1. *Given a Global Galois deformation problem, the full subcategory $\mathcal{D}F_N(\Lambda; G_{K,S}) \subset F_N(\Lambda; G_{K,S})$ satisfies conditions (1) and (2) of §23. It defines a deformation condition for $G_{K,S}$. The same is true, given a Global Galois deformation problem with fixed determinant δ, for the full subcategory $\mathcal{D}F_N(\Lambda; G_{K,S}; \delta) \subset F_N(\Lambda; G_{K,S}; \delta)$.*

Proof. A straightforward check of the definitions involved.

It follows that Global Galois deformation problems, with or without fixed determinant, give us "deformation conditions" and we may ask for a description of their associated Zariski tangent A-modules.

Proposition 2. *Let $\mathcal{D}$ be the deformation condition attached to a Global deformation problem for $G_{K,S}$ and for each $\lambda \in S$ let $\mathcal{D}_\lambda$ be the corresponding local deformation problem. Let A be a coefficient-Λ-algebra, and*

$$\rho : G_{K,S} \to \mathrm{GL}_N(A)$$

a continuous homomorphism. Let $V = V(\rho)$.

The Zariski tangent A-module $t_{\mathcal{D},\rho}$ fits into a cartesian square of A-modules

$$\begin{array}{ccc} t_{\mathcal{D},\rho} & \longrightarrow & H^1(G_{K,S}, \mathrm{End}(V)) \\ \downarrow & & \downarrow \\ \bigoplus_{\lambda \in S} H^1_{\mathcal{D}_\lambda}(G_{K_\lambda}, \mathrm{End}(V)) & \longrightarrow & \bigoplus_{\lambda \in S} H^1(G_{K_\lambda}, \mathrm{End}(V)). \end{array} \tag{7}$$

where the horizontal homomorphisms are the natural injections coming from the Proposition and surrounding discussion of §21, and the right-hand vertical homomorphism is induced from restriction of cohomology.

Proof. This again comes from a straightforward reduction of the definitions involved, together with the Proposition in §21 and its surrounding discussion.

If $\mathcal{D}$ is the deformation condition attached to a Global deformation problem with fixed determinant for $G_{K,S}$ then we have a very similar diagram as (7), but with one minor change, namely:

Proposition 3. *We have a cartesian diagram*

$$\begin{array}{ccc} t_{\mathcal{D},\rho} & \longrightarrow & H^1(G_{K,S}, \mathrm{End}(V)^\circ) \\ \downarrow & & \downarrow \\ \bigoplus_{\lambda \in S} H^1_{\mathcal{D}_\lambda}(G_{K_\lambda}, \mathrm{End}(V)) & \longrightarrow & \bigoplus_{\lambda \in S} H^1(G_{K_\lambda}, \mathrm{End}(V)). \end{array} \tag{8}$$

The "cartesian-ness" of diagram (7) is nothing more than the following "Selmer-like" description of the sub-A-module $H^1_{\mathcal{D}}(G_{K,S}, \mathrm{End}(V))$ in $H^1(G_{K,S}, \mathrm{End}(V))$. Namely,

Proposition 2′. *Under the hypotheses of Proposition 2,*

$$H^1_{\mathcal{D}}(G_{K,S}, \mathrm{End}(V))$$

consists of those cohomology classes in $H^1_{\mathcal{D}}(G_{K,S}, \mathrm{End}(V))$ *which, when restricted to* G_{K_λ}, *land in the submodule*

$$H^1_{\mathcal{D}_\lambda}(G_{K_\lambda}, \mathrm{End}(V)) \subset H^1(G_{K_\lambda}, \mathrm{End}(V))$$

for each $\lambda \in S$.

There is a similar paraphrase of Proposition 3.

In the special case when $\overline{\rho}$ is absolutely irreducible, the deformation problem for $\mathcal{D}$ is then representable and we have its corresponding universal deformation ring (Λ-algebra) $R_{\mathcal{D}}$. We may then use the proposition of §17 to give us an alternate description of $t_{\mathcal{D}}, \rho$ in terms of the module of Kähler differentials of the Λ-algebra $R_{\mathcal{D}}$. We recall this explicitly in the next proposition.

Proposition 4. *Under the hypotheses of Proposition 1, assume further that the residual representation* $\overline{\rho}$, *of which* ρ *is a lifting, is absolutely irreducible. We have natural A-module isomorphisms*

$$\mathrm{Hom}_A(\Omega_{R_{\mathcal{D}/\Lambda}} \hat{\otimes}_{R_{\mathcal{D}}} A, A) \cong t_{\mathcal{D},\rho} \cong H^1_{\mathcal{D}}(G_{K,S}, \mathrm{End}_A(V)).$$

We have a similar statement under the hypotheses of Proposition 3.

Proposition 4 has a number of slightly variant forms which come up and we give two of these in the next sections.

§27. Passage to the limit. Now let A be a coefficient-Λ-algebra which is p-torsion-free. Let $\rho : G_{K,S} \to \mathrm{GL}_N(A)$ be a continuous homomorphism whose associated residual representation is absolutely irreducible. Let

$$\rho_n : G_{K,S} \to \mathrm{GL}_N(A/p^n A)$$

be the "reduction mod p^n" of ρ. Denote $A/p^n A$ by A_n, for short. Assume that we have a Global Galois deformation problem $\mathcal{D}$, as in Proposition 4 of the previous section, where the ρ_n's are all of type $\mathcal{D}$. There is a completely analogous treatment of the case of Global Galois deformation problems with fixed determinant (obtained by replacing $\mathrm{End}(V)$'s by $\mathrm{End}^\circ(V)$'s) which we leave to the reader. By Proposition 4, we obtain commutative diagrams with the horizontal mappings isomorphisms, and the vertical mappings inclusions,

$$\begin{array}{ccc} \mathrm{Hom}_A(\Omega_{R_{\mathcal{D}/\Lambda}} \hat{\otimes}_{R_{\mathcal{D}}} A, A_n) & \cong & H^1_{\mathcal{D}}(G_{K,S}, \mathrm{End}_{A_n}(V_n)) \\ \subset \downarrow & & \subset \downarrow \\ \mathrm{Hom}_A(\Omega_{R/\Lambda} \hat{\otimes}_R A, A_n) & \cong & H^1(G_{K,S}, \mathrm{End}_{A_n}(V_n)). \end{array}$$

for each n.

Now consider the directed system of A-modules,

$$\cdots \to A_n \xrightarrow{i_n} A_{n+1} \to \cdots$$

where i_n is the mapping induced from multiplication by p viewed as endomorphism of the A-module A_{n+1}. The mappings i_n are injections because A is p-torsion-free. Consider also the directed system of $G_{K,S}$-modules

$$\cdots \to \mathrm{End}_{A_n}(V_n) \xrightarrow{j_n} \mathrm{End}_{A_{n+1}}(V_{n+1}) \to \cdots$$

where the j_n associates to an endomorphism

$$\varphi_n : V_n \to V_n$$

the endomorphism φ_{n+1} which is the composition

$$V_{n+1} \to V_n \xrightarrow{\varphi_n} V_n \xrightarrow{p} V_{n+1}$$

the unlabeled mapping being the natural projection. The direct limit of these systems are given, respectively, by

$$\varinjlim_{n\to\infty} A_n = A \otimes_{\mathbf{Z}_p} \mathbf{Q}_p/\mathbf{Z}_p$$

and

$$\varinjlim_{n\to\infty} \mathrm{End}_{A_n}(V_n) = \mathrm{End}_A(V) \otimes_{\mathbf{Z}_p} \mathbf{Q}_p/\mathbf{Z}_p.$$

We have

Proposition. *The diagram*

$$\begin{array}{ccc} \mathrm{Hom}_A(\Omega_{R/\Lambda} \hat{\otimes}_R A, A_n) & \cong & H^1(G_{K,S}, \mathrm{End}_{A_n}(V_n)) \\ {\scriptstyle i_n}\downarrow & & {\scriptstyle j_n}\downarrow \\ \mathrm{Hom}_A(\Omega_{R/\Lambda} \hat{\otimes}_R A, A_{n+1}) & \cong & H^1(G_{K,S}, \mathrm{End}_{A_{n+1}}(V_{n+1})). \end{array}$$

is commutative for each n, where the horizontal isomorphisms are those provided by Proposition 4 of §26.

Proof. One may, of course, just grind this out. But I find it useful, in thinking about this compatibility, to introduce a subring of $A_{n+1}[\epsilon]$, intermediate between $A_n[\epsilon]$ and $A_{n+1}[\epsilon]$ (call it $A_{n+1}[\epsilon]'$) given by

$$A_{n+1}[\epsilon]' = A_{n+1} \oplus \epsilon \cdot pA_{n+1} = A_{n+1} \oplus \epsilon \cdot A_n.$$

We have a little diagram of Λ-algebras

$$\begin{array}{ccccc} A_{n+1}[\epsilon]' & \longrightarrow & A_n[\epsilon] & \longrightarrow & 0 \\ \subset\downarrow & & & & \\ A_{n+1}[\epsilon]. & & & & \end{array}$$

Now if

$$\begin{array}{ccccc} \tilde{\rho}_n: & G_{K,S} & \to & \mathrm{GL}_N(A_n[\epsilon]) & = & \mathrm{GL}_N(A_n) \ltimes \{1+\epsilon\cdot M_N(A_n)\} \\ & x & \longrightarrow & & & (\rho_n(x),\gamma(x)) \end{array}$$

represents some element in t_{ρ_n}, we have a natural lifting, call it $\tilde{\rho}'_n$, of $\tilde{\rho}_n$ to $A_{n+1}[\epsilon]'$ defined in terms of the semi-direct product decomposition

$$\mathrm{GL}_N(A_{n+1}[\epsilon]) = \mathrm{GL}_N(A_{n+1})\{1+\epsilon\cdot M_N(A_n)\}$$

by the formula $\tilde{\rho}'_n(x) = (\rho_{n+1}(x), \gamma x))$ for $x \in G_{K,S}$.

The imbedding of $A_{n+1}[\epsilon]'$ in $A_{n+1}[\epsilon]$ allows us to view $\tilde{\rho}'_n$ as element, now, of $t_{\rho_{n+1}}$. This defines a mapping of A-modules, $k_n : t_{\rho_n} \to t_{\rho_{n+1}}$ and a somewhat more direct check, using the definitions involved, shows that each of the two diagrams,

$$\begin{array}{ccc} \mathrm{Hom}_A(\Omega_{R/\Lambda}\hat{\otimes}_R A, A_n) & \cong & t_{\rho_n} \\ {\scriptstyle i_n}\downarrow & & {\scriptstyle k_n}\downarrow \\ \mathrm{Hom}_A(\Omega_{R/\Lambda}\hat{\otimes}_R A, A_{n+1}) & \cong & t_{\rho_{n+1}} \end{array}$$

and

$$\begin{array}{ccc} t_{\rho_n} & \cong & H^1(G_{K,S}, \mathrm{End}_{A_n}(V_n)) \\ {\scriptstyle k_n}\downarrow & & {\scriptstyle j_n}\downarrow \\ t_{\rho_{n+1}} & \cong & H^1(G_{K,S}, \mathrm{End}_{A_{n+1}}(V_{n+1})). \end{array} \tag{10}$$

are commutative.

Since cohomology commutes with direct limits (as our modules are discrete Galois modules) we have that $H^1(G_{K,S}, \mathrm{End}_A(V)\otimes_{\mathbf{Z}_p}\mathbf{Q}_p/\mathbf{Z}_p)$ is the direct limit of the right-hand vertical direct system in (10) and we denote by

$$H^1_{\mathcal{D}}(G_{K,S}, \mathrm{End}_A(V)\otimes_{\mathbf{Z}_p}\mathbf{Q}_p/\mathbf{Z}_p) \subset H^1(G_{K,S}, \mathrm{End}_A(V)\otimes_{\mathbf{Z}_p}\mathbf{Q}_p/\mathbf{Z}_p)$$

the direct limit of the sub-system

$$\begin{array}{ccc} H^1_{\mathcal{D}}(G_{K,S}, \mathrm{End}_{A_n}(V_n)) & \subset & H^1(G_{K,S}, \mathrm{End}_{A_n}(V_n)) \\ {\scriptstyle j_n}\downarrow & & {\scriptstyle j_n}\downarrow \\ H^1_{\mathcal{D}}(G_{K,S}, \mathrm{End}_{A_{n+1}}(V_{n+1})) & \subset & H^1(G_{K,S}, \mathrm{End}_{A_{n+1}}(V_{n+1})). \end{array}$$

§28. The special case when $\Lambda = A$. Return to the situation of Proposition 4 of §26, and let

$$\rho : G_{K,S} \to \mathrm{GL}_N(\Lambda)$$

denote a lifting of $\overline{\rho}$, an absolutely irreducible residual representation. Assume that ρ is of type $\mathcal{D}$ (for $\mathcal{D}$ a given Global Galois deformation problem). From Proposition 4 of §26, we have isomorphisms

$$\mathrm{Hom}_\Lambda(\Omega_{R_\mathcal{D}/\Lambda} \otimes_{R_\mathcal{D}} \Lambda, \Lambda/J) \cong H^1{}_\mathcal{D}(G_{K,S}, \mathrm{End}_\Lambda(V) \otimes \Lambda/J),$$

for any ideal $J \subset \Lambda$ (by taking $A = \Lambda/J$).

The classifying mapping $R_\mathcal{D} \to \Lambda$ for the deformation ρ of $\overline{\rho}$ to Λ, being a homomorphism of Λ-algebras, is surjective. Denote its kernel by I, so that we have the canonical splitting as $R_\mathcal{D}$-modules,

$$R_\mathcal{D} = \Lambda \oplus I, \tag{13}$$

and there is a natural isomorphism of Λ-modules

$$I/I^2 \cong \Omega_{R_\mathcal{D}/\Lambda} \otimes_{R_\mathcal{D}} \Lambda.$$

We leave the proof of this as an exercise to the reader with the hint that the universal property of Kähler differentials may be the easiest way to see this. We then may paraphrase Proposition 4 of §26 as giving the isomorphism of Λ-modules

$$\mathrm{Hom}_\Lambda(I/I^2, \Lambda/J) \cong H^1_\mathcal{D}(G_{K,S}, \mathrm{End}_\Lambda(V) \otimes \Lambda/J)$$

for any ideal $J \subset \Lambda$, and we may similarly paraphrase Corollary 1 of §27 as giving, under the condition that Λ is p-torsion-free, the isomorphism of Λ-modules

$$H_\Lambda(I/I^2, \Lambda \otimes_{\mathbf{Z}_p} \mathbf{Q}_p/\mathbf{Z}_p) \cong H^1_\mathcal{D}(G_{K,S}, \mathrm{End}_\Lambda(V) \otimes_{\mathbf{Z}_p} \mathbf{Q}_p/\mathbf{Z}_p).$$

§29. A bestiary of local Galois deformation conditions. Let K_λ be a local field of characteristic ℓ. It remains to describe some useful local deformation conditions $\mathcal{D}_\lambda$. For each of these conditions, we must **(a)** check that they are indeed "deformation conditions" in the technical sense described above and **(b)** describe the sub A-modules

$$H^1_{\mathcal{D}_\lambda}(G_{K_\lambda}, \mathrm{End}_A(V)) \subset H^1(G_{K_\lambda}, \mathrm{End}_A(V))$$

concretely. We devote this section to the study of

Minimal ramification conditions for $\ell \neq p$ (in degree 2).
Let $\overline{\rho} : G_{K_\lambda} \to \mathrm{GL}_2(k)$ be a residual representation, with k a finite field of characteristic $p \neq \ell$.

Suppose either (**A**) or (**B**) below.
(**A**) The image of the inertia subgroup $I \subset G_{K_\lambda}$ under $\overline{\rho}$ is nontrivial, and is contained in a subgroup of $\mathrm{GL}_2(k)$ which is conjugate to

$$\begin{pmatrix} 1 & * \\ 0 & 1 \end{pmatrix}.$$

In this case say that a deformation ρ of $\overline{\rho}$ to a coefficient-ring A with residue field k is **minimally ramified** if the image under ρ of I is contained in a subgroup of $\mathrm{GL}_2(A)$ which is conjugate to

$$\begin{pmatrix} 1 & * \\ 0 & 1 \end{pmatrix}.$$

In this case, the action of the inertia group I factors through the "tame quotient," and more specifically, through the pro-p-completion of I, which is a free pro-p-group on one generator; cf. [Se3], §1.

Remark. To say that ρ is minimally ramified in the sense of (**A**) is equivalent to saying that we may choose a suitable topological generator γ of the pro-p-completion of I and we may modify ρ within its strict deformation class relative to $\overline{\rho}$ so that the restriction of ρ to the element γ is given by the matrix

$$\begin{pmatrix} 1 & 1 \\ 0 & 1 \end{pmatrix}.$$

Next, suppose that:
(**B**) The image of I under $\overline{\rho}$ is nonzero, and is contained in a subgroup of $\mathrm{GL}_2(k)$ which is conjugate to

$$\begin{pmatrix} * & 0 \\ 0 & 1 \end{pmatrix}.$$

In this case, say that a deformation ρ of $\overline{\rho}$ to a coefficient-ring A with residue field k is **minimally ramified** if the image under ρ of I is finite, and of order prime to p (equivalently: if $\rho(I)$ if finite and has the same order as $\overline{\rho}(I)$).

Note. Except for the conditions on the determinant, these conditions are the types A and B respectively on p. 458 of [W].

Proposition. *Fix Λ a coefficient-ring with residue field k a finite field of characteristic $p \neq \ell$. For A any coefficient-Λ-algebra, let $\mathcal{D}$ be the condition on a representation*

$$\rho : G_{K_\lambda} \to \mathrm{GL}_2(A)$$

that its associated residual representation

$$\bar{\rho} : G_{K_\lambda} \to \mathrm{GL}_2(k)$$

be minimally ramified, either in the sense of (**A**) *or of* (**B**) *above, and that* ρ *itself also be (correspondingly) minimally ramified. Then* $\mathcal{D}$ *is a deformation condition, and the sub-*A*-module*

$$H^1_{\mathcal{D}}(G_{K_\lambda}, \mathrm{End}_A(V)) \subset H^1(G_{K_\lambda}, \mathrm{End}_A(V))$$

is given by

$$H^1(G_{k_\lambda}, \mathrm{End}_A(V)^I) \subset H^1(G_{K_\lambda}, \mathrm{End}_A(V))$$

where k_λ *is the residue field of* K_λ*, where superscript* I *means the sub* A*-module of elements fixed by* I*, and where the inclusion above is the natural one.*

Proof. Let us first show that, in either of the two cases (**A**) or (**B**) the condition of being minimally ramified is a "deformation condition" in the sense that it satisfies Properties 1–3 of §23. This is immediate in case (**B**); so assume that we are in case (**A**). The following Lemma will be useful.

Lemma 1. *Let* A *be a commutative noetherian local ring with residue field* k*. Let* V *be a free* A*-module of rank two over* A*, and* $\gamma : V \to V$ *an* A*-linear homomorphism. These conditions are equivalent.*

(i) *There is an* A*-basis of* V *with respect to which the action of* γ *is given by the matrix:*

$$\begin{bmatrix} 1 & 1 \\ 0 & 1 \end{bmatrix}.$$

(ii) *The image of* $\gamma - 1 : V \to V$ *is equal to* $\ker(\gamma - 1)$*, and is a free* A*-module of rank 1, sitting as a direct-summand in* V*.*

(iii) *We have* $(\gamma - 1)^2 = 0$*, and* $\gamma \otimes k$ *is not the identity in* $V \otimes_A k$*.*

Proof. (i) implies (ii) which implies (iii). Now assume (iii), and let $\nu = \gamma - 1$, so that $\nu^2 = 0$ on V. Consider the mapping $\nu : V \to V$ which we break up as

$$V \to \nu \cdot V \subset V[\nu] \subset V. \tag{1}$$

Since $\nu \otimes_A k : V \otimes_A k \to V \otimes_A k$ is nonzero, and since, by (1) $\nu \otimes_A k$ factors through $V[\nu] \otimes_A k \to V \otimes_A k$, we have that the inclusion $V[\nu] \subset V$ induces a nontrivial homomorphism $V[\nu] \otimes_A k \to V \otimes_A k$. Therefore we can find an element x in $V[\nu] \subset V$ which reduces nontrivially in $V \otimes_A k$. By Nakayama's lemma (and the fact that V is free of rank 2 over A) we can find a element y in V such that x, y is a free A-basis for V. The matrix for ν in terms of the basis x, y is then

$$\begin{bmatrix} 0 & u \\ 0 & v \end{bmatrix}$$

where $u \cdot v = 0, v^2 = 0$, and u and v are not both in the maximal ideal of A. It follows that u is a unit, and consequently $v = 0$. Changing x to $u \cdot x$, we have found a basis for which the matrix for γ is:

$$\begin{bmatrix} 1 & 1 \\ 0 & 1 \end{bmatrix}$$

Lemma 1 is helpful in showing that the condition $\mathcal{D}$ of being "minimally ramified" is a deformation condition in the sense that it satisfies properties 1–3 of §23. First, by the remark just after the definition of case (**A**) it is clear that the condition of minimal ramification is functorial in couples (A, V), i.e., that Property 1 holds.

Proof of Property 2. Let A, B, C be artinian coefficient-Λ-algebras with coefficient-Λ-algebra homomorphisms $\alpha : A \to C$, and $\beta : B \to C$. Let ρ be a lifting of $\overline{\rho}$ to $A \times_C B$.

Clearly, if ρ satisfies condition (**A**) then so do its projections to A and to B. We must show the converse; i.e., that ρ satisfies condition (**A**) if its projections to A and B do. This boils down to showing that if an endomorphism γ (of a free rank two $A \times_C B$-module) satisfies the equivalent conditions of Lemma 1 when tensored with A and with B, then it satisfies those conditions even before such tensoring. But this is evident when you choose to look at condition (iii) of Lemma 1.

Proof of Property 3. We must show that if $A_0 \to A$ is an injective homomorphism of coefficient-Λ-algebras and if ρ_0 is a deformation of $\overline{\rho}$ to A_0 which "becomes" minimally ramified when you extend its scalars to A, then ρ_0 is already minimally ramified.

Let H_0 be $A_0 \times A_0$ viewed as A_0-module with G_{K_λ}-action via ρ_0 and $H = A \times_{A_0} H_0$ with its induced G_{K_λ}-action. Let γ_0 be the A_0-linear endomorphism of H_0 obtained by the action of a topological generator of the pro-p-completion of the inertia group $I \subset G_{K_\lambda}$ and let γ be the induced A-linear endomorphism of H. By assumption, we have that γ satisfies the three equivalent conditions of Lemma 1, and we choose to concentrate, again, on condition (iii). It follows that γ_0 also satisfies this condition because $H \otimes k \cong H_0 \otimes k$.

We now must establish the last assertion of our proposition. Let ρ be a minimally ramified lifting of $\overline{\rho}$ to the coefficient-Λ-algebra A, and denote by ρ_0 the canonical lifting of ρ to $A[\epsilon]$ (i.e., ρ_0 is the composition of ρ with the homomorphism $\mathrm{GL}_2(A) \to \mathrm{GL}_2(A[\epsilon])$ induced from the natural injection $A \to A[\epsilon]$). For any lifting ρ' of ρ to $A[\epsilon]$, we must show that ρ' is minimally ramified if and only if the "difference cocycle" $c_{\rho'}$ of §18 determines a cohomology class in the sub-A-module

$$H^1(G_{k_\lambda}, \mathrm{End}_A(V)^I) \subset H^1(G_{K_\lambda}, \mathrm{End}_A(V)),$$

or equivalently, if and only if the restriction of c'_ρ to the inertia subgroup I is cohomologous to zero. We do this in detail in case (**A**), the argument being more direct in case (**B**). Assume, then, that ρ is minimally ramified in the sense of (**A**) above.

Lemma 2. *ρ' is minimally ramified if and only if ρ_0 and ρ', when restricted to I, are in the same strict equivalence class relative to ρ.*

Proof. Let γ_0 and γ' denote the image of a topological generator of the pro-p-completion of I in $\mathrm{GL}_2(A[\epsilon])$ under the homomorphisms ρ_0 and ρ' respectively. Our assumptions give us that, after a suitable choice of topological generator and basis of $A \times A$ we may assume that γ_0 is the matrix

$$\begin{bmatrix} 1 & 1 \\ 0 & 1 \end{bmatrix},$$

that $\gamma' = \gamma_0 + \epsilon \cdot m$ for a matrix $m \in M_2(A)$, and that $(\gamma' - 1)^2 = 0$. These conditions give us, by straightforward calculation, that the matrix m is of the following form

$$m = \begin{bmatrix} a & b \\ 0 & -a \end{bmatrix}$$

and therefore γ' is the conjugate of γ_0 by the matrix $1 + \epsilon \cdot n$ where

$$n = \begin{bmatrix} b & 0 \\ -a & 0 \end{bmatrix}.$$

§30. The condition of being "ordinary." We keep to the local context and degree two representations, i.e., as in the previous section, we shall be considering residual representations

$$\overline{\rho} : G_{K_\lambda} \to \mathrm{GL}_2(k)$$

where K_λ is a local field, but now we suppose that ℓ, the residual characteristic of K_λ is equal to p, the characteristic of the finite field k. As before, denote by I the inertia subgroup of G_{K_λ}.

We say that $\overline{\rho}$ is "**ordinary and ramified**" if it is equivalent to a representation of the form

$$g \mapsto \begin{pmatrix} \overline{\chi}_1(g) & * \\ 0 & \overline{\chi}_2(g) \end{pmatrix}$$

where $\overline{\chi}_1$ is an unramified character and $\overline{\chi}_2$ is ramified.

Note. The condition of "ordinary-ness" found in [W] p. 457 also allows representations such as above, but which are unramified; but in that case, the deformation problem is posed differently.

Say that a deformation ρ of an ordinary and ramified residual representation $\overline{\rho}$ to a coefficient-Λ-algebra A is **an ordinary deformation** if it is equivalent to a representation of the form

$$g \mapsto \begin{pmatrix} \chi_1(g) & * \\ 0 & \chi_2(g) \end{pmatrix}$$

for characters χ_1 and $\chi_2 : G_{K_\lambda} \to A^*$, with χ_1 unramified.

Our discussion of "ordinariness" will use almost nothing concerning the nature of the local Galois group G_{K_λ} and its inertia subgroup (beyond the fact that the inertia subgroup is normal), and in fact will also apply mutatis mutandis to the case (**B**) treated in the previous section. To emphasize this, let G be any profinite group, and $I \subset G$ any closed, normal subgroup.

Proposition 1. *Let $\rho : G \to \mathrm{GL}_2(A)$ be a representation. These are equivalent:*
(i) *The representation ρ is equivalent to a representation of the form*

$$g \mapsto \begin{pmatrix} \chi_1(g) & * \\ 0 & \chi_2(g) \end{pmatrix}$$

for characters χ_1 and $\chi_2 : G \to A^*$ such that χ_1 trivial on I and the residual character $\overline{\chi}_2 : G \to k^*$ is nontrivial on I.
(ii) *The representation A-module $V(\rho) = V$ admits an I-stable filtration*

$$0 \to V^I \to V \to W \to 0$$

which splits noncanonically as a sequence of A-modules, where V^I is the A-module of I-invariant elements in V, where both V^I and W are free A-modules of rank 1, and finally where the I-representation space $\overline{W} = W \otimes_A k$ is not the trivial representation.
(iii) *The sub-A-module $V^I \subset V$ of I-invariant elements is free over A of rank 1; the natural mapping $V^I \to \overline{V}^I$ is surjective, where $\overline{V} = V \otimes_A k$, and $\overline{V}^I$ denotes the k-vector space of I-invariant elements in $\overline{V}$; the action of I on the quotient $\overline{W} = \overline{V}/\overline{V}^I$ is not trivial.*

Note. If these conditions hold we will say that the representation is I-**ordinary**, or if I is understood, we say that it is **ordinary**.

Proof. It is evident that (i) implies (ii) and that (ii) implies (iii). It suffices, then, to show that (iii) implies (i). Suppose (iii). Let α be an A-basis of the (free, rank one) A-module V^I and let β be any element in V whose projection to $\overline{V}$ does not lie in $\overline{V}^I$. Let $W = A \otimes A$ denote the free A-module of rank two, and consider the homomorphism $\varphi : W \to V$ which sends (x, y) to $x \cdot \alpha + y \cdot \beta$. Since φ projects surjectively to $\overline{V}$, by Nakayama's Lemma, φ is surjective. Since V is free over A, the A-homomorphism φ admits a right-inverse $\psi : V \to W$, giving that $W \cong V \oplus \ker(\varphi)$. But since

W is free of rank two over A, counting dimensions of $W \otimes_A k$ gives us that $\ker(\varphi) \otimes_A k$ vanishes, and therefore $\ker(\varphi)$, also vanishes, by Nakayama's Lemma. It follows that φ is an isomorphism. Writing ρ in terms of our choice of basis $W = A \otimes A$ we have that the image of I under ρ is contained in the semi-Borel subgroup

$$\begin{pmatrix} 1 & * \\ 0 & * \end{pmatrix}$$

Since I is normal in G, the image of ρ stabilizes V^I and therefore ρ is ordinary in the sense defined above.

Remark. It follows (e.g., from (ii)) that if ρ is ordinary and $V = V(\rho)$, then no nontrivial $A[I]$-subquotient of the (free, rank one) A-module $W = V/V^I$ has trivial I-action.

Given an I-ordinary representation $\rho : G \to \mathrm{GL}_2(A)$ we define

$$\mathrm{End}(V)_{\mathrm{ord}} \subset \mathrm{End}(V) = \mathrm{End}_A(V)$$

to be the rank two free sub A-module

$$\mathrm{Hom}_A(V/V^I, V) \subset \mathrm{Hom}_A(V, V) = \mathrm{End}_A(V).$$

Since the action of G on V stabilizes V^I the sub-module $\mathrm{End}(V)_{\mathrm{ord}} \subset \mathrm{End}(V)$ is stabilized under the adjoint action of G on $\mathrm{End}(V)$.

Proposition 2. *The homomorphism*

$$H^1(G, \mathrm{End}(V)_{\mathrm{ord}}) \to H^1(G, \mathrm{End}(V)) \tag{14}$$

induced by the inclusion $\mathrm{End}(V)_{\mathrm{ord}} \subset \mathrm{End}(V)$ *of* $A[G]$*-modules is an injection.*

Proof. Injectivity of (14) comes from consideration of the long exact sequence of G-cohomology coming from the exact sequence

$$0 \to \mathrm{End}_A(V)_{\mathrm{ord}} \to \mathrm{End}_A(V) \xrightarrow{j} \mathrm{Hom}_A(V^I, V) \to 0$$

once we show that the homomorphism j induces a surjection on G-invariant elements. But the A-module of G-invariant elements in $\mathrm{Hom}_A(V^I, V)$ is generated by the natural injection $V^I \to V$ and j maps the identity in $\mathrm{End}(V)$ to this natural injection.

Definition. The "**I-ordinary cohomology group,**"

$$H^1_{\mathrm{ord}}(G, \mathrm{End}_A(V)) \subset H^1(G, \mathrm{End}_A(V))$$

is the image of the injection (14). Thus

$$H^1_{\mathrm{ord}}(G, \mathrm{End}_A(V)) \cong H^1_{\mathrm{ord}}(G, \mathrm{End}(V)_{\mathrm{ord}})$$

Proposition 3. *Fix a profinite group G and a closed normal subgroup $I \subset G$. If $\mathcal{D}$ is the class of I-ordinary representations of G, then $\mathcal{D}$ is a "deformation condition" in the sense of §23, and if $\rho : G \to \mathrm{GL}_2(A)$ is ordinary, then the sub-A-module*

$$H^1_{\mathcal{D}}(G, \mathrm{End}_A(V) \subset H^1(G, \mathrm{End}_A(V))$$

is given by the I-ordinary cohomology submodule,

$$H^1_{\mathrm{ord}}(G, \mathrm{End}(V)) \subset H^1(G, \mathrm{End}_A(V))$$

Proof. Fixing the normal subgroup I, we shall refer to I-ordinary representations as simply "ordinary." We first must show that $\mathcal{D}$ satisfies Properties 1–3 of §23. Property 1 is evident from the definition of ordinariness. For the remaining two properties, it is convenient to put subscripts on our notation to indicate the coefficient-Λ-algebra we are dealing with: so if

$$\rho_A : G \to \mathrm{GL}_2(A)$$

is a representation and $V_A := V(\rho_A)$ its representation space, and if $A \to C$ is a homomorphism of coefficient-Λ-algebras, let ρ_C, V_C denote the induced representation and representation space. In particular, $V_C = V_A \otimes_A C$. We note that in this circumstance, we have:

Lemma 1.
(a) *Let ρ_A be ordinary, and $A \to C$ a homomorphism of coefficient-Λ-algebras. The natural homomorphism of C-modules*

$$(V_A)^I \otimes_A C \to (V_C)^I \tag{15}$$

is an isomorphism.
(b) *If we are given a diagram*

$$\begin{array}{ccc} A & & B \\ \alpha \searrow & & \swarrow \beta \\ & C & \end{array}$$

in $\mathcal{C}_\Lambda$ and an ordinary representation $\rho : G \to \mathrm{GL}_2(A \otimes_C B)$ with representation space V, we have natural isomorphisms

$$V \cong V_A \otimes_{V_C} V_B \tag{16}$$

and

$$V^I \cong (V_A)^I \times_{(V_C)^I} (V_B)^I. \tag{17}$$

Proof. (a) This follows directly from consideration of the (noncanonically) split short exact sequence of A-modules occurring in condition (ii) of Proposition 1.
(b) This is immediate.

Proof of Property 2. If ρ_A and ρ_B are ordinary, then $(V_A)^I$ and $(V_B)^I$ are free of rank one over A and B respectively, and therefore by part (a) of Lemma 1, $(V_C)^I$ is also free of rank one over C. Moreover, the mappings $(V_A)^I \to (V_C)^I$ and $(V_B)^I \to (V_C)^I$ come by tensoring $(V_A)^I$ and $(V_B)^I$ with $\alpha : A \to C$ and $\beta : B \to C$ respectively. By (17) we then have that V^I is free of rank one over $A \times_C B$, and by (15) we have that the natural mapping $V^I \otimes k \to \overline{V}^I$ is an isomorphism (the tensor product with k being taken over $A \times_C B$). Therefore V satisfies condition (iii) in Proposition 1, and $\rho : G \to \mathrm{GL}_2(A \times_C B)$ is ordinary.

Proof of Property 3. Now suppose given a representation $\rho_A : G \to \mathrm{GL}_2(A)$ and an injection of coefficient-Λ-algebras $i : A \to C$ such that ρ_C is ordinary.

We must show that ρ_A is ordinary. We use the notational conventions V_A and V_C for the corresponding representation spaces, and view V_A as sub-$A[I]$-module of V_C. We have that $W_C := V_C/(V_C)^I$ is a free C-module of rank one, and its I-action is via a character $\chi : I \to C^*$ which takes its values in A^* since χ may be identified with the determinant character of ρ_A. We have the further information that the residual character $\overline{\chi} : I \to k^*$ is nontrivial. Since i is an injection of coefficient-Λ-algebras, i induces an isomorphism of residue fields of A and C, and gives us a natural identification $\overline{V}_A = \overline{V}_C$.

Since

$$(V_A)^I = V_A \cap (V_C)^I$$

the natural mapping

$$W_A = V_A/(V_A)^I \to W_C = V_C/(V_C)^I \tag{18}$$

is injective. We have the exact sequence

$$(V_A)^I \otimes_A k \xrightarrow{j} \overline{V}_A \to W_A \otimes_A k \to 0. \tag{19}$$

From injectivity of (18) and the fact that W_A does not vanish, the action of I on W_A is via the character χ. In particular, since $\overline{\chi}$ is nontrivial, no nontrivial subquotient of $W_A \otimes_A k$ has trivial I-action. Since $\overline{V}_A = \overline{V}_C$, we then see from exactness of (19) that the k-dimension of $W_A \otimes_A k$ is equal to one. Now let us compare (19) with the corresponding exact sequence for C:

$$0 \to (V_C)^I \otimes_C k \to \overline{V}_C \to W_C \otimes_C k \to 0. \tag{20}$$

We get a commutative diagram of k-vector spaces

$$\begin{array}{ccccc} \overline{V}_A & \longrightarrow & W_A \otimes_A k & \longrightarrow & 0 \\ \| \downarrow & & \downarrow h & & \\ \overline{V}_C & \longrightarrow & W_C \otimes_C k & \longrightarrow & 0 \end{array}$$

from which we see that h is surjective (hence also injective since both domain and range are k-vector spaces of dimension 1). It follows from Nakayama's Lemma that $W_A \otimes_A C \to W_C$ is surjective (hence is an isomorphism since W_A is a cyclic A-module, and W_C is free as a C-module). It follows that the annihilator ideal of W_A in A is trivial. Therefore the cyclic A-module W_A is free over A of rank 1. We then get that the exact sequence of $A[I]$-modules

$$0 \to (V_A)^I \to V_A \to W_A \to 0$$

splits (noncanonically) as an exact sequence of A-modules. Since V_A is free of rank 2 over A, an application of Nakayama's Lemma then gives us that $(V_A)^I$ is a cyclic A-module, possessing trivial annihilator by (15) since $(V_C)^I$ is free of positive rank over C. That is, $(V_A)^I$ is also free of rank one over A, and we have shown all we need to show to conclude that ρ satisfies condition (iii) of Proposition 1.

To conclude the proof of Proposition 3 we must consider the "difference cocycle" c'_ρ of a lifting ρ' of ρ to $A[\epsilon]$, as in §18 and, by a direct calculation, show that ρ' is I-ordinary if and only if it takes its values in the sub A-module $\mathrm{Hom}_A(V/V^I, V) \subset \mathrm{End}_A(V)$. We leave this as an exercise.

§31. The condition of being "finite flat." Here we consider the case of $K_\lambda = \mathbf{Q}_\ell$, the field of ℓ-adic numbers, with $\ell \neq 2$. Let $\overline{\mathbf{Q}}_\ell$ be a choice of algebraic closure of $\mathbf{Q}_\ell$. As usual, $G_{\mathbf{Q}_\ell}$ denotes $\mathrm{Gal}(\overline{\mathbf{Q}}_\ell/\mathbf{Q}_\ell)$. If A is a coefficient-Λ-algebra, and

$$\rho : G_{\mathbf{Q}_\ell} \to \mathrm{GL}_N(A)$$

is a continuous representation, we say that ρ is **finite flat** (we probably should rather call it **pro-finite flat**) if for every artinian quotient A_0 of A the induced representation $\rho_0 : G_{\mathbf{Q}_\ell} \to \mathrm{GL}_N(A_0)$ has the property that its representation space $V(\rho_0)$, viewed as finite abelian group with $G_{\mathbf{Q}_\ell}$-action, is the $\mathbf{Q}_\ell$-Galois module of $\overline{\mathbf{Q}}_\ell$-rational points of some finite flat group scheme over $\mathrm{Spec}(\mathbf{Z}_\ell)$. If $\mathcal{D}$ is the condition of being "finite flat," then $\mathcal{D}$ satisfies Ramakrishna's "categorical conditions" (cf. §25) and therefore is indeed a "deformation condition." Also by Proposition 2 of §25, if A is artinian and ρ is finite flat coming from the finite flat group scheme M over $\mathrm{Spec}(\mathbf{Z}_\ell)$ we have

$$H^1_{\mathcal{D}}(G_{\mathbf{Q}_\ell}, \mathrm{End}(V)) = \mathrm{Ext}_{\mathrm{Spec}(\mathbf{Z}_\ell)}(M, M)$$

where V, the representation space of ρ, is the $G_{\mathbf{Q}_\ell}$-module $M(\overline{\mathbf{Q}}_\ell)$ and where $\mathrm{Ext}_{\mathrm{Spec}(\mathbf{Z}_\ell)}(-,-)$ means Ext in the category of finite flat group schemes over $\mathrm{Spec}(\mathbf{Z}_\ell)$. In contrast, however, with the other local conditions we have considered, the calculation of $\mathrm{Ext}_{\mathrm{Spec}(\mathbf{Z}_\ell)}(M,M)$ is more difficult and at present it has only been carried out in detail, in the specific case of interest; namely for $N=2$. This calculation depends on Fontaine's articles [Fo 1], [Fo 2]; see also the article by Fontaine-Laffaille [F-L]. See [W], and for a treatment of finite flat groups over finite field extensions of $\mathbf{Q}_\ell$ with ramification index $< \ell - 1$, see B. Conrad's Princeton Ph.D thesis, and his article [Co] in this volume.

A brief glossary of terminology and notation
(listed by the section in which they make their first appearance)

§1: $G_{K,S}$, p-finiteness condition, Galois representation.
§2: coefficient-ring, coefficient-ring homomorphism.
§4: residual representation $\overline{\rho}$.
§5: deformation of ρ to a coefficient-ring A, strict equivalence class, the categories $\mathcal{C}(A)$, $\hat{\mathcal{C}}(A)$, A-augmentation, the functor D_ρ, the deformation problem for ρ.
§9: universal deformation, the universal deformation ring $R(\overline{\rho})$ attached to a residual representation $\overline{\rho}$, universal deformation space.
§11: the functors $D_{\overline{V}}$, $D_{\overline{V},\Lambda}$.
§14: the categories $\mathcal{C}(A)$, $\hat{\mathcal{C}}(A)$, pro-representable, fiber product, Cartesian diagram, Mayer-Vietoris property.
§15: Zariski tangent (k-vector) space, t_R, t_D the "Tangent Space Hypothesis" $(\mathbf{T}_k)$.
§16: The Zariski tangent A-module, the "Tangent A-module Hypothesis" $(\mathbf{T}_A)$, $t_{D,A}$.
§17: Continuous Kähler differentials, $\Omega^{\mathrm{cont}}_{R/\Lambda} = \Omega_{R/\Lambda}$, $t_{D,A} = t_{D,\rho}$.
§18: Schlessinger's conditions (**H 1**)-(**H 4**), nearly representable, pro-representable hull, smooth morphism of functors.
§19: relatively representable subfunctor.
§20: weak near representability, strong near representability.
§23: deformation condition, $F_N(\Lambda;\Pi)$, $\mathcal{D}F_N(\Lambda;\Pi)$, type $\mathcal{D}$.
§24: "determinant δ."
§25: $\mathrm{Rep}_\Lambda(\Pi)$, $\mathcal{P}F_N$.
§26: Global Galois deformation problem.
§29: minimally ramified.
§20: ordinary, I-ordinary.

REFERENCES

[Az] Azumaya, G., *On maximally central algebras*, Nagoya Math. J. **2** (1951), 119–150.

[B 1] Boston, N., *Explicit deformation of Galois representations.*, Invent. Math. **103** (1991), 181–196.

[B 2] ———, *A refinement of the Faltings-Serre method,*, Number Theory (S. David, ed.), Cambridge Univ. Press, 1995, pp. 61–68.

[B-M] Boston, N., Mazur, B., *Explicit deformations of Galois representations*, Algebraic Number Theory — in honor of K. Iwasawa (Coates, Greenberg, Mazur, Satake, eds.), Advanced Studies in Pure Mathematics **17**, Academic Press, 1989, pp. 1–21.

[B-U] Boston, N., Ullom, S., *Representations related to CM elliptic curves*, Math. Proc. Camb. Phil. Soc. **113** (1993), 71–85.

[Bourb 1] Bourbaki, N., *Algebre*, Ch. VIII, Hermann, 1958.

[Bourb 2] ———, *Algebre commutative*, Ch. I–IV, Masson, 1985.

[By] Byrne, D., *Stop Making Sense*, Video, The Talking Heads.

[Ca] Carayol, H., *Formes modulaires et représentations Galoisiennes à valeurs dans un anneau local complet,*, p-adic Monodromy and the Birch-Swinnerton-Dyer Conjecture (Mazur and Stevens, eds.), Contemporary Mathematics **165**, Amer. Math. Soc., 1994, pp. 213–237.

[C] Coleman, R., *P-adic Banach spaces and families of modular forms*, Invent. math. (to appear).

[Co] Conrad, B., *The flat deformation functor*, (this volume).

[C-R] Curtis, C., Reiner, I., *Representation theory of finite groups and associative algebras*, Wiley, 1962.

[D-D-T] Darmon, H., Diamond, F., Taylor, R., *Fermat's Last Theorem*, Current Developments in Mathematics (Bott, Jaffe, Yau, eds.), (1995), International Press., 1995, pp. 1–108.

[D-W] Doran, C., Wong, S., *The deformation theory of Galois representations*, in preparation..

[Fo 1] Fontaine, J.-M., *Groupes p-divisibles sur les corps locaux*, Astérisque, Soc. Math. de France **47–48** (1977).

[Fo 2] ———, *Groupes finis commutatifs sur les vecteurs de Witt*, C.R. Acad. Sci. Paris **280** (1975), 1423–1425.

[F-L] Fontaine, J.-M., Laffaille, Construction de représentations p-adiques, Ann. Scient. E.N.S. **15** (1982), 547–608.

[F-M] Fontaine, J.-M., Mazur, B., *Geometric Galois representations*, Proceedings of the Conference on Elliptic Curves and Modular Forms (Coates, Yau, eds.), Hong Kong, Dec. 18–21 1993, International Press., 1995, pp. 47–78.

[Fr] Frobenius, G., *Über die Primfactoren der Gruppendeterminante*, Sitz. Preuss. Akad. Berlin (1896), 1343–1382 (pp. 38–77 in Ges. Abh. III).

[G] Gouvêa, F., *Arithmetic of p-adic modular forms*, Lecture Notes in Math. **1304**, Springer-Verlag, 1988.

[G-M] Gouvêa, F., Mazur, B., *Families of modular eigenforms*, Mathematics of Computation **58** (1992), 793–805.

[Gr] Grothendieck, A., *Éléments de Géométrie Algébrique IV (Étude locale des schémas et des morphismes de schémas)*, Publ. Math. IHES **20** (1964).

[H] Hida, H., *Elementary theory of L-functions and Eisenstein series*, London Math. Soc. (1993).

[Ha] Hartshorne, R., *Algebraic Geometry*, Springer-Verlag, 1977.

[K-O] Knus, M.-A., Ojanguren, M., *Théorie de la descente et algébres d'Azumaya*, Lecture Notes in Math **389**, Springer-Verlag, 1974.

[L-M] Lubotzky, A., Magid, A., *Varieties fo representations of finitely generated groups*, Memoirs of the A.M.S. **58** (1985).

[L-S] Lenstra, H., de Smit, B., *Explicit construction of universal deformation rings*, (this volume).

[Mat] Matsumura, H., *Commutative Algebra*, W.A. Benjamin Co., 1970.

[M 1] Mazur, B., *Deforming Galois representations*, Galois groups over **Q** (Ihara, Ribet, Serre, eds.), MSRI Publications **16**, Springer-Verlag, 1989, pp. 385–437.

[M 2] ———, *Two-dimensional p-adic Galois representations unramified away from p*, Compositio Math **74** (1990), 115–133.

[M 3] ———, *An "infinite fern" in the universal deformation space of Galois representations*, Proceedings of the Journées Arithmeétiques, held in Barcelona, July 1995. (to appear).

[M T] Mazur, B., Tilouine, J., *Représentations galoisiennes, différentielles de Kähler et conjectures principales*, Publ. Math. IHES **71** (1990), 65–103.

[Ny] Nyssen, L., *Pseudo-représentations*, preprint 1994.

[P 1] Procesi, C., *Rings with Polynoimal Identities*, Dekker, 1973.

[P 2] ———, *Representations of algebras*, Israel J. Math. **19** (1974), 169–182.

[Ram] Ramakrishna, R., *On a variation of Mazur's deformation functor*, Compositio Math **87** (1993), 269–286.

[Rouq] Rouquier, R., *Caractérisation des caractéres et pseudo-caractéres*, preprint 1995..

[Sch] Schlessinger, M., *Functors on Artin rings*, Trans. A.M.S. **130** (1968), 208–222.

[Se 1] Serre, J.-P., *Corps Locaux*, 2nd ed., Hermann, 1968.

[Se 2] ———, *Représentations linéaires sur des anneaux locaux (d'aprés Carayol)*, preprint, 1995.

[Se 3] ———, *Points d'ordre fini des courbes elliptiques*, Invent. Math. **15** (1972), 259–331.

[T] Taylor, R., *Galois representations associated to Siegel modular forms of low weight*, Duke Math. J. **63** (1991), 281–332.

[T-W] Taylor, R., Wiles, A., *Ring-theoretic properties of certain Hecke algebras*, Annals of Math. **141** (1995), 553–572.

[W] Wiles, A., *Modular elliptic curves and Fermat's Last Theorem*, Annals of Math. **141** (1995), 443–551.

EXPLICIT CONSTRUCTION OF UNIVERSAL DEFORMATION RINGS

Bart de Smit and Hendrik W. Lenstra, Jr.

1. Introduction

Let G be a profinite group and let k be a field. By a *k-representation* of G we mean a finite dimensional vector space over k with the discrete topology, equipped with a continuous k-linear action of G. If V is a k-representation of G and A is a complete local ring with residue field k, then a *deformation* of V in A is an isomorphism class of continuous representations of G over A that reduce to V modulo the maximal ideal of A; precise definitions are given in Section 2. We denote by $\mathrm{Def}(V, A)$ the set of such deformations.

Let V be an absolutely irreducible k-representation of G. The object of this chapter is to give a straight-forward construction of a ring R, the *universal deformation ring*, which represents the functor $\mathrm{Def}(V, -)$. In a purely algebraic setting, without considerations of continuity, a similar construction was already given by Procesi in the seventies [9, Chap. IV, Lemma 1.7; 10]. The existence of R in the present context was deduced first by Mazur [8] with Schlessinger's criteria for pro-representability [12]. An alternative construction was given recently by Faltings (see [5] and Section 7 below).

The main result of this chapter, formulated below as Theorem (2.3), is actually a little more general than Mazur's. Following Schlessinger, Mazur works only with noetherian rings, and this forces him to assume at the outset that a certain cohomology group is finite. For our argument, the noetherian condition is a hindrance, and we find it more convenient to follow Grothendieck [6] and work with not necessarily noetherian rings that are projective limits of artinian rings. This allows us to drop Mazur's cohomological condition; it reappears only at the end, as a necessary and sufficient condition for R to be noetherian.

Our construction of R proceeds in three steps. First we let G be finite, and we consider the functor that assigns to A a certain set of homomorphisms $G \to \mathrm{Gl}_n(A)$. Proving that this functor is representable is very easy: one just defines the corresponding 'universal' ring by generators and relations. Next, we take a projective limit and obtain a similar result for arbitrary profinite G (Proposition (2.5)). To conclude the construction, we pass to the closed subring generated by the traces of the elements of G; the proof that this ring has the required properties makes use of an argument of Serre [3, Théorème 2].

It is in the last step of the construction that the absolute irreducibility of V is crucially used. In Wiles's proof of Fermat's Last Theorem the existence of deformation rings is only needed for such V. Wiles also uses the fact that such deformation rings are generated by traces [13, pp. 509–512], so the approach above is particularly suitable for Wiles's applications.

It is, however, of interest to observe that the universal deformation ring also exists when V, instead of being absolutely irreducible, satisfies the weaker condition $\mathrm{End}_{k[G]}(V) = k$. In the noetherian case this was shown by Ramakrishna [11], as a consequence of Schlessinger's criteria. The general case is proved in Section 7. Instead of taking the subring generated by the traces we pass to the subring generated by a larger collection of elements, as suggested by an argument due to Faltings [5, Section 2.6]. We do not know whether a similar result holds in Procesi's purely algebraic setting.

Following Ramakrishna [11] we indicate in Section 6 how one can impose additional conditions on the deformations to obtain "ordinary" and "flat" deformation rings.

2. Main results

We denote the maximal ideal of a local ring A by $\mathfrak{m}_A$.

(2.1) Local complete rings. Let $\mathcal{O}$ be a noetherian local ring with residue field k. We denote by $\mathcal{C}$ the category of local topological $\mathcal{O}$-algebras A that satisfy the following two conditions: the natural map $\mathcal{O} \to A/\mathfrak{m}_A$ is surjective (so that k is also the residue field of A), and the map from A to the projective limit of its discrete artinian quotients is a topological isomorphism. Equivalently, the second condition asserts that A is complete and that its topology can be given by a collection of open ideals $\mathfrak{a}$ for which $A/\mathfrak{a}$ is artinian. Morphisms in $\mathcal{C}$ are continuous $\mathcal{O}$-algebra homomorphisms.

(2.2) Deformations. Let $\mathcal{O}$ and k be as above, let A be a ring in $\mathcal{C}$, and let G be a topological group. A *representation* of G over A, or an *A-representation* of G, is a finitely generated free A-module M with a continuous A-linear action; here we give M the product topology via an A-module isomorphism $M \cong_A A^n$, a topology that is independent of the choice of the isomorphism. Two A-representations M and M' are said to be *isomorphic* if there is an $A[G]$-module isomorphism $M \xrightarrow{\sim} M'$, and we denote this by $M \cong_{A[G]} M'$.

Let V be a k-representation of G. By a *deformation* of V in A we mean an isomorphism class of A-representations W of G for which $W \otimes_A k \cong_{k[G]} V$. The set of such deformations is denoted by $\mathrm{Def}(V, A)$. A morphism $f\colon A \to A'$ in $\mathcal{C}$ gives rise to a map $f_*\colon \mathrm{Def}(V, A) \to \mathrm{Def}(V, A')$ that sends the class of a representation W over A to the class of $W \otimes_A A'$.

Throughout the paper V is a representation of a profinite group G over the residue field k (with the discrete topology) of a noetherian local ring $\mathcal{O}$, and $\mathcal{C}$ is as above.

(2.3) Theorem. *If V is absolutely irreducible then*

(1) *there are a ring R in $\mathcal{C}$ and a deformation $D \in \mathrm{Def}(V, R)$ such that for all rings A in $\mathcal{C}$ we have a bijection $\mathrm{Hom}_{\mathcal{C}}(R, A) \xrightarrow{\sim} \mathrm{Def}(V, A)$ given by $f \mapsto f_*(D)$;*

(2) *the pair (R, D) is determined up to unique $\mathcal{C}$-isomorphism by the property in* (1);
(3) *the ring R is noetherian if and only if* $\dim_k H^1(G, \mathrm{End}_k(V)) < \infty$;
(4) *if R is noetherian then the following hold: R is $\mathfrak{m}_R$-adically complete and for each A in $\mathcal{C}$ we have a well-defined bijection*

$$\mathrm{Hom}_{\mathcal{O}\text{-Alg}}(R, A) \xrightarrow{\sim} \mathrm{Def}(V, A)$$

given by $f \mapsto f_*(D)$.

Recall that V is *absolutely irreducible* if $V \otimes_k K$ is a simple $K[G]$-module for every field extension K of k. The H^1 in (3) denotes the continuous cohomology group of the discrete G-module $\mathrm{End}_k(V)$, on which the G-action is given by $(g\varphi)(v) = g\varphi(g^{-1}v)$ for $\varphi \in \mathrm{End}_k(V)$ and $v \in V$. By "$\mathrm{Hom}_{\mathcal{O}\text{-Alg}}$" we denote the set of $\mathcal{O}$-algebra homomorphisms.

Statement (2) of the theorem follows from (1) by the standard uniqueness argument for universal objects. Statement (4) will follow immediately from (1) and the following proposition.

(2.4) Proposition. *Suppose A is a noetherian ring in $\mathcal{C}$. Then the topology on A is equal to the $\mathfrak{m}_A$-adic topology, and A is $\mathfrak{m}_A$-adically complete. Furthermore, every $\mathcal{O}$-algebra homomorphism $A \to A'$ with A' in $\mathcal{C}$ is continuous.*

The proof of (2.4) and the proof of part (3) of (2.3) are postponed to Section 5. By (2.4), the category $\mathcal{C}'$ whose objects are complete noetherian local $\mathcal{O}$-algebras with residue field k and whose morphisms are $\mathcal{O}$-algebra homomorphisms is a full subcategory of $\mathcal{C}$. We will use later that a closed sub-$\mathcal{O}$-algebra A' of a ring A in $\mathcal{C}$ is again in $\mathcal{C}$, which follows from the fact that a sub-$\mathcal{O}$-algebra of an artinian ring in $\mathcal{C}$ is again an artinian ring in $\mathcal{C}$. However, if A is in $\mathcal{C}'$ then A' need not be in $\mathcal{C}'$.

We will show (1) by an explicit construction, which starts by representing an easier functor. For this we will write representations as homomorphisms to matrix groups. Let V be any k-representation of G. If one chooses a k-basis $v_1, \ldots, v_n$ for V, then the G-action on V is given by a continuous homomorphism $\overline{\rho}\colon G \to \mathrm{Gl}_n(k)$. Now let W be a representation of G over some A in $\mathcal{C}$ such that $W/\mathfrak{m}_A W = W \otimes_A k \cong_{k[G]} V$. By Nakayama's lemma elements $w_1, \ldots, w_n \in W$ such that $w_i \mapsto v_i$ form an A-basis of W. The G-action on W is then given by a continuous group homomorphism $\rho\colon G \to \mathrm{Gl}_n(A)$ such that the composite map $G \to \mathrm{Gl}_n(A) \to \mathrm{Gl}_n(k)$ is $\overline{\rho}$. We denote the set of such maps ρ by $\mathrm{CHom}_{\overline{\rho}}(G, \mathrm{Gl}_n(A))$. Here "CHom" denotes the set of continuous homomorphisms, and the subscript $\overline{\rho}$ expresses the condition that the homomorphisms considered reduce to $\overline{\rho}$ over the residue field k of A.

(2.5) Proposition. *There are a ring R_b in $\mathcal{C}$ and a map*

$$\rho_b \in \mathrm{CHom}_{\overline{\rho}}(G, \mathrm{Gl}_n(R_b))$$

such that for each A in $\mathcal{C}$ we have a bijection

$$\mathrm{Hom}_{\mathcal{C}}(R_b, A) \xrightarrow{\sim} \mathrm{CHom}_{\overline{\rho}}(G, \mathrm{Gl}_n(A))$$

that sends a $\mathcal{C}$-morphism f to the composite map

$$G \xrightarrow{\rho_b} \mathrm{Gl}_n(R_b) \xrightarrow{f} \mathrm{Gl}_n(A).$$

The pair (R_b, ρ_b) is determined up to unique isomorphism by this property.

The ring R_b will be constructed in Section 3 as a projective limit over the discrete quotients of G of complete $\mathcal{O}$-algebras that are explicitly defined by generators and relations. The map ρ_b defines a representation $W_b = R_b^n$ of G in R_b such that $W_b \otimes_{R_b} k \cong_{k[G]} V$. We now let R be the smallest closed sub-$\mathcal{O}$-algebra of R_b that contains the traces of all matrices $\rho_b(g)$ with $g \in G$. Note that R is in $\mathcal{C}$ again. The following result asserts that we can define the representation W_b of G over the subring R. We let D be the $R[G]$-isomorphism class of this R-representation.

(2.6) Proposition. *Let W be a representation of G over some ring A in $\mathcal{C}$ and let $A' \subset A$ be an inclusion of rings in $\mathcal{C}$ so that A' has the induced topology of A. Suppose that A' contains the traces of all endomorphisms of W that are given by multiplication with an element of G, and suppose that $W \otimes_A A/\mathfrak{m}_A$ is absolutely irreducible. Then there is an A'-representation W' of G such that $W' \otimes_{A'} A \cong_{A[G]} W$.*

Proposition (2.6) is a variation of results due to Serre [3, Théorème 2] and Mazur [8, Proposition 4].

Let us assume (2.6) for the moment and prove that the pair (R, D) satisfies statement (1) of the theorem. Let W be a representation of G over a ring A in $\mathcal{C}$ for which $W \otimes_A k \cong_{k[G]} V$. Choosing a basis of W as in the argument before (2.5), one can give the G-action on W by a continuous homomorphism $\rho \in \mathrm{CHom}_{\overline{\rho}}(G, \mathrm{Gl}_n(A))$. By (2.5) there is a $\mathcal{C}$-morphism $f_b\colon R_b \to A$ such that the composite map $G \xrightarrow{\rho_b} \mathrm{Gl}_n(R) \xrightarrow{f_b} \mathrm{Gl}_n(A)$ is equal to ρ. Then the restriction $f\colon R \to A$ of f_b has the property that $f_*(D)$ is the $A[G]$-isomorphism class of W.

The trace of an element of G in some representation of G depends only on the representation up to isomorphism. Given $f_*(D)$ the map f is therefore uniquely determined on the traces of $\rho_b(g)$ for all $g \in G$. But the $\mathcal{O}$-algebra generated by these traces is dense in R, and f is continuous, so f is uniquely determined. This proves the universal property (1) in (2.3) once we know (2.5) and (2.6).

3. Lifting homomorphisms to matrix groups

In this section we prove (2.5). The last statement in (2.5) follows by the usual uniqueness argument.

Suppose first that G is finite, and denote its identity element by e. We define $\mathcal{O}[G,n]$ to be the commutative $\mathcal{O}$-algebra given by

generators: X_{ij}^{g} for $g \in G$ and $1 \le i,j \le n$;

relations: $X_{ij}^{e} = \begin{cases} 1 & \text{if } i = j, \\ 0 & \text{if } i \ne j; \end{cases}$

$$X_{ij}^{gh} = \sum_{l=1}^{n} X_{il}^{g} X_{lj}^{h} \quad \text{for } g,h \in G \text{ and } 1 \le i,j \le n.$$

For example, $\mathcal{O}[G,1]$ is just the group ring of the largest abelian quotient of G over $\mathcal{O}$.

For every $\mathcal{O}$-algebra A we have a canonical bijection

$$\text{(3.1)} \qquad \mathrm{Hom}_{\mathcal{O}\text{-Alg}}(\mathcal{O}[G,n], A) \cong \mathrm{Hom}(G, \mathrm{Gl}_n(A)),$$

where an $\mathcal{O}$-algebra homomorphism $f: \mathcal{O}[G,n] \to A$ corresponds to the group homomorphism ρ_f that sends $g \in G$ to the matrix $(f(X_{ij}^{g}))_{i,j}$.

By (3.1) the homomorphism $\overline{\rho}: G \to \mathrm{Gl}_n(k)$ gives rise to an $\mathcal{O}$-algebra homomorphism $\mathcal{O}[G,n] \to k$. Its kernel is a maximal ideal, which we denote by $\mathfrak{m}_{\overline{\rho}}$. Now let R_b be the completion of $\mathcal{O}[G,n]$ at $\mathfrak{m}_{\overline{\rho}}$. Certainly R_b is noetherian and lies in $\mathcal{C}$. The canonical map $\mathcal{O}[G,n] \to R_b$ gives by (3.1) a map $\rho_b: G \to \mathrm{Gl}_n(R_b)$ such that the diagram

$$\begin{array}{ccc} G & \xrightarrow{\rho_b} & \mathrm{Gl}_n(R_b) \\ \| & & \downarrow \\ G & \xrightarrow{\overline{\rho}} & \mathrm{Gl}_n(k) \end{array}$$

commutes.

To prove that the map in (2.5) is a bijection, let A be a ring in $\mathcal{C}$ and let $\rho \in \mathrm{CHom}_{\overline{\rho}}(G, \mathrm{Gl}_n(A))$. By (3.1), there is a unique $\mathcal{O}$-algebra homomorphism $f: \mathcal{O}[G,n] \to A$ such that $\rho_f = \rho$. The fact that ρ_f reduces to $\overline{\rho}$ modulo $\mathfrak{m}_A$ implies that $f(\mathfrak{m}_{\overline{\rho}}) \subset \mathfrak{m}_A$. The topology on A is given by open ideals $\mathfrak{a}$ for which $A/\mathfrak{a}$ is artinian, and the map $\mathcal{O}[G,n] \to A \to A/\mathfrak{a}$ is continuous for the $\mathfrak{m}_{\overline{\rho}}$-adic topology on $\mathcal{O}[G,n]$ for each such $\mathfrak{a}$. We therefore obtain a continuous $\mathcal{O}$-algebra homomorphism $\hat{f}: R_b \to A$ for which the diagram

$$\begin{array}{ccc} G & \xrightarrow{\rho_b} & \mathrm{Gl}_n(R_b) \\ \| & & \downarrow \hat{f} \\ G & \xrightarrow{\rho} & \mathrm{Gl}_n(A) \end{array}$$

commutes. Since the elements $\hat{f}(X_{ij}^g)$ are determined by ρ, and the X_{ij}^g generate a dense sub-$\mathcal{O}$-algebra of R_b, the map $\hat{f}$ is uniquely determined by the conditions that it be continuous and that the diagram commute. This finishes the proof of (2.5) in the case that G is finite.

For the general case, write G as $G = \varprojlim H$, with H ranging over those discrete quotients of G for which the representation $\overline{\rho}: G \to \mathrm{Gl}_n(k)$ factors through a map $\overline{\rho}_H: H \to \mathrm{Gl}_n(k)$. Each H is finite, so the construction above produces a ring R_H in $\mathcal{C}$ with a group homomorphism $H \to \mathrm{Gl}_n(R_H)$ that reduces to $\overline{\rho}_H: H \to \mathrm{Gl}_n(k)$. Using (2.5) for each H we get a projective system $(R_H)_H$ in $\mathcal{C}$.

Now let $R_b = \varprojlim R_H$. We have a continuous map $\rho_b: G \to \mathrm{Gl}_n(R_b)$ induced by the composite maps $G \to H \to \mathrm{Gl}_n(R_H)$. For fixed H, the images of the defining generators of $\mathcal{O}[H,n]$ generate each discrete artinian quotient of R_i over $\mathcal{O}$. But these images are contained in the image of R_b, so R_b surjects to each discrete artinian quotient of R_H. Moreover, each discrete artinian quotient of R_b arises in this way. In particular it follows that R_b lies in $\mathcal{C}$.

Let $A = \varprojlim A_i$ be a ring in $\mathcal{C}$ written as a projective limit of its discrete artinian quotients. We now have canonical isomorphisms

$$\begin{aligned}
\mathrm{CHom}_{\overline{\rho}}(G, \mathrm{Gl}_n(A)) &\cong \varprojlim_i \mathrm{CHom}_{\overline{\rho}}(G, \mathrm{Gl}_n(A_i)) \\
&\cong \varprojlim_i \varinjlim_H \mathrm{Hom}_{\overline{\rho}_H}(H, \mathrm{Gl}_n(A_i)) \\
&\cong \varprojlim_i \varinjlim_H \mathrm{CHom}_{\mathcal{O}\text{-Alg}}(R_H, A_i) \\
&\overset{(*)}{\cong} \varprojlim_i \mathrm{CHom}_{\mathcal{O}\text{-Alg}}(R_b, A_i) \\
&\cong \mathrm{CHom}_{\mathcal{O}\text{-Alg}}(R_b, A).
\end{aligned}$$

For $(*)$ we use that a continuous homomorphism $R_b \to A_i$ factors over some artinian quotient R' of R_b, and that R' can be chosen to be an artinian quotient of some R_H. This proves (2.5).

4. The condition of absolute irreducibility

In this section we show (2.6). Let $V = W \otimes_A k$. The G-action on V gives an $\mathcal{O}$-algebra homomorphism $\overline{\rho}: k[G] \to \mathrm{End}_k(V)$. The irreducibility of V implies that $D = \mathrm{End}_{k[G]}(V)$ is a division ring, and since V is absolutely irreducible, the tensor product $D \otimes_k K = \mathrm{End}_{K[G]}(V \otimes_k K)$ is also a division ring for any field extension K of k. This implies that $D = k$. By Wedderburn's theorem [7, chap. XVII, 3.5] one then deduces that $k[\overline{\rho}(G)] = \mathrm{End}_k(V)$.

Choosing a k-basis of V we may identify the k-algebra $\mathrm{End}_k(V)$ with the ring $M_n(k)$ of $n \times n$-matrices over k. Let $\bar{e}_1, \ldots, \bar{e}_{n^2}$ be a k-basis of $\mathrm{End}_k(V)$ for which each matrix $\bar{e}_i$ has exactly one non-zero entry. We denote the trace of an endomorphism f of a finitely generated free module over a ring R by $\mathrm{Tr}_R(f)$. An easy computation shows that the determinant of the matrix $(\mathrm{Tr}_k(\bar{e}_i\bar{e}_j))_{i,j} \in M_{n^2}(k)$ does not vanish.

Let B be the sub-A'-algebra of $\mathrm{End}_A(W)$ generated by the image of G. Denote the natural map $\mathrm{End}_A(W) \to \mathrm{End}_k(V)$ by φ. Then we have $\varphi(B) = k[\bar{\rho}(G)] = \mathrm{End}_k(V)$, so we can choose $e_i \in B$ such that $\varphi(e_i) = \bar{e}_i$. Since φ induces an isomorphism $\mathrm{End}_A(W) \otimes_A k \xrightarrow{\sim} \mathrm{End}_k(V)$, it follows from Nakayama's lemma that the e_i form an A-basis of $\mathrm{End}_A(W)$. We claim that they also form an A'-basis of B. Indeed, if we write an element $b \in B$ on this basis as $b = \sum_i a_i e_i$ with $a_i \in A$, then we have

$$\sum_{i=1}^{n^2} a_i \,\mathrm{Tr}_A(e_i e_j) = \mathrm{Tr}_A(b e_j) \in A',$$

because $\mathrm{Tr}_A(B) \subset A'$. The coefficient matrix $(\mathrm{Tr}_A(e_i e_j))_{ij} \in M_{n^2}(A')$ is invertible, because it is invertible modulo $\mathfrak{m}_{A'}$. Therefore all a_i lie in A', which proves our claim. It follows that $B \otimes_{A'} A = \mathrm{End}_A(W)$.

Choose an idempotent $\bar{\eta}$ in the ring $\mathrm{End}_k(V)$ that generates a minimal left-ideal; e.g., take a matrix with one diagonal entry equal to 1 and all other entries equal to 0. We claim that there exists $\eta \in B$ such that $\eta^2 = \eta$ and $\varphi(\eta) = \bar{\eta}$. If $x \in B$ and $l \geq 1$ are such that $x \equiv x^2 \bmod \mathfrak{m}_{A'}^l B$, then it is easy to check that $f(x) = 3x^2 - 2x^3$ satisfies $f(x) \equiv x \bmod \mathfrak{m}_{A'}^l B$ and $f(x)^2 \equiv f(x) \bmod \mathfrak{m}_{A'}^{2l} B$. Now choose any $\eta_0 \in B$ with $\varphi(\eta_0) = \bar{\eta}$ and consider the sequence $\eta_0, f(\eta_0), f(f(\eta_0)), \cdots$. This is clearly a Cauchy sequence for the $\mathfrak{m}_{A'}$-adic topology on B. But A' is a projective limit of artinian rings, so its $\mathfrak{m}_{A'}$-adic topology is at least as strong as the given topology on A', for which it is complete. This means that the sequence is a Cauchy sequence for the product topology on the free A'-module B, so that the sequence converges to a limit η in B. This η satisfies our conditions.

We have $B\eta \oplus B(1-\eta) = B$, and B is a free A'-module. It follows that the B-module $W' = B\eta$ is also free over A', and from $\varphi(\eta) = \bar{\eta}$ we see that its rank over A' equals $\dim_k(\mathrm{End}_k(V)\bar{\eta}) = n$. Choose an element w_0 of W whose image v_0 in V satisfies $\bar{\eta}v_0 \neq 0$. Then we have $\mathrm{End}_k(V)\bar{\eta}v_0 = V$, so Nakayama's lemma implies that the $\mathrm{End}_A(W)$-linear map $W' \otimes_{A'} A = \mathrm{End}_A(W)\eta \to W$ sending σ to σw_0 is surjective. By checking A-ranks one sees that it is an isomorphism. It follows that W and $W' \otimes_{A'} A$ are isomorphic over $B \otimes_{A'} A$, and in particular they are $A[G]$-isomorphic. It also follows that the G-action on W' is continuous. □

The following result will be needed for the proof of part (3) of (2.3).

(4.1) Lemma. *Let A be a local ring with residue field k and let G be a group. Let $\bar{\rho}: G \to \mathrm{Gl}_n(k)$ be a group homomorphism that makes k^n into an absolutely irreducible $k[G]$-module. Then two elements $\rho, \rho' \in \mathrm{Hom}_{\bar{\rho}}(G, \mathrm{Gl}_n(A))$ define isomorphic $A[G]$-module structures on A^n if and* *only if there is a matrix* $M \in \mathrm{Gl}_n(A)$ *reducing to the identity matrix in* $\mathrm{Gl}_n(k)$ *such that* $\rho(g) = M\rho'(g)M^{-1}$ *for all* $g \in G$.

Proof. The only non-trivial point is the following: if there exists $M \in \mathrm{Gl}_n(A)$ such that $\rho(g) = M\rho'(g)M^{-1}$ for all $g \in G$, then M can be chosen so that its reduction $\overline{M} \in \mathrm{Gl}_n(k)$ is the identity matrix. Note that $\overline{M}$ lies in $\mathrm{Aut}_{k[G]}(k^n)$, which by the first paragraph of the proof above is just k^*. But the scalar matrix $\overline{M}$ can then be lifted to a scalar matrix T in $\mathrm{Gl}_n(A)$, and we can now replace M by MT^{-1}. □

5. Projective limits

In this section we show (2.4) and statement (3) of (2.3).

Let A be a ring in $\mathcal{C}$ which is given as a projective limit $\varprojlim A_i$ of a collection of discrete artinian quotients, where i ranges over some directed index set. We let $\mathfrak{m}$ and $\mathfrak{m}_i$ be the maximal ideals of A and A_i.

(5.1) Lemma. *Suppose that we have a sequence of projective systems*

$$(M_i^1) \to (M_i^2) \to (M_i^3)$$

which for each i *is an exact sequence of finitely generated* A_i*-modules. Assume also that for each* $i' \leq i$ *and* $j = 1, 2, 3$, *the transition map* $M_i^j \to M_{i'}^j$ *is* A_i*-linear. Then the induced sequence*

$$\varprojlim_i M_i^1 \longrightarrow \varprojlim_i M_i^2 \xrightarrow{\varphi} \varprojlim_i M_i^3$$

is an exact sequence of A*-modules.*

Proof. The projective limits are A-modules by the condition on the transition maps. It is clear that the maps between them are A-linear, and that the composition of the two maps is zero.

Suppose that $(x_i)_i$ is an element in the kernel of φ. Let

$$E_i = \{x \in M_i^1 : x \mapsto x_i\}.$$

We need to show that $\varprojlim E_i$ is non-empty. In the case that k is finite one can see this by remarking that $\prod_i E_i$ is compact, and that $\varprojlim E_i$ is the intersection of a collection of closed subsets with the property that any finite subcollection has a non-empty intersection.

For the general case the reader is referred to the criterion for projective limits to be non-empty given in Bourbaki [2, III.7.4, Théorème 1]. To apply this criterion one lets $\mathfrak{S}_i$ be the set of subsets of E_i of the form $x + N$, where $x \in E_i$ and where N is a sub-A_i-module of the kernel of the map $M_i^1 \to M_i^2$ (see also [2, loc. cit., Exemple II]). □

(5.2) Remark. With a similar argument we will show the following, which will be used in Section 6. If X is a collection of open ideals I of A

which is closed under taking finite intersections, then the canonical map $\varphi\colon A \to A' = \varprojlim_{I\in X} A/I$ induces a topological isomorphism $A/F \xrightarrow{\sim} A'$, where $F = \bigcap_{I\in X} I$. Clearly, φ is continuous, and $\operatorname{Ker}\varphi = F$. Suppose first that k is finite. Then A and A' are compact and $\varphi(A)$ is a dense compact subset of A', so φ is surjective. A continuous bijection between compact Hausdorff spaces is a homeomorphism, so our claim follows.

Let us sketch the argument for general k. For $I \in X$ let A_i^I be the cokernel of the map $I \to A_i$. Since A_i is artinian, it surjects to $\varprojlim_I A_i^I$, and by (5.1) the ring A surjects to $\varprojlim_i \varprojlim_I A_i^I = \varprojlim_I \varprojlim_i A_i^I$. Since I is open we have $\varprojlim_i A_i^I = A/I$, and it follows that φ is surjective. In the same way one shows that the image in A' of any open ideal $\mathfrak{a}$ of A is $\varprojlim_I (\mathfrak{a}+I)/I$, which is open in A' because by (5.1) it is the kernel of the continuous map from A' to the discrete ring $\varprojlim_I A/(\mathfrak{a}+I)$. Thus, φ is an open map, and the map $A/F \to A'$ is a homeomorphism.

(5.3) Proposition. *The following two statements are equivalent:*

(1) *A is noetherian;*

(2) *$\dim_k(\mathfrak{m}_i/\mathfrak{m}_i^2)$ is a bounded function of i.*

If they hold, then the following are also true:

(3) *$\mathfrak{m}^a = \varprojlim \mathfrak{m}_i^a$ for all $a \geq 0$;*

(4) *the topology on A is the $\mathfrak{m}$-adic topology.*

This proposition implies (2.4). To obtain the last statement of (2.4), write $A' = \varprojlim A_i'$ with A_i' artinian and note that for each i the map $A \to A' \to A_i'$ is continuous in the $\mathfrak{m}$-adic topology on A. We already used this argument to show (2.5) in the case that G is finite.

Proof. Suppose that A is noetherian. Then $\mathfrak{m}$ can be generated as an A-ideal by a finite number d of elements of $\mathfrak{m}$. Since $\mathfrak{m}$ surjects to $\mathfrak{m}_i$ we have $\dim_k(\mathfrak{m}_i/\mathfrak{m}_i^2) \leq d$ for each i, so (1) implies (2).

Now assume that (2) holds. We need to show (1), (3) and (4). We start with (3). The statement is trivial for $a = 0$, and we will proceed by induction on a. Assume (3) holds for a and consider the sequence of projective systems

$$0 \longrightarrow \mathfrak{m}_i^{a+1} \longrightarrow \mathfrak{m}_i^a \longrightarrow \mathfrak{m}_i^a/\mathfrak{m}_i^{a+1} \longrightarrow 0.$$

Assumption (2) implies that $\mathfrak{m}_i^a/\mathfrak{m}_i^{a+1}$ also has bounded dimension, so the system on the right stabilizes, i.e., all transition maps for $j \geq i$ are isomorphisms if i is large enough. This implies that its limit is a finite dimensional k-vector space N. By (5.1) and the induction hypothesis we have a short exact sequence

$$(*) \qquad 0 \longrightarrow \varprojlim_i \mathfrak{m}_i^{a+1} \longrightarrow \mathfrak{m}^a \longrightarrow N \longrightarrow 0.$$

Choose elements $b_1, \ldots, b_l$ of $\mathfrak{m}^a$ whose images in N form a basis of N over k. For each i we have a surjection $A_i^l \to \mathfrak{m}_i^a$, sending $(x_1, \ldots, x_l)$ to $x_1 b_1 + \cdots + x_l b_l$. Taking limits we deduce from (5.1) and the induction hypothesis that $\mathfrak{m}^a$ is generated by $b_1, \ldots, b_l$ as an A-ideal. We now have $l \geq \dim_k(\mathfrak{m}^a/\mathfrak{m}^{a+1}) \geq \dim_k(N) = l$, so $\mathfrak{m}^{a+1}$ is equal to the kernel of the map $\mathfrak{m}^a \to N$. By the sequence $(*)$ above, this gives the induction step. This shows (3).

Applying (5.1) to the sequence

$$0 \longrightarrow \mathfrak{m}_i^a \longrightarrow A_i \longrightarrow A_i/\mathfrak{m}_i^a \longrightarrow 0$$

and using (3) we get $A/\mathfrak{m}^a = \varprojlim A_i/\mathfrak{m}_i^a$. Again with (2) one sees that this system stabilizes. But this means that the map $A \to A/\mathfrak{m}^a$ factors through A_i for some i, so that $\mathfrak{m}^a$ is open in A. We already mentioned in Section 4 that the $\mathfrak{m}$-adic topology on a ring in $\mathcal{C}$ is at least as strong as the given topology, so in this case the two topologies coincide. This shows (4).

We now know that A is $\mathfrak{m}$-adically complete, and that $\mathfrak{m}$ is a finitely generated A-ideal. To prove that A is noetherian we use a standard argument, which also goes into the proof that a completion of a noetherian ring is noetherian. The graded ring $\mathrm{G}(A) = \bigoplus_{m \geq 0} \mathfrak{m}^m/\mathfrak{m}^{m+1}$ is a finitely generated k-algebra, which is noetherian by Hilbert's basis theorem. By [1, (10.25)] this implies that A is noetherian. This shows (1). □

Proof of part (3) of (2.3). We consider deformations of V in the ring $A = k[\epsilon]$ with $\epsilon^2 = 0$. Write R as a projective limit of its discrete artinian quotients R_i. Let $\mathfrak{m}_i$ be the maximal ideal of R_i. One easily sees that

$$\begin{aligned} \mathrm{Hom}_{\mathcal{C}}(R, k[\epsilon]) &= \varinjlim_i \ \mathrm{Hom}_{\mathcal{O}\text{-Alg}}(R_i, k[\epsilon]) \\ &= \varinjlim_i \ \mathrm{Hom}_k(\mathfrak{m}_i/(\mathfrak{m}_i^2 + \mathfrak{m}_{\mathcal{O}} R_i), k). \end{aligned}$$

Let us denote the rightmost set by T, and note that T is a vector space over k. Recall that $\mathcal{O}$ is noetherian, so that the k-dimension d of $\mathfrak{m}_{\mathcal{O}}/\mathfrak{m}_{\mathcal{O}}^2$ is finite. Clearly $\dim_k(\mathfrak{m}_i/(\mathfrak{m}_i^2 + \mathfrak{m}_{\mathcal{O}} R_i))$ and $\dim_k(\mathfrak{m}_i/\mathfrak{m}_i^2)$ differ by at most d. Since the transition maps in the injective limit are injective, the dimension of T is finite if and only if the dimension of $\mathfrak{m}_i/\mathfrak{m}_i^2$ is bounded, which by (5.3) is equivalent to R being noetherian.

By part (1) of (2.3) the set $\mathrm{Def}(V, k[\epsilon])$ can be identified with T, so after choosing a basis of V over k one gets a surjection

$$\mathrm{CHom}_{\overline{\rho}}(G, \mathrm{Gl}_n(k[\epsilon])) \to T.$$

We have $\mathrm{Gl}_n(k[\epsilon]) = \mathrm{Gl}_n(k) \oplus M_n(k)\epsilon$, and one easily checks that the homomorphisms on the left are exactly the maps $g \mapsto (1 + c(g)\epsilon)\overline{\rho}(g)$ for which $c\colon G \to M_n(k)$ is a continuous 1-cocycle. Moreover, it follows from

(4.1) that two 1-cocycles give the same deformation in $k[\epsilon]$ if and only if they differ by a coboundary, so that we get a bijection $H^1(G, \mathrm{End}_k(V)) \xrightarrow{\sim} T$. In the case that k is finite, statement (3) follows at once. For the general case one checks that this bijection is k-linear, so that the same conclusion holds. □

6. Restrictions on deformations

In this section a class of additional properties of deformations is identified for which one gets a representable sub-functor of the deformation functor.

Suppose that for each ring A in $\mathcal{C}$ a subset $S(A)$ of $\mathrm{Def}(V, A)$ is given such that for each A in $\mathcal{C}$ and $D \in \mathrm{Def}(V, A)$ the following hold:

(1) we have $D \in S(A)$ if and only if $D/\mathfrak{a}D \in S(A/\mathfrak{a})$ for all open ideals $\mathfrak{a} \neq A$ in A;
(2) if $\mathfrak{a}$ and $\mathfrak{b}$ are open ideals $\neq A$ of A such that $D/\mathfrak{a}D \in S(A/\mathfrak{a})$ and $D/\mathfrak{b}D \in S(A/\mathfrak{b})$, then $D/(\mathfrak{a} \cap \mathfrak{b})D \in S(A/(\mathfrak{a} \cap \mathfrak{b}))$;
(3) if $A \subset A'$ is an inclusion of artinian rings in $\mathcal{C}$, then $D \in S(A)$ if and only if $D \otimes_A A' \in S(A')$.

(6.1) Proposition. *For any $\mathcal{C}$-morphism $f: A \to A'$ we have $f_*(S(A)) \subset S(A')$. If V is absolutely irreducible, then there is a closed ideal $\mathfrak{a}$ of the universal deformation ring R such that the map $\mathrm{Hom}_{\mathcal{C}}(R, A) \xrightarrow{\sim} \mathrm{Def}(V, A)$ in* (2.3) *induces a bijection $\mathrm{Hom}_{\mathcal{C}}(R/\mathfrak{a}, A) \xrightarrow{\sim} S(A)$.*

Proof. Let A be a ring in $\mathcal{C}$ and $D \in \mathrm{Def}(V, A)$. Using (5.2) one deduces from conditions (1) and (2) above that there is a unique closed ideal $\mathfrak{a}_S^D$ of A such that for every open ideal $\mathfrak{a}$ of A we have $D/\mathfrak{a}D \in S(A/\mathfrak{a})$ if and only $\mathfrak{a} \supset \mathfrak{a}_S^D$. By condition (1) we have $D \in S(A)$ if and only if $\mathfrak{a}_S^D = 0$.

Now let $f: A \to A'$ be a $\mathcal{C}$-morphism and put $D' = D \otimes_A A'$, where the tensor product is taken via f. Let $\mathfrak{a}'$ be an open A'-ideal and write $\mathfrak{a} = f^{-1}(\mathfrak{a}')$. By condition (3) we have $D'/\mathfrak{a}'D' \in S(A'/\mathfrak{a}')$ if and only if $D/\mathfrak{a}D \in S(A/\mathfrak{a})$. Therefore, $\mathfrak{a}_S^{D'} \subset \mathfrak{a}'$ if and only if $f(\mathfrak{a}_S^D) \subset \mathfrak{a}'$. In particular, $D' \in S(A')$ if and only if $\mathrm{Ker}\, f$ contains $\mathfrak{a}_S^D$.

The first statement of the proposition now follows at once, and by taking $\mathfrak{a} = \mathfrak{a}_S^D \subset R$, where D is the universal deformation, we obtain the second statement. □

(6.2) Ordinary deformations. Suppose that I is a closed subgroup of G. A 2-dimensional representation W of G over a ring A in $\mathcal{C}$ is said to be *ordinary* if the sub-A-module W^I of I-invariants is a direct summand of W of A-rank 1 (cf. [8, 1.7]). Suppose that V is 2-dimensional, absolutely irreducible, and ordinary. We want to show that the ordinary deformations form a representable functor on $\mathcal{C}$.

Using the fact that V is ordinary one can see that $D \in \mathrm{Def}(V, A)$ is ordinary if and only if the I-action on D is given by matrices $\left(\begin{smallmatrix} 1 & * \\ 0 & * \end{smallmatrix}\right)$ on a suitable A-basis of D, and if and only if D^I contains an element z not mapping to 0 in V. Now choose an element $g_0 \in I$ that does not act trivially on V. Then one checks that D is ordinary if and only if D is annihilated by the elements $(g-1)(g_0 - \det_D(g_0)) \in A[G]$ with $g \in I$ (for the if-part, choose $z = (g_0 - \det_D(g_0))y$ for suitable y). It is easy to verify that conditions (1)–(3) hold for this latter property.

(6.3) Flat deformations. Assume that k is a finite field of characteristic p. Let K be finite field extension of the field $\mathbb{Q}_p$ of p-adic numbers, let $\mathcal{O}_K$ be its ring of integers, and let $G = \mathrm{Gal}(\overline{K}/K)$, where $\overline{K}$ is an algebraic closure of K. We say that a $\mathbb{Z}[G]$-module of finite cardinality is *flat* if it G-isomorphic to the group of points in $\overline{K}$ of a finite flat group scheme over $\mathcal{O}_K$. The flatness property is preserved under passing to finite products, submodules, and quotients [11; 4]. Let us sketch the argument. For products it is clear. Suppose that $X' \subset X$ are $\mathbb{Z}[G]$-modules and that $X = \mathcal{G}(\overline{K})$ for a finite flat group scheme $\mathcal{G} = \operatorname{Spec} A$ over $\mathcal{O}_K$. Let I be the kernel of the map $A \to \prod_{x \in X'} \overline{K}$. The comultiplication $m^*\colon A \to A \otimes A$ induces a comultiplication on $A' = A/I$ and on $A'' = \{x \in A\colon m^*(x) \equiv x \otimes 1 \bmod A \otimes I\}$. Then $\mathcal{G}' = \operatorname{Spec} A'$ and $\mathcal{G}'' = \operatorname{Spec} A''$ are finite flat group schemes over $\mathcal{O}_K$ and one checks that $\mathcal{G}'(\overline{K}) \cong X'$ and $\mathcal{G}''(\overline{K}) \cong X/X'$.

A deformation of V in an artinian ring A in $\mathcal{C}$ is said to be flat if it is flat as a $\mathbb{Z}[G]$-module. Use condition (1) to define flatness for deformations to arbitrary rings A in $\mathcal{C}$. Then one easily checks (2) and (3). For (3) one notes that D' contains D as a sub-$\mathbb{Z}[G]$-module, and that D' is a quotient of a finite product of copies of D. Thus, the flat deformation functor on $\mathcal{C}$ is representable if V is absolutely irreducible and flat.

7. Relaxing the absolute irreducibility condition

In this section we will show that our main result already holds when $\mathrm{End}_{k[G]}(V) = k$. We saw in Section 4 that this is a weaker condition on V than absolute irreducibility. This improved result will not be needed in the rest of this book.

(7.1) Proposition. *If* $\mathrm{End}_{k[G]}(V) = k$ *then statements* (1)–(4) *of* (2.3) *hold.*

Proof. We will use the same construction as before, but we need to pass to a different subring of R_b: we may need more elements than the traces of the actions of the group elements. In order to describe a suitable set of elements we explain Faltings's notion of "well-placed" representations.

We choose a basis for V over k, so that the G-action on V is given by a continuous group homomorphism $\overline{\rho}\colon G \to \mathrm{Gl}_n(k)$. Since $M_n(k)$ is finite-dimensional over k, we can choose a finite number of elements $g_1, \ldots, g_r$ in G such that the only matrices in $M_n(k)$ commuting with all $\overline{\rho}(g_i)$ are the scalar matrices. Let a lift $E_i \in M_n(\mathcal{O})$ of each $\overline{\rho}(g_i)$ be chosen. For any ring A in $\mathcal{C}$ we let $M_n^0(A)$ be the matrix ring $M_n(A)$ modulo scalars; this is a free A-module of rank n^2-1. By Nakayama's lemma one sees that we have a split injection $i_A\colon M_n^0(A) \to M_n(A)^r$ given by $M \mapsto (ME_i - E_iM)_{i=1}^r$.

We now choose a splitting $\pi_{\mathcal{O}}$ of $i_{\mathcal{O}}$ once and for all. We have a canonical isomorphism $M_n^0(A) \cong M_n^0(\mathcal{O}) \otimes_{\mathcal{O}} A$, and $\pi_A = \pi_{\mathcal{O}} \otimes \mathrm{id}_A$ is a splitting of i_A. Consider the composite map

$$(7.2) \qquad \begin{array}{ccccc} \mathrm{CHom}_{\overline{\rho}}(G, \mathrm{Gl}_n(A)) & \longrightarrow & M_n(A)^r & \xrightarrow{\pi_A} & M_n^0(A). \\ \rho & \mapsto & (\rho(g_i))_{i=1}^r & & \end{array}$$

We say that ρ is *well-placed* if its image in $M_n^0(A)$ is $\pi_{\mathcal{O}}(E_1, \ldots, E_r) \otimes 1$.

(7.3) Lemma (Faltings). *For every $\rho \in \mathrm{CHom}_{\overline{\rho}}(G, \mathrm{Gl}_n(A))$ there is a matrix $M \in \mathrm{Gl}_n(A)$ reducing to $1 \in \mathrm{Gl}_n(k)$ so that $M\rho M^{-1}$ is well-placed. This matrix M is determined uniquely modulo $1 + \mathfrak{m}_A$.*

Proof. Put $\mathfrak{m} = \mathfrak{m}_A$. With induction to m we first show the lemma under the hypothesis that $\mathfrak{m}^m = 0$. For $m = 1$ this is clear. To make the induction step for $m \geq 2$ we can assume by the induction hypothesis that ρ is well-placed modulo $\mathfrak{m}^{m-1}$. We are done if we show that $(1+M)\rho(1+M)^{-1}$ is well-placed for a unique $M \in M_n^0(\mathfrak{m}^{m-1}) = \mathfrak{m}^{m-1}M_n^0(A)$, and this follows from the fact that the maps in (7.2) respect suitable actions of $M_n^0(\mathfrak{m}^{m-1})$: we let $M \in M_n^0(\mathfrak{m}^{m-1})$ act by conjugation with $1 + M$ on the leftmost set, by translation with $i_A(M)$ on the middle group, and by translation with M on $M_n^0(A)$.

To obtain the general case one refines the conjugating matrix modulo increasing powers of $\mathfrak{m}$ (recall that an $\mathfrak{m}$-adic Cauchy sequence in A converges to a unique limit in A even if A has a coarser topology). □

We apply the lemma to the deformation ρ_b of Proposition (2.5), and we let ρ be the well-placed conjugate of ρ_b. Define R to be the smallest closed sub-$\mathcal{O}$-algebra of R_b that contains all entries of $\rho(g)$ for all $g \in G$. Then ρ defines a deformation D of V in R, and we claim that properties (1)–(4) of Theorem (2.3) now hold. The map $\mathrm{Hom}_{\mathcal{C}}(R, A) \to \mathrm{Def}(V, A)$ in (1) is again surjective. To see injectivity, suppose that for $f_1, f_2 \in \mathrm{Hom}_{\mathcal{C}}(R, A)$ the well-placed composite maps

$$\rho_1, \rho_2 \colon G \xrightarrow{\rho} \mathrm{Gl}_n(R) \xrightarrow{f_1, f_2} \mathrm{Gl}_n(A)$$

give the same deformation of V in A. By the argument of (4.1) together with the uniqueness statement in (7.3) it follows that $\rho_1 = \rho_2$, and by the definition of R this implies that $f_1 = f_2$. The proofs of (2) and (4) are as before. For (3) we just remark that the argument at the end of Section 5 showing that $H^1(G, \mathrm{End}_k(V)) \cong T$, only uses that $\mathrm{End}_{k[G]}(V) = k$. This proves (7.1). □

ACKNOWLEDGEMENT. The authors thank N. Boston, E. Das, B. Edixhoven, and C. Procesi for their assistance. The first author was employed by the Erasmus Universiteit Rotterdam while most of this paper was being written. Both authors received support from the Nederlandse Organisatie voor Wetenschappelijk Onderzoek. The second author was supported by the National Science Foundation under Grant No. DMS 9224205.

References

1. M. F. Atiyah and I. G. Macdonald, *Introduction to commutative algebra*, Addison-Wesley, Reading, Mass., 1969.
2. N. Bourbaki, *Théorie des ensembles*, Hermann, Paris, 1970.
3. H. Carayol, *Formes modulaires et représentations galoisiennes à valeurs dans un anneau local complet*, pp. 213–237 in: B. Mazur and G. Stevens (eds), *p-adic monodromy and the Birch and Swinnerton-Dyer conjecture*, Contemp. Math. **165**, Amer. Math. Soc., Providence, 1994.
4. B. Conrad, *The flat deformation functor*, Chapter XIV in this volume.
5. H. Darmon, F. Diamond, and R. Taylor, *Fermat's Last Theorem*, pp. 1–107 in: R. Bott, A. Jaffe, and S. T. Yau (eds), *Current developments in mathematics, 1995*, International Press, Cambridge, MA, 1995.
6. A. Grothendieck, *Technique de descente et théorèmes d'existence en géométrie algébrique, II*, Sém. Bourbaki **12** (1959/60), n° 195.
7. S. Lang, *Algebra*, 3rd ed., Addison-Wesley, Reading, Mass., 1993.
8. B. Mazur, *Deforming Galois representations*, pp. 385–437 in: Y. Ihara, K. Ribet, and J-P. Serre (eds), *Galois groups over* $\mathbb{Q}$, MSRI Publications **16**, Springer-Verlag, New York, 1989.
9. C. Procesi, *Rings with polynomial identities*, Marcel Dekker, New York, 1973.
10. C. Procesi, *Deformations of representations*, preprint, December 1995.
11. R. Ramakrishna, *On a variation of Mazur's deformation functor*, Compositio Math. **87** (1993), 269–286.
12. M. Schlessinger, *Functors of Artin rings*, Trans. Amer. Math. Soc. **130** (1968), 208–222.
13. A. Wiles, *Modular elliptic curves and Fermat's Last Theorem*, Ann. of Math. (2) **141** (1995), 443–551.

HECKE ALGEBRAS AND THE GORENSTEIN PROPERTY

Jacques Tilouine

Université de Paris-Nord

The goal of this paper is to show the importance of the Gorenstein property for the Hecke algebra and its relation with the local freeness of the cohomology of modular curves as a module over the Hecke algebra.

Hence the text revolves around Section 2.1 of [W4]. The main theorem of this part of Wiles' paper (Theorem 2.1 of Section 2.1 in [W4]) is used crucially in the proof of the Shimura-Taniyama-Weil conjecture for semistable curves in two instances:

(i) On line 2, page 559 of Taylor-Wiles paper [TW], to insure that it is enough to prove that a cohomology group is free over the group algebra $\mathcal{O}[\Delta]$ in order to conclude that the Hecke algebra is free over $\mathcal{O}[\Delta]$.

(ii) In the proof of Proposition 2.15 of [W4] (page 507) , which in turn is used to prove Theorem 3.1.

The Gorenstein property has evolved in the 40's–50's, from the study of singular plane curves and in particular of their duality theory. It had been used since then in classical algebraic geometry until B. Mazur had the idea of its relevance in the study of congruences between modular forms or equivalently, of local components of the Hecke algebra. It is first introduced as an important tool in [M1].

There, among other things, it is shown that if $\mathfrak{M}$ is a p-ordinary maximal ideal of the Hecke algebra $\mathbb{T}$ acting on the jacobian J of $X_0(N)$ (for N prime), then:

(A) $J[\mathfrak{M}]$ is of dimension two; it is also shown there (by a different method) that the completion $\mathbb{T}_{\mathfrak{M}}$ is a complete intersection, hence:

(B) $\mathbb{T}_{\mathfrak{M}}$ is Gorenstein.

Moreover, by an elementary argument, given the basic context of [M1], one can show directly that (A) is equivalent to (B) [see the Appendix of this paper]; therefore, in effect, one has two routes for showing both (A) and (B).

However, in [M1] already, B. Mazur also noticed that the other direction was interesting, in the "non-Eisenstein" case. In fact, in that case, one can often directly establish that $J[\mathfrak{M}]$ is 2-dimensional. One must (i) first cut it into two pieces, then (ii) bound by one the dimension of one piece, and (iii) finally deduce by some trick that the other as well is one dimensional; for these steps one uses the following tools:

- If $\mathfrak{M}$ is supersingular, one gets (i) by the theory of Dieudonné modules.
- If $\mathfrak{M}$ is ordinary with potential good reduction, one gets (i) by the étale-connected dévissage of finite flat group schemes.
- If $\mathfrak{M}$ is ordinary semistable, one gets (i) by Raynaud's theory of semistable abelian varieties.
- Then, by various ingenious arguments, (ii) is brought down to the question of "weak multiplicity one" in characteristic p. More precisely, in each of the points above, one piece of $J[\mathfrak{M}]$ is related to $H^0(X_0(N) \otimes \bar{\mathbb{F}}_p, \Omega)[\mathfrak{M}]$ so that one needs to see that this space is one-dimensional. After q-expansion, this amounts to the above-mentioned weak multiplicity one question.
- In all cases, one gets (iii) by various concluding arguments (using the Brauer-Nesbitt or Krull-Schmidt-Akizuki theorems).

Actually, this scheme of proof has been used in several papers, such as [W1], [H], [MW], [Ti], [Mri], [W4].

In the present paper, we shall start by briefly recalling some properties of Gorenstein rings which are relevant to us (Section 1); then, we shall focus on the modular situation, defining the local Hecke algebra we want to study (Section 2). After these preliminaries we state the Main theorem and its corollaries (Section 3). Then we explain the strategy of the proof (Section 4). Finally, (Section 5), we give the proof itself according to the lines of Section 4. The author wishes to thank B. Mazur and K. Ribet for correspondence and discussions which clarified several questions. Finally, the author is also thankful to Mrs. C. Simon from Université de Paris-Nord who prepared the final form from a rough manuscript.

1. The Gorenstein property

Let $(R, \mathfrak{M}, k)$ be a triple consisting of a noetherian local ring R with maximal ideal $\mathfrak{M}$ and residue field k. For any R-module M and any ideal $\mathfrak{A}$ of R, we denote by $M[\mathfrak{A}]$ the submodule of M consisting of vectors annihilated by $\mathfrak{A}$. Let $d = \dim R$ be the Krull dimension of R.

Definition 1.1. We say that R is *Gorenstein of dimension* d (briefly, R is a G_d-ring) if:

$$(*) \qquad \mathrm{Ext}^i_R(k, R) \cong \begin{cases} 0 & \text{if } i < d, \\ k & \text{if } i = d. \end{cases}$$

Comment 1. By [Mat] (Theorem 2.6 and 16.A), condition $(*)$ implies that a Gorenstein ring is Cohen-Macaulay; moreover by [B], Theorem 4.1, a

Cohen-Macaulay ring is a G_d-ring if and only if the injective envelope E of its residue field k is a "dualizing module for artinian R-modules M"; that is, there are functorial isomorphisms

$$\mathrm{Ext}^d_R(M,R) \xrightarrow{\sim} \mathrm{Hom}_R(M,E)$$

for all finite length R-modules M.

Comment 2. In fact, one can get a better picture of the significance of this condition by using Grothendieck's local duality theory [Gr].

Proposition and Definition 1.2. *(See* [Gr] *Propositions 4.9 and 4.10.) A dualizing module for R is an injective R-module I such that one of the following conditions is satisfied:*
(i) *For each artinian R-module M,* $\mathrm{Hom}_R(M,I)$ *is finitely generated and the canonical homomorphism*

$$M \longrightarrow \mathrm{Hom}(\mathrm{Hom}(M,I),I)$$

is an isomorphism.
(ii) *$I[\mathfrak{M}]$ is one-dimensional.*
(iii) *I is an injective hull of the residue field $k = R/\mathfrak{M}$.*

Example. For $R = \mathbb{Z}_p$, one has $I = \mathbb{Q}_p/\mathbb{Z}_p$.

Grothendieck defines the local cohomology of a finitely generated R-module M by

$$H^i_{\mathfrak{M}}(M) = \varinjlim_{n\mapsto\infty} \mathrm{Ext}^i_R(R/\mathfrak{M}^n, M)$$

for any $i = 0, 1, \ldots, d$. Let us put

$$I = H^i_{\mathfrak{M}}(M) = \varinjlim_{n\mapsto\infty} \mathrm{Ext}^i_R(R/\mathfrak{M}^n, M).$$

One has the proposition (Proposition 4.14 of [Gr]):

Proposition 1.2. *R is a G_d-ring if and only if R is Cohen-Macaulay and $I = H^d_{\mathfrak{M}}(R)$ is dualizing.*

If R is regular, or more generally, if it is locally a complete intersection, then the theory of Koszul complexes shows that R is Gorenstein. (See [Ku], Proposition 3.22, Chapter VI, or [Gr] page 67.) In particular, if $R = \mathbb{Z}_p[\eta]$ is generated by one element η, then it is Gorenstein. Finally, Grothendieck's duality theorem ([Gr], Theorem 6.3, page 85) for Gorenstein rings reads as follows:

Theorem 1.3. *Let R be a complete noetherian local G_d-ring, and let $I = H^d_{\mathfrak{M}}(R)$. For any finitely generated R-module N, denote by $D(N) = \mathrm{Hom}_R(N, I)$. Yoneda's pairing*

$$H^i_{\mathfrak{M}}(M) \times \mathrm{Ext}^{d-i}_R(M, R) \longrightarrow I$$

is "perfect" in the sense that for any i, it induces

(1) *an isomorphism $H^i_{\mathfrak{M}}(M) \cong D(\mathrm{Ext}^{d-i}_R(M, R))$,*
(2) *an isomorphism $\mathrm{Ext}^{d-i}_R(M, R) \cong D(H^i_{\mathfrak{M}}(M))$.*

In this paper, we shall be concerned only with local rings R which are finite flat $\mathbb{Z}_p$-algebras. These rings are complete noetherian local of dimension 1. We put $\bar{R} = R/pR$. It is an artinian local ring. We denote its maximal ideal by $\bar{\mathfrak{M}}$.

Proposition 1.4. *Let R be a finite flat local $\mathbb{Z}_p$-algebra. The following statements are equivalent:*
(i) *R is a G_1-ring.*
(ii) *$\bar{R}$ is a G_0-ring.*
(iii) *$\bar{R}[\mathfrak{M}]$ is one-dimensional over k.*
(iii)′ *$\bar{R}^* = \mathrm{Hom}_{\mathbb{F}_p\text{-lin}}(\bar{R}, \mathbb{F}_p)$ is free (of rank 1) over $\bar{R}$.*
(iv) *$\mathrm{Hom}_{\mathbb{Z}_p\text{-lin}}(R, \mathbb{Z}_p)$ is free (of rank 1) over R.*

Proof. To show the equivalence of (i), (ii), and (iii), we consider the short exact sequence of R-modules

$$0 \longrightarrow R \xrightarrow{p} R \longrightarrow \bar{R} \longrightarrow 0.$$

We apply the functors $\mathrm{Ext}^0_R(k, -)$ and we observe

$$\mathrm{Ext}^1_R(k, R) = \mathrm{Hom}_R(k, \bar{R}) = \mathrm{Ext}^0_{\bar{R}}(k, \bar{R}) = \bar{R}[\mathfrak{M}].$$

To prove the equivalence of (iii), (iii)′ and (iv), one applies Nakayama's lemma for $\mathfrak{M}$, respectively pR. We leave the details to the reader

2. Hecke algebras

Let $p > 2$ be a rational prime, and let $N \geq 1$ an integer such that $\mathrm{ord}(N) \leq 1$. We consider the curve $X_1(N)_{\mathbb{Q}}$ which represents the moduli problem of generalized elliptic curves over $\mathbb{Q}$ with a $\mathbb{Q}$-rational point of exact order N (see [K-M] chapter 3, 3.2, and [D-R]).

It is a geometrically connected smooth projective curve defined over $\mathbb{Q}$. Observe that in this model, the ∞-cusp is $\mathbb{Q}(\zeta_N)$-rational but not rational; on the other hand, the 0-cusp is $\mathbb{Q}$-rational. Let $H \subset (\mathbb{Z}/N\mathbb{Z})^\times$ be a subgroup,

$$X_{\mathbb{Q}} = X_1(N)_{\mathbb{Q}}/H, \quad \text{the quotient by } H,$$
$$J_{\mathbb{Q}} = \mathrm{Alb}(X_{\mathbb{Q}}), \quad \text{its jacobian variety,}$$

viewed for the covariant functoriality for algebraic correspondences.

We introduce the Hecke algebra $\mathbb{T}$, defined as the subring of $\mathrm{End}(J_{\mathbb{Q}})$ generated by the Hecke operators T_p and the operators $\langle a \rangle$ with $a \in (\mathbb{Z}/N\mathbb{Z})^\times/H$. If p divides N, we use the Atkin-Lehner notation U_p instead of T_p, in order to emphasize that p divides the level. It is well-known that $\mathbb{T}$ is a finite flat $\mathbb{Z}$-algebra.

We assume we are given a maximal ideal $\mathfrak{M}$ of $\mathbb{T}$ satisfying the following conditions:

(1) The residue field $k = \mathbb{T}/\mathfrak{M}$ has characteristic p.

(2) There exists a continuous representation

$$\rho_{\mathfrak{M}} : \mathrm{Gal}(\bar{\mathbb{Q}}/\mathbb{Q}) \longrightarrow \mathrm{GL}_2(k)$$

which is unramified outside Np, and such that for any ℓ relatively prime to Np,

$$\begin{cases} \mathrm{Tr}\, \rho_{\mathfrak{M}}(\mathrm{Frob}_\ell) \equiv T_\ell \pmod{\mathfrak{M}}, \\ \det \rho_{\mathfrak{M}}(\mathrm{Frob}_\ell) \equiv \ell\langle \ell \rangle \pmod{\mathfrak{M}}. \end{cases}$$

(3) $\rho_{\mathfrak{M}}$ is absolutely irreducible.

Remark 1. Note that condition (2) implies that if c is a complex conjugation, then $\det \rho_{\mathfrak{M}}(c) = -1$. This can be seen by choosing $\ell \equiv -1 \pmod{Np}$ (so that $\ell\langle -1 \rangle = -1$).

Remark 2. Condition 3 will be referred to in the sequel as the *non-Eisenstein condition.*

From now on, we assume that $\mathfrak{M}$ is a non-Eisenstein maximal ideal satisfying (1), (2), (3). Let $\mathbb{T}_{\mathfrak{M}}$ be the completion of $\mathbb{T}$ at $\mathfrak{M}$. It is a local finite flat $\mathbb{Z}_p$-algebra. Let $\mathbb{T} = \mathbb{T} \otimes \mathbb{Z}_p$; then $\mathbb{T}_{\mathfrak{M}}$ is a direct factor of the semi-local ring $\mathbb{T}_p$. Hence it is a projective $\mathbb{T}_p$-algebra.

3. THE MAIN THEOREM

Let $\mathfrak{M}$ be a maximal ideal of $\mathbb{T}$ satisfying (1), (2), (3) as above.

Definition 3.1. We say that $\mathfrak{M}$ is *ordinary* if

$$\begin{cases} T_p \notin \mathfrak{M} \quad (\text{if } p \nmid N), \\ \qquad \text{or} \\ U_p \notin \mathfrak{M} \quad (\text{if } p | N). \end{cases}$$

Let $G_p \subset \mathrm{Gal}(\bar{\mathbb{Q}}/\mathbb{Q})$ be a decomposition group at p and $I_p \subset G_p$ its inertia group. We recall an important result.

Theorem 3.2. *If $\mathfrak{M}$ is ordinary, then*

$$\rho_{\mathfrak{M}}|_{G_p} \sim \begin{pmatrix} \chi_1 & * \\ 0 & \chi_2 \end{pmatrix},$$

where $\chi_2(I_p) = 1$ and if $p|N$, we have

$$\chi(\mathrm{Frob}_p) \equiv U_p \pmod{\mathfrak{M}}.$$

Comment. This fact was known to Deligne and Serre in the early 70's (letter from Deligne to Serre) and is proven in [H] proposition 4.4. Note that the presentation in [W2] theorem 2.2 treats also the p-adic case. We postpone its proof until section 5 below.

Definition 3.3. A maximal ideal $\mathfrak{M}$ of $\mathbb{T}$ satisfying conditions (1), (2), (3) is called *G_p-distinguished* if it is ordinary with $\chi_1 \neq \chi_2$.

Let $\mathfrak{M}$ be a maximal ideal of $\mathbb{T}$ satisfying (1), (2), (3). Let $R = \mathbb{T}_{\mathfrak{M}}$ and $\bar{R} = R/pR$. The main theorem of this paper is the following.

Theorem 3.4. *If $p \nmid N$, or if $p|N$ and $\mathfrak{M}$ is G_p-distinguished, then*

$$V = J[p](\bar{\mathbb{Q}})_{\mathfrak{M}}$$

is free of rank 2 over $\bar{R}$, and $\bar{R}$ is a G_0-ring.

Let us draw some corollaries.

Corollaries.

(1) *R is a G_1-ring, and*

$$R \cong \mathrm{Hom}_{\mathbb{Z}_p\text{-lin}}(R, \mathbb{Z}_p)$$

as an R-module.

(2) *$M = \mathrm{Ta}_p(J_{\mathbb{Q}})_{\mathfrak{M}}$ and $M^* = \mathrm{Hom}_{\mathbb{Z}_p\text{-lin}}(M, \mathbb{Z}_p)$ are free if rank 2 over R.*

(3) *$H_1(X(\mathbb{C}), \mathbb{Z}_p)_{\mathfrak{M}}$ is free of rank 2 over $\mathbb{T}_{\mathfrak{M}}$.*

Comment. The analogue of statement (3) above is not known in the following situations:

- (i) before localization at a maximal prime,
- (ii) after localization at an Eisenstein prime (i.e., such that $\rho_{\mathfrak{M}}$ is reducible),
- (iii) after localization at an ordinary maximal ideal which isn't G_p-distinguished. Actually, the question of whether such an ideal exists in level Np is still open (see [Bu]).

Proof of the corollaries. (1) This is clear from Proposition 1.2.
(2) We have $M/pM = V$, hence by Nakayama's lemma, we have a surjective R-linear homomorphism $R^2 \twoheadrightarrow M$. It must be injective because it induces an isomorphism after tensoring with $\mathbb{Q}_p$ and because R is flat over $\mathbb{Z}_p$.

For M^*, one recall that the Weil pairing induces a perfect Galois-equivariant pairing

$$\langle \cdot\,, \cdot \rangle : \mathrm{Ta}_p(J_{\mathbb{Q}}) \times \mathrm{Ta}_p(J_{\mathbb{Q}}) \longrightarrow \mathbb{Z}_p(1)$$

satisfying

$$\langle tx, y \rangle = \langle x, t^* y \rangle \quad \text{for all } x, y \in \mathrm{Ta}_p(J_{\mathbb{Q}}) \text{ and } t \in \mathbb{T},$$

where t^* is the image of t by the Rosati involution.

On the other hand, consider the Atkin-Lehmer automorphism w_ζ (for a fixed primitive N^{th} root of unity ζ) of $X_1(N)$ defined by

$$w_\zeta : (E, P) \longmapsto (E/\langle P \rangle, \bar{Q}),$$

where Q is an N-torsion point on the elliptic curve $E/\langle P \rangle$, such that

$$e_{E,N}(P, Q) = \zeta$$

for the Weil pairing $e_{E,N} : E[N] \times E[N] \to \mu_N$ on $E[N]$. It has the property that

$$w_\xi \circ t \circ w_\zeta = t^*.$$

Therefore if we introduce the new pairing

$$[x, y] = \langle w_\zeta x, y \rangle,$$

we obtain, by localization at $\mathfrak{M}$, a perfect R-bilinear pairing

$$M \times M \longrightarrow \mathbb{Z}_p.$$

Therefore, as R-modules, we have

$$M \cong \mathrm{Hom}_{\mathbb{Z}_p\text{-lin}}(M, \mathbb{Z}_p) = M^*.$$

Hence $M^* \cong R^2$.
(3) The transcendental description of the Albanese variety,

$$H_1(X(\mathbb{C}), \mathbb{Z}_p) \cong \mathrm{Ta}_p(J_{\mathbb{Q}}),$$

is $\mathbb{T}_p$-equivariant. Therefore after localizing at $\mathfrak{M}$, we obtain (3).

STRATEGY OF THE PROOF OF THEOREM 3.4

There are several cases to examine. They are, chronologically,

1. $p \nmid N$, (the first case studied, see [Ma1] section 14, proposition 14.2).
2. $p|N$, $\det \rho_{\mathfrak{M}}\big|_{I_p} \neq \omega\big|_{I_p}$, and $\mathfrak{M}$ is G_p distinguished.
3. $p|N$ and $\det \rho_{\mathfrak{M}}\big|_{I_p} = \omega\big|_{I_p}$.

In this last case, note that G_p-distinguishability is automatic, since χ_2 is unramified while $\chi_1\big|_{I_p} = \omega\big|_{I_p}$ is ramified. Keeping in mind the application to Fermat's last theorem, we need only study cases (1) and (3). (The so-called flat, respectively ordinary, cases.)

Remark. Note that in case (1), we have also

$$\det \rho_{\mathfrak{M}}\big|_{I_p} = \omega\big|_{I_p},$$

and that case (3) does not prevent $\rho_{\mathfrak{M}}$ from being flat — this is what happens if we start with a modular elliptic curve E with multiplicative reduction at p, but such that

$$p|\operatorname{ord}_p(q_E).$$

We shall focus on case (3). As for case (1), no change is needed from Mazur's original treatment.

Let us write $N = N'p$ with $p \nmid N'$.

Step 0. We can replace $X_{\mathbb{Q}}$ by

$$X_1(N',p) = X_1(N)/H_0$$

for $H_0 = (\mathbb{Z}/p\mathbb{Z})^\times \subset (\mathbb{Z}/N\mathbb{Z})^\times$.

Recall $X_{\mathbb{Q}}$ is defined as $X_1(N)_{\mathbb{Q}}/H$ for some subgroup $H \subset (\mathbb{Z}/N\mathbb{Z})^\times$. The reason why we can do this reduction is that the maps between the jacobian induced by

$$\alpha : X_1(N)_{\mathbb{Q}} \longrightarrow X_{\mathbb{Q}} \quad \text{and} \quad \beta : X_1(N)_{\mathbb{Q}} \longrightarrow X_1(N',p)_{\mathbb{Q}}$$

give rise to isomorphisms on the p-divisible groups localized at $\mathfrak{M}$. Let $J_1(N',p) = \operatorname{Jac}(X_1(N',p))$. Then

$$\alpha : J_1(N)[p^\infty]_{\mathfrak{M}} \longrightarrow J[p^\infty]_{\mathfrak{M}}$$

and

$$\beta : J_1(N)[p^\infty]_{\mathfrak{M}} \longrightarrow J_1(N',p)[p^\infty]_{\mathfrak{M}}$$

are isomorphisms by the theory of the fundamental group. The kernel of α (respectively β) is a quotient of H or of $(\mathbb{Z}/p\mathbb{Z})^\times$, but it must be a p-group with Galois action whose Jordan-Hölder factors are isomorphic to $\rho_{\mathfrak{M}}$. This is absurd (unless it is 0), since H has multiplicative-type Galois action.

What is the gain of this reduction? The scheme $X_1(N',p)_{\mathbb{Z}_p}$ defined as the moduli space for triples (E,P,C), where E is an elliptic curve, P is a point of order exactly N', and C is a cyclic subgroup of E of order exactly p, has the advantage of being "almost semistable;" that is , its minimal regular model (which can be obtained by blowing-up some points in the special fiber) is semistable. By this we mean not only that the special fiber has ordinary double points, but that the singularities are of type A_0:

$$\mathbb{Z}_p[\![X,Y]\!]/(XY-p).$$

Step 1. (which implies Theorem 3.2.) If $\mathfrak{M}$ is ordinary, then there exists a short exact sequence of $\bar{R}[G_p]$-modules,

$$0 \longrightarrow V^0 \longrightarrow V \longrightarrow V^E \longrightarrow 0 \tag{$\dagger$}$$

such that V^0 has all its irreducible subquotients ramified, while V^E has all its irreducible subquotients unramified. Moreover,

$$V^E \cong \mathrm{Hom}(V^0, \mu_p)$$

as $\bar{R}[I_p]$-modules.

Step 2. Assume that V^0 is free of $\bar{R}$. Then V is also free. Hence $V \cong \bar{R}^2$ and $\bar{R}$ is a G_0-ring.

Step 3. We may assume by Step 0 that

$$X = X_1(N',p) \quad \text{and} \quad J = J_1(N',p).$$

Let $J_{\mathbb{Z}_p}$ be the Néron model of J over $\mathbb{Z}_p$, and let $J_{\mathbb{Z}_p}[p]^t$ be the largest finite flat subgroup scheme of $J_{\mathbb{Z}_p}[p]$ which is of multiplicative type. Let $\mathcal{V} = J_{\mathbb{Z}_p}[p]^t(\bar{\mathbb{Q}}_p)_{\mathfrak{M}}$. Then $\mathcal{V} = V^0$.

Step 4. $\mathcal{V}$ is free of rank 1 over $\bar{R}$.

5. Sketch of the proof

Step 1. One shows actually that the whole p-divisible group

$$\mathcal{D} = J[p^\infty](\bar{\mathbb{Q}})_{\mathfrak{M}}$$

admits such a decomposition into p-divisible groups $\mathcal{D}^0$ and $\mathcal{D}^E$, as $\bar{R}[G_p]$-modules. Then one concludes by putting

$$V^0 = \mathcal{D}^0[p] \quad \text{and} \quad V^E = \mathcal{D}^E[p].$$

We observe first that there is an isogeny

$$\mathcal{D} \sim \prod_f A_f[\mathfrak{P}^\infty],$$

where the product runs over the eigenforms $f \in S_2(\Gamma_1(N))$, and the primes $\mathfrak{P}|p$ in the field of eigenvalues K_f are such that the eigensystem modulo $\mathfrak{P}$ defined by f factors through $\mathbb{T}_{\mathfrak{M}}$.

Recall that the abelian variety A_f is defined as a quotient of $J_1(N',p)$ on which Hecke operators act by the eigenvalues of f (there is a canonical embedding of K_f in $\mathrm{End}(A_f) \otimes \mathbb{Q}$). Therefore, it is enough to deal separately with each $A_f[\mathfrak{P}^\infty]$.

(a) Suppose first that p divides the conductor of f. Note that the p-part of the character of f is trivial, and $\det \rho_{f,\beta} = \chi\varepsilon$, where χ is the p-cyclotomic character and ε is a character unramified at p, as can be seen by reduction mod $\mathfrak{P}$. Let us show that this implies that A_f has purely multiplicative reduction $\mathfrak{P}$.

We consider the two coverings

$$X_{\mathbb{Q}} \rightrightarrows X_1(N')$$

induces by $\tau \mapsto \tau$ and $\tau \mapsto p\tau$ on the upper half-plane. By Albanese functoriality, they induce a morphism

$$J_{\mathbb{Q}} \xrightarrow{\phi} J_1(N')^2.$$

Let A be the neutral component of $\mathrm{Ker}\,\phi$. By comparing the cotangent spaces, we see that A_f is a quotient of A.

By analyzing the Néron model of A over $\mathbb{Z}_p$, we see it has purely toric reduction, hence so has A_f. The ingredients for this verification are:

- Theorem 2.5 of [Ra2], which identifies the connected component of the Néron model $J_{\mathbb{Z}_p}$ of $J_{\mathbb{Q}}$ over $\mathbb{Z}_p$ as $\mathrm{Pic}^0(M_{\mathbb{Z}_p})$, where $M_{\mathbb{Z}_p}$ is the regular minimal model of $X_{\mathbb{Z}_p}$.
- The rigidity of tori as in [Ra1] or [M2], to obtain a decomposition of the group-scheme $A_f[\mathfrak{P}^\infty]_{\mathbb{Z}_p}$ as

$$0 \longrightarrow T_{\mathbb{Z}_p} \longrightarrow A_f[\mathfrak{P}^\infty]_{\mathbb{Z}_p} \longrightarrow E_{\mathbb{Z}_p} \longrightarrow 0,$$

where T and E are p-divisible groups of multiplicative type, respectively étale over $\mathbb{Z}_p$.

By taking the $\bar{\mathbb{Q}}_p$-points, we obtain the desired decomposition of $\rho_{f,\mathfrak{P}}$.

(b) If p does not divide the conductor of f, then A_f is a quotient of $J_1(N')$; it has good reduction at p over $\mathbb{Z}_p$. If f is ordinary at $\mathfrak{P}$, we consider the connected-étale decomposition of the p-divisible group

$$0 \longrightarrow C \longrightarrow A_f[\mathfrak{P}^\infty]_{\mathbb{Z}_p} \longrightarrow E \longrightarrow 0.$$

The corank over $\mathcal{O}_{K_f,\mathfrak{P}}$ of the middle term is 2, so one has to show that E has corank 1.

To show that corank $E \leq 1$, the method of Mazur ([Ma1], proposition 14.7) is to use the Cartier map δ which defines the Hecke-equivariant injection

$$\delta : J_1(N')[p](\bar{\mathbb{F}}_p) \otimes \bar{\mathbb{F}}_p \hookrightarrow H^0(X_1(N')_{/\bar{\mathbb{F}}_p}, \Omega^1).$$

It is defined, for a linear class $[D]$ of a divisor D such that $pD = (g)$, by

$$\delta([D] \otimes 1) = \frac{dg}{g}.$$

Then considering the q-expansion at ∞ of a global section of Δ^1, and by the very definition of Hecke operators, we see that

$$H^0(X_1(N')_{/\bar{\mathbb{F}}_p}, \Omega^1)[\mathfrak{M}]$$

is one-dimensional (see [Ma1], proposition 9.3). To show that the corank of E is exactly one, one uses the Eichler-Shimura relations as in [H], proposition 4.4 (4.17).

This takes care of the ordinary case. If f is supersingular at p, one makes use of the Dieudonné module exactly as in [Ma1], proof of proposition 14.2, case 1, pages 114–116.

Step 1. For the duality statement, we consider the modified Weil pairing,

$$[\cdot, \cdot] : V \times V \longrightarrow \mu_p.$$

Observe that it is only $\bar{R}[I_p]$-equivariant because there may be an unramified character ε such that

$$[x^\sigma, y^\sigma] = [x, y]^{\omega\varepsilon(\sigma)}.$$

Nevertheless, $\mathrm{Hom}(V^0, \mu_p)$ is the maximal quotient of $\mathrm{Hom}(V^0, \mu_p) = V$ with unramified irreducible subquotients. It must therefore by V^E.

Step 2. Let us show that the short exact sequence (†) splits over $\bar{R}$ (not as s sequence of Galois-modules!).

Let $\sigma \in I_p$ such that $\omega(\sigma)$ generates $\mathbb{F}_p^\times$. Let $\tilde{\omega}(\sigma)$ be the Teichmüller lifting in $\mathbb{Z}_p^\times$ of $\omega(\sigma)$. Since V is finite, there exists an integer $h > 0$ such that

$$V = \ker(\sigma - \tilde{\omega}(\sigma))^n \oplus \ker(\sigma - 1)^h.$$

One can then easily check that the first summand is V^0, while the second is isomorphic to V^E. By assumption, we have

$$V^0 \cong \bar{R},$$

and we have proven that, as $\bar{R}$-modules,

$$V^E \cong \mathrm{Hom}(V^0, \mathbb{F}_p),$$

so

$$V^E \cong \bar{R}^* = \mathrm{Hom}_{\mathbb{F}_p\text{-lin}}(\bar{R}, \mathbb{F}_p).$$

On the other hand, since $p > 2$, the complex conjugation c induces another decomposition of $\bar{R}$-modules:

$$V = V^+ \oplus V^-.$$

So by Krull-Schmidt-Akizuki theorem (see Curtis-Reiner, volume I, theorem 14.5), we conclude:

$$V^\pm \cong \bar{R} \qquad \text{and} \qquad V^\mp \cong \bar{R}^*$$

as $\bar{R}$-modules. (If a module is a sum of indecomposable submodules in two different ways, then up to permutation the indecomposable factors are isomorphic; this, over any ring).

We then observe that

$$\dim_k V^+[\mathfrak{M}] = \dim_k V^-[\mathfrak{M}],$$

because $V[\mathfrak{M}]^{\text{s.s.}}$ is a direct sum of copies of $\rho_{\mathfrak{M}}$ and because $\rho_{\mathfrak{M}}$ is odd, as already noticed. Hence both eigenvalues ± 1 have the same multiplicities in $V_{\mathfrak{M}}$.

We therefore conclude that

$$\dim_k \bar{R}[\mathfrak{M}] = 1.$$

This is a criterion for $\bar{R}$ to be a G_0-ring; and so $\bar{R}^* \cong R$. The theorem follows.

Step 3. We have defined in the proof of Step 1 an abelian variety A of $J_1(N', p)$,

$$A = \left(\ker(J_1(N', p) \xrightarrow{\phi} J(N')^2)\right)^0.$$

It is defined over $\mathbb{Q}$ and it is stable under the action of $\mathbb{T}$.

Let $B = J/A$; it is defined over $\mathbb{Q}$ as well and carries an action of $\mathbb{T}$. Moreover, it has good reduction over $\mathbb{Z}_p$, whereas A has toric reduction over $\mathbb{Z}_p$.

Since $p > 2$, we have by [Ma2], Proposition 1.3, a diagram whose rows are short exact sequences of connected finite flat group schemes:

$$(\mathcal{C}) \qquad \begin{array}{ccccccccc} 0 & \longrightarrow & A[p]^t_{\mathbb{Z}_p} & \longrightarrow & J[p]^t_{\mathbb{Z}_p} & \longrightarrow & B[p]^t_{\mathbb{Z}_p} & \longrightarrow & 0 \\ & & {\scriptstyle\alpha}\downarrow\subset & & \downarrow\subset & & {\scriptstyle\beta}\downarrow\subset & & \\ 0 & \longrightarrow & A[p]^0_{\mathbb{Z}_p} & \longrightarrow & J[p]^0_{\mathbb{Z}_p} & \longrightarrow & B[p]^0_{\mathbb{Z}_p} & \longrightarrow & 0 \end{array}$$

The exactness can be checked on generic fibers by Raynaud's theorem ([Ra3]) and the assumption $p > 2$ (hence $1 = e < p - 1$).

Since all irreducible subquotients of $\mathcal{V} = J[p]^t(\bar{\mathbb{Q}}_p)_{\mathfrak{M}}$ are ramified, we have

$$\mathcal{V} \subset \mathcal{V}^0.$$

To check the reverse inclusion, we take the $\bar{\mathbb{Q}}_p$-points of $(\mathcal{C})$ and localize at $\mathfrak{M}$:

(1) From the fact that $B[p]_{\mathfrak{M}}$ is ordinary, we see that the map induced by β is an isomorphism.
(2) By finiteness and flatness, we check that α induces an isomorphism simply by comparing the ranks; it is enough to to do that on the geometric special fiber; there it is obvious since $A_{/\bar{\mathbb{F}}_p}$ is a split torus.

Hence $\mathcal{V}^0 = \mathcal{V}$.

Step 4. We pass to the tangent spaces and tangent maps in the first row of $(\mathcal{C})$. Again from [Ma2], Proposition 1.3, we have a diagram with exact rows:

$$(\mathcal{C}) \qquad \begin{array}{ccccccccc} 0 & \longrightarrow & t_{A[p]^t_{\bar{\mathbb{F}}_p}} & \longrightarrow & t_{J[p]^t_{\bar{\mathbb{F}}_p}} & \longrightarrow & t_{B[p]^t_{\bar{\mathbb{F}}_p}} & \longrightarrow & 0 \\ & & \alpha\downarrow & & \gamma\downarrow & & \beta\downarrow & & \\ 0 & \longrightarrow & t_{A/\bar{\mathbb{F}}_p} & \longrightarrow & t_{J/\bar{\mathbb{F}}_p} & \longrightarrow & t_{B/\bar{\mathbb{F}}_p} & \longrightarrow & 0 \end{array}$$

Let us localize at $\mathfrak{M}$.

(1) $\alpha_{\mathfrak{M}}$ is an isomorphism because $A_{/\bar{\mathbb{F}}_p}$ is a torus.
(2) $\beta_{\mathfrak{M}}$ is an isomorphism because $B[p]_{\mathfrak{M}}$ is ordinary, so $j_{\mathfrak{M}}$ is an isomorphism.

At this final stage, we make use again of the weak multiplicity one theorem in characteristic p to prove:

$$(**) \qquad (t_{J/\mathbb{F}_p})_{\mathfrak{M}} \cong \bar{R}$$

(as $\bar{R}$-modules).

By Nakayama's lemma, one must check

$$\dim_k(t_{J/\mathbb{F}_p}/\mathfrak{M} \cdot t_{J/\mathbb{F}_p}) = 1.$$

Since $X_1(N', p)$ has ordinary double points we find that Ω is the sheaf of regular differentials on this curve, there is an $\mathbb{F}_p$-duality:

$$t_{J/\mathbb{F}_p}/\mathfrak{M} \cdot t_{J/\mathbb{F}_p} \longleftrightarrow H^0(X_1(N', p)_{\mathbb{F}_p}, \Omega)[\mathfrak{M}].$$

The right-hand side is one-dimensional by Mazur's argument.

To conclude that (∗∗) holds we actually go up to $\mathbb{Z}_p$ and notice that

$$R \twoheadrightarrow (t_{J/\mathbb{Z}_p})_{\mathfrak{M}}$$

is an isomorphism when tensored with $\mathbb{Q}_p$, hence it is injective.

We have thus proven that

$$t_{J[p]^t_{\bar{\mathbb{F}}_p}} \cong \bar{R} \otimes_{\mathbb{F}_p} \bar{\mathbb{F}}_p.$$

We observe then that for any finite flat multiplicative group scheme there is a functorial isomorphism

$$t_{G_{\bar{\mathbb{F}}_p}} \cong G(\bar{\mathbb{Q}}_p) \otimes \bar{\mathbb{F}}_p$$

(see (2.9), page 488 of [W4]).

So we see that

$$\mathcal{V} \otimes \bar{\mathbb{F}}_p \cong \bar{R} \otimes \bar{\mathbb{F}}_p$$

as $\bar{R}$-modules.

We leave it as an exercise of Galois descent that this implies that $\mathcal{V}$ is free over $\bar{R}$.

Appendix by B. Mazur (letter to K. Ribet and to the author) Gorenstein-ness and the multiplicity one theorem in Sections II.15 and II.16 of [M1]

The best manner to correct the erroneous proof of Lemma II.16.3 of [M1] (which simply refers the reader to Lemma II.15.1) is to give a more comprehensive proof of Lemma II.15.1. This new version of the lemma shows that Corollary II.16.2 implies Corollary II.16.3. Actually, in Lemma II.15.1 one shouldn't make any distinction between "Eisenstein maximal ideal" and any "ordinary, good reduction" maximal ideal. It should be stated in the following degree of generality:

Let $\mathbb{T}$ be any finite flat (commutative) $\mathbb{Z}_p$-algebra such that $\mathbb{T} \otimes \mathbb{Q}_p$ is etale (or just Gorenstein!) and J any p-divisible group (over $\operatorname{Spec} \mathbb{Z}_p$, say) which is "ordinary" in the sense that it is an extension of an etale p-divisible group J^{et} by a multiplicative type p-divisible group J^{mt}. Suppose that $\mathbb{T}$ is a ring of endomorphisms of the p-divisible group J over $\operatorname{Spec} \mathbb{Z}_p$. Let ∗ denote Pontrjagin dual.

Make these further assumptions:

(1) $J^{\mathrm{et}*}$ is a free $\mathbb{T}$-module (rank $d > 0$, say).

(2) J is self-dual as a p-divisible group ("Cartier" self-dual) and the action of $\mathbb{T}$ is "Hermitian" with respect to this self-duality, "Hermitian" meaning of course "self-adjoint."

That's it, I believe, for assumptions. Then these things are equivalent:

(i) $J^{\mathrm{et}}[\mathfrak{M}]$ is of dimension d over the residue field.
(ii) $J[\mathfrak{M}]$ is of dimension $2d$ over the residue field.
(iii) J^{mt^*} is free of rank d over $\mathbb{T}$.
(iv) J^* is free of rank $2d$ over $\mathbb{T}$.
(v) $\mathbb{T}$ is Gorenstein.

Proof. Since the self-duality produces a duality between etale and multiplicative type parts we get that the $\mathbb{T}$-modules J^{et^*} and J^{mt^*} are $\mathbb{Z}_p$-duals. Since $J^{\mathrm{et}^*} \otimes \mathbb{Q}_p$ is free of rank d over $\mathbb{T} \otimes \mathbb{Q}_p$, we get (by Gorenstein-ness of $\mathbb{T} \otimes \mathbb{Q}_p$) that $J^{\mathrm{mt}^*} \otimes \mathbb{Q}_p$ is also free of rank d over $\mathbb{T} \otimes \mathbb{Q}_p$.

Now, I guess we can begin to tote up the equivalences. In dealing with property (iv) it will be useful to remember that $\mathbb{T}$ is Gorenstein if and only if the quotient $\mathbb{T}$-module

$$\mathrm{Hom}(\mathbb{T}, \mathbb{Z}_p)/\mathfrak{M}\,\mathrm{Hom}(\mathbb{T}, \mathbb{Z}_p)$$

is of dimension one over the residue field.

I claim that (i) is equivalent to (iii) because $J[\mathfrak{M}]$ is the Pontrjagin dual of $J^*/\mathfrak{M}J^*$. Also, (i) implies (iii), for by Nakayama, given (i), we can find a surjective $\mathbb{T}$-homomorphism from $\mathbb{T}^d$ to J^{mt^*} which must have trivial kernel, as can be seen by counting the ranks over $\mathbb{Q}_p$ of these modules tensored with $\mathbb{Q}_p$, since otherwise $J^{\mathrm{mt}^*} \otimes \mathbb{Q}_p$ could not be free of rank d over $\mathbb{T} \otimes \mathbb{Q}_p$, which it is. By the same argument, (iv) implies (ii). By what we "remembered" above, counting dimensions over the residue field, we see that (v) is equivalent to (i) and to (ii), noting that J^{mt^*} is the $\mathbb{Z}_p$-dual of a free $\mathbb{T}$-module of rank d, and noting that J^* is self-$\mathbb{Z}_p$-dual.

REFERENCES

[A-K] A. Altman, S. Kleinman, *Introduction to Grothendieck duality theory*, SLN, Springer.

[B] H. Bass, *On the ubiquity of Gorenstein rings*, Math. Zeit. **82** (1963), 8–28.

[Bu] K. Buzzard, *The levels of modular representations*, Ph.D. thesis, Cambridge, 1995.

[D-R] P. Deligne, M. Rapoport, *Les Schémas de modules de courbes elliptiequs*, Modular functions of one variable, SLN 349, Springer, 1973.

[Gr] A. Grothendieck, *Local cohomology*, SLN 41, Springer, 1967.

[H] H. Hida, *On congruence divisors of cusp forms as special values of their zeta functions*, Invent. Math. **64** (1981), 221–262.

[Ku] E. Kunz, *Introduction to commutative algebra and algebraic geometry*, Birkhaüser, 1985.

[Mat] H. Matsumura, *Commutative algebra*, Benjamin, 1970.

[M1] B. Mazur, *Modular curves and the Eisenstein ideal*, Publ. Math. IHES **47** (1977), 133–186.

[M2] ———, *Rational isogenics of prime degree*, Invent. Math. **44** (1978), 129–162.

[M-Ri] B. Mazur, K. Ribet, *Two dimensional representations in the arithmetic of modular curves*, in Courbes modulaires et courbes de Shimura, Astérisque **196–197** (1991).

[M-T] B. Mazur, J. Tilouine, *Représentationns galoisiennes différentielles de Kähler et conjectures principales*, Publ Math. IHES **71** (1990), 65–103.

[M-W] B. Mazur, A. Wiles, *Class-fields of abelian extensions of* $\mathbb{Q}$, Invent. Math. **76** (1984), 179–330.

[Ra1] M. Raynaud, *Variétés abéliennes et géométrie rigide*, Actes Congres Int. Math., vol. 1, 1970, pp. 473–477.

[Ra2] ———, *Spécialisation du foncteur de Picard*, Publ. Math. IHES **38** (1970), 27–76.

[Ra3] ———, *Schémas en groupes de type* $(p, p, \ldots, p)$, Bull. Soc. Math. France **102** (1974), 241–280.

[Ta] J. Tate, *p-divisible groups*, Proceedings of a conference on local fields (Driebergen 1966), Springer, 1967, pp. 158–183.

[T-W] R. Taylor, A. Wiles, *Ring theoretic properties of certain Hecke algebras*, Ann. Math. **141** (1995), 553–572.

[Ti] J. Tilouine, *Un sous-groupe p-divisible de la jacobienne de* $X_1(N_p^r)$ *comme module sur l'algébre de Hecke*, Bull. Soc. Math. France **115** (1987), 329–360.

[W1] A. Wiles, *Modular curves and the class group of* $\mathbb{Q}(\mu_p)$, Invent. Math. **58** (1980), 407–456.

[W2] ———, *On p-adic representations for totally real fields*, Ann. Math. **123** (1986), 407–456.

[W3] ———, *The Iwasawa conjecture for totally real fields*, Ann. Math. **131** (1990), 493–540.

[W4] ———, *Modular elliptic curves and Fermat's last theorem*, Ann. Math. **142** (1995), 443–551.

CRITERIA FOR COMPLETE INTERSECTIONS

Bart de Smit, Karl Rubin, and René Schoof

Introduction

In this paper we discuss two results in commutative algebra that are used in A. Wiles's proof that all semi-stable elliptic curves over $\mathbf{Q}$ are modular [11].

We first fix some notation that is used throughout this paper. Let $\mathcal{O}$ be a complete Noetherian local ring with maximal ideal $\mathfrak{m}_{\mathcal{O}}$ and residue field $k = \mathcal{O}/\mathfrak{m}_{\mathcal{O}}$. Suppose that we have a commutative triangle of surjective homomorphisms of complete Noetherian local $\mathcal{O}$-algebras:

$$\begin{array}{ccc} R & \xrightarrow{\varphi} & T \\ & {\scriptstyle \pi_R} \searrow \quad \swarrow {\scriptstyle \pi_T} & \\ & \mathcal{O}. & \end{array}$$

Assume that T is a finite flat $\mathcal{O}$-algebra, i.e., that T is finitely generated and free as an $\mathcal{O}$-module. In the applications in Wiles's proof $\mathcal{O}$ is a discrete valuation ring, R is a deformation ring, T is a Hecke algebra and π_T is the homomorphism associated to a certain eigenform.

We show two distinct criteria, formulated as Criterion I and Criterion II below, which give sufficient conditions to conclude that φ is an isomorphism and that R and T are complete intersections. We say that a local $\mathcal{O}$-algebra that is finitely generated as an $\mathcal{O}$-module is a *complete intersection* over $\mathcal{O}$ if it is of the form

$$\mathcal{O}[[X_1,\ldots,X_n]]/(f_1,\ldots,f_n), \qquad \text{with } f_1,\ldots,f_n \in \mathcal{O}[[X_1,\ldots,X_n]].$$

We first state Criterion I. We put $I_R = \ker \pi_R$ and $I_T = \ker \pi_T$. The *congruence ideal* of T is defined to be the $\mathcal{O}$-ideal $\eta_T = \pi_T \operatorname{Ann}_T(I_T)$.

Criterion I. *Suppose that $\mathcal{O}$ is a complete discrete valuation ring and that $\eta_T \neq 0$. Then*

$$\operatorname{length}_{\mathcal{O}}(I_R/I_R^2) \;\geq\; \operatorname{length}_{\mathcal{O}}(\mathcal{O}/\eta_T).$$

Moreover, equality holds if and only if φ is an isomorphism between complete intersections over $\mathcal{O}$.

Wiles used a slightly weaker form of this criterion, where T is assumed to be Gorenstein, to show that certain "non-minimal" deformation rings are isomorphic to Hecke algebras [4]. The present version, without the Gorenstein condition, is due to H.W. Lenstra [6]. In Section 3 we give an alternative argument for Criterion I that was found by the first and the third author. Criterion I is an easy consequence of the following result, which holds without any conditions on $\mathcal{O}$ or η_T.

Theorem. *The map φ is an isomorphism between complete intersections over $\mathcal{O}$ if and only if $\varphi \operatorname{Fit}_R(I_R) \not\subset \mathfrak{m}_{\mathcal{O}} T$.*

Here $\operatorname{Fit}_R(I_R)$ denotes the R-Fitting ideal of I_R. Fitting ideals are instrumental in the proof of Criterion I. We recall their definition and basic properties in Section 1.

A crucial special case of the theorem can already be found in a 1969 paper of H. Wiebe [10]; see also [1, Thm. 2.3.16]. More precisely, Wiebe's result covers the case that $\mathcal{O} = k$ is a field, and φ is the identity on $R = T$. The statement is then that T is a complete intersection over k if and only if the Fitting ideal of its maximal ideal is non-zero.

For the proof of Criterion I we need some properties of complete intersections that go back to J.T. Tate [8]. In Section 2 we formulate Tate's result and prove it using Koszul complexes. These are discussed in Section 1. As a consequence we find that complete intersections have the Gorenstein property. The Gorenstein property does not occur in our proof of Criterion I, but we briefly discuss its significance in our context at the end of Section 2.

In order to formulate Criterion II, assume that $\operatorname{char}(k) = p > 0$, and let $n \geq 1$. The ring $\mathcal{O}[[S_1, \ldots, S_n]]$ is filtered by the ideals J_m, with $m \geq 0$, given by $J_m = (\omega_m(S_1), \ldots, \omega_m(S_n))$, where $\omega_m(S)$ denotes the polynomial $(1+S)^{p^m} - 1$. Note that $J_0 = (S_1, \ldots, S_n)$.

Criterion II. *Suppose that for every $m > 0$ there is a commutative diagram of $\mathcal{O}$-algebras*

$$\begin{array}{ccccc} \mathcal{O}[[S_1, \ldots, S_n]] & \longrightarrow & R_m & \xrightarrow{\varphi_m} & T_m \\ & & \downarrow & & \downarrow \\ & & R & \xrightarrow{\varphi} & T \end{array}$$

with the properties:

(i) there is a surjection of $\mathcal{O}$-algebras $\mathcal{O}[[X_1, \ldots, X_n]] \longrightarrow R_m$;
(ii) the map $\varphi_m \colon R_m \longrightarrow T_m$ is surjective;
(iii) the vertical arrows induce isomorphisms

$$R_m / J_0 R_m \xrightarrow{\sim} R \qquad \text{and} \qquad T_m / J_0 T_m \xrightarrow{\sim} T.$$

(iv) the quotient ring $T_m / J_m T_m$ is finite flat over $\mathcal{O}[[S_1, \ldots, S_n]]/J_m$;

Then $\varphi \colon R \longrightarrow T$ is an isomorphism between complete intersections over $\mathcal{O}$.

Criterion II, with the additional condition that k be a finite field, first appeared in the paper by R. Taylor and A. Wiles [9] with an improvement due to G. Faltings. It is used by Wiles for the "minimal" deformation problem [2]. In Section 4 we present a proof due to the second author. It is independent of the proof of Criterion I. Our approach avoids the original non-canonical limiting process, and it works for arbitrary complete Noetherian local rings $\mathcal{O}$.

1. Preliminaries.

In this section we first recall the definition and basic properties of Fitting ideals. Then we do the same for Koszul complexes following [3]. For more details see [5, Sections XIX.2, XXI.4].

Fitting ideals. Let A be a ring and let M be a finitely generated A-module with generators $m_1, \ldots, m_n$. Let $f\colon A^n \twoheadrightarrow M$ be the surjective A-homomorphism defined by $f(e_i) = m_i$ for $i = 1, \ldots, n$. Here e_i denotes the ith standard basis vector of A^n. The *Fitting ideal* $\mathrm{Fit}_A(M)$ of M is the ideal generated by the determinants $\det(v_1, \ldots, v_n)$ for which the column vectors $v_1, \ldots, v_n$ lie in $\ker f$. Clearly, $\mathrm{Fit}_A(M)$ is already generated by the elements $\det(v_1, \ldots, v_n)$ with $v_1, \ldots, v_n$ in a fixed set of A-module generators of $\ker f$.

The Fitting ideal does not depend on the choice of the generators m_i. To see this, let $m_{n+1} = \sum_{i=1}^{n} \alpha_i m_i$, with $\alpha_i \in A$, be an additional generator of M. The kernel of the surjective homomorphism $\psi\colon A^{n+1} \twoheadrightarrow M$ given by $\psi(e_i) = m_i$ for $i = 1, \ldots, n+1$, is generated by the vector $(\alpha_1, \ldots, \alpha_n, -1)$ and the vectors $(v, 0)$ with $v \in \ker f$. It follows at once that the Fitting ideal does not change when we replace the generators $m_1, \ldots, m_n$ by $m_1, \ldots, m_n, m_{n+1}$. Inductively, this implies that any two generating sets $m_1, \ldots, m_n$ and $m'_1, \ldots, m'_{n'}$ give rise to the same Fitting ideal as their union $m_1, \ldots, m_n, m'_1, \ldots, m'_{n'}$.

The following proposition contains the properties of the Fitting ideal that we will use.

Proposition 1.1. *Let A be a ring and let M be a finitely generated A-module. Then*

(i) we have $\mathrm{Fit}_A(M) \subset \mathrm{Ann}_A(M)$;
(ii) for any A-algebra B we have $\mathrm{Fit}_B(M \otimes_A B) = \mathrm{Fit}_A(M) \cdot B$;
(iii) for any ideal $\mathfrak{a} \subset A$ we have $\mathrm{Fit}_A(A/\mathfrak{a}) = \mathfrak{a}$;
(iv) for every A-module N we have $\mathrm{Fit}_A(M \times N) = \mathrm{Fit}_A(M)\mathrm{Fit}_A(N)$.

Proof. We sketch the proof. If $v_1, \ldots, v_n$ are in the kernel of $A^n \xrightarrow{f} M$, then the matrix σ with columns $v_1, \ldots, v_n$ has the property that the composite map $A^n \xrightarrow{\sigma} A^n \xrightarrow{f} M$ is equal to zero. By multiplying first with the adjoint matrix of σ, we see that $\det(\sigma) \cdot A^n \subset \ker f$. Since f is surjective, this implies that $\det(\sigma) \in \mathrm{Ann}_A(M)$, and (*i*) follows. Part (*ii*) follows from the fact that taking the tensor product with B is right exact. Part (*iii*) is immediate from the definition if we take $n = 1$. We leave part (*iv*) to the reader.

If A is a principal ideal domain, then, by the theory of elementary divisors, every finitely generated A-module M is of the form

$$M \cong A/\mathfrak{a}_1 \times \ldots \times A/\mathfrak{a}_s$$

for certain ideals $\mathfrak{a}_i \subset A$. By (*iii*) and (*iv*), we see that $\mathrm{Fit}_A(M) = \mathfrak{a}_1 \cdots \mathfrak{a}_s$. If A is a discrete valuation ring with maximal ideal $\mathfrak{m}_A$, then we see that

$$\mathrm{Fit}_A(M) = \mathfrak{m}_A^{\mathrm{length}_A(M)},$$

with the convention that $\mathfrak{m}_A^\infty = 0$.

Example. Let $\mathcal{O}$ be any ring and let $A = \mathcal{O}[[X_1, \ldots, X_n]]/J$ with $J = (f_1, \ldots, f_r)$ an ideal contained in $I = (X_1, \ldots, X_n)$. We put $I_A = I/J$. Suppose that $g_{ij} \in \mathcal{O}[[X_1, \ldots, X_n]]$ satisfy

$$f_i = \sum_{j=1}^{n} g_{ij} X_j \qquad \text{for } i = 1, \ldots, r.$$

Then the Fitting ideal $\mathrm{Fit}_A(I_A)$ contains the determinants, taken modulo J, of the $n \times n$ submatrices of the matrix (g_{ij}). Actually, it can be shown that these determinants generate $\mathrm{Fit}_A(I_A)$ by applying Proposition 1.3 below with $i = 1$ to the sequence $X_1, \ldots, X_n$ in $\mathcal{O}[[X_1, \ldots, X_n]]$. This will not be used in the sequel. By a different argument, we will obtain a special case in Proposition 2.1.

Koszul complexes. Let A be a ring, let $V = A^n$ and let $f = (f_1, \ldots, f_n) \in V$. For any A-module M and $m \geq 0$ we set

$$K_m(f, M) = \mathrm{Hom}_A(\textstyle\bigwedge_A^m V, M).$$

For $\varphi \in K_m(f, M)$ we define $d\varphi \in K_{m-1}(f, M)$ by $d\varphi(x) = \varphi(f \wedge x)$. Since $d^2 = 0$, we obtain a complex $K_\bullet(f, M)$, which we call the *Koszul complex* of f on M:

$$0 \longrightarrow K_n(f, M) \xrightarrow{d} \cdots \xrightarrow{d} K_1(f, M) \xrightarrow{d} K_0(f, M) \longrightarrow 0.$$

Note that $K_\bullet(f, M) = K_\bullet(f, A) \otimes_A M$ and that $K_m(f, A)$ is a free A-module of rank $\binom{n}{m}$. The m-th homology group of $K_\bullet(f, M)$ is denoted by $H_m(f, M)$. We have $H_0(f, M) = M/IM$, where I is the A-ideal generated by the f_i.

Lemma 1.2. *The homology groups $H_m(f, M)$ are annihilated by I.*

Proof. Let $\varphi \in K_m(f, M)$ with $d\varphi = 0$. For each generator f_i of I we must show that there is $\psi \in K_{m+1}(f, M)$ with $d\psi = f_i\varphi$. To see this, write $V = Ae_i \times V'$ where e_i is the ith standard basis vector of V over A, and V' is generated by the other standard basis vectors. Then every $x \in \bigwedge^{m+1} V$ can be written as $x = e_i \wedge x' + x''$ for unique $x' \in \bigwedge^m V'$ and $x'' \in \bigwedge^{m+1} V'$. Now define $\psi \in K_{m+1}(f, M)$ by $\psi(x) = \varphi(x')$. From $d\varphi = 0$ one deduces that $d\psi = f_i\varphi$, as required.

We say that a sequence of elements $p_1, \ldots, p_n$ in A is *M-regular*, if for $i = 1, \ldots, n$ the multiplication by p_i on $M/(p_1, \ldots, p_{i-1})M$ is an injective map. The following proposition can also be found in [1, Thm. 1.6.16].

Proposition 1.3. *Let $f = (f_1, \ldots, f_n) \in A^n$ and let M be an A-module. If the A-ideal I generated by $f_1, \ldots, f_n$ contains an M-regular sequence of length n, then $H_i(f, M) = 0$ for $i \geq 1$.*

Proof. Let $p_1, \ldots, p_n \in A$ be an M-regular sequence in I. For any integer j with $0 \leq j \leq n$ we claim that $H_i(f, M/(p_1, \ldots, p_j)M) = 0$ for all $i \geq j+1$. For $j = n$ this is trivial, and for $j = 0$ this is the content of the proposition. We prove the claim by induction on j, decreasing j by 1 in each step.

Assume that the claim holds for some integer j with $1 \leq j \leq n$. We put $M' = M/(p_1, \ldots, p_{j-1})M$. Since the sequence $p_1, \ldots, p_n$ is M-regular, there is an exact sequence

$$0 \longrightarrow M' \xrightarrow{p_j} M' \longrightarrow M'/p_jM' \longrightarrow 0.$$

For each m we apply the exact functor $\mathrm{Hom}_A(\bigwedge^m V, -)$. This gives us a short exact sequence of complexes

$$0 \longrightarrow K_\bullet(f, M') \xrightarrow{p_j} K_\bullet(f, M') \longrightarrow K_\bullet(f, M'/p_jM') \longrightarrow 0.$$

By Lemma 1.2 the homology groups of $K_\bullet(f, M')$ are annihilated by I and therefore by p_j. This implies that the long exact homology sequence breaks up into short exact sequences. For every $i \geq 1$ we obtain an exact sequence

$$0 \longrightarrow H_i(f, M') \longrightarrow H_i(f, M'/p_jM') \longrightarrow H_{i-1}(f, M') \longrightarrow 0.$$

The induction hypothesis implies that the middle group is zero for $i \geq j+1$. This implies that $H_i(f, M') = 0$ for $i \geq j$, which is the claim for $j - 1$.

2. Complete intersections.

This section is devoted to the proof of the following result, which goes back to Tate [8].

Proposition 2.1. *Let $\mathcal{O}$ be a complete Noetherian local ring. Let A be a finite flat $\mathcal{O}$-algebra of the form $A = \mathcal{O}[[X_1, \ldots, X_n]]/(f_1, \ldots, f_n)$ with $(f_1, \ldots, f_n) \subset (X_1, \ldots, X_n)$. Write $f_i = \sum_{j=1}^n g_{ij}X_j$, let d be the image of $\det(g_{ij})$ in A, and let I_A be the A-ideal $I_A = (X_1, \ldots, X_n)/(f_1, \ldots, f_n)$. Then we have*

(i) $\mathrm{Fit}_A(I_A) = \mathrm{Ann}_A(I_A) = (d)$;

(ii) the A-ideal (d) is a direct $\mathcal{O}$-summand of A of $\mathcal{O}$-rank 1.

Proof. Let $P = \mathcal{O}[[X_1, \ldots, X_n]]$, and let f be the vector $(f_1, \ldots, f_n) \in P^n$. Multiplication by the matrix (g_{ij}) gives an P-linear map $P^n \longrightarrow P^n$ sending the vector $X = (X_1, \ldots, X_n)$ to f. It induces a morphism of Koszul complexes

$$K_\bullet(f, P) \longrightarrow K_\bullet(X, P).$$

Since A is finitely generated as an $\mathcal{O}$-module, the P-ideal $(f_1, \ldots, f_n)$ contains a monic polynomial $p_i \in \mathcal{O}[X_i]$ for each i. The sequence $p_1, \ldots, p_n$ is $\mathcal{O}[X_1, \ldots, X_n]$-regular and by exactness of completion it is also P-regular. By Proposition 1.3 the homology groups of both Koszul complexes vanish and we obtain the following commutative diagram with exact rows

$$\begin{array}{ccccccccccccc}
0 & \longrightarrow & P & \xrightarrow{(f_1,\ldots,f_n)} & P^n & \longrightarrow & \cdots & \longrightarrow & P^n & \xrightarrow{(f_1,\ldots,f_n)} & P & \longrightarrow & A & \longrightarrow & 0 \\
 & & \downarrow{\scriptstyle \det(g_{ij})} & & \downarrow & & & & \downarrow{\scriptstyle (g_{ij})} & & \| & & \downarrow{\scriptstyle \pi_A} & & \\
0 & \longrightarrow & P & \xrightarrow{(X_1,\ldots,X_n)} & P^n & \longrightarrow & \cdots & \longrightarrow & P^n & \xrightarrow{(X_1,\ldots,X_n)} & P & \longrightarrow & \mathcal{O} & \longrightarrow & 0.
\end{array}$$

Here π_A is the $\mathcal{O}$-algebra map $A \longrightarrow \mathcal{O}$ with kernel I_A. We now tensor the whole diagram *on the right* with the P-module A. Since the rows are P-free resolutions of A and $\mathcal{O}$, the homology groups of the rows become $\mathrm{Tor}_j^P(A, A)$ and $\mathrm{Tor}_j^P(\mathcal{O}, A)$ respectively. Hence, we obtain a commutative diagram with exact rows:

$$\begin{array}{ccccccc}
0 & \longrightarrow & \mathrm{Tor}_n^P(A, A) & \longrightarrow & A & \xrightarrow{0} & A^n \\
 & & \downarrow{\scriptstyle \pi_{A*}} & & \downarrow{\scriptstyle d} & & \downarrow \\
0 & \longrightarrow & \mathrm{Tor}_n^P(\mathcal{O}, A) & \longrightarrow & A & \xrightarrow{(X_1,\ldots,X_n)} & A^n.
\end{array}$$

It follows that $\mathrm{Tor}_n^P(\mathcal{O}, A) \cong \mathrm{Ann}_A(I_A)$. In order to determine this Tor-group and the image of π_{A*}, we tensor the P-resolution $K_\bullet(f, P)$ of A *on the left* with the P-module map $\pi_A \colon A \longrightarrow \mathcal{O}$. This gives a map between two complexes with homology groups $\mathrm{Tor}_j^P(A, A)$ and $\mathrm{Tor}_j^P(\mathcal{O}, A)$ respectively. Since one can compute Tor-functors using resolutions of either argument

[5, Chap. XX, Prop. 8.2′], the same map π_{A*} then makes the following diagram with exact rows commute:

$$\begin{array}{ccccccc} 0 & \longrightarrow & \mathrm{Tor}_n^P(A,A) & \longrightarrow & A & \xrightarrow{0} & A^n \\ & & \downarrow{\scriptstyle \pi_{A*}} & & \downarrow{\scriptstyle \pi_A} & & \downarrow{\scriptstyle (\pi_A,\ldots,\pi_A)} \\ 0 & \longrightarrow & \mathrm{Tor}_n^P(\mathcal{O},A) & \longrightarrow & \mathcal{O} & \xrightarrow{0} & \mathcal{O}^n. \end{array}$$

In particular we see that π_{A*} is surjective, so that $(d) = \mathrm{Ann}_A(I_A)$ and (d) is free of rank 1 as an $\mathcal{O}$-module. On the other hand, we have

$$(d) \subset \mathrm{Fit}_A(I_A) \subset \mathrm{Ann}_A(I_A),$$

and therefore equality holds everywhere. By applying what we have already proved to the complete intersection $A \otimes_{\mathcal{O}} k$ over k we see that $d \otimes 1 \neq 0$ in $A \otimes_{\mathcal{O}} k$, so that $d \notin \mathfrak{m}_{\mathcal{O}} A$. By Nakayama's lemma we can therefore make the element d part of an $\mathcal{O}$-basis of A, so that the inclusion $(d) \subset A$ splits as an $\mathcal{O}$-linear map. This proves the proposition.

Corollary 2.2. *If in the situation of Proposition 2.1 the ring $\mathcal{O}$ is a field, then (d) is the unique minimal non-zero ideal of A.*

Proof. Proposition 2.1 says that (d) has dimension 1 over $\mathcal{O} = k$, so (d) contains no smaller non-zero ideals. On the other hand, every minimal ideal $\mathfrak{a}$ is annihilated by the maximal ideal I_A of A, and by Proposition 2.1 we have $\mathrm{Ann}_A(I_A) = (d)$, so $\mathfrak{a} \subset (d)$.

Corollary 2.3. *Let A be a finite flat $\mathcal{O}$-algebra with a section $\pi_A : A \longrightarrow \mathcal{O}$ and let $I_A = \ker \pi_A$. If A is a complete intersection over $\mathcal{O}$, then $\mathrm{Fit}_A(I_A) = \mathrm{Ann}_A(I_A)$, and this ideal is a non-zero direct $\mathcal{O}$-summand of A.*

Proof. Suppose that $A = \mathcal{O}[[X_1, \ldots, X_n]]/(f_1, \ldots, f_n)$. Since $\mathcal{O}$ is a complete local ring, we can replace each variable X_i by $X_i - \pi_A(X_i)$. This ensures that $(f_1, \ldots, f_n) \subset (X_1, \ldots, X_n)$. The result then follows from Proposition 2.1.

We conclude this section with some remarks that will not be used in the rest of this paper.

The Gorenstein condition. Let A be a finite flat $\mathcal{O}$-algebra. Then the $\mathcal{O}$-linear dual $A^\vee = \mathrm{Hom}_{\mathcal{O}}(A, \mathcal{O})$ of A has an A-module structure given by $(af)(x) = f(ax)$ for $f \in A^\vee$ and $a, x \in A$. The algebra A is called *Gorenstein* over $\mathcal{O}$ if $A^\vee$ is a free A-module of rank 1.

It follows from Proposition 2.1 (*ii*) that for A of the form

$$\mathcal{O}[[X_1, \ldots, X_n]]/(f_1, \ldots, f_n) \quad \text{with} \quad (f_1, \ldots, f_n) \subset (X_1, \ldots, X_n),$$

there exists an $\mathcal{O}$-linear map $t\colon A \longrightarrow \mathcal{O}$ with $t(d) = 1$. This homomorphism t generates $A^\vee$ as an A-module, so that A is Gorenstein over $\mathcal{O}$. To see this when $\mathcal{O}$ is a field, one notes that $(d) \not\subset \mathrm{Ann}_A(t)$, so that $\mathrm{Ann}_A(t) = 0$ by Corollary 2.2. With Nakayama's lemma the general case then follows as well.

In general, suppose that A is Gorenstein, so there exists an A-module isomorphism $s\colon A^\vee \xrightarrow{\sim} A$. Assume further that there is a section $\pi_A\colon A \longrightarrow \mathcal{O}$ and put $I_A = \ker \pi_A$. Then the image of the composite map

$$\mathcal{O} \cong \mathcal{O}^\vee \xrightarrow{\pi_A^\vee} A^\vee \underset{s}{\xrightarrow{\sim}} A$$

is $\mathrm{Ann}_A(I_A)$. To see this, one notes that the image of $\pi_A^\vee$ is

$$\mathcal{O} \cdot \pi_A = \{f \in A^\vee : f(I_A) = 0\},$$

and that

$$f(I_A) = 0 \iff I_A \cdot f = 0 \iff s(f) \in \mathrm{Ann}_A(I_A).$$

Applying π_A, we see that the congruence ideal $\eta_A = \pi_A \,\mathrm{Ann}_A(I_A)$ is equal to the $\mathcal{O}$-ideal generated by $\pi_A \circ s \circ \pi_A^\vee(1)$. It is this property that Wiles uses to *define* the congruence ideal in the Gorenstein case.

More general complete intersections. The statement that finite complete intersection algebras are Gorenstein holds over much more general base rings, and it also holds if there is no section $A \longrightarrow \mathcal{O}$. Moreover, one can omit the flatness condition on A in Proposition 2.1, because it follows from the other assumptions. More precisely, if $\mathcal{O}$ is any ring and the ring $A = \mathcal{O}[X_1, \ldots, X_n]/(f_1, \ldots, f_n)$ is finitely generated as an $\mathcal{O}$-module, then one can show with Koszul complexes that A is projective as an $\mathcal{O}$-module [3]. An argument of Tate [7, appendix] then implies that $A^\vee$ is free of rank 1 over A. For Noetherian $\mathcal{O}$ the class of finite $\mathcal{O}$-algebras of the form $\mathcal{O}[[X_1, \ldots, X_n]]/(f_1, \ldots, f_n)$ is a subclass of the class of finite algebras of the form $\mathcal{O}[X_1, \ldots, X_n]/(f_1, \ldots, f_n)$; see [3]. In particular, these algebras are also projective and Gorenstein over $\mathcal{O}$.

3. Proof of Criterion I

In this section we first prove the theorem in the introduction and then show Criterion I. Using Nakayama's lemma we first show that the question whether φ is an isomorphism reduces to the case that $\mathcal{O}$ is a field.

Lemma 3.1. *Let $f\colon A \longrightarrow B$ be a surjective homomorphism of Noetherian local $\mathcal{O}$-algebras for which B is finite flat over $\mathcal{O}$. Suppose that the induced map $\overline{f}\colon A \otimes_{\mathcal{O}} k \longrightarrow B \otimes_{\mathcal{O}} k$ is an isomorphism. Then f is an isomorphism.*

Proof. By applying Nakayama's lemma to B as an $\mathcal{O}$-module we see that f is surjective. Since B is $\mathcal{O}$-free, $(\ker f) \otimes_{\mathcal{O}} k$ is the kernel of $\overline{f}$, which is zero. The ring A is Noetherian, so $\ker f$ is finitely generated as an A-module. Since $\mathfrak{m}_{\mathcal{O}}$ is contained in the maximal ideal of A we can apply Nakayama's lemma to the A-module $\ker f$ and conclude that $\ker f = 0$.

Now we give the proof of the theorem stated in the introduction. Recall that we have a commutative triangle of surjective homomorphisms of complete Noetherian local $\mathcal{O}$-algebras with T finite and flat over $\mathcal{O}$:

$$\begin{array}{ccc} R & \xrightarrow{\varphi} & T \\ & {\scriptstyle \pi_R} \searrow \quad \swarrow {\scriptstyle \pi_T} & \\ & \mathcal{O}. & \end{array}$$

We let $I_R = \ker \pi_R$ and $I_T = \ker \pi_T$.

Theorem. *The map φ is an isomorphism between complete intersections over $\mathcal{O}$ if and only if $\varphi \operatorname{Fit}_R(I_R) \not\subset \mathfrak{m}_{\mathcal{O}} T$.*

Proof. In order to show "only if", we note that by Corollary 2.3, $\operatorname{Fit}_T(I_T)$ is a non-zero direct $\mathcal{O}$-summand of T and in particular

$$\varphi \operatorname{Fit}_R(I_R) = \operatorname{Fit}_T(I_T) \not\subset \mathfrak{m}_{\mathcal{O}} T.$$

To show "if", suppose first that $\mathcal{O} = k$ is a field. Since R is complete and Noetherian, we can write $R = k[[X_1, \ldots, X_n]]/J_R$ where J_R is a $k[[X_1, \ldots, X_n]]$-ideal. Since T is a finite dimensional k-vector space, we can do this in such a way that the elements $\varphi(X_i \bmod J_R)$ generate I_T as a k-vector space. The kernel J_T of the composite map

$$k[[X_1, \ldots, X_n]] \twoheadrightarrow R \xrightarrow{\varphi} T$$

is contained in the ideal $I = (X_1, \ldots, X_n)$. We assume that $\varphi \operatorname{Fit}_R(I_R) \neq 0$, which means that there are polynomials $g_{ij} \in k[[X_1, \ldots, X_n]]$ so that $\sum_j g_{ij} X_j \in J_R$ for $i = 1, \ldots, n$ and $\det(g_{ij}) \notin J_T$.

Since the elements X_i generate I/J_T as a k-vector space, the monomials $X_i X_j$ generate I^2/IJ_T as a k-vector space. This implies that every element of the quotient ring $k[[X_1, \ldots, X_n]]/IJ_T$ is represented by a polynomial of total degree at most 2. Therefore, we can, for $i = 1, \ldots, n$, find polynomials p_i and q_i of total degree at most 2, so that

$$p_i \equiv \sum_j g_{ij} X_j \pmod{IJ_T},$$

$$q_i \equiv X_i^3 \pmod{IJ_T}.$$

We now let the polynomials $f_1, \ldots, f_n$ be

$$f_i = X_i^3 - q_i + p_i \qquad \text{for } i = 1, \ldots, n.$$

Note first that $f_i \in IJ_T + J_R \subset J_T$ and that $f_i = \sum_j G_{ij}X_j$ with $G_{ij} \equiv g_{ij} \bmod J_T$.

The k-algebra $B = k[X_1, \ldots, X_n]/(f_1, \ldots, f_n)$ has finite dimension as a k-vector space, because every element in B is represented by a polynomial of degree at most 2 in each variable. Therefore, B is Artinian and it is a finite product of local Artinian rings. Hence, the completion $\hat{B} = k[[X_1, \ldots, X_n]]/(f_1, \ldots, f_n)$ of B at $(X_1, \ldots, X_n)$ is a factor of B, so it is also finite dimensional over k. By Corollary 2.2 the $\hat{B}$-ideal generated by $\det(G_{ij})$ is the unique minimal non-zero ideal of $\hat{B}$. Since $\det(G_{ij}) \equiv \det(g_{ij}) \not\equiv 0 \pmod{J_T}$, this minimal ideal does not map to 0 in T. It follows that the map $\hat{B} \longrightarrow T$ is an isomorphism. Thus, T is a complete intersection over k, and $J_T = (f_1, \ldots, f_n) \subset IJ_T + J_R$. By Nakayama's lemma we must have $J_T = J_R$ so that φ is an isomorphism. This completes the proof in the case that $\mathcal{O} = k$.

We now prove the "if" part for general $\mathcal{O}$. The map π_R: $R \longrightarrow \mathcal{O}$ is an $\mathcal{O}$-split surjection, so the induced map $R \otimes_{\mathcal{O}} k \longrightarrow k$ has kernel $I_R \otimes_{\mathcal{O}} k$. Since $\mathrm{Fit}_k(I_R \otimes_{\mathcal{O}} k)$ is the image in $R \otimes_{\mathcal{O}} k$ of $\mathrm{Fit}_R(I_R)$, the case that we proved already implies that the map $R \otimes_{\mathcal{O}} k \longrightarrow T \otimes_{\mathcal{O}} k$ is an isomorphism between complete intersections over k. Lemma 3.1 implies that φ is an isomorphism. Moreover, we can lift any k-algebra isomorphism

$$k[[X_1, \ldots, X_n]]/(f_1, \ldots, f_n) \xrightarrow{\sim} T \otimes_{\mathcal{O}} k.$$

to a surjective $\mathcal{O}$-algebra homomorphism ψ: $\mathcal{O}[[X_1, \ldots, X_n]] \longrightarrow T$. The kernel of ψ contains lifts $\tilde{f}_i$ of the elements f_i, and by Lemma 3.1 the induced map

$$\mathcal{O}[[X_1, \ldots, X_n]]/(\tilde{f}_1, \ldots, \tilde{f}_n) \twoheadrightarrow T.$$

is an isomorphism. This proves the theorem.

Proof of Criterion I. First we show the inequality. By Proposition 1.1 (*i*) we have $\mathrm{Fit}_R(I_R) \subset \mathrm{Ann}_R(I_R)$. Since the map $I_R \xrightarrow{\varphi} I_T$ is surjective, we have $\varphi \, \mathrm{Ann}_R(I_R) \subset \mathrm{Ann}_T(I_T)$. Hence we see that

$$\pi_R \, \mathrm{Fit}(I_R) = \pi_T \varphi \, \mathrm{Fit}_R(I_R) \; \subset \; \pi_T \, \mathrm{Ann}_T(I_T) = \eta_T = \mathfrak{m}_{\mathcal{O}}^{\mathrm{length}_{\mathcal{O}}(\mathcal{O}/\eta_T)}.$$

Viewing $\mathcal{O}$ as an R-algebra via π_R: $R \longrightarrow \mathcal{O}$ we have $I_R \otimes_R \mathcal{O} = I_R/I_R^2$. By Proposition 1.1 (*ii*) this implies that

$$\pi_R \, \mathrm{Fit}_R(I_R) = \mathrm{Fit}_{\mathcal{O}}(I_R/I_R^2) = \mathfrak{m}_{\mathcal{O}}^{\mathrm{length}_{\mathcal{O}}(I_R/I_R^2)},$$

and it follows that $\mathrm{length}_{\mathcal{O}}(I_R/I_R^2) \geq \mathrm{length}_{\mathcal{O}}(\mathcal{O}/\eta_T)$. Moreover, if φ is an isomorphism between complete intersections, then by Corollary 2.3 we have $\varphi \, \mathrm{Fit}_R(I_R) = \mathrm{Ann}_T(I_T)$, and therefore the two lengths are equal.

To show the converse, assume that the two lengths are equal, so that $\pi_R \, \mathrm{Fit}_R(I_R) = \pi_T \, \mathrm{Ann}_T(I_T)$. We first show that $I_T \cap \mathrm{Ann}_T(I_T) = 0$. Since

$\eta_T \neq 0$ there is an element $y \in \mathrm{Ann}_T(I_T)$ for which $\pi_T(y) \neq 0$. For any element $x \in I_T \cap \mathrm{Ann}_T(I_T)$ we clearly have

$$xy = 0 \qquad \text{and} \qquad x(y - \pi_T(y)) = 0.$$

But then $\pi_T(y)x = 0$, and since T is free as a module over the discrete valuation ring $\mathcal{O}$ this implies that $x = 0$. This shows that $I_T \cap \mathrm{Ann}_T(I_T) = 0$.

It follows that the map $\pi_T\colon \mathrm{Ann}_T(I_T) \longrightarrow \eta_T$ is an isomorphism. Since

$$\pi_T\varphi \,\mathrm{Fit}_R(I_R) = \pi_R \,\mathrm{Fit}_R(I_R) = \pi_T \,\mathrm{Ann}_T(I_T),$$

we conclude that $\varphi \,\mathrm{Fit}_R(I_R) = \mathrm{Ann}_T(I_T)$. This non-zero $\mathcal{O}$-submodule of T cannot be contained in $\mathfrak{m}_{\mathcal{O}}T$ because $T/\mathrm{Ann}_T(I_T)$ injects canonically to $\mathrm{End}_{\mathcal{O}}(I_T)$, which is torsion free as an $\mathcal{O}$-module. By the theorem this can only happen if φ is an isomorphism of complete intersections. This proves Criterion I.

Remark. If T is Gorenstein over $\mathcal{O}$ (see the end of Section 2), or if $\mathcal{O}$ is a complete discrete valuation ring, then it is not hard to show that $\mathrm{Ann}_T(I_T)$ is a non-zero direct $\mathcal{O}$-summand of T. By Corollary 2.3 the condition $\varphi \,\mathrm{Fit}_R(I_R) \not\subset \mathfrak{m}_{\mathcal{O}}T$ in the theorem can then be replaced by $\varphi \,\mathrm{Fit}_R(I_R) = \mathrm{Ann}_T(I_T)$. This may fail for other $\mathcal{O}$ and T. For instance, let k be a field, and let $\mathcal{O} = k[\varepsilon]$ with $\varepsilon^2 = 0$. The ring $T = \mathcal{O}[[X,Y]]/(X^2, Y^2, XY - \varepsilon X - \varepsilon Y)$, with $I_T = (X, Y)$, is a finite flat $\mathcal{O}$-algebra with $\mathrm{Fit}_T(I_T) = \mathrm{Ann}_T(I_T) = (\varepsilon X, \varepsilon Y)$, but T is not a complete intersection over $\mathcal{O}$.

4. Proof of Criterion II

In this section we prove Criterion II. Just as in Section 3, we first give the argument over a field, and then apply Nakayama's lemma.

Lemma 4.1. *Let k be a field and let $n \geq 1$. Suppose we have k-algebra homomorphisms*

$$k[[S_1, \ldots, S_n]] \longrightarrow k[[X_1, \ldots, X_n]] \overset{f}{\twoheadrightarrow} A$$

with f surjective, and suppose that the k-algebra $A/(S_1, \ldots, S_n)A$ has finite dimension d as a vector space over k. Assume that for some $N > n^{n-1}d^n$, the induced map

$$k[[S_1, \ldots, S_n]]/(S_1^N, \ldots, S_n^N) \overset{g}{\longrightarrow} A/(S_1^N, \ldots, S_n^N)A$$

is injective. Then f induces an isomorphism of k-algebras

$$k[[X_1, \ldots, X_n]]/(S_1, \ldots, S_n) \overset{\sim}{\longrightarrow} A/(S_1, \ldots, S_n)A.$$

Proof. The ring $k[[X_1, \ldots, X_n]]$ is a local ring with maximal ideal

$$I = (X_1, \ldots, X_n).$$

Since $A/(S_1,\ldots,S_n)A$ has length d as a module over $k[[X_1,\ldots,X_n]]$ it is annihilated by I^d. Writing $J = \ker f$ this means that

$$I^d \subset J + (S_1,\ldots,S_n),$$

where $(S_1,\ldots,S_n)$ denotes the ideal of $k[[X_1,\ldots,X_n]]$ generated by the S_i. We will show that $J \subset I^{d+1}$ by assuming that we can find $\alpha \in J$ with $\alpha \notin I^{d+1}$, and deriving a contradiction. Consider the multiplication by α map:

$$0 \longrightarrow \ker \longrightarrow k[[X_1,\ldots,X_n]]/I^{ndN} \xrightarrow{\alpha} k[[X_1,\ldots,X_n]]/I^{ndN} \longrightarrow \mathrm{cok} \longrightarrow 0.$$

Since $k[[X_1,\ldots,X_n]]/I^{ndN}$ has finite dimension over k, we have

$$\dim_k(\ker) = \dim_k(\mathrm{cok}).$$

We give estimates for these two dimensions. We have inclusions of ideals in the ring $k[[X_1,\ldots,X_n]]$,

$$I^{ndN} \subset (J + (S_1,\ldots,S_n))^{nN} \subset J + (S_1^N,\ldots,S_n^N),$$

so the cokernel $\mathrm{cok} = k[[X_1,\ldots,X_n]]/(I^{ndN}+(\alpha))$ now maps surjectively to the quotient ring $k[[X_1,\ldots,X_n]]/(J+(S_1^N,\ldots,S_n^N)) = A/(S_1^N,\ldots,S_n^N)A$. Since g is injective this gives

$$\begin{aligned} \dim_k \mathrm{cok} &\geq \dim_k A/(S_1^N,\ldots,S_n^N)A \\ &\geq \dim_k k[[S_1,\ldots,S_n]]/(S_1^N,\ldots,S_n^N) \\ &= N^n. \end{aligned}$$

On the other hand, since $\alpha \notin I^{d+1}$, we have $\ker \subset I^{ndN-d}/I^{ndN}$, so that the $\dim_k(\ker)$ is at most the number of monomials of degree δ with $ndN - d \leq \delta < ndN$. For such a monomial we have at most ndN choices for the exponent of each of the variables $X_1,\ldots,X_{n-1}$, and then at most d choices for the exponent of X_n. Therefore

$$\dim_k \ker \leq d(ndN)^{n-1}.$$

Combining the two estimates we see that $N^n \leq d(ndN)^{n-1}$, which contradicts the assumption that $N > n^{n-1}d^n$. This proves that $J \subset I^{d+1}$.

To finish the proof of the lemma, consider the inclusions

$$I^d \subset J + (S_1,\ldots,S_n) \subset I^{d+1} + (S_1,\ldots,S_n).$$

By Nakayama's lemma we see that $I^d \subset (S_1,\ldots,S_n)$, so that

$$\ker f = J \subset I^{d+1} \subset (S_1,\ldots,S_n).$$

Since f induces an isomorphism $k[[X_1,\ldots,X_n]]/J \xrightarrow{\sim} A$, the lemma follows.

We now return to the setting in which Criterion II is formulated: we let $\mathcal{O}$ be a complete Noetherian local ring and suppose that its residue field k has characteristic $p > 0$. Let $n \geq 1$ and for $m \geq 0$ let J_m be the $\mathcal{O}[[S_1,\ldots,S_n]]$-ideal $(\omega_m(S_1),\ldots,\omega_m(S_n))$, where $\omega_m(S)$ denotes the polynomial $(1+S)^{p^m} - 1$.

Corollary 4.2. *Suppose we have $\mathcal{O}$-algebra homomorphisms*

$$\mathcal{O}[[S_1,\ldots,S_n]] \longrightarrow \mathcal{O}[[X_1,\ldots,X_n]] \xrightarrow{f} A$$

with f surjective, and $A/(S_1,\ldots,S_n)A$ free of rank $d>0$ over $\mathcal{O}$. If, for some m with $p^m > n^{n-1}d^n$ the quotient ring A/J_mA is free as a module over $\mathcal{O}[[S_1,\ldots,S_n]]/J_m$, then the induced map

$$h\colon \mathcal{O}[[X_1,\ldots,X_n]]/(S_1,\ldots,S_n) \longrightarrow A/(S_1,\ldots,S_n)A$$

is an isomorphism between complete intersections over $\mathcal{O}$.

Proof. Taking everything modulo $\mathfrak{m}_\mathcal{O}$ we see that for the k-algebra $\overline{A} = A\otimes_\mathcal{O} k$, the quotient ring $\overline{A}/(S_1^{p^m},\ldots,S_n^{p^m})\overline{A}$ is a non-zero free module over $k[[S_1,\ldots,S_n]]/(S_1^{p^m},\ldots,S_n^{p^m})$. By Lemma 4.1 we see that h is an isomorphism modulo $\mathfrak{m}_\mathcal{O}$, and Lemma 3.1 then implies that h is an isomorphism. In particular we see that $\mathcal{O}[[X_1,\ldots,X_n]]/(S_1,\ldots,S_n)$ is finitely generated as an $\mathcal{O}$-module, so that it is a complete intersection. This shows 4.2.

Proof of Criterion II. Let d denote the $\mathcal{O}$-rank of T, and let m be so large that $p^m > n^{n-1}d^n$. By property (i) there is a surjection

$$\mathcal{O}[[X_1,\ldots,X_n]] \twoheadrightarrow R_m.$$

We now lift the homomorphism $\mathcal{O}[[S_1,\ldots,S_n]] \longrightarrow R_m$ to an $\mathcal{O}$-algebra homomorphism $\mathcal{O}[[S_1,\ldots,S_n]] \longrightarrow \mathcal{O}[[X_1,\ldots,X_n]]$ and we apply Corollary 4.2 with $A = T_m$. We conclude that the composite map

$$\mathcal{O}[[X_1,\ldots,X_n]]/(S_1,\ldots,S_n) \twoheadrightarrow R_m/(S_1,\ldots,S_n)R_m \twoheadrightarrow T_m/(S_1,\ldots,S_n)T_m$$

is an isomorphism between complete intersections. It follows from property (iii) that φ is an isomorphism between complete intersections as well.

Bibliography

[1] Bruns, W. and Herzog, J.: *Cohen-Macaulay rings*, Cambridge University Press, Cambridge 1993.

[2] De Shalit, E.: Hecke rings and universal deformation rings, this volume.

[3] De Smit, B. and Lenstra, H.W., Jr.: Finite complete intersection algebras and the completeness radical, *J. Algebra* **196** (1997), 520–531.

[4] Diamond, F. and Ribet, K.: l-adic modular deformations and Wiles's "Main Conjecture", this volume.

[5] Lang, S.: Algebra, 3rd ed., Addison-Wesley, Reading, Mass., 1993.

[6] Lenstra, H.W., Jr.: Complete intersections and Gorenstein rings, in J. Coates and S.T. Yau: *Elliptic curves, modular forms and Fermat's Last Theorem*, International Press, Cambridge 1995.

[7] Mazur, B. and Roberts, L.: Local Euler characteristics, *Invent. Math.* **9** (1970), 201–234.
[8] Tate, J.T.: Homology of Noetherian rings and local rings, *Illinois Math. Journal* **1** (1957), 14–27.
[9] Taylor, R. and Wiles, A.: Ring theoretic properties of certain Hecke algebras, *Annals of Math.* **141** (1995), 553–572.
[10] Wiebe, H.: Über homologische Invarianten lokaler Ringe, *Math. Annalen* **179** (1969), 257–274.
[11] Wiles, A.: Modular elliptic curves and Fermat's Last Theorem, *Annals of Math.* **141** (1995), 443–551.

ℓ-ADIC MODULAR DEFORMATIONS AND WILES'S "MAIN CONJECTURE"

FRED DIAMOND AND KENNETH A. RIBET

1. Introduction

Let E be an elliptic curve over $\mathbf{Q}$. The Shimura-Taniyama conjecture asserts that E is modular, i.e., that there is a weight-two newform f such that $a_p(f) = a_p(E)$ for all primes p at which E has good reduction. Let ℓ be a prime, choose a basis for the Tate module $T_\ell(E)$ and consider the ℓ-adic representation

$$\rho_{E,\ell} : G_{\mathbf{Q}} \to \mathrm{Aut}(T_\ell(E)) \cong \mathbf{GL}_2(\mathbf{Z}_\ell).$$

Then E is modular if and only if $\rho_{E,\ell}$ is modular, i.e., if and only if $\rho_{E,\ell}$ is equivalent over $\mathbf{Q}_\ell$ to the representation $\rho_{f,\ell}$ for some f (see [22]).

We aim to prove a stronger result which characterizes ℓ-adic representations arising from modular forms. Since the coefficients of a newform lie in the ring of integers of a number field which is not necessarily $\mathbf{Q}$, we are led to consider representations

$$\rho : G_{\mathbf{Q}} \to \mathbf{GL}_2(A) \tag{1}$$

where A is the ring of integers of a finite extension of $\mathbf{Q}_\ell$, and ask which of these arise from modular forms. In fact, it turns out to be convenient to consider representations ρ as in (1) where now A is a complete local Noetherian ring with finite residue field k, and formulate a notion of modularity for such a representation. Roughly speaking, the main theorem of Wiles in [26] supposes we are given such a representation which is "plausibly modular" and concludes that under certain technical hypotheses:

$$\bar{\rho} \text{ modular} \Longrightarrow \rho \text{ modular}, \tag{2}$$

where $\bar{\rho} : G_{\mathbf{Q}} \to \mathbf{GL}_2(k)$ is the reduction of ρ.

Returning to the elliptic curve E, suppose we know that

$$\bar{\rho}_{E,\ell} : G_{\mathbf{Q}} \to \mathbf{GL}_2(\mathbf{F}_\ell)$$

is modular (by Langlands-Tunnell [16], for example, if $\ell = 3$ and the representation is irreducible). A result of the form (2) then implies $\rho_{E,\ell}$ is modular (assuming it is "plausibly modular" and satisfies the technical hypotheses), hence so is E!

To orient the reader, we recall the connection with Fermat's Last Theorem, which begins with an idea of G. Frey (see [14]): Suppose that we have a non-trivial solution $a^n + b^n = c^n$, where $n \geq 5$ is prime, b is even and $a \equiv -1 \bmod 4$. Consider the elliptic curve

$$E : Y^2 = X(X - a^n)(X + b^n).$$

The modularity of E implies in particular that the mod n representation $\bar{\rho}_{E,n}$ is modular. The main result of [21] (see [12]) then shows that $\bar{\rho}_{E,n}$ arises from a (non-zero) modular form of weight two and level two, a contradiction as there are no such forms.

2. STRATEGY

We begin our formal discussion by returning to the phrase "plausibly modular," which we used in connection with representations like (1). For simplicity, assume again that A is the ring of integers of a finite extension of $\mathbf{Q}_\ell$. Consider representations $\rho : G_{\mathbf{Q}} \to \mathbf{GL}_2(A)$ which arise from weight-two eigenforms on $\Gamma_0(N)$, where N is allowed to vary. The idea is to exhibit a list of conditions satisfied by these representations which encapsulates their modularity. Those ρ which arise from weight-two forms on $\Gamma_0(N)$ are irreducible with cyclotomic determinant. Further, they are unramified outside a finite set of prime numbers (namely, those not dividing ℓN). It is tempting to guess that any ρ satisfying these simple conditions is likely to be modular. In fact, however, one also needs to impose a further condition on the restriction of ρ to a decomposition group at ℓ. A sufficient condition in this direction is conjectured by Fontaine and Mazur in [13], but with the current technology we need to impose a stronger one in order to obtain results. We will introduce a condition which is both convenient to work with and sufficient for applications to elliptic curves with semistable reduction at ℓ. Namely, suppose now that the prime number ℓ is odd. Then we will assume that the representation ρ is "semistable" [2], i.e., that it is either finite flat or ordinary at ℓ in the terminology of [19].

Let us fix an irreducible mod ℓ representation

$$\bar{\rho} : G_{\mathbf{Q}} \to \mathbf{GL}_2(k)$$

whose determinant is the mod ℓ cyclotomic character. Denote by $\mathcal{R}$ the set of isomorphism classes of "plausibly modular" ρ as above with reduction $\bar{\rho}$. We denote by $\mathcal{T}$ the set of (genuinely) modular isomorphism classes in $\mathcal{R}$. Thus $\mathcal{T}$ is a set of modular forms giving rise to $\bar{\rho}$, and the goal is to prove $\mathcal{R} = \mathcal{T}$.

We suppose that $\mathcal{T}$ is not empty; we can paraphrase this condition by the statement that $\bar{\rho}$ is modular and semistable (locally at the prime ℓ). It is known that $\mathcal{T}$ is infinite [20], so that $\mathcal{R} \supseteq \mathcal{T}$ is infinite. This circumstance makes $\mathcal{R}$ rather unwieldy, so that we are led to filter $\mathcal{R}$ as follows: For each finite set of primes Σ, we will let $\mathcal{R}_\Sigma$ be the set of ρ in $\mathcal{R}$ which are of "type Σ," meaning they are well-behaved outside Σ, and set $\mathcal{T}_\Sigma = \mathcal{T} \cap \mathcal{R}_\Sigma$. Before making precise the notion of being "well-behaved," we remark that $\mathcal{R}$ will be the union of the $\mathcal{R}_\Sigma$ over all finite sets of primes Σ. Therefore, to prove that $\mathcal{R} = \mathcal{T}$, it will suffice to prove

$$\mathcal{R}_\Sigma = \mathcal{T}_\Sigma \tag{3}$$

for all Σ. With the definition we give, each set $\mathcal{R}_\Sigma$ corresponds, at least conjecturally, to an easily described finite set of modular forms.

We now give a preliminary definition of "type Σ" in terms of the conductors[1] $N(\rho)$ and $N(\bar{\rho})$: We say that ρ is *type* Σ if Σ contains the set of prime divisors of $N(\rho)/N(\bar{\rho})$ (an integer by [1] or [18]).

We shall assume below that $N(\bar{\rho})$ is square-free, in which case this preliminary definition of type Σ turns out to be suitable (cf. remark 3.3), but we shall have to extend it (along with the notions of plausibly modular and modular) to representations ρ as in (1) where A is a complete local Noetherian ring with finite residue field. The purpose is to work in the context of Mazur's deformation theory [19], thereby introducing more structure into the problem and enabling us to use tools from commutative algebra. The desired equality (3) is subsumed by Wiles's "Main Conjecture," a precise version of (2) which takes the following form: A certain ring homomorphism

$$R_\Sigma \to \mathbf{T}_\Sigma \tag{4}$$

is an isomorphism, where

- R_Σ is a universal deformation ring which parametrizes representations of type Σ with a fixed residual representation $\bar{\rho}$;
- $\mathbf{T}_\Sigma$ is a Hecke algebra which parametrizes the newforms of weight two giving rise to such representations.

Wiles's strategy is to prove this first in the case $\Sigma = \emptyset$, and then deduce the result for arbitrary Σ. The aim of this article is to explain the statement of the conjecture and the reduction to the case $\Sigma = \emptyset$.

3. The "Main Conjecture"

3.1. The Hecke algebra. Once again, we fix a representation

$$\bar{\rho} : G_{\mathbf{Q}} \to \mathbf{GL}_2(k)$$

where k is a finite field of characteristic ℓ. We assume:

(a) ℓ is odd;
(b) $\bar{\rho}$ is irreducible;

---

[1] Use the following ad hoc definition of the exponent of ℓ in the conductor of a representation which is semistable at ℓ and has cyclotomic determinant: it is trivial if the representation is finite flat and 1 otherwise.

(c) the restriction of $\bar{\rho}$ to a decomposition group at ℓ is finite flat or ordinary;
(d) $\bar{\rho}$ has cyclotomic determinant;
(e) $\bar{\rho}$ has square-free conductor.

Remark 3.1. While hypotheses (a)–(c) are needed for the existing methods, the last two are made to simplify the exposition. Their presence causes no problem in the application to Fermat's Last Theorem, since they are satisfied by the mod ℓ representations coming from semistable elliptic curves. See [8] for a discussion of how to work without them—hypothesis (d) is not so serious; hypothesis (e) was removed in [7].

For a reason fundamental to Wiles's method, we must make the further assumption that $\bar{\rho}$ is modular. Roughly speaking, this means that $\bar{\rho}$ is equivalent to the reduction of a representation $\rho_{f,\lambda}$ arising from a modular form. To make this assumption precise, let us fix embeddings $\overline{\mathbf{Q}} \hookrightarrow \overline{\mathbf{Q}}_\ell$, $\overline{\mathbf{Q}} \hookrightarrow \mathbf{C}$. We also fix an embedding of k in $\overline{\mathbf{F}}_\ell$, where $\overline{\mathbf{F}}_\ell$ is the residue field of the ring of integers of $\overline{\mathbf{Q}}_\ell$. Suppose that f is a newform of weight two, level N_f and trivial character, and let K_f denote the number field generated by its coefficients $a_n(f)$. The chosen embeddings determine a prime λ of $\mathcal{O}_f$, the ring of integers of K_f. We write simply ρ_f for $\rho_{f,\lambda}$ (see [22] where this representation is denoted ρ_λ and defined in the course of the proof of theorem 4). Thus ρ_f is the absolutely irreducible representation

$$G_{\mathbf{Q}} \to \mathbf{GL}_2(K_{f,\lambda})$$

characterized up to isomorphism by the following property:

$$\text{(5)} \quad \begin{array}{c} \text{If } p \text{ is a prime not dividing } \ell N_f, \text{ then } \rho_f \text{ is unramified at } p, \\ \operatorname{tr}\rho_f(\mathrm{Frob}_p) = a_p(f) \text{ and } \det\rho_f(\mathrm{Frob}_p) = p. \end{array}$$

One can choose a basis so that the image of ρ_f is contained in $\mathbf{GL}_2(\mathcal{O}_{f,\lambda})$, and the reduction

$$\bar{\rho}_f : G_{\mathbf{Q}} \to \mathbf{GL}_2(\mathcal{O}_f/\lambda)$$

is well-defined up to semi-simplification.

We assume $\bar{\rho}$ is equivalent over $\overline{\mathbf{F}}_\ell$ to $\bar{\rho}_f$ for some f as above. It turns out that if this assumption holds, then in fact there are infinitely many f to choose from. (As we mentioned above, this follows from the results of [20].) Given a finite set of primes Σ, we can then ask which of these f give rise to representations of type Σ, in the sense that $\rho_{f,\lambda}$ is semistable at ℓ and Σ contains the set of primes dividing $N(\rho_{f,\lambda})/N(\bar{\rho})$. A sufficient condition is that the level of f divide N_Σ where

$$N_\Sigma = N(\bar{\rho}) \prod_{p \in \Sigma} p^{m_p}$$

and the m_p are defined as follows:

- $m_p = 2$ if p does not divide $\ell N(\bar{\rho})$;
- $m_p = 1$ if $p \neq \ell$ and p divides $N(\bar{\rho})$;
- $m_\ell = 1$ if $\bar{\rho}$ is finite flat and ordinary at ℓ;
- $m_\ell = 0$ otherwise.

The motivation for this definition of N_Σ is that this condition is known to be necessary as well as sufficient, as long as we restrict our attention to forms f with trivial character and level not divisible by ℓ^2. (The proof of the necessity relies on the Deligne-Langlands-Carayol theorem, an analysis of possible values of $N(\rho)/N(\bar{\rho})$ which is due independently to Carayol and Livné, and well-known results on the reduction of modular curves, abelian varieties and ℓ-divisible groups. We shall not, however, make use of this fact; indeed it turns out to be a consequence of what follows.)

Let Φ_Σ denote the set of newforms f of weight two, trivial character and level dividing N_Σ. As explained in [12], it follows from the results of [21] and others on Serre's conjecture that the set $\Phi_\emptyset$ is non-empty. The analogous statement holds *a fortiori* for each Φ_Σ. We can then consider the ring

$$\tilde{\mathbf{T}}_\Sigma := \prod_{f \in \Phi_\Sigma} \mathcal{O}_{f,\lambda}.$$

Recall that for each f, the prime λ of K_f is determined by our choices of embeddings and note that $\tilde{\mathbf{T}}_\Sigma$ is semilocal and finitely generated as a $\mathbf{Z}_\ell$-module. For each prime p not in Σ, we let T_p denote the element $(a_p(f))_{f\in\Phi_\Sigma}$. We define the Hecke algebra $\mathbf{T}_\Sigma$ as the $\mathbf{Z}_\ell$-subalgebra of $\tilde{\mathbf{T}}_\Sigma$ generated by the elements T_p for p not in $\Sigma \cup \{\ell\}$.

We can give another description of $\mathbf{T}_\Sigma$ in terms of the subring $\mathbf{T}$ of the ring of endomorphisms of $S = S_2(\Gamma_0(N_\Sigma))$ generated by the operators T_p for all primes p. We suppose that f is in Φ_Σ and we define f_Σ as a certain $\mathbf{T}$-eigenform in S for which f is the associated newform. The eigenform f_Σ is characterized by this together with the properties:

- if p is in $\Sigma \setminus \{\ell\}$, then $a_p(f_\Sigma) = 0$;
- if ℓ divides N_Σ, then $a_\ell(f_\Sigma)$ is an ℓ-adic unit.

The map sending T_p to the reduction of $a_p(f_\Sigma)$ defines a homomorphism $\mathbf{T} \to \overline{\mathbf{F}}_\ell$, and we write $\mathbf{T}_\mathfrak{m}$ for the completion of $\mathbf{T}$ at the kernel $\mathfrak{m}$ of this homomorphism. We then have the lemma:

Lemma 3.2. *If ℓ is not in Σ, then the element T_ℓ of $\tilde{\mathbf{T}}_\Sigma$ is in $\mathbf{T}_\Sigma$. For arbitrary Σ there is an isomorphism $\mathbf{T}_\Sigma \xrightarrow{\sim} \mathbf{T}_\mathfrak{m}$ such that $T_p \mapsto T_p$ for all p not in Σ.*

We refer the reader to section 2.3 of [26] or section 4.2 of [2] for the verification, which is tedious, unilluminating, and ultimately unnecessary (see the discussion in section 5 below).

3.2. **The universal deformation ring.** We appeal to Mazur's deformation theory of Galois representations, discussed in [4] and [19], to define a certain universal deformation ring R_Σ.

Keep the technical hypotheses imposed on $\bar{\rho}$ at the beginning of the preceding section, and consider deformations of $\bar{\rho}$ of the form

$$\rho : G_{\mathbf{Q}} \to \mathbf{GL}_2(A),$$

where A is a complete local Noetherian ring with residue field k. We say such a deformation is of type Σ if the following statements are true:

(a) $\det \rho$ is cyclotomic.
(b) ρ is finite flat or ordinary at ℓ.
(c) let p be a prime not in Σ. Then:
- if $p \neq \ell$ and $\bar{\rho}$ is unramified at p, then so is ρ;
- if $p \neq \ell$ and $\bar{\rho}$ is of type A at p, then so is ρ;
- if $p = \ell$ and $\bar{\rho}$ is finite flat, then so is ρ.

Remark 3.3. Suppose that A is the ring of integers of a finite extension of $\mathbf{Q}_\ell$ and the first two conditions are satisfied. In that case, one can check that condition (c) is equivalent to the equality $\mathrm{ord}_p N(\rho) = \mathrm{ord}_p N(\bar{\rho})$. We shall not use this fact.

The results discussed in [4] and [19] furnish a universal deformation ring R_Σ and a universal deformation

$$\rho_\Sigma^{\mathrm{univ}} : G_{\mathbf{Q}} \to \mathbf{GL}_2(R_\Sigma)$$

of $\bar{\rho}$ of type Σ.

Suppose now that $\bar{\rho}$ is modular. For a newform f in Φ_Σ, we let A_f denote the subring of $\mathcal{O}_{f,\lambda}$ consisting of those elements whose reduction mod λ is in k. One checks that with respect to some basis, we have

$$\rho_f : G_{\mathbf{Q}} \to \mathbf{GL}_2(A_f),$$

a deformation of $\bar{\rho}$ of type Σ. The universal property of R_Σ therefore furnishes a unique homomorphism $\pi_{f,\Sigma} : R_\Sigma \to A_f$ such that the composite

$$G_{\mathbf{Q}} \to \mathbf{GL}_2(R_\Sigma) \to \mathbf{GL}_2(A_f)$$

is equivalent to ρ_f. Since R_Σ is topologically generated by the traces of $\rho_\Sigma^{\mathrm{univ}}(\mathrm{Frob}_p)$ for p not in $\Sigma \cup \{\ell\}$, we conclude that the image of the map

$$\begin{array}{ccc} R_\Sigma & \longrightarrow & \tilde{\mathbf{T}}_\Sigma \\ r & \longmapsto & (\pi_{f,\Sigma}(r))_{f \in \Phi_\Sigma} \end{array}$$

is precisely $\mathbf{T}_\Sigma$. We define ϕ_Σ to be the resulting surjective ring homomorphism $R_\Sigma \to \mathbf{T}_\Sigma$.

3.3. **Statement of the conjecture.** We keep the hypotheses on $\bar{\rho}$ imposed in section 3.1 (the hypothesis of modularity as well as the technical conditions listed at the beginning of the section). We suppose that Σ is a finite set of primes and we consider the map ϕ_Σ defined at the end of the section 3.2. In our setting, Wiles's conjecture 2.16 in [26] becomes:

Conjecture 3.4. *The map ϕ_Σ is an isomorphism.*

We briefly recall how the conjecture implies the Shimura-Taniyama Conjecture for semistable elliptic curves (see also [3]).

Theorem 3.5. *Suppose conjecture 3.4 holds and E is an elliptic curve with square-free conductor N_E. If there is an odd prime ℓ such that $\bar{\rho}_{E,\ell}$ is irreducible and modular, then E is modular.*

To prove this, one checks that under these hypotheses, the mod ℓ representation $\bar{\rho}_{E,\ell}$ satisfies the technical conditions of §3.1. Moreover the ℓ-adic representation $\rho_{E,\ell}$ is a deformation of $\bar{\rho}_{E,\ell}$ of type Σ for some Σ; for example, one can take Σ to be the set of primes dividing $N_E\ell$. On taking $\mathcal{O} = \mathbf{Z}_\ell$, we obtain from the universal property of R_Σ a homomorphism $\theta : R_\Sigma \to \mathbf{Z}_\ell$ where

$$\operatorname{tr}(\rho_\Sigma^{\text{univ}}(\text{Frob}_p)) \longmapsto a_p(E) = p+1 - \#E(\mathbf{F}_p)$$

for $p \neq \ell$ not in Σ. Now the conjecture implies ϕ_Σ is an isomorphism, so we may consider the composite

$$\mathbf{T}_\Sigma \xrightarrow{\phi_\Sigma^{-1}} R_\Sigma \xrightarrow{\theta} \mathbf{Z}_\ell.$$

One sees from the definition of $\mathbf{T}_\Sigma$ that such a homomorphism is necessarily a projection $\pi_{f,\Sigma}$ for some f in Φ_Σ. It follows that $\rho_{E,\ell}$ is isomorphic to ρ_f, or equivalently that

$$a_p(f) = a_p(E) \quad \text{for all } p \notin \Sigma.$$

Therefore E is modular.

If $\bar{\rho}_{E,3}$ is irreducible, then the Langlands-Tunnell theorem [16] shows that $\bar{\rho}_{E,3}$ is modular, hence conjecture 3.4 implies that E is modular. If $\bar{\rho}_{E,3}$ is reducible, then Wiles argues (see [23]) that $\bar{\rho}_{E,5}$ is isomorphic to $\bar{\rho}_{E',5}$ for some semistable E' with irreducible $\bar{\rho}_{E',3}$. Since E' is modular, so is $\bar{\rho}_{E',5} \approx \bar{\rho}_{E,5}$, so we may apply the preceding theorem with $\ell = 5$.

4. REDUCTION TO THE CASE $\Sigma = \emptyset$

4.1. **Commutative algebra.** Suppose now that $f = \sum a_n q^n$ is a newform in $\Phi_\emptyset$ (recall from [12] that such an f exists), hence in Φ_Σ for every finite set of primes Σ. Consider the commutative triangle

$$\begin{array}{ccc} R_\Sigma & \xrightarrow{\phi_\Sigma} & \mathbf{T}_\Sigma \\ & \searrow & \downarrow \\ & & A_f \end{array}$$

where the downwards arrow is $\pi_{f,\Sigma}$ and the diagonal one corresponds to ρ_f via the universal property of R_Σ.

In order to apply the commutative algebra results explained in [5], it will be convenient to work with $\mathcal{O}$-algebras where $\mathcal{O} = \mathcal{O}_{f,\lambda}$. Note that ϕ_Σ is an isomorphism if and only if

$$\phi_\Sigma \otimes_{W(k)} \mathcal{O} : R_\Sigma \otimes_{W(k)} \mathcal{O} \to \mathbf{T}_\Sigma \otimes_{W(k)} \mathcal{O}$$

is an isomorphism. From now on, we replace ϕ_Σ, R_Σ and $\mathbf{T}_\Sigma$ by their tensor products over $W(k)$ with $\mathcal{O}$. The resulting representation $G_{\mathbf{Q}} \to \mathbf{GL}_2(R_\Sigma)$ is universal for type-Σ deformations of $\bar{\rho}$ as in (1), where now A is a complete local Noetherian $\mathcal{O}$-algebra with residue field $\mathcal{O}/\lambda$. Writing $\mathcal{O}'_g$ for the $\mathcal{O}$-subalgebra of $\overline{\mathbf{Q}}_\ell$ generated by the Fourier coefficients of f, we may identify $\mathbf{T}_\Sigma$ with the $\mathcal{O}$-subalgebra of

$$\prod_{g \in \Phi_\Sigma} \mathcal{O}'_g$$

generated by the operators T_p for p not in $\Sigma \cup \{\ell\}$. Our commutative triangle becomes

$$\begin{array}{ccc} R_\Sigma & \xrightarrow{\phi_\Sigma} & \mathbf{T}_\Sigma \\ & \searrow & \downarrow \\ & & \mathcal{O} \end{array} .$$

Write π_Σ for the map $\mathbf{T}_\Sigma \to \mathcal{O}$, let $\mathfrak{p}_\Sigma$ denote the kernel of $R_\Sigma \to \mathcal{O}$, and let η_Σ denote the ideal $\pi_\Sigma(\mathrm{Ann}_{\mathbf{T}_\Sigma} \ker \pi_\Sigma)$ (which is non-zero). According to Criterion I of [5], we have

$$\text{length}_{\mathcal{O}}(\mathfrak{p}_\Sigma/\mathfrak{p}_\Sigma^2) \geq \text{length}_{\mathcal{O}}(\mathcal{O}/\eta_\Sigma) \tag{6}$$

and the following are equivalent:

- ϕ_Σ is an isomorphism between complete intersections over $\mathcal{O}$.
- Equality holds in (6).

It is actually the following strengthening of conjecture 3.4 whose proof we reduce to the case of $\Sigma = \emptyset$.

Conjecture 4.1. *The surjection ϕ_Σ is an isomorphism between complete intersections over $\mathcal{O}$.*

In the remaining sections we sketch the proof that if

$$\text{length}_{\mathcal{O}}(\mathfrak{p}_\Sigma/\mathfrak{p}_\Sigma^2) = \text{length}_{\mathcal{O}}(\mathcal{O}/\eta_\Sigma),$$

then

$$\text{length}_{\mathcal{O}}(\mathfrak{p}_{\Sigma'}/\mathfrak{p}_{\Sigma'}^2) \leq \text{length}_{\mathcal{O}}(\mathcal{O}/\eta_{\Sigma'})$$

where $\Sigma' = \Sigma \cup \{p\}$. This implies:

Theorem 4.2. *If conjecture 4.1 holds in the case $\Sigma = \emptyset$, then it holds for all Σ.*

The proof of conjecture 4.1 in the case $\Sigma = \emptyset$ is explained in [3].

4.2. **Selmer groups.** Recall from Chapter VI of [19] (see also §2.7 of [2]) that the $\mathcal{O}$-module $\mathfrak{p}_\Sigma/\mathfrak{p}_\Sigma^2$ has a natural description in terms of Galois cohomology. Write M_f for $\mathcal{O}^2$ with Galois action defined by ρ_f and E_f for $\text{ad}^0 M_f$, the $\mathcal{O}[G_{\mathbf{Q}}]$-module of trace-zero $\mathcal{O}$-endomorphisms of M_f. Let $E_{f,n}$ denote $E_f \otimes_{\mathcal{O}} \lambda^{-n}\mathcal{O}/\mathcal{O}$ and set

$$E_{f,\infty} = \varinjlim_n E_{f,n} \cong E_f \otimes_{\mathcal{O}} K/\mathcal{O} \cong E_f \otimes_{\mathbf{Z}_\ell} \mathbf{Q}_\ell/\mathbf{Z}_\ell.$$

According to §28 of [19] (or more precisely its analogue in the case of fixed determinant; cf. Proposition 3 of §26), we have a canonical isomorphism

$$\text{Hom}(\mathfrak{p}_\Sigma/\mathfrak{p}_\Sigma^2, K/\mathcal{O}) \cong H^1_{\mathcal{D}}(G_{\mathbf{Q},\Sigma\cup\{\ell\}}, E_{f,\infty})$$

where $\mathcal{D}$ is the type-Σ deformation condition. Appealing to the descriptions in §§29–31 of the resulting conditions on local cohomology classes, we find that the latter $\mathcal{O}$-module is the "generalized Selmer group"

$$\varinjlim_n H^1_{\mathcal{L}}(\mathbf{Q}, E_{f,n})$$

in the notation of §6 of [25] (with the appropriate choice of local condition at ℓ from §7).

The analogous statements hold with Σ replaced by Σ' (in which case we write $\mathcal{D}'$ and $\mathcal{L}'$), so to compare the lengths of $\mathfrak{p}_\Sigma/\mathfrak{p}_\Sigma^2$ and $\mathfrak{p}_{\Sigma'}/\mathfrak{p}_{\Sigma'}^2$ we consider the cokernel of the natural inclusion

$$H^1_{\mathcal{D}}(G_{\mathbf{Q},\Sigma\cup\{\ell\}}, E_{f,\infty}) \to H^1_{\mathcal{D}'}(G_{\mathbf{Q},\Sigma\cup\{\ell\}}, E_{f,\infty}).$$

Suppose now that p does not divide $\ell N(\bar{\rho})$. From the definitions of the generalized Selmer groups, we see that our cokernel embeds naturally in

$$H^1(I_p, E_{f,\infty})^{G_p/I_p}.$$

Since the action of I_p on $E_{f,\infty}$ is trivial, the cohomology group can be identified with

$$\text{Hom}(I_p, E_{f,\infty}) \cong \text{Hom}(\mathbf{Z}_\ell(1), E_{f,\infty})$$

as a module for G_p/I_p. We are therefore reduced to computing the length of

$$H^0\left(G_p/I_p, E_{f,\infty}(-1)\right).$$

This is just the kernel of the endomorphism $1 - \text{Frob}_p^{-1}$ of

$$E_f(-1) \otimes_{\mathcal{O}} K/\mathcal{O}.$$

This kernel is finite if and only if $1 - \mathrm{Frob}_p^{-1}$ defines an automorphism of the K-vector space $E_f(-1) \otimes_{\mathcal{O}} K$, in which case its $\mathcal{O}$-length is simply the valuation of the determinant. We compute this determinant using the fact that the characteristic polynomial of Frob_p on $M_f \otimes_{\mathcal{O}} K$ is $X^2 - a_pX + p$ (see (5)). The result is that our determinant is

$$(1-p)\left((1+p)^2 - a_p^2\right) \tag{7}$$

(non-zero by [22], theorem 5), so we conclude that

$$\mathrm{length}_{\mathcal{O}}(\mathfrak{p}_{\Sigma'}/\mathfrak{p}_{\Sigma'}^2) \leq \mathrm{length}_{\mathcal{O}}(\mathfrak{p}_{\Sigma}/\mathfrak{p}_{\Sigma}^2) + v_\lambda(c_p) \tag{8}$$

where c_p is given by (7).

The cohomological calculation above is a special case of the following general result, whose proof is left as an exercise.

Proposition 4.3. *Suppose that $p \neq \ell$ and X a finitely generated free $\mathcal{O}$-module with a continuous action of G_p. Let $X_\infty = X \otimes_{\mathcal{O}} K/\mathcal{O}$. Then the Fitting ideal of the $\mathcal{O}$-module $H^1(I_p, X_\infty)^{D_p/I_p}$ is generated by the determinant of the endomorphism $1 - \mathrm{Frob}_p^{-1}p$ of $(X \otimes_{\mathcal{O}} K)_{I_p}$.*

Applying it in the case that $p \neq \ell$, but p divides $N(\bar{\rho})$, we find that the space $(E_f \otimes_{\mathcal{O}} K)_{I_p}$ is one-dimensional, Frob_p acts by the inverse of the cyclotomic character, and (8) holds with $c_p = 1 - p^2$.

Finally, in the case $p = \ell$, the groups H^1_Σ and $H^1_{\Sigma'}$ coincide unless $\bar{\rho}$ is ordinary and finite flat. In that case, bounding the cokernel is more subtle and one uses the calculations in [25] to show that the length increases by at most the valuation of $1-\alpha_\ell^2$, where α_ℓ is the unit root of $X^2 - a_\ell X + \ell = 0$.

Summing up, we have

Lemma 4.4. *In all of the cases above,*

$$\mathrm{length}_{\mathcal{O}}(\mathfrak{p}_{\Sigma'}/\mathfrak{p}_{\Sigma'}^2) \leq \mathrm{length}_{\mathcal{O}}(\mathfrak{p}_{\Sigma}/\mathfrak{p}_{\Sigma}^2) + v_\lambda(c_p)$$

where

$$c_p = \begin{cases} (1-p)\left((1+p)^2 - a_p^2\right) & \textit{if } p \nmid N(\bar{\rho}) \\ 1 - p^2 & \textit{otherwise.} \end{cases}$$

4.3. **Congruence modules.** Now we have to prove the inequality

$$\mathrm{length}_{\mathcal{O}}(\mathcal{O}/\eta_{\Sigma'}) \geq \mathrm{length}_{\mathcal{O}}(\mathcal{O}/\eta_{\Sigma}) + v_\lambda(c_p),$$

or equivalently:

$$\eta_{\Sigma'} \subset c_p \eta_\Sigma, \tag{9}$$

where c_p is defined above. Before sketching the proof, we describe the general strategy and consider an example.

The first observation to make is that the ideal η_Σ measures congruences between f and other forms of level N_Σ. Suppose for simplicity that $\mathcal{O}$ contains the coefficients of all newforms in Φ_Σ. Then $\mathbf{T}_\Sigma$ is contained in a product of copies of $\mathcal{O}$ indexed by these newforms and π_Σ is the projection onto the coordinate corresponding to f. The ideal η_Σ consists of those x such that

$$(x, 0, 0, \dots, 0) \in \mathbf{T}_\Sigma,$$

where the first coordinate corresponds to f. If there are just two forms, f and g, in Φ_Σ, then η_Σ is the ideal generated by $a_p(f) - a_p(g)$ for p not dividing $N_\Sigma \ell$. So η_Σ measures how congruent are the coefficients of f and g at "good" primes. More generally, one finds that $\eta_\Sigma \subset \lambda^n$ if $f \equiv g \bmod \lambda^n$ (in the sense above) for some g which is a linear combination over $\mathcal{O}$ of the forms different from f.

Consider for example the unique newform f of weight 2 and level 11. Its Fourier coefficients are rational, and the associated L-function is that of (the isogeny class of) the elliptic curve E defined by

$$Y^2 + Y = X^3 - X.$$

The 3-adic representation attached to f is equivalent to

$$\rho_{E,3} : G_{\mathbf{Q}} \to \mathrm{Aut}(\mathcal{T}_3(E)) \cong \mathbf{GL}_2(\mathbf{Z}_3),$$

and we let $\bar{\rho}$ denote the reduction. (We leave it to the reader to verify that $\bar{\rho}$ satisfies our running hypotheses.) Since $\Phi_\emptyset$ is the singleton set $\{f\}$ (as $N_\emptyset = 11$), we have $\eta_\emptyset = \mathcal{O}$. Take $\Sigma = \{3\}$. Then $N_\Sigma = 33$, and Φ_Σ consists of f and the unique newform g of (weight 2, trivial character and) level 33. To prove this statement, one needs to check that f and g are congruent mod 3; for this, it suffices to compare the Galois action on the 3-division points of the corresponding elliptic curves. Therefore, $\eta_\Sigma = 3\mathcal{O}$ is indeed generated by c_3.

Now consider the problem of comparing η_Σ and $\eta_{\Sigma'} = \eta_\Sigma \cup \{p\}$ when $p = 7$. Then $N_{\Sigma'} = 1617$ and c_p is 9 times a unit. So we could verify this case of the desired inclusion by finding congruent newforms among the levels 77, 231, 539 and 1617, and then writing down a linear combination of g with these forms which is congruent to f mod $27\mathcal{O}$.

This type of problem ("raising the level") was first addressed in [20], where the general strategy is as follows: Rather than produce such congruences directly, one detects them using the cohomology of modular curves[2], or in our case, their Jacobians. The problem of comparing these "cohomological congruences" at different levels is then reduced to studying the kernel of a certain homomorphism of Jacobians induced by degeneracy maps on the curves. This last issue is then resolved by a result of Ihara.

The method and results of [20] were sharpened and generalized in various ways in such articles as [6], [11], [26] and [7]. We now sketch Wiles's proof ([26], section 2.2) of (9), with some modifications taken from [2].

---

[2]This approach is suggested by work of Hida [17], which also establishes a relation between congruences and certain values of L-functions; see also section 4.4 of [2] and [9].

Let $\mathfrak{m}$ denote the maximal ideal of $\mathbf{T} \otimes \mathcal{O}$ containing the kernel of the homomorphism to $\mathcal{O}$ determined by f_Σ. Define M_Σ as the localization at $\mathfrak{m}$ of $\mathcal{T}_\ell(J_0(N)) \otimes_{\mathbf{Z}_\ell} \mathcal{O}$. One deduces from lemma 3.2 that the $\mathcal{O}$-algebra $\mathbf{T}_\Sigma$ is isomorphic to $(\mathbf{T} \otimes \mathcal{O})_\mathfrak{m}$, so we may regard M_Σ as a module for $\mathbf{T}_\Sigma$, hence R_Σ. Recall from [24] that Wiles proves the following generalization of results of Mazur and others:

Theorem 4.5. *The $\mathbf{T}_\Sigma$-module M_Σ is free of rank two.*

The module M_Σ is equipped with an alternating pairing, $\langle\, , \rangle_\Sigma$ that induces an isomorphism

$$M_\Sigma \cong \mathrm{Hom}_\mathcal{O}(M_\Sigma, \mathcal{O})$$

of $\mathbf{T}_\Sigma$-modules. Since M_Σ is free of rank two over $\mathbf{T}_\Sigma$, the submodule $M_\Sigma[\mathfrak{p}_\Sigma]$ (the set of elements annihilated by every t in $\mathfrak{p}_\Sigma$) is free of rank two over $\mathbf{T}_\Sigma/\mathfrak{p}_\Sigma = \mathcal{O}$. On combining these facts, one shows easily that if $\{x, y\}$ is a basis for $M_\Sigma[\mathfrak{p}_\Sigma]$, then

$$\eta_\Sigma = \langle x, y \rangle_\Sigma.$$

(See [2] for the details.)

To compare η_Σ and $\eta_{\Sigma'}$, one defines a $\mathbf{T}_{\Sigma'}$-equivariant map

$$M_{\Sigma'} \to M_\Sigma.$$

Its definition employs the degeneracy maps $X_0(N_{\Sigma'}) \to X_0(N_\Sigma)$ induced by $\tau \mapsto p^i\tau$ for $i \leq m_p$. These induce by Albanese functoriality maps $J_0(N_{\Sigma'}) \to J_0(N_\Sigma)$, hence maps

$$\mathcal{T}_\ell(J_0(N_{\Sigma'})) \otimes_{\mathbf{Z}_\ell} \mathcal{O} \to \mathcal{T}_\ell(J_0(N_\Sigma)) \otimes_{\mathbf{Z}_\ell} \mathcal{O} \to M_\Sigma$$

which we denote δ_i. A suitable $\mathbf{T}_\Sigma$-linear combination of these, namely

- $\delta_0 - p^{-1}T_p\delta_1 + p^{-1}\delta_2$ if $p \neq \ell$ and $m_p = 2$;
- $\delta_0 - p^{-1}T_p\delta_1$ if $p \neq \ell$ and $m_p = 1$;
- $\delta_0 - \tau_\ell^{-1}\delta_1$ if $p = \ell$ and $m_\ell = 1$, where τ_ℓ is the unit root in $\mathbf{T}_\Sigma$ of
$$X^2 - T_\ell X + \ell = 0;$$
- δ_0 if $p = \ell$ and $m_\ell = 0$;

induces the desired homomorphism

$$\beta : M_{\Sigma'} \to M_\Sigma$$

of $\mathbf{T}_\Sigma$-modules (see p. 119 of [2]).

Write β' for the adjoint of β with respect to the pairings $\langle\, , \rangle_\Sigma$ and $\langle\, , \rangle_{\Sigma'}$. A straightforward computation shows that the composite $\beta\beta'$ is an endomorphism in $\mathbf{T}_\Sigma$ which is a unit times

- $(1-p)\left((1+p)^2 - T_p^2\right)$ if $p \nmid N(\bar\rho)$,
- $1 - p^2$ otherwise

(see p. 121 of [2]). Note that this operator is c_p mod $\mathfrak{p}_\Sigma$. (Moreover this holds with $\mathfrak{p}_\Sigma$ replaced by any minimal prime of $\mathbf{T}_\Sigma$ and c_p replaced by its analogue defined using the corresponding newform. Since these are non-zero, $\beta\beta'$, and hence β', is injective.) To obtain (9), it suffices to prove that β' has torsion-free cokernel, for then a basis $\{x,y\}$ for M_Σ yields a basis $\{\beta'(x),\beta'(y)\}$ for $M_{\Sigma'}$ and we conclude that

$$\eta_{\Sigma'} = (\langle\beta'(x),\beta'(y)\rangle_{\Sigma'}) = c_p(\langle x,y\rangle)_\Sigma = c_p\eta_\Sigma.$$

Since β' is injective, it has torsion-free cokernel if and only if β is surjective (or equivalently, β' mod λ is injective). Recall that β is defined using the maps on homology (or Tate modules) induced by the degeneracy maps $X_0(N_{\Sigma'}) \to X_0(N_\Sigma)$. We therefore wish to analyze the cokernel of the homomorphism

$$H_1(X_0(N'_\Sigma),\mathcal{O}) \to H_1(X_0(N_\Sigma),\mathcal{O})^{m_p+1}$$

gotten from the degeneracy maps. This map is not surjective in general, but it is enough to prove:

Lemma 4.6. *Suppose that $N_\Sigma > 3p$ if p divides N_Σ. Then the map*

$$H_1(X_0(N'_\Sigma),\mathbf{Z}_\ell) \to H_1(X_0(N_\Sigma),\mathbf{Z}_\ell)^{m_p+1}_{\mathfrak{m}'}$$

is surjective, where $\mathfrak{m}'$ is the intersection of $\mathfrak{m}$ with the subalgebra of $\mathbf{T}\otimes\mathbf{Z}_\ell$ generated by the operators T_r for primes $r \neq p$.

We sketch the proof of the lemma in the generic (and most difficult) case of $m_p = 2$ and then explain what changes are needed to treat the remaining cases.

First recall that a similar problem is solved in [20]. Let $X_1(N_\Sigma, p)$ be the modular curve associated to $\Gamma_1(N_\Sigma)\cap\Gamma_0(p)$ and consider the map

$$(\pi_{1,*},\pi_{2,*}) : H_1(X_1(N_\Sigma,p),\mathbf{Z}_\ell) \to H_1(X_1(N_\Sigma),\mathbf{Z}_\ell)^2, \tag{10}$$

where here and below, $\pi_{1,*}$ (resp. $\pi_{2,*}$) will denote the map induced by $\tau \mapsto \tau$ (resp. $\tau \mapsto p\tau$). In section 4 of [20], the surjectivity of this map is proved by a group-theoretic argument using results of Ihara. The surjectivity of (10) is a key ingredient in the proof of the lemma.

Next consider the sequence of homology groups of non-compact modular curves

(11)

$$\begin{array}{ccccc} H_1(Y_1(N_\Sigma p,p^2),\mathbf{Z}_\ell) & \to & H_1(Y_1(N_\Sigma p),\mathbf{Z}_\ell)^2 & \to & H_1(Y_1(N_\Sigma),\mathbf{Z}_\ell) \\ x & \mapsto & (\pi_{1,*}x,\pi_{2,*}x); \quad (y,z) & \mapsto & \pi_{2,*}y - \pi_{1,*}z. \end{array}$$

Wiles proves the exactness of this sequence by an elementary group-theoretic argument (see lemma 2.5 of [26]). If we could replace X_1 and Y_1 with X_0 in (10) and (11), we could now deduce that

$$H_1(X_0(N_\Sigma p^2),\mathbf{Z}_\ell) \to H_1(X_0(N_\Sigma),\mathbf{Z}_\ell)^3$$

is surjective. A minor complication arises when we try to make this change, and one finds instead that the cokernel is supported only at "Eisenstein" maximal ideals of the Hecke algebra. (See the last half of section 4.5 of [2] for more details). The irreducibility hypothesis on $\bar{\rho}$ ensures that $\mathfrak{m}'$ is not Eisenstein, from which we deduce the lemma.

Suppose now that $m_p = 1$. If $p \neq \ell$ (in which case p divides N_Σ), then one uses the exactness of (11) with N_Σ replaced by N_Σ/p. If $p = \ell$, then one just uses the surjectivity of (10). In either case, the lemma follows as above since $\mathfrak{m}'$ is not Eisenstein, using also that $\mathfrak{m}'$ is not in the support of $H_1(X_0(N_\Sigma/p), \mathbf{Z}_\ell)$ in the case $p \neq \ell$. There is nothing to prove if $m_p = 0$.

5. Epilogue

Some parts of the exposition above, especially the last section, draw from [2]. As explained there, theorem 4.5 is actually only used in the case $\Sigma = \emptyset$.

In a recent article [10], the first author has presented a modification of the method of Taylor-Wiles-Faltings which makes no appeal to lemma 3.2 and theorem 4.5.[3] Instead, in the modified approach, one deduces these two results as *by-products* of the proof of the Main Conjecture. The new idea is to use tools from commutative algebra to prove directly, without any initial reference to the Hecke algebra $\mathbf{T}_\emptyset$, that $M_\emptyset$ is free over $R_\emptyset$.

References

[1] H. Carayol, *Sur les représentations galoisiennes modulo ℓ attachées aux formes modulaires,* Duke Math. J. **59** (1989), 785–801.

[2] H. Darmon, F. Diamond, R. Taylor, *Fermat's Last Theorem,* in Current Developments in Mathematics, 1995, International Press, 1–154.

[3] E. de Shalit, *Hecke rings and universal deformation rings,* this volume.

[4] B. de Smit, H. Lenstra, *Explicit construction of universal deformation rings,* this volume.

[5] B. de Smit, K. Rubin, R. Schoof, *Criteria for complete intersections,* this volume.

[6] F. Diamond, *On congruence modules associated to Λ-adic forms,* Comp. Math. **71** (1989), 49–83.

[7] F. Diamond, *On deformation rings and Hecke rings,* Annals of Math. **144** (1996), 137–166.

[8] F. Diamond, *An extension of Wiles' results,* this volume.

[9] F. Diamond, *Congruences between modular forms: Raising the level and dropping Euler factors,* to appear in Proc. NAS.

[10] F. Diamond, *The Taylor-Wiles construction and multiplicity one,* to appear in Invent. Math.

[11] F. Diamond, R. Taylor, *Non-optimal levels of mod ℓ modular representations,* Invent. Math. **115** (1994) 435–462.

[12] B. Edixhoven, *Serre's conjecture,* this volume.

[3] A similar method was found independently by Fujiwara [15].

[13] J.-M. Fontaine, B. Mazur, *Geometric Galois representations,* in: Elliptic Curves, Modular Forms and Fermat's Last Theorem, International Press, Cambridge (1995), 41–78.

[14] G. Frey, *On ternary equations of Fermat type and relations with elliptic curves,* this volume.

[15] K. Fujiwara, *Deformation rings and Hecke algebras in the totally real case,* preprint.

[16] S. Gelbart, Three lectures on the modularity of $\bar{\rho}_{E,3}$ and the Langlands reciprocity conjecture, this volume.

[17] H. Hida, *Congruences of cusp forms and special values of their zeta functions,* Inv. Math. **63** (1981), 225–261.

[18] R. Livné, *On the conductors of mod ℓ representations coming from modular forms,* J. Number Theory **31** (1989), 133–141.

[19] B. Mazur, *An introduction to the deformation theory of Galois representations,* this volume.

[20] K. Ribet, *Congruence relations between modular forms,* Proc. ICM, 1983, 503–514.

[21] K. Ribet, *On modular representations of* $\mathrm{Gal}(\overline{\mathbf{Q}}/\mathbf{Q})$ *arising from modular forms,* Inv. Math. **100** (1990), 431–476.

[22] D. Rohrlich, *Modular functions and modular curves,* this volume.

[23] K. Rubin, *Modularity of mod 5 representations,* this volume.

[24] J. Tilouine, *Hecke algebras and the Gorenstein property,* this volume.

[25] L. Washington, *Galois cohomology,* this volume.

[26] A. Wiles, *Modular elliptic curves and Fermat's Last Theorem,* Annals of Math. **141** (1995), 443–551.

THE FLAT DEFORMATION FUNCTOR

Brian Conrad

Introduction

Let $E_{/\mathbf{Q}}$ be a semistable elliptic curve and p a prime. In the proof that E is modular, properties of the local representation

$$\rho_{E,p}|_{D_p} : D_p = \mathrm{Gal}(\overline{\mathbf{Q}}_p/\mathbf{Q}_p) \to \mathrm{Aut}(\mathrm{Ta}_p(E(\overline{\mathbf{Q}}_p))) \simeq \mathrm{GL}_2(\mathbf{Z}_p)$$

play an essential role. If E has either ordinary or multiplicative reduction at p, then one can describe the essential properties of $\rho_{E,p}|_{D_p}$ quite explicitly in representation-theoretic terms. However, if E has supersingular reduction at p, then the situation is more subtle. More precisely, the residual representation $\overline{\rho}_{E,p}|_{D_p} = \rho_{E,p}|_{D_p} \bmod p$ is *absolutely irreducible* (which already distinguishes this case from the other two cases) and for all $n \geq 1$, the finite discrete D_p-module $\rho_{E,p}|_{D_p} \bmod p^n$ has the property that the corresponding $\mathbf{Q}_p$-group scheme arises as the generic fiber of a finite *flat* $\mathbf{Z}_p$-group scheme (necessarily commutative, with order p^{2n}).

In order to study the supersingular case, Wiles considers a 'flat' deformation problem which makes critical use of the theory of finite flat group schemes. The representability of the deformation functor in this case, as well as the (abstract) 'computation' of the corresponding representing ring (when $p \neq 2$), were worked out by Ramakrishna in his thesis [27]. The central tool he uses is Fontaine's work on finite flat group schemes [13] (made more 'explicit' by the work of Fontaine-Laffaille [14, §9]). For $p \neq 2$, Fontaine constructed an equivalence of categories between finite flat commutative $\mathbf{Z}_p$-group schemes with p-power order and a category consisting of finite-length $\mathbf{Z}_p$-modules with certain extra structures. This allows one to reformulate the study of certain group schemes over $\mathbf{Z}_p$ as the study of certain modules. Since it is far easier to manipulate and construct modules than it is to manipulate and construct group schemes, this reformulation is very useful.

We begin with an explanation of why 'flat' representations are a natural thing to consider and outline the role they play in Wiles' proof. After relating flat representations to local Galois cohomology, we formulate and partially prove the main results (due to Ramakrishna). We then review certain fundamental results due to Fontaine [13]. With these in our toolbox, we complete the proof of Ramakrishna's theorem concerning the 'explicit' structure of the deformation ring which represents the 'flat deformation functor.' This yields a theorem about Galois cohomology; we also prove a similar cohomological result in the residually split ordinary case (where the deformation functor is *not* known to be representable). Throughout, we assume a familiarity with the basic notions in the deformation theory of Galois representations, as given in [23], and the theory of finite flat commutative group schemes, as given in [34]. Our discussion provides a 'concrete' application of these ideas.

Much of what we say over $\mathbf{Z}_p$, $\mathbf{Q}_p$, and $\mathbf{F}_p$ is valid somewhat more generally, but we will stick with the more concrete setting when it is convenient to do so. Also, though we will frequently use Néron models of elliptic curves, we do not use anything from the theory of Néron models; all we need is that an integral Weierstrass model in $\mathbf{P}^2_{\mathbf{Z}_p}$ with a smooth closed fiber (and therefore a smooth generic fiber) admits the structure of a group scheme extending the canonical group scheme structure on the generic fiber. This can be proven directly [32, Ch IV, Remark 5.4.1].

§0. Notation

We fix algebraic closures $\overline{\mathbf{Q}}$ and $\overline{\mathbf{Q}}_\ell$ (for all places ℓ of $\mathbf{Q}$), as well as embeddings $\iota_\ell : \overline{\mathbf{Q}} \hookrightarrow \overline{\mathbf{Q}}_\ell$. For any field K with a fixed separable closure K_s, let $G_K = \mathrm{Gal}(K_s/K)$. Let $D_\ell \subseteq G_{\mathbf{Q}}$ denote the decomposition group at ℓ arising from ι_ℓ. When $\ell \neq \infty$, let $I_\ell \subseteq D_\ell$ denote the inertia subgroup. All finite extensions of $\mathbf{Q}_\ell$ are understood to lie inside of $\overline{\mathbf{Q}}_\ell$ and for $\ell \neq \infty$, all finite fields of characteristic ℓ are understood to be subfields of the residue field $\overline{\mathbf{F}}_\ell$ naturally attached to $\overline{\mathbf{Q}}_\ell$. These choices are not important, but making them at the start simplifies the exposition below.

For Σ a finite set of places of $\mathbf{Q}$, let $\mathbf{Q}_\Sigma \subseteq \overline{\mathbf{Q}}$ denote the maximal subextension unramified outside of Σ, with $G_\Sigma = \mathrm{Gal}(\mathbf{Q}_\Sigma/\mathbf{Q})$ the corresponding quotient of $G_{\mathbf{Q}}$. Finally, when a prime p is fixed under discussion, let $\epsilon : D_p \to \mathbf{Z}_p^\times$ denote the p-adic cyclotomic character and let

$$\omega = \epsilon \bmod p : D_p \to \mathbf{F}_p^\times$$

denote its reduction modulo p. We allow for the possibility $p = 2$ unless we explicitly say otherwise. Finally, we adopt a common abuse of terminology: when we say 'finite flat S-group scheme,' it is always understood that we mean 'finite flat *commutative* S-group scheme' (of finite presentation over S — but our bases are always noetherian). We sometimes omit reference to S when the context makes it clear. When $S = \mathrm{Spec}(L)$ with L a field, we will abbreviate this to 'finite L-group scheme.' The notation $X_{/S}$ denotes an S-scheme X. If $S = \mathrm{Spec}(A)$, then this is usually written $X_{/A}$ instead. Also, for certain standard group schemes G over $\mathbf{Z}$ or $\mathbf{F}_p$, such as $\mathbf{G}_m$, α_p, and μ_p, we write $G_{/T}$ to denote the base extension to T, and we sometimes omit reference to T when the context makes it clear.

§1. MOTIVATION AND FLAT REPRESENTATIONS

Fix a semistable elliptic curve $E_{/\mathbf{Q}}$ and a prime p. Suppose p is *odd*, the restriction of $\overline{\rho}_{E,p}$ to $\mathrm{Gal}(\overline{\mathbf{Q}}/\mathbf{Q}(\sqrt{(\frac{-1}{p})p}))$ is absolutely irreducible, and $\overline{\rho}_{E,p}$ is modular (e.g., $p=3$ and $\overline{\rho}_{E,3}|_{\mathrm{Gal}(\overline{\mathbf{Q}}/\mathbf{Q}(\sqrt{-3}))}$ is absolutely irreducible [38, pp. 541-542]). Wiles' proof provides an inductive procedure to show that for successively larger and larger n, the representations $\rho_{E,p} \bmod p^n$ are 'modular' (in an appropriate sense). Moreover, sufficiently tight control is kept on the level (as well as other properties) of the modular forms which are constructed so that one can 'pass to the limit' and conclude that $\rho_{E,p}$ is 'modular.' In order to carry out this procedure, there is an extremely delicate balancing act to handle, with (abstract) deformation rings on one side and (concrete) Hecke rings on the other side. The latter provide a link to modular forms and representations 'coming from modular forms,' whereas the former provide a link to the particular representation of interest, $\rho_{E,p}$, which we want to prove 'comes from a modular form.' The relation between the two different types of rings — leading to the proof that they're isomorphic — is supplied by a numerical criterion from commutative algebra [9]. The hard part is to check that this numerical criterion actually can be applied! In order to do this, one has to prove highly non-obvious theorems about the commutative algebra properties of the rings in question. This requires a very detailed understanding of both the deformation rings and the Hecke rings.

Refined knowledge about Hecke rings can be obtained via methods of algebraic geometry [36], essentially because Hecke rings act on algebro-geometric objects such as Jacobians of modular curves. However, refined knowledge about deformation rings is supplied by completely different methods, namely techniques from Galois cohomology. The problem of estimating the sizes of certain 'global' Galois cohomology groups is the central issue, and is the means by which the 'flat deformation functor' enters into Wiles' proof. The cohomology groups whose order must be bounded are of the form $H^1_{\mathcal{D}}(G_\Sigma, \mathrm{ad}(\rho))$, where $\rho : G_\Sigma \to \mathrm{GL}_2(A)$ is a continuous representation, with A a suitable local artin ring having finite residue field k of characteristic p. Here, $\mathcal{D}$ is a well-chosen G_Σ-deformation problem for $\overline{\rho} = \rho \bmod \mathfrak{m}_A$. In practice, $\overline{\rho}$ is essentially $\overline{\rho}_{E,p}$, up to extension of scalars on the residue field k. The deformation problem $\mathcal{D}$ imposes 'local' conditions on G_Σ-deformations of $\overline{\rho}$ at the finite set of places Σ, including p and ∞. In the application to elliptic curves, the other places in Σ consist of the places where $E_{/\mathbf{Q}}$ has bad reduction (though in his arguments, Wiles also needs to consider other sets Σ' which include auxiliary primes used in the study of the corresponding 'minimal' deformation problem). The study of 'type $\mathcal{D}$' deformations of $\overline{\rho}$ translates cohomologically into studying elements of $H^1(G_\Sigma, \mathrm{ad}(\rho))$ whose local restrictions at places in Σ satisfy certain conditions; these special elements constitute the A-submodule $H^1_{\mathcal{D}}(G_\Sigma, \mathrm{ad}(\rho))$.

As is explained in [37] and [38, Prop 1.6], for any finite discrete G_Σ-module X with p-power order (e.g., $X = E[p^n](\overline{\mathbf{Q}})$), the problem of estimating the size of $H^1_{\mathcal{D}}(G_\Sigma, X)$ reduces essentially to the problem of estimating the size of certain D_ℓ-cohomology groups for the places $\ell \in \Sigma$. We say 'essentially' because there is another 'dual' group $H^1_{\mathcal{D}^*}(G_\Sigma, X^*)$ that must be considered, too. But at a critical stage in what Wiles calls the 'minimal deformation problem,' this extra factor is not hard to handle (this essentially follows from the injectivity of ε_Q in [38, (3.1)]).

Among the sizes of the various D_ℓ-cohomology groups to be estimated, the only one which turns out to present serious difficulties is the case $\ell = p$. Wiles considers two types of D_p-cohomology groups, both modeled on *semistability* of E at p and based on an attempt to capture the essential properties of $\rho_{E,p}|_{D_p}$ in terms of deformation theory and Galois cohomology. Particularly in view of recent work of Diamond [10], these local deformation-theoretic/Galois-cohomological conditions at p are the central reason that semistability conditions enter into Wiles' proof. Any attempt to prove the full Taniyama-Shimura Conjecture will almost certainly have to first involve the formulation of local deformation conditions at p which accurately describe cases with additive reduction.

Before precisely defining the D_p-cohomology group and the deformation functor that will concern us, we will first prove some preliminary results. We now allow $p = 2$. In order to state the first result, we recall the notion of a *fundamental character* (this will occur repeatedly below, albeit in a limited context). The tame inertia group I_p^t of $\mathbf{Q}_p$ admits a canonical isomorphism

$$I_p^t \simeq \varprojlim \mathbf{F}_{p^n}^\times,$$

where the (surjective) transition maps are given by taking norms (recall our convention about finite fields in the Notation section above). This follows readily from the structure theorem for tamely ramified finite extensions of local fields. See [30, §1] for more details and generality. The projection

$$\psi_n : I_p \twoheadrightarrow I_p^t \twoheadrightarrow \mathbf{F}_{p^n}^\times$$

is called the *fundamental character of level* n (for $\mathbf{Q}_p$). Explicitly, if $\pi \in \mathbf{Z}_p$ is a uniformizer and $\alpha_n \in \overline{\mathbf{Q}}_p^\times$ satisfies $\alpha_n^{p^n-1} = \pi$, then for $g \in I_p$,

$$\psi_n(g) = \frac{g(\alpha_n)}{\alpha_n}.$$

For example, $\psi_1 = \omega|_{I_p}$ (check!). We remind the reader that the notion of a fundamental character is not functorial. That is, if U is an open subgroup of D_p with fixed field K, then we can define fundamental characters for K in a similar manner, but $\psi_n|_U$ is usually *not* a fundamental character for K (even if K has residue field $\mathbf{F}_p$).

Upon choosing a basis for $\mathbf{F}_{p^2}$ over $\mathbf{F}_p$, we get an injection of groups

$$j_2 : \mathbf{F}_{p^2}^\times \hookrightarrow \mathrm{GL}_2(\mathbf{F}_p).$$

Observe that as a 2-dimensional $\mathbf{F}_p$-representation for I_p, $j_2 \circ \psi_2$ is semisimple and therefore irreducible (as the character ψ_2 does not take all of its values in $\mathbf{F}_p^\times$). This representation remains irreducible under any extension of scalars of odd degree over $\mathbf{F}_p$. However, once we extend scalars so that $\mathbf{F}_{p^2}$ lies in the field of scalars, the representation decomposes into a direct sum of the two *distinct* characters, ψ_2 and ψ_2^p (viewed as taking values in the field of scalars).

Theorem 1.1. *Let $E_{/\mathbf{Q}_p}$ be a semistable elliptic curve. If the reduction type is either ordinary or multiplicative, then*

$$\rho_{E,p} \simeq \begin{pmatrix} \epsilon\chi & * \\ 0 & \chi^{-1} \end{pmatrix},$$

with $\chi : D_p \to \mathbf{Z}_p^\times$ a continuous unramified character. In particular,

$$\overline{\rho}_{E,p} \simeq \begin{pmatrix} \omega\overline{\chi} & * \\ 0 & \overline{\chi}^{-1} \end{pmatrix},$$

with $\overline{\chi} : D_p \to \mathbf{F}_p^\times$ a continuous unramified character.

If the reduction type is good supersingular, then the 2-dimensional $\mathbf{F}_p$-representation

$$\overline{\rho}_{E,p}|_{I_p} : I_p \to \mathrm{Aut}(E[p](\overline{\mathbf{Q}}_p)) \simeq \mathrm{GL}_2(\mathbf{F}_p)$$

is isomorphic to the fundamental character of level 2. Moreover, $\overline{\rho}_{E,p}$ is absolutely irreducible and $\mathbf{Q}_p \otimes_{\mathbf{Z}_p} \rho_{E,p}$ is irreducible.

REMARKS. The irreducibility features of the supersingular case are quite a contrast with the relatively simple (and 'reducible') representation-theoretic description of the other cases. The supersingular case therefore requires a completely different treatment in Wiles' proof, and that is the purpose of the 'flat' deformation functor, as we shall see. The convenience of the general description of the representations in Theorem 1.1 (when combined with Theorem 1.2) is why semistable cases are much more accessible (at present) than cases with additive reduction.

The results in Theorem 1.1 for $\overline{\rho}_{E,p}$ in cases of good reduction were first proven by Serre [30, §1.11, Prop 11, 12] using formal groups. Since group schemes will be fundamental in our work below, we give a different proof that uses the theory of group schemes, particularly general theorems of Raynaud [28, §3] which were motivated by (but do not logically depend upon) the results of Serre.

We will use Raynaud's results to prove (in Theorem 1.8) a generalization of the good reduction cases under the hypothesis $p \neq 2$; however, the case of elliptic curves merits special treatment since in the 'reducible' cases we can interpret the unramified quotient χ^{-1} quite concretely, as is shown in the proof below.

PROOF. First we consider good ordinary and bad multiplicative reduction. In both cases, we wish to produce an unramified rank 1 quotient of $\rho_{E,p}$. If we call the character on such a quotient χ^{-1} and recall that $\det \rho_{E,p} = \epsilon$, the rest is then clear. Suppose the reduction is bad multiplicative. In this case, the theory of Tate models supplies us with a D_p-module isomorphism $E(\overline{\mathbf{Q}}_p) \simeq (\overline{\mathbf{Q}}_p^\times / q^{\mathbf{Z}})(\chi)$ for some $q \in p\mathbf{Z}_p$ and some continuous unramified character $\chi : D_p \to \langle -1 \rangle$ [32, Ch V, Thm 5.3]. Choosing consistent p^nth roots of 1 in $\overline{\mathbf{Q}}_p^\times$ then gives rise to a rank 1 subrepresentation inside of $\rho_{E,p}$ and the quotient by this is visibly free of rank 1 and unramified.

Now suppose the reduction is good ordinary. Observe that the Néron model $\mathcal{E}_{/\mathbf{Z}_p}$ of $E_{/\mathbf{Q}_p}$ is an elliptic curve over $\mathbf{Z}_p$ whose p-divisible group has a closed fiber with non-trivial connected and étale factors (this is essentially the definition of what it means for the closed fiber $\overline{E}_{/\mathbf{F}_p}$ of the Néron model to be an ordinary elliptic curve). Thus, the p-divisible group of $\mathcal{E}_{/\mathbf{Z}_p}$ has non-trivial connected and étale factors (since formation of the connected-étale sequence of a finite flat group scheme over a henselian local base is compatible with base change by a *local* map of base rings, such as $\mathbf{Z}_p \to \mathbf{F}_p$). Passing to the generic fiber over $\mathbf{Q}_p$ and then to $\overline{\mathbf{Q}}_p$-points, the non-trivial connected-étale sequence over $\mathbf{Z}_p$ gives rise to the desired decomposition of $\rho_{E,p}$, since a finite étale cover of $\mathrm{Spec}(\mathbf{Z}_p)$ has a generic fiber $\mathrm{Spec}(L)$, with L a finite product of finite *unramified* extensions of $\mathbf{Q}_p$ (and base change preserves the exactness property of a short exact sequence of finite flat group schemes and thus of p-divisible groups). In particular, we can interpret the unramified quotient of $\rho_{E,p} = \mathrm{Ta}_p(E(\overline{\mathbf{Q}}_p))$ as precisely $\mathrm{Ta}_p(\overline{E}(\overline{\mathbf{F}}_p))$, via the reduction map on points.

Lastly, consider the case in which E has good supersingular reduction at p. Assume that we have established the desired description of $\overline{\rho}_{E,p}|_{I_p}$ in terms of the fundamental character of level 2. Let us see how to prove the other assertions. First of all, $\overline{\rho}_{E,p}$ must be absolutely irreducible. This is because if it has a stable line over $\overline{\mathbf{F}}_p$, then the eigencharacter along this line gives a character on D_p whose restriction to I_p is either ψ_2 or

ψ_2^p (the two characters occurring in the decomposition of the semisimple $\overline{\mathbf{F}}_p \otimes_{\mathbf{F}_p} \overline{\rho}_{E,p}|_{I_p}$). But conjugation by a Frobenius element on I_p interchanges ψ_2 and ψ_2^p (check!), so neither ψ_2 nor ψ_2^p extends to an $\overline{\mathbf{F}}_p^\times$-valued character on D_p. This proves that $\overline{\rho}_{E,p}$ is (absolutely) irreducible. If $\mathbf{Q}_p \otimes_{\mathbf{Z}_p} \rho_{E,p}$ is reducible, then we can scale a generator of a D_p-stable line so that it is part of a basis for the lattice $\rho_{E,p}$. Reducing this mod p then would contradict the irreducibility of $\overline{\rho}_{E,p}$.

It remains to prove that $\overline{\rho}_{E,p}|_{I_p}$ has the asserted form. We postpone the proof of this until after Theorem 1.7, as it will require some results from the theory of finite flat group schemes. The careful reader can check that this does not lead to circular reasoning. ■

Theorem 1.1 shows that from the point of view of Galois representations, it is natural to combine the cases of ordinary and multiplicative reduction into one case, called 'ord' by Wiles, and to treat the case of supersingular reduction separately; this latter case is called 'fl' by Wiles, as its study involves the theory of finite flat group schemes in order to circumvent the lack of upper triangular representations in the supersingular case. We want to formulate 'abstract' properties satisfied by the deformation $\rho_{E,p}$ of $\overline{\rho}_{E,p}$ (and we will require these for *all* deformations of $\overline{\rho}_{E,p}$ that we study). The 'ord' condition has an obvious analogue for a representation into $\mathrm{GL}_2(R)$, where R is *any* topological $\mathbf{Z}_p$-algebra. However, it is not clear how to formulate a representation-theoretic condition which both captures the supersingular Tate module representations and also makes sense in deformation theory. For example, 'irreducibility' of a $\mathrm{GL}_2(R)$-representation is a bad notion when R is not a field. The correct condition is supplied by the theory of finite flat group schemes, and it is a very subtle condition from a representation-theoretic viewpoint.

It is worth pointing out that the finite flat group scheme techniques we will use to study the supersingular case are also needed at a critical technical step in the 'minimal' deformation problem for some 'ord' cases (essentially because of [38, (3.1)]; also see Example 1.3(i) below). Thus, what we are about to do is not only needed in the supersingular cases.

The result which makes finite flat group schemes a relevant notion for us is the following fundamental fact, whose most 'natural' proof requires a serious use of algebraic geometry to handle the case of torsion levels divisible by p, which is the case of interest to us here. In fact, Theorem 1.2 below was implicitly invoked in the proof of Theorem 1.1 when we used the theory of p-divisible groups to treat the ordinary reduction case. If in Theorem 1.2 one only cared about torsion levels which are prime to p, an argument using just the criterion of Néron-Ogg-Shafarevich (for elliptic curves) and Galois descent of rings [3, §6.2, Ex B] would suffice.

Theorem 1.2. *Let $E_{/\mathbf{Q}_p}$ have good reduction. Then for all $m \neq 0$, the*

$\mathbf{Q}_p$-group scheme $E[m]$ which is canonically attached to the D_p-representation $E[m](\overline{\mathbf{Q}}_p)$ is the generic fiber of a finite flat *group scheme over $\mathbf{Z}_p$ (in fact, the m-torsion on the Néron model over $\mathbf{Z}_p$ is such a finite flat $\mathbf{Z}_p$-group scheme).*

PROOF. We will first give a 'natural' proof that uses general principles in algebraic geometry. At the end we will give an ad hoc proof which is not at all 'natural,' but uses only the theory of elliptic curves over fields and basic facts from commutative algebra and the theory of schemes.

Let $\mathcal{E}_{/\mathbf{Z}_p}$ denote the Néron model of $E_{/\mathbf{Q}_p}$. This is an abelian scheme over $\mathbf{Z}_p$ with relative dimension 1. It suffices to prove that if $A \to \mathrm{Spec}(R)$ is any abelian scheme of relative dimension d over an affine noetherian base (i.e., a proper smooth group scheme with d-dimensional connected fibers, so necessarily geometrically connected [16, IV, 4.5.13]), then the multiplication map $m_A : A \to A$ is finite and flat. Base extension by the identity section of $A_{/R}$ then yields the desired conclusion that $A[m] \to \mathrm{Spec}(R)$ is finite and flat. The general principle is to reduce to the case in which R is an algebraically closed field and then to appeal to the 'classical' theory of abelian varieties (or just elliptic curves). We give the argument in the case of arbitrary relative dimension for conceptual clarity.

Here is the proof that m_A is finite and flat. Certainly m_A is proper. It is also quasi-finite, as can be checked on geometric fibers, in which case it is a standard result in the theory of abelian varieties over algebraically closed fields [24, p. 62] (or for elliptic curves in our setting of interest [31, Ch III, Prop 4.2(a)]). It follows that $m_A : A \to A$ is a quasi-finite and proper map between noetherian schemes, so by [16, IV, 8.11.1] it is necessarily finite (one can avoid noetherian and even finite presentation hypotheses [16, IV, 18.12.4], but more work is then needed). The essential content of this statement is that f is *affine.* Even in the 'concrete' setting of an elliptic curve $\mathcal{E}$ over $\mathbf{Z}_p$ given by an explicit integral Weierstrass model in $\mathbf{P}^2_{\mathbf{Z}_p}$, it is not at all a priori obvious that the 'multiplication by m' map on $\mathcal{E}$ is actually affine (but we will see below that one can prove by elementary ad hoc means that $\mathcal{E}[m]$ is affine).

For flatness, we observe that since $A \to \mathrm{Spec}(R)$ is flat and of finite presentation, by the fiber-by-fiber criterion for flatness [16, IV, 11.3.10(2)] it suffices to check the flatness of $m_A : A \to A$ along fibers over $\mathrm{Spec}(R)$, so we may suppose that R is a field, even algebraically closed (by faithful flatness of field extensions). By the classical theory of abelian varieties [24, p. 62], m_A is a map between two smooth, irreducible varieties of the same dimension d over an algebraically closed field, with fibers over all closed points of dimension $d - d = 0$. It is then a consequence of the fundamental 'local criterion for flatness' that such a map is necessarily flat [21, Cor, Thm 23.1].

Now we sketch an 'elementary' proof that $E[m]$ is the generic fiber of a finite flat $\mathbf{Z}_p$-group scheme (this proof will work for any good reduction elliptic curve over the fraction field of a discrete valuation ring). Let $\mathcal{E} \hookrightarrow \mathbf{P}^2_{\mathbf{Z}_p}$ denote an integral 'good' Weierstrass model for $E_{/\mathbf{Q}_p}$. One can check without appealing to the general theory of Néron models that $\mathcal{E}$ has the structure of a *group scheme* over $\mathbf{Z}_p$ which extends that on the generic fiber $E_{/\mathbf{Q}_p}$ [32, Ch IV, Remark 5.4.1]. It remains to check that the $\mathbf{Z}_p$-group scheme $\mathcal{E}[m]$ is finite and flat over $\mathbf{Z}_p$. By the classical theory of elliptic curves over *fields*, we know that the generic and closed fibers are both finite schemes (over $\mathbf{Q}_p$ and $\mathbf{F}_p$ respectively) with the same rank, namely m^2. Note that the closed fiber might not be reduced!

Assume for the moment that we know $\mathcal{E}[m]$ is an affine scheme, say of the form $\mathrm{Spec}(R)$ for a $\mathbf{Z}_p$-algebra R. Since $\mathrm{Spec}(R) \to \mathrm{Spec}(\mathbf{Z}_p)$ is universally closed, by [1, Exer. 35] (this is the essential commutative algebra input) we see that R is integral over $\mathbf{Z}_p$. As it is of finite type by construction, R is finite over $\mathbf{Z}_p$. The equality of the generic and closed fiber dimensions then yields freeness over $\mathbf{Z}_p$ (by the structure theorem for finite $\mathbf{Z}_p$-modules), and hence flatness.

It remains to show that $\mathcal{E}[m]$ is an affine scheme. Since it is at least a closed subscheme of $\mathbf{P}^2_{\mathbf{Z}_p}$, if we can find an open affine in $\mathbf{P}^2_{\mathbf{Z}_p}$ which contains $\mathcal{E}[m]$, then we will have exhibited $\mathcal{E}[m]$ as a closed subscheme of an affine scheme, and so $\mathcal{E}[m]$ must be affine. The closed fiber of $\mathcal{E}[m]$ is finite over $\mathbf{F}_p$, so we can find a homogeneous polynomial $\overline{f} \in \mathbf{F}_p[X,Y,Z]$ such that $V(\overline{f}) \subseteq \mathbf{P}^2_{\mathbf{F}_p}$ does not meet the closed fiber of $\mathcal{E}[m]$. Choose a lifting of $\overline{f}$ to a homogeneous polynomial $f \in \mathbf{Z}_p[X,Y,Z]$. Since $V(f)$ and $\mathcal{E}[m]$ are closed subschemes of $\mathbf{P}^2_{\mathbf{Z}_p}$, if their intersection is non-empty, it contains a closed point x. Such a point maps to the closed point of $\mathrm{Spec}(\mathbf{Z}_p)$ by the properness of $\mathbf{P}^2_{\mathbf{Z}_p} \to \mathrm{Spec}(\mathbf{Z}_p)$. Our assumption on $V(\overline{f})$ therefore rules out the existence of such a point x, so $\mathcal{E}[m]$ is a closed subscheme of the open affine scheme $\mathrm{Spec}(\mathbf{Z}_p[X,Y,Z]_{(f)}) = \mathbf{P}^2_{\mathbf{Z}_p} \setminus V(f)$ inside of $\mathbf{P}^2_{\mathbf{Z}_p}$. ∎

We note in passing that it is not enough in the proof of Theorem 1.2 to have a proper, smooth, integral model for $E_{/\mathbf{Q}_p}$ (e.g., a 'good' projective integral Weierstrass model); it is critical to know that such a model can be chosen that is a *group scheme* (compatible with the group scheme structure on the generic fiber).

The converse to Theorem 1.2 is also true in a strong sense, though we will not need it in any of our proofs. More precisely, if $E[p^r]$ is the generic fiber of a finite flat group scheme over $\mathbf{Z}_p$ for all $r \geq 1$, then $E_{/\mathbf{Q}_p}$ has good reduction. This is a special case of a more general theorem of Grothendieck's on p-divisible groups [18, IX, Cor 5.10], which says in particular that an abelian variety A over $\mathbf{Q}_p$ has good reduction if and only if its p-divisible group Γ has 'good reduction' in the sense that it is the generic fiber of a p-divisible group over $\mathbf{Z}_p$. This is the correct analogue of the Néron-Ogg-Shafarevich criterion 'at p.' In order to apply this, we need to invoke Raynaud's theorem [28, 2.3.1] which asserts that Grothendieck's criterion is equivalent to the a priori weaker condition that each torsion level of Γ is the generic fiber of a finite flat group scheme over $\mathbf{Z}_p$ (i.e., we do not need to assume that there is any compatibility between the $\mathbf{Z}_p$-group schemes for each p-power torsion level). Keep in mind that it is essential for the converse of Theorem 1.2 that we have a condition on large torsion levels. This is illustrated by the first part of the following example.

Example 1.3.
(i) It is possible that $E_{/\mathbf{Q}_p}$ can have bad reduction (and so $E[p^n]$ is not the generic fiber of a finite flat $\mathbf{Z}_p$-group scheme for some large n) while $E[p]$ has good reduction (i.e., it is the generic fiber of a finite flat $\mathbf{Z}_p$-group scheme). For example, if $E_{/\mathbf{Q}_p}$ has bad split multiplicative reduction with non-integral $j(E)$ a pth power in $\mathbf{Q}_p^\times$, then the theory of Tate models shows

$$\overline{\rho}_{E,p} \simeq \omega \oplus 1,$$

which is certainly the D_p-representation attached to the generic fiber of a finite flat group scheme over $\mathbf{Z}_p$, namely $\mu_p \times_{\mathbf{Z}_p} \underline{\mathbf{Z}/p}$.

For $E_{/\mathbf{Q}}$ such that $E_{/\mathbf{Q}_p}$ has bad reduction and $E[p]_{/\mathbf{Q}_p}$ has good reduction, Wiles' proof studies $\rho_{E,p}$ as an 'ord' deformation of $\overline{\rho}_{E,p}$. But in the critical associated 'minimal' deformation problem, the 'flat' methods of this article are needed; see [38, (3.1)].

(ii) If $E_{/\mathbf{Q}_p}$ has good ordinary reduction of the form

$$\overline{\rho}_{E,p} = \begin{pmatrix} \omega & * \\ 0 & 1 \end{pmatrix},$$

then the p-torsion on the Néron model $\mathcal{E}_{/\mathbf{Z}_p}$ of $E_{/\mathbf{Q}_p}$ is a finite flat group scheme over $\mathbf{Z}_p$ with order p^2, though it is not easy to describe this $\mathbf{Z}_p$-scheme explicitly. It can be shown that $\mathcal{E}[p]$ fits into a short exact sequence of finite flat commutative $\mathbf{Z}_p$-group schemes. Such a sequence *must* have the form

$$0 \to \mu_p \to \mathcal{E}[p] \to \underline{\mathbf{Z}/p} \to 0,$$

except when $\mathcal{E}[p] \simeq \mu_p \times_{\mathbf{Z}_p} \underline{\mathbf{Z}/p}$, in which case there is also a sequence corresponding to the other way of splitting $\mathcal{E}[p]$. Theorems 1.6 and 1.7 supply the results needed to justify this claim. We leave this as an exercise (it will not be needed).

Recall that for a field F of characteristic 0, all finite F-group schemes are étale over F [34], and so with a choice of an algebraic closure $\overline{F}$, we get an order-preserving equivalence of categories between finite flat group schemes over F and finite discrete modules over $G_F = \mathrm{Gal}(\overline{F}/F)$. This allows us to single out certain Galois modules as special.

Definition 1.5. Let K be a field of characteristic 0 that is the fraction field of a henselian (e.g., complete) discrete valuation ring R with residual characteristic p. Fix an algebraic closure $\overline{K}$. We say that a continuous representation

$$\rho\colon G_K \to \mathrm{Aut}(M)$$

on a finite abelian p-group M is *R-flat* (or just *flat* when R is understood from the context) if there exists a finite flat group scheme $H_{/R}$ such that the G_K-representation $H(\overline{K})$ is isomorphic to ρ (or, equivalently, such that the finite K-group scheme canonically attached to ρ is the generic fiber of a finite flat group scheme over R).

The 'flatness' refers to the essential property of $H \to \mathrm{Spec}(R)$. The representation ρ is not being required to be flat over anything. Also, it should be emphasized that flatness as defined above is really a property of *finite* Galois modules (though in Definition 2.1 we will extend it in a formal way to more general Galois modules, such as Tate modules of elliptic curves). Note that in the setting of the above definition, the H which arises necessarily has order equal to the size of the abelian group M (since by R-flatness, this can be checked after passage to a geometric fiber over the generic point).

If there were many different choices for H, the notion of 'flatness' as defined above would not be a natural one to use. In addition, if the representation ρ had extra structure such as that of a vector space over a large finite field k (with the k-action commuting with the G_K-action of ρ), then we would like such extra endomorphism structure to extend to H, or else the concept of a 'flat' representation would likely be too weak to use. Fortunately, we have the following two fundamental results due to Raynaud.

Theorem 1.6. (Raynaud) *Let R, K, $\overline{K}$, and G_K be as above. Consider the covariant functor*

$$F_R : H \rightsquigarrow H(\overline{K})$$

from finite flat p-power order R-group schemes to flat G_K-modules. When $e(R) < p-1$ (so $p \neq 2$), F_R is an equivalence of categories. Moreover, the category of finite flat p-power order R-group schemes is an abelian category via the usual scheme-theoretic kernel and quotient constructions.

Suppose $e(R) = p-1$ and V is a continuous k-linear representation of G_K on a one-dimensional vector space over a finite field k of characteristic p. Assume that the $\mathbf{F}_p[G_K]$-module underlying V is simple (i.e., the natural *map* $\mathbf{F}_p[G_K] \to k$ *is surjective). Then either there is a unique (up to canonical isomorphism) finite flat R-group scheme H with* $H(\overline{K}) \simeq V$*, or else there are two such R-group schemes, one étale and the other multiplicative (i.e., with an étale Cartier dual).*

The essential image of the functor F_R *is stable under passage to subrepresentations, to quotients, and to (finite) direct sums without restriction on* $e(R)$.

PROOF. The proof of full faithfulness consists of reducing to the case in which $H(\overline{K})$ is a simple representation, and then one writes down the most general form of H that could possibly give rise to the given representation (see Theorem 1.7 and the discussion preceding it). We discuss the quasi-inverse functor below. The proof of the stability of the essential image under various constructions is much simpler and is proved by the method of 'scheme-theoretic closure.' See [28, 2.1, 3.3.2(3), 3.3.6] or [34, section 4] for further details.

Though [28, 2.1] refers to the fppf topology for formation of quotients, in our setting it is possible to get away with a more naive construction of quotients that exploits Cartier duality for finite flat group schemes. One essentially defines the quotient by a closed subgroup scheme to be the Cartier dual of a suitable kernel. We omit the details of this alternative construction (though there is some work needed to verify flatness), as this would be too much of a digression here, except we note that such an alternative construction supplies an elementary proof that base change preserves short exactness and that applying Cartier duality to a short exact sequence of finite flat group schemes preserves the property of being short exact. See [5, §2.2] for further details. ■

One can describe the 'quasi-inverse' to the functor in Theorem 1.6 when $e(R) < p-1$. In down-to-earth terms, consider a flat representation ρ with a representation space V_ρ whose underlying group is a finite abelian p-group. With our choice of $\overline{K}$, there is (via Galois descent) canonically attached to V_ρ a finite K-group scheme $\mathrm{Spec}(K_\rho)$ whose representation on $\overline{K}$-points is V_ρ. The essential content of Theorem 1.6 is that inside of K_ρ there is a *unique* finite R-subalgebra $\mathcal{O}_\rho$ which has generic fiber K_ρ and which is 'stable' under the group law (or, rather, K-Hopf algebra) morphisms on K_ρ. This $\mathcal{O}_\rho$ is the affine R-algebra for the unique finite flat R-group scheme whose generic fiber representation is V_ρ. The functor $\rho \rightsquigarrow \mathrm{Spec}(\mathcal{O}_\rho)$ is the sought-after 'quasi-inverse' functor.

The idea behind Raynaud's proof is to build everything up from an analysis of finite flat R-group schemes with generic fiber representations that are *simple* of p-power order. This generalizes earlier results of Oort-Tate [26, Thm 2] in the case of finite flat group schemes of order p. Consider a (possibly non-simple) finite discrete G_K-representation V with p-power order. How many (if any) finite flat R-group schemes H admit V as a generic fiber representation (i.e., $V \simeq H(\overline{K})$ as G_K-modules)? By the theory of Galois descent of schemes [3, §6.2, Ex B], or else by an ad hoc modification of the proof of classical Galois descent for fields [31, Ch II, Lemma 5.8.1], knowledge of such an H is equivalent to knowledge of V and of $H \times_R R^{\mathrm{sh}}$, with R^{sh} the strict henselization of R (e.g., when $R = \mathbf{Z}_p$, $R^{\mathrm{sh}} = \mathbf{Z}_p^{\mathrm{un}}$; see [3, §2.3, Prop 10, 11]). Thus, the essential case is when R is strictly henselian, so we can replace R by R^{sh} (and G_K by the inertia subgroup $I_K = G_{K^{\mathrm{sh}}}$).

For R strictly henselian and V *simple*, $k = \mathrm{End}_{G_K}(V)$ is a finite division ring and so by Wedderburn's Theorem is a finite field with characteristic p (here, we temporarily abandon our usual convention that such a field is equipped with an embedding into $\overline{\mathbf{F}}_p$). Hence, V is canonically an irreducible $k[G_K]$-module with finite k-dimension. Because R is a strictly henselian discrete valuation ring with residue characteristic p, G_K has a pro-p normal subgroup (wild inertia) with a pro-prime-to-p abelian quotient (tame inertia). To justify this structure for G_K, note that the integral closure of R in a finite (necessarily separable) extension of K is finite as an R-module. Since R is a strictly henselian discrete valuation ring, such an integral closure is also a (strictly henselian) discrete valuation ring; thus, we can easily pass to the completion of R in place of R and then use [30, §1] to analyze G_K.

We conclude from the structure of G_K that V is a tame (and therefore abelian) representation [29, Ch IX, Lemma 2], so from the structure theory of semisimple rings we see that $\dim_k(V) = 1$. That is, V is given by a continuous character

$$\psi : G_K \to k^\times.$$

Raynaud succeeded in describing precisely the ψ which can arise in this way. The description is in terms of fundamental characters. Since we have only discussed fundamental characters for $\mathbf{Q}_p$, we will state the result in that limited context with $R = \mathbf{Z}_p^{\mathrm{un}}$ (which suffices for our purposes).

Theorem 1.7. (Raynaud) *Consider a continuous character*

$$\psi : G_{\mathbf{Q}_p^{\mathrm{un}}} = I_p \to k^\times,$$

with k a finite field of characteristic p. Choose an embedding $k \hookrightarrow \overline{\mathbf{F}}_p$. Let V be a one-dimensional k-vector space with I_p-action given by ψ. Then V is $\mathbf{Z}_p^{\mathrm{un}}$-flat if and only if

$$\psi = \psi_n^e,$$

where $|k| = p^n$, ψ_n is the fundamental character of level n, and $e = e_0 + e_1 p + \cdots + e_{n-1}p^{n-1}$, with $0 \le e_i \le e(\mathbf{Z}_p^{\mathrm{un}}) = 1$.

PROOF. See [28, 3.4.3] or [34, section 4] for details. The proof of 'only if' proceeds as follows. Let $H_{/\mathbf{Q}_p^{\mathrm{un}}}$ be the finite p-power order commutative $\mathbf{Q}_p^{\mathrm{un}}$-group scheme associated to V, so the k-action on the (one-dimensional) flat I_p-representation space V can be viewed as a k-action on H. The key step is to prove that this action *necessarily extends* to a k-action on certain finite flat $\mathbf{Z}_p^{\mathrm{un}}$-group schemes which have generic fiber H [28, 3.3.1]. We remark in passing that one can exploit this to show that there is a unique finite flat $\mathbf{Z}_p^{\mathrm{un}}$-group scheme with generic fiber representation V when $p > 2$ and there are at most two such $\mathbf{Z}_p^{\mathrm{un}}$-group schemes when $p = 2$ and V is a simple $\mathbf{F}_p[I_p]$-module; this is used in the proof of Theorem 1.6.

Note that the choice of embedding $k \hookrightarrow \overline{\mathbf{F}}_p$ is harmless, since changing this merely has the effect of cyclically permuting the 'digits' e_i. ■

The example of μ_2 and $\underline{\mathbf{Z}/2}$ over $\mathbf{Z}_2$ shows that the full faithfulness in Theorem 1.6 is false when $p = 2$, since the generic fibers of μ_2 and $\underline{\mathbf{Z}/2}$ over $\mathbf{Q}_2$ are isomorphic, but the closed fibers over $\mathbf{F}_2$ are not (one is reduced, the other is not), so μ_2 and $\underline{\mathbf{Z}/2}$ are not $\mathbf{Z}_2$-isomorphic (as schemes, let alone as group schemes). Note that this also explains why Theorem 1.6 treats the case $e(R) = p - 1$ separately.

Theorem 1.7 and the discussion preceding it show that for $\chi : D_p \to \mathbf{F}_p^\times$ an unramified continuous character, $\omega^i\chi : D_p \to \mathbf{F}_p^\times$ is flat if and only if $i \equiv 0, 1 \bmod p-1$. The flatness condition is a therefore a very severe constraint on a representation. Though Theorem 1.2 shows that flat representations arise naturally from elliptic curves, the general problem of describing *all* flat representations (not just the ones as in Theorem 1.7) is very subtle. An indirect answer to this question was given by Fontaine in a special case, and we will discuss his theory in §4.

Gathering together our results so far, we can complete the proof of Theorem 1.1 and in the supersingular reduction case can give a description of properties of $\rho_{E,p}$ which make sense for representations into $\mathrm{GL}_2(R)$, where R is any complete local noetherian ring with a finite residue field of characteristic p. This will be the starting point for the deformation-theoretic study of the supersingular case. First, we complete the proof of Theorem 1.1.

PROOF. (of Theorem 1.1, continued) By Theorem 1.7 and the discussion preceding it, it suffices to check that in the supersingular case, $\overline{\rho}_{E,p}|_{I_p}$ is irreducible (and then we can apply Theorem 1.7 with $k = \mathbf{F}_{p^2}$). If $\overline{\rho}_{E,p}|_{I_p}$ is reducible, then by Theorem 1.6, the diagonal characters are $\mathbf{Z}_p^{\mathrm{un}}$-flat. Since $\det \overline{\rho}_{E,p} = \omega$, we see from Theorem 1.7 and the fact that the level 1 fundamental character is $\psi_1 = \omega|_{I_p}$ that there exists a short exact sequence of the form

$$0 \to \chi_1 \to \overline{\rho}_{E,p}|_{I_p} \to \chi_2 \to 0,$$

where $\{\chi_1, \chi_2\} = \{\omega|_{I_p}, 1\}$. By Theorem 1.6, it follows that there exists a short exact sequence of finite flat $\mathbf{Z}_p^{\mathrm{un}}$-group schemes

$$0 \to G_1 \to \mathcal{E}^{\mathrm{un}}[p] \to G_2 \to 0$$

which on $\overline{\mathbf{Q}}_p$-points realizes the given decomposition of $\overline{\rho}_{E,p}|_{I_p}$. Here, $\mathcal{E}^{\mathrm{un}}$ is the Néron model of the good reduction curve $E_{/\mathbf{Q}_p^{\mathrm{un}}}$. It happens to be the case that $\mathcal{E}^{\mathrm{un}}$ is the base extension of the Néron model of $E_{/\mathbf{Q}_p}$ [3, §7.2, Thm 1(*ii*)], but we will not need this. What we *will* need is that $\mathcal{E}^{\mathrm{un}}$ has a supersingular closed fiber. This is clear naively, since $j(E_{/\mathbf{Q}_p^{\mathrm{un}}}) = j(E_{/\mathbf{Q}_p})$ is integral, and the image of this in $\overline{\mathbf{F}}_p$ is the j-invariant of the closed fiber. Since the reduction type is determined by the j-invariant, we see that $\mathcal{E}^{\mathrm{un}}$ has supersingular reduction.

Now assume $p \neq 2$ (so $\omega|_{I_p}$ is non-trivial). Let G denote whichever of G_1 or G_2 has a trivial generic fiber representation. We claim that G is étale. If this were true, then passing to the closed fiber would prove that the étale factor of the finite flat group scheme $\mathcal{E}^{\mathrm{un}}[p] \times_{\mathbf{Z}_p^{\mathrm{un}}} \overline{\mathbf{F}}_p$ is non-trivial, and so there exist non-trivial p-torsion geometric points on the closed fiber, contradicting the supersingularity condition.

Hence, for the case of $p \neq 2$ we are reduced to checking the claim that if $H_{/\mathbf{Z}_p^{\mathrm{un}}}$ is a finite flat group scheme with p-power order and $H(\overline{\mathbf{Q}}_p)$ has trivial inertial action, then $H \to \mathrm{Spec}(\mathbf{Z}_p^{\mathrm{un}})$ is étale. The generic fiber representation of $G_{\mathbf{Q}_p^{\mathrm{un}}} = I_p$ is trivial, so the full faithfulness in Theorem 1.6 implies that that $H_{/\mathbf{Z}_p^{\mathrm{un}}}$ is a constant group scheme and so it is clearly étale over $\mathbf{Z}_p^{\mathrm{un}}$. This completes the proof of Theorem 1.1 for odd p.

Now consider the case $p = 2$. This will require (at the end) more advanced results from algebraic geometry; since the case $p = 2$ is only being included for completeness and is not used in Wiles' proof, the reader can skip this case.

Note that each G_i has trivial inertial generic fiber representation. It is clear that μ_2 and $\underline{\mathbf{Z}/2}$ are two non-isomorphic $\mathbf{Z}_2^{\mathrm{un}}$-group schemes with trivial generic fiber representations, so by Theorem 1.6 we see that each G_i is $\mathbf{Z}_2^{\mathrm{un}}$-isomorphic to μ_2 or $\underline{\mathbf{Z}/2}$. Since $\mathcal{E}^{\mathrm{un}}[p]$ is connected, G_1 and G_2 are connected. Thus, each G_i is isomorphic to μ_2.

Applying Cartier duality to the short exact sequence

$$0 \to \mu_2 \to \mathcal{E}^{\mathrm{un}}[p] \to \mu_2 \to 0,$$

we get another short exact sequence. Granting that $\mathcal{E}^{\mathrm{un}}[p]$ is self-dual, the middle term in the Cartier-dualized short exact sequence is $\mathcal{E}^{\mathrm{un}}[p]$. However, the outer terms now are isomorphic to $\underline{\mathbf{Z}/2}$ over $\mathbf{Z}_2^{\mathrm{un}}$, so we have a contradiction.

In order to check the self-duality of $\mathcal{E}^{\mathrm{un}}[p]$, it is not enough to know that the generic fiber is self-dual (via the Weil pairing), since we cannot apply

Theorem 1.6 (with $p = 2$) to the non-simple generic fiber representation. The essential problem is that we do not know naively that the Cartier dual of $\mathcal{E}^{\mathrm{un}}[p]$ is still connected (otherwise results of Raynaud and Fontaine extending Theorem 1.6 for $p = 2$ could be used). This connectedness is only known *after* we prove the stronger assertion of self-duality. Self-duality follows from the Cartier-Nishi duality theorem [25, Cor 1.3(*ii*)], applied to the morphism $p : \mathcal{E}^{\mathrm{un}} \to \mathcal{E}^{\mathrm{un}}$, and the canonical autoduality of elliptic curves [20, 2.1.2]. ■

When $E_{/\mathbf{Q}_p}$ is an elliptic curve with good supersingular reduction, we can give a list of special properties of the representation $\overline{\rho} = \overline{\rho}_{E,p}$ and its deformation class to $\mathrm{GL}_2(\mathbf{Z}_p)$ represented by $\rho = \rho_{E,p}$. Namely, $\det \overline{\rho} = \omega$, $\overline{\rho}$ is flat (Theorem 1.2) and absolutely irreducible (Theorem 1.1), while $\det \rho = \epsilon$ and for every $n \geq 1$, the finite p-power order discrete D_p-module $\rho \bmod p^n$ is *flat* (Theorem 1.2). The absolute irreducibility of $\overline{\rho}$ indicates that we are not in the 'ord' case. Keeping in mind the converse to Theorem 1.2, these conditions will motivate our definition (in the next section) of the 'flat deformation functor' as a deformation problem for D_p-representations. The remarkable fact that we will be able to explicitly describe the deformation ring associated to this deformation problem (Theorem 3.5) will allow us to reverse the usual process 'H^1 tells us about the deformation ring' to get information about a local H^1 term from knowledge about a deformation ring. This ring-theoretic knowledge will be obtained by methods based on the work of Fontaine and Ramakrishna. As we said earlier, this local H^1 data is then pieced together with other Galois cohomological data (at other places in Σ) to tell us about the size of a global H^1 term, and thereby gives us a grip on a global deformation ring (e.g., allows us to show it is isomorphic to a Hecke ring after a lot more work).

Using Raynaud's results (Theorems 1.6, 1.7), we can somewhat generalize the 'good reduction' parts of Theorem 1.1 when $p \neq 2$. This is useful insofar as it will allow us to state the Main Theorem 3.3 with fewer hypotheses (and it is an aesthetically pleasing result too).

Theorem 1.8. *Let R be a complete local noetherian ring with finite residue field k of characteristic $p \neq 2$. Choose a* flat *representation*

$$\overline{\rho} : D_p \to \mathrm{GL}_2(k)$$

with $\det \overline{\rho}|_{I_p} = \omega|_{I_p}$ *and a continuous lift* $\rho : D_p \to \mathrm{GL}_2(R)$ *which gives rise to an element of* $\mathcal{D}^{\mathrm{fl}}_{\overline{\rho}}(R)$ *(see Definition* 2.1, *with* $\mathcal{O} = W(k)$ *there).*

Let

$$\varphi_2 : I_p \xrightarrow{\psi_2} \mathbf{F}_{p^2}^{\times} \hookrightarrow \mathrm{GL}_2(\mathbf{F}_p) \hookrightarrow \mathrm{GL}_2(k)$$

be the map arising from a choice of an $\mathbf{F}_p$*-basis of* $\mathbf{F}_{p^2}$.

(i) If $\overline{\rho}$ is reducible, then there exist continuous unramified characters $\overline{\chi}_i : D_p \to k^\times$ such that

$$\overline{\rho} \simeq \begin{pmatrix} \omega\overline{\chi}_1 & * \\ 0 & \overline{\chi}_2 \end{pmatrix}.$$

Otherwise $\overline{\rho}$ is absolutely irreducible and $\overline{\rho}|_{I_p} \simeq \varphi_2$ (and so $\overline{\rho}|_{I_p} \otimes_k \overline{\mathbf{F}}_p \simeq \psi_2 \oplus \psi_2^p$).

(ii) If $\overline{\rho}$ is reducible, then there exist unique continuous unramified characters $\chi_i : D_p \to R^\times$ such that

$$\rho \simeq \begin{pmatrix} \epsilon\chi_1 & * \\ 0 & \chi_2 \end{pmatrix}.$$

In particular, $\det \rho|_{I_p} = \epsilon|_{I_p}$ *and* $\chi_i \bmod \mathfrak{m}_R = \overline{\chi}_i$.

(iii) If $\overline{\rho}$ is irreducible, then $\det \rho|_{I_p} = \epsilon|_{I_p}$.

REMARKS. It is essential in Theorem 1.8(*i*) that we assume a condition on $\det \overline{\rho}|_{I_p}$. Otherwise one could construct counterexamples in the reducible case using Raynaud's Theorem 1.7 and in the irreducible case using an unramified character $D_p \twoheadrightarrow k_2^\times$, with k_2 the quadratic extension of k.

In Theorem 1.8(*ii*),(*iii*) the main examples of such R to keep in mind are $\mathcal{O}/\lambda^n$ and $(\mathcal{O}/\lambda^n)[\varepsilon] = (\mathcal{O}/\lambda^n)[T]/(T^2)$, with $\mathcal{O}$ and λ as defined in the beginning of §2.

The proof we give for (*iii*) involves Raynaud's work on determinants of p-divisible groups and so ultimately relies on the Zariski-Nagata theorem on purity of the branch locus [17, p. 118]. With extra work, the proof of (*iii*) can be extended to the case $p = 2$ also. The reader should appreciate that there is a substantial amount of algebraic geometry lying behind the assertion in (*iii*). An alternative proof (valid at least for $p \geq 5$ and perhaps also for $p = 3$) can probably be obtained by a brute-force calculation using Fontaine's ideas from §4, together with [14, §6, §9]. But this alternative procedure is not very insightful.

PROOF. (*i*) Suppose $\overline{\rho}$ is reducible, say with diagonal characters $\eta_1, \eta_2 : D_p \to k^\times$. By local class field theory for $\mathbf{Q}_p$, $\eta_j|_{I_p} = \omega^{n_j}|_{I_p}$ as $k^\times$-valued characters, for suitable $n_j \in \mathbf{Z}/(p-1)$. Using the observations after Theorem 1.7, together with the fact that

$$\eta_1\eta_2|_{I_p} = \det \overline{\rho}|_{I_p} = \omega|_{I_p},$$

we see that $\{n_1, n_2\} = \{0, 1\}$. It remains to show that if

$$\overline{\rho} \simeq \begin{pmatrix} \eta_1 & * \\ 0 & \eta_2 \end{pmatrix}$$

with η_1 unramified and $\eta_2|_{I_p} = \omega|_{I_p}$, then $\overline{\rho}$ splits. Equivalently, we must show that $\overline{\rho}$ has a non-zero unramified quotient.

Consider the connected-étale sequence (see [34, section 3.7]) of the finite flat $\mathbf{Z}_p$-group scheme G with generic fiber representation $\overline{\rho}$,

$$0 \to G^0 \to G \to G^{\text{et}} \to 0.$$

Recall the following:

Unramifiedness criterion (U): For $p \neq 2$, a flat D_p-representation is unramified if and only if the corresponding $\mathbf{Z}_p$-group scheme is *étale* over $\mathbf{Z}_p$.

To prove this, we can make a base change to $\mathbf{Z}_p^{\text{un}}$, and then use the argument which proved the irreducibility of $\overline{\rho}_{E,p}$ in the supersingular case in Theorem 1.1 (this amounts to the fact that for *odd* p, constant finite group schemes over $\mathbf{Q}_p^{\text{un}}$ admit only constant extensions to finite flat group schemes over $\mathbf{Z}_p^{\text{un}}$). If $p = 2$ this critical fact is not true (consider μ_2 over $\mathbf{Z}_2^{\text{un}}$).

Combining (U), Theorem 1.6, and the universal properties characterizing G^0 and G^{et} in the category of finite flat $\mathbf{Z}_p$-group schemes, it follows that the generic fiber representation of G^{et} is precisely the maximal unramified quotient of $\overline{\rho}$ as an $\mathbf{F}_p[D_p]$-module. However, by functoriality of the connected-étale sequence, we see that the canonical k-action on G gives rise to compatible canonical k-actions on G^0 and G^{et}. In other words, the generic fiber representation of G^{et} is also the maximal unramified quotient of $\overline{\rho}$ as a $k[D_p]$-module. Thus, if $\overline{\rho}$ is *not* split as a $k[D_p]$-module, then G^{et} is trivial and so G is a connected group scheme. But all subrepresentations of $\overline{\rho}$ arise as generic fiber representations of finite flat closed subgroup schemes of G, and these are all necessarily connected, too. Hence, such representations must be ramified, by (U). This contradicts the existence of the unramified subrepresentation η_1.

Now consider the case in which $\overline{\rho}$ is irreducible. Since $\det \overline{\rho}|_{I_p} = \omega|_{I_p}$ and p is odd, it follows that $\overline{\rho}$ is absolutely irreducible (the argument is a local analogue of the global theorem that for $p \neq 2$, an odd continuous irreducible two-dimensional representation of $G_{\mathbf{Q}}$ over k is absolutely irreducible). The normality of I_p in D_p then implies that $\overline{\rho}|_{I_p}$ is semisimple and therefore tame [29, Ch IX, Lemma 2]. Thus, $\overline{\rho}|_{I_p}$ is abelian, so

$$\overline{\rho}|_{I_p} \otimes_k \overline{\mathbf{F}}_p \simeq \chi_1 \oplus \chi_2$$

with continuous characters $\chi_i : I_p \to \overline{\mathbf{F}}_p^{\times}$. These eigencharacters must take their values in the quadratic extension k_2 of k and $\chi_1 \neq \chi_2$ since $\overline{\rho}$ is absolutely irreducible (and therefore non-abelian). The absolute irreducibility of $\overline{\rho}$ then forces a Frobenius element of D_p to interchange the two inertial lines, so from the well-known explicit description of the conjugation action of Frobenius on tame inertia, we obtain

$$\chi_2 = \chi_1^p, \quad \chi_1 = \chi_2^p.$$

Thus, $\chi_1 : I_p \to \mathbf{F}_{p^2}^\times$ and $\chi_1(I_p)$ does not lie in $\mathbf{F}_p^\times$.

The flatness of $\overline{\rho}$ implies (by Theorem 1.6) that Theorem 1.7 can be applied to χ_1, so χ_1 is equal to either ψ_2 or ψ_2^p. This proves that the semisimple $k[I_p]$-modules $\overline{\rho}|_{I_p}$ and φ_2 are isomorphic over $\overline{k} = \overline{\mathbf{F}}_p$. By the Brauer-Nesbitt Theorem, $\overline{\rho}|_{I_p} \simeq \varphi_2$.

(*ii*) Our argument will follow the method suggested by the brief sketch given in the proof of [8, Lemma 2.19(*b*)]. Note that we can immediately reduce to the case in which R is artinian (and therefore *finite*), so ρ arises as the generic fiber of the finite flat $\mathbf{Z}_p$-group scheme G. By Theorem 1.6 ($p \neq 2$), R acts on G. Let

$$0 \to G^0 \to G \to G^{\text{et}} \to 0$$

denote the connected-étale sequence of G. By the universal properties of the connected-étale sequence, it follows that R acts on G^0 and G^{et}, so the generic fibers give rise to a short exact sequence of $R[D_p]$-modules, with ρ in the middle. We will show that both G^0 and G^{et} have generic fiber R-modules which are free of rank 1 and that the characters on these 'lines' are of the desired type.

Now comes the critical step where we exploit the theory of finite flat group schemes. We claim that all $\mathbf{Z}_p[D_p]$-module Jordan-Hölder factors of the generic fiber representation of G^0 are ramified. The reason is quite simple. All such representations are flat and the corresponding finite flat $\mathbf{Z}_p$-group scheme fits into a 'decomposition series' for the connected group scheme G^0. Since all closed subgroup schemes and quotients of a connected object in the category of finite flat $\mathbf{Z}_p$-group schemes are again *connected* (as connectedness can be determined on the closed fiber and base change commutes with formation of short exact sequences of finite flat group schemes). A non-trivial connected finite flat $\mathbf{Z}_p$-group scheme is not étale and its generic fiber representation must therefore be ramified (by the criterion (U) above). This proves the claim concerning Jordan-Hölder factors of the generic fiber representation of G^0.

Let V_ρ be the representation space underlying ρ and let V_ρ^{et} denote the (non-zero) maximal unramified abelian group quotient. This is the generic fiber representation of G^{et} and so it has a canonical R-module structure. Due to the form of $\overline{\rho}$, $V_{\overline{\rho}} = V_\rho/\mathfrak{m}_R \to V_\rho^{\text{et}}/\mathfrak{m}_R$ is surjective with $V_\rho^{\text{et}}/\mathfrak{m}_R$ non-zero, so $V_\rho^{\text{et}}/\mathfrak{m}_R$ is exactly 1-dimensional over k. Hence, V_ρ^{et} has a single generator as an R-module. In order to show that it is a free R-module of rank 1, it suffices to check that it has the same R-length as R. We will check this below. On the other hand, the kernel V_ρ^0 of $V_\rho \twoheadrightarrow V_\rho^{\text{et}}$

is the generic fiber representation of G^0 and consequently all of its Jordan-Hölder factors (either as an $R[D_p]$-module or even as a $\mathbf{Z}_p[D_p]$-module) are *ramified*. Since formation of the connected-étale sequence is compatible with the local base change $\mathbf{Z}_p \to \mathbf{Z}_p^{\mathrm{un}}$, similar reasoning proves the same 'ramified' Jordan-Hölder factor property for V_ρ^0 when viewed as a $\mathbf{Z}_p[I_p]$ or $R[I_p]$-module.

Since the R-action and D_p-action on V_ρ commute, the form of $\overline{\rho}$ shows that all $R[I_p]$-module Jordan-Hölder factors of the rank 2 free R-module V_ρ are 1-dimensional k-vector spaces of the form $k(1)$ or k. Note that since $p \neq 2$, k and $k(1)$ are not isomorphic as D_p-modules. Also, clearly all $R[I_p]$-module Jordan-Hölder factors of V_ρ^{et} have the form k, and all such factors of V_ρ^0 have the form $k(1)$. Because V_ρ is free of rank 2 as an R-module, and the quotient V_ρ^{et} has a single R-module generator, we see that the number of unramified Jordan-Hölder factors is less than or equal to the number of ramified ones.

We claim that it is enough to show that $V_\rho^0/\mathfrak{m}_R$ is 1-dimensional over k. Indeed, this implies that V_ρ^0 and V_ρ^{et} are both quotients of R as R-modules, so since V_ρ is free of rank 2 over R, comparing lengths proves that V_ρ^{et} and V_ρ^0 are free of rank 1 over R. Set $\rho' = \mathrm{Hom}_R(\rho, R(1))$. Injecting $R(1)$ into a finite product of copies of $(\mathbf{Q}_p/\mathbf{Z}_p)(1)$ gives an injection of ρ' into a finite product of copies of the flat Cartier dual $\mathrm{Hom}_{\mathbf{Z}_p}(\rho, (\mathbf{Q}_p/\mathbf{Z}_p)(1))$, so ρ' is a flat representation (by Theorem 1.6). A consideration of Jordan-Hölder factors then shows that $\rho' \bmod \mathfrak{m}_R$ is reducible with cyclotomic inertial determinant (so (i) may be applied) and we readily conclude that $V_\rho^0 = \mathrm{Hom}_R(V_{\rho'}^{\mathrm{et}}, R(1))$. Thus, the inertial actions on V_ρ^{et} and V_ρ^0 are respectively trivial and cyclotomic, as desired. The uniqueness of the χ_i and the equality $\overline{\chi}_i = \chi_i \bmod \mathfrak{m}_R$ are clear.

We now verify that the right exact sequence

$$0 \to V_\rho^0/\mathfrak{m}_R \to V_\rho/\mathfrak{m}_R \to V_\rho^{\mathrm{et}}/\mathfrak{m}_R \to 0$$

is exact, from which the desired 1-dimensionality of $V_\rho^0/\mathfrak{m}_R$ follows. Choose a finite set of generators $X = \{x_i\}$ for $\mathfrak{m}_R$ and consider the obvious $R[D_p]$-linear map

$$\bigoplus_{x \in X} \rho \to \rho$$

with cokernel $\rho \bmod \mathfrak{m}_R$. If we extend this to a diagram arising from the connected-étale sequences associated to each side, we can apply the snake lemma in the abelian category of flat D_p-representations. The resulting coboundary map must be 0 because any map between a connected and an étale finite flat group scheme over a local base ring must be trivial. This implies that $V_\rho^0/\mathfrak{m}_R \to V_\rho/\mathfrak{m}_R$ is injective.

(iii) The proof we give is taken from [7, Thm 13.1(i)], which considers a slightly more general setting. The main point to note is that the proof uses

Ramakrishna's Theorem 3.5 below, so when reading the proof of Theorem 3.5, note that the present result is never used, and thus there is no circular reasoning. The idea of the proof is to use an observation of Faltings' [38, pp.457-8] and an auxiliary trick to reduce to the special case where $\mathcal{O} = \mathbf{Z}_p$. In that case we can use Raynaud's fundamental work on determinants of p-divisible groups [28, Thm 4.2.1].

Using $\mathcal{O} = W(k)$ in Theorem 3.5 below, we know that a universal 'flat' $W(k)$-deformation ring $R_{\overline{\rho}}^{\mathrm{fl}}$ exists and that it is isomorphic to $W(k)[\![T_1, T_2]\!]$. If we could show that the associated universal 'flat' representation has inertial determinant $\epsilon|_{I_p}$, then we'd be done. The observation of Faltings mentioned above ensures that we can always extend the field k without loss of generality. Thus, assume k is large enough so that there exists a continuous unramified character $\chi : D_p \to k^\times$ such that $\omega^{-1} \det \overline{\rho} = \chi^2$ (this is possible since $D_p/I_p \simeq \widehat{\mathbf{Z}}$ has no non-trivial 2-torsion). Recall that flatness of a D_p-representation is unaffected by unramified twisting (due to Galois descent, as mentioned after Theorem 1.6). Thus, twisting by the unramified Teichmüller lift of χ^{-1} gives (via Yoneda's Lemma) an isomorphism between the universal flat deformation rings of $\overline{\rho}$ and $\overline{\rho}\chi^{-1}$. This reduces us to the case in which $\det \overline{\rho} = \omega$.

It is relatively straightforward to check that up to isomorphism, there is only one continuous representation $D_p \to \mathrm{GL}_2(k)$ which has determinant ω and which is isomorphic to φ_2 on I_p (and in particular, this D_p-representation is self-dual). In addition, this unique representation is defined over $\mathbf{F}_p$. Now applying (i) and Faltings' observation 'in reverse,' we reduce to the case $k = \mathbf{F}_p$, so $R_{\overline{\rho}}^{\mathrm{fl}} = \mathbf{Z}_p[\![T_1, T_2]\!]$. Because this universal ring has such a special form, we see that to prove that the universal flat deformation has inertial determinant $\epsilon|_{I_p}$, it is enough to check this on all $\mathbf{Z}_p$-valued points. But a $\mathbf{Z}_p$-valued point is the same thing as a p-divisible group over $\mathbf{Z}_p$ with p-torsion representation $\overline{\rho}$.

We are therefore reduced to checking that if $\rho : D_p \to \mathrm{GL}_2(\mathbf{Z}_p)$ is the generic fiber representation of a p-divisible group Γ over $\mathbf{Z}_p$ with $\rho \bmod p \simeq \overline{\rho}$, then $\det \rho|_{I_p} = \epsilon|_{I_p}$. By [28, Thm 4.2.1], it is enough to check that Γ has dimension 1. Since $\overline{\rho}$ is ramified, clearly Γ is not étale and so $\dim \Gamma \geq 1$. The dual Γ^* has generic fiber p-torsion representation $\overline{\rho}^* \simeq \overline{\rho}$, so $\dim \Gamma^* \geq 1$. By [33, §2.3, Prop 3],

$$\dim \Gamma + \dim \Gamma^* = \mathrm{height}(\Gamma) = 2,$$

so $\dim \Gamma = 1$, as desired. ■

Since Ramakrishna's Theorem 3.5 is true in the residually ramified and irreducible case when $p = 2$ (though we omit the proof of this), Theorem 1.8(iii) is true when $p = 2$. This is not needed in Wiles' proof.

§2. Defining the Functor

For the remainder of this article, we will fix the following notation. Let $\mathcal{O}$ denote the valuation ring of a finite extension of $\mathbf{Q}_p$, with λ a uniformizer. Let $k = \mathcal{O}/\lambda$ denote the finite residue field of characteristic p. Choose a continuous representation

$$\overline{\rho} : D_p \to \mathrm{GL}_N(k)$$

such that $\overline{\rho}$ is flat. Later on we will specialize to the case $N = 2$, $p \neq 2$, and $\det \overline{\rho}|_{I_p} = \omega|_{I_p}$, but for now this is not necessary. Also, we remind the reader that even though we are ultimately interested in the case $\overline{\rho} = \overline{\rho}_{E,p}|_{D_p}$, the D_p-representation arising from the p-torsion of an elliptic curve E over $\mathbf{Q}$, for technical reasons in Wiles' method it is necessary to extend the field of scalars from $\mathbf{F}_p$ to a finite extension k which contains all of the eigenvalues of the finite subgroup $\overline{\rho}_{E,p}(G_{\mathbf{Q}}) \subseteq \mathrm{GL}_2(\mathbf{F}_p)$. This is one reason why it is critical that we work in the level of generality fixed above.

Let $\mathcal{D}_{\overline{\rho}}^{\mathrm{univ}}$ denote the universal deformation functor attached to $\overline{\rho}$, on the category $\widehat{\mathcal{C}}_{\mathcal{O}}$ of complete local noetherian $\mathcal{O}$-algebras with residue field k. If $\overline{\rho}$ is absolutely irreducible (e.g., if $\overline{\rho} = \overline{\rho}_{E,p}$ for $E_{/\mathbf{Q}_p}$ an elliptic curve with supersingular reduction), then $\mathcal{D}_{\overline{\rho}}^{\mathrm{univ}}$ is representable [23, §20]. We let $R_{\overline{\rho}}^{\mathrm{univ}}$ denote the representing ring. Recall, as was mentioned earlier, that our methods below will be needed to handle certain types of 'ord' cases (e.g., those arising from Example 1.3), so it would be too restrictive for us to assume at the outset that $\overline{\rho}$ is (absolutely) irreducible. Nevertheless, the absolutely irreducible case is good to keep in mind, since in this case the universal deformation functor is known to be representable.

The essential problem which we need to solve for input into Wiles' proof is to compute the orders of D_p-cohomology groups attached to certain deformations of $\overline{\rho}$. In terms of deformation functors, this will amount to computing the size of certain 'distinguished' subsets of $\mathcal{D}_{\overline{\rho}}^{\mathrm{univ}}(A)$ for suitable artinian objects A in $\widehat{\mathcal{C}}_{\mathcal{O}}$ (e.g., $A = \mathcal{O}/\lambda^n$). If a representing ring $R_{\overline{\rho}}^{\mathrm{univ}}$ exists (e.g., if $\overline{\rho}$ is absolutely irreducible), then

$$\mathcal{D}_{\overline{\rho}}^{\mathrm{univ}}(A) = \mathrm{Hom}_{\widehat{\mathcal{C}}_{\mathcal{O}}}(R_{\overline{\rho}}^{\mathrm{univ}}, A),$$

so if we could determine the structure of $R_{\overline{\rho}}^{\mathrm{univ}}$, then we would be on the right track. This is the basic reason for interest in deformation rings. In principle, it does not matter what the deformation ring looks like, so long as we can *somehow* compute the size of certain 'distinguished' subsets of $\mathcal{D}_{\overline{\rho}}^{\mathrm{univ}}(A)$ (and these subsets will make sense even when the universal deformation ring is not known to exist). However, in many cases (e.g., supersingular cases) the computations we will need to make are consequences of a much more precise structure theorem for a 'restricted' universal deformation ring, and so we will aim to prove this structure theorem and then will derive its consequences for the 'restricted' deformation problem of interest. This also seems to be a more natural way to proceed, since the existence of a representing ring begs the question of figuring out what it is. So let us start by first defining the 'flat deformation functor.'

Definition 2.1. For A in $\widehat{\mathcal{C}}_{\mathcal{O}}$, define $\mathcal{D}^{\mathrm{fl}}_{\overline{\rho}}(A)$ to be the subset in $\mathcal{D}^{\mathrm{univ}}_{\overline{\rho}}(A)$ consisting of those deformations of $\overline{\rho}$ to $\mathrm{GL}_N(A)$ whose representative liftings

$$\rho : D_p \to \mathrm{GL}_N(A)$$

have the property that for all $n \geq 1$, the *finite* D_p-module $\rho \bmod \mathfrak{m}_A^n$ is flat.

A few comments are in order concerning this definition. First of all, note that the condition on the lifting ρ can be checked on any single representative in the same deformation class of $\overline{\rho}$, as flatness is a property of the isomorphism class of a finite discrete D_p-module. Also, by Theorem 1.6, we see that it suffices to check the 'flatness' constraint on $\rho \bmod \mathfrak{m}_A^n$ just for sufficiently large n. If A is artinian, this is the same as saying that the finite D_p-module ρ is flat, in the sense of Definition 1.5. Of course, for any continuous representation $\rho : D_p \to \mathrm{GL}_N(A)$, $\rho \bmod \mathfrak{m}_A^n$ is a *finite* D_p-module with p-power order because A is a noetherian local ring with a *finite* residue field of characteristic p. Lastly, in [38, p. 457], the definition analogous to Definition 2.1 is given in terms of a condition on $\rho \bmod \mathfrak{a}$ for *all* open ideals $\mathfrak{a}$ in A, not just the ideals $\mathfrak{m}_A^n$ for all $n \geq 1$. However, since the ideals $\mathfrak{m}_A^n$ give a base of opens around 0 in A, it follows that every $\rho \bmod \mathfrak{a}$ is a quotient of some $\rho \bmod \mathfrak{m}_A^n$, so by Theorem 1.6 it follows that the above definition is equivalent to that given in [38].

Example 2.2. Let $E_{/\mathbf{Q}_p}$ have good reduction and $\overline{\rho} = \overline{\rho}_{E,p}$. Let $\mathcal{O} = \mathbf{Z}_p$ and $A = \mathcal{O}$. Then $\rho = \rho_{E,p}$ represents an element in $\mathcal{D}^{\mathrm{fl}}_{\overline{\rho}}(A)$. This uses Theorem 1.2 and the fact that $\rho \bmod \mathfrak{m}_A^n$ is nothing other than $E[p^n](\overline{\mathbf{Q}}_p)$. This is the primordial example to keep in mind and is the main reason why we care about $\mathcal{D}^{\mathrm{fl}}_{\overline{\rho}}$ here.

Though Example 2.2 shows that the sets $\mathcal{D}^{\mathrm{fl}}_{\overline{\rho}}(\mathcal{O}/\lambda^n)$ are the main sets in which we are interested, we also see that at this point we have done nothing of mathematical substance with these sets other than define them. It might even appear hopeless to say anything about such abstractly defined sets. We need a technique for understanding and constructing flat deformations. There is a theory due to Fontaine which will enable us to actually construct (albeit in a somewhat indirect manner) many flat deformations of $\overline{\rho}$. When combined with the fact that $\mathcal{D}^{\mathrm{fl}}_{\overline{\rho}}$ is a representable functor in certain cases (something we will prove shortly), we will be able to get enough of a hold on the representing ring that we can say exactly what it is! This will then make it essentially an easy exercise to answer the questions we will have about flat deformations (when $\mathcal{D}_{\overline{\rho}}^{\mathrm{fl}}$ is a representable functor).

To begin, we need to justify a couple of facts about $\mathcal{D}_{\overline{\rho}}^{\mathrm{fl}}$. For example, at this point it is not even clear if this is a functor! To make this clearer, consider the simpler (and essential) case of a map $\varphi : A \to B$ in $\widehat{\mathcal{C}}_{\mathcal{O}}$ with A and B both artinian. We then have a natural map of sets

$$\mathcal{D}_{\overline{\rho}}^{\mathrm{univ}}(\varphi) : \mathcal{D}_{\overline{\rho}}^{\mathrm{univ}}(A) \to \mathcal{D}_{\overline{\rho}}^{\mathrm{univ}}(B),$$

which, in concrete terms, just takes a representation into $\mathrm{GL}_N(A)$ and applies φ to the matrix entries to give a representation into $\mathrm{GL}_N(B)$. What we would most like is $\mathcal{D}_{\overline{\rho}}^{\mathrm{univ}}(\varphi)$ to map the subset $\mathcal{D}_{\overline{\rho}}^{\mathrm{fl}}(A)$ into $\mathcal{D}_{\overline{\rho}}^{\mathrm{fl}}(B)$. In other words, given a flat representation into $\mathrm{GL}_N(A)$, if we map the matrix entries into B, then we want the resulting representation into $\mathrm{GL}_N(B)$ to be flat. This is not obvious (until one sees the proof)! That this is the case is part of the next result.

Theorem 2.3. (Ramakrishna) *The association $A \rightsquigarrow \mathcal{D}_{\overline{\rho}}^{\mathrm{fl}}(A)$ from $\widehat{\mathcal{C}}_{\mathcal{O}}$ to* **Set** *is a subfunctor of $\mathcal{D}_{\overline{\rho}}^{\mathrm{univ}}$. That is, given any morphism $\varphi : A \to B$ in $\widehat{\mathcal{C}}_{\mathcal{O}}$,*

$$\mathcal{D}_{\overline{\rho}}^{\mathrm{univ}}(\varphi)(\mathcal{D}_{\overline{\rho}}^{\mathrm{fl}}(A)) \subseteq \mathcal{D}_{\overline{\rho}}^{\mathrm{fl}}(B).$$

If $\overline{\rho}$ has trivial centralizer (e.g., if it is absolutely irreducible), then $\mathcal{D}_{\overline{\rho}}^{\mathrm{fl}}$ is representable, by an object $R_{\overline{\rho}}^{\mathrm{fl}}$ in $\widehat{\mathcal{C}}_{\mathcal{O}}$. The resulting natural map

$$R_{\overline{\rho}}^{\mathrm{univ}} \to R_{\overline{\rho}}^{\mathrm{fl}}$$

is surjective.

PROOF. Since $\overline{\rho}$ is assumed to be flat, we see that $\mathcal{D}_{\overline{\rho}}^{\mathrm{fl}}(k)$ is a one element set; that is,

$$\mathcal{D}_{\overline{\rho}}^{\mathrm{fl}}(k) \subseteq \mathcal{D}_{\overline{\rho}}^{\mathrm{univ}}(k) = \{\overline{\rho}\}$$

is not empty. By Theorem 1.6, the property of a finite D_p-module being flat is preserved under passage to subrepresentations, quotients, and (finite) direct sums. It therefore follows from [23, Prop 1, §25; Cor, §23] that $\mathcal{D}_{\overline{\rho}}^{\mathrm{fl}}$ is a subfunctor of $\mathcal{D}_{\overline{\rho}}^{\mathrm{univ}}$ as claimed and that when $\overline{\rho}$ is absolutely irreducible, $\mathcal{D}_{\overline{\rho}}^{\mathrm{fl}}$ is represented by a quotient ring of $R_{\overline{\rho}}^{\mathrm{univ}}$.

If $\overline{\rho}$ has trivial centralizer (but is not assumed to be absolutely irreducible), and if we can show that all lifts of $\overline{\rho}$ have a trivial centralizer, then the method of proof in [23] can still be applied (i.e., the only property of residual absolute irreducibility which is needed is the fact that lifts have trivial centralizer).

So we are left with the problem of proving that if

$$\overline{\rho} : G \to \mathrm{GL}_N(k)$$

is a representation of any group G (with k any field) and $\overline{\rho}$ has trivial centralizer, then for any complete local noetherian ring A with residue field k and any lift

$$\rho : G \to \mathrm{GL}_N(A)$$

of $\overline{\rho}$, ρ has trivial centralizer. Passing to the limit, we are reduced to the case in which A is a local artin ring and we induct on the length of A as an A-module, the case of length 1 being our original hypothesis on $\overline{\rho}$.

Choose non-zero $x \in \mathfrak{m}_A$, so by induction $\rho \bmod x$ has trivial centralizer. Let $c \in \mathrm{M}_N(A)$ commute with the action of ρ, so $c \equiv a \bmod x\mathrm{M}_N(A)$ for some $a \in A$. Replacing c by $c - a$, we can assume $c = xc'$ for some $c' \in \mathrm{M}_N(A)$. Since c centralizes ρ, we see that c' centralizes $\rho \bmod \mathrm{ann}(x)$. By our inductive length assumption and the fact that $\mathrm{ann}(x) \subsetneq A$, c' is congruent to a scalar matrix modulo the annihilator of x, so we conclude as desired that $c = xc'$ is a scalar matrix. ■

Note that the only input from the theory of finite flat group schemes in the above proof occurs when we invoke [23], which uses only the fact that the 'flatness' condition on a finite D_p-module has certain formal properties; namely, it is preserved under passage to (finite) direct sums, subrepresentations, and quotients. This does not make any deep use of the theory of finite flat group schemes (and in particular, this part of Theorem 1.6 is not the hard part; far from it, in fact).

§3. Local Galois Cohomology and Deformation Theory

We now review some results given in [23] and formulate them in a way that will be convenient for our purposes. We then will state the main result concerning the orders of local H^1's. When $\overline{\rho}$ has trivial centralizer (so $R^{\mathrm{fl}}_{\overline{\rho}}$ is known to exist!), we will give an interpretation of this main result as a statement about the structure of $R^{\mathrm{fl}}_{\overline{\rho}}$. The proofs of these results can be reduced to constructing 'enough' flat representations. The actual construction of such representations will be carried out in §5, as it will first require a review of Fontaine's approach to the theory of finite flat group schemes (to be discussed in §4).

Pick an artinian object A in $\widehat{\mathcal{C}}_{\mathcal{O}}$ and a flat lifting ρ of $\overline{\rho}$ to $\mathrm{GL}_N(A)$, so the deformation given by ρ is an element of $\mathcal{D}^{\mathrm{fl}}_{\overline{\rho}}(A)$. There are two examples to keep in mind, with $A = \mathcal{O}/\lambda^n$, $N = 2$, and $p \neq 2$. These are

$$\rho = (\mathcal{O} \otimes_{\mathbf{Z}_p} \rho_{E,p}) \bmod \lambda^n \quad \text{and} \quad \rho = \rho_{f,\lambda} \bmod \lambda^n.$$

We have $\overline{\rho} = k \otimes_{\mathbf{F}_p} \overline{\rho}_{E,p}$ and $E_{/\mathbf{Q}_p}$ an elliptic curve with good reduction in the first example and $\overline{\rho} = \rho_{f,\lambda} \bmod \lambda$ in the second, with f a weight 2 newform having level prime to p and $\mathcal{O}$ the completion of the integer ring $\mathcal{O}_f \subseteq \mathbf{C}$ at a prime above p.

Consider $H^1(D_p, \mathrm{ad}(\rho))$ in our general setting. It is shown in [23, §22] that as an A-module, this is naturally identified with $\mathrm{Ext}^1_{A[D_p]}(\rho, \rho)$. This is an A-module that classifies isomorphism classes of exact sequences of $A[D_p]$-modules

$$0 \to \rho \to X \to \rho \to 0$$

in which X, necessarily an $A[D_p]$-module with finite A-length, has a D_p-action which is trivial on an open subgroup of D_p. That is, the D_p-action must be continuous with respect to the discrete topology on the finite set underlying X. Inside of $\mathrm{Ext}^1_{A[D_p]}(\rho, \rho)$, there is a natural subset $\mathrm{Ext}^{1,\mathrm{fl}}_{A[D_p]}(\rho, \rho)$ consisting of those elements with a representative short exact sequence in which X is a *flat* representation (note that X automatically has p-power order). This condition, of course, then holds for any choice of representative short exact sequence.

From the formal properties of the 'flat' condition on representations (Theorem 1.6), it follows that

$$\mathrm{Ext}^{1,\mathrm{fl}}_{A[D_p]}(\rho, \rho) \subseteq \mathrm{Ext}^1_{A[D_p]}(\rho, \rho)$$

is not just a subset, but is an A-submodule [23]. Via the A-module isomorphism

$$\mathrm{Ext}^1_{A[D_p]}(\rho, \rho) \simeq H^1(D_p, \mathrm{ad}(\rho)),$$

we can define a corresponding A-submodule

$$H^1_{\mathrm{fl}}(D_p, \mathrm{ad}(\rho)) \subseteq H^1(D_p, \mathrm{ad}(\rho)).$$

See [23, §25] for a more detailed explanation of all of this. Beware that, despite what the notation may suggest, $H^1_{\mathrm{fl}}(D_p, \mathrm{ad}(\rho))$ is best thought of as a functor of ρ, *not* of $\mathrm{ad}(\rho)$. We will give some ad hoc definitions of $H^1_{\mathrm{fl}}(D_p, *)$ for other $*$'s, but there is not a general definition of an H^1_{fl} functor in our situation. Note that $H^1_{\mathrm{fl}}(D_p, \mathrm{ad}(\rho))$ is exactly the H^1 term whose size Wiles needs to tightly control in his 'local-to-global' Galois cohomology estimate (for suitable A and ρ).

Let's give an easy (but important!) example of some elements in the A-module $H^1_{\mathrm{fl}}(D_p, \mathrm{ad}(\rho))$.

Example 3.1 For an unramified continuous additive homomorphism

$$\chi : D_p \to A,$$

we can consider X_χ which has the 'block matrix form'

$$X_\chi \simeq \begin{pmatrix} \rho & \chi\rho \\ 0 & \rho \end{pmatrix}.$$

When restricted to I_p, this is just $\rho|_{I_p} \oplus \rho|_{I_p}$, so $X_\chi|_{I_p}$ is certainly the generic fiber representation of a finite flat group scheme G over $\mathbf{Z}_p^{\text{un}}$; namely, if ρ is the generic fiber representation of a finite flat $\mathbf{Z}_p$-group scheme H, then we can take

$$G = (H \times_{\mathbf{Z}_p} \mathbf{Z}_p^{\text{un}}) \times_{\mathbf{Z}_p^{\text{un}}} (H \times_{\mathbf{Z}_p} \mathbf{Z}_p^{\text{un}}).$$

It follows from standard facts about Galois descent of rings that G canonically descends to a finite flat $\mathbf{Z}_p$-group scheme whose generic fiber representation is X_χ. This is the same thing that we needed in the discussion of descent following Theorem 1.6.

Though Example 3.1 gives an A-line of 'flat' elements in $H^1(D_p, \text{ad}(\rho))$, we can construct even more. However, we should first point out a subtlety in the definition of $H^1_{\text{fl}}(D_p, \text{ad}(\rho))$ which is easy to overlook but which will slightly complicate our life. Assume for now that p does not divide N. In the case of interest, $N = 2$ and p is odd, so this is not a problematic assumption. Since N is then invertible in A, if we let $\text{ad}^0(\rho) \subseteq \text{ad}(\rho)$ denote the A-submodule consisting of elements with trace 0, then there is a canonical isomorphism of $A[D_p]$-modules

$$\text{ad}^0(\rho) \oplus A \simeq \text{ad}(\rho).$$

The D_p-action on the scalar line A is of course trivial. We have a canonical isomorphism of A-modules

$$H^1(D_p, \text{ad}^0(\rho)) \oplus H^1(D_p, A) \simeq H^1(D_p, \text{ad}(\rho)).$$

Before connecting this up with H^1_{fl}, we should mention that the reason for interest in $\text{ad}^0(\rho)$ is that under the identification of $H^1(D_p, \text{ad}(\rho))$ with the set of 'infinitesimal' deformations of ρ to $\text{GL}_N(A[\varepsilon])$, $H^1(D_p, \text{ad}^0(\rho))$ corresponds to those deformations whose determinant is the 'same' as $\det \rho$ (via the canonical inclusion $A^\times \hookrightarrow A[\varepsilon]^\times$). This is explained in [23, §24] and is of interest because of the fact that we are primarily interested in studying deformations of $\overline{\rho}_{E,p}$ with a 'fixed' determinant, namely the cyclotomic character. A cyclotomic determinant condition is but one of several deformation-theoretic conditions satisfied by the Tate module deformation $\rho_{E,p}$. The reason it is useful to impose such conditions on the deformations we consider is that they 'cut down' on the size of the corresponding deformation ring — if it exists — and so we can expect the deformation ring to encode more and more refined information about the Tate module representation $\rho_{E,p}$.

Now we return to a general setting. We define

$$H^1_{\text{fl}}(D_p, \text{ad}^0(\rho)) = H^1(D_p, \text{ad}^0(\rho)) \cap H^1_{\text{fl}}(D_p, \text{ad}(\rho))$$

and

$$H^1_{\mathrm{fl}}(D_p, A) = H^1(D_p, A) \cap H^1_{\mathrm{fl}}(D_p, \mathrm{ad}(\rho)),$$

so we have the A-linear inclusion

$$\iota_\rho : H^1_{\mathrm{fl}}(D_p, \mathrm{ad}^0(\rho)) \oplus H^1_{\mathrm{fl}}(D_p, A) \hookrightarrow H^1_{\mathrm{fl}}(D_p, \mathrm{ad}(\rho)).$$

We might expect this map to be an isomorphism. But at the present time it isn't known how to prove this in general. Nevertheless, for the type of ρ which arise in Wiles' proof, we can prove the result. It should be emphasized that this is actually not needed; without this, the Main Theorem 3.3 below would merely be an inequality of the form $\leq$, which is adequate for Wiles' needs. But it seems like a good policy to prove results in as strong a form as possible (who knows what will be useful in the future?), so we give here a proof that ι_ρ is sometimes an isomorphism.

Lemma 3.2. *Let A and ρ be as above, with $N = 2$, $p \neq 2$, and $\det \overline{\rho}|_{I_p} = \omega|_{I_p}$. Then ι_ρ is an isomorphism and*

$$\mathrm{Hom}_{\mathrm{cont}}(D_p/I_p, A) \hookrightarrow H^1_{\mathrm{fl}}(D_p, A)$$

is an isomorphism (i.e., $|H^1_{\mathrm{fl}}(D_p, A)| = |A|$).

REMARK. The above result applies, in particular, if $A = \mathcal{O}/\lambda^n$. This is the only case which arises in Wiles' method. Also, the proof in the residually irreducible case ultimately relies on Theorem 1.8(*iii*), whose proof requires a vast amount of algebraic geometry. Thus, the reader can skip Lemma 3.2 in the residually irreducible case and insert inequalities where appropriate in our later arguments (i.e., Theorem 3.3) and Wiles' method will still go through (and will yield equalities after the method succeeds).

PROOF. Consider the composite map

$$H^1_{\mathrm{fl}}(D_p, \mathrm{ad}(\rho)) \hookrightarrow H^1(D_p, \mathrm{ad}(\rho)) \to H^1(D_p, A).$$

The lemma is readily seen to be equivalent to the assertion that the image is exactly the A-submodule $\mathrm{Hom}_{\mathrm{cont}}(D_p/I_p, A)$ inside of $H^1_{\mathrm{fl}}(D_p, A)$. Note that it is *not* a priori clear that the image even lies in $H^1_{\mathrm{fl}}(D_p, A)$. Since $\mathrm{Hom}_{\mathrm{cont}}(D_p/I_p, A)$ is the kernel of the natural restriction map

$$H^1(D_p, A) \to H^1(I_p, A),$$

the lemma is equivalent to the statement that the natural map

$$H^1_{\mathrm{fl}}(D_p, \mathrm{ad}(\rho)) \to H^1(I_p, \mathrm{ad}(\rho))$$

has image inside of $H^1(I_p, \mathrm{ad}^0(\rho))$.

Under the identification [23, §23]

$$H^1_{\text{fl}}(D_p, \text{ad}(\rho)) \simeq \mathcal{D}^{\text{fl}}_{\rho}(A[\varepsilon]),$$

our assertion is equivalent to the statement that if

$$\rho_1 : D_p \to \text{GL}_2(A[\varepsilon])$$

is a flat deformation of ρ, then $\det \rho_1|_{I_p} = \det \rho|_{I_p}$. Since both ρ and ρ_1 are flat deformations of $\overline{\rho}$, both have the same inertial determinant, namely $\epsilon|_{I_p}$ (treat the cases in which $\overline{\rho}$ is reducible and irreducible separately, using Theorem 1.8). ■

There does not (at the present time) exist a satisfactory theory of any sort of H^1_{fl} *functor* in our setting. For us, $H^1_{\text{fl}}(D_p, \text{ad}(\rho))$ is primarily an artifice for singling out a special submodule of 'distinguished' cohomology classes, and we need to find out how many of them there are in certain special cases.

We now are in a position to state an important result that Wiles needs. This is the main result of the present article.

Main Theorem 3.3. *Let* $\overline{\rho} : D_p \to \text{GL}_2(k)$ *be flat with* $\det \overline{\rho}|_{I_p} = \omega|_{I_p}$ *and* $p \neq 2$. *Let*

$$\rho : D_p \to \text{GL}_2(\mathcal{O}/\lambda^n)$$

be a flat lifting of $\overline{\rho}$. *Then*

$$|H^1_{\text{fl}}(D_p, \text{ad}^0(\rho))| = |H^0(D_p, \text{ad}^0(\rho))| \cdot |\mathcal{O}/\lambda^n|.$$

REMARKS. Using Theorem 1.8(*ii*),(*iii*), the hypotheses imply that

$$\det \rho|_{I_p} = \epsilon|_{I_p}.$$

However, in the residually irreducible case the proof requires sophisticated methods. The condition is automatic in the application to elliptic curves, so the reader can insert this as an extra hypothesis in Theorem 3.3 and thereby bypass Theorem 1.8(*iii*) without affecting the applicability of Theorem 3.3 to the study of semistable elliptic curves over $\mathbf{Q}$.

Since Lemma 3.2 in the residually irreducible case requires Theorem 1.8(*iii*), the reader who would prefer to avoid Theorem 1.8(*iii*) should insert $\leq$ in place of $=$ in Theorem 3.3 above. Such an inequality suffices for the successful application of Wiles' methods to elliptic curves.

Before addressing the proof of Theorem 3.3, we make a few observations about what it says. The H^0 term is nothing other than the number of trace 0 matrices in $\text{M}_2(\mathcal{O}/\lambda^n)$ which commute with the action of ρ. For example, if $\overline{\rho}$ is not split, then by Theorem 1.8(i) we see that $\overline{\rho}$ has trivial centralizer. We saw in the proof of Theorem 2.3 that this implies that ρ must also have trivial centralizer. In particular, the only trace 0 matrix commuting with ρ is the zero matrix, so $|H^0(D_p, \mathrm{ad}^0(\rho))| = 1$ in these cases.

Now we prove Theorem 3.3. By Lemma 3.2, the injection ι_ρ is an isomorphism and

$$|H^1_{\mathrm{fl}}(D_p, \mathcal{O}/\lambda^n)| = |\mathcal{O}/\lambda^n|,$$

so we have

$$|H^1_{\mathrm{fl}}(D_p, \mathrm{ad}^0(\rho))| = |\mathcal{O}/\lambda^n|^{-1}|H^1_{\mathrm{fl}}(D_p, \mathrm{ad}(\rho))|.$$

Therefore, it suffices to prove

Lemma 3.4. *Under the hypotheses in Theorem* 3.3, *we have*

$$|H^1_{\mathrm{fl}}(D_p, \mathrm{ad}(\rho))| = |H^0(D_p, \mathrm{ad}^0(\rho))| \cdot |\mathcal{O}/\lambda^n|^2.$$

Since Theorem 1.8(iii) is true for $p = 2$ and Theorem 4.5 below can be extended to the case $p = 2$ with some connectedness hypotheses, Lemma 3.4 can be extended to the case $p = 2$ when $\overline{\rho}$ is irreducible. We omit the arguments needed for this case, as it is not needed in Wiles' proof.

The idea of the proof of Lemma 3.4 is quite simple: we will write down every flat representation corresponding to an element in $H^1_{\mathrm{fl}}(D_p, \mathrm{ad}(\rho)) \simeq \mathcal{D}^{\mathrm{fl}}_\rho((\mathcal{O}/\lambda^n)[\varepsilon])$ and just count how many we have! This is not the most conceptually satisfying way to proceed, and in the case where $\overline{\rho}$ has trivial centralizer, Fontaine has said that he can give a more 'pure thought' proof (with a small amount of computation required). We will say more about this below. We use a computational 'brute force' proof because it involves a really clever idea, and also because it is needed (at present) to handle the cases with non-trivial centralizer at the residual level. Regardless of how one proves Lemma 3.4, Fontaine's work on finite flat group schemes is indispensable.

Before proving Lemma 3.4, we make some remarks in the case when $p \neq 2$, $\overline{\rho}$ as above has trivial centralizer, and $\rho = \rho_0 \bmod \lambda^n$ for some

$$\rho_0 : D_p \to \mathrm{GL}_2(\mathcal{O})$$

lifting $\overline{\rho}$, with ρ_0 corresponding to an element in $\mathcal{D}^{\mathrm{fl}}_{\overline{\rho}}(\mathcal{O})$. In practice, ρ_0 is the local restriction of a modular lifting of $\overline{\rho} = \overline{\rho}_{E,p} \otimes_{\mathbf{F}_p} k$, but we will see in Theorem 3.5 below that some ρ_0 as above always exists. The following analysis serves as a good indication of what Lemma 3.4 really means.

Since ρ has a trivial centralizer, Lemma 3.4 amounts to the assertion

$$|H^1_{\mathrm{fl}}(D_p, \mathrm{ad}(\rho))| = |\mathcal{O}/\lambda^n|^2.$$

By Theorem 2.3, the functor $\mathcal{D}_{\overline{\rho}}^{\mathrm{fl}}$ is represented by some $R_{\overline{\rho}}^{\mathrm{fl}}$ in $\widehat{\mathcal{C}}_{\mathcal{O}}$. Let us interpret Lemma 3.4 as an assertion about $R_{\overline{\rho}}^{\mathrm{fl}}$. In the case $n = 1$, Lemma 3.4 says that the k-vector space $H^1_{\mathrm{fl}}(D_p, \mathrm{ad}(\overline{\rho}))$ is 2-dimensional. Recalling the fundamental isomorphism relating Galois cohomology and deformation theory [23, §28]

$$\mathrm{Hom}_k(\mathfrak{m}/(\mathfrak{m}^2, \lambda), k) \simeq H^1_{\mathrm{fl}}(D_p, \mathrm{ad}(\overline{\rho})),$$

with $\mathfrak{m}$ the maximal ideal of $R_{\overline{\rho}}^{\mathrm{fl}}$, we see that choosing a k-basis of $\mathfrak{m}/(\mathfrak{m}^2, \lambda)$ gives rise to a surjection of rings

$$\pi : \mathcal{O}[\![T_1, T_2]\!] \twoheadrightarrow R_{\overline{\rho}}^{\mathrm{fl}}.$$

We now consider whether π is an isomorphism.

Let $\mathfrak{p}$ denote the kernel of the natural map $R_{\overline{\rho}}^{\mathrm{fl}} \twoheadrightarrow \mathcal{O}$ induced by ρ_0. By suitable change of coordinates, we can assume $\mathfrak{p}$ is the image of (T_1, T_2) under π, so $\mathfrak{p}$ has two generators. The method of proof of [38, Prop 1.2] produces an $\mathcal{O}$-linear isomorphism

$$H^1_{\mathrm{fl}}(D_p, \mathrm{ad}(\rho)) \simeq \mathrm{Hom}_{\mathcal{O}}(\mathfrak{p}/\mathfrak{p}^2, \mathcal{O}/\lambda^n)$$

(this construction uses the choice of the representation ρ_0 corresponding to $\mathfrak{p}$). Hence, Lemma 3.4 says that the $\mathcal{O}/\lambda^n$-module $\mathfrak{p}/(\mathfrak{p}^2, \lambda^n\mathfrak{p})$, which has two generators, is in fact free of rank 2. Passing to the limit (using $\rho_0 \bmod \lambda^m$ with $m \geq n$), this says that $\mathfrak{p}/\mathfrak{p}^2$ is a *free* $\mathcal{O}$-module of rank 2. This fact is not at all obvious. Moreover, passing to the direct limit on cohomology gives

$$\begin{aligned} H^1_{\mathrm{fl}}(D_p, \mathrm{ad}(\rho_0) \otimes_{\mathcal{O}} (K/\mathcal{O})) &\stackrel{\mathrm{def}}{=} \varinjlim H^1_{\mathrm{fl}}(D_p, \mathrm{ad}(\rho_0 \bmod \lambda^m)) \\ &\simeq (K/\mathcal{O}) \oplus (K/\mathcal{O}), \end{aligned}$$

so this cohomology module is actually p-divisible as a group. This is all very important in [38, Prop 1.9(v)]. What sort of structure must $R_{\overline{\rho}}^{\mathrm{fl}}$ have if $\mathfrak{p}/\mathfrak{p}^2$ is free of rank 2? One possibility is that the surjection π above is an isomorphism. This is *much* stronger than the statement that $\mathfrak{p}/\mathfrak{p}^2 \simeq \mathcal{O} \oplus \mathcal{O}$ (e.g., a priori, we could have $R_{\overline{\rho}}^{\mathrm{fl}} \simeq \mathcal{O}[\![T_1, T_2]\!]/(T_1, T_2)^2$, which has a unique $\mathcal{O}$-valued point $\mathfrak{p}$ and $\mathfrak{p}/\mathfrak{p}^2 \simeq \mathcal{O} \oplus \mathcal{O}$). Nevertheless, if π is an isomorphism, this is a 'good' explanation for the $\mathcal{O}$-module freeness. This may seem like a lot to ask for, but Ramakrishna came up with a very clever way to prove that indeed, π is an isomorphism.

Theorem 3.5. (Ramakrishna) *Assume $p \neq 2$. For a flat representation $\overline{\rho} : D_p \to \mathrm{GL}_2(k)$ with trivial centralizer and with* $\det \overline{\rho}|_{I_p} = \omega|_{I_p}$,

$$R_{\overline{\rho}}^{\mathrm{fl}} \simeq \mathcal{O}[\![T_1, T_2]\!].$$

In particular, the natural map $\mathcal{D}^{\mathrm{fl}}_{\overline{\rho}}(\mathcal{O}) \to \mathcal{D}^{\mathrm{fl}}_{\overline{\rho}}(\mathcal{O}/\lambda^n)$ *is surjective for all* $n \geq 1$.

PROOF. The centralizer hypothesis ensures that $R^{\mathrm{fl}}_{\overline{\rho}}$ exists. Let $\mathfrak{m}$ be the maximal ideal of $R^{\mathrm{fl}}_{\overline{\rho}}$. Using Fontaine's theory from §4 (this requires $p \neq 2$), we will write down in §5 all short exact sequences of *flat* $k[D_p]$-modules

$$0 \to \overline{\rho} \to X \to \overline{\rho} \to 0$$

and count that there are $|k|^2$ of them (up to equivalence). Since we have k-linear isomorphisms

$$\mathrm{Hom}_k(\mathfrak{m}/(\mathfrak{m}^2, \lambda), k) \simeq H^1_{\mathrm{fl}}(D_p, \mathrm{ad}(\overline{\rho})) \simeq \mathrm{Ext}^{1,\mathrm{fl}}_{k[D_p]}(\overline{\rho}, \overline{\rho}),$$

we obtain a surjection $\mathcal{O}[\![T_1, T_2]\!] \twoheadrightarrow R^{\mathrm{fl}}_{\overline{\rho}}$. Let I denote the kernel.

A natural approach to proving $I = 0$ is to try to use methods from commutative algebra. But our knowledge of the commutative algebra properties $R^{\mathrm{fl}}_{\overline{\rho}}$ is (right now) quite minimal. Instead, we will exploit the defining property of $R^{\mathrm{fl}}_{\overline{\rho}}$ in the following remarkable manner. Choose $f \in I$. If we can show that the induced injective map on $\mathcal{O}/\lambda^n$-valued points

$$\mathrm{Hom}_{\hat{\mathcal{C}}_{\mathcal{O}}}(R^{\mathrm{fl}}_{\overline{\rho}}, \mathcal{O}/\lambda^n) \hookrightarrow \mathrm{Hom}_{\hat{\mathcal{C}}_{\mathcal{O}}}(\mathcal{O}[\![T_1, T_2]\!], \mathcal{O}/\lambda^n)$$

is a bijection, then it follows that $f(t_1, t_2) \in \lambda^n\mathcal{O}$ for all $t_1, t_2 \in \lambda\mathcal{O}$ and all $n \geq 1$. Thus, f vanishes on the open λ-adic unit disc and so by a basic result from non-archimedean analysis we may conclude that $f = 0$!

To show that the injection on $\mathcal{O}/\lambda^n$-valued points is a bijection, all we have to show is that both sides have the same size. Since the right side trivially has size $|\lambda\mathcal{O}/\lambda^n\mathcal{O}|^2 = |k|^{2(n-1)}$, what we need to show is

$$|\mathrm{Hom}_{\hat{\mathcal{C}}_{\mathcal{O}}}(R^{\mathrm{fl}}_{\overline{\rho}}, \mathcal{O}/\lambda^n)| = |k|^{2(n-1)}.$$

But $|\mathrm{Hom}_{\hat{\mathcal{C}}_{\mathcal{O}}}(R^{\mathrm{fl}}_{\overline{\rho}}, \mathcal{O}/\lambda^n)|$ actually means something: it is the number of flat deformations of $\overline{\rho}$ to $\mathrm{GL}_2(\mathcal{O}/\lambda^n)$. We again use Fontaine's theory ($p \neq 2$) to simply write down all such possible deformations (this will be done in §5) and count the number of possibilities; it turns out to be exactly what we want. ■

Note that the above proof shows that *any* surjection $\mathcal{O}[\![T_1, T_2]\!] \twoheadrightarrow R^{\mathrm{fl}}_{\overline{\rho}}$ is an isomorphism (which is to be expected, since any surjection from a noetherian ring to itself must be an isomorphism). One may ask if the isomorphism in Theorem 3.5 can be chosen 'naturally.' That is, can one 'interpret' what the parameters T_1 and T_2 actually mean? In the proof of Theorem 3.5, we chose arbitrarily a basis of the reduced Zariski cotangent space $\mathfrak{m}/(\mathfrak{m}^2, \lambda)$ and then arbitrarily lifted these to elements of $\mathfrak{m}$. Is there a natural way to make these choices? Recent work of Fontaine and Mazur suggests that natural choices might exist. In the special case of Theorem 3.5, when $\mathcal{O} = \mathbf{Z}_p$, $p \geq 5$, and $\overline{\rho}$ is absolutely irreducible, they give an explicit parameterization of $\mathrm{Hom}_{\widehat{\mathcal{C}}_{\mathcal{O}}}(R^{\mathrm{fl}}_{\overline{\rho}}, \mathcal{O})$ which assigns explicit meaning to T_1 and T_2 [15, Thm B2(*ii*)]. One of the ingredients for this is an analogue of the ideas to be described in the next section.

The fact that $R^{\mathrm{fl}}_{\overline{\rho}}$ is a power series ring should have a cohomological interpretation as a vanishing condition on an appropriately defined H^2 (compare with [22, §1.6, Prop 2] in the 'unrestricted' case). Fontaine has said that he can prove such a vanishing condition directly, using [2, Lemma 4.4] and a 'cohomological dimension' argument, thereby giving a conceptual proof that $R^{\mathrm{fl}}_{\overline{\rho}}$ is a power series ring. One then needs to do the calculation that $\dim_k \mathrm{Ext}^{1,\mathrm{fl}}_{k[D_p]}(\overline{\rho}, \overline{\rho}) = 2$ as above in order to prove that the number of variables is 2. This procedure has the advantage of working for GL_N and thereby highlights the role of GL_2 as an artifact needed only for the residual Ext^1 calculation (which requires Theorem 1.8(*i*)). Also, one can use the theory of Fontaine-Laffaille to bypass the residual Ext^1 calculation and to compute directly the k-dimension of $H^1_{\mathrm{fl}}(D_p, \mathrm{ad}(\overline{\rho}))$. This produces the number 2 via a completely different (but much more complicated) calculation that explains more conceptually where the '2' comes from. See [4, pp. 6-11, esp. Thm 5] for further details.

It is fairly easy to use Theorem 3.5 to deduce Lemma 3.4 in the case of a residually trivial centralizer. Indeed, Theorem 3.5 proves that for any flat lift $\rho : D_p \to \mathrm{GL}_2(\mathcal{O}/\lambda^n)$ of $\overline{\rho}$, there exists $\rho_0 : D_p \to \mathrm{GL}_2(\mathcal{O})$ lifting ρ with ρ_0 giving an element in $\mathcal{D}^{\mathrm{fl}}_{\overline{\rho}}(\mathcal{O})$. Hence, we can apply the discussion preceding Theorem 3.5, and combining this with $R^{\mathrm{fl}}_{\overline{\rho}} \simeq \mathcal{O}[\![T_1, T_2]\!]$, Lemma 3.4 follows if $\overline{\rho}$ has a trivial centralizer. In §5, we will give a direct proof of Lemma 3.4 in all residually reducible cases, as well as complete the unfinished steps in the proof of Theorem 3.5.

We conclude this section by showing how imposing extra deformation conditions cuts down quite a lot on the 'size' of the deformation ring. For example, suppose

$$\overline{\rho} : D_p \to \mathrm{GL}_2(k)$$

is flat and irreducible, with cyclotomic determinant on inertia. Also assume $p \neq 2$. Ramakrishna showed using Tate local duality that

$$R^{\mathrm{univ}}_{\overline{\rho}} \simeq \mathcal{O}[\![X_1, X_2, X_3, X_4, X_5]\!],$$

where $5 = \dim_k \mathrm{ad}(\overline{\rho}) + h^0(D_p, \mathrm{ad}(\overline{\rho}))$ [27, Thm 4.1], [38, pp. 457-8] (remark: the cohomological calculations used to prove [27, Thm 4.1] can be simplified by the use of a k-linear version of Tate local duality rather than just an $\mathbf{F}_p$-linear version). By Theorem 3.5, we see that imposing a flatness condition on the deformations of $\overline{\rho}$ yields the quotient $R^{\mathrm{fl}}_{\overline{\rho}} \simeq \mathcal{O}[\![T_1, T_2]\!]$ of

$R_{\overline{\rho}}^{\mathrm{univ}}$. Moreover, by [7, Thm 13.1(*ii*)], it follows that if we impose the added deformation condition that the determinant be some fixed $\mathcal{O}^\times$-valued character χ lifting $\det \overline{\rho}$, with $\chi|_{I_p} = \epsilon|_{I_p}$, then the resulting deformation ring is isomorphic to a quotient of the form $\mathcal{O}[\![T]\!]$. From an intuitive viewpoint, if we required the D_p-representations to actually come from *global* representations of a suitably restricted type, then this would cut down on the deformation ring even more. Hence, one way to think about why the global deformation rings considered by Wiles are so 'small' is that the functors they represent involve a lot of (local) constraints!

There is one interesting property of all of the Galois deformation rings which have been computed, whether local or global: they are flat over $\mathcal{O}$. Does this fact have any deep meaning?

§4. Fontaine's Approach to Finite Flat Group Schemes.

The classical theory of complex (or real) Lie groups can be 'linearized' insofar as the theory of Lie algebras often allows one to translate theorems and constructions concerning (connected) Lie groups into issues concerning Lie algebras. Since the Lie algebra only perceives tangential information at the origin, an attempt to construct a theory of Lie algebras in the context of algebraic groups is reasonable as long as the group schemes are *reduced* (since tangent spaces can't distinguish between a scheme and the underlying reduced subscheme). In particular, everything is fine in characteristic 0. However, in characteristic p there are many group schemes which are not reduced. This is the most important fact about the theory of group schemes in characteristic p. Any attempt at constructing a 'Lie algebra' theory for non-reduced group schemes over a field of positive characteristic must use more subtle infinitesimal information than that detected at the level of tangent spaces.

Let k be any field of characteristic $p > 0$. For the study of finite k-group schemes (recall the commutativity hypotheses) there is a theory of *Dieudonne modules* which serves as a good analogue to the theory of Lie algebras, up to the fact that the Dieudonne theory is contravariant (like a cotangent space rather than a tangent space). In fact, the theory of Dieudonne modules covers a much wider class of commutative k-group schemes than the finite ones, but we restrict ourselves to this case, as it is all that we will need. Before discussing the basic ingredients of this theory, we mention that Fontaine's idea (following Grothendieck) is that a finite flat group scheme G over $\mathbf{Z}_p$ should be classified by specifying its closed fiber (a finite $\mathbf{F}_p$-group scheme, or equivalently, a 'Dieudonne module'), together with some additional 'lifting data.' In other words, he proposed a refinement of the theory of Dieudonne modules which would create an analogue to the theory of Lie algebras for finite flat $\mathbf{Z}_p$-group schemes. Fontaine's theory actually applies to finite flat group schemes over any base $W(k)$ with k a perfect field of characteristic p. We will only discuss the case $k = \mathbf{F}_p$, as this case is all we shall need and it avoids some technical Frobenius-semilinearity issues (thereby making computations simpler). Omitting this Frobenius-semilinearity may create some mistaken impressions about the shape of the more general theory, so what follows should be taken as a slightly skewed perspective on Fontaine's theory.

The reason that it is preferable to classify a finite flat $\mathbf{Z}_p$-group scheme by its closed fiber, together with 'extra data,' rather than by its generic fiber, together with 'extra data,' is that closed fibers can be classified by the very explicit linear-algebraic notion of a Dieudonne module. The generic fibers, on the other hand, constitute the entire theory of finite discrete modules over D_p. Since the structure of D_p is a still quite a mystery, using Theorem 1.6 to 'classify' finite flat $\mathbf{Z}_p$-group schemes via flat representations is not a very useful 'classification.'

We now give the fundamental classification of finite $\mathbf{F}_p$-group schemes via Dieudonne modules. An essentially self-contained development of the general theory of Dieudonne modules is given in [12, Ch II]. Define the 'Dieudonne ring' $D = \mathbf{Z}_p[F, V]/(FV - p)$ to be a ring with the variables F ('Frobenius') and V ('Verschiebung'). By a *finite D-module*, we will mean a D-module with finite $\mathbf{Z}_p$-length (and so the underlying abelian group is finite). The category of finite D-modules is an abelian category in an evident way. The category of finite $\mathbf{F}_p$-group schemes with p-power order is also an abelian category, using scheme-theoretic kernels and quotients. This is one of the essential ingredients in the proof of

Theorem 4.1. (Dieudonne-Cartier) *There exists a contravariant additive anti-equivalence of abelian categories* $\mathbf{M} : G \rightsquigarrow \mathbf{M}(G)$ *from the category of finite* $\mathbf{F}_p$*-group schemes of p-power order to the category of finite D-modules. Moreover, the order of G is equal to the order of* $\mathbf{M}(G)$ *(i.e.,* $p^{\ell_{\mathbf{Z}_p}(\mathbf{M}(G))}$*).*

PROOF. See [12, Ch III] for Fontaine's proof, where it is obtained from more general considerations in the setting of certain formal commutative group schemes over an arbitrary perfect field of characteristic p. ■

For a finite $\mathbf{F}_p$-group scheme G, the *Dieudonne module* of G is the finite D-module $\mathbf{M}(G)$ from Theorem 4.1. Just to give a hint as to where the construction of $\mathbf{M}(G)$ comes from, let us look at finite abelian p-groups, which are the same thing as finite $\mathbf{C}$-group schemes of p-power order. Cartier duality $G \rightsquigarrow G^*$, with

$$G^*(T) = \mathrm{Hom}_T(G \times_{\mathbf{C}} T, \mathbf{G}_{m/T})$$

for $\mathbf{C}$-schemes T, becomes on $\mathbf{C}$-points just classical duality of finite abelian p-groups:

$$G \rightsquigarrow G^* = \mathrm{Hom}(G, \mathbf{C}^\times) = \mathrm{Hom}(G, \mathbf{Q}/\mathbf{Z}) = \mathrm{Hom}(G, \mathbf{Q}_p/\mathbf{Z}_p).$$

The $\mathbf{Q}_p/\mathbf{Z}_p$ term is a direct limit of the groups $\mathbf{Z}/p^n = W_n(\mathbf{F}_p)$, where W_n is the 'Witt ring scheme of length n.' Thus, if we could somehow exploit the group/ring-scheme theoretic properties of the W_n's, we could try to construct a 'schemified version' of $\mathbf{Q}_p/\mathbf{Z}_p$ as a formal scheme. Considering group scheme maps into such an object would give a reasonable candidate for a 'linear algebra' object attached to G. The actual construction of $\mathbf{M}(G)$ requires some care (e.g., W_n is a finite type scheme, not finite or even formal), but the above gives the flavor of the basic idea.

The essential content of Theorem 4.1 is that one can pass from 'linear algebra data' such as a finite D-module and produce something as subtle as a (possibly non-reduced) $\mathbf{F}_p$-group scheme. By considering how the functor $\mathbf{M}$ is constructed as a sort of dual and trying to mimic 'double duality' to get a quasi-inverse to $\mathbf{M}$, the general proof of Theorem 4.1 uses a finite D-module M to define a functor (analogous to a 'double dual') from finite $\mathbf{F}_p$-algebras to abelian groups. Then a general (and essentially formal) 'pro-representability' theorem of Grothendieck's is invoked. See [5, §1.4] for a precise formulation and proof of this 'pro-representability' theorem in the form needed. In this way, one gets a commutative formal $\mathbf{F}_p$-group scheme whose affine ring R_M is an inverse limit of finite $\mathbf{F}_p$-algebras over an enormous index set. One then has to show that R_M is actually finite over $\mathbf{F}_p$ (and that the finite $\mathbf{F}_p$-group scheme $G_M = \mathrm{Spec}(R_M)$ has $\mathbf{M}(G_M) \simeq M$ naturally in M). But at least R_M *is* a ring to work with, albeit an abstract one (so Theorem 4.1 is not a complete black hole).

There are more refined versions of Theorem 4.1 which translate various notions from the theory of finite $\mathbf{F}_p$-group schemes over into the language of finite D-modules. We give a limited sampling that is all we shall need.

Lemma 4.2 *Let G be a finite $\mathbf{F}_p$-group scheme with p-power order, $M = \mathbf{M}(G)$. Then G is étale if and only if $F(M) = M$ (or, equivalently, $F : M \to M$ is a $\mathbf{Z}_p$-linear isomorphism) and G is connected if and only if the action of F is nilpotent. Define $M^* = \mathrm{Hom}_{\mathbf{Z}_p}(M, \mathbf{Q}_p/\mathbf{Z}_p)$ and let F and V act on M^* as duals to the actions of V and F on M respectively. In this way, M^* has the structure of a finite D-module and there is a natural isomorphism of finite D-modules*

$$\varphi_G : \mathbf{M}(G^*) \to M^*,$$

with G^* the Cartier dual of G.

The étale criterion in Lemma 4.2 corresponds to the fact that a finite $\mathbf{F}_p$-algebra R is étale if and only if $r \mapsto r^p$ is an automorphism of R. The connectedness criterion corresponds to the fact [12, Ch I, Rem 9.5.2] that a finite connected $\mathbf{F}_p$-group scheme always has an affine ring $R = \mathbf{F}_p[X_1, \dots, X_m]/(X_j^{p^{n_j}})$, so it necessarily has p-power order, and for $n \geq$

$\max(n_j)$ the nth iterate of $r \mapsto r^p$ kills the augmentation ideal of R. The isomorphism φ_G is constructed in [12, Ch II, §5] in a very indirect manner. In particular, it is not at all clear from the definition that φ_G is a 'symmetric' duality pairing in the sense that the natural diagram

$$\begin{array}{ccc} \mathbf{M}(G)^{**} & \xrightarrow{\varphi_G^*} & \mathbf{M}(G^*)^* \\ \simeq\uparrow & & \uparrow \varphi_{G^*} \\ \mathbf{M}(G) & \xleftarrow{\mathbf{M}(\alpha_G)} & \mathbf{M}(G^{**}) \end{array}$$

commutes, where $\alpha_G : G \simeq G^{**}$ is the canonical isomorphism. This is a complicated technical point which we will not need, so we will not discuss it. What we will need is the existence of the (natural) isomorphism φ_G.

Example 4.3 By Theorem 4.1, $\mathbf{M}(\mathbf{Z}/p)$, $\mathbf{M}(\mu_p)$, and $\mathbf{M}(\alpha_p)$ are all 1-dimensional over $\mathbf{F}_p$. The actions of F and V on these are given as follows:
$G = \mathbf{Z}/p$: $F = 1$, $V = 0$,
$G = \mu_p$: $F = 0$, $V = 1$,
$G = \alpha_p$: $F = V = 0$.
By Lemma 4.2, the only thing which remains to be checked is that $V = 1$ when $G = \mu_p$. This follows from [12, Ch I, §8.7; Ch III, Prop 4.3].

In general, it is possible that non-isomorphic finite flat $\mathbf{Z}_p$-group schemes can have isomorphic closed fibers. In the notation of [26, Rem 5, pp.15-16], if $\pi^3 = p$ and we work in the category of finite flat group schemes over $\mathbf{Z}_p[\pi]$, then $G^{\pi^2}_{\pi,\mathbf{Z}_p[\pi]}$ and $G^{\pi}_{\pi^2,\mathbf{Z}_p[\pi]}$ are non-isomorphic, yet they have isomorphic closed fiber α_p. With more work, one can construct examples over $\mathbf{Z}_p$ as well. In other words, the analogue to Theorem 1.6 for passage to the closed fiber is false: the functor $G \rightsquigarrow G \times_{\mathbf{Z}_p} \mathbf{F}_p$ is not fully faithful. The emphasis here is on the 'fully' part; it is shown in the course of the proof of Theorem 4.5 below that this functor is faithful for $p \neq 2$.

Thus, if one wishes to describe finite flat $\mathbf{Z}_p$-group schemes in terms of 'linear algebra data,' then one needs to find some 'extra structure' within $\mathbf{M}(G \times_{\mathbf{Z}_p} \mathbf{F}_p)$ that encodes the lifting G to $\mathbf{Z}_p$. Using the affine ring of $G_{/\mathbf{Z}_p}$, Fontaine constructs a $\mathbf{Z}_p$-submodule of 'logarithms'

$$\mathcal{L}(G) \subseteq \mathbf{M}(G \times_{\mathbf{Z}_p} \mathbf{F}_p)$$

which is *not necessarily stable under* F *and* V but which satisfies the following two properties:
(1) $V|_{\mathcal{L}(G)} : \mathcal{L}(G) \to \mathbf{M}(G \times_{\mathbf{Z}_p} \mathbf{F}_p)$ is injective
(2) The natural $\mathbf{Z}_p$-linear composite map

$$\mathcal{L}(G)/p \to \mathbf{M}(G \times_{\mathbf{Z}_p} \mathbf{F}_p)/p \twoheadrightarrow \mathbf{M}(G \times_{\mathbf{Z}_p} \mathbf{F}_p)/F$$

is an isomorphism.
This is valid even if $p = 2$.

For a finite $\mathbf{F}_p$-group scheme H, $\mathbf{M}(H)/F$ can be canonically identified with the cotangent space of H at its origin [12, Ch III, Prop 4.4(*ii*)], so the second condition above shows that $\mathcal{L}(G)$ provides some sort of 'minimal basis' lifting of the cotangent space, partially explaining the name 'logarithm.' [12, Ch IV] develops the ideas which motivate the construction of $\mathcal{L}(G)$, as well as the techniques which are needed to actually construct it. The brief article [13] gives an outline of the actual construction, whose justification relies on the theory of p-divisible groups; nearly the entire contents of the book [12] are needed for this! For a more detailed explanation of [13], see [7, §1].

It is now reasonable to make the following definition, following Fontaine.

Definition 4.4 The category SH^f of *finite Honda systems* (over $\mathbf{Z}_p$) consists of pairs (L, M) with M a finite D-module and L a $\mathbf{Z}_p$-submodule satisfying the properties (1) and (2) above. The notion of a morphism is defined in the obvious manner.

One can show directly that SH^f is an abelian category, using the obvious candidates for kernel and cokernel as the kernel and cokernel objects. See [14, §1, §9], noting that [14, Prop 9.10] provides a translation of SH^f into the language of finite filtered modules $\underline{MF}^f_{\text{tor}}$ as is used in [14, Prop 1.8]. In addition, the construction of $\mathcal{L}(G)$ is sufficiently natural for a finite flat $\mathbf{Z}_p$-group scheme G so that

$$LM : G \rightsquigarrow (\mathcal{L}(G), \mathbf{M}(G \times_{\mathbf{Z}_p} \mathbf{F}_p))$$

is an additive contravariant functor from the category of finite flat $\mathbf{Z}_p$-group schemes to the category SH^f. We have the following fundamental fact, whose proof makes essential use of the theory of p-divisible groups:

Theorem 4.5 (Fontaine [13]) *For $p \neq 2$, the additive contravariant functor LM is an anti-equivalence of abelian categories.*

The idea of the proof is to simultaneously show that all finite flat $\mathbf{Z}_p$-group schemes embed into p-divisible groups over $\mathbf{Z}_p$ and to invoke Fontaine's classification theory for such p-divisible groups (when $p \neq 2$) in terms of a 'finite free' analogue of the notion of a finite Honda system. The special fact about p-divisible groups used in all of this is that over a field k of characteristic p (where the Dieudonne theory classifies various commutative group schemes when the field is perfect), a connected p-divisible group is the same thing as a finite-dimensional formal Lie group Γ such that for all finite k-algebras R, every element of the abelian group $\Gamma(R)$ has p-power order.

Theorem 4.5, in conjunction with a good understanding of the construction of LM (e.g., the theory of Dieudonne modules, Lemma 4.2, etc.), enables us to translate questions concerning the construction and study of finite flat $\mathbf{Z}_p$-group schemes (for $p \neq 2$) into analogous questions in the setting of SH^f, where (in principle) everything is just a matter of 'linear algebra'! There are more refined questions one could ask, such as whether one can give an explicit description of the functor that passes from a finite Honda system to the generic fiber D_p-representation of the associated finite flat $\mathbf{Z}_p$-group scheme. This can be done and is the essential point of [14, §9]. We will not need this. However, we note in passing that this point has led to the misunderstanding that [14] is critical to the proof of Ramakrishna's theorem and Wiles' proof of the modularity of semistable elliptic curves over $\mathbf{Q}$. This is not true. Everything which Ramakrishna and Wiles need is contained in Theorem 4.5, which was proven by Fontaine long before [14] was written. Nevertheless, we remind the reader that [14, Prop 9.12] can still be useful in the present setting; the main issue that [14, Prop 9.12] enables us to handle is the analysis of generic fiber representation 'tensor' constructions such as 'determinant'; see also [14, §6.13(*b*)] and its corrected form in [6, §7.11]. Generic fiber representation 'tensor' constructions are difficult to study solely from the point of view of group schemes because the representation-theoretic notion of a tensor product has no analogue in the context of group schemes.

Before using Theorem 4.5 to construct $\mathbf{Z}_p$-flat representations of D_p in the next section, we give a modified formulation which is used to handle the cases in which some finite extension $\mathcal{O}$ of $\mathbf{Z}_p$ acts on the representation space. Keeping in mind Raynaud's full faithfulness result (Theorem 1.6), it is not hard to deduce a variant of Theorem 4.5 in the following manner.

Let A be a finite local $\mathbf{Z}_p$-algebra and let $D_A = A[F,V]/(FV-p)$. We can define the notion of a finite D_A-module in the obvious way; the category of such objects forms an A-linear subcategory of the category of finite D-modules. In a similar way, we can define a finite A-Honda system (L, M) by replacing $\mathbf{Z}_p$ by A in the definition of a finite Honda system (still using L/p and not $L/\mathfrak{m}_A$ in condition (2)). We then get a category SH_A^f which one readily checks is an abelian subcategory of SH^f, with the 'forgetful functor' $SH_A^f \to SH^f$ exact. We then have

Corollary 4.6 *For A as above and $p \neq 2$, the functor LM induces an A-linear anti-equivalence of abelian categories between the category of finite flat $\mathbf{Z}_p$-group schemes with an A-action on the generic fiber representation and the category SH_A^f. If $A = \mathcal{O}$ or $\mathcal{O}/\lambda^n$ and G is a finite flat $\mathbf{Z}_p$-group scheme with an A-action on its generic fiber representation ρ, then ρ and $\mathbf{M}(G \times_{\mathbf{Z}_p} \mathbf{F}_p)$ are non-canonically isomorphic as A-modules.*

PROOF. The only point we need to check is the final one. For a finite-length $\mathcal{O}$-module N, it is clear that the abstract $\mathcal{O}$-module structure of N is determined by the $\mathcal{O}$-module structure of λN and the value of $\dim_k(N[\lambda])$. Arguing par devissage and using the fact that the dimension of a k-vector space is determined by the underlying $\mathbf{F}_p$-vector space, we are reduced to the case $\mathcal{O} = \mathbf{Z}_p$ and objects killed by p. It now suffices to invoke the fact that $p^{\ell_{\mathbf{Z}_p}(\mathbf{M}(G\times_{\mathbf{Z}_p}\mathbf{F}_p))}$ is equal to the order of $G \times_{\mathbf{Z}_p} \mathbf{F}_p$, which is equal to the order of G. ■

§5. Applications to Flat Deformations

We now apply Fontaine's theory to complete the proof of Lemma 3.4 and Ramakrishna's Theorem 3.5. For the proof of Lemma 3.4 in the residually reducible (i.e., 'ord') cases, we give a variant on the arguments in [27] and [8, §2.5]. Our argument is different insofar as we make direct use of the results in §4, rather than work in the language of Fontaine-Laffaille modules. This makes the role of the Dieudonne theory more explicit and clarifies the role of the theory of finite flat group schemes in the calculations.

After finishing the proof of Lemma 3.4 in the residually reducible case, we will carry out the unfinished calculations from our earlier sketch of the proof of Theorem 3.5 (which, as we have already seen, implies Lemma 3.4 in the residually irreducible case).

Let $\overline{\rho}$ and ρ be as in Lemma 3.4, with $p \neq 2$ and $\overline{\rho}$ possibly irreducible. By Theorem 1.6, there is a canonical finite flat $\mathbf{Z}_p$-group scheme $G(\rho)$ with generic fiber representation ρ. By Corollary 4.6, we obtain an object

$$LM(\rho) = (L(\rho), M(\rho)) \stackrel{\text{def}}{=} (\mathcal{L}(G(\rho)), \mathbf{M}(G(\rho) \times_{\mathbf{Z}_p} \mathbf{F}_p))$$

in $SH^f_{\mathcal{O}}$. As an $\mathcal{O}$-module, $M(\rho)$ is free of rank 2 over $\mathcal{O}/\lambda^n$, by Corollary 4.6. Observe that we have natural $\mathcal{O}$-linear isomorphisms

$$\begin{aligned} H^1_{\mathrm{fl}}(D_p, \mathrm{ad}(\rho)) &\simeq \mathcal{D}^{\mathrm{fl}}_{\rho}((\mathcal{O}/\lambda^n)[\varepsilon]) \\ &\simeq \mathrm{Ext}^{1,\mathrm{fl}}_{(\mathcal{O}/\lambda^n)[D_p]}(\rho, \rho) \\ &\simeq \mathrm{Ext}^1_{SH^f_{\mathcal{O}/\lambda^n}}(LM(\rho), LM(\rho)). \end{aligned}$$

Using the above chain of isomorphisms, the following gives a more precise version of Lemma 3.4.

Theorem 5.1. *As $\mathcal{O}/\lambda^n$-modules, there is a non-canonical isomorphism*

$$\mathrm{Ext}^1_{SH^f_{\mathcal{O}/\lambda^n}}(LM(\rho), LM(\rho)) \simeq (\mathcal{O}/\lambda^n) \oplus (\mathcal{O}/\lambda^n) \oplus H^0(D_p, \mathrm{ad}^0(\rho)).$$

Proof. For *any* finite $\mathcal{O}$-Honda system (L, M), the composite map

$$L/p \to M/p \to M/FM$$

is an isomorphism of $\mathcal{O}$-modules. If we apply $\otimes_{\mathcal{O}} k$, we get a composite isomorphism

$$L/\lambda \to M/\lambda \to (M/FM) \otimes_{\mathcal{O}} k,$$

so the map $L/\lambda \to M/\lambda$ is injective. Thus, L is necessarily an $\mathcal{O}$-module direct summand of M.

Consider the pair $(L(\rho), M(\rho))$. We claim that $L(\rho) \neq 0$ and $L(\rho) \neq M(\rho)$, so therefore $L(\rho)$ is a rank 1 free $\mathcal{O}/\lambda^n$-module direct summand of $M(\rho)$. If $L(\rho) = 0$, then $M(\rho)/F(M(\rho)) = 0$, so by Lemma 4.2, $G(\rho) \times_{\mathbf{Z}_p} \mathbf{F}_p$ is étale. But then $G(\rho)$ is étale over $\mathbf{Z}_p$, in which case ρ is unramified. This contradicts the fact that $\overline{\rho} = \rho \bmod \lambda$ has a ramified determinant.

If $L(\rho) = M(\rho)$, then the injectivity of V on $L(\rho)$ implies that V is an automorphism on $M(\rho)$. Hence, by Lemma 4.2, the Cartier dual of $G(\rho)$ is a finite étale $\mathbf{Z}_p$-group scheme, so the Cartier dual ρ^* of ρ is unramified. Thus, the Cartier dual $\overline{\rho}^*$ of $\overline{\rho}$ is unramified, since $\overline{\rho}^* \simeq \rho^*[\lambda]$. This is inconsistent with the classification of possibilities for $\overline{\rho}$ in Theorem 1.8(i).

Choose a basis e_1, e_2 of $M(\rho)$ over $\mathcal{O}/\lambda^n$ such that e_2 is a basis for the direct summand $L(\rho)$. Consider an extension (L, M) of $LM(\rho)$ by itself in the abelian category $SH^f_{\mathcal{O}/\lambda^n}$. By the very nature of the abelian category structure of $SH^f_{\mathcal{O}/\lambda^n}$ (i.e., the construction of kernels and cokernels), it follows that M must be free of rank 4 as an $\mathcal{O}/\lambda^n$-module and L must be free of rank 2 as an $\mathcal{O}/\lambda^n$-module. Now choose an abstract rank 4 free $\mathcal{O}/\lambda^n$-module M with a chosen basis m_1, m_2, m_3, m_4 and we set L to be the submodule spanned by m_2 and m_4. We fix a short exact sequence of $\mathcal{O}/\lambda^n$-modules

$$0 \to M(\rho) \xrightarrow{j} M \xrightarrow{h} M(\rho) \to 0$$

determined by $j(e_1) = m_1$, $j(e_2) = m_2$ and $h(m_3) = e_1$, $h(m_4) = e_2$. Our problem is to count the number of ways (up to equivalence respecting L) we can impose a D-module structure on M compatible with the D-module structure on $M(\rho)$ via j and h. The main point is that since $LM(\rho)$ is an object in $SH^f_{\mathcal{O}/\lambda^n}$, any such D-module structure on (L, M) structure would have to make (L, M) an object in $SH^f_{\mathcal{O}/\lambda^n}$ (and so the resulting sequence would be a short exact sequence in $SH^f_{\mathcal{O}/\lambda^n}$). We will check this below.

We will only consider the case in which $\overline{\rho}$ is reducible, since otherwise we have seen in §3 that Theorem 5.1 follows from Theorem 3.5, whose proof we will finish later. Since $\overline{\rho}$ is reducible, by Theorem 1.8(ii) there exist continuous unramified characters $\chi_i : D_p \to (\mathcal{O}/\lambda^n)^\times$ and a short exact sequence of $(\mathcal{O}/\lambda^n)[D_p]$-modules

$$0 \to \epsilon\chi_1 \to \rho \to \chi_2 \to 0.$$

Lemma 4.2 and the contravariance of LM enable us to modify our choice of e_1 so that $(0, (\mathcal{O}/\lambda^n)e_1)$ is a subobject of $LM(\rho)$, corresponding to the unramified quotient χ_2. In addition, we have $F(e_1) = ve_1$ and $V(e_1) = v^{-1}pe_1$ for some $v \in (\mathcal{O}/\lambda^n)^\times$.

Set $V(e_2) = ae_1 + ue_2$ and $F(e_2) = be_1 + ce_2$. Since the subrepresentation $\epsilon\chi_1$ of ρ has an unramified Cartier dual, it follows that $u \in (\mathcal{O}/\lambda^n)^\times$. Finally, the conditions $FV = VF = p$ force $c = u^{-1}p$ and $a = -v^{-1}ub$, so with respect to the ordered basis $\{e_1, e_2\}$ of $M(\rho)$, we have matrices

$$F_{M(\rho)} = \begin{pmatrix} v & b \\ 0 & u^{-1}p \end{pmatrix}, \quad V_{M(\rho)} = \begin{pmatrix} v^{-1}p & -v^{-1}ub \\ 0 & u \end{pmatrix}.$$

Since we can write the actions of F and V on M in the 'block matrix' form

$$F_M = \begin{pmatrix} F_{M(\rho)} & X \\ 0 & F_{M(\rho)} \end{pmatrix}, \quad V_M = \begin{pmatrix} V_{M(\rho)} & -Y \\ 0 & V_{M(\rho)} \end{pmatrix},$$

the conditions $FV = VF = p$ yield the matrix equations

$$XV_{M(\rho)} = F_{M(\rho)}Y, \quad V_{M(\rho)}X = YF_{M(\rho)}.$$

We claim that with such data, (L, M) will necessarily be an object in $SH^f_{\mathcal{O}/\lambda^n}$. Since $V|_{L(\rho)}$ is injective, $V|_L$ is injective. In order to deduce that $L/p \to M/FM$ is an isomorphism from the corresponding fact for $L(\rho)$ and $M(\rho)$, the crux of the argument is to check that the sequence

$$0 \to M(\rho)/F(M(\rho)) \to M/F(M) \to M(\rho)/F(M(\rho)) \to 0$$

is actually exact on the left. In terms of the explicit basis for $M(\rho)$, this reduces to the statement that if $m \in M(\rho)$ is of the form $X(m')$ with $F_{M(\rho)}(m') = 0$, then $m \in F_{M(\rho)}(M(\rho))$. Using length considerations, it is a straightforward consequence of the axioms for a finite Honda system that the sequence

$$0 \to M(\rho)/V(M(\rho)) \xrightarrow{F} M(\rho)/p \to M(\rho)/F(M(\rho)) \to 0$$

is exact (and not just right exact). Thus, $F_{M(\rho)}(m') = 0$ implies that $m' = V_{M(\rho)}(m_0)$, so

$$m = X(m') = XV_{M(\rho)}(m_0) = F_{M(\rho)}Y(m_0) \in F_{M(\rho)}(M(\rho)),$$

as desired.

Define the $\mathcal{O}/\lambda^n$-module

$$E(\rho) = \{(X, Y) \in \mathrm{M}_2(\mathcal{O}/\lambda^n) \times \mathrm{M}_2(\mathcal{O}/\lambda^n) \mid XV_{M(\rho)} = F_{M(\rho)}Y,\ V_{M(\rho)}X = YF_{M(\rho)}\}.$$

Inside of here is a submodule of 'commutators'

$$C(\rho) = \{([F_{M(\rho)}, A], -[V_{M(\rho)}, A]) \mid A = (a_{ij}) \in \mathrm{M}_2(\mathcal{O}/\lambda^n),\ a_{12} = 0\}$$

(the $a_{12} = 0$ condition corresponds to preserving the line $L(\rho) = (\mathcal{O}/\lambda^n)e_2$ inside of $M(\rho)$). It follows from [19, Ch III] that we have an $\mathcal{O}/\lambda^n$-module isomorphism

$$\mathrm{Ext}^1_{SH^f_{\mathcal{O}/\lambda^n}}(LM(\rho), LM(\rho)) \simeq E(\rho)/C(\rho).$$

We now will determine the $\mathcal{O}/\lambda^n$-module structures of $E(\rho)$ and $C(\rho)$. Choose $X = (x_{ij})$ and $Y = (y_{ij})$ in $\mathrm{M}_2(\mathcal{O}/\lambda^n)$. The condition that $(X, Y) \in E(\rho)$ is easily checked to be equivalent to the simultaneous constraints

$$x_{21} = u^{-1}vy_{21},\ x_{22} = u^{-1}(by_{21} + u^{-1}py_{22}),\ y_{11} = v^{-1}(v^{-1}px_{11} - by_{21}),$$

and

$$y_{12} = v^{-1}(-v^{-1}ubx_{11} + ux_{12} - by_{22}),$$

with x_{11}, x_{12}, y_{21}, and y_{22} arbitrarily chosen in $\mathcal{O}/\lambda^n$. Thus,

$$E(\rho) \simeq (\mathcal{O}/\lambda^n)^{\oplus 4}.$$

In order to determine the $\mathcal{O}/\lambda^n$-module structure of $C(\rho)$, we note that there is an obvious surjection

$$q : \{A \in \mathrm{M}_2(\mathcal{O}/\lambda^n) \mid a_{12} = 0\} \twoheadrightarrow C(\rho)$$

and the kernel is identified (as an $\mathcal{O}/\lambda^n$-module) with all $\mathcal{O}/\lambda^n$-module endomorphisms of $M(\rho)$ which stabilize $L(\rho)$ and commute with the actions of F and V — in other words, we have an $\mathcal{O}$-linear isomorphism

$$\ker(q) \simeq \mathrm{End}_{SH^f_{\mathcal{O}/\lambda^n}}(LM(\rho)) \simeq H^0(D_p, \mathrm{ad}(\rho)),$$

where the second isomorphism is $\mathcal{O}$-linear because of the linearity properties of the functor in Corollary 4.6. It is straightforward to compute that for $A = (a_{ij}) \in \mathrm{M}_2(\mathcal{O}/\lambda^n)$ with $a_{12} = 0$, $A \in \ker(q)$ if and only if $a_{21} = 0$ and $b(a_{11} - a_{22}) = 0$, so

$$H^0(D_p, \mathrm{ad}(\rho)) \simeq (\mathcal{O}/\lambda^n) \oplus (\mathcal{O}/\lambda^n)[b].$$

Thus, we obtain

$$C(\rho) \simeq (\mathcal{O}/\lambda^n)^{\oplus 3}/((\mathcal{O}/\lambda^n) \oplus (\mathcal{O}/\lambda^n)[b]) \simeq (\mathcal{O}/\lambda^n) \oplus (b \cdot (\mathcal{O}/\lambda^n))$$

and

$$H^0(D_p, \mathrm{ad}^0(\rho)) \simeq (\mathcal{O}/\lambda^n)[b] \simeq (\mathcal{O}/\lambda^n)/b.$$

Since

$$E(\rho)/C(\rho) \simeq (\mathcal{O}/\lambda^n)^{\oplus 2} \oplus (\mathcal{O}/\lambda^n)/b,$$

we are done. ■

We now complete the proof of Theorem 3.5 (which requires $\overline{\rho}$ to have trivial centralizer). We will handle the residually reducible and irreducible cases separately. First, we need to check that $|\operatorname{Ext}^1_{SH^f_k}(LM(\overline{\rho}), LM(\overline{\rho}))| = |k|^2$ in order to obtain a surjection

$$\pi : \mathcal{O}[\![T_1, T_2]\!] \twoheadrightarrow R^{\mathrm{fl}}_{\overline{\rho}}.$$

In the residually reducible case, this is just the calculation in the proof of Theorem 5.1, with $n = 1$. Now consider the case in which $\overline{\rho}$ is irreducible. By Theorem 1.8(i), $\overline{\rho}|_{I_p}$ is self-dual. In particular, the closed fiber of $G(\overline{\rho})$ is connected with a connected dual, so by Lemma 4.2, F and V act in a nilpotent manner on $M(\overline{\rho})$. Thus, $\ker(V) \neq 0$, so we have a natural k-linear isomorphism

$$\ker(V) \oplus L(\overline{\rho}) \simeq M(\overline{\rho}),$$

giving two natural lines in $M(\overline{\rho})$. Let $L(\overline{\rho}) = ke_2$ and $\ker(V) = ke_1$, so with respect to this basis, we get the matrix

$$V_{M(\overline{\rho})} = \begin{pmatrix} 0 & \overline{b} \\ 0 & 0 \end{pmatrix},$$

with $\overline{b} \neq 0$. The condition $FV = VF = p = 0$ on $M(\overline{\rho})$ yields

$$F_{M(\overline{\rho})} = \begin{pmatrix} 0 & \overline{a} \\ 0 & 0 \end{pmatrix},$$

for some $\overline{a} \in k$. Applying the above reasoning to the irreducible flat 'connected' dual $\overline{\rho}^*$, we see that $\overline{a} \neq 0$ by Lemma 4.2.

Thus, $V = \overline{c}F$, with $\overline{c} = \overline{a}^{-1}\overline{b} \in k^\times$. It won't matter for us what the value of $\overline{c} \in k^\times$ is, but we mention for completeness that $\overline{c}$ is determined by the unramified character

$$\omega^{-1} \det \overline{\rho} : D_p \to k^\times,$$

with $\overline{c} = -1$ when $\det \overline{\rho} = \omega$ (see [7, Lemma 6.1] for more details). Defining $E(\overline{\rho})$ and $C(\overline{\rho})$ as in the proof of Theorem 5.1, we compute that $E(\overline{\rho})$ is 4-dimensional over k, while $C(\overline{\rho})$ is 2-dimensional over k, so

$$\operatorname{Ext}^1_{SH^f_k}(LM(\overline{\rho}), LM(\overline{\rho})) \simeq E(\overline{\rho})/C(\overline{\rho})$$

is 2-dimensional over k, as desired.

As we saw in the sketch of the proof of Theorem 3.5 earlier, it remains to check that

$$|\mathcal{D}^{\mathrm{fl}}_{\overline{\rho}}(\mathcal{O}/\lambda^n)| = |k|^{2(n-1)}.$$

In fact, the surjection π already gives us $\leq$, so it is enough to simply construct $|k|^{2(n-1)}$ distinct flat deformations of $\overline{\rho}$ to $\mathrm{GL}_2(\mathcal{O}/\lambda^n)$. First consider the residually irreducible case. We will list $|k|^{2(n-1)}$ objects $X = (L_X, M_X)$ in $SH^f_{\mathcal{O}/\lambda^n}$ with M_X free of rank 2 over $\mathcal{O}/\lambda^n$ and

$$X[\lambda] \simeq LM(\overline{\rho})$$

in $SH^f_{\mathcal{O}}$ (recall that LM is contravariant). Fix $a \in \mathcal{O}/\lambda^n$ lifting $\overline{a} \in k^\times$ (where $\overline{a}$ is defined via a matrix for $F_{M(\overline{\rho})}$ as above). Choose any $\alpha, \beta \in \mathcal{O}/\lambda^n$ with $\alpha \equiv 0 \bmod \lambda$ and $\beta \bmod \lambda = \overline{b}$ (recall $\overline{b} \in k^\times$ from above). Define M_X to be free with basis e_1, e_2 and define $L_X = (\mathcal{O}/\lambda^n)e_2$. Also, define

$$F_{M_X} = \begin{pmatrix} \alpha & a \\ p\beta^{-1}a^{-1} & 0 \end{pmatrix}, \quad V_{M_X} = \begin{pmatrix} 0 & \beta a \\ pa^{-1} & -\alpha\beta \end{pmatrix}.$$

It is easy to check that for each of the $|k|^{2(n-1)}$ different choices of (α, β), the corresponding (L_X, M_X) is an object in $SH^f_{\mathcal{O}/\lambda^n}$ and that different pairs (α, β) give rise to non-isomorphic flat deformations of $\overline{\rho}$ of the desired type. When $\mathcal{O} = W(k)$, [27] gives a non-explicit direct proof that $|\mathcal{D}^{\mathrm{fl}}_{\overline{\rho}}(\mathcal{O}/\lambda^n)| = |k|^{2(n-1)}$ for $\overline{\rho}$ irreducible or reducible as in Theorem 3.5.

Before finishing off the residually reducible case in Theorem 3.5 (with trivial centralizer), note that what we have done so far completes the proof of Lemma 3.4 in all cases, which is what is needed in Wiles' method.

Back to the reducible flat $\overline{\rho}$ in Theorem 3.5. By the argument used to prove Theorem 5.1, we can choose a k-basis $\{e_1, e_2\}$ for $M(\overline{\rho})$ with $L(\overline{\rho}) = ke_2$ and

$$F_{M(\overline{\rho})} = \begin{pmatrix} \overline{v} & \overline{b} \\ 0 & 0 \end{pmatrix}, \quad V_{M(\overline{\rho})} = \begin{pmatrix} 0 & -\overline{v}^{-1}\overline{u}\overline{b} \\ 0 & \overline{u} \end{pmatrix},$$

with $\overline{u}, \overline{v} \in k^\times$ and $\overline{b} \in k$. Fix $b \in \mathcal{O}/\lambda^n$ lifting $\overline{b}$. For each of the $|k|^{2(n-1)}$ pairs (u, v) with $u, v \in \mathcal{O}/\lambda^n$ lifting $\overline{u}$ and $\overline{v}$ respectively, define

$$F_M = \begin{pmatrix} v & b \\ 0 & u^{-1}p \end{pmatrix}, \quad V_M = \begin{pmatrix} v^{-1}p & -v^{-1}ub \\ 0 & u \end{pmatrix}.$$

It is easy to check that in this way, (L, M) acquires the structure of an object in $SH^f_{\mathcal{O}/\lambda^n}$ with λ-torsion isomorphic to $LM(\overline{\rho})$, so the corresponding $(\mathcal{O}/\lambda^n)[D_p]$-module gives rise to a flat deformation of $\overline{\rho}$ to $\mathrm{GL}_2(\mathcal{O}/\lambda^n)$. Moreover, different pairs (u, v) are readily checked to give rise to non-isomorphic deformations. This concludes the proof of Theorem 3.5.

References

[1] M. Atiyah, I. MacDonald, *Introduction to Commutative Algebra*, Addison-Wesley Publishing Company, 1969.

[2] S. Bloch, K. Kato, *L-functions and Tamagawa Numbers of Motives*, The Grothendieck Festschrift, vol. I, Birkhäuser, 1990, pp. 333-400.

[3] S. Bosch, W. Lütkebohmert, M. Raynaud, *Néron Models*, Springer-Verlag, 1980.

[4] B. Conrad, *Assorted Extras and Tidbits*, notes for lectures at Princeton, Spring, 1995.

[5] B. Conrad, *Background Notes on p-divisible Groups over Local Fields*, notes for lectures in Seminar on Assorted Topics, Princeton, Fall, 1995.

[6] B. Conrad, *Filtered Modules, Galois Representations, and Big Rings*, notes for lectures at Princeton, Fall, 1994.

[7] B. Conrad, *Finite Honda Systems and Supersingular Elliptic Curves*, thesis, Princeton University, 1996.

[8] H. Darmon, F. Diamond, R. Taylor, *Fermat's Last Theorem*, preprint.

[9] B. de Smit, K. Rubin, R Schoof, *Criteria for Complete Intersections*, this volume.

[10] F. Diamond, *On Deformation Rings and Hecke Rings*, preprint.

[11] F. Diamond, K. Ribet, *ℓ-adic Modular Deformations and Wiles's "Main Conjecture"*, this volume.

[12] J.-M. Fontaine, *Groupes p-divisible sur les corps locaux*, Astérisque 47-48, Soc. Math. de France, 1977.

[13] J.-M. Fontaine, *Groupes finis commutatiffs sur les vecteurs de Witt*, C.R. Acad. Sci. **280** (1975), pp. 1423-1425.

[14] J.-M. Fontaine, G. Laffaille, *Construction de représentations p-adiques*, Ann. Sci. E.N.S. (1982), pp. 547-608.

[15] J.-M. Fontaine, B. Mazur, *Geometric Galois Representations* in Elliptic Curves, Modular Forms, and Fermat's Last Theorem, International Press, 1995, pp. 41-78.

[16] A. Grothendieck, *Éléments de Géométrie Algébrique*, Math Publ. IHES.

[17] A. Grothendieck, *Séminaire de Géométrie Algébrique* 2.

[18] A. Grothendieck, *Séminaire de Géométrie Algébrique* 7.

[19] P. Hilton, U. Stammbach, *A Course in Homological Algebra*, GTM 4, Springer-Verlag, 1970.

[20] N. Katz, B. Mazur, *Arithmetic Moduli of Elliptic Curves*, Princeton Univ. Press, Princeton, 1985.

[21] H. Matsumura, *Commutative Ring Theory*, Cambridge Univ. Press, 1986.

[22] B. Mazur, *Deforming Galois Representations* in Galois Groups over **Q**, pp. 385-437.

[23] B. Mazur, *An Introduction to the Deformation Theory of Galois Representations*, this volume.

[24] D. Mumford, *Abelian Varieties*, Oxford University Press, 1970.

[25] T. Oda, *The first DeRham Cohomology and Dieudonne Modules*, Ann. Sci. E.N.S., 5e série t.2, 1969, pp. 63-135.

[26] F. Oort, J. Tate, *Group Schemes of Prime Order*, Ann. Sci. E.N.S. (1970), pp. 1-21.

[27] R. Ramakrishna, *On a Variation of Mazur's Deformation Functor*, Compositio Math. (3) **87** (1993), pp. 269-286.

[28] M. Raynaud, *Schémas en groupes de type* $(p, p, \ldots, p)$, Bull. Soc. Math. France **102** (1974), pp. 241-280.

[29] J-P. Serre, *Local Fields*, GTM 67, Springer-Verlag, 1979.

[30] J-P. Serre, *Propriétés galoisiennes des points d'ordre fini des courbes elliptiques*, Inv. Math. **15** (1972), pp. 259-331.

[31] J. Silverman, *The Arithmetic of Elliptic Curves*, Springer-Verlag, 1986.

[32] J. Silverman, *Advanced Topics in the Arithmetic of Elliptic Curves*, Springer-Verlag, 1993.

[33] J. Tate, *p-divisible groups* in Proceedings of a Conference on Local Fields (Driebergen), pp. 158-183, 1966.

[34] J. Tate, *Finite Flat Group Schemes*, this volume.

[35] R. Taylor, A. Wiles, *Ring-theoretic Properties of Certain Hecke Algebras*, Annals of Mathematics (3) **141** (1995), pp. 553-572.

[36] J. Tilouine, *Hecke Algebras and the Gorenstein Property*, this volume.

[37] L. Washington, *Galois Cohomology*, this volume.

[38] A. Wiles, *Modular Elliptic Curves and Fermat's Last Theorem*, Annals of Mathematics (3) **141** (1995), pp. 443-551.

HECKE RINGS AND UNIVERSAL DEFORMATION RINGS

EHUD DE SHALIT

1. Introduction

Wiles' proof of the Shimura-Taniyama-Weil conjecture for semi-stable elliptic curves is based on the "modularity" of certain universal deformation rings.

Fix an odd irreducible representation

$$\bar{\rho} : G_{\mathbb{Q}} \to GL_2(k) \tag{1.1}$$

from the absolute Galois group of $\mathbb{Q}$ to the group of 2×2 invertible matrices over a finite field k, and a *deformation type* $\mathcal{D}$ (see section 2 for precise definitions). One constructs then a certain complete noetherian local ring, the universal deformation ring $R_{\mathcal{D}} = R_{\mathcal{D}}(\bar{\rho})$, and a universal deformation $\rho_{\mathcal{D}}^{\text{univ}} : G_{\mathbb{Q}} \to GL_2(R_{\mathcal{D}})$, whose specializations give all the deformations of $\bar{\rho}$ of type $\mathcal{D}$, up to strict equivalence (see [M2], [M3]). Here it is implicitly assumed that $\bar{\rho}$ itself is of type $\mathcal{D}$, to begin with.

If we assume in addition that $\bar{\rho}$ is *modular* (in the sense that it comes from reduction mod λ of the λ-adic representation associated to some cusp form, see 2.1 below), one can also associate to $\mathcal{D}$ another complete noetherian local ring, the Hecke algebra $T_{\mathcal{D}}$, and a canonical homomorphism $\varphi_{\mathcal{D}} : R_{\mathcal{D}} \to T_{\mathcal{D}}$ of local rings. $T_{\mathcal{D}}$ is the (p-adic completion of the) Hecke algebra acting on *all* the modular forms whose associated λ-adic representation is of type $\mathcal{D}$, and lifts $\bar{\rho}$. The modularity of $\bar{\rho}$ is needed to assure that there is at least one such form. The homomorphism $\varphi_{\mathcal{D}}$ is derived from the universality of $(R_{\mathcal{D}}, \rho_{\mathcal{D}}^{\text{univ}})$. Similarly, any deformation ρ of $\bar{\rho}$ with values in a complete local ring R defines a homomorphism $\varphi : R_{\mathcal{D}} \to R$ bringing $\rho_{\mathcal{D}}^{\text{univ}}$ to ρ, and ρ is called *modular* if and only if φ factors through $\varphi_{\mathcal{D}}$. The assertion that every deformation of type $\mathcal{D}$ is modular is therefore equivalent to the assertion that $\varphi_{\mathcal{D}}$ is an isomorphism. In this set-up the main theorem to be proved is the following.

Theorem 1. *Assume that $\mathcal{D}$ is a* minimal *deformation type. Then* (i) *$\varphi_{\mathcal{D}}$ is an isomorphism* (ii) *$T_{\mathcal{D}}$ is a local complete intersection (l.c.i.).*

We shall follow the proof of this theorem given by R. Taylor and A. Wiles in the appendix to their paper[1]. The proof gives (i), and the complete

[1] Our exposition benefitted at several points also from the excellent survey paper by Darmon, Diamond and Taylor [D-D-T].

intersection property (ii) comes as a by-product, rather than a prerequisite, to (i). In the earlier proof of (i), given by Wiles in chapter 3 of [W], (ii) had to be known in advance, and its independent proof made up the main body of [T-W]. It was observed by Faltings that a slight modification of the arguments of Taylor and Wiles yields both (i) and (ii) simultaneously, and this observation was incorporated into the appendix of [T-W].

The *minimality* condition on $\mathcal{D}$ (see remark 2 in section 2 below for a precise definition) is essential for the proof to work. It means, roughly speaking, that a deformation of type $\mathcal{D}$ has "no more ramification than what $\bar{\rho}$ forces it to have." While the main theorem remains true without the minimality assumption, different ideas are needed to pass from minimal to non-minimal $\mathcal{D}$. Very roughly, one measures by how much $R_{\mathcal{D}}$ and $T_{\mathcal{D}}$ change when $\mathcal{D}$ is modified, and proves that the equality $R_{\mathcal{D}} = T_{\mathcal{D}}$ propagates from the minimal deformation type to any $\mathcal{D}$. The Gorenstein-ness of $T_{\mathcal{D}}$ plays a crucial role in establishing a tool to measure the "change in $T_{\mathcal{D}}$." The passage from the minimal case to the general case is treated in Ribet's article in this volume [Ri2].

We shall not work in greatest generality. For example, we shall assume throughout that the determinant of ρ is the cyclotomic character, and at $l \neq p$ (p is the characteristic of k) we shall assume that $\mathcal{D}$ is of "type A" in Wiles' terminology (Wiles himself considered types "B" and "C" as well, and Diamond [Di2] [Di3] completed the picture by allowing the restriction of ρ to the decomposition group at l to be arbitrary). However, this will be enough for the application to the Shimura-Taniyama-Weil conjecture, in the minimal case.

Before we turn to a detailed outline of the proof of theorem 1, let us explain how it implies the Shimura-Taniyama-Weil conjecture for semi-stable elliptic curves. By abuse of language one calls an elliptic curve E over $\mathbb{Q}$ semi-stable if it has good or semi-stable (multiplicative) reduction everywhere. This is equivalent to the fact that the conductor of E is a square-free integer. Let E be a semi-stable elliptic curve defined over $\mathbb{Q}$, take $p = 3$, and *assume* that $\bar{\rho} = \bar{\rho}_{E,3}$, the representation of $G_{\mathbb{Q}}$ on the 3-division points, is irreducible. (The reducible case is handled via a trick of Wiles which involves $\bar{\rho}_{E,5}$ as well. See chapter 5 of [W] and [Ru].) Then in fact $\bar{\rho}$ satisfies all the technical conditions listed below, in section 2.1. Let us quickly check them.

First, $GL_2(\mathbb{F}_3)$ can be lifted to a subgroup of $GL_2(A)$ for some ring of integers A in a number field in which 3 splits completely, and therefore $\bar{\rho}$ can be viewed as a complex representation. Since $\mathrm{PGL}_2(\mathbb{F}_3) \cong S_4$, $\bar{\rho}$ is modular by the Langlands-Tunnell theorem. Second, the determinant of $\bar{\rho}$ is the cyclotomic character mod 3 thanks to the Weil pairing. Third, $\bar{\rho}$ remains absolutely irreducible when restricted to the absolute Galois group of $L = \mathbb{Q}(\sqrt{-3})$. If $\bar{\rho}|G_L$ were reducible, it would have two invariant lines (in $\bar{\mathbb{F}}_3^2$), for if it had only one such line, that line would be invariant under $G_{\mathbb{Q}}$ as well. Thus $\bar{\rho}(G_L)$ is a torus, the splitting field of $\bar{\rho}$ is an abelian extension M of L, and $[M : L]$ is relatively prime to 3. Since E is semi-stable, M/L would be unramified outside 3. Class field theory tells us that there are no abelian extensions of L of degree relatively prime to 3 which are unramified outside 3, so $M = L$, contradicting the irreducibility of $\bar{\rho}$. Finally, $\bar{\rho}$ is "type A" at every $l \neq 3$ and "Selmer" or "flat" at 3 by the semi-stability of E, as follows from the l-adic analytic model of E as a Tate curve (see below).

For any p, $\rho = \rho_{E,p}$, the representation of $G_{\mathbb{Q}}$ on the p-adic Tate module of E, is a deformation of $\bar{\rho} = \bar{\rho}_{E,p}$, the representation on the p-division points. It is a *minimal* deformation if and only if for every prime of bad reduction (including possibly p) the order of the minimal discriminant Δ_E of E at that prime is not divisible by p. This follows easily from the Tate parametrization of E at the bad primes. At a prime l of split multiplicative reduction one has $E(\bar{\mathbb{Q}}_l) \cong \bar{\mathbb{Q}}_l^{\times} / \langle q_{E,l} \rangle$ as a Galois module. In particular one has a short exact sequence of Galois modules

$$0 \to \mu_p \to E[p] \to \mathbb{Z}/p\mathbb{Z} \to 0 \tag{1.2}$$

and the splitting field of $E[p]$ is $\mathbb{Q}_l(\mu_p, q_{E,l}^{1/p})$. If the order of Δ_E, hence also of the Tate period $q_{E,l}$, *is* divisible by p, $\bar{\rho}$ is unramified (if $l \neq p$) or flat (if $l = p$), but ρ is ramified or non-flat (resp.). On the other hand if the order of $q_{E,l}$ is not divisible by p, already $\bar{\rho}$ is ramified (if $l \neq p$) or Selmer *non*-flat (if $l = p$). If l is a prime of non-split multiplicative reduction then the above analysis applies to the unramified quadratic twist of E. Since the notion of being unramified or flat is invariant under unramified twists, the same conclusion holds.

Now let $\mathcal{D}$ be the minimal deformation type described in section 2.2 below, and assume that the order of Δ_E at primes of bad reduction is not divisible by 3. Then $\rho = \rho_{E,3}$ is a deformation of type $\mathcal{D}$ of $\bar{\rho}$. The main theorem therefore implies that ρ factors through $T_{\mathcal{D}}$, so there exists a ring homomorphism $h : T_{\mathcal{D}} \to \mathbb{Z}_3$, such that $h \circ \varphi_{\mathcal{D}}$ brings $\rho_{\mathcal{D}}^{\text{univ}}$ to ρ. But h defines a $\mathbb{Z}_3$-valued weight-2 newform with trivial nebentypus, and for all but finitely many primes l, $\varphi_{\mathcal{D}}(\text{tr}(\rho_{\mathcal{D}}^{\text{univ}}(\text{Frob}_l))) = T_l$ is the l-Hecke operator. It follows that $\text{tr}(\rho(\text{Frob}_l)) = h(T_l)$. Since also $\det(\rho(\text{Frob}_l)) = l$, and ρ is irreducible, ρ *is* the representation associated to the modular form h. If the level of h is N then the Isogeny Theorem (due in this case to Serre and in general to Faltings) implies that there exists a non-constant morphism from the modular curve $X_0(N)$ to E. However, for many applications, such as the analytic continuation and functional equation of $L(E/\mathbb{Q}, s)$, it is enough to know that ρ is associated to a modular form.

Example 1. Take for E the curve $y^2 + xy = x^3 - x^2 - x$, $p = 3$, and $\bar{\rho} = \bar{\rho}_{E,3}$. Then

- $\text{Im}(\bar{\rho}) = GL_2(\mathbb{F}_3)$ (exercise !), $\det(\bar{\rho}) = \omega$
- $\bar{\rho}$ is flat non-ordinary at 3 (E has good, supersingular reduction there)
- $\bar{\rho}$ is semi-stable *ramified* at 73 (E is a Tate curve at 73 and $\Delta_E = 73$)

- E has good reduction outside 73.

The first point implies the irreducibility of $\bar{\rho}$. The last two points imply that $\rho_{E,3}$ is a minimal deformation of $\bar{\rho}$. Thus the main theorem applies as it stands, and proves that E is modular. Indeed, it is the curve labelled 73A in the Antwerp tables.

2. An outline of the proof

2.1. **Set-up.** Let $p \geq 3$ be a prime number, k a finite field of characteristic p, and $\bar{\rho} : G_{\mathbb{Q}} \to GL_2(k)$ an odd irreducible continuous representation. Assume

- $\bar{\rho}$ is *modular* — there exist (a) a newform f of weight κ, level N and nebentypus ψ (for some κ, N and ψ), (b) a prime λ in the field K_f generated over $\mathbb{Q}$ by the Fourier coefficients of f, dividing p, (c) an embedding of $\mathcal{O}_{K_f}/\lambda$ in $\bar{k}$, such that for every l not dividing N, $\bar{\rho}$ is unramified at l and

$$\det(X - \bar{\rho}(\mathrm{Frob}_l)) \equiv X^2 - a_l(f)X + \psi(l)l^{\kappa-1} \mod \lambda \tag{2.1}$$

- $\det(\bar{\rho}) = \omega$, the cyclotomic character mod p
- $\bar{\rho}|G_L$ is absolutely irreducible, where $L = \mathbb{Q}\left(\sqrt{\left(\frac{-1}{p}\right)p}\right)$
- $\bar{\rho}|G_l \sim \begin{pmatrix} \omega\chi^{-1} & * \\ & \chi \end{pmatrix}$ and χ is unramified, for $l \neq p$ (G_l is the decomposition group at l)
- $\bar{\rho}|G_p$ is *either flat* — it is the Galois module attached to the generic fiber of a finite flat k-vector group scheme over $\mathbb{Z}_p$, *or* it is not flat but *Selmer* — $\bar{\rho}|G_p \sim \begin{pmatrix} \omega\psi^{-1} & * \\ & \psi \end{pmatrix}$ for an unramified ψ. (In such a case, the $*$ has to be "trés ramifié" in the language of [Se], otherwise $\bar{\rho}|G_p$ would be both Selmer *and* flat.)

Remark 1. (i) $N(\bar{\rho})$, the prime-to-p Artin conductor of $\bar{\rho}$, is square free, and $\bar{\rho}$ is ramified at p too.

(ii) In the Selmer non-flat case (at p), and in the ramified case (at $l \neq p$), the unramified character ψ (resp. χ) is trivial or quadratic. This observation is due to Diamond ([Di1], 6.1 and 6.2).

For example, at $l \neq p$, the proof of (ii) goes as follows: $\bar{\rho}|G_l$ factors through a two-step solvable extension of $\mathbb{Q}_l$, with an unramified quotient and a tamely ramified submodule which is isomorphic to k. The unramified quotient acts by conjugation on the ramified submodule through the character $\omega\chi^{-2}$. On the other hand, we know from the structure theory of tamely ramified extensions, that this action should be through ω, hence $\chi^2 = 1$.

2.2. **Deformations.** Let $\mathcal{O}$ be the ring of integers in a finite extension of $\mathbb{Q}_p$ such that, denoting by λ the prime ideal of $\mathcal{O}$, $\mathcal{O}/\lambda = k$. Let Σ be the set of primes where $\bar{\rho}$ is ramified ($p \in \Sigma$). The key technical point in the proof of theorem 1 is the introduction of an *auxiliary set of primes* $Q = \{q_1, ..., q_r\}$ satisfying:

- $Q \cap \Sigma = \emptyset$, for all $q \in Q$ with $q \equiv 1 \bmod p$.
- For all $q \in Q$, $\bar{\rho}(\mathrm{Frob}_q)$ has distinct eigenvalues $\{\alpha_q, \beta_q\}$ contained in k.

If k is too small to contain the eigenvalues of some Frobenius, replace it by its quadratic extension, and change $\mathcal{O}$ and λ accordingly. More assumptions on Q will be imposed along the way. The set Q will vary, but ultimately our interest lies in $Q = \emptyset$. By abuse of notation we shall write Q also for the product of the primes in Q. Let $\mathbb{Q}_S$ (where S is a set of primes) be the maximal extension of $\mathbb{Q}$ which is unramified outside the primes in S, and $G_S = \mathrm{Gal}(\mathbb{Q}_S/\mathbb{Q})$. We are now ready to define the deformation type.

Definition 1. A *deformation of type* $\mathcal{D}_Q$ of $\bar{\rho}$ is a continuous representation $\rho : G_{\Sigma \cup Q} \to GL_2(R)$, where R is a local complete noetherian $\mathcal{O}$-algebra, with maximal ideal $\mathfrak{m}_R$ and residue field $R/\mathfrak{m}_R = k$, satisfying

- $\rho \bmod \mathfrak{m}_R = \bar{\rho}$
- $\det(\rho) = \epsilon$, the cyclotomic character
- $\rho|G_l \sim \begin{pmatrix} \epsilon\chi^{-1} & * \\ & \chi \end{pmatrix}$ and χ is unramified, for $l \in \Sigma$, $l \neq p$ (ρ is "type A")
- If $\bar{\rho}|G_p$ is *flat*, so is ρ — meaning that for every ideal $I \subseteq R$ with R/I finite, $\rho \bmod I$ is the Galois module attached to the generic fiber of a finite flat group scheme over $\mathbb{Z}_p$, endowed with an action of R/I making it free of rank 2 over R/I.

  If $\bar{\rho}$ is not flat but *Selmer* then $\rho|G_p \sim \begin{pmatrix} \omega\psi^{-1} & * \\ & \psi \end{pmatrix}$ for an unramified ψ.

Remark 2. (i) As above, in the Selmer non-flat case (at p), and in the ramified case (at $l \in \Sigma$, $l \neq p$), the unramified character ψ (resp. χ) is trivial or quadratic. Since $p \neq 2$, it follows that it is the same character as the one figuring in $\bar{\rho}$.

(ii) When $Q = \emptyset$, $\mathcal{D}$ ($= \mathcal{D}_\emptyset$) is *minimal:* if $\bar{\rho}$ is unramified (at $l \neq p$) or flat (at p), so is ρ.

Although no condition is imposed at $q \in Q$, we have

Lemma 2. ([T-W], appendix, lemma 7) *If $q \in Q$, then*

$$\rho|G_q \sim \begin{pmatrix} \phi_1 & \\ & \phi_2 \end{pmatrix}.$$

Proof: It is enough to prove the lemma when R is Artinian. Since $\bar{\rho}$ is unramified at q, and R is a p-adic ring, $\rho(I_q)$ is a pro-p group. Since

$q \neq p$, ρ is tamely ramified at q. Let f and t be topological generators of $\mathrm{Gal}(\mathbb{Q}_q^{\mathrm{tr}}/\mathbb{Q}_q)$ such that f restricts to the Frobenius automorphism on the maximal unramified extension $\mathbb{Q}_q^{\mathrm{nr}}$ of $\mathbb{Q}_q$, and t fixes $\mathbb{Q}_q^{\mathrm{nr}}$. Using the fact that $\alpha_q \neq \beta_q$ choose a basis for the space of ρ in which $\rho(f) = \begin{pmatrix} a & \\ & b \end{pmatrix}$ is diagonal. Since $\bar{\rho}$ is unramified at q, $\rho(t) \equiv 1 \bmod \mathfrak{m}_R$. Now suppose that $\rho(t) = \begin{pmatrix} \mu_1 & \\ & \mu_2 \end{pmatrix}(1+N) = M(1+N)$ and $N \equiv 0 \bmod \mathfrak{m}_R^n$ $(n \geq 1)$. Since M and N commute modulo $\mathfrak{m}_R^{n+1}$ we have

$$\rho(t)^q \equiv \begin{pmatrix} \mu_1^q & \\ & \mu_2^q \end{pmatrix}(1+qN) \bmod \mathfrak{m}_R^{n+1}$$

Using the relation $ftf^{-1} = t^q$ one gets

$$\begin{pmatrix} a & \\ & b \end{pmatrix}\begin{pmatrix} \mu_1 & \\ & \mu_2 \end{pmatrix}(1+N)\begin{pmatrix} a^{-1} & \\ & b^{-1} \end{pmatrix} \equiv \begin{pmatrix} \mu_1^q & \\ & \mu_2^q \end{pmatrix}(1+qN) \bmod \mathfrak{m}_R^{n+1}$$

which implies (since $q \equiv 1 \bmod p$) that N is diagonal $\bmod \mathfrak{m}_R^{n+1}$, and the desired result follows by induction on the nilpotency degree of $\mathfrak{m}_R$. □

Definition 2. Let Δ_q be the p-Sylow subgroup of $(\mathbb{Z}/q\mathbb{Z})^\times$, and let $\chi_q : G_{\mathbb{Q}} \to \Delta_q$ be the composition of the cyclotomic character mod q and the projection from $(\mathbb{Z}/q\mathbb{Z})^\times$ to Δ_q. (We call χ_q the *nebentypus character of conductor* q.) Further, let

$$\Delta_Q = \prod_{q \in Q} \Delta_q, \quad \Lambda_Q = \mathcal{O}[\Delta_Q] \text{ (a local ring)}, \quad \text{and} \quad \chi_Q = \prod_{q \in Q} \chi_q.$$

Let us also distinguish between ϕ_1 and ϕ_2 using the convention that ϕ_1 mod $\mathfrak{m}_R$ sends Frob_q to α_q (this is possible since $\alpha_q \neq \beta_q$).

Corollary 3. *$\phi_1|I_q = (\phi_2|I_q)^{-1}$ factors through $\chi_q|I_q$: $\phi_1|I_q = \tilde{\phi}_1 \circ \chi_q|I_q$ for a unique character $\tilde{\phi}_1 : \Delta_q \to (1+\mathfrak{m}_R)$.*

Proof : The first equality follows simply from the fact that $\epsilon = \phi_1\phi_2$ is unramified at q. Since ϕ_1 mod $\mathfrak{m}_R$ is unramified too, $\phi_1(I_q) \subset 1+\mathfrak{m}_R$. Now ϕ_1 factors through the maximal abelian extension of $\mathbb{Q}_q$, which, by the local Kronecker-Weber theorem, is generated by roots of unity. In particular $\phi_1|I_q$ factors through the cyclotomic character into $\mathbb{Z}_q^\times$, and since $1+\mathfrak{m}_R$ is a pro-p group, through the cyclotomic character mod q, and eventually through χ_q. □

The local conditions defining the deformation type $\mathcal{D}_Q$ are "conditions" in the technical sense of [M3]. It follows from the irreducibility of $\bar{\rho}$ that the deformation problem is representable. Thus there exist a *universal deformation ring* R_Q, and a *universal deformation* $\rho_Q^{\mathrm{univ}} : G_{\mathbb{Q}} \to GL_2(R_Q)$, such that every deformation $\rho : G_{\Sigma \cup Q} \to GL_2(R)$ of type $\mathcal{D}_Q$ is strictly equivalent to a unique specialization of ρ_Q^{univ} under a unique homomorphism $R_Q \to R$.

Next, apply the above discussion, on the shape of $\rho|I_q$, to the universal deformation ρ_Q^{univ}. Denote the $\tilde{\phi}_1$ of the corollary in this case by $\tilde{\phi}_1^{\text{univ}}$. We shall give R_Q the structure of a Λ_Q-algebra by mapping Δ_Q to it via the homomorphism $(\tilde{\phi}_1^{\text{univ}})^2$ (the reason for the square will become clear soon; note that R_Q is already an $\mathcal{O}$-algebra). Let $\mathfrak{a}_Q$ be the augmentation ideal in Λ_Q. Then from lemma 2 and corollary 3, ρ_Q^{univ} mod $\mathfrak{a}_Q R_Q$ is unramified at the q's dividing Q, hence it is of type $\mathcal{D}$. The universality of $R\,(=R_\emptyset)$ implies that there is a unique homomorphism $R \to R_Q/\mathfrak{a}_Q R_Q$ bringing $\rho_\emptyset^{\text{univ}}$ to ρ_Q^{univ} mod $\mathfrak{a}_Q R_Q$. On the other hand $\rho_\emptyset^{\text{univ}}$ is clearly of type $\mathcal{D}_Q$, so the universality of R_Q implies that there is a unique homomorphism $R_Q \to R$ bringing ρ_Q^{univ} to $\rho_\emptyset^{\text{univ}}$. We conclude that $R_Q/\mathfrak{a}_Q R_Q$ can be canonically identified with R.

2.3. **The Hecke ring.** Put $N = N(\bar\rho)$ if $\bar\rho|G_p$ is flat, and $N = N(\bar\rho)p$ if $\bar\rho|G_p$ is non-flat but Selmer. (This will turn out to be the minimal level at which one can find a weight 2 newform whose associated Galois representation lifts $\bar\rho$.) Let

$$(2.2)\qquad \Gamma_Q = \left\{ \begin{pmatrix} a & b \\ c & d \end{pmatrix} \in \Gamma_0(NQ) \,\middle|\, \begin{array}{c} \text{the order of } d \bmod Q \text{ in} \\ (\mathbb{Z}/Q\mathbb{Z})^\times \text{ is prime to } p \end{array} \right\}$$

so that $\Gamma_0(NQ)/\Gamma_Q = \Delta_Q$. Let $S_2(\Gamma_Q, \mathcal{O})$ be the space of $\mathcal{O}$-valued weight-2 cusp-forms on Γ_Q, and let $T(\Gamma_Q)$ be the subalgebra of $\mathrm{End}(S_2(\Gamma_Q, \mathcal{O}))$ generated over $\mathcal{O}$ by the Hecke operators T_l and $\langle l \rangle$ for primes l not dividing NQ, and U_l for $l|NQ$. Since $\langle l \rangle$ depends only on the image of l in Δ_Q, the diamond operators make $T(\Gamma_Q)$ a Λ_Q-algebra.

Let $\mathfrak{m}_Q$ be the ideal of $T(\Gamma_Q)$ generated by λ, $T_l - \mathrm{tr}(\bar\rho(\mathrm{Frob}_l))$ and $l\langle l \rangle - \det(\bar\rho(\mathrm{Frob}_l))$ for $(l, NQ) = 1$, $U_l - \chi(\mathrm{Frob}_l)$ for $l|N(\bar\rho)$, $U_p - \psi(\mathrm{Frob}_p)$ if $\bar\rho$ is Selmer non-flat, and $U_q - \beta_q$ for $q|Q$. Here χ and ψ are the unramified characters figuring in section 2.1, and β_q is the eigenvalue of $\bar\rho(\mathrm{Frob}_q)$ which coincides with $\tilde{\phi}_2^{\text{univ}}(\mathrm{Frob}_q)$ (recall that we chose $\tilde{\phi}_1^{\text{univ}}$ to define the action of Δ_Q on R_Q). The expressions of the form $X - x$, where $X \in T(\Gamma_Q)$ and $x \in k$, are shorthand for $X - \tilde{x}$, where $\tilde{x}$ is a lifting of x to $\mathcal{O}$. Since $\lambda \in \mathfrak{m}_Q$, it does not matter which lifting we choose. If $\mathfrak{m}_Q$ is a proper ideal (i.e., not equal to the whole ring), it is clearly maximal, and the homomorphism from $T(\Gamma_Q)$ to k defined by it *is* a k-valued eigenform, whose associated representation is $\bar\rho$.

The following theorem is very deep and contains Ribet's theorem on "lowering the level" [Ri1], as well as improvements due to Carayol, Gross, Coleman-Voloch, Edixhoven, Wiles and Diamond (although under the circumstances considered here, some of these names may be dispensable). (See [Di1] theorem 6.4 and [EG]).

Theorem 4. *$\mathfrak{m}_Q$ is a* proper *maximal ideal in $T(\Gamma_Q)$.*

Note that $\mathfrak{m}_Q$, if proper, contains (at least one) minimal prime, which corresponds to a weight 2 newform on Γ_Q whose associated Galois representation lifts $\bar{\rho}$. Thus the theorem is equivalent to the statement that $\bar{\rho}$ is *modular of weight 2 and the minimal possible level.* Note that this is a *variant* of Serre's "Conjecture ϵ," because Serre stipulated a level which was always prime-to-p. In the Selmer non-flat case considered here (where $p|N$) Serre would "pay" for the omission of p from the level in allowing the newform to be of weight $p+1$ rather than 2. Note also that here, and only here, is the crucial hypothesis that $\bar{\rho}$ is modular (of *some* weight and level), being used. Sketching the proof of this theorem would take us outside the scope of our survey. Let us indicate only how it follows, in the form stated here, from [Di1], theorem 6.4.

Consider first the case $Q = \emptyset$. It then follows from [Di1], theorem 6.4, that there is a newform f of weight 2 and level N, a prime λ' dividing p in the field K_f, and an embedding of $\mathcal{O}_{K_f}/\lambda'$ in $\bar{k}$ such that for every prime l not dividing Np,

$$a_l(f) \equiv \mathrm{tr}(\bar{\rho}(\mathrm{Frob}_l)) \bmod \lambda'.$$

Parts (2) and (3) of theorem 6 below, describing the restriction of the λ'-adic representation associated to f to the decomposition groups at bad l (those dividing $N(\bar{\rho})$) or at $l = p$, imply that $a_l(f) \equiv \chi(\mathrm{Frob}_l) \bmod \lambda'$ for $l|N(\bar{\rho})$, and $a_p(f) \equiv \psi(\mathrm{Frob}_p) \bmod \lambda'$ in the Selmer non-flat case. The nebentypus of f has order prime to p, so can be read from the determinant of $\bar{\rho} = \rho_{f,\lambda'} \bmod \lambda'$. But $\det(\bar{\rho}) = \omega$, and the weight is 2, so the nebentypus is trivial, and f is on $\Gamma_0(N)$. The kernel of the homomorphism of the Hecke algebra (into $\bar{k}$) defined by $f \bmod \lambda'$ is the desired maximal ideal.

When Q is not empty, one resorts to the theory of old-forms. For simplicity assume that $Q = \{q\}$ consists of a single prime (the general case is handled similarly). If α and β are the two eigenvalues of $\rho_{f,\lambda'}(\mathrm{Frob}_q)$, then $\alpha\beta = q$ and $\alpha + \beta = a_q(f)$. It follows that in the two-dimensional space of old-forms spanned by $f_1(z) = f(z)$ and $f_2(z) = f(qz)$, the matrix of U_q is $\begin{pmatrix} a_q & -1 \\ q & 0 \end{pmatrix}$, and there exists a unique linear combination g of f_1 and f_2 satisfying $U_q g = \beta g$. This g is an old eigenform on $\Gamma_0(NQ)$, hence on Γ_Q, and defines a homomorphism from $T(\Gamma_Q)$ to $\bar{k}$ whose kernel is the desired $\mathfrak{m}_Q$. This completes the proof. An important point to bear in mind is that since $q \equiv 1 \bmod p$, there are other *new*forms on Γ_Q whose λ-adic representation lifts $\bar{\rho}$, i.e., which are *congruent* to g. These newforms have non-trivial nebentypus factoring through Δ_Q. This will become clear once we prove, in section 3, the main theorem about the structure of T_Q. □

Definition 3. Let T_Q be the localization of $T(\Gamma_Q)$ in $\mathfrak{m}_Q$. Let X_Q be the modular curve over $\mathbb{Q}$ corresponding to the congruence group Γ_Q, and $J_Q = \mathrm{Jac}(X_Q)$.

The Hecke operators act as correspondences on X_Q, and as endomorphisms on J_Q. Let $\mathrm{Ta}_p(J_Q)$ be the p-adic Tate module of J_Q, which becomes a $T(\Gamma_Q)$-module after we tensor it over $\mathbb{Z}_p$ with $\mathcal{O}$. We can therefore localize it at $\mathfrak{m}_Q$. The resulting module, $\mathrm{Ta}_p(J_Q)_{\mathfrak{m}_Q}$, is a T_Q -module. Thanks to the assumption that the residual representation of $\mathfrak{m}_Q$, namely $\bar{\rho}$, is irreducible, we know the following result, which again is very deep, and is related to the Gorenstein-ness of T_Q (see [M1] and [Ti]).

Theorem 5. $\mathrm{Ta}_p(J_Q)_{\mathfrak{m}_Q}$ *is free of rank 2 over* T_Q. □

Definition 4. Let $\rho = \rho_Q^{\mathrm{mod}} : G_{\mathbb{Q}} \to GL_2(T_Q)$ be the Galois representation on $\mathrm{Ta}_p(J_Q)_{\mathfrak{m}_Q}$.

Remark 3. It is easy to see that $\mathrm{Ta}_p(J_Q)_{\mathfrak{m}_Q} \otimes_{\mathbb{Z}_p} \mathbb{Q}_p$ is free of rank 2 over $T_Q \otimes_{\mathbb{Z}_p} \mathbb{Q}_p$. One therefore obtains ρ, but with entries in $T_Q \otimes_{\mathbb{Z}_p} \mathbb{Q}_p$, by "gluing" the p-adic representations associated to the individual newforms. From there it is possible to find ρ with entries in T_Q by the method of "pseudo-representations" (see the discussion following [W], 2.1). However, theorem 5 will be needed later again.

The first part of the next theorem is "classical" — it is the Eichler-Shimura relation. Parts (2) and (3) are more recent.

Theorem 6. (1) (Eichler-Shimura-Igusa) *ρ is unramified outside $\Sigma \cup Q$, and for $(l, NQ) = 1$, $l \neq p$, the characteristic polynomial of $\rho(\mathrm{Frob}_l)$ is* $X^2 - T_l X + \langle l \rangle l$.
(2) (Carayol, following Deligne and Langlands) *For $l | N(\bar{\rho})$*

$$\rho | G_l \sim \begin{pmatrix} \chi^{-1}\chi_Q \epsilon & * \\ 0 & \chi \end{pmatrix}$$

with an unramified character χ, $\chi(\mathrm{Frob}_l) = U_l$, and $\chi^2 = \chi_Q$. The $*$ in the upper right corner is ramified.

For $q|Q$

$$\rho | G_q \sim \begin{pmatrix} \chi^{-1}\chi_Q \epsilon & * \\ 0 & \chi \end{pmatrix}$$

with an unramified character χ, $\chi(\mathrm{Frob}_q) = U_q$.
(3) (Unpublished correspondence of Fontaine-Serre, Wiles) *If $N(\bar{\rho}) = N$, ρ is flat at p. If, on the other hand, $p|N$, then*

$$\rho | G_p \sim \begin{pmatrix} \psi^{-1}\chi_Q \epsilon & * \\ 0 & \psi \end{pmatrix}$$

with an unramified character ψ, and $\psi(\mathrm{Frob}_p) = U_p$. □

In case (2), the automorphic representation of $GL_2(\mathbb{A}_{\mathbb{Q}})$ associated to any eigenform of level NQ whose λ-adic representation lifts $\bar{\rho}$, is "special" at $l|N(\bar{\rho})$, and "principal series" at $q|Q$.

Corollary 7. *The representation* $\rho'_Q = \rho_Q^{\text{mod}} \otimes \chi_Q^{-1/2}$ *is of type* $\mathcal{D}_Q$, *and there exists a unique surjective homomorphism of* Λ_Q*-algebras* $R_Q \to T_Q$, *bringing* ρ_Q^{univ} *to* ρ'_Q.

Proof: After twisting, $\det(\rho'_Q) = \epsilon$. Parts (2) and (3) of the theorem now imply that ρ'_Q is of type $\mathcal{D}_Q$. By the universality of R_Q we obtain the desired map to T_Q. Since this map brings $\phi_1^{\text{univ}}|I_q$ to $\sqrt{\chi_q|I_q}$, it respects the Δ_q action (this finally justifies the peculiar square in the definition of the Δ_q-action on R_Q). It remains to prove that the map is surjective, or that every Hecke operator is contained in the image. That T_l and $\langle l \rangle$ are in the image ($(l, NQ) = 1$) follows from the relations $\text{tr}(\rho_Q^{\text{mod}}(\text{Frob}_l)) = T_l$, and $\det(\rho_Q^{\text{mod}}(\text{Frob}_l)) = l\langle l \rangle$. Parts (2) and (3) of the theorem show directly that U_l ($l|N(\bar{\rho})$), U_q ($q|Q$) and U_p (if $\bar{\rho}$ is Selmer non-flat) are in the image. □

When $Q = \emptyset$ we drop the subscript and write R (resp. T) for $R_\emptyset$ (resp. $T_\emptyset$). The following is the main theorem (theorem 1).

Theorem 8. *The map* $R \to T$ *is an isomorphism, and* T *is a l.c.i.*

2.4. **The Taylor-Wiles-Faltings criterion.** The following commutative-algebra criterion lies at the basis of the proof of the main theorem. See [T-W], appendix, or [dSRS][2].

Lemma 9. *Let* $\phi : R \to T$ *be a surjective homomorphism of local complete noetherian* $\mathcal{O}$*-algebras, and assume that* T *is finite and flat over* $\mathcal{O}$. *Suppose that for some* $r \geq 1$ *and for every* $n \geq 1$ *there exist local complete noetherian* $\mathcal{O}$*-algebras* R_Q *and* T_Q *and a commutative diagram*

$$\begin{array}{ccccc} \mathcal{O}[[S_1, ..., S_r]] & \to & R_Q & \twoheadrightarrow & R \\ & & \downarrow & & \downarrow \\ & & T_Q & \twoheadrightarrow & T \end{array} \tag{2.3}$$

where all four maps in the square on the right of the diagram are surjective, and

(i) $(S_1, ..., S_r)R_Q = \text{Ker}(R_Q \to R)$

(ii) $(S_1, ..., S_r)T_Q = \text{Ker}(T_Q \to T)$

(iii) $\mathfrak{b} = \text{Ker}(\mathcal{O}[[S_1, ..., S_r]] \to T_Q) \subseteq ((1 + S_1)^{p^n} - 1, ..., (1 + S_r)^{p^n} - 1)$, *and* T_Q *is free of finite rank over* $\mathcal{O}[[S_1, ..., S_r]]/\mathfrak{b}$.

(iv) R_Q *is topologically generated as an* $\mathcal{O}$*-algebra by* r *elements.*

Then $R \to T$ *is an isomorphism, and they are l.c.i.* □

One should note that there are two types of assumptions here. Points (ii)–(iii) mean that *the* $\{T_Q\}$ *are large* and "regularly controlled" by r parameters, namely the S_i's. As n increases these parameters consume an increasingly significant part of T_Q. The subscript Q here has no meaning, and simply hints for the upcoming application. The R_Q and T_Q are not

[2] Recent improvements of the criterion by Rubin and Schoof eliminate the need to pass to R_∞ and T_∞ (see below). Here we stick to the original presentation.

canonical, and are not assumed even to relate to each other as we vary n. Indeed, the first step in the proof of the lemma is to show, by a "Mittag-Leffler" argument (after we reduce the picture mod λ), that it is possible to make an inverse system of diagrams as above, and pass to a limit

$$\begin{array}{ccccc} \mathcal{O}[[S_1,...,S_r]] & \rightarrow & R_\infty & \twoheadrightarrow & R \\ & & \downarrow & & \downarrow \\ & & T_\infty & \twoheadrightarrow & T \end{array}$$

where T_∞ is now finite free over $\mathcal{O}[[S_1,...,S_r]]$.

On the other hand, point (iv) means that *the R_Q are uniformly small*, so in the limit diagram R_∞ will still be generated topologically as an $\mathcal{O}$-algebra by r elements. Considerations of Krull dimension then yield $R_\infty = T_\infty \cong \mathcal{O}[[X_1,...,X_r]]$, from which the desired equality $R = T = \mathcal{O}[[X_1,...,X_r]]/(S_1,...S_r)$ is finally deduced.

To apply the lemma Taylor and Wiles find an r depending only on $\bar{\rho}$, and for every $n \geq 1$ they choose a set Q as above, containing precisely r primes, all congruent to 1 modulo p^n (and not merely p), so that (iv) will hold. The R_Q, the T_Q and the maps between them are chosen as in the previous sections. The map $\mathcal{O}[[S_1,...,S_r]] \rightarrow R_Q$ is taken to be the map that sends $1+S_i$ to a fixed generator of Δ_{q_i}. If we let p^{n_i} be the order of Δ_{q_i}, the p-Sylow subgroup of $(\mathbb{Z}/q_i\mathbb{Z})^\times$ (so that $n_i \geq n$), then Λ_Q is identified with $\mathcal{O}[[S_1,...,S_r]]/\mathfrak{b}$, where $\mathfrak{b} = ((1+S_1)^{p^{n_1}}-1,...,(1+S_r)^{p^{n_r}}-1)$, and the augmentation ideal $\mathfrak{a}_Q$ is identified with $(S_1,...,S_r)/\mathfrak{b}$. Point (i) was already noticed at the end of section 2.2. Points (ii) and (iii) are guaranteed by proposition 10 below, to be proved in section 3, and by the fact that all the $q_i \equiv 1 \bmod p^n$. The proof of proposition 10 is essentially topological. By an extension of a result of Mazur, a certain piece of the (singular) cohomology of the modular curve, with coefficients in $\mathcal{O}$, is free of rank one over T_Q. This allows to replace the question on the structure of T_Q over Λ_Q by a similar question on the structure of the cohomology.

The greatest difficulty lies in point (iv). We have to choose r and the Q's so that *the number r of primes in Q (which is the number of parameters S_i) is just the minimal number of generators of R_Q*. This is guaranteed by proposition 11 below, to be proved in section 4. The proof uses in a deep way Tate's global duality theorem in Galois cohomology. As will become clear, this is where the assumption that the deformation type is *minimal* enters.

We are thus left with the task of proving the following two propositions. Together with the commutative-algebra criterion of lemma 9, they imply the main theorem.

Proposition 10. *Let Q be a set of r primes as in section 2.2. Then* (i) *T_Q is finite and free over Λ_Q* (ii) *$rank_{\Lambda_Q} T_Q = rank_{\mathcal{O}} T$. Equivalently, $T_Q/\mathfrak{a}_Q T_Q = T$.*

Proposition 11. *There exists an r (namely, $\dim_k H^1_{\mathcal{D}*}(G_\Sigma, \mathrm{Symm}^2\bar{\rho})$, see section 4) such that for every $n \geq 1$ there exists a set Q of r primes, disjoint from Σ, satisfying*

- *For every $q \in Q$, $q \equiv 1 \bmod p^n$*
- *For every $q \in Q$, $\bar{\rho}(\mathrm{Frob}_q)$ has distinct eigenvalues contained in k*
- *R_Q can be topologically generated as an $\mathcal{O}$-algebra by r elements.*

3. Proof of proposition 10 — on the structure of the Hecke algebra[3]

3.1. **New and old.** Let $\tilde{\Gamma}_Q = \Gamma_0(NQ)$, and define the Hecke algebra $\tilde{T}_Q$ as in section 2.3, but with respect to $\tilde{\Gamma}_Q$ instead of Γ_Q, omitting the diamond operators. In other words, we first let $T(\tilde{\Gamma}_Q)$ be the Hecke algebra on $S_2(\tilde{\Gamma}_Q, \mathcal{O})$ (over $\mathcal{O}$). The inclusion $S_2(\tilde{\Gamma}_Q, \mathcal{O}) \subset S_2(\Gamma_Q, \mathcal{O})$ induces a surjective homomorphism $T(\Gamma_Q) \twoheadrightarrow T(\tilde{\Gamma}_Q)$. The image of the maximal ideal $\mathfrak{m}_Q$ in $T(\tilde{\Gamma}_Q)$ is also a proper maximal ideal $\tilde{\mathfrak{m}}_Q$ (cf the proof of theorem 4), and we set $\tilde{T}_Q = T(\tilde{\Gamma}_Q)_{\tilde{\mathfrak{m}}_Q}$. Localizing the above restriction map between the two Hecke algebras we get a surjective homomorphism $T_Q \twoheadrightarrow \tilde{T}_Q$. Note that since $T(\tilde{\Gamma}_Q)$ is pro-artinian, it is a product of its localizations at maximal ideals, and every $T(\tilde{\Gamma}_Q)$-module is the direct sum of its localizations.

The next theorem (see also theorem 5) was first proved, for $\Gamma_0(N)$ and N prime, by Mazur, and then generalized by Tilouine, Ribet, Gross, Edixhoven, and Wiles (see [W], theorem 2.1, and [Ti]). It is a "multiplicity one" result for certain finite Hecke modules (killed by p). The Gorenstein-ness of the Hecke algebra, which is a consequence of the complete intersection property in theorem 1, is known to follow from it. However, the Gorenstein property itself is not directly used in the Taylor-Wiles proof of theorem 1 in the minimal case (although it is heavily used in the passage from minimal to non-minimal $\mathcal{D}$). Let $Y_Q \subset X_Q$ be the open modular curve which is obtained from X_Q when we delete the cusps.

Theorem 12. *The following modules are finite free over T_Q:*

- $\mathrm{Ta}_p(J_Q)_{\mathfrak{m}_Q} = (\mathrm{Ta}_p(J_Q) \otimes_{\mathbb{Z}_p} \mathcal{O}) \otimes_{T(\Gamma_Q)} T_Q$ *— free of rank 2*
- $H^1(X_Q, \mathcal{O})^{\pm}_{\mathfrak{m}_Q} = H^1(Y_Q, \mathcal{O})^{\pm}_{\mathfrak{m}_Q}$ *($\pm$ refers to the action of complex conjugation on the modular curve) — each free of rank 1.*

Similar statements hold for $\tilde{T}_Q$. □

The theorem, as well as the fact that we can replace the closed curve X by the open curve Y, rely on the irreducibility of $\bar{\rho}$ (so $\mathfrak{m}_Q$ is not an "Eisenstein" prime).

[3] We follow the proof in [TW]. F. Diamond kindly informed us that in the final version of [DDT] there is a new argument that directly uses the q-expansion principle instead of "multiplicity one."

Recall that $T = T_\emptyset = \tilde{T}_\emptyset$ is the localization of $T(\Gamma_0(N))$ at $\mathfrak{m} = \mathfrak{m}_\emptyset$. While there is no map from $T(\Gamma_0(NQ))$ to $T(\Gamma_0(N))$ (the U_q for $q|Q$ do not preserve $S_2(\Gamma_0(N), \mathcal{O}) \subset S_2(\Gamma_0(NQ), \mathcal{O})$), the next lemma shows that after we localize at $\tilde{\mathfrak{m}}_Q$ and $\mathfrak{m}$, such a map exists, and in fact is an isomorphism.

Lemma 13. *There exists an isomorphism $\tilde{T}_Q \cong T$, mapping the Hecke operators T_l $(l \nmid N)$ and U_l $(l|N)$ in $\tilde{T}_Q$ to the corresponding operators in T.*

Proof: (1) In the first step one uses the fact that $\alpha_q \neq \beta_q$ and $q \equiv 1 \bmod p$, to show that $\mathrm{Ta}_p(\tilde{J}_Q)_{\tilde{\mathfrak{m}}_Q}$ is "Q-old." By "Q-old" we mean that $\mathrm{Ta}_p(\tilde{J}_Q)_{\tilde{\mathfrak{m}}_Q}$, which is a direct summand of $\mathrm{Ta}_p(\tilde{J}_Q) \otimes_{\mathbb{Z}_p} \mathcal{O}$, is contained in the Tate module of the Q-old subvariety of $\tilde{J}_Q$. This Q-old subvariety is isogenous to a product of 2^r copies of $J = J_0(N)$.

One can prove (1) by computing the module of fusion between the Q-old and Q-new parts of the Jacobian as in [W], proposition 2.4' (p. 503, see the remark at the end of the section there). For simplicity let us illustrate the proof when $Q = \{q\}$ consists of one prime (the proof of the general case is the same, proceeding one q at a time). Let $\tilde{J}_Q^{\mathrm{old}}$ and $\tilde{J}_Q^{\mathrm{new}}$ be the old and new subvarieties of $\tilde{J}_Q$. Thus, if $\mu : J^2 \to \tilde{J}_Q$ is the map derived by Pic functoriality from the two degeneracy maps $X_0(Nq) \to X_0(N)$, and $\hat{\mu} : \tilde{J}_Q \to J^2$ is the dual map, $\tilde{J}_Q^{\mathrm{old}}$ is the image of μ, and $\tilde{J}_Q^{\mathrm{new}}$ is the connected component of $\mathrm{Ker}(\hat{\mu})$. Both $\tilde{J}_Q^{\mathrm{new}}$ and $\tilde{J}_Q^{\mathrm{old}}$ are stable under the Hecke algebra, including the U_q-operator. Consider $J[p^\infty]_\mathfrak{m} = (J[p^\infty] \otimes_{\mathbb{Z}_p} \mathcal{O}) \otimes_{T(\Gamma)} T$ (recall $T = T(\Gamma)_\mathfrak{m}$ is the localization of the Hecke algebra at $\mathfrak{m}$) and suppose that

$$J[p^\infty]_\mathfrak{m}^2 \cap \mathrm{Ker}(\hat{\mu} \circ \mu) = \{0\}.$$

Then the relevant module of fusion

$$\mathcal{F} = \tilde{J}_Q^{\mathrm{old}}[p^\infty]_{\tilde{\mathfrak{m}}_Q} \cap \tilde{J}_Q^{\mathrm{new}}[p^\infty]_{\tilde{\mathfrak{m}}_Q} = \{0\},$$

for it is contained in $\mu(J[p^\infty]_\mathfrak{m}^2) \cap \mathrm{Ker}(\hat{\mu})$. In particular

$$\tilde{J}_Q^{\mathrm{old}}[\tilde{\mathfrak{m}}_Q] \cap \tilde{J}_Q^{\mathrm{new}}[\tilde{\mathfrak{m}}_Q] = \{0\}$$

(where $J[\mathfrak{m}]$ is the kernel of $\mathfrak{m}$ in $J[p] \otimes k$). However, it follows from theorem 12 that $\dim_k \tilde{J}_Q[\tilde{\mathfrak{m}}_Q] = 2$. Equivalently, the multiplicity of $\bar{\rho}$ in the semisimplification of this Galois module is 1. Since $\bar{\rho}$ is a constituent (i.e., a subquotient Galois module) of $\tilde{J}_Q^{\mathrm{old}}[p^\infty]_{\tilde{\mathfrak{m}}_Q}$, hence of $\tilde{J}_Q^{\mathrm{old}}[\tilde{\mathfrak{m}}_Q]$, it is *not* a constituent of $\tilde{J}_Q^{\mathrm{new}}[\tilde{\mathfrak{m}}_Q]$ or of $\tilde{J}_Q^{\mathrm{new}}[p^\infty]_{\tilde{\mathfrak{m}}_Q}$, which means that $\tilde{J}_Q^{\mathrm{new}}[p^\infty]_{\tilde{\mathfrak{m}}_Q} = 0$, as desired.

Clearly $J[p^\infty]_\mathfrak{m}^2 \cap \mathrm{Ker}(\hat{\mu} \circ \mu)$ is a finite group, and has a decomposition series all of whose factors are killed by $\mathfrak{m}$, so to prove $J[p^\infty]_\mathfrak{m}^2 \cap \mathrm{Ker}(\hat{\mu} \circ \mu) = \{0\}$, it will be enough to show that $J[\mathfrak{m}]^2 \cap \mathrm{Ker}(\hat{\mu} \circ \mu) = 0$. An explicit calculation gives

$$\hat{\mu}\circ\mu = \begin{pmatrix} q+1 & T_q \\ T_q & q+1 \end{pmatrix} \in \mathrm{End}(J^2) = M_2(\mathrm{End}(J)).$$

Since $q \equiv 1 \bmod \mathfrak{m}$ and $T_q \equiv \alpha_q + \beta_q \bmod \mathfrak{m}$, the matrix representing $\hat{\mu}\circ\mu$ on $J[\mathfrak{m}]^2$ is $\begin{pmatrix} 2 & t \\ t & 2 \end{pmatrix}$ where $t = \alpha_q + \beta_q$. This matrix is non-singular since $\alpha_q\beta_q = q = 1$ and $\alpha_q \neq \beta_q$ imply $t \neq \pm 2$.

An alternative proof goes as follows. A newform g which is new at q must be "special" there (because q divides the level to the first power, compare theorem 6 (2)). The local representation $\rho_{g,\lambda}|G_q$ therefore has the form described in theorem 6. Since $q \equiv 1 \bmod p$, the character ω is trivial on G_q, and we find that $\bar{\rho}_{g,\lambda}|G_q \sim \begin{pmatrix} \chi & * \\ & \chi \end{pmatrix}$. On the other hand if g belongs to $\tilde{\mathfrak{m}}_Q$ then $\bar{\rho}_{g,\lambda} = \bar{\rho}$, and the assumption was that the two eigenvalues of $\bar{\rho}(\mathrm{Frob}_q)$ are distinct. This contradiction proves that all the newforms belonging to $\tilde{\mathfrak{m}}_Q$ are Q-old.

(2) On the Q-old forms the Hecke algebra is simply

$$T(\Gamma_0(N))[u_q]/(u_q^2 - T_q u_q + \langle q\rangle q)$$

(one variable for each q in Q). Since the roots of $u_q^2 - T_q u_q + \langle q\rangle q$ are distinct modulo $\mathfrak{m}_\emptyset$, and since $U_q - \beta_q$ is contained in $\tilde{\mathfrak{m}}_Q$, Hensel's lemma shows that after we localize at $\tilde{\mathfrak{m}}_Q$ we get $T_\emptyset$. □

3.2. **Freeness of the Hecke algebra over Λ_Q.** Let $\tilde{Y}_Q$ be the open modular curve associated to $\tilde{\Gamma}_Q$. As we shall prove below, $H^1(Y_Q,\mathcal{O})^-$ is a free Λ_Q-module of rank equal to the $\mathcal{O}$-rank of $H^1(\tilde{Y}_Q,\mathcal{O})^-$. It follows from theorem 12 that T_Q is a free Λ_Q-module, whose rank is equal to the $\mathcal{O}$-rank of $\tilde{T}_Q$, and invoking lemma 13 we get proposition 10.

Proposition 14. *$H^1(Y_Q,\mathcal{O})^-$ is a free Λ_Q-module of rank equal to the $\mathcal{O}$-rank of $H^1(\tilde{Y}_Q,\mathcal{O})^-$.*

Proof: First, let us *pretend* that $\tilde{\Gamma}_Q$ had no elliptic elements. Then $\tilde{\Gamma}_Q = \pi_1(\tilde{Y}_Q)$ is free as the fundamental group of the incomplete curve. By Shapiro's lemma one may identify

$$H^1(Y_Q,\mathcal{O}) = H^1(\Gamma_Q,\mathcal{O}) = H^1(\tilde{\Gamma}_Q,\mathcal{O}[\Delta_Q]). \tag{3.1}$$

The action of Δ_Q, the deck-transformation group of $Y_Q/\tilde{Y}_Q$, on $H^1(Y_Q,\mathcal{O})$, corresponds in this isomorphism to the action of $\tilde{\Gamma}_Q/\Gamma_Q$ on $H^1(\Gamma_Q,\mathcal{O})$ through conjugation. Via the Shapiro isomorphism this, in turn, gets translated to the action of Δ_Q on the coefficients of $H^1(\tilde{\Gamma}_Q,\mathcal{O}[\Delta_Q])$. Since $\tilde{\Gamma}_Q$ is free, the module of 1-cocycles $Z^1(\tilde{\Gamma}_Q,\mathcal{O}[\Delta_Q])$ is a free $\mathcal{O}[\Delta_Q]$-module. In fact, choosing g free generators for $\tilde{\Gamma}_Q$, say $\gamma_1,\ldots,\gamma_g$, the map

$$c \longmapsto (c(\gamma_1),\ldots,c(\gamma_g)) \tag{3.2}$$

is an isomorphism of $Z^1(\tilde{\Gamma}_Q, \mathcal{O}[\Delta_Q])$ onto $\mathcal{O}[\Delta_Q]^g$, because a cocycle can be fixed arbitrarily on the generators, and then it is determined on any word.

The action of complex conjugation on $H^1(Y_Q, \mathcal{O})$ is deduced from the map $z \mapsto -\bar{z}$ on the upper half plane. If $\gamma \in \Gamma_Q$ gets mapped in $\pi_1(Y_Q)$ to the path $[\gamma]$ which is the projection (in Y_Q) of the geodesic $\{i, \gamma(i)\}$ connecting i to $\gamma(i)$ in the upper half plane, then the geodesic $\{-\bar{i}, -\overline{\gamma(i)}\} = \{i, \xi\gamma\xi^{-1}(i)\}$ projects to the path $[\xi\gamma\xi^{-1}]$, which is its complex conjugate. Here $\xi = \begin{pmatrix} -1 & \\ & 1 \end{pmatrix}$, and as a base point for the fundamental group we took the image of i. The action of complex conjugation on $H^1(Y_Q, \mathcal{O})$ therefore corresponds to the action of ξ on $H^1(\Gamma_Q, \mathcal{O})$ through conjugation. Via the Shapiro isomorphism it gets translated to an action of ξ on $H^1(\tilde{\Gamma}_Q, \mathcal{O}[\Delta_Q])$, where ξ is still acting by conjugation on the group, and trivially on the coefficients. This action can be defined already at the level of 1-cocycles and 1-coboundaries. Clearly $Z^1(\tilde{\Gamma}_Q, \mathcal{O}[\Delta_Q])^-$ is free over $\mathcal{O}[\Delta_Q]$, as a direct factor of a free module. About $B^1(\tilde{\Gamma}_Q, \mathcal{O}[\Delta_Q])$ we need not worry because the 1-coboundaries are in the + eigenspace for the action of ξ (since $\xi \begin{pmatrix} a & b \\ c & d \end{pmatrix} \xi^{-1} = \begin{pmatrix} a & -b \\ -c & d \end{pmatrix}$). It follows that $H^1(\tilde{\Gamma}_Q, \mathcal{O}[\Delta_Q])^-$ is free over $\Lambda_Q = \mathcal{O}[\Delta_Q]$, and

(3.3)
$$H^1(\tilde{\Gamma}_Q, \mathcal{O}[\Delta_Q])^- / \mathfrak{a}_Q H^1(\tilde{\Gamma}_Q, \mathcal{O}[\Delta_Q])^- = H^1(\tilde{\Gamma}_Q, \mathcal{O})^- = H^1(\tilde{Y}_Q, \mathcal{O})^-$$

from where the statement about the rank follows.

Unfortunately, $\tilde{\Gamma}_Q$ might have elliptic elements. This is remedied by introducing an auxiliary prime $R > 3$, and replacing Γ_Q and $\tilde{\Gamma}_Q$ by $\Gamma'_Q = \Gamma_Q \cap \Gamma_1(R)$ and $\tilde{\Gamma}'_Q = \tilde{\Gamma}_Q \cap \Gamma_1(R)$ respectively. We contend ourselves with a brief sketch of how one modifies the arguments above, which now apply *verbatim* to Γ'_Q and $\tilde{\Gamma}'_Q$. To deduce proposition 10 for the original Γ_Q and $\tilde{\Gamma}_Q$, we need to find a maximal ideal $\mathfrak{m}'_Q$ (resp. $\tilde{\mathfrak{m}}'_Q$) in $T(\Gamma'_Q)$ (resp. $T(\tilde{\Gamma}'_Q)$) such that localization at $\mathfrak{m}'_Q$ yields an *isomorphism* $T'_Q \cong T_Q$ (resp. $\tilde{T}'_Q \cong \tilde{T}_Q$). In the language of congruences between modular forms we want to make sure that there are no congruences between R-new and R-old forms that belong to $\mathfrak{m}'_Q$. The computations of Ribet and Wiles, similar to those in lemma 13, show that this will be the case as long as

- $\{1, \alpha_R, \beta_R, \alpha_R\beta_R = R\}$ are 4 distinct elements of k, where as before $\{\alpha_R, \beta_R\}$ are the two eigenvalues of Frob_R.

At this point there is a slight inaccuracy in [T-W], where lemma 3 of [D-T] is misquoted (this was pointed out to us by F. Diamond). One way to overcome the difficulty is to change R to R^2 (see [D-D-T] for details). However, for the purpose of proving the Shimura-Taniyama-Weil conjecture, it is enough to consider the case $k = \mathbb{F}_p, p \geq 3$. The following lemma (combined with the Čebotarev density theorem) therefore serves our purpose.

Lemma *Let $\bar{\rho} : G_{\mathbb{Q}} \to GL_2(\mathbb{F}_p), p \geq 3$ satisfy the assumptions of Section 2.1. Then there exists a $\sigma \in G_{\mathbb{Q}}$ such that if α and β are the two eigenvalues of $\bar{\rho}(\sigma), \{1, \alpha, \beta, \alpha\beta = \omega(\sigma)\}$ are all distinct.*

Proof: If $p = 3$, a case-by-case analysis reveals that $\mathrm{Im}(\bar{\rho})$ contains a matrix of determinant -1 and trace ± 1 satisfying the requirements of the lemma.

If $p > 3$ let $R \in \mathbb{F}_p^\times$ be a primitive root mod p and choose σ with $\omega(\sigma) = R$. If α, β are the two eigenvalues of $\bar{\rho}(\sigma)$ and they are not in $\mathbb{F}_p$ then we are done, because they must be distinct as conjugates of each other. If they are in $\mathbb{F}_p$, again they must be distinct because R is not a square. Suppose $\{\alpha, \beta\} = \{1, R\}$. Then in a suitable basis $\bar{\rho}(\sigma) = \begin{pmatrix} 1 & 0 \\ 0 & R \end{pmatrix}$. Let $L = \mathbb{Q}(\zeta_p)$ and pick $\bar{\rho}(\tau) = \begin{pmatrix} a & b \\ c & d \end{pmatrix} \in \bar{\rho}(G_L)$ with $bc \neq 0$. If this is not possible, $\bar{\rho}(G_L)$ is contained in Borel subgroup, so either $\bar{\rho}(G_{\mathbb{Q}})$ is reducible, or it is (generalized) dihedral, and both cases are ruled out by our assumptions. Conjugating $\bar{\rho}(\tau)$ by a power of $\bar{\rho}(\sigma)$ we get in $\mathrm{Im}(\bar{\rho})$ every matrix of the form $\begin{pmatrix} a & xb \\ x^{-1}c & d \end{pmatrix}, x \in \mathbb{F}_p^\times$ (since R was a primitive root modulo p), hence $\mathrm{Im}(\bar{\rho})$ contains

$$u = \begin{pmatrix} 1 & 0 \\ 0 & R \end{pmatrix} \begin{pmatrix} a & b \\ c & d \end{pmatrix} \begin{pmatrix} a & xb \\ x^{-1}c & d \end{pmatrix} = \begin{pmatrix} a^2 + bx^{-1}c & * \\ * & R(d^2 + cxb) \end{pmatrix}$$

and $\mathrm{tr}(u) = a^2 + bcx^{-1} + R(d^2 + bcx)$. Since $p > 3$ we can find an x such that $\mathrm{tr}(u) \neq 1 + R$ and this finishes the proof of the lemma. □

4. Proof of Proposition 11 — On the structure of the universal deformation ring

4.1. The Selmer group. For the concepts defined in this section, see also Washington's paper in this volume [Wa]. Let $\bar{\rho}$ be as in 2.1, and define

(4.1) $\quad W = \mathrm{ad}^0(\bar{\rho})$ (trace-0 matrices in the adjoint representation of $\bar{\rho}$)

(4.2) $$W^* = \mathrm{Hom}(W, \mu_p) \cong W(1) \cong \mathrm{Symm}^2(\bar{\rho})$$

(the last isomorphism stems from

$$\mathrm{ad}(\bar{\rho}) \cong \bar{\rho} \otimes \bar{\rho}^\vee \cong \bar{\rho} \otimes \bar{\rho}(-1) \cong \mathrm{Symm}^2(\bar{\rho})(-1) \oplus k,$$

hence $W \cong \mathrm{Symm}^2(\bar{\rho})(-1)$). If $\bar{\rho}$ is "Selmer," U is the space of $\bar{\rho}$, and $0 \to U^0 \to U \to U^1 \to 0$ is the filtration defining the Selmer condition at p (so that U^i are modules for $\bar{\rho}|G_p$ and U^0 is the ω-eigenspace for I_p), put

(4.3) $$W^0 = \mathrm{Hom}(U/U^0, U^0) \subset W^1 = \mathrm{Hom}^0((U, U^0), (U, U^0)) \subset W.$$

Thus W^1 is the submodule of upper-triangular, trace-0 matrices, and W^0 consists of those whose diagonal vanishes. Note that G_p acts on W^0 through $\omega\psi^{-2} = \omega$ (see remark 2.1(ii)), trivially on W^1/W^0, and through $\omega^{-1}\psi^2 = \omega^{-1}$ on W/W^1.

We shall next define the "local conditions," which are subgroups L_v of $H^1(G_v, W)$ for the various decomposition groups G_v. We shall later consider global cohomology classes whose restriction to every G_v falls into L_v. Since $p \geq 3$, $H^1(G_\infty, W) = 0$, and we only have to define these local conditions at the finite places. Let

- $L_l = H^1(G_l/I_l, W^{I_l})$ if $l|N(\bar\rho)$
- $L_p = H^1_{Se}(G_p, W) := \mathrm{Ker}(H^1(G_p, W) \to H^1(I_p, W/W^0))$ if $\bar\rho$ is Selmer but not flat; and $L_p = H^1_{\mathrm{fl}}(G_p, W)$ if $\bar\rho$ is flat.
- $L_q = H^1(G_q, W)$ if $q|Q$.

Some explanations are in order. First, we did not define L_l for primes $l \notin \Sigma \cup Q$, but they could be defined there too as $H^1(G_l/I_l, W^{I_l}) = H^1(G_l/I_l, W)$. Instead, we shall assume in (4.5) below that our cohomology classes are unramified outside $\Sigma \cup Q$, which, of course, is a synonimous condition.

At $l|N(\bar\rho)$ our local condition confirms with [W], p.461, because

$$H^1(G_l/I_l, W^{I_l}) = \mathrm{Ker}(H^1(G_l, W) \to H^1(G_l, W/W^{I_l})).$$

To check this use the exact diagram

$$(4.4)\quad \begin{array}{ccccc} H^1(G_l/I_l, W^{I_l}) & \hookrightarrow & H^1(G_l, W^{I_l}) & \to & H^1(I_l, W^{I_l}) \\ \| & & \downarrow & & \downarrow 0 \\ H^1(G_l/I_l, W^{I_l}) & \hookrightarrow & H^1(G_l, W) & \to & H^1(I_l, W) \\ & & \downarrow & & \downarrow \\ & & H^1(G_l, W/W^{I_l}) & \to & H^1(I_l, W/W^{I_l}) \end{array}$$

To justify the arrow labeled with a 0 note that it is part of the long exact sequence

$$0 \to W^{I_l} \simeq W^{I_l} \to (W/W^{I_l})^{I_l} \to H^1(I_l, W^{I_l}) \to H^1(I_l, W)$$

and both $(W/W^{I_l})^{I_l}$ and $H^1(I_l, W^{I_l}) = \mathrm{Hom}(I_l, k)$ are one-dimensional over k.

At the auxiliary $q's$ we simply took L_q to be the whole H^1, which means that there is no local condition imposed there.

The most subtle condition is the one imposed at p. If $\bar\rho$ is Selmer non-flat the definition is analogous to that of L_l for $l|N(\bar\rho)$. If $\bar\rho$ is flat, then $H^1_{\mathrm{fl}}(G_p, W)$ has the same meaning as in [Co]. Let us recall that there is a canonical isomorphism between $H^1(G_p, \mathrm{ad}(\bar\rho))$ and $\mathrm{Ext}^1_{k[G_p]}(\bar\rho, \bar\rho)$, the Ext group being computed in the category of $k[G_p]$-modules. Since $\bar\rho|G_p$ is assumed to be the representation attached to a finite flat group scheme Γ (say), one may ask whether a specific Galois module, which is an extension of $\bar\rho$ by $\bar\rho$, in fact arises from a finite flat group scheme over $\mathbb{Z}_p$ which is an extension of Γ by Γ in the category of group schemes over $\mathbb{Z}_p$. This need not be the case in general, and one denotes $H^1_{\mathrm{fl}}(G_p, \mathrm{ad}(\bar{\rho}))$ those cohomology classes whose corresponding extension arises in such a manner. Then $H^1_{\mathrm{fl}}(G_p, W) = H^1_{\mathrm{fl}}(G_p, \mathrm{ad}(\bar{\rho})) \cap H^1(G_p, W)$. While this definition of the local condition at p is appealing intuitively, it is not directly amenable to computations in Galois cohomology. For computational purposes it is important to have a "linear-algebra" criterion for an extension class to be "flat," and this is supplied by Fontaine's work on Honda systems (as explained in [Co]), or by later developments due to Fontaine and Lafaille (as used in [W]).

Definition 5. The "Selmer group" is

$$H^1_{\mathcal{D}_Q}(\mathbb{Q}, W) = \mathrm{Ker}\left(H^1(G_{\Sigma\cup Q}, W) \to \prod_{v\in\Sigma\cup Q} H^1(G_v, W)/L_v \right) \tag{4.5}$$

In other words, it is the group of cohomology classes which at every place v satisfy the "local condition" L_v.

The importance of the Selmer group stems from the fact that it is canonically isomorphic to the *reduced tangent space* of the deformation problem. Recall that we denoted by R_Q the universal deformation ring for deformations of $\bar{\rho}$ of type $\mathcal{D}_Q$, and that, for notational convenience, we drop the subscript Q if Q is empty.

Theorem 15. ([M2], [M3]) *There is a canonical isomorphism*

$$H^1_{\mathcal{D}}(\mathbb{Q}, W) \cong \mathrm{Hom}(\mathfrak{m}_R/(\lambda, \mathfrak{m}_R^2), k). \tag{4.6}$$

A similar identity holds with R_Q and $\mathcal{D}_Q$. In particular, R_Q can be topologically generated as an $\mathcal{O}$-algebra by $r(Q) = \dim_k H^1_{\mathcal{D}_Q}(\mathbb{Q}, W)$ elements, and this is the minimal number of generators.

Proof: The last assertion, about the number of generators needed to generate R_Q as a topological $\mathcal{O}$-algebra, is a consequence of (4.6) and Nakayama's lemma. Let us prove (4.6). A k-linear homomorphism from $\mathfrak{m}_R/(\lambda, \mathfrak{m}_R^2)$ to k defines a local $\mathcal{O}$-algebra homomorphism from R to the ring of dual numbers $k[\epsilon]$ ($\epsilon^2 = 0$), and vice versa. It therefore corresponds to ρ, an infinitesimal deformation of $\bar{\rho}$ (i.e., a deformation with values in $GL_2(k[\epsilon])$), which is unique up to strict equivalence. Writing

$$\rho(g)\bar{\rho}(g)^{-1} = 1 + \epsilon c_\rho(g), \tag{4.7}$$

$c_\rho(g)$ becomes a 1-cocycle in W for the adjoint action, namely

$$c_\rho(gh) = c_\rho(g) + \mathrm{ad}(\bar{\rho})(g)c_\rho(h) \tag{4.8}$$

($c_\rho(g)$ has trace zero because $\det(\rho) = \det(\bar{\rho}) = \omega$). Replacing ρ by a strictly equivalent deformation has the effect of changing c_ρ by a coboundary, and every cohomology class is obtained in such a way from an infinitesimal deformation of $\bar{\rho}$. This is the construction which associates to an infinitesimal deformation of $\bar{\rho}$ a class $[c_\rho]$ in $H^1(\mathbb{Q}, W)$. It remains to check that ρ is of type $\mathcal{D}$ if and only if $[c_\rho] \in H^1_{\mathcal{D}}(\mathbb{Q}, W)$.

Assume first that $l|N(\bar{\rho})$, and ρ is of type $\mathcal{D}$. Then for $g \in I_l$ we have $c_\rho(g) \in \left\{ \begin{pmatrix} 0 & b \\ 0 & 0 \end{pmatrix} \middle| b \in k \right\} = W^{I_l}$, so the image of $[c_\rho]$ in $H^1(I_l, W/W^{I_l})$ is 0. In view of (4.4) (the injectivity of $H^1(I_l, W) \to H^1(I_l, W/W^{I_l})$) , $[c_\rho]$ satisfies the local condition at l. Conversely, if this condition is satisfied $\rho|I_l$ is upper triangular, hence so is $\rho|G_l$ (I_l being normal in G_l), and it follows that it has the shape of definition 2.1.

The argument at p, in the case that $\bar{\rho}$ is Selmer non-flat, is entirely analogous, W^0 replacing W^{I_l}. In the flat case, the definition of $H^1_{\text{fl}}(G_p, W)$ makes the desired result a tautology. Finally at $l \notin \Sigma$, ρ is unramified if and only if $[c_\rho]$ is unramified.

A similar proof works when Q is not necessarily empty. □

4.2. The dual Selmer group. By local Tate duality, at each place v, $H^1(G_v, W)$ and $H^1(G_v, W^*)$ are dual abelian groups. We let L_v^* be the exact annihilator of L_v. At $l|N(\bar{\rho})$ we have $L_l^* = H^1(G_l/I_l, W^{*I_l})$, and at $q|Q$ we simply have $L_q^* = 0$. The dual Selmer group is defined by

(4.9)

$$H^1_{\mathcal{D}_Q*}(\mathbb{Q}, W^*) = \mathrm{Ker}\left(H^1(G_{\Sigma\cup Q}, W^*) \to \prod_{v \in \Sigma \cup Q} H^1(G_v, W^*)/L_v^* \right).$$

The central observation concerning the Selmer group and the dual Selmer group is that while their orders can be pretty hard to compute, their ratios are expressible as a product of local terms. The next theorem is a corollary of Tate's global duality (the Tate-Poitou exact sequence). See [Gr] and [Wa] in this volume for its derivation.

Theorem 16. *Notation as above*

$$\frac{\#H^1_{\mathcal{D}}(\mathbb{Q}, W)}{\#H^1_{\mathcal{D}*}(\mathbb{Q}, W^*)} = \frac{\#H^0(\mathbb{Q}, W)}{\#H^0(\mathbb{Q}, W^*)} \prod_{v \in \Sigma\infty} \frac{\#L_v}{\#H^0(G_v, W)} \tag{4.10}$$

A similar result holds with $\mathcal{D}_Q$ replacing $\mathcal{D}$, and $\Sigma \cup Q$ replacing Σ. Wiles' idea was that with a cleverly chosen set Q one can make $H^1_{\mathcal{D}_Q*}(\mathbb{Q}, W^*)$ vanish, and so control $H^1_{\mathcal{D}_Q}(\mathbb{Q}, W)$, hence $r(Q)$. To this end we have to compute the local terms that intervene in the right hand side of the formula.

4.3. Computation of the local terms.

- $\#L_p = \#H^0(G_p, W) \cdot \#k$

In the flat case this computation is quite delicate. The main point is that if $\bar{\rho}|G_p$ is absolutely irreducible, $\dim H^1_{\text{fl}}(G_p, \mathrm{ad}(\bar{\rho})) = 2$. This was proved by Ramakrishna in his thesis [Ra] using the theory of Fontaine-Lafaille (alternatively, one can use Fontaine's theory of "Honda systems," see [Co]). It follows that $L_p = H^1_{\text{fl}}(G_p, \mathrm{ad}^0(\bar{\rho}))$ is 1-dimensional, which is the desired result, since if $\bar{\rho}|G_p$ is irreducible, $H^0(G_p, W)$ vanishes. See [Ra], [Co] or [D-D-T] (section 2.5) for full details, and a treatment of the case where $\bar{\rho}|G_p$ is reducible as well.

In the Selmer non-flat case we have the filtration $0 \subset W^0 \subset W^1 \subset W$ as explained in 4.1 above, so $H^0(G_p, W) = 0$, and we have to show that $L_p = \mathrm{Ker}(H^1(G_p, W) \to H^1(I_p, W/W^0))$ is one-dimensional. The key point is that $L_p = \mathrm{Ker}(H^1(G_p, W) \to H^1(G_p, W/W^0))$ (compare the situation at $l|N(\bar{\rho})$, diagram (4.4)). Denote for the moment the latter group by $L'_p \subset L_p$. L_p is the tangent space to the (local) deformation problem classifying deformations of $\bar{\rho}|G_p$ of the form $\begin{pmatrix} \omega\tilde{\psi}^{-1} & * \\ & \tilde{\psi} \end{pmatrix}$ where $\tilde{\psi}$ is an unramified character of G_p. On the other hand L'_p is the tangent space to the (local) deformation problem classifying deformations of $\bar{\rho}|G_p$ of the form $\begin{pmatrix} \omega\psi^{-1} & * \\ & \psi \end{pmatrix}$ where ψ is the character appearing in $\bar{\rho}$. As explained in remark 2.2(i), the assumption that $\bar{\rho}$ is *non-flat* implies that in any deformation $\tilde{\psi} = \psi$. (In the language of [W], any Selmer deformation is strict, see [W], proposition 1.1, p.459, and [Dil], 6.1.) Hence the two tangent spaces are the same, and $L'_p = L_p$. Alternatively, one can use a diagram similar to (4.4)

$$
(4.11)\qquad
\begin{array}{ccc}
H^1(G_p, W^0) & \to & H^1(I_p, W^0)^{G_p/I_p} \\
\downarrow & & \downarrow \\
H^1(G_p, W) & \to & H^1(I_p, W)^{G_p/I_p} \\
\downarrow & & \downarrow \\
H^1(G_p, W/W^0) & \to & H^1(I_p, W/W^0).
\end{array}
$$

Since $W^{I_p} = 0$, the horizontal maps on the top and in the middle are isomorphisms. This immediately implies $L_p = L'_p$. Now compute

$$
\begin{aligned}
H^0(G_p, W/W^0) &= W^1/W^0 \cong k, \\
H^1(G_p, W^0) &= H^1(G_p, \mu_p \otimes k) = \mathbb{Q}_p^\times/\mathbb{Q}_p^{\times p} \otimes k \cong k^2,
\end{aligned}
$$

and the desired result comes out from the long exact sequence in cohomology associated to $0 \to W^0 \to W \to W/W^0 \to 0$.

- $\# L_\infty = \# H^0(G_\infty, W)/\# k = 1$

Here $L_\infty = H^1(G_\infty, W) = (0)$ by definition. Since the eigenvalues of complex conjugation on $\bar{\rho}$ are ± 1, the eigenvalues on W are $\{-1, -1, +1\}$, and the second equality follows.

- $\# L_l = \# H^0(G_l, W)$ for $l|N(\bar{\rho})$

This follows from the exact sequence

(4.12)
$$0 \to H^0(G_l, W) \to H^0(I_l, W) \to H^0(I_l, W) \to H^1(G_l/I_l, W^{I_l}) \to 0$$

where the arrow in the middle is $\mathrm{Frob}_l - 1 : W^{I_l} \to W^{I_l}$.

- $H^0(\mathbb{Q}, W) = H^0(\mathbb{Q}, W^*) = 0$

For W this follows from the absolute irreducibility of $\bar{\rho}$, since by Schur's lemma the only endomorphisms commuting with the Galois action are the scalars, which are missing from W.

For W^* the argument is a little different. A Galois-invariant vector in W^* is, by (4.2), an invariant symmetric bilinear form. If this bilinear form is degenerate, then its kernel is invariant, contradicting the irreducibility of $\bar{\rho}$. If it is non-degenerate, this means that the image of $\bar{\rho}$ is contained in some orthogonal group. This already contradicts the fact that $\det(\bar{\rho}) = \omega$, unless $p = 3$ (when $\omega^2 = 1$), but in this case it contradicts the absolute irreducibility of $\bar{\rho}|G_L$, L as in 2.1. Indeed, over $\bar{k}$ (which is of odd characteristic) the invariant quadratic form could be taken to be the standard one $x^2 + y^2$, and its orthogonal group the group

$$O(2, \bar{k}) = \left\{ \begin{pmatrix} a & b \\ -b & a \end{pmatrix} \middle| a^2 + b^2 = 1 \right\} \cup \left\{ \begin{pmatrix} a & b \\ b & -a \end{pmatrix} \middle| a^2 + b^2 = 1 \right\}.$$

Then $\bar{\rho}(G_L) \subset SO(2, \bar{k}) = \left\{ \begin{pmatrix} a & b \\ -b & a \end{pmatrix} \middle| a^2 + b^2 = 1 \right\}$, which is diagonalizable.

- $\#H^0(G_q, W) = \#k$ for $q \in Q$

The eigenvalues of Frob_q on W are $\alpha_q \beta_q = q = 1$ (since $q \equiv 1 \bmod p$), α_q^2 and β_q^2. The latter two are not equal to 1 by our assumption that $\alpha_q \neq \beta_q$.

- $\#H^1(G_q, W) = \#k^2$

Here, at $q \in Q$, $H^1(G_q/I_q, W) = W/(\mathrm{Frob}_q - 1)W$ is one-dimensional (by the assumptions on α_q and β_q),

$$\begin{aligned} H^1(I_q, W)^{G_q/I_q} &= \mathrm{Hom}(\mathbb{Z}_p(1), W)^{\mathrm{Frob}_q} \\ &= W[\mathrm{Frob}_q - q] = W[\mathrm{Frob}_q - 1] = W^{\mathrm{Frob}_q} \end{aligned}$$

is again one-dimensional, and $H^2(G_q/I_q, W) = 0$ since $G_q/I_q \cong \hat{\mathbb{Z}}$. The desired result follows from the inflation-restriction exact sequence.

These computations allow us to deduce the following corollary from theorem 16.

Corollary 17. (i) $r(Q) = \dim_k H^1_{\mathcal{D}_Q}(\mathbb{Q}, W) = \dim_k H^1_{\mathcal{D}_Q *}(\mathbb{Q}, W^*) + \#(Q)$.
(ii) *When we add a new prime q as above to a given set of primes Q,* either $\dim_k H^1_{\mathcal{D}_Q *}(\mathbb{Q}, W^*)$ *drops by 1,* or $\dim_k H^1_{\mathcal{D}_Q}(\mathbb{Q}, W)$ *grows by 1.* □

To prove proposition 11 it is therefore enough to prove the following:

Proposition 18. *Let $r = \dim_k H^1_{\mathcal{D}*}(\mathbb{Q}, W^*)$. Then for every n it is possible to choose a set Q of r primes, disjoint from Σ, satisfying*

- *For every $q \in Q$, $q \equiv 1 \bmod p^n$.*
- *For every $q \in Q$, $\bar{\rho}(\mathrm{Frob}_q)$ has distinct eigenvalues in k.*
- *$H^1_{\mathcal{D}_Q *}(\mathbb{Q}, W^*) = 0$.*

It will then follow from the corollary that $r = r(Q)$, and from theorem 4.1, that R_Q can be topologically generated as an $\mathcal{O}$-algebra by r elements.
Discussion. The "miracle" occuring in the minimal case is that all the local terms in the product expressing the ratio $\#H^1_{\mathcal{D}}(\mathbb{Q}, W)/\#H^1_{\mathcal{D}*}(\mathbb{Q}, W^*)$ are 1, hence we can achieve an equality $r = r(Q)$. In the numerical criterion of lemma 9 it is crucial to know that r, the number of variables S_i (which turns out to be the number of primes in Q, see the discussion preceding proposition 10), is actually equal to $r(Q)$, the minimal number of generators of R_Q, and is not smaller. In the non-minimal case some of the local terms in (4.10) would be bigger than 1, and corollary 17(i) would look like

$$r(Q) = \dim_k H^1_{\mathcal{D}_Q}(\mathbb{Q}, W) = \dim_k H^1_{\mathcal{D}_Q*}(\mathbb{Q}, W^*) + t + \#(Q)$$

for some $t > 0$, so we would never be able to achieve $r(Q) = r$.

What remains to be checked is that we can "kill" the dual Selmer group by a clever choice of the set Q, while maintaining the other restrictions imposed on the q's in Q. The proof of this fact, given in the next section, is group-theoretical in nature, based on the Čebotarev density theorem.

5. Conclusion of the proof : some group theory

5.1. A reduction step. Since

$$H^1_{\mathcal{D}_Q*}(\mathbb{Q}, W^*) = \mathrm{Ker}\left(H^1_{\mathcal{D}*}(\mathbb{Q}, W^*) \to \prod_{q \in Q} H^1(G_q/I_q, W^*)\right),$$

it is enough to prove that for every cohomology class $0 \neq [\psi] \in H^1_{\mathcal{D}*}(\mathbb{Q}, W^*)$ (ψ is a cocycle representing the cohomology class) there are infinitely many primes q not in Σ such that

- $q \equiv 1 \bmod p^n$
- the two eigenvalues of $\bar{\rho}(\mathrm{Frob}_q)$ are distinct
- $\mathrm{res}_q([\psi]) \neq 0$.

We can then choose the q's successively, each time decreasing the dimension of $H^1_{\mathcal{D}_Q*}(\mathbb{Q}, W^*)$ by 1, until we annihilate it completely after r steps.

By Čebotarev's density theorem it is enough to find a $\sigma \in G_{\mathbb{Q}}$ such that

- $\sigma|\mathbb{Q}(\zeta_{p^n}) = 1$,
- the eigenvalues of $\bar{\rho}(\sigma)$ are distinct,
- $\psi_\sigma \notin (\sigma - 1)W^*$.

If we find such a σ, let q be a prime with $\sigma|L = \left(\frac{L/\mathbb{Q}}{\mathfrak{Q}}\right)$, where L is a Galois extension of $\mathbb{Q}$ containing $\mathbb{Q}(\zeta_{p^n})$ and the splitting fields of $\bar{\rho}$ and ψ, and $\mathfrak{Q}$ is a suitable prime above q in L. Then σ belongs to the decomposition group of q in L, so the last condition clearly implies $\mathrm{res}_q([\psi]) \neq 0$, and the first two obviously imply the other two restrictions imposed on q.

Consider the following diagram of fields

$$\left.\begin{array}{ll} F & = \text{extension of } \mathbb{Q}(\zeta_{p^n}) \text{ cut by } \operatorname{Ker}(\bar{\rho}) \\ \Big| & \\ K & = \text{extension of } \mathbb{Q}(\zeta_{p^n}) \text{ cut by } \operatorname{Ker}(\operatorname{ad}^0(\bar{\rho})) \\ \Big| H & \\ \mathbb{Q}(\zeta_{p^n}) & \end{array}\right\} \tilde{H} \subset \tilde{G} = \operatorname{Im}(\bar{\rho})$$

Since $\operatorname{Ker}\big(\operatorname{Im}(\bar{\rho}) \to \operatorname{Im}(\operatorname{ad}^0(\bar{\rho}))\big)$ are the scalars in $\operatorname{Im}(\bar{\rho})$, we have

$$H = \tilde{H}k^\times/k^\times \subset G = \operatorname{Im}(\operatorname{ad}^0(\bar{\rho})) \subset \operatorname{PGL}_2(k).$$

Lemma 19. $H^1(K/\mathbb{Q}, W^*) = 0$.

Let us postpone the proof of the lemma for the moment, and see how to use it to find a σ as above. From the lemma,

$$0 \neq \psi|G_K \in \operatorname{Hom}(G_K, W^*)^{\operatorname{Gal}(K/\mathbb{Q})}.$$

But $\psi(G_K)$ is a $G_{\mathbb{Q}}$-submodule of W^*, hence by the irreducibility of $\bar{\rho}|G_L$, which implies the irreducibility of W^* (see the argument in section 4.3), it is all of W^*.

Next we claim that it is possible to find $\sigma_0 \in G_{\mathbb{Q}(\zeta_{p^n})}$ such that $\bar{\rho}(\sigma_0)$ has distinct eigenvalues $\{\alpha, \beta\}$. If this is not the case, $\bar{\rho}(G_{\mathbb{Q}(\zeta_{p^n})})$ is contained in the upper-triangular matrices, by an elementary computation with 2×2 matrices. But then, either $\bar{\rho}(G_{\mathbb{Q}(\zeta_{p^n})})$ is contained in a torus, in which case $\bar{\rho}$ is dihedral, and therefore $\bar{\rho}|G_L$ is not irreducible, or $\bar{\rho}|G_{\mathbb{Q}(\zeta_{p^n})}$ has a unique invariant line, in which case it must be invariant under $\bar{\rho}$, contradicting its irreducibility.

The eigenvalues of σ_0 on W are α/β, 1, and β/α. These are also its eigenvalues on W^* because σ_0 fixes ζ_p. Thus $(\sigma_0 - 1)W^* \neq W^*$, and $\psi(G_K) \not\subset (\sigma_0 - 1)W^*$.

Now look for $\sigma = \tau\sigma_0$ with $\tau \in G_K$. Note that τ acts trivially on W and W^*. Since $\bar{\rho}(\tau)$ is a scalar matrix, $\bar{\rho}(\sigma)$ will still have distict eigenvalues, and clearly σ leaves the p^n roots of unity invariant. All that remains to check is that

$$\psi_\sigma = \tau\psi_{\sigma_0} + \psi_\tau = \psi_{\sigma_0} + \psi_\tau \notin (\sigma - 1)W^* = (\sigma_0 - 1)W^*. \tag{5.1}$$

If $\sigma = \sigma_0$ is not good, then by the above we can find a $\tau \in G_K$ such that adding ψ_τ will move ψ_{σ_0} out of $(\sigma_0 - 1)W^*$. □

Proof of the lemma: We need the following classification theorem, due to Dickson ([Hu], II.8.27).

Theorem 20. *Any finite subgroup of* $\operatorname{PGL}_2(\bar{k})$ *is one of the following:*
(1) *a subgroup of a Borel group,*
(2) *conjugate to* $\operatorname{PGL}_2(k')$ *or* $\operatorname{PSL}_2(k')$ *for a finite field* k',

(3) *isomorphic to the dihedral group* D_{2n} *for* $(n,p)=1$,
(4) *isomorphic to* A_4, S_4, *or* A_5. □

Let $\tilde{G} = \mathrm{Im}(\bar{\rho}) = \mathrm{Gal}(M/\mathbb{Q})$, where M is the splitting field of $\bar{\rho}$. Let $G = \mathrm{Im}(\mathrm{ad}^0(\bar{\rho})) = \mathrm{Gal}(N/\mathbb{Q})$, where N is the splitting field of $\mathrm{ad}^0(\bar{\rho})$. Let $Z = \mathrm{Ker}(\tilde{G} \to G) = \mathrm{Gal}(M/N)$, the scalar matrices in $\mathrm{Im}(\bar{\rho})$.

Case 1. $Z \neq \{\pm 1\}$. Then $\det(Z) \neq 1$, so $W^Z = W$ implies that $W^{*Z} = 0$. Now Z is cyclic of order prime to p, and F/M is a p-extension since $\mathbb{Q}(\zeta_p) \subset M \subset F = M(\zeta_{p^n})$, so Z lifts to a subgroup (still denoted Z) of $\mathrm{Gal}(F/\mathbb{Q})$. From the inflation-restriction exact sequence we get in this case $H^1(F/\mathbb{Q}, W^*) = H^1(F^Z/\mathbb{Q}, W^{*Z}) = 0$. A fortiori $H^1(K/\mathbb{Q}, W^*) = 0$.

Case 2. $Z = \{\pm 1\}$, $p > 3$. Then Z fixes $\mathbb{Q}(\zeta_p)$ and therefore $\Delta = \mathrm{Gal}(\mathbb{Q}(\zeta_p)/\mathbb{Q})$ is a quotient of $G \subset \mathrm{PGL}_2(k)$. Using the classification theorem we see that G is contained in a Borel, or has order prime to p (the other groups don't have a cyclic quotient of order $p-1$). The first option contradicts the irreducibility of $\bar{\rho}$. The second implies that $H = \mathrm{Gal}(K/\mathbb{Q}(\zeta_{p^n}))$ also has order prime to p, so $H^1(K/\mathbb{Q}, W^*) = H^1(\mathbb{Q}(\zeta_{p^n})/\mathbb{Q}, W^{*H})$. But $W^{*H} = 0$, or else W^* is reducible, contradicting the absolute irreducibility of $\bar{\rho}|G_L$.

Case 3. $Z = \{\pm 1\}$, $p = 3$. We may assume that $3|\#G$, and G is not contained in a Borel, otherwise we finish the proof as in step 2. Again, $\Delta = \mathrm{Gal}(\mathbb{Q}(\zeta_3)/\mathbb{Q})$ is a quotient of G, so this rules out $G \cong A_5$, and we are left with the possibility $G = \mathrm{PGL}_2(k')$ ($\mathrm{PSL}_2(k')$ is simple if $k' \neq \mathbb{F}_3$, and for $\mathbb{F}_3$ it is A_4, which does not have a normal subgroup of index 2. Also $S_4 = \mathrm{PGL}_2(\mathbb{F}_3)$).

If $k' = \mathbb{F}_3$ then $S_4 = \mathrm{PGL}_2(\mathbb{F}_3) = \mathrm{Gal}(N/\mathbb{Q})$ has a normal subgroup $V_4 \subset \mathrm{Gal}(N/L)$ (its unique 2-Sylow subgroup). Since $\mathrm{Gal}(K/N)$ is a 3-group, this V_4 lifts to a normal subgroup of $\mathrm{Gal}(K/L)$. Since $\bar{\rho}|G_L$ is absolutely irreducible, $W^{*V_4} = 0$, and inflation-restriction finishes the proof as before.

There remains the case where $k' = \mathbb{F}_{3^n}$ and $n > 1$. Then

$$H = \mathrm{Gal}(K/\mathbb{Q}(\zeta_{p^n})) \cong \mathrm{Gal}(N/\mathbb{Q}(\zeta_p)) \cong \mathrm{PSL}_2(k'),$$

and $W = W^*$ is the standard adjoint representation of $\mathrm{PSL}_2(k')$ on trace-0 matrices. Wiles relies here on a result of Cline Parshall and Scott [CPS] that $H^1(H, W^*) = 0$. With this the lemma is settled, and with it are concluded also the proofs of propositions 18 and 11, and of the main theorem. □

Bibliography

[Co] B. Conrad: *The Flat Deformation Functor*, this volume.

[CPS] E. Cline, B. Parshall, L. Scott: *Cohomology of finite groups of Lie type I*, Publ. Math. IHES **45** (1975), 169-191.

[DDT] H. Darmon, F. Diamond, R. Taylor: *Fermat's Last Theorem*, Current Developments in Mathematics, 1995, International Press, Boston (preliminary version).

[dSRS] B. de Smit, K. Rubin, R. Schoof: *Criteria for Complete Intersections*, this volume.

[D1] F. Diamond: *The refined conjecture of Serre*, in: Elliptic curves, modular forms and Fermat's last theorem, J. Coates and S.T. Yau (edts.), International Press, Boston (1995) 22-37.

[D2] F. Diamond: *On deformation rings and Hecke rings,* to appear.

[D3] F. Diamond: *An Extension of Wiles' Results*, this volume.

[DR] F. Diamond, K. Ribet: *ℓ-adic Modular Deformations and Wiles's "Main Conjecture"*, this volume.

[DT] F. Diamond, R. Taylor: *Lifting modular mod l representations*, Duke Math. J. **74** (1994), 253-269.

[EG] B. Edixhoven: *Serre's Conjectures*, this volume.

[Gr] R. Greenberg: *Iwasawa theory for p-adic representations*, in: Algebraic Number Theory, in honor of K. Iwasawa, Adv. Studies in Pure Math. **17**, Academic Press, San-Diego (1989), 97-137.

[Hu] B. Huppert: Endliche Gruppen I, Grundlehren Math. Wiss. **134**, Springer-Verlag, New-York, Berlin, Heidelberg (1983).

[M1] B. Mazur: *Modular curves and the Eisenstein ideal*, Publ. Math. IHES **47** (1977), 133-186.

[M2] B. Mazur: *Deforming Galois representations*, in: Galois groups over $\mathbb{Q}$, Y. Ihara, K. Ribet, J.-P. Serre edts. MSRI publications **16** Springer-Verlag, New-York, Berlin, Heidelberg (1989) 385-437.

[M3] B. Mazur: *An Introduction to the Deformation Theory of Galois Representations*, this volume.

[Ra] R. Ramakrishna: *On a variation of Mazur's deformation functor*, Comp. Math **87** (1993), 269-286.

[R1] K. Ribet: *On modular representations of* $\mathrm{Gal}(\bar{\mathbb{Q}}/\mathbb{Q})$ *arising from modular forms*, Inv. Math. **100** (1990), 431-476.

[Ru] K. Rubin: *Modularity of Mod 5 Representations*, this volume.

[Se] J.-P. Serre: *Sur les représentations modulaires de degré 2 de* $\mathrm{Gal}(\bar{\mathbb{Q}}/\mathbb{Q})$, Duke math. J. **54** (1987), 179-230.

[TW] R. Taylor, A. Wiles: *Ring-theoretic properties of certain Hecke algebras*, Ann. Math. **141** (1995), 553-572.

[Ti] J. Tilouine: *Hecke Algebras and the Gorenstein Property*, this volume.

[Wa] L. Washington: *Galois Cohomology*, this volume.

[Wi] A. Wiles: *Modular elliptic curves and Fermat's Last Theorem*, Ann. Math. **141** (1995), 443-551.

EXPLICIT FAMILIES OF ELLIPTIC CURVES WITH PRESCRIBED MOD N REPRESENTATIONS

ALICE SILVERBERG

INTRODUCTION

In Part 1 we explain how to construct families of elliptic curves with the same mod 3, 4, or 5 representation as that of a given elliptic curve over $\mathbf{Q}$. In §4 we give equations for the families in the mod 4 case. The mod 3 and mod 5 cases were given in [9] (see also [8]). The results remain true (with the same proofs) with the field of rational numbers replaced by any field whose characteristic does not divide the level.

In Part 2 we use the work of Wiles, Taylor-Wiles, and Diamond to give explicit equations for infinite families of modular elliptic curves. In §7 (see Theorem 7.3) we show how to find infinite families of modular elliptic curves with the same mod 4 representation. In §8 we prove that if E is an elliptic curve over $\mathbf{Q}$, and the torsion subgroup of $E(\mathbf{Q})$ is not cyclic of order 1, 2, 3, 6, or 9 (i.e., the torsion subgroup is cyclic of order 4, 5, 7, 8, 10, or 12 or is of the form $\mathbf{Z}/2\mathbf{Z} \times \mathbf{Z}/2N\mathbf{Z}$ for $N = 1$, 2, 3 or 4), then E is modular (see Theorem 8.1 and Corollary 8.10).

The proofs of the results in §4 use symbolic computer computations, which were done using the programs Pari and Mathematica. I would like to thank Ken Ribet and Karl Rubin for useful conversations, and the IHES for its hospitality.

Notation. Let $\mathbf{Z}$, $\mathbf{Q}$, and $\mathbf{C}$ denote, respectively, the integers, rational numbers, and complex numbers.

We will suppose that N is a positive integer, and $N \geq 3$. If E is an elliptic curve over a field k with algebraic closure $\bar{k}$, let $E[N]$ denote the kernel of multiplication by N on $E(\bar{k})$, and let $j(E)$ denote the j-invariant of E. If $F \subseteq \bar{\mathbf{Q}}$ is a number field, let $G_F = \mathrm{Gal}(\bar{\mathbf{Q}}/F)$. Let $\boldsymbol{\mu}_N$ be the $G_{\mathbf{Q}}$-module of N-th roots of unity, and let

$$e_N : E[N] \times E[N] \to \boldsymbol{\mu}_N$$

denote the Weil pairing. Let $\mathfrak{H}$ denote the complex upper half plane, and let

$$\Gamma(N) = \{\left(\begin{smallmatrix} a & b \\ c & d \end{smallmatrix}\right) \in \mathrm{SL}_2(\mathbf{Z}) : \left(\begin{smallmatrix} a & b \\ c & d \end{smallmatrix}\right) \equiv \left(\begin{smallmatrix} 1 & 0 \\ 0 & 1 \end{smallmatrix}\right) \pmod{N}\}.$$

Let I denote the 2×2 identity matrix.

Part 1. Elliptic curves with the same mod N representation

1. MODULAR CURVES AND ELLIPTIC MODULAR SURFACES OF LEVEL N

Let

$$V_N = \mathbf{Z}/N\mathbf{Z} \times \boldsymbol{\mu}_N,$$

a $G_{\mathbf{Q}}$-module, and define a $G_{\mathbf{Q}}$-equivariant pairing

$$\eta_N : V_N \times V_N \to \boldsymbol{\mu}_N$$

by

$$\eta_N((a_1, \zeta_1), (a_2, \zeta_2)) = \zeta_2^{a_1}/\zeta_1^{a_2}.$$

Denote by Y_N the (non-compact) modular curve over $\mathbf{Q}$ which parametrizes triples (E, P, C) where E is an elliptic curve, P is a point of exact order N on E, C is a cyclic subgroup of order N on E, and C and P generate $E[N]$. Let $Y(N)$ denote the modular curve which parametrizes elliptic curves with full level N structure (see [12]). If ζ is a fixed primitive N-th root of unity in $\bar{\mathbf{Q}}$, then the map

$$(E, P, C) \mapsto (E, P, Q),$$

where Q is the unique point in C such that $e_N(P, Q) = \zeta$, induces an isomorphism (defined over $\mathbf{Q}(\zeta)$) from Y_N onto one connected component of $Y(N)$. Thus $Y_N(\mathbf{C})$ is isomorphic to $\mathfrak{H}/\Gamma(N)$. Let X_N denote the compactification of Y_N. Then X_N has genus 0 if and only if $N \leq 5$ (see p. 23 of [12]).

Lemma 1.1. *The curve Y_N parametrizes isomorphism classes of pairs (E, ϕ), where E is an elliptic curve and*

$$\phi : V_N \to E[N]$$

is a group isomorphism with the property that for all u, $v \in V_N$,

$$\eta_N(u, v) = e_N(\phi(u), \phi(v)).$$

Proof. Given (E, P, C), define ϕ by $\phi(a, \zeta) = aP + Q$ for the unique $Q \in C$ such that $e_N(P, Q) = \zeta$. Conversely, given (E, ϕ), let $P = \phi(1, 1)$ and let $C = \phi(0 \times \boldsymbol{\mu}_N)$. □

There is a quasi-projective surface W_N defined over $\mathbf{Q}$, with a projection morphism

$$\pi_N : W_N \to Y_N$$

and a zero-section $Y_N \to W_N$, both defined over $\mathbf{Q}$, with N^2 sections defined over $\bar{\mathbf{Q}}$ of order dividing N, and such that the fibers of π_N correspond to the triples (E, P, C) classified by Y_N. (Note that this notation differs from that of [9], where W_N denoted a compactification.) The variety W_N can be viewed as the universal elliptic curve with level structure as above.

See [13] for the theory of elliptic modular surfaces of level N. Analytically, we have

$$W_N(\mathbf{C}) \cong (\mathfrak{H} \times \mathbf{C})/(\Gamma(N) \ltimes \mathbf{Z}^2).$$

If $\tau \in \mathfrak{H}$, then the equivalence class of τ in $\mathfrak{H}/\Gamma(N)$ corresponds to the $\mathbf{C}$-isomorphism class of the triple $(\mathbf{C}/\mathbf{Z}\tau + \mathbf{Z}, \tau/N, \langle 1/N \rangle)$. Let $W_N[N]$ denote the N^2 sections of π_N of order dividing N, viewed as a $G_{\mathbf{Q}}$-module.

2. TWISTS OF MODULAR CURVES AND ELLIPTIC MODULAR SURFACES

Let $\mathrm{Aut}(V_N, \eta_N)$ denote the group of automorphisms of V_N which preserve η_N. Suppose V is a free rank-2 module over $\mathbf{Z}/N\mathbf{Z}$ with a continuous and linear $G_{\mathbf{Q}}$-action and suppose

$$\eta : V \times V \to \boldsymbol{\mu}_N$$

is a non-degenerate alternating $G_{\mathbf{Q}}$-equivariant pairing. Fix a group isomorphism $\varphi : V_N \to V$ under which the pairing η_N corresponds to the pairing η. Then $\tau \mapsto \varphi^{-1} \circ \tau(\varphi)$ defines a cocycle on $G_{\mathbf{Q}}$ with values in $\mathrm{Aut}(V_N, \eta_N)$. By the universal property of W_N, there is a natural injective $G_{\mathbf{Q}}$-equivariant homomorphism

$$\mathrm{Aut}(V_N, \eta_N) \hookrightarrow \mathrm{Aut}(W_N).$$

There is also a natural $G_{\mathbf{Q}}$-equivariant homomorphism

$$\mathrm{Aut}(W_N) \to \mathrm{Aut}(Y_N).$$

Therefore, the above cocycle induces cocycles c and c_0 on $G_{\mathbf{Q}}$ with values in $\mathrm{Aut}(W_N)$ and in $\mathrm{Aut}(Y_N)$, respectively. Let W (respectively, Y) denote the twist of W_N (respectively, Y_N) by the cocycle c (respectively, c_0) (see [10]). Then W and Y are quasi-projective varieties defined over $\mathbf{Q}$. Up to isomorphism, W and Y are independent of the choice of φ. We obtain isomorphisms

$$\psi : W \to W_N \qquad \text{and} \qquad \psi_0 : Y \to Y_N$$

defined over $\bar{\mathbf{Q}}$, and a projection morphism $\pi : W \to Y$ defined over $\mathbf{Q}$, such that the diagram

$$\begin{array}{ccc} W & \xrightarrow{\psi} & W_N \\ \downarrow \pi & & \downarrow \pi_N \\ Y & \xrightarrow{\psi_0} & Y_N \end{array}$$

commutes, and such that for every $\tau \in G_{\mathbf{Q}}$,

$$c(\tau) = \psi \circ \tau(\psi)^{-1} \quad \text{and} \quad c_0(\tau) = \psi_0 \circ \tau(\psi_0)^{-1}.$$

It follows from the definition of W that if $t \in Y(\mathbf{C})$ and E_t is the elliptic curve $\pi^{-1}(t)$, then $E_t[N]$ and V are isomorphic as $\mathrm{Gal}(\overline{\mathbf{Q}(t)}/\mathbf{Q}(t))$-modules.

Theorem 2.1. *Suppose $N = 3$, 4, or 5, and E is an elliptic curve over $\mathbf{Q}$. Then there are infinitely many elliptic curves E' over $\mathbf{Q}$ such that $E[N]$ and $E'[N]$ are isomorphic as $G_{\mathbf{Q}}$-modules.*

Proof. Let $V = E[N]$ and let $\eta = e_N$, the Weil pairing on $E[N]$. Let W and Y be the varieties constructed as above, for V and η, and let X denote the compactification of Y. Since E is defined over $\mathbf{Q}$, $Y(\mathbf{Q})$ is nonempty. Since $N \leq 5$, X has genus 0. Therefore, X is isomorphic to $\mathbf{P}^1$, and $X(\mathbf{Q})$ and $Y(\mathbf{Q})$ are infinite. The points of $Y(\mathbf{Q})$ correspond to the desired elliptic curves E'. □

3. Models

Suppose now that $N = 3$, 4, or 5 and E is an elliptic curve over $\mathbf{Q}$ with Weierstrass model $y^2 = x^3 + ax + b$, with $a, b \in \mathbf{Q}$. We will construct a model for W, where Y and W are the twists of Y_N and W_N as in §2, with $V = E[N]$ and $\eta = e_N$. For $N = 3$, 4, and 5, let $m = 1$, 2, and 5, respectively. Then $12m = \#\mathrm{PSL}_2(\mathbf{Z}/N\mathbf{Z})$.

We can find a model for W_N (see (1) and (3) of [9] for $N = 3$ and $N = 5$, respectively, and (5) below for $N = 4$) such that for $u \in Y_N$, the fiber $E_u = \pi_N^{-1}(u)$ is of the form

$$E_u : y^2 = x^3 + a_4(u)x + a_6(u) \tag{1}$$

where $a_4(u), a_6(u) \in \mathbf{Q}[u]$ and $\deg(a_j) = jm$ for $j = 4, 6$. Let u_0 be an algebraic number such that E_{u_0} is isomorphic (over $\bar{\mathbf{Q}}$) to E. The isomorphism $\psi_0 : Y \to Y_N$ extends to an isomorphism $\psi_0 : X \to X_N$ on the compactifications. Since X and X_N are isomorphic to $\mathbf{P}^1$, the isomorphism ψ_0 can be given by a linear fractional transformation, which can be normalized so that 0 is sent to u_0. Let

$$A = \begin{pmatrix} \alpha & u_0 \\ \gamma & 1 \end{pmatrix} \in \mathrm{GL}_2(\bar{\mathbf{Q}})$$

be such a transformation. Since ψ takes a fiber $\mathcal{E}_t = \pi^{-1}(t)$ in W isomorphically onto a fiber $E_u = \pi_N^{-1}(u)$ in W_N, the isomorphism ψ takes a point $(t, x, y) \in W \subseteq \mathbf{P}^1 \times \mathbf{P}^2$ to a point of the form $(A(t), h(t)^{-2}x, h(t)^{-3}y) \in W_N \subseteq \mathbf{P}^1 \times \mathbf{P}^2$, for appropriate $h(t)$. Therefore, $(h(t)^{-2}x, h(t)^{-3}y)$ lies on $E_{A(t)}$. Using (1), it follows that (t, x, y) satisfies

$$y^2 = x^3 + h(t)^4 a_4(A(t))x + h(t)^6 a_6(A(t)). \tag{2}$$

When $t = 0$, we would like (2) to be an equation for the elliptic curve E. We will solve for $h(t)^2$, α, and γ so that

$$h(t)^4 a_4(A(t)) \quad \text{and} \quad h(t)^6 a_6(A(t)) \tag{3}$$

are in $\mathbf{Q}[t]$ (i.e., so that (2) is a model over $\mathbf{Q}$ for W) and take on the values a and b, respectively, when $t = 0$. From

$$h(0)^4 a_4(A(0)) = a \quad \text{and} \quad h(0)^6 a_6(A(0)) = b, \tag{4}$$

we can solve for $h(0)^2$. Write $h(t)^2 = h(0)^2(\gamma t + 1)^{2m}$ and substitute into (3). Then the expressions in (3) become polynomials in $\bar{\mathbf{Q}}[t]$, and have constant terms a and b, respectively. In particular, (2) is E when $t = 0$. Take any ordered pair of rational numbers (r, s) which is not a rational multiple of $(4a, 6b)$, set the coefficients of t in the polynomials in (3) equal to r and s, respectively, and solve for α and γ. With these values, (2) is a model over $\mathbf{Q}$ for W. Different choices of the pair (r, s) give rise to $\mathbf{Q}$-isomorphic elliptic surfaces.

Let $J = j(E)/1728$. The resulting model for W is of the form

$$y^2 = x^3 + af_4(J,t)x + bf_6(J,t)$$

where $f_4, f_6 \in \mathbf{Z}[J, t]$, f_4 and f_6 depend only on N, and $\deg_t(f_j) = jm$ for $j = 4, 6$.

4. LEVEL 4

We begin by writing down a model for the elliptic modular surface W_4, following [14]. We can view W_4 as a surface over $\mathbf{Q}$ or as an elliptic curve A_u over a function field in one variable. Define

$$A_u : y^2 = x(x-1)\left(x - \frac{(u+u^{-1})^2}{4}\right).$$

If $u \in \mathbf{C}$ and $u \notin \{0, 1, -1, i, -i\}$, then A_u is an elliptic curve over $\mathbf{Q}(u)$, and a Weierstrass model for A_u is given by

$$E_u : y^2 = x^3 - 27(u^8 + 14u^4 + 1)x - 54(u^{12} - 33u^8 - 33u^4 + 1). \tag{5}$$

We have

$$j(A_u) = j(E_u) = \frac{16(u^8 + 14u^4 + 1)^3}{u^4(u^4 - 1)^4}. \tag{6}$$

Let

$$P_u = \left(\frac{u^2+1}{2u^2}, \frac{1-u^4}{4u^3}\right) \in A_u[4].$$

Let C_u be the $\mathrm{Gal}(\overline{\mathbf{Q}(u)}/\mathbf{Q}(u))$-invariant cyclic subgroup of $A_u[4]$ generated by the point (of order 4)

$$\left(\frac{u^2+1}{2u}, \frac{i(u^2+1)(u-1)^2}{4u^2}\right).$$

The map $u \mapsto (A_u, P_u, C_u)$ induces a morphism $f : \mathbf{P}^1 \to X_4$ defined over $\mathbf{Q}$. The morphism $j : X_4 \to \mathbf{P}^1$ induced by $(E, P, C) \mapsto j(E)$ has degree $\#\mathrm{PSL}_2(\mathbf{Z}/4\mathbf{Z}) = 24$. By (6), the degree of the composition $j \circ f$ is 24, which is the same as the degree of j. Therefore, f is an isomorphism. Identify X_4 with $\mathbf{P}^1$ via f.

Next, we give models for the twisted surfaces W.

Theorem 4.1. *Fix an elliptic curve over* $\mathbf{Q}$*:*

$$E: \quad y^2 = x^3 + ax + b,$$

with $a, b \in \mathbf{Q}$*, and let* $J = j(E)/1728 = 4a^3/(4a^3 + 27b^2)$*. Let* $\mathcal{E}_t$ *be*

$$\mathcal{E}_t: \quad y^2 = x^3 + a(t)x + b(t), \tag{7}$$

where

$$\begin{aligned} a(t) &= ((J-1)^4(144J^2 - 56J - 7)t^8 - 48(J-1)^4(4J+1)t^7 + \\ &\quad 28(J-1)^3(4J+5)t^6 + 224(J-1)^3t^5 + 42(J-1)^2(4J-5)t^4 - \\ &\quad 112(J-1)^2t^3 + 28(J-1)t^2 + 1)a, \\ b(t) &= ((J-1)^6(1728J^3 - 144J^2 + 116J + 1)t^{12} - \\ &\quad 12(J-1)^5(288J^3 - 128J^2 + 82J + 1)t^{11} + \\ &\quad 66(J-1)^5(48J^2 - 56J - 1)t^{10} - \\ &\quad 44(J-1)^4(208J^2 - 176J - 5)t^9 - \\ &\quad 99(J-1)^4(48J^2 - 104J - 5)t^8 + 792(J-1)^3(8J^2 - 10J - 1)t^7 - \\ &\quad 924(J-1)^3(4J+1)t^6 + 792(J-1)^2t^5 - 99(4J-5)(J-1)^2t^4 + \\ &\quad 44(J-1)(6J-5)t^3 - 66(J-1)t^2 + 12t + 1)b. \end{aligned}$$

Then for every rational number t *such that* $\mathcal{E}_t$ *is nonsingular,* $\mathcal{E}_t[4]$ *is isomorphic as a* $G_{\mathbf{Q}}$*-module to* $E[4]$*. If* $ab \neq 0$*, then (7) is a model for* W *over* $\mathbf{Q}$*, where* W *is constructed as in §2 from* $V = E[4]$*,* $\eta = e_4$*.*

Proof. If $a = 0$ then $\mathcal{E}_t$ is $y^2 = x^3 + (t+1)^{12}b$ and if $b = 0$ then $\mathcal{E}_t$ is $y^2 = x^3 + ax$. In both cases the elliptic surface is isotrivial, and $\mathcal{E}_t[4]$ is isomorphic as a $G_{\mathbf{Q}}$-module to $E[4]$. Now assume $ab \neq 0$. Let $j = j(E)$. Using (6), a computation shows that

$$j(E_u) - j(E) = \frac{16(u^{24}+1) + (672-j)(u^{20}+u^4) + (9456+4j)(u^{16}+u^8) + (45248-6j)u^{12}}{u^4(u^4-1)^4}.$$

Let u_0 be a root of the numerator. Then $j(E_{u_0}) = j(E)$. Following the algorithm in §3, we deduce from (4) that

$$h(0)^2 = \frac{a_4(u_0)b}{a_6(u_0)a} = \frac{b(u_0^8 + 14u_0^4 + 1)}{2a(u_0^{12} - 33u_0^8 - 33u_0^4 + 1)} \in \bar{\mathbf{Q}}^\times.$$

Now solve for α and γ so that the coefficients of t in the polynomials in (3) are 0 and $12b$, respectively. (This choice $(r, s) = (0, 12b)$ leads to the relatively simple polynomials $a(t)$ and $b(t)$ in the statement of the theorem.) We obtain

$$\alpha = \frac{(7u_0^4+1)b}{2^2 3^5 u_0^3 (1-u_0^4)^3 h(0)^6}, \qquad \gamma = \frac{(u_0^4+7)b}{2^2 3^5 (u_0^4-1)^3 h(0)^6}.$$

With these values, (2) is a model over $\mathbf{Q}$ for W, and (2) is (7) with the stated $a(t)$ and $b(t)$. This elliptic surface is not isotrivial. □

Theorem 4.2. *Fix a nonzero integer D and define $\mathcal{E}_t$ by*

$$y^2 = x^3 + a(t)x + b(t)$$

where

$$a(t) = D(81D^2t^4 + 6Dt^2 + 1)(81D^2t^4 - 90Dt^2 + 1),$$
$$b(t) = 8D^2t(9Dt^2 + 1)(9D^2t^4 - 2Dt^2 + 1)(729D^2t^4 - 18Dt^2 + 1).$$

If $t \in \mathbf{Q}$ and $9Dt^2 \neq 1$, then $\mathcal{E}_t$ is an elliptic curve over $\mathbf{Q}$ and $\mathcal{E}_t[4]$ is isomorphic as a $G_{\mathbf{Q}}$-module to $E[4]$, where E is the elliptic curve

$$y^2 = x^3 + Dx.$$

Proof. Using (6), a computation shows that

$$j(E_u) - j(E) = \frac{2^4(u^2-2u-1)^2(u^2+2u-1)^2(u^4+1)^2(u^4+6u^2+1)^2}{u^4(u^4-1)^4}.$$

Let $u_0 = 1+\sqrt{2}$, a root of $u^2 - 2u - 1$. Now follow the algorithm in §3. We obtain

$$h(0)^4 = \frac{D}{a_4(u_0)} = \frac{(12\sqrt{2}-17)D}{2^4 3^4}.$$

Let

$$h(0)^2 = \frac{(2\sqrt{2}-3)\sqrt{-D}}{36}, \quad r = 0, \quad \text{and} \quad s = 8D^2.$$

Then

$$\alpha = 3\sqrt{-D}, \qquad \gamma = -3(1+\sqrt{2})\sqrt{-D},$$

and we obtain $\mathcal{E}_t$ as in the statement of the theorem. The discriminant of $\mathcal{E}_t$ is

$$\Delta(\mathcal{E}_t) = -2^6 D^3 (9Dt^2 - 1)^4 (81D^2t^4 + 54Dt^2 + 1)^4.$$

Thus if $t \in \mathbf{P}^1(\mathbf{Q})$ and $9Dt^2 \neq 1$, then $\mathcal{E}_t$ is an elliptic curve. The j-invariant of $\mathcal{E}_t$ is

$$j(\mathcal{E}_t) = \frac{1728(81D^2t^4 + 6Dt^2 + 1)^3(81D^2t^4 - 90Dt^2 + 1)^3}{(9Dt^2 - 1)^4(81D^2t^4 + 54Dt^2 + 1)^4}.$$

□

Theorem 4.3. *Fix a nonzero integer D and define $\mathcal{E}_t$ by*

$$y^2 = x^3 - 12Dt(8Dt^3 - 1)(Dt^3 + 1)x - D(8D^2t^6 + 88Dt^3 - 1)(8D^2t^6 + 1).$$

For every rational number t, $\mathcal{E}_t$ is an elliptic curve over $\mathbf{Q}$ and $\mathcal{E}_t[4]$ is isomorphic as a $G_{\mathbf{Q}}$-module to $E[4]$, where E is the elliptic curve

$$y^2 = x^3 + D.$$

Proof. We have $j(E) = 0$ and

$$j(E_u) = \frac{2^4(u^4 - 2u^3 + 2u^2 + 2u + 1)^3(u^4 + 2u^3 + 2u^2 - 2u + 1)^3}{u^4(u^4 - 1)^4}.$$

Let u_0 be a root of $u^4 - 2u^3 + 2u^2 + 2u + 1$. Let

$$\beta = u_0(1 + u_0 - u_0^2) \qquad \text{and} \qquad \lambda = \frac{(39\beta + 18)^{2/3}D^{1/3}}{6}.$$

Applying the algorithm of §3, we have

$$h(0)^6 = \frac{D(13\beta - 84)}{2^6 3^5}.$$

Let

$$h(0)^2 = \frac{[3D(13\beta - 84)]^{1/3}}{36}, \qquad r = 12D, \qquad \text{and} \qquad s = 0.$$

Then

$$\alpha = (11u_0^3 - 33u_0^2 + 49u_0 - 11)\lambda, \quad \gamma = (3\beta - 19)\lambda,$$

and we obtain $\mathcal{E}_t$ as in the statement of the theorem. The discriminant and j-invariant of $\mathcal{E}_t$ are given by

$$\Delta(\mathcal{E}_t) = -2^4 3^3 D^2(8D^2t^6 - 20Dt^3 - 1)^4,$$

$$j(\mathcal{E}_t) = \frac{-2^{14}3^3 Dt^3(8Dt^3 - 1)^3(Dt^3 + 1)^3}{(8D^2t^6 - 20Dt^3 - 1)^4}.$$

Since $\Delta(\mathcal{E}_t)$ has no rational roots, the theorem follows. □

Part 2. Explicit families of modular elliptic curves

5. MODULAR j-INVARIANTS

If E and E' are elliptic curves over $\mathbf{Q}$, and E and E' are isomorphic over $\mathbf{C}$, then E is modular if and only if E' is modular. It therefore makes sense to talk about modular j-invariants, i.e., the rational numbers which are j-invariants of modular elliptic curves. Before the work of Wiles, it was not known that there are infinitely many modular j-invariants. Using the results of Wiles [19], Taylor-Wiles [18], and Diamond [2], it is now very easy to write down infinite families of modular j-invariants.

We begin by stating Diamond's improvement of the results of Wiles and Taylor-Wiles. While Theorems 7.3 and 8.1 below follow easily from this statement, in fact such results generally follow from the theorems stated in [19], with some additional work.

Theorem 5.1. *If E is an elliptic curve over $\mathbf{Q}$ which has semistable reduction at 3 and at 5, then E is modular.*

Proof. See Theorem 1.2 of [2]. □

6. Semistable reduction

We next state some results which will be used in the proofs of Propositions 7.1 and 8.4. If F is a number field, and v is a prime ideal of F, let $\mathcal{I}_v$ denote the inertia subgroup of G_F corresponding to an extension to $\bar{\mathbf{Q}}$ of the v-adic valuation on F.

Theorem 6.1. *If E is an elliptic curve over $\mathbf{Q}$, then there is a number field over which E has everywhere semistable reduction.*

Proof. See Proposition 3.6 of [4]. See also Proposition 5.4 on p. 181 of [17]. □

Theorem 6.2. *If E is an elliptic curve over $\mathbf{Q}$, p and ℓ are distinct prime numbers, $\rho_p : G_{\mathbf{Q}} \to \mathrm{GL}_2(\mathbf{Z}_p)$ is the p-adic representation associated to E, and $\tau \in \mathcal{I}_\ell$, then the characteristic polynomial of $\rho_p(\tau)$ has integer coefficients which are independent of p.*

Proof. See Theorem 4.3 of [4]. □

Theorem 6.3. *Suppose E is an elliptic curve over a number field F, p and ℓ are distinct prime numbers, v is a prime ideal of F dividing ℓ, and $\rho_p : G_F \to \mathrm{GL}_2(\mathbf{Z}_p)$ is the p-adic representation associated to E. Then E has good reduction at v if and only if $\rho_p(\mathcal{I}_v)$ is trivial.*

Proof. See Theorem 1 of [11]. □

Theorem 6.4. *Suppose E is an elliptic curve over a number field F, p and ℓ are distinct prime numbers, v is a prime ideal of F dividing ℓ, and $\rho_p : G_F \to \mathrm{GL}_2(\mathbf{Z}_p)$ is the p-adic representation associated to E. Then the following are equivalent:*

(i) *E has semistable reduction at v,*
(ii) *for every $\tau \in \mathcal{I}_v$, all the eigenvalues of $\rho_p(\tau)$ are 1 (i.e., $\mathcal{I}_v$ acts unipotently on the p-adic Tate module of E),*
(iii) *for every $\tau \in \mathcal{I}_v$, $(\rho_p(\tau) - I)^2 = 0$.*

Proof. See Proposition 3.5 and Corollaire 3.8 of [4]. □

Let $\bar{\mathbf{Z}}$ denote the ring of algebraic integers.

Theorem 6.5. *If α is a root of unity in $\bar{\mathbf{Z}}$, and either*

(a) $5 \leq N \in \mathbf{Z}$ *and* $(\alpha - 1)^2 \in N\bar{\mathbf{Z}}$, *or*
(b) $3 \leq N \in \mathbf{Z}$ *and* $\alpha - 1 \in N\bar{\mathbf{Z}}$,

then $\alpha = 1$.

Proof. Part (b) is well-known. See Theorem 3.1 of [16] for proofs of (a) and (b). □

Lemma 6.6. *Suppose that N is a positive integer, and for each prime divisor p of N we have a matrix $A_p \in M_2(\mathbf{Z}_p)$ such that the characteristic polynomials of the A_p have integral coefficients independent of p, and such that $(A_p - I)^2 \in NM_2(\mathbf{Z}_p)$. Then for every eigenvalue α of A_p, $(\alpha-1)/\sqrt{N}$ satisfies a monic polynomial with integer coefficients.*

Proof. See Lemma 5.2 of [15]. □

7. MOD 4 REPRESENTATIONS

Proposition 7.1. *Suppose E and E' are elliptic curves over $\mathbf{Q}$, N is a positive integer, $E[N]$ and $E'[N]$ are isomorphic as $G_{\mathbf{Q}}$-modules, and ℓ is a prime number which does not divide N. If*

(a) *$N \geq 5$ and E has semistable reduction at ℓ, or*
(b) *$N = 3$ or 4 and E has good reduction at ℓ,*

then E' has semistable reduction at ℓ.

Proof. We give a proof in the spirit of [15]. Let $\mathcal{I}_\ell$ denote the inertia subgroup of $G_{\mathbf{Q}}$ corresponding to an extension $\bar{\lambda}$ to $\bar{\mathbf{Q}}$ of the ℓ-adic valuation on $\mathbf{Q}$. Suppose $\tau \in \mathcal{I}_\ell$, p is a prime divisor of N, and $\rho_{E,p}$ and $\rho_{E',p}$ are the p-adic representations of $G_{\mathbf{Q}}$ associated to E and E', respectively. Suppose α is an eigenvalue of $\rho_{E',p}(\tau)$. There is a number field F such that E has semistable reduction at the restriction λ of $\bar{\lambda}$ to F (by Theorem 6.1). Therefore, $\tau^m \in \mathcal{I}_\lambda$ for some positive integer m. By Theorem 6.4,

$$(\rho_{E',p}(\tau)^m - I)^2 = 0.$$

Thus, $(\alpha^m - 1)^2 = 0$, so $\alpha^m = 1$.

Since E has semistable reduction at ℓ,

$$(\rho_{E,p}(\tau) - I)^2 = 0$$

by Theorem 6.4. Since $E[N] \cong E'[N]$, we have

$$\rho_{E,p}(\tau) - \rho_{E',p}(\tau) \in N\mathrm{M}_2(\mathbf{Z}_p)$$

for appropriate choices of bases for the p-adic Tate modules of E and of E'. Therefore,

$$(\rho_{E',p}(\tau) - I)^2 \in N\mathrm{M}_2(\mathbf{Z}_p).$$

The characteristic polynomial of $\rho_{E',p}(\tau)$ is independent of the choice of prime divisor p of N (by Theorem 6.2). By Lemma 6.6, $(\alpha - 1)^2 \in N\bar{\mathbf{Z}}$. Suppose $N \geq 5$. By Theorem 6.5a, $\alpha = 1$. Therefore, $\mathcal{I}_\ell$ acts on the Tate module of E' by unipotent operators. By Theorem 6.4, E' has semistable reduction at ℓ. Now suppose $N = 3$ or 4 and E has good reduction at ℓ. By Theorem 6.3, $\rho_{E,p}(\tau) = I$. Therefore, $\rho_{E',p}(\tau) - I \in N\mathrm{M}_2(\mathbf{Z}_p)$, $\alpha - 1 \in N\bar{\mathbf{Z}}$, and $\alpha = 1$ (using Lemma 6.6 and Theorem 6.5b). By Theorem 6.4, E' has semistable reduction at ℓ. □

Examples 7.2. To see that Proposition 7.1a fails for $N = 3$ or 4, let E be the elliptic curve $y^2 = x^3 + x + 1$. The conductor of E is $2^4 \cdot 31$, so E has multiplicative reduction at 31. Consider Theorem 4.1 with $a = b = 1$ and let E' be the elliptic curve obtained by letting $t = 1$ (letting $t = 0$ gives E). Then $E[4] \cong E'[4]$, and it is easy to check that E' has additive reduction at 31. Theorem 4.1 of [9] is the analogue of Theorem 4.1 of this paper, with $N = 3$ instead of $N = 4$. Consider Theorem 4.1 of [9] with $a = b = 1$, and let E'' be the elliptic curve obtained by letting $t = 1$ (again, $t = 0$ gives E). Then $E[3] \cong E''[3]$, and it is easy to check that E'' has additive reduction at 31.

To see that Proposition 7.1b fails for $N = 2$, let ℓ be an odd prime, let E be the elliptic curve $y^2 = x^3 - x$, and let E' be the elliptic curve $y^2 = x^3 - \ell^2 x$. Then E has good reduction at ℓ (the conductor of E is 2^5), E' has additive reduction at ℓ (the conductor of E' is $2^5\ell^2$), and $E[2] \cong E'[2] \cong \mathbf{Z}/2\mathbf{Z} \times \mathbf{Z}/2\mathbf{Z}$.

Theorem 7.3. *If E is an elliptic curve over $\mathbf{Q}$ which has good reduction at 3 and at 5, and E' is an elliptic curve over $\mathbf{Q}$ such that $E[4]$ and $E'[4]$ are isomorphic as $G_{\mathbf{Q}}$-modules, then E' is modular.*

Proof. By Proposition 7.1b with $N = 4$, E' has semistable reduction at 3 and at 5. Therefore E' is modular by Theorem 5.1. □

Therefore, the explicit families of §4 give infinite families of modular elliptic curves, as long as one of the elliptic curves in the family has good reduction at 3 and at 5.

8. TORSION SUBGROUPS

Theorem 8.1. *If E is an elliptic curve over $\mathbf{Q}$ which has:*

(1) *all its points of order* 2,
(2) *a cyclic subgroup of order* 4,
(3) *a point of order* 5, *or*
(4) *a point of order* 7,

defined over $\mathbf{Q}$, then E is modular.

We prove Theorem 8.1 in a series of lemmas.

Lemma 8.2. *If E is an elliptic curve over $\mathbf{Q}$, and $E(\mathbf{Q}) \supseteq \mathbf{Z}/2\mathbf{Z} \times \mathbf{Z}/2\mathbf{Z}$, then E is modular.*

Proof. It is easy to see that E is isomorphic over $\mathbf{C}$ to an elliptic curve E' of the form

$$y^2 = x(x-A)(x-B)$$

where A and B are relatively prime integers. Suppose p is an odd prime. Since the right hand side does not have a triple root modulo p, E' has semistable reduction at p. In other words, one can twist away any additive reduction on E at odd primes. The lemma now follows from Theorem 5.1. □

Lemma 8.2 was an observation made with K. Rubin. The main theorem of [3] shows that Lemma 8.2 follows from the results of [19] and [18], without using [2].

Lemma 8.3. *If E is an elliptic curve over $\mathbf{Q}$, and E has a cyclic subgroup of order 4 defined over $\mathbf{Q}$, then E is modular.*

Proof. Suppose C is a rational cyclic subgroup of E of order 4. Let $D = C \cap E[2]$. Then D is a subgroup of E of order 2, and D is defined over $\mathbf{Q}$. Let $E' = E/D$, an elliptic curve over $\mathbf{Q}$. The quotient map $\varphi : E \to E'$ is an isogeny defined over $\mathbf{Q}$. Therefore to show that E is modular, it suffices to show that E' is modular. Fix a generator x of C and fix $y \in E[2] - D$. Then $\varphi(x)$ generates C/D, which is a rational subgroup of E' of order 2. Therefore, $\varphi(x)$ is defined over $\mathbf{Q}$. Similarly, $\varphi(y)$ generates $E[2]/D$, a rational subgroup of E' of order 2, so $\varphi(y)$ is defined over $\mathbf{Q}$. Since $x - y \notin C$, we have $x - y \notin D$, so $\varphi(x) \neq \varphi(y)$. Therefore, E' has all its points of order 2 defined over $\mathbf{Q}$. By Lemma 8.2, E' is modular. □

Proposition 8.4 (Néron, Frey, Flexor-Oesterlé,...).
If E is an elliptic curve over $\mathbf{Q}$, $5 \leq N \in \mathbf{Z}$, and $E(\mathbf{Q}) \supseteq \mathbf{Z}/N\mathbf{Z}$, then E has semistable reduction at every prime which does not divide N.

Proof. We give a proof from [15] (see Theorem 6.2). Suppose ℓ is a prime which does not divide N. Let $\mathcal{I}_\ell$ denote the inertia subgroup of $G_{\mathbf{Q}}$ corresponding to an extension $\bar{\lambda}$ to $\bar{\mathbf{Q}}$ of the ℓ-adic valuation on $\mathbf{Q}$. Suppose $\tau \in \mathcal{I}_\ell$. Since ℓ does not divide N, $\mathbf{Q}(\zeta_N)$ is unramified at ℓ, so $\mathcal{I}_\ell$ acts as the identity on the N-th roots of unity. Suppose p is a prime divisor of N, and let $\rho_p : G_{\mathbf{Q}} \to \mathrm{GL}_2(\mathbf{Z}_p)$ denote the p-adic representation associated to E. Since $E(\mathbf{Q})$ has a point of order N,

$$\rho_p(\tau) \equiv \begin{pmatrix} 1 & * \\ 0 & 1 \end{pmatrix} \pmod{N\mathrm{M}_2(\mathbf{Z}_p)},$$

for an appropriate choice of basis for the p-adic Tate module of E. Therefore, $(\rho_p(\tau) - I)^2 \in N\mathrm{M}_2(\mathbf{Z}_p)$. There is a number field F such that E has semistable reduction at the restriction λ of $\bar{\lambda}$ to F (by Theorem 6.1). Then $\tau^m \in \mathcal{I}_\lambda$ for some positive integer m. By Theorem 6.4, $(\rho_p(\tau)^m - I)^2 = 0$.

Let α be an eigenvalue of $\rho_p(\tau)$. Then $(\alpha^m - 1)^2 = 0$, so $\alpha^m = 1$. By Lemma 6.6 and Theorem 6.2, $(\alpha - 1)^2 \in N\mathrm{M}_2(\bar{\mathbf{Z}})$. By Theorem 6.5a, $\alpha = 1$. Therefore, $\mathcal{I}_\ell$ acts on the Tate module of E by unipotent operators. By Theorem 6.4, E has semistable reduction at ℓ. □

Examples 8.5. Proposition 8.4 fails for $N < 5$. The point $(48, -15)$ is a point of order 4 on the elliptic curve $y^2+xy = x^3-x^2-7056x+229905$, and this curve has additive reduction at 3 (it is the curve 63A5 in Cremona's tables [1]). The point $(1, 1)$ is a point of order 3 on the elliptic curve $y^2 = x^3 + x^2 - x$, and this curve has additive reduction at 2 (it is 20A2 in [1]).

Lemma 8.6. *If E is an elliptic curve over $\mathbf{Q}$, and $E(\mathbf{Q}) \supseteq \mathbf{Z}/7\mathbf{Z}$, then E is modular.*

Proof. By Proposition 8.4 with $N = 7$, E has semistable reduction at 3 and at 5. By Theorem 5.1, E is modular. □

Theorem 8.7. *If E is an elliptic curve over $\mathbf{Q}$, p is an odd prime number, E has semistable reduction at p, the mod p representation $\bar{\rho}_{E,p}$ for E is modular, and the restriction of $\bar{\rho}_{E,p}$ to $G_{\mathbf{Q}(\sqrt{-1^{(p-1)/2}p})}$ is absolutely irreducible, then E is modular.*

Proof. This is Theorem 5.3 of [2], applied to the p-adic representation associated to the elliptic curve E. □

Proposition 8.8. *If E is an elliptic curve over $\mathbf{Q}$, the mod 5 representation for E is reducible, and the restriction to $G_{\mathbf{Q}(\sqrt{-3})}$ of the mod 3 representation for E is not absolutely irreducible, then E is modular.*

Proof. See Proposition 13 of [7] (by work of J. E. Cremona, the elliptic curves over $\mathbf{Q}$ whose j-invariants are given in Proposition 13 of [7] are all modular), p. 544 of [19], or the proof of Theorem 5.4 of [2]. □

Lemma 8.9. *If E is an elliptic curve over $\mathbf{Q}$, and $E(\mathbf{Q}) \supseteq \mathbf{Z}/5\mathbf{Z}$, then E is modular.*

Proof. Since $E(\mathbf{Q}) \supseteq \mathbf{Z}/5\mathbf{Z}$, the mod 5 representation for E is reducible. By Proposition 8.8, we may assume that the restriction to $G_{\mathbf{Q}(\sqrt{-3})}$ of the mod 3 representation for E is absolutely irreducible. The mod 3 representation for E is modular by the Langlands-Tunnell Theorem. By Proposition 8.4, E has semistable reduction at 3. By Theorem 8.7, E is modular. □

Theorem 8.1 follows from Lemmas 8.2, 8.3, 8.6, and 8.9.

By [6], if E is an elliptic curve over $\mathbf{Q}$, then the torsion subgroup of $E(\mathbf{Q})$ is isomorphic to one of the following 15 groups:

$$\begin{array}{ll} \mathbf{Z}/N\mathbf{Z} & \text{for } N = 1, \ldots, 10 \text{ or } 12, \\ \mathbf{Z}/2\mathbf{Z} \times \mathbf{Z}/2N\mathbf{Z} & \text{for } N = 1, 2, 3, \text{ or } 4. \end{array}$$

This and Theorem 8.1 immediately imply the following result.

Corollary 8.10. *If E is an elliptic curve over $\mathbf{Q}$, and the torsion subgroup of $E(\mathbf{Q})$ is not a cyclic group of order 1, 2, 3, 6, or 9, then E is modular.*

Given a model for an elliptic curve E over $\mathbf{Q}$, the Nagell-Lutz Theorem (see Corollary 7.2 of Chapter VIII of [17]) provides an algorithm for computing the torsion subgroup of $E(\mathbf{Q})$.

9. Explicit families of modular elliptic curves

By Theorem 8.1, the elliptic curves below are modular. See Table 3 on p. 217 of [5] for such parametrizations of elliptic curves over $\mathbf{Q}$ with points of finite order.

Example 9.1 (rational 2-torsion).
Elliptic curves over $\mathbf{Q}$ with all points of order 2 defined over $\mathbf{Q}$ are given by

$$Dy^2 = x(x-1)(x-\lambda)$$

with $D, \lambda \in \mathbf{Q}^\times, \lambda \neq 1$. The corresponding (modular) j-invariants are all the numbers of the form

$$\frac{2^8(\lambda^2 - \lambda + 1)^3}{\lambda^2(\lambda - 1)^2} \quad \text{with} \quad \lambda \in \mathbf{Q} - \{0, 1\}.$$

Example 9.2 (rational cyclic subgroup of order 4).
The family of elliptic curves with a rational cyclic subgroup of order 4 is given by

$$y^2 = x^3 + D(1-4b)x^2 - 8D^2bx + 16D^3b^2$$

with $b, D \in \mathbf{Q}^\times, b \neq -\frac{1}{16}$. The rational cyclic subgroup of order 4 is

$$\{0, (4bD, 0), (0, \pm 4bD\sqrt{D})\}$$

and the j-invariant is

$$\frac{(16b^2 + 16b + 1)^3}{(16b + 1)b^4}.$$

Example 9.3 (rational points of order 5).
The family of elliptic curves over $\mathbf{Q}$ with a rational point of order 5 is given by

$$y^2 + (1-b)xy - by = x^3 - bx^2$$

with $b \in \mathbf{Q}^\times$. The point $(0, 0)$ has order 5.

Example 9.4 (rational points of order 7).
The family of elliptic curves over $\mathbf{Q}$ with a rational point of order 7 is given by

$$y^2 + (1 - d(d-1))xy - d^2(d-1)y = x^3 - d^2(d-1)x^2$$

with $d \in \mathbf{Q} - \{0, 1\}$. The point $(0,0)$ has order 7.

REFERENCES

[1] J. E. Cremona, Algorithms for modular elliptic curves, Cambridge Univ. Press, Cambridge, 1992.

[2] F. Diamond, *On deformation rings and Hecke rings*, preprint.

[3] F. Diamond, K. Kramer, *Modularity of a family of elliptic curves*, Math. Research Letters **2** (1995), 299–305.

[4] A. Grothendieck, *Modèles de Néron et monodromie*, in Groupes de monodromie en géometrie algébrique, SGA7 I, A. Grothendieck, ed., Lecture Notes in Math. **288**, Springer, Berlin-Heidelberg-New York, 1972, pp. 313–523.

[5] D. S. Kubert, *Universal bounds on the torsion of elliptic curves*, Proc. London Math. Soc. **33** (1976), 193–237.

[6] B. Mazur, *Modular curves and the Eisenstein ideal*, Publ. Math. IHES **47** (1977), 133–186.

[7] K. Rubin, *Modularity of mod* 5 *representations*, this volume.

[8] K. Rubin, A. Silverberg, *A report on Wiles' Cambridge lectures*, Bull. Amer. Math. Soc. (N. S.) **31**, no. 1 (1994), 15–38.

[9] K. Rubin, A. Silverberg, *Families of elliptic curves with constant mod p representations*, in Conference on Elliptic Curves and Modular Forms, Hong Kong, December 18–21, 1993, Intl. Press, Cambridge, Massachusetts, 1995, pp. 148–161.

[10] J-P. Serre, *Cohomologie galoisienne*, Lecture Notes in Mathematics **5**, Springer-Verlag, Berlin-New York, 1965.

[11] J-P. Serre, J. Tate, *Good reduction of abelian varieties*, Ann. of Math. **88** (1968), 492–517.

[12] G. Shimura, Introduction to the arithmetic theory of automorphic functions, Princeton Univ. Press, Princeton, 1971.

[13] T. Shioda, *On elliptic modular surfaces*, J. Math. Soc. Japan **24** (1972), 20–59.

[14] T. Shioda, *On rational points of the generic elliptic curve with level N structure over the field of modular functions of level N*, J. Math. Soc. Japan **25** (1973), 144–157.

[15] A. Silverberg, Yu. G. Zarhin, *Semistable reduction and torsion subgroups of abelian varieties*, Ann. Inst. Fourier **45** (1995), 403–420.

[16] A. Silverberg, Yu. G. Zarhin, *Variations on a theme of Minkowski and Serre*, J. Pure and Applied Algebra **111** (1996), 285–302.

[17] J. Silverman, The Arithmetic of Elliptic Curves, Springer-Verlag, New York-Berlin-Heidelberg-Tokyo, 1986.

[18] R. Taylor, A. Wiles, *Ring-theoretic properties of certain Hecke algebras*, Ann. Math. **141** (1995), 553–572.

[19] A. Wiles, *Modular elliptic curves and Fermat's Last Theorem*, Ann. Math. **141** (1995), 443–551.

MODULARITY OF MOD 5 REPRESENTATIONS

KARL RUBIN

Introduction

The aim of this paper is to tie everything together to prove the following theorem.

Theorem A (Wiles [19] + Taylor & Wiles [17] + Diamond [3]).
If E is an elliptic curve defined over $\mathbf{Q}$ and E is semistable at 3 and 5, then E is modular.

Our starting point will be Theorems B and C below. If E is an elliptic curve defined over a field F of characteristic zero, let $G_F = \mathrm{Gal}(\bar{F}/F)$ and for every prime p, let

$$\bar{\rho}_{F,E,p} : G_F \to \mathrm{GL}_2(\mathbf{F}_p)$$

be (the isomorphism class of) the representation of G_F on the p-torsion $E[p]$ in $E(\bar{F})$. We will write simply $\bar{\rho}_{E,p}$ for $\bar{\rho}_{\mathbf{Q},E,p}$.

Theorem B (Wiles [19] + Taylor & Wiles [17] + Diamond [3]).
Suppose E is an elliptic curve over $\mathbf{Q}$, and p is an odd prime, such that

- *E is semistable at p,*
- *$\bar{\rho}_{E,p}$ restricted to $\mathrm{Gal}(\bar{\mathbf{Q}}/\mathbf{Q}(\sqrt{(-1)^{(p-1)/2}p}))$ is absolutely irreducible, and*
- *$\bar{\rho}_{E,p}$ is modular.*

Then E is modular.

Theorem C (Langlands [6] + Tunnell [18]). *Suppose $\rho : G_{\mathbf{Q}} \to \mathrm{GL}_2(\mathbf{F}_3)$ is a continuous representation satisfying*

- *ρ is irreducible, and*
- $\det(\rho(\textit{complex conjugation})) = -1$.

Then ρ is modular.

Partially supported by an NSF grant. The author also thanks the Institut des Hautes Etudes Scientifiques and the Institute for Advanced Study for their hospitality and support.

Theorem B (for $p = 3$) and Theorem C together prove Theorem A under the extra hypothesis that $\bar{\rho}_{E,3}$ restricted to $G_{\mathbf{Q}(\sqrt{-3})}$ is absolutely irreducible. When $\bar{\rho}_{E,3}$ restricted to $G_{\mathbf{Q}(\sqrt{-3})}$ is *not* absolutely irreducible we will complete the proof of Theorem A by applying Theorem B with $p = 5$. We need to show that the hypotheses of Theorem B are satisfied, and this will follow from the following two theorems:

Theorem 1. *Suppose E is an elliptic curve over $\mathbf{Q}$ such that*

- *$\bar{\rho}_{E,3}$ restricted to $G_{\mathbf{Q}(\sqrt{-3})}$ is not absolutely irreducible,*
- *E is semistable at 5, and*
- *the j-invariant $j(E)$ is not $11^3/2^3$ or $-29^3 41^3/2^{15}$.*

Then $\bar{\rho}_{E,5}$ restricted to $G_{\mathbf{Q}(\sqrt{5})}$ is absolutely irreducible.

Theorem 2. *Suppose E is an elliptic curve over $\mathbf{Q}$ such that*

- *E is semistable at 3, and*
- *$\bar{\rho}_{E,5}$ is irreducible.*

Then $\bar{\rho}_{E,5}$ is modular.

Theorem A is immediate from Theorems B, C, 1 and 2, together with the observation that the exceptional curves in Theorem 1 are modular. (More precisely, these curves are twists of the curves listed as 338E1 and 338E2 in [2]. Although it is stated in §2.14 of [2] that these models are not proved to be modular, John Cremona informs me that due to subsequent progress the computations in [2] do suffice to prove that these curves are the modular curves they appear to be.)

The fundamental hypothesis in Theorem B is that $\bar{\rho}_{E,p}$ is modular. When $p = 3$ this is known because of Theorem C, and when $p = 5$ we will use Theorem 2. To prove Theorem 2 we must have at our disposal a large collection of modular forms. The modular forms we use will be the ones produced by Theorems B and C when $p = 3$.

The proofs of Theorems 1 and 2 are essentially due to Wiles, and can be found with varying degrees of generality and detail in [19] Chapter 5 and in [3], proof of Theorem 5.4.

If F is an arbitrary field, let $G_F = \mathrm{Gal}(F^s/F)$ where F^s is a separable closure of F. If p is a prime not equal to the characteristic of F let $\chi_p : G_F \to \mathrm{Aut}(\boldsymbol{\mu}_p) \xrightarrow{\sim} \mathbf{F}_p^\times$ be the cyclotomic character.

The following statement was first pointed out to me by Richard Taylor, and is proved in [11]. Although not necessary for the proof of Theorem A, it is related to Theorem 2 (see Theorem 4 below) and we will prove it in §5.

Theorem 3. *Let p be 3 or 5, and let F be a field of characteristic different from p. Suppose $\rho : G_F \to \mathrm{GL}_2(\mathbf{F}_p)$ is a representation such that $\det(\rho) = \chi_p$. Then there is an elliptic curve E defined over F such that $\rho \cong \bar{\rho}_{F,E,p}$.*

Combining Theorems 2 and 3 (with $F = \mathbf{Q}$) gives the following.

Theorem 4. *Suppose $\rho : G_{\mathbf{Q}} \to \mathrm{GL}_2(\mathbf{F}_5)$ is a representation satisfying*

- *ρ is semistable at 3 (i.e., the image of an inertia group at 3 is unipotent),*
- *ρ is irreducible, and*
- $\det(\rho) = \chi_5$.

Then ρ is modular.

Remark. The proof of Theorem 1 depends in a crucial way on the fact that the genus of the modular curve $X_0(3{\cdot}5)$ is greater than zero, and the proof of Theorem 2 on the fact that the genus of $X(5)$ is equal to zero. For primes p (see [13] §1.6),

$$\mathrm{genus}(X_0(3p)) > 0 \Leftrightarrow p \geq 5,$$
$$\mathrm{genus}(X(p)) = 0 \Leftrightarrow p \leq 5.$$

Thus 5 is the only prime that will work for Wiles' argument.

Acknowledgment. I would like to thank Alice Silverberg for helpful conversations.

1. PRELIMINARIES: GROUP THEORY

Lemma 5. *Suppose E is an elliptic curve over a field F and $p > 2$ is prime. If F has an embedding into $\mathbf{R}$ and $\bar{\rho}_{F,E,p}$ is irreducible, then $\bar{\rho}_{F,E,p}$ is absolutely irreducible.*

Proof. Fix a complex conjugation $\tau \in G_F$ and write $E[3]^+$ (resp. $E[3]^-$) for the subspace of $E[3]$ where τ acts via $+1$ (resp. -1). The characteristic polynomial of $\bar{\rho}_{F,E,p}(\tau)$ is $x^2 - 1$, so $\dim_{\mathbf{F}_3} E[3]^+ = \dim_{\mathbf{F}_3} E[3]^- = 1$.

Suppose $\bar{\rho}_{F,E,p}$ is not absolutely irreducible, i.e., there is a one-dimensional subspace W of $E[3] \otimes \bar{\mathbf{F}}_p$ which is stable under G_F. Then

$$W = E[3]^{\pm} \otimes \bar{\mathbf{F}}_p$$

for some choice of sign, since no other one-dimensional subspace is stable under τ. Hence $E[3]^{\pm} = W \cap E[3]$ is stable under G_F, so $\bar{\rho}_{F,E,p}$ is reducible. □

Proposition 6. *Suppose E is an elliptic curve over $\mathbf{Q}$ such that*

- *$\bar{\rho}_{E,3}$ is irreducible, and*
- *the restriction of $\bar{\rho}_{E,3}$ to $G_{\mathbf{Q}(\sqrt{-3})}$ is not absolutely irreducible.*

Then there are distinct subgroups C_3, C_3' of order 3 of $E[3]$ such that the unordered set $\{C_3, C_3'\}$ is fixed by $G_{\mathbf{Q}}$.

Proof. Fix a complex conjugation $\tau \in G_{\mathbf{Q}}$. Our hypotheses on $\bar{\rho}_{E,3}$ imply (using Lemma 5) that there is a one-dimensional subspace W of $E[3] \otimes \bar{\mathbf{F}}_3$ which is stable under $G_{\mathbf{Q}(\sqrt{-3})}$ but not under $G_{\mathbf{Q}}$. Then W^τ is also stable under $G_{\mathbf{Q}(\sqrt{-3})}$, and $E[3] \otimes \bar{\mathbf{F}}_3 = W \oplus W^\tau$.

It follows that $\bar{\rho}_{E,3}(G_{\mathbf{Q}(\sqrt{-3})})$ is a cyclic subgroup of $\mathrm{SL}_2(\mathbf{F}_3)$, of order prime to 3, and so the image of $G_{\mathbf{Q}(\sqrt{-3})}$ in $\mathrm{PGL}_2(\mathbf{F}_3)$ has order at most 2. Let $\mathcal{C}$ denote the set of 4 subgroups of order 3 of $E[3]$, and let $E[3]^+ \in \mathcal{C}$ be the subgroup of points fixed by complex conjugation. Then $G_{\mathbf{Q}}$ acts on $\mathcal{C}$ by a subgroup of S_4 of order at most 4. Since τ fixes $E[3]^+$ but acts nontrivially on $\mathcal{C}$, the $G_{\mathbf{Q}}$-orbit of $E[3]^+$ has order at most 2. The assumption that $\bar{\rho}_{E,3}$ is irreducible implies that this orbit has order exactly 2, so this orbit is the desired set $\{C_3, C_3'\}$. □

Proposition 7. *Suppose E is an elliptic curve over $\mathbf{Q}$ such that*

- *$\bar{\rho}_{E,5}$ is irreducible and*
- *E is semistable at* 5.

Then the restriction of $\bar{\rho}_{E,5}$ to $G_{\mathbf{Q}(\sqrt{5})}$ is absolutely irreducible.

Proof. Let $I_5 \subset G_{\mathbf{Q}}$ denote the inertia group of some prime above 5, and fix $\sigma \in I_5$ such that $\sigma \notin G_{\mathbf{Q}(\sqrt{5})}$ (possible since 5 ramifies in $\mathbf{Q}(\sqrt{5})/\mathbf{Q}$). Suppose that E satisfies the hypotheses of the proposition but the restriction of $\bar{\rho}_{E,5}$ to $G_{\mathbf{Q}(\sqrt{5})}$ is not absolutely irreducible. By Lemma 5, this restriction is then reducible, so there is a subgroup C_5 of $E[5]$ which is stable under $G_{\mathbf{Q}(\sqrt{5})}$ but not under $G_{\mathbf{Q}}$. Then C_5^σ is also stable under $G_{\mathbf{Q}(\sqrt{5})}$, and $E[5] = C_5 \oplus C_5^\sigma$. With a basis of $E[5]$ chosen from C_5 and C_5^σ, since

$$\det(\bar{\rho}_{E,5}(G_{\mathbf{Q}(\sqrt{5})})) = \chi_5(G_{\mathbf{Q}(\sqrt{5})}) = \{\pm 1\} \quad \text{and}$$
$$\det(\bar{\rho}_{E,5}(\sigma)) = \chi_5(\sigma) \notin \{\pm 1\}$$

we see that

$$\bar{\rho}_{E,5}(G_{\mathbf{Q}(\sqrt{5})}) \subset \{\left(\begin{smallmatrix} a & 0 \\ 0 & \pm a \end{smallmatrix}\right) : a \in \mathbf{F}_5^\times\} \quad \text{and}$$
$$\bar{\rho}_{E,5}(\sigma G_{\mathbf{Q}(\sqrt{5})}) \subset \{\left(\begin{smallmatrix} 0 & a \\ \pm 2a & 0 \end{smallmatrix}\right) : a \in \mathbf{F}_5^\times\}.$$

Case I: E is ordinary or multiplicative at 5. Since $\sigma \in I_5$, σ stabilizes a proper subgroup of $E[5]$ (see [9], Proposition 13 and the remarks before Proposition 11). But the description above shows this is impossible.

Case II: E is supersingular at 5. Proposition 12 of [9] shows that $\bar{\rho}_{E,5}(I_5)$ is cyclic of order 24. Again, the description above shows that $\bar{\rho}_{E,5}(G_{\mathbf{Q}})$ has no such subgroup. □

2. Preliminaries: modular curves

Write $\mathcal{H}$ for the complex upper half plane. If Γ is a congruence subgroup of $\mathrm{SL}_2(\mathbf{Z})$, $\hat{\mathcal{H}}/\Gamma$ will denote the compactification of $\mathcal{H}/\Gamma$, obtained by adjoining a finite set of cusps ([13] §1.3).

We will need the congruence groups

$$\Gamma(N) = \{ \left(\begin{smallmatrix} a & b \\ c & d \end{smallmatrix}\right) \in \mathrm{SL}_2(\mathbf{Z}) : \left(\begin{smallmatrix} a & b \\ c & d \end{smallmatrix}\right) \equiv \left(\begin{smallmatrix} 1 & 0 \\ 0 & 1 \end{smallmatrix}\right) \pmod N \}$$

$$\Gamma_0(N) = \{ \left(\begin{smallmatrix} a & b \\ c & d \end{smallmatrix}\right) \in \mathrm{SL}_2(\mathbf{Z}) : c \equiv 0 \pmod N \}$$

$$\Gamma_{\mathrm{split}}(N) = \{ \left(\begin{smallmatrix} a & b \\ c & d \end{smallmatrix}\right) \in \mathrm{SL}_2(\mathbf{Z}) : \text{either } b \equiv c \equiv 0 \text{ or } a \equiv d \equiv 0 \pmod N \}.$$

Lemma 8. *As in [13] §1.6 define*

$\mu(\Gamma) = [\mathrm{SL}_2(\mathbf{Z}) : \{\pm 1\}\Gamma]$,

$\nu_\infty(\Gamma) =$ *number of cusps of* $\mathcal{H}/\Gamma$,

$\nu_i(\Gamma) =$ *number of elliptic points of order* i *of* $\mathcal{H}/\Gamma$, $i = 2$ *or* 3.

Then we have the following table:

Γ	$\mu(\Gamma)$	$\nu_\infty(\Gamma)$	$\nu_2(\Gamma)$	$\nu_3(\Gamma)$	*genus of* $\mathcal{H}/\Gamma$
$\Gamma_0(15)$	24	4	0	0	1
$\Gamma_0(5) \cap \Gamma_{\mathrm{split}}(3)$	36	4	4	0	1
$\Gamma(5) \cap \Gamma_0(3)$	240	24	0	0	9
$\Gamma(5) \cap \Gamma_{\mathrm{split}}(3)$	360	24	0	0	19

Proof. Let Γ denote one of the four groups in the table. Then $\Gamma(15) \subset \Gamma \subset \mathrm{SL}_2(\mathbf{Z})$, so $\mu(\Gamma)$ is the index of the image of Γ in $\mathrm{PSL}_2(\mathbf{Z}/15\mathbf{Z})$, which is easily computed.

The cusps of $\mathcal{H}/\Gamma(15)$ are described in [13] §1.6: there are 96 of them, and they can be identified with the coset space $\mathrm{PSL}_2(\mathbf{Z}/15\mathbf{Z})/U$ where U is the subgroup generated by $\left(\begin{smallmatrix} 1 & 1 \\ 0 & 1 \end{smallmatrix}\right)$. From this it is not difficult to work out the number of cusps for each of the four groups (for $\Gamma_0(15)$ it is done in Proposition 1.43 of [13]).

For elliptic points, Propositions 1.43 and 1.39 of [13] cover all of the 0 entries in the table, and $\nu_2(\Gamma_0(5) \cap \Gamma_{\mathrm{split}}(3))$ can be worked out with some effort using the methods of [13] §1.6.

The last column follows from Proposition 1.40 of [13], which says that the genus of $\mathcal{H}/\Gamma$ is $1 + \mu(\Gamma)/12 - \nu_2(\Gamma)/4 - \nu_3(\Gamma)/3 - \nu_\infty(\Gamma)/2$. □

2.1. **The modular curve $X_0(15)$.** Let $X_0(15)$ be the modular curve over $\mathbf{Q}$ which parametrizes isomorphism classes of pairs (E, C_{15}) where E is an elliptic curve and C_{15} is a subgroup of E of order 15.

Lemma 9. (i) $X_0(15)$ *is a curve of genus one with four cusps, all of which are rational over* $\mathbf{Q}$.

(ii) $\#(X_0(15)(\mathbf{Q})) = 8$.

(iii) *The four rational points of* $X_0(15)$ *which are not cusps correspond to four pairs* $(E_i, C_{15}^{(i)})$ *where*

$$j(E_i) \in \{-25/2, -5^2 241^3/2^3, -5{\cdot}29^3/2^5, 5{\cdot}211^3/2^{15}\}.$$

(iv) *If* E *is an elliptic curve over* $\mathbf{Q}$ *with a rational subgroup of order* 15, *then* E *is not semistable at* 5.

Proof. Since $X_0(15)(\mathbf{C}) = \hat{\mathcal{H}}/\Gamma_0(15)$, the genus and the number of cusps of $X_0(15)$ are given in Lemma 8. The fact that the four cusps are rational follows from the action of $G_{\mathbf{Q}}$ on the cusps (see for example [5] §VI.5).

From the tables [1] 15C or [2] 15A1 one can read off that $X_0(15)$ has 8 rational points. In the same tables one finds 4 elliptic curves over $\mathbf{Q}$ (50A1, 50A2, 50A3, and 50A4 in [2]), with invariants

$$j = -25/2,\ -5^2 241^3/2^3,\ -5{\cdot}29^3/2^5,\ \text{and}\ 5{\cdot}211^3/2^{15},$$

respectively, each isogenous to $E_0 : y^2 + xy + y = x^3 - x - 2$ and having a rational subgroup of order 15. Thus these four curves represent all the non-cusp rational points of $X_0(15)$. This proves (iii).

Suppose E is defined over $\mathbf{Q}$ and has a rational subgroup C_{15} of order 15. Then $(E, C_{15}) \in X_0(15)(\mathbf{Q})$, so by (iii), E is isogenous to a quadratic twist of E_0. The equation above for E_0 is minimal and is not semistable at 5. Since it has discriminant $-2{\cdot}5^4$, this equation remains minimal over any quadratic extension of $\mathbf{Q}$ (see for example [16] Exercise 7.1). It follows that E is not semistable at 5. □

2.2. **The modular curve $X_{0,\mathrm{split}}(5,3)$.** Let $X_{0,\mathrm{split}}(5,3)$ be the modular curve over $\mathbf{Q}$ which parametrizes triples $(E, C_5, \{C_3, C_3'\})$ where E is an elliptic curve, C_5 is a subgroup of order 5, and $\{C_3, C_3'\}$ is an unordered set of distinct subgroups of order 3.

Lemma 10. (i) *$X_{0,\mathrm{split}}(5,3)$ is a curve of genus one with four cusps, all of which are rational over $\mathbf{Q}$.*
(ii) $\#(X_{0,\mathrm{split}}(5,3)(\mathbf{Q})) = 8$.
(iii) *The four rational points of $X_{0,\mathrm{split}}(5,3)$ which are not cusps correspond to four triples $(E_i, \dots)$ where $j(E_i) \in \{11^3/2^3, -29^3 41^3/2^{15}\}$.*

Proof. The classical theory of modular curves shows that

$$X_{0,\mathrm{split}}(5,3)(\mathbf{C}) = \hat{\mathcal{H}}/(\Gamma_0(5) \cap \Gamma_{\mathrm{split}}(3)).$$

Thus the genus and the number of cusps of $X_{0,\mathrm{split}}(5,3)$ are given in Lemma 8. The fact that the four cusps are rational follows from the action of $G_{\mathbf{Q}}$ on the cusps (see for example [5] §VI.5).

Define a map f from $X_{0,\mathrm{split}}(5,3)$ to the jacobian $J_0(15)$ of $X_0(15)$ by

$$f : (E, C_5, \{C_3, C_3'\}) \mapsto (E, C_5 + C_3) + (E, C_5 + C_3') - 2[\infty]$$

where $[\infty]$ denotes the infinity cusp on $X_0(15)$. Then f is not constant, because if E is an elliptic curve without complex multiplication by $\mathbf{Q}(i)$ and C_3, C_3', C_3'' are three distinct subgroups of order 3, then

$$f(E, C_5, \{C_3, C_3'\}) - f(E, C_5, \{C_3, C_3''\}) = (E, C_5 + C_3') - (E, C_5 + C_3'')$$

which is not zero in $J_0(15)$ since $(E, C_5 + C_3'), (E, C_5 + C_3'')$ represent distinct points of $X_0(15)$ and $X_0(15)$ has genus 1 (Lemma 9(i)). Therefore

$X_{0,\mathrm{split}}(5,3)$ is isogenous to $X_0(15)$ and so $\#(X_{0,\mathrm{split}}(5,3)(\mathbf{Q})) \leq 8$ (see [1] or [2]).

In [2], one finds that the elliptic curve denoted 338E1, $E : y^2+xy+y = x^3+x^2+3x-5$ has a rational subgroup of order 5. Changing variables to the model $Y^2 = X^3+5X^2+56X-304$, one computes that the X-coordinates of the points of order 3 are the roots of $(X^2+12X+192)(3X^2-16X-48)$. Thus E gives rise to two of the non-cusp rational points of $X_{0,\mathrm{split}}(5,3)$, corresponding to the two choices of $G_{\mathbf{Q}}$-stable pairs of subgroups of order 3. (These two points are distinct because $\mathrm{Aut}(E) = \pm 1$.) One computes easily that $j(E) = 11^3/2^3$. The quotient of E by its rational subgroup of order 5 (denoted 338E2 in [2]) gives rise to the other two non-cusp points of $X_{0,\mathrm{split}}(5,3)(\mathbf{Q})$ and has j-invariant $-29^3 41^3/2^{15}$. $\square$

2.3. **Twists of modular curves.** Suppose $p > 2$ is a prime. As in [15] §1 or [7] §1, there is an open modular curve Y_p over $\mathbf{Q}$ whose points correspond to isomorphism classes of pairs (E, ϕ) where E is an elliptic curve and $\phi : E[p] \to \mathbf{F}_p \times \boldsymbol{\mu}_p$ is an isomorphism such that $\det(\phi) : \wedge^2 E[p] \to \boldsymbol{\mu}_p$ is the Weil pairing. Define X_p to be the compactification of Y_p. When $p = 3$ or 5 (which are the only cases we will use) X_p has genus zero, and explicit models are given in [7]. Over $\mathbf{Q}(\boldsymbol{\mu}_p)$, a choice of primitive p-th root of unity induces an isomorphism from X_p to one component of the usual modular curve $X(p)$.

More generally, suppose that V is a 2-dimensional $\mathbf{F}_p$-vector space with an action of $G_{\mathbf{Q}}$ and $\eta : \wedge^2 V \xrightarrow{\sim} \boldsymbol{\mu}_p$ is a non-degenerate alternating pairing. As explained in [7] Remark 2.4 or [15] §2, there is an open curve Y_V defined over $\mathbf{Q}$ whose points parametrize isomorphism classes of pairs (E, ϕ) where E is an elliptic curve and $\phi : E[5] \xrightarrow{\sim} V$ is an isomorphism such that

$$\eta \circ \det(\phi) : \wedge^2 E[5] \to \wedge^2 V \to \boldsymbol{\mu}_p$$

is the Weil pairing. Define X_V to be the compactification of Y_V. In particular $X_p = X_{\mathbf{F}_p \times \boldsymbol{\mu}_p}$.

Now suppose $p = 5$, E is an elliptic curve, $V = E[5]$, and η is the Weil pairing. We will write Y_E and X_E for Y_V and X_V, and we have the following explicit result ([7] §5).

Proposition 11. *Suppose E is an elliptic curve over $\mathbf{Q}$ and $p = 5$. There is an isomorphism $\psi : \mathbf{P}^1 \xrightarrow{\sim} X_E$ defined over $\mathbf{Q}$ and two polynomials $f_E(t), g_E(t) \in \mathbf{Q}[t]$ of degrees 20 and 30, respectively, such that if $t \in \mathbf{Q}$ is not in the finite set $\psi^{-1}(X_E(\mathbf{Q}) - Y_E(\mathbf{Q}))$ then $\psi(t) \in Y_E(\mathbf{Q})$ is represented by the elliptic curve*

$$E_t : y^2 = x^3 + f_E(t)x + g_E(t).$$

In particular for every such t, $\bar{\rho}_{E_t,5} \cong \bar{\rho}_{E,5}$.

Similarly there are open curves Y'_E and Y''_E which parametrize isomorphism classes of triples (E', ϕ, C_3) and $(E', \phi, \{C_3, C'_3\})$, respectively, where

- E' is an elliptic curve,
- $\phi : E'[5] \xrightarrow{\sim} E[5]$ is an isomorphism taking the Weil pairing on $E'[5]$ to that on $E[5]$,
- C_3 and C'_3 are (distinct) subgroups of order 3 of $E'[3]$.

These curves come equipped with natural (forgetful) maps to Y_E.

Lemma 12. *With notation as above, $Y'_E(\mathbf{Q})$ and $Y''_E(\mathbf{Q})$ are finite.*

Proof. Let X'_E and X''_E denote the compactifications of Y'_E and Y''_E, respectively. The classical theory of modular curves shows that $X'_E(\mathbf{C}) = \hat{\mathcal{H}}/(\Gamma(5) \cap \Gamma_0(3))$ and $X''_E(\mathbf{C}) = \hat{\mathcal{H}}/(\Gamma(5) \cap \Gamma_{\mathrm{split}}(3))$. Thus by Lemma 8, X'_E and X''_E are both curves of genus greater than 1, and the lemma follows from Faltings' Theorem (the Mordell Conjecture). □

3. Proof of the irreducibility theorem (Theorem 1)

Proposition 13. *Suppose E is an elliptic curve over $\mathbf{Q}$ such that*

- *$\bar{\rho}_{E,3}$ restricted to $G_{\mathbf{Q}(\sqrt{-3})}$ is not absolutely irreducible,*
- *$\bar{\rho}_{E,5}$ is reducible.*

Then

$$j(E) \in \{-25/2, -5^2 241^3/2^3, -5{\cdot}29^3/2^5, 5{\cdot}211^3/2^{15}, 11^3/2^3, -29^3 41^3/2^{15}\}.$$

Proof. Suppose C_5 is a nontrivial subgroup of $E[5]$ stable under $G_{\mathbf{Q}}$.

Case I: $\bar{\rho}_{E,3}$ is reducible. In this case $E[3]$ has a proper $G_{\mathbf{Q}}$-stable subgroup C_3, and then $(E, C_3 + C_5)$ represents a rational point of $X_0(15)$. By Lemma 9(iii), $j(E) \in \{-25/2, -5^2 241^3/2^3, -5{\cdot}29^3/2^5, 5{\cdot}211^3/2^{15}\}$.

Case II: $\bar{\rho}_{E,3}$ is irreducible. By Proposition 6, $E[3]$ has two distinct subgroups C_3, C'_3 of order 3 such that $(E, C_5, \{C_3, C'_3\})$ represents a rational point of $X_{0,\mathrm{split}}(5,3)$. By Lemma 10(iii), $j(E) \in \{11^3/2^3, -29^3 41^3/2^{15}\}$. □

Proof of Theorem 1. Theorem 1 is now immediate from Propositions 7 and 13 and Lemma 9(iv). □

4. Proof of the modularity theorem (Theorem 2)

Fix an elliptic curve E over $\mathbf{Q}$ which is semistable at 3 and such that $\bar{\rho}_{E,5}$ is irreducible. We want to show that $\bar{\rho}_{E,5}$ is modular.

Let the notation be as in §2.3 with $p = 5$. Define a subset B of $X_E(\mathbf{Q})$ by

$$B = (X_E(\mathbf{Q}) - Y_E(\mathbf{Q})) \cup \mathrm{image}(Y'_E(\mathbf{Q}) \to Y_E(\mathbf{Q})) \cup \mathrm{image}(Y''_E(\mathbf{Q}) \to Y_E(\mathbf{Q}))$$

and let $T = \psi^{-1}(B) \subset \mathbf{P}^1(\mathbf{Q})$. By Lemma 12, T is a finite set.

Suppose $t \in \mathbf{Q} - T$, and let E' be the elliptic curve $y^2 = x^3 + f_E(t)x + g_E(t)$ from Proposition 11. Since $\psi(t)$ is not the image of a rational point of Y'_E or of Y''_E, Proposition 6 shows that $\bar{\rho}_{E',3}$ restricted to $G_{\mathbf{Q}(\sqrt{-3})}$ is absolutely irreducible. Further, since E is semistable at 3, Proposition 7.1 of [15] shows that E' is semistable at 3 as well. Therefore by Theorems B and C applied with $p = 3$, E' is modular, and so $\bar{\rho}_{E',5}$ is modular. By Proposition 11, $\bar{\rho}_{E',5} \cong \bar{\rho}_{E,5}$, so $\bar{\rho}_{E,5}$ is modular. □

5. MOD 5 REPRESENTATIONS AND ELLIPTIC CURVES

Fix a prime $p > 2$ and a representation $\rho : G_{\mathbf{Q}} \to \mathrm{GL}_2(\mathbf{F}_p)$ such that $\det(\rho)$ is the cyclotomic character χ_p. Equivalently, fix a 2-dimensional $\mathbf{F}_p$-vector space $V(\rho)$ with a $G_{\mathbf{Q}}$-action such that $\wedge^2 V(\rho)$ is isomorphic to $\boldsymbol{\mu}_p$ as a $G_{\mathbf{Q}}$-module.

Let $V_p = \mathbf{Z}/p\mathbf{Z} \times \boldsymbol{\mu}_p$. There is a canonical identification $\wedge^2 V_p \cong \boldsymbol{\mu}_p$. Define

$$\mathrm{SL}(V_p) = \{\psi \in \mathrm{Aut}(V_p) : \det(\psi) = 1 \text{ on } \wedge^2 V_p\}.$$

Fix an $\mathbf{F}_p$-vector space isomorphism $\varphi : V_p \xrightarrow{\sim} V(\rho)$.

Lemma 14. *The map $\sigma \mapsto \varphi^{-1} \circ \varphi^{\sigma}$ is a one-cocycle on $G_{\mathbf{Q}}$ with values in $\mathrm{SL}(V_p)$. The class $c \in H^1(\mathbf{Q}, \mathrm{SL}(V_p))$ of this cocycle depends only on the isomorphism class of ρ, and not on the choice of φ.*

Proof. The fact that $\varphi^{-1} \circ \varphi^{\sigma} \in \mathrm{SL}(V_p)$ follows from the assumption on $\det(\rho)$. The other assertions of the lemma are all easy computations. (Note that we are using non-commutative Galois cohomology [8].) □

Let X_p be the modular curve defined in §2.3. Let $\mathcal{E}_p$ be the universal elliptic curve over X_p (see for example [12] or [14]), and $\iota : X_p \to \mathcal{E}_p$ the zero section. Finally, define the line bundle $\mathcal{L}_p$ to be the pull-back along ι of the cotangent bundle of $\mathcal{E}_p$ over X_p.

Define $\mathrm{Aut}_0(\mathcal{E}_p)$ to be the isomorphisms from $\mathcal{E}_p$ to itself which map the image of the zero section $\iota(X_p)$ to itself. There are natural $G_{\mathbf{Q}}$-equivariant maps

$$\mathrm{SL}(V_p) \to \mathrm{Aut}_0(\mathcal{E}_p) \to \mathrm{Aut}(\mathcal{L}_p) \to \mathrm{Aut}(X_p)$$

and we apply these maps to the class c of Lemma 14

$$(1) \quad \begin{array}{ccccc} H^1(\mathbf{Q}, \mathrm{SL}(V_p)) & \to & H^1(\mathbf{Q}, \mathrm{Aut}_{\bar{\mathbf{Q}}}(\mathcal{L}_p)) & \to & H^1(\mathbf{Q}, \mathrm{Aut}_{\bar{\mathbf{Q}}}(X_p)) \\ c & \mapsto & c_{\mathcal{L}} & \mapsto & c_X. \end{array}$$

Define $\eta_\rho = \det(\varphi^{-1}) : \wedge^2 V(\rho) \to \boldsymbol{\mu}_p$. Let $X(\rho)$ denote the modular curve $X_{V(\rho)}$ of §2.3. Then $X(\rho)$ contains an open curve $Y(\rho)$ whose points correspond to isomorphism classes of pairs (E, ϕ) where E is an elliptic curve and $\phi : E[p] \xrightarrow{\sim} V(\rho)$ is an isomorphism such that $\eta_\rho \circ \det(\phi) : \wedge^2 E[p] \to \wedge^2 V(\rho) \to \boldsymbol{\mu}_p$ is the Weil pairing. In particular (see for example

Proposition 2.1 of [7]) a point of $Y(\rho)(\mathbf{Q})$ gives an elliptic curve E over $\mathbf{Q}$ such that $\bar{\rho}_{E,p} \cong \rho$.

As explained in [15] §2 and [7] Remark 2.4, $X(\rho)$ is the twist of X_p by c_X in the sense of [8]. Now suppose $p = 3$ or 5, so $X_p \cong \mathbf{P}^1$ over $\mathbf{Q}$. To prove Theorem 4, we need only show that $Y(\rho)(\mathbf{Q})$ is nonempty. We will do this by showing that $c_X = 0$, so $X(\rho) \cong X_p \cong \mathbf{P}^1$ over $\mathbf{Q}$. For a related discussion see [10].

Lemma 15. *Suppose $p = 3$ or 5, and fix an isomorphism $X_p \cong \mathbf{P}^1$ over $\mathbf{Q}$.*

(i) $\mathrm{Aut}_{\bar{\mathbf{Q}}}(X_p) \cong \mathrm{PGL}_2(\bar{\mathbf{Q}}) \cong \mathrm{PSL}_2(\bar{\mathbf{Q}})$.

(ii) $\mathrm{Aut}_{\bar{\mathbf{Q}}}(\mathcal{L}_p) \cong \mathrm{GL}_2(\bar{\mathbf{Q}})/\mu_m$, *where*

$$m = \begin{cases} 1 & \text{if } p = 3 \\ 5 & \text{if } p = 5 \end{cases}$$

and μ_m lies inside $\mathrm{GL}_2(\bar{\mathbf{Q}})$ as scalars.

Proof. Assertion (i) is well-known.

By [4] §12.1, the degree of the line bundle $\mathcal{L}_p$ is $\#(\mathrm{PSL}_2(\mathbf{F}_p))/12 = m$. Since $X_p \cong \mathbf{P}^1$, $\mathcal{L}_p$ is determined by its degree so we can identify $H^0(X_p, \mathcal{L}_p)$, the space of global sections of $\mathcal{L}_p$, with the space of homogeneous polynomials of degree m in two variables. Further $\mathcal{L}_p$ is generated by its global sections, so this identification gives

$$\mathrm{Aut}(\mathcal{L}_p) \cong \mathrm{Aut}(H^0(X_p, \mathcal{L}_p)) \cong \mathrm{GL}_2/\mu_m$$

where $\left(\begin{smallmatrix} a & b \\ c & d \end{smallmatrix}\right) \in \mathrm{GL}_2$ sends a polynomial $f(x, y)$ to $f(ax + by, cx + dy)$. □

Proof of Theorem 3. To simplify notation we give the proof in the case that $F = \mathbf{Q}$. The case of an arbitrary field of characteristic different from p is identical.

Let $p = 3$ or 5 and $m = 1$ or 5, respectively, as in Lemma 15(ii). Let $\rho : G_{\mathbf{Q}} \to \mathrm{GL}_2(\mathbf{F}_p)$ be a representation whose determinant is χ_p and let c_X and $c_{\mathcal{L}}$ be as above. By Lemma 15, there are exact sequences

$$0 \to \pm 1 \to \mathrm{SL}_2(\bar{\mathbf{Q}}) \to \mathrm{Aut}_{\bar{\mathbf{Q}}}(X_p) \to 0,$$
$$0 \to \mu_m \to \mathrm{GL}_2(\bar{\mathbf{Q}}) \to \mathrm{Aut}_{\bar{\mathbf{Q}}}(\mathcal{L}_p) \to 0.$$

Since $H^1(\mathbf{Q}, \mathrm{GL}_2(\bar{\mathbf{Q}})) = H^1(\mathbf{Q}, \mathrm{SL}_2(\bar{\mathbf{Q}})) = 0$ (see [8] III Lemme 1), the induced long exact sequences from Galois cohomology show that $c_X \in H^1(\mathbf{Q}, \mathrm{Aut}_{\bar{\mathbf{Q}}}(X_p))$ is killed by 2 and $c_{\mathcal{L}} \in H^1(\mathbf{Q}, \mathrm{Aut}_{\bar{\mathbf{Q}}}(\mathcal{L}_p))$ is killed by m. But m is odd, and c_X is the image of $c_{\mathcal{L}}$ under (1), so $c_X = 0$. Therefore $X(\rho)$ is isomorphic over $\mathbf{Q}$ to $\mathbf{P}^1$, so $Y(\rho)(\mathbf{Q})$ is infinite and there are (infinitely many) elliptic curves E over $\mathbf{Q}$ with $\bar{\rho}_{E,p}$ isomorphic to ρ. □

Remark . The above proof of Theorem 3 relies heavily on the fact that X_p has genus zero, which is true for $p = 3$ and 5 but false for $p > 5$. However, one can replace p by 4 and define X_4 corresponding to $V_4 = \mathbf{Z}/4\mathbf{Z} \times \mu_4$ (see [15]). Then X_4 has genus zero, and one can define $\mathcal{E}_4$ and $\mathcal{L}_4$ and apply the method of the proof above. But the corresponding integer m of Lemma 15(ii) is $\#(\mathrm{PSL}_2(\mathbf{Z}/4\mathbf{Z}))/12 = 2$, so we cannot conclude that $c_X = 0$.

In fact, it is not true that every representation $\rho : G_{\mathbf{Q}} \to \mathrm{GL}_2(\mathbf{Z}/4\mathbf{Z})$ with determinant equal to the mod 4 cyclotomic character is isomorphic to the representation of $G_{\mathbf{Q}}$ on $E[4]$ for some elliptic curve E over $\mathbf{Q}$. For example, let ρ be the nontrivial representation of $G_{\mathbf{Q}}$ into $\{1, (\begin{smallmatrix} 1 & 2 \\ 2 & 3 \end{smallmatrix})\} \subset \mathrm{GL}_2(\mathbf{Z}/4\mathbf{Z})$ factoring through $\mathrm{Gal}(\mathbf{Q}(i)/\mathbf{Q})$. Then $\det(\rho)$ is the mod 4 cyclotomic character, but if ρ came from an elliptic curve E we would have $E(\mathbf{R})[4] \cong \mathbf{Z}/2\mathbf{Z} \times \mathbf{Z}/2\mathbf{Z}$ which is impossible.

REFERENCES

[1] B. Birch, W. Kuyk, eds., *Modular functions of one variable IV.* Lecture Notes in Math. **476** Springer-Verlag, Berlin-New York (1975) 74–144.

[2] J. Cremona, Algorithms for modular elliptic curves, Cambridge Univ. Press, Cambridge (1992).

[3] F. Diamond, On deformation rings and Hecke rings. To appear in *Annals of Math.*

[4] F. Diamond, J. Im, Modular forms and modular curves. In: *Seminar on Fermat's Last Theorem.* Canadian Math. Soc. Conf. Proc. **17** (1995) 39-133.

[5] P. Deligne, M. Rapoport, Les schémas de modules de courbes elliptiques. In: *Modular functions of one variable II.* Lecture Notes in Math. **349** (1973) 143–316.

[6] R. Langlands, *Base change for* GL(2). Annals of Math. Studies **96**, Princeton University Press, Princeton (1980).

[7] K. Rubin, A. Silverberg, Families of elliptic curves with constant mod p representations. In: *Elliptic curves, modular forms, and Fermat's Last Theorem (Hong Kong, December 1994),* Coates and Yau, eds. International Press, Cambridge (1995) 148–161.

[8] J-P. Serre, *Cohomologie galoisienne.* Lecture Notes in Math. **5**, Springer-Verlag, Berlin-New York (1964).

[9] ———, Propriétés galoisiennes des points d'ordre fini des courbes elliptiques. *Invent math.* **15** (1972) 259–331.

[10] ———, Extensions icosaédriques (Letter to J.D. Gray). In: *Collected Papers* Vol. III, Springer-Verlag, Berlin-New York (1986) 550–554.

[11] N.I. Shepherd-Barron, R. Taylor, Mod 2 and mod 5 icosahedral representations. *J. Amer. Math. Soc.* **10** (1997) 283–298.

[12] G. Shimura, Moduli and fibre systems of abelian varieties. *Annals of Math* **83** (1966) 294–338.

[13] ———, *Introduction to the arithmetic theory of automorphic functions.* Princeton Univ. Press, Princeton (1971).

[14] T. Shioda, On elliptic modular surfaces. *J. Math. Soc. Japan* **24** (1972) 20–59.

[15] A. Silverberg, Explicit families of elliptic curves with prescribed mod N representations. This volume.

[16] J. Silverman, *The arithmetic of elliptic curves,* Graduate Texts in Math. **106**, Springer-Verlag, Berlin-New York (1986).

[17] R. Taylor, A. Wiles, Ring theoretic properties of certain Hecke algebras. *Annals of Math.* **141** (1995) 553–572.

[18] J. Tunnell, Artin's conjecture for representations of octahedral type. *Bull. A.M.S.* **5** (1981) 173–175.

[19] A. Wiles, Modular elliptic curves and Fermat's Last Theorem. *Annals of Math.* **141** (1995) 443–551.

AN EXTENSION OF WILES' RESULTS

FRED DIAMOND

1. Introduction

Suppose that E is an elliptic curve defined over $\mathbf{Q}$. We wish to prove that E is modular, or equivalently, that the associated ℓ-adic representation

$$\rho : G_{\mathbf{Q}} \to GL_2(\mathbf{Z}_\ell)$$

is modular for some prime ℓ.

If we are assuming that E is semistable, i.e., has square-free conductor, then we can impose some convenient hypotheses on the local behavior of the Galois representations ρ we consider. By "local behavior," we mean the behavior of the representation

$$\theta : G_p \to GL_2(\mathbf{Z}_\ell)$$

defined by restricting ρ to a decomposition group at p.

Recall that if E has good reduction at p, then θ is unramified. If E has multiplicative reduction at p, then a convenient description of θ results from the Tate parametrization of E (§17 of [S]). In particular, we see that

$$\theta|_{I_p} \sim \begin{pmatrix} 1 & * \\ 0 & 1 \end{pmatrix}.$$

To consider elliptic curves with additive reduction at some primes $p \neq \ell$, we must allow more general types of θ. We can actually consider representations ρ with arbitrary local behavior at primes $p \neq \ell$. This is carried out in [D] where, building on the work of Wiles [W] and Taylor-Wiles [TW], we prove a result of the form

$$(1) \qquad \bar{\rho} \text{ modular} \Rightarrow \rho \text{ modular},$$

and deduce

Theorem 1.1. *If E has good or multiplicative reduction at 3 and 5, then E is modular.*

The details of the proof can be found in [D]. Here we give an exposition which we hope is more motivated and systematic. We often follow [DDT], admitting results which are straightforward generalizations of those there or elsewhere in this volume. For the proofs of some of the key lemmas, we refer completely to [D].

This article is structured as follows:

Some background is given in §2 on local Galois representations

$$\sigma : G_p \to GL_2(k),$$

where $p \neq \ell$ and k is an algebraic closure of $\mathbf{F}_\ell$. A classification of the possible σ, though not logically necessary for the proof of theorem 1.1, helps provide some insight into local Galois representations and their deformations. (The appendix with K. Kramer determines precisely how local Galois representations arising from elliptic curves fit into this classification.)

In §3 we explain what it means for a deformation of σ to be "minimally ramified" at p.

Suppose that ℓ is odd and

$$\bar{\rho} : G_{\mathbf{Q}} \to GL_2(k)$$

is an irreducible representation which is semistable at ℓ. We formulate in §4 a certain deformation problem for each finite set of primes Σ. This deformation problem turns out to be representable by a ring R_Σ whose tangent space is described in terms of Galois cohomology (see [M]).

Suppose now that $\bar{\rho}$ is modular. The goal of §5 is to define a corresponding Hecke algebra $\mathbf{T}_\Sigma$, and modular deformation

$$\tau : G_{\mathbf{Q}} \to GL_2(\mathbf{T}_\Sigma)$$

arising from a homomomorphism $\phi_\Sigma : R_\Sigma \to \mathbf{T}_\Sigma$ (see [Ri2]).

If we can show that it is an isomorphism, then we obtain a result of the form (1) as a corollary. The main results are stated in §6.

To prove that ϕ_Σ is an isomorphism, we must modify some of the techniques used in [W] and [TW]. In particular, the analysis of the Hecke algebras becomes more difficult. In our sketch of the proof in §7, we indicate where the complications arise, but give only a rough idea of how they are dealt with in [D].

2. Local representations mod ℓ

Suppose that p is a prime. We let $G_p = \mathrm{Gal}(\bar{\mathbf{Q}}_p/\mathbf{Q}_p)$ and let I_p denote the inertia subgroup of G_p. Suppose that ℓ is an odd prime different from p and consider continuous representations

$$\sigma : G_p \to \mathrm{GL}_2(k),$$

where k is an algebraic closure of $\mathbf{F}_\ell$. We let $\tilde{\sigma}$ denote the associated projective representation.

We let χ denote the cyclotomic character $G_p \to k^\times$. Note that χ is nontrivial if and only if $p \not\equiv 1 \bmod \ell$. In that case, we write sp_2 for the representation

$$\begin{pmatrix} \chi & u \\ 0 & 1 \end{pmatrix}, \tag{2}$$

where u is a cocycle representing the image of a uniformizer under the Kummer map

$$\mathbf{Q}_p^\times \to \mathbf{Q}_p^\times/(\mathbf{Q}_p^\times)^\ell \to H^1(G_p, k(1)).$$

The equivalence class is independent of the choice of uniformizer and cocycle. (See the proof of proposition 2.2 below.)

If ψ is a character $G_p \to k^\times$, then we write $k(\psi)$ for the one-dimensional vector space over k on which G_p acts via ψ. Recall that two representations σ_1 and σ_2 are called twist-equivalent if σ_1 is equivalent to $\psi \otimes \sigma_2$ for some character $\psi : G_p \to k^\times$.

We classify σ according to the following four types of behavior (principal, special, vexing or harmless).

P : σ is reducible and $\sigma|_{I_p}$ is decomposable.
S : σ is reducible and $\sigma|_{I_p}$ is indecomposable.
V : σ is irreducible and $\sigma|_{I_p}$ is reducible.
H : σ is irreducible and $\sigma|_{I_p}$ is irreducible.

Proposition 2.1. *The following are equivalent:*

1. *σ is reducible and $\sigma|_{I_p}$ is decomposable.*
2. *σ is twist-equivalent to a representation either of the form*
   (a) $\begin{pmatrix} \psi & 0 \\ 0 & 1 \end{pmatrix}$ *for some character ψ, or*
   (b) $\begin{pmatrix} 1 & \phi \\ 0 & 1 \end{pmatrix}$ *for some additive unramified character ϕ.*
3. *Either $\tilde{\sigma}(G_p)$ is cyclic of order not divisible by ℓ, or it has order ℓ and $\tilde{\sigma}(I_p)$ is trivial.*
4. *$\tilde{\sigma}(G_p)$ is cyclic and the order of $\tilde{\sigma}(I_p)$ is not divisible by ℓ.*

Proof: Suppose 1 holds. Then σ is twist-equivalent to a representation of the form

$$\begin{pmatrix} \psi & u \\ 0 & 1 \end{pmatrix}$$

for some character ψ, where u is a cocycle representing a class

$$x \in H^1(G_p, k(\psi)).$$

If σ is indecomposable, then x is nontrivial. On the other hand, the image of x in $H^1(I_p, k(\psi))$ vanishes, so x is in the image of $H^1(G_p/I_p, k(\psi)^{I_p})$. This last group is trivial unless ψ is trivial, so 2 follows.

The implications $2 \Rightarrow 3 \Rightarrow 4 \Rightarrow 1$ are clear.

Proposition 2.2. *The following are equivalent:*

1. *σ is reducible and $\sigma|_{I_p}$ is indecomposable.*
2. *$p \equiv 1 \bmod \ell$ and σ is twist-equivalent to a representation of the form*
$$\begin{pmatrix} 1 & \phi \\ 0 & 1 \end{pmatrix}$$
for some additive ramified character ϕ, or $p \not\equiv 1 \bmod \ell$ and σ is twist-equivalent to sp_2.
3. *$p \equiv 1 \bmod \ell$, $\tilde{\sigma}(I_p)$ has order ℓ and $\tilde{\sigma}(G_p)$ has order dividing ℓ^2, or $p \not\equiv 1 \bmod \ell$, $\tilde{\sigma}(I_p)$ has order ℓ and $\tilde{\sigma}(G_p)$ has order $d\ell$ where d is the order of p in $\mathbf{F}_\ell^\times$.*
4. *$\tilde{\sigma}(I_p)$ is cyclic of order divisible by ℓ.*

Proof: Suppose 1 holds. Then σ is twist-equivalent to a representation of the form
$$\begin{pmatrix} \psi & u \\ 0 & 1 \end{pmatrix}$$
for some character ψ where u is a cocycle representing a class
$$x \in H^1(G_p, k(\psi)).$$
Since $\sigma|_{I_p}$ is indecomposable, the image of x in $H^1(I_p, k(\psi))^{G_p}$ does not vanish. This group is isomorphic to $\mathrm{Hom}_{G_p}(\mu_\ell(\bar{\mathbf{Q}}_p), k(\psi))$, which vanishes unless $\psi = \chi$. Moreover if χ is non-trivial, then $H^1(G_p, k(1))$ is one-dimensional over k, so 2 follows.

The implications $2 \Rightarrow 3 \Rightarrow 4$ are clear. If 4 holds, then 1 follows from the fact that $\sigma(G_p)$ is contained in the normalizer of the ℓ-Sylow subgroup of $\sigma(I_p)$.

Proposition 2.3. *The following are equivalent:*

1. *σ is irreducible and $\sigma|_{I_p}$ is reducible.*
2. *σ is equivalent to a representation of the form $\mathrm{Ind}_{G_M}^{G_p} \xi$, where M is the unramified quadratic extension of $\mathbf{Q}_p$ and ξ is a character of G_M not equal to its conjugate under the action of $\mathrm{Gal}(M/\mathbf{Q}_p)$.*
3. *$\tilde{\sigma}(I_p)$ is cyclic of order not divisible by ℓ, and $\tilde{\sigma}(G_p)$ is dihedral of twice that order.*
4. *$\tilde{\sigma}(I_p)$ is cyclic of order not divisible by ℓ, and $\tilde{\sigma}(G_p)$ is not cyclic.*

Proof: Suppose that 1 holds. Consider the action of G_p on $\mathbf{P}^1(k)$ gotten from $\tilde{\sigma}$. Note first that $\tilde{\sigma}(I_p)$ is nontrivial. Let S denote the set of elements in $\mathbf{P}^1(k)$ fixed by I_p. Since $\sigma|_{I_p}$ is reducible, S is not empty. Since σ is irreducible, S has no elements fixed by G_p and it follows that S has exactly two elements. Moreover G_p acts transitively on S via the unramified quadratic character, so 2 holds.

Suppose next that 2 holds. Then $\tilde{\sigma}(G_p)$ is a dihedral group in which $\tilde{\sigma}(G_M)$ is a cyclic subgroup of index two and order not divisible by ℓ. Since

$M^\times = \mathbf{Q}_p^\times \mathcal{O}_M^\times$, we see from local class field theory that

$$\xi^{-1}\xi'(G_M) = \xi^{-1}\xi'(I_p),$$

where ξ' is the conjugate of ξ, and 3 follows. The implication $3 \Rightarrow 4$ is clear, and $4 \Rightarrow 1$ follows from the converse of the corresponding one in Proposition 2.1.

Proposition 2.4. *The following are equivalent:*

1. $\sigma|_{I_p}$ *is irreducible.*
2. p *is odd and* σ *is equivalent to a representation of the form* $\mathrm{Ind}_{G_M}^{G_p}\xi$, *where* M *is a ramified quadratic extension of* $\mathbf{Q}_p$ *and* ξ *is a character of* G_M *whose restriction to* I_M *is not equal to its conjugate under the action of* $\mathrm{Gal}(M/\mathbf{Q}_p)$, *or* $p = 2$ *and the restriction of* σ *to the wild inertia subgroup of* G_p *is irreducible.*
3. $\tilde{\sigma}(I_p)$ *is dihedral of order* $2p^r$ *for some* $r \geq 1$ *and* $\tilde{\sigma}(G_p)$ *is dihedral of order dividing* $4p^r$, *or* $p = 2$, $\tilde{\sigma}(I_p)$ *(respectively* $\tilde{\sigma}(G_p)$*) is isomorphic to* D_4 *(respectively* A_4*),* A_4 *(respectively* A_4*) or* A_4 *(respectively* S_4*).*
4. $\tilde{\sigma}(I_p)$ *is not cyclic.*

Proof: Suppose that 1 holds and furthermore that $\sigma|_{P_p}$ is irreducible, where P_p is the wild inertia subgroup of I_p. Consider the action of G_p on $\mathbf{P}^1(k)$ gotten from $\tilde{\sigma}$. Since $\tilde{\sigma}(I_p)$ is not cyclic, we see that $\tilde{\sigma}(P_p)$ is nontrivial. Let S denote the set of elements in $\mathbf{P}^1(k)$ fixed by P_p. Then S is not empty and has no elements fixed by I_p. It follows that S has exactly two elements and that I_p acts transitively. Therefore p is odd and G_p acts transitively on S via a ramified quadratic character. We deduce that 2 holds, where M is the corresponding quadratic extension of $\mathbf{Q}_p$. (We have that $\xi \neq \xi'$ on P_p, hence on I_M.)

Suppose now that 2 holds. First consider the case of odd p. Then $\tilde{\sigma}(G_p)$ (respectively, $\tilde{\sigma}(I_p)$) is dihedral, and $\tilde{\sigma}(G_M)$ (respectively, $\tilde{\sigma}(I_M)$) is a cyclic subgroup of index two. Letting U denote the kernel of the reduction map on $\mathcal{O}_M^\times$, we have $\mathbf{Q}_p^\times U = \mathbf{Q}_p^\times \mathcal{O}_M^\times$ has index two in $M^\times$. From local class field theory it follows that $\tilde{\sigma}(I_M) = \tilde{\sigma}(P_M)$ has p-power order and index at most two in $\tilde{\sigma}(G_M)$. We conclude that 3 holds.

In the case of $p = 2$, we see that $D = \tilde{\sigma}(P_p)$ is dihedral, since it is not cyclic and is a finite subgroup of $PGL_2(k)$ of 2-power order. Furthermore $\tilde{\sigma}(G_p)$ is contained in the normalizer of D. If D has order greater than 4, the normalizer is dihedral and we may use the same argument as in the case of odd p. If D has order 4, then the normalizer is isomorphic to S_4, and 4 follows.

The implication $3 \Rightarrow 4$ is clear, as is $4 \Rightarrow 1$ (in view of the preceding propositions).

3. Minimally ramified liftings

For a fixed representation

$$\sigma : G_p \to \mathrm{GL}_2(k),$$

we consider liftings

$$\theta : G_{\mathbf{Q}} \to \mathrm{GL}_2(R)$$

of σ, where R is a complete local Noetherian $W(k)$-algebra with residue field k. We shall now say what it means for θ to be minimally ramified. We use $\tilde{}$ to denote composition with the Teichmüller lift

$$k^\times \to W(k)^\times \to R^\times.$$

Definition 3.1.

1. If σ is of type **P** or **V** , then

$$\sigma|_{I_p} \sim \begin{pmatrix} \xi_1 & 0 \\ 0 & \xi_2 \end{pmatrix},$$

and we say θ is *minimally ramified* if

$$\theta|_{I_p} \sim \begin{pmatrix} \tilde{\xi}_1 & 0 \\ 0 & \tilde{\xi}_2 \end{pmatrix}.$$

2. If σ is of type **S** , then

$$\sigma|_{I_p} \sim \xi \otimes \begin{pmatrix} 1 & * \\ 0 & 1 \end{pmatrix},$$

and we say θ is *minimally ramified* if

$$\theta|_{I_p} \sim \tilde{\xi} \otimes \begin{pmatrix} 1 & * \\ 0 & 1 \end{pmatrix}.$$

3. If σ is of type **H** , then we say θ is *minimally ramified* if $\det \theta|_{I_p}$ is the Teichmüller lift of $\det \sigma|_{I_p}$.

Remark 3.2. First note that if χ is a character of $G_p \to k^\times$, then θ is a minimally ramified lifting of σ if and only if $\tilde{\chi} \otimes \theta$ is a minimally ramified lifting of $\chi \otimes \sigma$.

Remark 3.3. If σ is of type **P** , then it has a twist which is either unramified or of type B in the terminology of [W]. Note that if σ is unramified, then θ is minimally ramified if and only if θ is unramified.

Remark 3.4. If σ is of type **S** , then it has a twist of type A in the terminology of [W]. Recall that if θ arises from the ℓ-adic Tate module of an elliptic curve E over $\mathbf{Q}_p$ with split multiplicative reduction, then $\theta|_{I_p}$ is equivalent to a representation of the form $\begin{pmatrix} 1 & * \\ 0 & 1 \end{pmatrix}$. It is minimally ramified if and only if σ is ramified if and only if $v_p(\Delta_E)$ is divisible by ℓ.

Remark 3.5. Suppose now that σ is type **V** . If $p \not\equiv -1 \bmod \ell$, then σ is type C in the terminology of [W]. Suppose instead that $p \equiv -1 \bmod \ell$ and write $\sigma = \mathrm{Ind}_{G_M}^{G_p} \xi$ as in proposition 2.3. Let $\mu : G_M \to \mathcal{O}^\times$ be a ramified character of G_M of ℓ-power order. Then

$$\theta = \mathrm{Ind}_{G_M}^{G_p} \tilde{\xi}\mu \tag{3}$$

is a lifting of σ which is *not* minimally ramified.

Remark 3.6. Now consider σ of type **H** . Suppose that

$$\sigma|_{I_p} \sim \mathrm{Ind}_{I_M}^{I_p} \xi$$

as in proposition 2.4. Then θ is minimally ramified if and only if

$$\theta|_{I_p} \sim \mathrm{Ind}_{I_M}^{I_p} \tilde{\xi}.$$

Remark 3.7. Suppose that $\theta : G_p \to \mathrm{GL}_2(\mathcal{O})$ is a minimally ramified lifting of σ, where $\mathcal{O}$ is the ring of integers of a finite extension of the field of fractions of $W(k)$. Then $\det \theta|_{I_p}$ is the Teichmüller lift of $\det \sigma|_{I_p}$ and the Artin conductors of θ and σ coincide. In [W] and [TW] a technical hypothesis is imposed to ensure that a partial converse holds. This hypothesis rules out the existence of liftings as in (3) and facilitates the characterization of the modular forms which give rise to minimally ramified liftings. The main contribution of [D] is to dispense with that hypothesis.

4. UNIVERSAL DEFORMATION RINGS

Now consider an irreducible representation

$$\bar{\rho} : G_{\mathbf{Q}} \to \mathrm{GL}_2(k).$$

For each prime p we fix an embedding of $\bar{\mathbf{Q}}$ in $\bar{\mathbf{Q}}_p$ and regard G_p as a decomposition group in $G_{\mathbf{Q}}$. We suppose that $\bar{\rho}|_{G_\ell}$ is semistable in the sense of [DDT], section 2.4.

Suppose that K is a finite extension of the field of fractions of $W(k)$. Let $\mathcal{O}$ denote the integral closure of $W(k)$ in K; thus $\mathcal{O}$ is a complete discrete valuation ring with residue field k. We consider liftings of $\bar{\rho}$ of the form

$$\rho : G_{\mathbf{Q}} \to \mathrm{GL}_2(R),$$

where R is in the category **C** of local complete $\mathcal{O}$-algebras with residue field k. A deformation of $\bar{\rho}$ is an isomorphism class of such liftings (see [dSL] (2.1), (2.2)).

If Σ is a finite set of primes, we say that ρ is type Σ if

1. $\chi_\ell^{-1} \det \rho$ has finite order not divisible by ℓ;
2. ρ is minimally ramified outside Σ;
3. ρ is semistable at ℓ in the sense of [DDT].

The notion depends only on the isomorphism class of ρ and is independent of the choice of embeddings of $\bar{\mathbf{Q}}$ in $\bar{\mathbf{Q}}_p$.

Consider the functor which associates to R the set of deformations of $\bar{\rho}$ of type Σ. The type Σ restriction satisfies the conditions listed at the beginning of §6 of [dSL] (see also §29 of [M] and §2.4 of [DDT]). From [dSL] (2.4) and (6.1) we conclude that the functor is represented by a complete local $\mathcal{O}$-algebra R_Σ, the identity map of R_Σ corresponding to the universal deformation of type Σ:

$$\rho_\Sigma^{\mathrm{univ}} : G_{\mathbf{Q}} \to \mathrm{GL}_2(R_\Sigma).$$

Suppose now that we are given a lifting

$$\rho : G_{\mathbf{Q}} \to \mathrm{GL}_2(\mathcal{O})$$

of type Σ. The universal property of R_Σ yields a surjective morphism

$$\pi : R_\Sigma \to \mathcal{O}$$

such that ρ is equivalent to the pushforward of $\rho_\Sigma^{\mathrm{univ}}$. Let $\mathfrak{p}$ denote the kernel of π. We define the group

$$H^1_\Sigma(G_{\mathbf{Q}}, (\mathrm{ad}^0\rho) \otimes (K/\mathcal{O}))$$

as in §2.7 of [DDT]. A generalization of results of Mazur (see §23–25 of [M]) yields a canonical isomorphism

$$\mathrm{Hom}_{\mathcal{O}}(\mathfrak{p}/\mathfrak{p}^2, K/\mathcal{O}) \cong H^1_\Sigma(G_{\mathbf{Q}}, (\mathrm{ad}^0\rho) \otimes_{\mathcal{O}} (K/\mathcal{O})). \tag{4}$$

5. HECKE ALGEBRAS

Recall that given a newform

$$f(\tau) = \sum a_n(f) e^{2\pi i n\tau}$$

of weight 2, level N_f and character ψ_f, a construction of Eichler and Shimura (see [Ro]) associates to f a continuous representation

$$\rho_f : G_{\mathbf{Q}} \to \mathrm{GL}_2(\bar{\mathbf{Q}}_\ell),$$

where we have fixed embeddings $\bar{\mathbf{Q}} \to \mathbf{C}$ and $\bar{\mathbf{Q}} \to \bar{\mathbf{Q}}_\ell$. The representation ρ_f is characterized up to isomorphism by the following property: For all primes p not dividing $N_f\ell$, ρ_f is unramified at p and the characteristic polynomial of $\rho_f(\mathrm{Frob}_p)$ is

$$X^2 - a_p(f)X + \psi_f(p)p.$$

We wish to continue working over the ring $\mathcal{O}$ introduced above, so we also fix an embedding $\bar{\mathbf{Q}}_\ell \to \bar{K}$ and view ρ_f as taking values in $\mathrm{GL}_2(K_f)$, where K_f is the subfield of $\bar{K}$ generated by K and the Fourier coefficents of f. We denote the ring of integers $\mathcal{O}_f$, which we regard as an object of $\mathbf{C}$. Define

$$\bar{\rho}_f : G_{\mathbf{Q}} \to \mathrm{GL}_2(k)$$

as the semisimplification of the reduction of ρ_f.

We assume that our fixed representation $\bar{\rho}$ is modular in the sense that it is isomorphic to $\bar{\rho}_f$ for some weight 2 newform f. We let Φ_Σ denote the set of newforms g such that ρ_g is a deformation of $\bar{\rho}$ of type Σ and N_g is not divisible by ℓ^2.

Theorem 5.1. *If $\bar{\rho}$ is modular, then $\Phi_\emptyset \neq \emptyset$.*

This is a refinement of Serre's ϵ-conjecture for which a crucial ingredient is Ribet's theorem [Ri1] (see [E]). The result stated here is a consequence of [D] which builds on the work Ribet and many others.

For each g in Φ_Σ, we consider the map $R_\Sigma \to \mathcal{O}_g$ corresponding to ρ_g. We then define

$$\mathbf{T}_\Sigma \subset \prod_{g \in \Phi_\Sigma} \mathcal{O}_g$$

as the image of R_Σ. Since R_Σ is topologically generated by traces, we may also regard $\mathbf{T}_\Sigma$ as the $\mathcal{O}$-subalgebra generated by the elements

$$T_p = (a_p(g))_{g \in \Phi_\Sigma}$$

for primes p not dividing $N\ell$. We wish to prove that the surjective map

$$\phi_\Sigma : R_\Sigma \to \mathbf{T}_\Sigma$$

is an isomorphism. Note that Φ_Σ gives rise to a type Σ deformation

$$\rho_\Sigma^{\mathrm{mod}} : G_{\mathbf{Q}} \to \mathrm{GL}_2(\mathbf{T}_\Sigma)$$

of $\bar{\rho}$, such that for each $g \in \Phi$, the composition with the projection to $\mathrm{GL}_2(\mathcal{O}_g)$ is equivalent to ρ_g.

For finite sets of primes $\Sigma \supset \Theta$, there is a natural surjective homomorphism $R_\Sigma \to R_\Theta$ defined by regarding $\rho_\Theta^{\mathrm{univ}}$ as a deformation of $\bar{\rho}$ of type Σ. We have also the natural surjection $\mathbf{T}_\Sigma \to \mathbf{T}_\Theta$ so that the diagram

$$\begin{array}{ccc} R_\Sigma & \stackrel{\phi_\Sigma}{\to} & \mathbf{T}_\Sigma \\ \downarrow & & \downarrow \\ R_\Theta & \stackrel{\phi_\Theta}{\to} & \mathbf{T}_\Theta \end{array} \tag{5}$$

commutes.

6. THE MAIN RESULTS

Recall our assumption that ℓ is odd and $\bar{\rho}$ is semistable at ℓ. We let $L = \mathbf{Q}(\sqrt{\varepsilon\ell})$, where $\varepsilon = (-1)^{(\ell-1)/2}$. We suppose that Σ is an arbitrary finite set of primes. The main result is the following:

Theorem 6.1. *If $\bar{\rho}|_{G_L}$ is irreducible, then ϕ_Σ is an isomorphism and $\mathbf{T}_\Sigma$ is a complete intersection.*

We shall sketch the proof below referring to [D] for the full details.

Corollary 6.2. *Suppose that $\rho : G_{\mathbf{Q}} \to \mathrm{GL}_2(\mathcal{O})$ is continuous and unramified outside a finite set of primes. Suppose that ρ is semistable at ℓ and $\bar{\rho}|_{G_L}$ is irreducible. If $\bar{\rho}$ is modular, then ρ is modular.*

Applying the Langlands-Tunnell theorem (see [G]) as in [Ru], we conclude:

Corollary 6.3. *Suppose that E is an elliptic curve over $\mathbf{Q}$ with good or multiplicative reduction at 3, and that $[\mathbf{Q}(E[3]) : \mathbf{Q}] = 16$ or 48. Then E is modular.*

We refer to [Ru] for the deduction of theorem 1.1 from corollaries 6.2 and 6.3.

7. Sketch of proof

7.1. **Vague principle.** A formulation of the problem such as theorem 6.1 enables us to use tools from commutative algebra. We shall use information about the vertical maps in (5) and one of the horizontal maps to prove that the other horizontal map is an isomorphism. The information about $R_\Sigma \to R_\Theta$ comes from the description of tangent spaces in terms of Galois cohomology (4); the information about $\mathbf{T}_\Sigma \to \mathbf{T}_\Theta$ comes from the connection with congruences between modular forms.

7.2. **Some preparation.** We begin with two reduction steps and a definition.

One can check that if ψ is a character $G_{\mathbf{Q}} \to k^\times$ unramified outside ℓ, then theorem 6.1 holds for $\bar{\rho}$ if and only if it holds for $\bar{\rho}' = \bar{\rho} \otimes \chi$. Indeed if we define $\phi'_\Sigma : R'_\Sigma \to \mathbf{T}'_\Sigma$ using $\bar{\rho}'$ instead of $\bar{\rho}$, then we obtain a natural commutative diagram

$$\begin{array}{ccc} R_\Sigma & \xrightarrow{\sim} & R'_\Sigma \\ \downarrow{\scriptstyle \phi_\Sigma} & & \downarrow{\scriptstyle \phi'_\Sigma} \\ \mathbf{T}_\Sigma & \xrightarrow{\sim} & \mathbf{T}'_\Sigma . \end{array}$$

We can therefore assume that for each prime $p \neq \ell$ such that $\bar{\rho}|_{G_p}$ is reducible (i.e., **P** or **S**), we have $\bar{\rho}^{I_p} \neq 0$.

We also find that theorem 6.1 is well-behaved under extension of scalars. More precisely, suppose that K' is a finite extension of K. Defining

$$\phi'_\Sigma : R'_\Sigma \to \mathbf{T}'_\Sigma$$

using K' instead of K, we find that there is a natural commutative diagram

$$\begin{array}{ccc} R_\Sigma \otimes_{\mathcal{O}} \mathcal{O}' & \xrightarrow{\sim} & R'_\Sigma \\ \downarrow{\scriptstyle \phi_\Sigma \otimes 1} & & \downarrow{\scriptstyle \phi'_\Sigma} \\ \mathbf{T}_\Sigma \otimes_{\mathcal{O}} \mathcal{O}' & \xrightarrow{\sim} & \mathbf{T}'_\Sigma , \end{array}$$

where $\mathcal{O}'$ is the ring of integers of K'. One deduces from this that if theorem 6.1 holds for some K, then it holds for all K. In particular, we may assume that there is an $\mathcal{O}$-algebra homomorphism $\mathbf{T}_\emptyset \to \mathcal{O}$.

In view of remark 3.7, we must exercise extra care with primes p such that $\bar{\rho}|_{G_p}$ is of type $\mathbf{V}$. We denote by P the set of such vexing primes. (In [W] and [TW], it is assumed that P consists only of primes which are not congruent to $-1 \bmod \ell$.)

7.3. **The case $\Sigma = \emptyset$.** Recall that the strategy of Wiles and Taylor-Wiles in the "minimal case" is to choose, for each $n \geq 1$, a certain set $Q = Q_n$ consisting of primes congruent to 1 mod ℓ^n. These sets Q are chosen so that R_Q and $R_\emptyset$ can be topologically generated as an $\mathcal{O}$-algebra by r elements, where r is the cardinality of Q. Moreover the choice is made so that $\mathbf{T}_Q$ and $\mathbf{T}_\emptyset$ can be related using their natural structure as algebras over a group ring where the group is generated by r elements. One then proves $\phi_\emptyset$ is an isomorphism using the arguments of §3 of [TW] and Chapter 3 of [W], or using the Taylor-Wiles-Faltings criterion ([TW], Appendix or [DDT], §3.4). Alternatively, using Rubin's simplification of the isomorphism criterion (see [dSRS]), it suffices to choose a single set $Q = Q_n$ as in [TW], where n is made explicit.

Our strategy is the same, but the set P introduces several complications. A minor complication is that we use a version over $\mathcal{O}$ of the isomorphism criterion (see §5 of [D]). We shall now state such a version along the lines of Rubin's simplification, leaving it as an exercise to make the necessary modifications to the proof of Criterion II of [dSRS].

We fix an integer $r \geq 0$ and consider power series rings

$$\mathcal{O}[[S]] = \mathcal{O}[[S_1, \dots, S_r]] \qquad \text{and} \qquad \mathcal{O}[[X]] = \mathcal{O}[[X_1, \dots, X_r]].$$

Let $\mathfrak{m}$ denote the maximal ideal of $\mathcal{O}[[S]]$. Recall that the polynomial

$$f(x) = \prod_{i=0}^{r} (x+i)$$

satisfies $f(n)/(r+1)! = \text{length}_{\mathcal{O}}(\mathcal{O}[[S]]/\mathfrak{m}^n)$ for all integers $n \geq 1$. We also fix $\mathcal{O}$-algebra homomorphisms

$$\mathcal{O}[[S]] \to \mathcal{O}[[X]] \to R \to T \tag{6}$$

with $\mathcal{O}[[X]] \to R$ and $R \to T$ surjective. Suppose that $T/(S_1, \dots, S_r)T$ is finitely generated as an $\mathcal{O}$-module; let s denote its rank and t the $\mathcal{O}$-length of its torsion.

Theorem 7.1. *Suppose that there are positive integers d and N such that*

1. $d \geq st + s + t$,
2. $f(N) + f(dN - d) - f(dN) > 0$,
3. $\mathcal{O}[[S]]/\mathfrak{m}^N \to T/\mathfrak{m}^N T$ *is injective.*

Then

- $R/(S_1,\dots,S_r)R \to T/(S_1,\dots,S_r)T$ *is an isomorphism,*
- $T/(S_1,\dots,S_r)T$ *is a local complete intersection,*
- $s > 0$ *and* $t = 0$.

We shall apply the criterion with $R = R_Q$ and $\mathbf{T} = \mathbf{T}_Q$ for a certain set Q as in [TW]. We shall explain below how r, d, N and Q are to be chosen, and the maps in (6) are to be defined.

For arbitrary Σ, let I_Σ denote the kernel of the map $R_\Sigma \to R_\emptyset$. One can check that the kernel of the natural surjection

$$\mathbf{T}_\Sigma/I_\Sigma\mathbf{T}_\Sigma \to \mathbf{T}_\emptyset$$

is torsion. In particular, the rank of $\mathbf{T}_\Sigma/I_\Sigma\mathbf{T}_\Sigma$ is independent of Σ, and we denote it s'. We denote by t' the $\mathcal{O}$-length of the torsion submodule of $\mathbf{T}_P/I_P\mathbf{T}_P$. We set $d = s't' + s' + t'$, $r = \dim_k H^1_\emptyset(G_\mathbf{Q}, \mathrm{ad}^0\bar\rho(1))$ and choose N so that the inequality of theorem 7.12 is satisfied. (Note that $f(N) + f(dN - d) - f(dN)$ is a polynomial with leading term N^{r+1}.)

By the same Galois cohomology argument as in §4 of [TW] (or see [dSh] or [DDT]), we choose a finite set of primes Q such that

- $\#Q = r$,
- R_Q can be topologically generated as an $\mathcal{O}$-algebra by r elements,
- if $q \in Q$, then the following hold:
  - $q \equiv 1 \bmod \ell^N$;
  - $\bar\rho$ is unramified at q;
  - $\bar\rho(\mathrm{Frob}_q)$ has distinct eigenvalues.

Since R_Q is generated by r elements as an $\mathcal{O}$-algebra, we can define a surjective homomorphism $\mathcal{O}[[X]] \to R_Q$. Let G denote the maximal quotient of $\prod_{q\in Q}(\mathbf{Z}/q\mathbf{Z})^\times$ of ℓ-power order. We endow $R_{P\cup Q}$, hence R_Q, with the structure of an $\mathcal{O}[G]$-algebra as in [TW], appendix (or see [dSh] or [DDT]). Choosing generators $g_1,\dots,g_r$ for g, we define a surjection

$$\begin{array}{ccc} \mathcal{O}[[S]] & \to & \mathcal{O}[G] \\ S_i & \mapsto & g_i - 1 \end{array}$$

whose kernel is contained in $\mathfrak{m}^N$. We then define the $\mathcal{O}$-algebra homomorphism $\mathcal{O}[[S]] \to \mathcal{O}[[X]]$ so that the diagram

$$\begin{array}{ccc} \mathcal{O}[[S]] & \longrightarrow & \mathcal{O}[[X]] \\ \downarrow & & \downarrow \\ \mathcal{O}[G] & \longrightarrow & R_Q \end{array}$$

commutes.

The verification of hypothesis 3 can be viewed as the main obstacle in improving the methods of [TW] and [W] to cover the setting of theorem 6.1. Recall that Taylor and Wiles use a method of de Shalit to prove that (under their hypotheses), $\mathbf{T}_Q$ is free over $\mathcal{O}[G]$ and $\mathbf{T}_Q/I_Q\mathbf{T}_Q \xrightarrow{\sim} \mathbf{T}_\emptyset$ (see §2 of [TW] or [dSh], or see §4.3 of [DDT] for an alternative argument using the q-expansion principle). The key observation made in [D] is that it suffices to prove the following:

Lemma 7.2. *There exists a nonzero* $\mathbf{T}_Q$*-module which is free over* $\mathcal{O}[G]$.

The proof of the lemma is very technical and is related to the methods of [DT]. We refer the reader to §4 of [D] for details, mentioning here only that it relies on the Jacquet-Langlands correspondence and a cohomological construction. We also point out that to prove the lemma and other results used below on the fine structure of the algebras $\mathbf{T}_\Sigma$, one first realizes them as completions of Hecke algebras acting on spaces of modular forms. (See for example §4.1 and §4.2 of [DDT].)

To verify that the hypothesis 1 of the theorem is satisfied, one uses that $\mathcal{O}[G] \to R_Q$ was defined so that the augmentation ideal of $\mathcal{O}[G]$ maps onto $I_Q = \ker(R_Q \to R_\emptyset)$. Thus we have

$$s = \operatorname{rank}_{\mathcal{O}}(\mathbf{T}_Q/I_Q\mathbf{T}_Q) = \operatorname{rank}_{\mathcal{O}}\mathbf{T}_\emptyset = \operatorname{rank}_{\mathcal{O}}(\mathbf{T}_P/I_Q\mathbf{T}_P) = s'.$$

The arguments of [TW] discussed (or [DDT] §4.3) can be used to show that the natural map

$$\mathbf{T}_{P\cup Q}/(S_1, \dots, S_r)\mathbf{T}_{P\cup Q} \to \mathbf{T}_P \tag{7}$$

is an isomorphism (see [D], lemma 3.3). One then deduces that

$$\mathbf{T}_{P\cup Q}/I_{P\cup Q}\mathbf{T}_{P\cup Q} \xrightarrow{\sim} \mathbf{T}_P/I_P,$$

from which it follows that $t \le t'$ and $d \le d'$.

We now apply theorem 7.1 to conclude that

$$R_Q/I_Q \to \mathbf{T}_Q/I_Q\mathbf{T}_Q$$

is an isomorphism, and these rings are complete intersections and torsion-free over $\mathcal{O}$. From this follows theorem 6.1 in the case $\Sigma = \emptyset$.

7.4. **The case of arbitrary** Σ**.** Our situation now is that we have a commutative diagram of surjective $\mathcal{O}$-algebra homomorphisms

$$\begin{array}{ccc} R_\Sigma & \xrightarrow{\phi_\Sigma} & \mathbf{T}_\Sigma \\ \downarrow & & \downarrow \\ R_\emptyset & \xrightarrow{\phi_\emptyset} & \mathbf{T}_\emptyset; \end{array}$$

we know that the bottom row is an isomorphism and the rings are local complete intersections, and we wish to prove this holds for the top row.

Recall that we have assumed the existence of a map $\mathbf{T}_\emptyset \to \mathcal{O}$ of $\mathcal{O}$-algebras. Such a homomorphism necessarily corresponds to newform f with coefficients in $\mathcal{O}$ such that ρ_f is a deformation of $\bar{\rho}$ of type $\emptyset$.

For arbitrary Θ, we write $\mathfrak{p}_\Theta$ for the kernel of $R_\Theta \to \mathcal{O}$, and $\mathfrak{P}_\Theta$ for the kernel of $\pi_\Theta : \mathbf{T}_\Theta \to \mathcal{O}$. We consider the $\mathcal{O}$-module $\Phi_\Theta = \mathfrak{p}_\Theta/\mathfrak{p}_\Theta^2$ and the $\mathcal{O}$-ideal $\eta_\Theta = \pi_\Theta(\operatorname{Ann}_{\mathbf{T}_\Theta}\mathfrak{P}_\Theta)$. We omit the subscript Θ when $\Theta = \emptyset$. According to the Wiles-Lenstra criterion, Criterion I of [dSRS], we know that

$$\operatorname{length}_{\mathcal{O}}(\Phi) = \operatorname{length}_{\mathcal{O}}(\mathcal{O}/\eta),$$

and we wish to prove that

$$\operatorname{length}_{\mathcal{O}}(\Phi_\Sigma) \leq \operatorname{length}_{\mathcal{O}}(\mathcal{O}/\eta_\Sigma).$$

Using (4), one obtains as in §4.2 of [Ri2]

$$\operatorname{length}_{\mathcal{O}}(\Phi_\Sigma) \leq \operatorname{length}_{\mathcal{O}}(\Phi) + \sum_{p\in\Sigma} d_p, \tag{8}$$

where d_p is the length of

- $H^0(G_p, \operatorname{ad}^0\rho_f \otimes_{\mathcal{O}} K/\mathcal{O}(1))$ if $p \neq \ell$;
- $\mathcal{O}/(a_\ell(f)^2 - \psi_f(\ell))$ if $p = \ell$ does not divide N_f;
- 0 otherwise.

(We have used here that ρ_f is of type $\emptyset$.)

Using that $\mathbf{T}_\Sigma$ is Gorenstein for $\Sigma \supset P$ (Wiles' generalization of results of Mazur and others discussed in [Ti]), together with Wiles' calculations of the change in η discussed in §4.3 of [Ri2], we find that

$$\operatorname{length}_{\mathcal{O}}(\mathcal{O}/\eta_\Sigma) \geq \operatorname{length}_{\mathcal{O}}(\mathcal{O}/\eta_P) + \sum_{p\in\Sigma-P} d_p, \tag{9}$$

(provided $\Sigma \supset P$). We complement this with the inequality

$$\operatorname{length}_{\mathcal{O}}(\mathcal{O}/\eta_P) \geq \operatorname{length}_{\mathcal{O}}(\mathcal{O}/\eta) + \sum_{p\in P} d_p \tag{10}$$

established by lemma 3.6 of [D].

Applying the Wiles-Lenstra criterion together with (8), (9) and (10), we conclude that Φ_Σ is an isomorphism if $\Sigma \supset P$. (This is all that is proved in [D] and all that is needed for the corollaries.) We leave it as an exercise for the reader to treat the case of arbitrary Σ by showing that if $P_0 \subset P \subset \Sigma$, then $\prod_{p\in P_0}(p+1)$ is an element of $\pi_\Sigma(J)$, where J is the annihilator in $\mathbf{T}_\Sigma$ of the kernel of $\mathbf{T}_\Sigma \to \mathbf{T}_{\Sigma-P_0}$.

Acknowledgement. The author was supported by the United Kingdom's EPSRC (#GR/J4761) while this paper was written.

REFERENCES

[DDT] H. Darmon, F. Diamond, R. Taylor, *Fermat's Last Theorem,* in Current Developments in Mathematics, 1995, International Press, 1–154.

[dSh] E. de Shalit, *Hecke rings and universal deformation rings,* this volume.

[dSL] B. de Smit, H. Lenstra, *Explicit construction of unversal deformation rings,* this volume.

[dSRS] B. de Smit, K. Rubin, R. Schoof, *Criteria for complete intersections,* this volume.

[D] F. Diamond, *On deformation rings and Hecke rings,* Annals of Math. **144** (1996), 137-166.

[Ri2] F. Diamond, K. Ribet, *p-adic modular deformations and Wiles' "Main Conjecture,"* this volume.

[DT] F. Diamond, R. Taylor, *Lifting modular mod l representations,* Duke Math J. **74** (1994), 253–269.

[E] B. Edixhoven, *Serre's conjecture,* this volume.

[G] S. Gelbart, *Three lectures on the modularity of $\bar{\rho}_{E,3}$ and the Langlands reciprocity conjecture,* this volume.

[M] B. Mazur, *An introduction to the deformation theory of Galois representations,* this volume.

[Ri1] K. Ribet, *On modular representations of* $\mathrm{Gal}(\bar{\mathbf{Q}}/\mathbf{Q})$ *arising from modular forms,* Inv. Math. **100** (1990), 431–476.

[Ro] D. Rohrlich, *Modular functions and modular curves,* this volume.

[Ru] K. Rubin, *Modularity of mod 5 representations,* this volume.

[S] J.H. Silverman, *A survey of the arithmetic theory of elliptic curves,* this volume.

[TW] R. Taylor, A. Wiles, *Ring theoretic properties of certain Hecke algebras,* Annals of Math. **141** (1995), 553–572.

[Ti] J. Tilouine, *Hecke algebras and the Gorenstein property,* this volume.

[W] A. Wiles, *Modular elliptic curves and Fermat's Last Theorem,* Annals of Math. **141** (1995), 443–551.

APPENDIX: CLASSIFICATION OF $\overline{\rho}_{E,\ell}$ BY THE j-INVARIANT OF E

FRED DIAMOND AND KENNETH KRAMER

Let K be a finite extension of $\mathbf{Q}_p$ with ring of integers $\mathcal{O}_K$ and valuation v_K. Suppose that E is an elliptic curve over K with absolute invariant j and minimal discriminant Δ_E. Assume throughout that ℓ is a prime different from p. Let $G_K = \mathrm{Gal}(\overline{K}/K)$ and consider the mod-ℓ representation

$$\overline{\rho}_{E,\ell} : G_K \to \mathrm{Aut}(E[\ell]) \approx \mathrm{GL}_2(\mathbf{F}_\ell).$$

In so far as possible, we wish to describe the representation type of $\overline{\rho}_{E,\ell}$ as defined in section 2 of [Di] and the conductor of E in terms of j. We rely on observations of Serre [Se2] and on Tate's algorithm [Ta]. For extensive tables of Kodaira reduction types in terms of congruences on the coefficients of a generalized Weierstrass model for E, see [Pa].

As motivation for this exercise, note that certain calculations of conductor or representation type have been used to study various Diophantine equations ([Da], [Rib]) or to prove the modularity of elliptic curves A defined over $\mathbf{Q}$ ([DK],[LR]). For example, if A is semistable at 3 and if the wild ramification group G_1 corresponding to the j-invariant of A in Table 3 below is isomorphic to the quaternions $\mathfrak{Q}$ then A is modular by [Di, Cor. 6.3].

In case E has good reduction, $\overline{\rho}_{E,\ell}$ is unramified with cyclic image, and therefore of type $\mathbf{P}$. If E is potentially multiplicative (i.e., if $v_K(j) < 0$), then E acquires multiplicative reduction over at most a quadratic extension of K, and the twist of E by this extension is semistable. Then the parametrization of E by p-adic theta functions may be used to determine the representation of G_K on the ℓ-adic Tate module. (See for example [Si2, Chap. V, Prop. 6.1 and ex. 5.13].) In particular, $\overline{\rho}_{E,\ell}$ is of type $\mathbf{P}$ or type $\mathbf{S}$ according to whether or not $v_K(\Delta_E)$ is divisible by ℓ.

We now restrict our attention to elliptic curves E with potential good reduction. Let $\ell' = 4$ if $\ell = 2$ and $\ell' = \ell$ otherwise. Then E acquires good reduction over any of the division fields $K(E[\ell'])$. Indeed, the kernel of $\overline{\rho}_{E,\ell'}$ restricted to the inertia subgroup I_K of G_K is independent of ℓ. We denote by G_0 the abstract group defined by the image $\overline{\rho}_{E,\ell'}(I_K)$. The following lemma and the conditions of [Di, Prop. 2.1–2.4], may be used to check the extent to which the representation type of $\overline{\rho}_{E,\ell}$ also is independent of ℓ.

Lemma 0.1. *Write $\sigma = \overline{\rho}_{E,\ell}$ and $\tilde{\sigma}$ for the associated projective representation. Then*

1. *$\tilde{\sigma}(I_K)$ is cyclic if and only if G_0 is cyclic.*
2. *Suppose that G_0 is cyclic. Then $|\tilde{\sigma}(I_K)|$ is divisible by ℓ if and only if $|G_0|$ is divisible by ℓ'.*
3. *Suppose that G_0 is cyclic and $|G_0|$ is not divisible by ℓ'. Then σ is reducible over $\overline{\mathbf{F}}_\ell$ if and only if E acquires good reduction over a cyclic extension of K.*

For $j \neq 0$, 1728, one choice of model with absolute invariant j is given by

$$y^2 = x^3 - 3cx - 2c, \tag{0.2}$$

with $c = j/(j-1728)$ and discriminant $12^6 j^2/(j-1728)^3$. Upon twisting E by a quadratic character ψ, the discriminant varies by a sixth power. Furthermore, $\mathbf{f}(E^\psi) \leq \max\{\mathbf{f}(E), 2\,\mathbf{f}(\psi)\}$, with equality if $\mathbf{f}(E) \neq 2\,\mathbf{f}(\psi)$. The representation type is not affected. If $p \neq 2$ and ψ is ramified, such a twist changes Kodaira symbols I_ν, II, III or IV to I_ν^*, IV*, III* or II* respectively. For a discussion of the effect of a quadratic twist on Kodaira symbol and conductor when $p = 2$, see [Co].

As for the elliptic curves with many automorphisms, if $j = 0$, the model $y^2 + y = x^3$ has good reduction for $p \neq 3$. If $j = 1728$, the model $y^2 = x^3 - x$ has good reduction for $p \neq 2$. Note that for $j = 0$ (resp. 1728), twisting by a ramified cubic or sextic (resp. quartic) field may affect the representation type.

Definition. Let $j \in \mathcal{O}_K$ be given. We say that E is j-minimal if E has the minimal conductor exponent among elliptic curves with absolute invariant equal to j.

For $p \geq 5$, we extract the table below from the well-known table of reduction types [Si2, Table 4.1]. The classification of representation types follows from tame ramification theory [Se2, §1.3–1.5].

$v_K(j-1728)$ mod 2	$v_K(j)$ mod 3	$v_K(\Delta_E)$	G_0	$\mathbf{f}(E)$
0	0	0	0	0
1	0	3 or 9	$\mathbf{Z}/4$	2
0	± 1	4 or 8	$\mathbf{Z}/3$	2
0	± 1	2 or 10	$\mathbf{Z}/6$	2

TABLE 1. j-minimal curves over p-adic fields K $p \geq 5$, $j \neq 0$, 1728

Proposition 0.3. *Suppose $p \neq 2$ and $G_0 \approx \mathbf{Z}/4$. Then $\overline{\rho}_{E,2}$ is of type* **S**. *For $\ell \neq 2$, $\overline{\rho}_{E,\ell}$ is of type* **P** *or type* **V** *according to whether or not $\boldsymbol{\mu}_4 \subset K$.*

Suppose $p \neq 3$ and $G_0 \approx \mathbf{Z}/3$ or $\mathbf{Z}/6$. Then $\overline{\rho}_{E,3}$ is of type **S**. *For $\ell \neq 3$, $\overline{\rho}_{E,\ell}$ is of type* **P** *or type* **V** *according to whether or not $\boldsymbol{\mu}_3 \subset K$.*

RESIDUE CHARACTERISTIC $p = 3$

The table of j-minimal curves over $\mathbf{Q}_3$ given below may be constructed by applying Tate's algorithm to the generic curve (0.2). For the determination of representation types, including cases of more general 3-adic base fields, see the subsequent remarks. To reduce the number of entries, we have imposed the following convention.

Convention. In each family of j-minimal curves for fixed j, the table includes only the curves E with minimal valuation of discriminant.

In Table 2, we write v for $v_{\mathbf{Q}_3}$ and j^* for $j - 1728$. By **P-V** we mean representation type **P** if $j^* \in Q_3^{\times 2}$ or type **V** otherwise. By **V-S** we mean type **V** if $\ell \geq 5$ or type **S** if $\ell = 2$.

$v(j)$		Kodaira Symbol	min $v(\Delta_E)$	$\mathbf{f}(E)$	type of $\overline{\rho}_{E,\ell}$
0		I_0	0	0	**P**
1		IV	5	3	**H**
2		II	4	4	**P-V**
3,	$j^*/27 \equiv 1\,(9)$	III	3	2	**V-S**
	$j^*/27 \equiv 4, 7\,(9)$	II	3	3	**H**
	$v(j^*) = 4$	IV	6	4	**P-V**
	$v(j^*) = 5$	II	3	3	**H**
	$v(j^*) \geq 6$, even	I_0	0	0	**P**
	$v(j^*) \geq 7$, odd	III	3	2	**V-S**
$3n+1$,	$n \geq 1$	II	5	5	**H**
$3n+2$,	$n \geq 1$	IV	7	5	**H**
$3n$, $n \geq 2$,	$j/3^{3n} \equiv \pm 4\,(9)$	III	3	2	**V-S**
	$j/3^{3n} \equiv \pm 1, \pm 2\,(9)$	II	3	3	**H**

TABLE 2. j-minimal curves over $\mathbf{Q}_3$ with $j \neq 0, 1728$

Suppose that K is a general 3-adic field. If E is a j-minimal curve with $\mathbf{f}(E) = 2$, then $G_0 \approx \mathbf{Z}/4$ and the representation type of $\overline{\rho}_{E,\ell}$ is given by Proposition 0.3. For $K = \mathbf{Q}_3$, this includes the 0-minimal curve $y^2 = x^3 + 1$.

The representation $\overline{\rho}_{E,\ell}$ is wildly ramified precisely when $\overline{\rho}_{E,\ell}(P_K) \approx \mathbf{Z}/3$, where P_K is the wild ramification subgroup of I_K. Equivalently, $\mathbf{f}(E) \geq 3$.

Proposition 0.4. *The following are equivalent:*

1. $\mathbf{f}(E)$ *is odd and at least 3,*
2. $\overline{\rho}_{E,\ell}$ *is of type* $\mathbf{H}$ *for some* ℓ*, or equivalently for all* ℓ*,*
3. $\overline{\rho}_{E,2}(I_K) = \mathrm{GL}_2(\mathbf{F}_2)$*,*
4. G_0 *is isomorphic to the semi-direct product* $\mathbf{Z}/3 \ltimes \mathbf{Z}/4$*.*

Proof. Write $\mathbf{b}(\overline{\rho}_{E,2})$ for the wild conductor exponent ([Se3, §4.9]). Recall that $\mathbf{f}(E) = 2 + \mathbf{b}(\overline{\rho}_{E,2})$ when E has additive reduction. Furthermore $\mathbf{f}(E) = \mathbf{f}(\overline{\rho}_{E,2})$ when the restriction $\overline{\rho}_{E,2} \mid I_K$ has no non-zero fixed points.

If (1) is true, $\mathbf{b}(\overline{\rho}_{E,2})$ is odd. Then $\overline{\rho}_{E,2} \mid I_K$ must be irreducible, so that (2) holds. Otherwise, after possible extension of scalars, we have

$$\overline{\rho}_{E,2} | I_K \sim \begin{pmatrix} \psi & * \\ 0 & \psi^{-1} \end{pmatrix}$$

and $\mathbf{b}(\overline{\rho}_{E,2}) = \mathbf{b}(\psi) + \mathbf{b}(\psi^{-1}) = 2\,\mathbf{b}(\psi)$ is even.

Conversely, if (2) holds, then $\overline{\rho}_{E,2}(G_K) = \overline{\rho}_{E,2}(I_K) = \mathrm{GL}_2(\mathbf{F}_2)$. Thus $\overline{\rho}_{E,2}$ is induced from a one-dimensional representation ψ of G_F, where $F = K(\Delta^{1/2})$ is the quadratic extension of K inside $M = K(E[2])$. By the inductive property of conductors [Se1, VI, §2, Cor. to Prop. 4], we have $\mathbf{f}(E) = \mathbf{f}(\overline{\rho}_{E,2}) = \mathbf{f}(\psi) + 1$.

Let us verify that $\mathbf{f}(\psi)$ is even. According to the notion of conductor in abelian class field theory, there is a unit $u \in F$ with the property that u is not a norm from M and $v_F(u-1) = \mathbf{f}(\psi) - 1$. The Artin symbol $s = (u, M/F)$ provides a generator for $\mathrm{Gal}(M/F)$. If $\mathbf{f}(\psi)$ were odd, we could find an element $\theta \in K$ such that

$$v_K(\theta) > 0 \quad \text{and} \quad v_F(u - 1 - \theta)) \geq \mathbf{f}(\psi).$$

But then $s = (1+\theta, M/F)$, so that $\mathrm{Gal}(F/K)$ acts trivially on s, a contradiction. The equivalence of (1) and (2) is proved.

One easily checks the equivalence of conditions (2), (3) and (4). ∎

The situation is somewhat more complicated when $\mathbf{f}(E)$ is even. By examining the 2-division field of the generic curve (0.2), we get a classification which may be useful when the size of $\mathbf{f}(E)$ is known.

Corollary 0.5. *Suppose that* E *has absolute invariant* $j \in \mathcal{O}_K$ *with* $j \neq 0, 1728$.

1. *If* $v_K(j - 1728)$ *is odd, then either* $\mathbf{f}(E) = 2$, $G_0 \approx \mathbf{Z}/4$ *and* $\overline{\rho}_{E,\ell}$ *is described by Proposition 0.3, or else* $\mathbf{f}(E)$ *is odd,* $\mathbf{f}(E) \geq 3$ *and* $\overline{\rho}_{E,\ell}$ *is of type* $\mathbf{H}$.
2. *If* $v_K(j - 1728)$ *is even, then either* E *achieves good reduction by at most a quadratic twist and* $\overline{\rho}_{E,\ell}$ *is of type* $\mathbf{P}$*, or else* $\mathbf{f}(E)$ *is even,* $\mathbf{f}(E) \geq 4$ *and* $\overline{\rho}_{E,\ell}$ *is of type* $\mathbf{P}$ *or type* $\mathbf{V}$ *according to whether or not* $j - 1728 \in K^2$.

At least for $j \in \mathcal{O}_K$ with $v_K(j) \not\equiv 0 \pmod 3$, wild ramification is assured in the ℓ-division fields of E. Indeed, we then have $\mathbf{f}(E) = v_K(j-1728)+2$.

RESIDUE CHARACTERISTIC $p = 2$

The table of j-minimal curves over $\mathbf{Q}_2$ given below may be constructed by using Tate's algorithm on the curve (0.2). Our convention on minimal discriminant within a j-minimal family also holds for this table. We denote by G_1 the 2-Sylow (wild ramification) subgroup of the inertia group G_0. Then G_1 is isomorphic to a subgroup of the quaternion group $\mathfrak{Q}$ of order 8. We explain the determination of G_1 and representation type appearing in Table 3, as well as other information for more general 2-adic base fields, after giving the table.

In Table 3, we write v for $v_{\mathbf{Q}_2}$ and j^* for $j - 1728$. By **P-V** we mean representation type **P** if $j/64 \equiv 1 \pmod 8$ or type **V** if $j/64 \equiv 5 \pmod 8$. By **V-S** we mean type **V** if $\ell \geq 5$ or type **S** if $\ell = 3$. The special cases labeled (a) and (b) involve $j^* = 2^{2n}u$ with $n \geq 4$. Case (a) is defined by $u \equiv 1 \pmod 4$ and also includes $j = 1728$. Case (b) is defined by $u \equiv 3 \pmod 4$.

We now assume that K is a general 2-adic field and that E has potential good reduction. To study the image of inertia under $\overline{\rho}_{E,\ell}$ it suffices to examine the 3-division field $L = K(E[3])$. The subfield $L_X = K(X(E[3]))$ of L obtained by adjoining to K the x-coordinates of points of order 3 in a Weierstrass model contains the field $K_1 = K(\mu_3, \Delta_E^{1/3})$. Furthermore $\mathrm{Gal}(L_X/K_1) \hookrightarrow \mathbf{Z}/2 \oplus \mathbf{Z}/2$. Note that $\mathrm{Gal}(L_X/K) \approx \tilde{\sigma}(G_K)$, where $\tilde{\sigma}$ is the projective representation derived from $\overline{\rho}_{E,3}$.

We have $G_0 = G_1$ if $v_K(\Delta_E) \equiv 0 \pmod 3$ and $G_0/G_1 \approx Z/3$ otherwise. In particular, the representations $\overline{\rho}_{E,\ell}$ are tamely ramified, i.e., $\mathbf{f}(E) = 2$, precisely when $G_0 \approx Z/3$. In that case, the representation type is given by Proposition 0.3. The representations $\overline{\rho}_{E,\ell}$ are of type **H** precisely when $G_1 \approx \mathfrak{Q}$ and this is equivalent to L_X/K_1 being totally ramified of degree 4. Furthermore, type **H** is guaranteed when $\mathbf{f}(E) \geq 3$ and odd by the argument in the proof of Proposition 0.4.

The next lemma, which is valid over any field of characteristic not 2 or 3, may be checked by consideration of generic j as in [Ig, p. 456, p. 461, Thm. 3, Thm. 6] or by direct computation. We shall use it to examine ramification in L_X/K_1. For another approach, including a detailed study of various Galois extensions of $\mathbf{Q}_2$, see the thesis of A. Rio [Rio]. In particular, the representations of type **V** occurring over $\mathbf{Q}_2$ have also been determined in [LR].

Lemma 0.6. *Let ζ be a primitive cube root of unity. For $j \neq 1728$, the field L_X may be obtained from K_1 by adjoining the square roots of the*

$v(j)$		G_1	Kodaira Symbol	min $v(\Delta_E)$	$\mathbf{f}(E)$	type of $\overline{\rho}_{E,\ell}$
0		0	I_0	0	0	**P**
1		$\mathfrak{Q}$	II^*	11	3	**H**
2		$\mathfrak{Q}$	III^*	10	3	**H**
3		$\mathfrak{Q}$	I_0^*	9	5	**H**
4,	$j/16 \equiv 1\,(4)$	$\mathfrak{Q}$	I_1^*	8	3	**H**
	$j/16 \equiv 3\,(4)$	0	IV^*	8	2	**V-S**
5		$\mathfrak{Q}$	II	7	7	**H**
6,	$v(j^*) = 7$	$\mathbf{Z}/4$	III	9	8	**P-V**
	case (a) above	$\mathfrak{Q}$	III	6	5	**H**
	case (b) above	$\mathfrak{Q}$	I_3^*	12	5	**H**
	$v(j^*) \geq 9$, odd	$\mathfrak{Q}$	III	9	8	**H**
7		$\mathfrak{Q}$	III	8	7	**H**
8,	$j/256 \equiv 1\,(4)$	$\mathfrak{Q}$	III	4	3	**H**
	$j/256 \equiv 3\,(4)$	0	IV	4	2	**V-S**
9		$\mathfrak{Q}$	III^*	12	5	**H**
10		$\mathfrak{Q}$	I_1^*	8	3	**H**
11		$\mathfrak{Q}$	III	4	3	**H**
$3n$,	$n \geq 4$, or $j = 0$	0	I_0	0	0	**P**
$3n+1$,	$n \geq 4$	0	IV^*	8	2	**V-S**
$3n+2$,	$n \geq 4$	0	IV	4	2	**V-S**

TABLE 3. j-minimal curves over $\mathbf{Q}_2$

Kummer generators

$$\begin{aligned}\kappa_0 &= (j^{1/3} - 12\zeta)(j^{1/3} - 12\zeta^2),\\ \kappa_1 &= (j^{1/3} - 12)(j^{1/3} - 12\zeta),\\ \kappa_2 &= (j^{1/3} - 12)(j^{1/3} - 12\zeta^2).\end{aligned}$$

Note that the relation $\kappa_0\kappa_1\kappa_2 = (j-1728)^2 \in K^2$ holds.

Proposition 0.7. *Let E be a j-minimal curve with $j \neq 0$. Write $e_K = v_K(2)$.*

1. *If $v_K(j) = 0$, then E has good reduction.*
2. *Suppose that $v_K(j) \geq 12e_K$. If $v_K(j) \equiv 0 \pmod 3$, then E has good reduction. Otherwise, $\overline{\rho}_{E,\ell}$ is tamely ramified.*
3. *If $0 < v_K(j) < 12e_K$ and $v_K(j)$ is odd, then $G_1 \approx \mathfrak{Q}$ and $\overline{\rho}_{E,\ell}$ is of type* **H**.

Proof. For the special case $v_K(j) = 12e_K$, the element $a = 64(1728-j)/j$ is a unit in K. The curve $y^2 + ay = x^3 + 3ax$ has absolute invariant equal to j and good reduction.

If $v_K(j) = 0$, consider the model $y^2 + xy = x^3 - 4bx - b$, with $\Delta = b(1-64b)^2$ and absolute invariant $(1+192b)^3/\Delta$. Hensel's lemma shows that if $j \in \mathcal{O}_K$ is a unit, then there exists a unit root $r \in K$ of the equation $(X+192)^3 - j(X-64)^2 = 0$. Setting $b = 1/r$ provides a curve with good reduction and absolute invariant equal to j.

In the remaining cases, one determines the ramification in L_X/K_1 by examining the Kummer generators of Lemma 0.6. We omit details here, but give a similar argument in the next proposition. ∎

Proposition 0.8. *An elliptic curve E has maximal conductor exponent, namely $\mathbf{f}(E) = 6e_K + 2$, in precisely the following cases:*

1. *$j = 1728J$ with $J \in \mathcal{O}_K$ and $v_K(J-1)$ odd,*
2. *$j = 1728$ and E has a model of the form $y^2 = x^3 - ax$ with $v_K(a)$ odd.*

Proof. It follows from [Se1, Chap. IV, §2, ex. 3] and [BK, Prop. 3.7] that $\mathbf{f}(E) = 6e_K + 2$ precisely when there exists an element with odd valuation among the Kummer generators for L_X/K_1. By examining the generators of Lemma 0.6, we find that this forces $v_K(j) = 6e_K$, in which case we may write $j = 1728J$ for some unit $J \in \mathcal{O}_K$. Assume for the moment that $J \neq 1$. Then our Kummer generators have even valuation $4e_K$ unless $v_K(J-1) > 0$. Suppose that indeed $v_K(J-1) > 0$. It follows that J is a cube, say $J = (1+s)^3$ with $s \in K$ and $v_K(s) \geq 1$. Then we have

$$(0.9) \qquad \begin{aligned} \kappa_1 &= 144s[(1-\zeta)+s] = 144\zeta^2 s[1+(s+2)\zeta], \\ \kappa_0 &= 144(3+3s+s^2). \end{aligned}$$

Clearly $v_K(\kappa_0) = 4e_K$ is even. Furthermore, $v_K(\kappa_1)$ is odd if and only if $v_K(s)$ is odd. Case (1) arises in this way.

The most general curve with $j = 1728$ has a model $y^2 = x^3 - ax$, with non-zero $a \in K$, for which $L_X = K(\mu_{12}, \sqrt{a(1+2\zeta)})$. But then the extension $L_X/K(\mu_3)$ is generated by square roots of units of $K(\mu_3)$ unless $v_K(a)$ is odd. Case (2) arises in this way. ∎

It remains to find the representation type for j-minimal curves E having even conductor exponent $\mathbf{f}(E) \geq 4$. When the base field is $\mathbf{Q}_2$, it then follows from the list of conductors in Table 3, as computed by Tate's algorithm, that $\mathbf{f}(E) = 8$ and $j = 1728J$ with $v(J-1)$ being odd. Let G be the image of $\overline{\rho}_{E,3}$ and write $\mathbf{D}_8$ for the dihedral group of order 8.

Lemma 0.10. *Suppose $j \in \mathbf{Z}_2$ has the form $j = 1728J$ with $v(J-1)$ odd.*

1. *If $v(J-1) \geq 3$, then $G_1 \approx \mathfrak{Q}$ and $\overline{\rho}_{E,\ell}$ is of type $\mathbf{H}$.*

2. *If* $v(J-1)=1$, *then* $G_1 \approx \mathbf{Z}/4$. *Furthermore, either* $G \approx \mathbf{Z}/8$ *and* $\overline{\rho}_{E,\ell}$ *is of type* $\mathbf{P}$, *or else* $G \approx \mathbf{D}_8$ *and* $\overline{\rho}_{E,\ell}$ *is of type* $\mathbf{V}$ *according as* $J \equiv 3 \pmod 8$ *or* $J \equiv 7 \pmod 8$.

Proof. Under these hypotheses, J is a cube, say $J=(1+s)^3$ as in the notation of the proof of Proposition 0.8. Then $K_1 = \mathbf{Q}_2(\boldsymbol{\mu}_3, j^{1/3}) = \mathbf{Q}_2(\boldsymbol{\mu}_3)$. In case (1), we have $v(s) \geq 3$. From the generators in (0.9) we see that L_X/K_1 is totally ramified of degree 4. Thus $G_1 \approx \mathfrak{Q}$.

In case (2), the following congruences hold modulo 8: either

$$s \equiv 2 \quad \text{and} \quad \kappa_0/144 \equiv -3, \qquad \text{or else} \qquad s \equiv -2 \quad \text{and} \quad \kappa_0/144 \equiv 1,$$

according as $J \equiv 3$ or $J \equiv 7$. For either possibility, we have $L_X = K_1(\sqrt{\kappa_1})$. The extension L_X/K_1 clearly is ramified. It follows that $G_1 \approx \mathbf{Z}/4$. If ϕ denotes a generator for $\mathrm{Gal}(K_1/\mathbf{Q}_2)$, we have $\kappa_1\kappa_1^{\phi} \in \kappa_0(\mathbf{Q}_2^{\times})^2$. Thus $\mathrm{Gal}(L_X/\mathbf{Q}_2) \approx \mathbf{Z}/2 \oplus \mathbf{Z}/2$ or $\mathbf{Z}/4$ depending on whether or not $\kappa_0 \in (\mathbf{Q}_2^{\times})^2$. Then the claimed structure of G follows accordingly. ∎

REFERENCES

[BK] A. Brumer and K. Kramer, *The conductor of an abelian variety*, Comp. Math. **92** (1994), 227–248.

[Co] Comalada, *Twists and reduction of an elliptic curve*, J. No. Theory, **49** (1994), 45–62.

[Da] H. Darmon, *The equations* $x^n + y^n = z^2$ *and* $x^n + y^n = z^3$, IMRN **10** (1993), 263–274.

[Di] F. Diamond, *An extension of Wiles' results*, this volume.

[DK] F. Diamond and K. Kramer, *Modularity of a family of elliptic curves*, Math. Research Letters, **2** (1995), 299–304.

[Ig] J.-I. Igusa, *Fiber systems of Jacobian varieties*, (III. *Fiber systems of elliptic curves*), Am. J. Math. **81** (1959), 453–476.

[LR] J.-C. Lario and A. Rio, *Elliptic modularity for octahedral Galois representations*, Math. Research Letters, **3** (1996), 329–342.

[Pa] I. Papadopoulos, *Sur la classification de Néron des courbes elliptiques en caractéristique residuelle 2 et 3*, J. No. Theory, **44(2)** (1993), 119–152.

[Rib] K. Ribet, *On the equation* $a^p + 2^\alpha b^p + c^p = 0$, preprint.

[Rio] A. Rio, *Representacions de Galois octaèdriques*, Tesi Doctoral, Universitat de Barcelona, 1995.

[Se1] J.-P. Serre, Local Fields, GTM **67**, Springer-Verlag, New York, 1979.

[Se2] J.-P. Serre, *Propriétés galoisiennes des points d'ordre fini des courbes elliptiques*, Invent. Math. **15** (1972), 259–331.

[Se3] J.-P. Serre, *Sur les représentations modulaires de degré 2 de* $\mathrm{Gal}(\overline{\mathbf{Q}}/\mathbf{Q})$, Duke Math. J. **54** (1987), 179–230.

[Si1] J. Silverman, The Arithmetic of Elliptic Curves, GTM **106**, Springer-Verlag, New York, 1986.

[Si2] J. Silverman, Advanced Topics in the Arithmetic of Elliptic Curves, GTM **151**, Springer-Verlag, New York, 1994.

[Ta] J. T. Tate, *Algorithm for determining the type of a singular fiber in an elliptic pencil.* In: Modular Functions of One Variable IV. Lecture Notes in Math. **476**, Springer-Verlag, New York 1975.

CLASS FIELD THEORY AND THE FIRST CASE OF FERMAT'S LAST THEOREM

H. W. LENSTRA, JR. AND P. STEVENHAGEN

For a prime number p, the *first case* of Fermat's last theorem for exponent p asserts that for any three integers x, y, z with $x^p + y^p + z^p = 0$, at least one of x, y, z is divisible by p. In the present chapter we use class field theory to prove several classical results concerning the first case. Our treatment is based on Hasse's exposition [6, Section 22], but whereas Hasse applied explicit reciprocity laws, our proofs depend only on general properties of power and norm residue symbols.

Theorem 1. *The first case of Fermat's last theorem with exponent p is correct for each prime number p for which $2p+1$ is prime.*

This theorem is due to Sophie Germain (1823).

For a positive integer k, we define $N_k = \prod_{\eta,\vartheta} (1 + \eta + \vartheta)$, the product ranging over all kth roots of unity η and ϑ in an algebraic closure of the field $\mathbf{Q}$ of rational numbers. It is easy to see that N_k is a rational integer for each k, and that N_k vanishes if and only if k is divisible by 3.

Theorem 2. *Let p be a prime number, and suppose that there exists a positive integer k not divisible by p for which $kp+1$ is a prime number not dividing N_k. Then the first case of Fermat's last theorem with exponent p is correct.*

This result, which is similar to a theorem of Wendt (1894), is taken from [1]. The integer k is necessarily even and not divisible by 3.

Let k be a positive integer, and let T_k be the set of odd primes p for which p divides k or $kp+1$ is a prime factor of N_k. By Theorem 2, the first case of Fermat's last theorem is correct for exponent p if p is a prime number not in T_k for which $kp+1$ is prime. When k is not divisible by 3, the estimate $|N_k| \leq 3^{k^2}$ shows that the exceptional set T_k has cardinality at most $k^2 + \log k$.

In 1985, Adleman, Heath-Brown, and Fouvry [1, 4] deduced from Theorem 2 that the first case is valid for infinitely many p, as follows. Using sieve methods, Fouvry showed that there exists $c > 0$ with the following property: for all sufficiently large t, there are at least $c \cdot t/\log t$ prime numbers $q \leq t$ with $q \equiv 2 \bmod 3$ for which $q-1$ has a prime factor $p > t^{0.6687}$. Each pair q, p gives rise to an integer $k = (q-1)/p$ that is less than $u = t^{0.3313}$.

The inequality $c \cdot t/\log t > u \cdot (u^2 + \log u)$, which is valid when t is large enough, shows that some value of k must arise for more than $k^2 + \log k$ pairs q, p. For at least one of these pairs the number p is outside T_k, so that the first case holds for p.

From $N_2 = -3$ one finds that T_2 is empty, so Theorem 1 follows from Theorem 2, with $k = 2$. In general, when k is a given positive integer that is not divisible by 3, then it is usually easy to deduce from Theorem 2 that the first case of Fermat's last theorem is correct for each prime exponent p for which $kp + 1$ is prime. For example, from

$$N_4 = -3 \cdot 5^3, \qquad N_8 = -3^7 \cdot 5^3 \cdot 17^3, \qquad N_{10} = -3 \cdot 11^9 \cdot 31^3$$

one finds $T_4 = T_8 = \emptyset$ and $T_{10} = \{3, 5\}$. Since Theorem 1 applies to $p = 3$ and to $p = 5$, one concludes that the first case is true for p if $4p + 1$, $8p+1$, or $10p+1$ is a prime number. This result is due to Legendre (1823). Exceptional primes p that may arise for other values of k are generally easily dealt with by means of the following theorem.

Theorem 3. *Let p be a prime number, and suppose that the first case of Fermat's last theorem for exponent p is false. Then we have*
(a) $2^{p-1} \equiv 1 \bmod p^2$,
(b) $3^{p-1} \equiv 1 \bmod p^2$.

These two results are due to Wieferich (1909) and Mirimanoff (1910), respectively.

There is an efficient algorithm that for a given prime number p tests the validity of (a) and (b). It is believed that there is not a single prime p satisfying both (a) and (b), so that this algorithm, combined with Theorem 3, could be used to prove the first case of Fermat's last theorem for any given prime exponent. This belief is borne out by numerical experiments. In fact, of all primes for which (a) has ever been tested—and this includes all primes less than $4 \cdot 10^{12}$ (see [3])—only $p = 1093$ and $p = 3511$ satisfy (a), and neither of these primes satisfies (b). (The only primes $p < 2^{32} \approx 4.3 \cdot 10^9$ satisfying (b) are $p = 11$ and $p = 1{,}006{,}003$, see [8].)

It is an amusing consequence of (a) that the first case of Fermat's last theorem holds for exponents that are Mersenne or Fermat primes.

Several mathematicians proved, with the same hypotheses as in Theorem 3, that for various other small prime numbers q one has $q^{p-1} \equiv 1 \bmod p^2$. The best result of this nature, prior to the work of Wiles and Taylor, was obtained by Granville and Monagan [5], who covered all prime numbers $q \leq 89$. If it had been possible to replace 89 by an expression that tends sufficiently rapidly to infinity with p, such as $4 \cdot (\log p)^2$, then the first case of Fermat's last theorem would have followed for all p, by [7]; but this could apparently not be achieved by the method of [5]. However, by a theorem of Gunderson (1948) the bound 89 is good enough to imply the first case for all p up to the limit in the title of [5]. Tanner and Wagstaff [9] improved upon Gunderson's work and raised the limit to 156,442,236,847,241,729.

In the proofs, we let p be a prime number, and we let ζ be a primitive pth root of unity in an extension field of $\mathbf{Q}$. We denote by $(\overline{-})$ the pth power residue symbol for the cyclotomic field $\mathbf{Q}(\zeta)$, and by $\mathfrak{p} = (\zeta - 1)$ the unique prime of $\mathbf{Q}(\zeta)$ lying over p. The properties of power and norm residue symbols that we use can all be found in [2, pp. 348–353].

Let it now be supposed that x, y, z are integers not divisible by p that satisfy $x^p + y^p + z^p = 0$. Clearly, p is odd. Removing a greatest common divisor, we may assume that x, y, z are pairwise coprime. We have $\prod_{i=0}^{p-1}(x + y\zeta^i) = x^p + y^p = -z^p$, and from $\gcd(x, y) = \gcd(p, z) = 1$ it follows that the factors $x + y\zeta^i$ are pairwise coprime. Hence each factor generates an ideal that is a pth ideal power.

Lemma 1. *Let n be an integer that is coprime to p and z. Then we have $\left(\frac{x+y\zeta}{n}\right) = \left(\frac{\zeta}{n}\right)^{-y/z}$, where the exponent $-y/z$ is computed modulo p.*

Proof. With $\alpha = (x + y\zeta)\zeta^{y/z}$, the assertion reads $\left(\frac{\alpha}{n}\right) = 1$. Note that (α) is a pth ideal power that is coprime to n, so the definition of the power residue symbol gives $\left(\frac{n}{\alpha}\right) = 1$. The general power reciprocity law (see [2, p. 352, Exercise 2.10]) asserts in this case that $\left(\frac{\alpha}{n}\right)\left(\frac{n}{\alpha}\right)^{-1}$ equals the $\mathfrak{p}$-adic pth power norm residue symbol $(n, \alpha)_{\mathfrak{p}}$. Hence it suffices to prove $(n, \alpha)_{\mathfrak{p}} = 1$. We do this by a computation in the ring of integers of the local field at $\mathfrak{p}$. The units of that ring taken modulo $\mathfrak{p}^2$ are of the form $a + b(\zeta - 1)$, where $a, b \in \mathbf{Z}/p\mathbf{Z}$, $a \neq 0$. They form a group of order $(p-1)p$, which is the direct product of a group of order $p - 1$, consisting of the elements with $b = 0$, and a group of order p, consisting of the elements with $a = 1$; the latter group is generated by ζ, since $\zeta^b \equiv 1 + b(\zeta - 1) \bmod (\zeta - 1)^2$. A general element $a + b(\zeta - 1)$ is decomposed as $a \cdot \zeta^{b/a}$. Applying this to $x + y\zeta \pmod{\mathfrak{p}^2}$, which has $a = x + y \equiv -z \bmod p$ and $b = y$, we find that the $\langle\zeta\rangle$-component of $x + y\zeta \pmod{\mathfrak{p}^2}$ equals $\zeta^{-y/z}$. The other component must then be $(x + y\zeta)/\zeta^{-y/z} = \alpha$. Therefore the order of $\alpha \pmod{\mathfrak{p}^2}$ divides $p - 1$, and $\alpha^{p-1} = 1 - \beta$ with $\beta \in \mathfrak{p}^2$. Also, n^{p-1} is of the form $1 - \gamma$, with $\gamma \in (p) = \mathfrak{p}^{p-1}$. From $\beta\gamma \in \mathfrak{p}^{p+1}$ it follows that $1 - \beta\gamma = \delta^p$ for some non-zero δ in the $\mathfrak{p}$-adic field (cf. [2, p. 353, Exercise 2.12]). Using the bimultiplicativity of the norm residue symbol and the fact that $(1 - \gamma, \gamma)_{\mathfrak{p}} = 1$ we find

$$(n, \alpha)_{\mathfrak{p}} = (n^{p-1}, \alpha^{p-1})_{\mathfrak{p}} = (1 - \gamma, 1 - \beta)_{\mathfrak{p}} = (1 - \gamma, (1 - \beta)\gamma)_{\mathfrak{p}} = 1,$$

the last step because $(1 - \gamma) + (1 - \beta)\gamma = \delta^p$ (see [2, p. 351, Exercise 2.5]). This proves Lemma 1.

From Lemma 1, we obtain the following result of Furtwängler (1912).

Lemma 2. *We have $q^{p-1} \equiv 1 \bmod p^2$ for every prime number q that satisfies one of the following conditions:*

(i) *q divides x, y, or z;*

(ii) *one of the differences $x-y$, $y-z$, $z-x$ is divisible by q but not by p.*

Proof. Suppose first that q is a prime number dividing y. Then q does not divide p or z, so we can apply Lemma 1 with $n=q$ to find $\left(\frac{x}{q}\right)=\left(\frac{x+y\zeta}{q}\right)=\left(\frac{\zeta}{q}\right)^{-y/z}$. As $\left(\frac{x}{q}\right)$ is a Galois-invariant pth root of unity, it equals 1. Also $-y/z\not\equiv 0 \bmod p$, so we have $\left(\frac{\zeta}{q}\right)=1$. The formula $\left(\frac{\zeta}{q}\right)=\zeta^{(q^{p-1}-1)/p}$ from [2, p. 349, Exercise 1.6] now implies $q^{p-1}\equiv 1 \bmod p^2$.

Next, suppose that q is a prime number dividing $x-y$, and that $x-y$ is not divisible by p. Clearly, we may assume that q does not divide z. From the equality $\left(\frac{x+y\zeta}{q}\right)=\left(\frac{y+x\zeta}{q}\right)$ it follows, by another application of Lemma 1, that $\left(\frac{\zeta}{q}\right)^{-y/z}$ and $\left(\frac{\zeta}{q}\right)^{-x/z}$ are equal. As $-y/z$ and $-x/z$ are not congruent modulo p, this implies $\left(\frac{\zeta}{q}\right)=1$. As before, we obtain $q^{p-1}\equiv 1 \bmod p^2$. This proves Lemma 2.

We derive Theorem 3 from Lemma 2. By the assumption of the theorem, there exist x, y, z as above. As one of x, y, z is even, condition (i) holds for $q=2$. This yields (a). To prove (b), we first note that by (a) we have $p\neq 3$. It suffices to show that one of the conditions in Lemma 2 is met by $q=3$. If 3 divides one of x, y, z, then (i) holds. Otherwise, the congruence $x^p+y^p+z^p\equiv 0 \bmod 3$ shows that 3 divides all differences $x-y$, $y-z$, $z-x$; but from $3x^p\not\equiv 0 \bmod p$ it follows that these differences are not all divisible by p, so (ii) holds. This completes the proof of Theorem 3.

We next prove Theorem 2. Let k be a positive integer for which $q=kp+1$ is prime. It suffices to show that if x, y, z are as above, then p divides k or q divides N_k. We distinguish two cases. First suppose that one of x, y, z is divisible by q. From Lemma 2 it follows that $q^{p-1}\equiv 1 \bmod p^2$, so we have $1+kp=q\equiv q^p=(1+kp)^p\equiv 1 \bmod p^2$. Thus, in this case p divides k. Next, suppose that none of x, y, z is divisible by q. From $p=(q-1)/k$ we see that each of x^p, y^p, z^p, when taken modulo q, is a kth root of unity in the finite field $\mathbf{Z}/q\mathbf{Z}$. Hence there are, in the ring of q-adic integers, kth roots of unity ϵ, $\epsilon\eta$, $\epsilon\vartheta$ (say) that are congruent to x^p, y^p, and z^p, respectively, modulo q. From $x^p+y^p+z^p=0$ we find $1+\eta+\vartheta\equiv 0 \bmod q$, so that now q divides N_k. This proves Theorem 2.

Above we saw already that Theorem 1 follows from Theorem 2.

REFERENCES

1. L. M. Adleman, D. R. Heath-Brown, *The first case of Fermat's last theorem*, Invent. math. **79** (1985), 409–416.
2. J. W. S. Cassels, A. Fröhlich (eds), *Algebraic number theory*, Proceedings of an Instructional Conference, Academic Press, 1967.
3. R. Crandall, K. Dilcher, C. Pomerance, *A search for Wieferich and Wilson primes*, Math. Comp. **66** (1997), no. 217, 433–449.
4. E. Fouvry, *Théorème de Brun-Titchmarsh; application au théorème de Fermat*, Invent. math. **79** (1985), 383–407.

5. A. Granville, M. B. Monagan, *The first case of Fermat's last theorem is true for all prime exponents up to* 714,591,416,091,389, Trans. Amer. Math. Soc. **306** (1988), 329–359.
6. H. Hasse, *Bericht über neuere Untersuchungen und Probleme aus der Theorie der algebraischen Zahlkörper, Teil II: Reziprozitätsgesetz*, Jber. Deutsch. Math.-Verein. Ergänzungsband **6** (1930), 1–204.
7. H. W. Lenstra, Jr., *Miller's primality test*, Inform. Process. Lett. **8** (1979), 86–88.
8. P. Montgomery, *New solutions of* $a^{p-1} \equiv 1 \pmod{p^2}$, Math. Comp. **61** (1993), 361–363.
9. J. W. Tanner, S. S. Wagstaff, Jr., *New bound for the first case of Fermat's last theorem*, Math. Comp. **53** (1989), 743–750.

Acknowledgment.
The assistance of Andrew Granville and Wilfrid Keller is gratefully acknowledged. The first author was supported by NSF grant No. DMS–9224205.

REMARKS ON THE HISTORY OF FERMAT'S LAST THEOREM 1844 TO 1984

MICHAEL ROSEN

Introduction

It is arguably true that Fermat's Last Theorem (FLT) has been the most famous of all mathematical problems for at least three centuries. There has been debate about whether it is a serious and important problem or merely a curiosity, but there can be no denying its popularity. Generations of mathematicians, both professional and amateur, have tried their hand at solving it. These efforts have resulted in a mighty body of theory with many deep and important results. Nevertheless, until 1984, when G. Frey connected the problem in an intimate way with the arithmetic theory of elliptic curves and the conjecture of Taniyama-Shimura-Weil (after earlier work in the same direction by Y. Helloguarch), a solution seemed as far away as ever.

We will attempt to review some of the highlights among the results obtained in the period from 1844 to 1984 (and a few that came after).

As is well known the study of Fermat's last theorem for exponent n reduces rapidly to the cases where $n = 4$ or n is equal to an odd prime p. Fermat himself proved the theorem for $n = 4$ and claimed to have proven it for $p = 3$. By 1844 the theorem was known only for $p = 3, 5$, and 7 with proofs by L. Euler, L. Dirichlet and A. Legendre, and G. Lamé, respectively. The greatest contributions to our subject up to the last few years were made by E. Kummer. We begin our discussion with the year 1844, because that was the year Kummer published his results on the theory of ideal numbers in the field $\mathbf{Q}(\zeta_p)$. There is some controversy over whether Kummer's primary motivation for creating this theory was his interest in higher reciprocity laws or his interest in FLT. He was clearly interested in both although he thought the reciprocity laws were more important. In any case, three years later in 1847 he completed the proof of his great theorem: FLT is true for regular primes p, i.e., primes p which do not divide the class number of $\mathbf{Q}(\zeta_p)$. To do this he had to define the class group, prove it is finite, analyze it as a product of two factors $h_p = h_p^+ h_p^-$ (the real class number times the relative class number), and do a profound investigation of the unit group. Beyond this he was able to relate the regularity of the prime p to divisibility by p of certain Bernoulli numbers. We will review this work and give a sketch of his proof of FLT for regular primes. Kummer claimed that FLT is true for regular primes p even if we allow entries from $\mathbf{Q}(\zeta_p)$. Kummer's proof of this latter claim contains an error, but his error was later patched up by Hilbert. It is interesting to ask if FLT is true for all odd primes p if we allow entries from $\mathbf{Q}(\zeta_p)$. The modular proof over $\mathbf{Z}$ does not seem to extend easily to this case.

Kummer returned to FLT in 1857 when he published a paper dealing with certain cases of irregular primes p for which FLT can be proven. There are some problems with this paper. The results were put on a firm foundation and further refined many years later (1929) by H.S.Vandiver. We will mention some of this work which is the foundation for calculations which have shown FLT to be true up to very large bounds. This culminated in a 1993 paper of J. Buhler-R. Crandall-R. Ernvall-T. Metsänkylä which showed FLT to be true for all primes up to four million. This may be of limited interest now that we know FLT is true without restriction, but these calculations have bearing on other issues as well, for example Vandiver's conjecture that p does not divide h_p^+ (already conjectured by Kummer many years earlier). The truth or falsity of this is still unknown, but it is true for all primes up to four million.

Irregular primes and Vandiver's conjecture both relate to the interesting question of the structure of A_p, the p-primary part of the class group of $\mathbf{Q}(\zeta_p)$. Kummer made a fundamental contribution to this by proving a special case of Stickelberger's theorem on the prime decomposition of Gauss sums. We will discuss the consequences of this, among them the important theorem of J. Herbrand (1932). It's converse was proved by K. Ribet in 1976. Ribet's proof used the arithmetic theory of modular forms and Galois representation theory and can thus be seen as an early breakthrough demonstrating the power of the methods which eventually led to a proof of FLT. We will also discuss subsequent work of A. Wiles and Mazur-Wiles on the structure of A_p.

The first case of FLT, i.e., when $x^p + y^p = z^p$ and it is assumed that $(xyz, p) = 1$, will be dealt with elsewhere in this volume [LS]. We will concentrate on the second case. With a subject so vast it is inevitable that much of value will be left out. We hope that the material to be included will be sufficient to give a good sense of the fascinating and intricate history of this old conjecture which is finally a theorem.

Before beginning, we point out that we will assume the reader is familiar with elementary algebraic number theory and the elements of p-adic numbers. We will not attempt to use the language and notation of the nineteenth century, but will use more or less standard modern notation. What is lost in historical flavor by this process is made up for by an increase in clarity (or so we hope).

Section 1. Fermat's last theorem for polynomials.
We begin our discussion, somewhat a-historically, by considering $x^n + y^n = z^n$, $n > 2$, over a polynomial ring $A = k[T]$ where k is a field of characteristic prime to n. We claim there are no non-zero solutions in A except possibly for constant solutions, i.e., when x, y, and z are in k [Gr]. Suppose there is a solution $x, y, z \in A$ where x, y, and z are non-zero and the maximum of the three degrees is $d > 0$. We will show that there is a solution x', y', and z' all non-zero and with the maximum of the three degrees d' satisfying $d > d' > 0$. We could then repeat the process indefinitely. Of course, this constitutes a contradiction since there are only finitely many positive integers less than d. This is Fermat's method of "infinite descent."

Before we begin, note that it is no restriction to assume that k is algebraically closed. Further, since A is a unique factorization domain, a moment's reflection shows that we may assume that x, y, and z are relatively prime. Now, consider the identity

$$\prod_{i=0}^{n-1} (x + \zeta^i y) = -z^n. \tag{1}$$

Here, ζ is a primitive n'th root of unity in k^*. The factorization in equation (1) is the basis of all attempts to prove FLT before the elliptic curve/modular function approach. Every pair of factors on the left is relatively prime since if w divided the i'th and j'th factor it would also divide $(\zeta^j - \zeta^i)y$ and $(\zeta^{j-i} - 1)x$ and these two polynomials are relatively prime. It follows easily that each factor is itself a constant times an n'th power. Since we have assumed that k is algebraically closed, each constant is also an n'th power. Thus, there are polynomials u, v, and w such that

$$x + y = u^n, \quad x + \zeta y = v^n, \quad x + \zeta^2 y = w^n.$$

Eliminating x and y from these three equations, we find

$$w^n + \zeta u^n = (1 + \zeta) v^n.$$

Again using that k is algebraically closed we set $x' = w$, $y' = \sqrt[n]{\zeta}\, u$, and $z' = \sqrt[n]{1 + \zeta}\, v$. Then, x', y', and z' constitute a new solution with the required properties. In fact, $d' \leq d/n$.

It may be asked where the assumption $n > 2$ was used? The point is that if $n = 2$, then $\zeta = -1$ and so $1 + \zeta = 0$ and $z' = 0$. This shows that the proof breaks down if $n = 2$ but is sound for all $n > 2$.

Another interesting point is that the proof applies equally well to polynomials in several variables over a field.

Section 2. Kummer's work on cyclotomic fields.

The proof of FLT for the polynomial ring $A = k[T]$ is very short and sweet. It contains the main ideas of the early attempts to prove Fermat's assertion, but is much easier. Why is that? To begin with $\mathbf{Z}$, the ring of rational integers, contains no roots of unity except ± 1. To compensate for this the relevant roots of unity were added to $\mathbf{Z}$ and the arithmetic of the resulting rings investigated. Letting $i = \sqrt{-1}$. C.F. Gauss investigated the ring $\mathbf{Z}[i]$ in his celebrated pair of papers on biquadratic reciprocity [G]. Let $\omega = e^{2\pi i/3}$. The ring $\mathbf{Z}[\omega]$ was investigated by C.G. Jacobi and independently by G. Eisenstein in the course of formulating and proving the law of cubic reciprocity. Gauss had investigated the same ring in an unpublished paper proving FLT for $p = 3$. These authors, and others, also experimented with other roots of unity. For small primes p it turns out that $\mathbf{Z}[\zeta_p]$ is a unique factorization domain. In fact, this is true for all primes p less than 23. G. Lamé, perhaps led astray by this fact, announced in 1847 that he had a proof of FLT. Liouville almost immediately pointed out that he was implicitly assuming unique factorization. Soon thereafter, Liouville received a letter from Kummer which pointed out that not only is the assumption without proof, it is not correct. Kummer had shown three years earlier that $\mathbf{Z}[\zeta_{23}]$ is not a unique factorization domain. As we shall see in a moment, Kummer had done much, much more than that.

There is a story, told by K. Hensel in an address given on the hundredth anniversary of Kummer's birth, that Kummer himself had once constructed a proof of FLT assuming unique factorization and that the error had been pointed out to him by Dirichlet. Although this story has been widely retold in subsequent works on number theory, it is probably incorrect. This was pointed out, with fairly convincing evidence, by H. Edwards in a pair of papers [Ed2,Ed3] which are interesting reading both because of this issue and also for a general historical account of the events of 1844 to 1847 which bear on FLT.

Kummer not only noticed the failure of unique factorization, he invented his theory of ideal numbers to restore this immensely useful property for the rings $\mathbf{Z}[\zeta_p]$. Later, R. Dedekind extended Kummer's work by discussing general rings of algebraic numbers and by reinterpreting Kummer's ideal numbers by means of ideals. Because of the fact that Dedekind's language is so much more familiar to modern readers we will use it rather than Kummer's. The interested reader can consult Edward's book [Ed1] for a discussion of Kummer's point of view.

For the rings under consideration, Kummer proved that every ideal is the product of prime ideals in a unique way. He also defined the usual equivalence relation on ideals, proved the equivalence classes form a group, and that this group is finite. Let's call the ideal class group of $\mathbf{Z}[\zeta_p]$,

Cl_p, and its order, h_p, the class number. We also refer to h_p as the class number of the field $\mathbf{Q}(\zeta_p)$. Unique factorization holds for elements if and only if $h_p = 1$.

Another complication arises when $p \geq 5$. Namely, the unit group will contain elements of infinite order. As is well known the units in $\mathbf{Z}[i]$ are the fourth roots of unity and those in $\mathbf{Z}[\omega]$ are the sixth roots of unity. For $p \geq 5$ we define the elements (set $\zeta_p = \zeta$)

$$\xi_k = \sqrt{\frac{\zeta^k - 1}{\zeta - 1}\,\frac{\zeta^{-k} - 1}{\zeta^{-1} - 1}} = \frac{\sin(k\pi/p)}{\sin(\pi/p)} \quad \text{for } k = 2, 3, \ldots, \frac{p-1}{2}. \tag{2}$$

These elements are easily seen to be units, called cyclotomic units, in $\mathbf{Z}[\zeta_p]$. Kummer shows that they are independent; i.e., they generate a free abelian group of rank $\frac{p-3}{2}$. Let C_p be the subgroup of the unit group E_p generated by the ξ_k and the roots of unity. C_p is called the group of cyclotomic units. Kummer shows that C_p is of finite index in E_p and gives the following beautiful interpretation of the index.

Theorem 2.1. *The index $[E_p : C_p] = h_p^+$, where h_p^+ is the class number of $\mathbf{Q}(\zeta_p + \zeta_p^{-1}) = \mathbf{Q}(\zeta_p)^+$, the maximal real subfield of $\mathbf{Q}(\zeta_p)$.*

Further, it turns out that h_p^+ divides h_p, so that we can define the integer $h_p^- = h_p/h_p^+$, called the relative class number. In the older literature h_p^- and h_p^+ are referred to as the first and second factors of the class number respectively. Kummer also gives a beautiful and important formula for h_p^- which we shall now proceed to explain.

Let χ be a Dirichlet character modulo p. We say that χ is odd if $\chi(-1) = -1$ and even if $\chi(-1) = 1$. Define $B_{1,\chi} = p^{-1} \sum_{a=1}^{p-1} \chi(a)a$. It is easy to see that $B_{1,\chi} = 0$ if χ is even and not trivial. On the other hand:

Theorem 2.2.

$$h_p^- = 2p \prod_{\chi \text{ odd}} \left(-\frac{1}{2}\right) B_{1,\chi}.$$

The numbers $B_{1,\chi}$ are sometimes called generalized Bernoulli numbers. We will relate them to ordinary Bernoulli numbers a little later. First, we mention another of Kummer's important theorems about the class number.

Theorem 2.3. *If p divides h_p^+, then p divides h_p^- .*

Definition. A prime number p is called *regular* if p does not divide h_p, otherwise it is called *irregular*.

As was mentioned in the introduction, Kummer's greatest contribution to FLT was to show that it is true for regular primes. We will sketch a proof of this in Section 3. For the remainder of this section we discuss Kummer's criterion for regularity which allows one to actually compute whether a given prime is regular or not. We begin by recalling the definition of the Bernoulli numbers and some of their properties (see [BS], [IR], [Ri], or [W]).

The simplest way to define the Bernoulli numbers B_n is by way of the power series expansion

$$\frac{t}{e^t - 1} = \sum_{n=0}^{\infty} B_n \frac{t^n}{n!}.$$

From this it is easy to derive the formula

$$(m+1)B_m = -\sum_{k=0}^{m-1} \binom{m+1}{k} B_k.$$

The Bernoulli numbers may now be computed recursively. One finds $B_1 = -\frac{1}{2}, B_2 = \frac{1}{6}, B_3 = 0, B_4 = -\frac{1}{30}, B_5 = 0, B_6 = \frac{1}{42}$, etc. For $n > 1$ and odd one has $B_n = 0$. The even numbered Bernoulli numbers grow quite rapidly. In fact, $|B_{2m}| > 2(m/\pi e)^{2m}$.

The Bernoulli numbers are rational numbers whose denominators are known thanks to the theorem of Von-Staudt and Claussen. This asserts that the denominator of B_n is the product of those primes p such that $p - 1$ divides n. In particular, if $p > 3$ the numbers $B_2, B_4, B_6, \ldots, B_{p-3}$ are p-integral.

Let $\mathbf{Z}_p$ denote the p-adic numbers. It is well known that the unit group $\mathbf{Z}_p^*$ contains the $(p-1)^{\text{st}}$ roots of unity, and that there is a character $\omega : (Z/pZ)^* \to Z_p^*$ with the property that $\omega(a) \equiv a \pmod{p}$ for all rational integers a prime to p. ω is an odd character of order $p-1$. Using Theorem 2.2, one easily derives the following p-adic version.

Theorem 2.2 P.

$$h_p^- = 2p \prod_{\substack{i=1 \\ i \text{ odd}}}^{p-2} \left(-\frac{1}{2} B_{1,\omega^i}\right)$$

The equality here takes place inside $\mathbf{Z}_p$.

To get Kummer's criterion from this we need the following congruence which is proved in [W, Cor.5.15] using p-adic L-functions and in [Lg1, Theorem 2.5] using the theory of p-adic distributions. Because of its importance we give another, more elementary proof, in the appendix to this paper.

Proposition 2.3. *If n is odd, and $p - 1$ does not divide $n + 1$, we have*

$$B_{1,\omega^n} \equiv \frac{B_{n+1}}{n+1} \pmod{p}.$$

We are now in a position to demonstrate the following wonderful result of Kummer.

Theorem 2.4. *A prime number p is regular if and only if it does not divide the numerator of any of the Bernoulli numbers $B_2, B_4, \ldots, B_{p-3}$.*

PROOF. By Theorem 2.3 it follows that p divides h_p if and only if p divides h_p^-. We will use the expression for h_p^- given by Theorem 2.2 P.

First consider the case $i = p-2$. We have,

$$pB_{1,\omega^{p-2}} = \sum_{a=1}^{p-1} \omega^{p-2}(a)a \equiv \sum_{a=1}^{p-1} a^{p-1} \equiv p-1 \equiv -1 \pmod{p} .$$

Thus,

$$h_p^- \equiv \prod_{\substack{i=1 \\ i \text{ odd}}}^{p-4} \left(-\frac{1}{2}B_{1,\omega^i}\right) \equiv \prod_{\substack{i=1 \\ i \text{ odd}}}^{p-4} \left(-\frac{1}{2}\frac{B_{i+1}}{i+1}\right) \pmod{p} .$$

The result follows directly from this congruence.

By using this theorem it is possible to check a given prime for regularity. Kummer showed that the only irregular primes less than 100 are 37, 59, and 67. He later checked all the primes up to 164 and found that 101, 103, 131, 149, and 157 are the only additional irregular primes. At one point he thought that he had a proof that there are infinitely many regular primes. In 1915, K.L. Jensen proved that there are infinitely many irregular primes [Ri; Lecture VI, Section 4]. However, to this day it is not known if there are infinitely many regular primes. It is possible to give a probabilistic argument to show that over 60% of the primes should be regular.

Let's assume that the probability that an even indexed Bernoulli number B_{2m} be divisible by p is $\frac{1}{p}$. If this is so, the probability that none of the numbers $B_2, B_4, \ldots, B_{p-3}$ be divisible by p is

$$\left(1 - \frac{1}{p}\right)^{(p-3)/2} \approx e^{-1/2} \approx .6065 .$$

This estimate for the percentage of regular primes agrees very well with the experimental evidence [BCEM]. It would be nice to have a rigorous proof.

We end this section with two more of Kummer's results. These two concern units. The first is fairly easy, but very useful. The second is quite deep, and is crucial to Kummer's proof of FLT for regular primes.

Recall some simple facts about $\mathbf{Q}(\zeta_p)$. The prime p is totally ramified in this field and the prime lying above (p) in $\mathbf{Z}[\zeta_p]$ is (λ), where $\lambda = \zeta_p - 1$. Note that $\zeta_p \equiv 1 \pmod{\lambda}$ and $\mathbf{Z}[\zeta_p]/(\lambda) \cong \mathbf{Z}/p\mathbf{Z}$.

Proposition 2.5. *Let u be a unit in $\mathbf{Z}[\zeta_p]$. Then u can be written in a unique way as $\pm\zeta_p^i e$, where i is determined modulo p, and e is real and positive.*

PROOF. Let bar denote, as usual, complex conjugation. Then, $u/\bar{u}$ is a unit such that it and all its conjugates have absolute value 1. By a well known result, $u/\bar{u}$ is a root of unity. By ramification theoretic considerations the only roots of unity in $\mathbf{Z}[\zeta_p]$ are $\pm\zeta_p^i$ for $0 \le i \le p-1$. If $u/\bar{u} = \zeta_p^i$, find an integer j such that $2j \equiv i \pmod{p}$. Then one easily checks $\zeta_p^{-j}u$ is equal to its own conjugate, i.e., is real. The result follows in this case by adjusting the sign.

Suppose $u/\bar{u} = -\zeta^i$. We will show this leads to a contradiction. Choosing j as above, and setting $w = \zeta_p^{-j}u$, we find that w is a unit such that $\bar{w} = -w$. Thus $w^2 = -w\bar{w} = -v$. It follows that $-v \in \mathbf{Z}[\zeta_p]^+$ is a negative unit and so w generates the extension $\mathbf{Q}(\zeta_p)/\mathbf{Q}(\zeta_p)^+$. It would follow that this extension is unramified at p (recall $p \neq 2$). However, it is ramified at p, so we have reached a contradiction.

Corollary. *Let E_p^+ be the real positive units in E_p and C_p^+ be the subgroup of C_p generated by the cyclotomic units ξ_k (see equation (2) above). Then, $E_p^+/C_p^+ \cong E_p/C_p$ and both groups have order h_p^+.*

PROOF. This follows directly from Theorem 2.1 and Proposition 2.5.

We remark that there is another way to finish the proof of the proposition. Suppose w is a unit such that $w = -\bar{w}$. By the remarks preceding the proposition, there is a rational integer M such that $w \equiv M \pmod{\lambda}$. It follows that $M \equiv -M \pmod{\lambda}$ and so $2M \equiv 0 \pmod{\lambda}$. Thus, λ divides M and so also w. However, w is a unit so this is a contradiction.

The final result we need is known as Kummer's lemma. It is simple to state. Any unit in $\mathbf{Q}(\zeta_p)$ which is congruent to a rational integer modulo p is a pth power. The usual proof is by means of a close analysis of the unit group and especially of the local unit group in the completion of $\mathbf{Q}(\zeta_p)$ at the prime (λ). This involves the p-adic logarithm, the local expansion of the p-adic logarithm of the cyclotomic units, etc. The analysis is nicely carried out in Section 6 of Chapter 5 in [B-S]. We give an alternative approach, suggested by Hilbert in his Zahlbericht [H], which depends on the study of ramification in Kummer extensions. This approach brings out more clearly the underlying role played by class field theory.

Lemma 2.6. *Let $\beta \in \mathbf{Z}[\zeta_p]$. Suppose λ does not divide β and that $x^p \equiv \beta \pmod{\lambda^p}$ is solvable. Let K be the field obtained by adjoining $A = \sqrt[p]{\beta}$ to $\mathbf{Q}(\zeta_p)$. Then the extension $K/\mathbf{Q}(\zeta_p)$ is unramified at p.*

PROOF. Suppose $\alpha^p \equiv \beta \pmod{\lambda^p}$ and set $\tau = (A - \alpha)/\lambda$. Then τ is a root of the monic polynomial

$$f(x) = \frac{(\lambda x + \alpha)^p - \beta}{\lambda^p}.$$

Using the fact that $p = u\lambda^{p-1}$ where u is a unit, we see that all the coefficients of $f(x)$ are algebraic integers. Thus, τ is an algebraic integer. A short computation shows that all the coefficients of $f'(x)$ are in (λ) except the constant term $u\alpha^p$ which is prime to (λ). Thus, $f'(\tau)$ is prime to (λ) and it follows that the relative discriminant is prime to (λ). The proof is complete.

Theorem 2.7. *Let p be a regular prime and e a unit in $\mathbf{Q}(\zeta_p)$ which is congruent to a pth power modulo λ^p. Then, e is the pth power of a unit.*

PROOF. Consider the extension $L = \mathbf{Q}(\zeta_p, \sqrt[p]{e})$ of $\mathbf{Q}(\zeta_p)$. This extension is cyclic of degree 1 or p. Suppose the degree is p. By Lemma 2.6, the extension is unramified at p. Since e is a unit it is easy to see it is unramified at every other prime as well. Thus, it is an unramified, abelian extension of degree p and it follows by class field theory that $p|h_p$, which contradicts the assumption that p is regular. Thus, $\sqrt[p]{e} \in \mathbf{Q}(\zeta_p)$ and the theorem follows.

The phrase "it follows by class field theory" can be avoided. The result needed follows from Theorem 94 of Hilbert's Zahlbericht [H; page 155]. We sketch a short cohomological proof of this in the Appendix.

The usual form of Kummer's Lemma can be deduced from Theorem 2.7 as follows. If $e \equiv a \pmod{p}$ with $a \in \mathbf{Z}$ we have $e = a + p\alpha$ with $\alpha \in \mathbf{Z}_p$. Now, $\alpha \equiv b \pmod{\lambda}$ with $b \in \mathbf{Z}$. Thus, $e \equiv a + bp \pmod{\lambda^p}$. Let σ be an element of $Gal(\mathbf{Q}(\zeta_p)/\mathbf{Q})$. Then, $e^\sigma/e \equiv 1 \pmod{\lambda^p}$ and is thus a pth power by Theorem 2.7: $e^\sigma = e\eta(\sigma)^p$. Taking the product over all σ and remembering that the norm of a unit is ± 1, we find $\pm 1 = e^{p-1}\eta^p$ for a suitable η and finally, $e = (\pm e\eta)^p$.

Section 3. Fermat's last theorem for regular primes and certain other cases.

Having assembled a number of the powerful tools forged by Kummer, we will now give part of his proof that FLT is true for regular primes. Here is the statement of the full theorem.

Theorem 3.1. *Let p be a regular prime. Then, the equation $x^p + y^p = z^p$ has no solution with $x, y, z \in \mathbf{Z}[\zeta_p]$ and $xyz \neq 0$.*

The proof is usually broken up into two parts. The first case is when λ does not divide xyz and the second case is when λ does divide xyz (recall, $\lambda = \zeta_p - 1$). The first case is easier. If one confines one's attention to $\mathbf{Z}$ rather than $\mathbf{Z}[\zeta_p]$, proofs of the first case can be found in [BS,IR,W] and many other places. For a proof of the first case in $\mathbf{Z}[\zeta_p]$ see [La] or [H]. We will concentrate on the second case.

It is interesting that Kummer made a simple error at the beginning of his proof of the second case. He asserts that it is no loss of generality to assume that x, y, z are pairwise relatively prime. This is certainly true over $\mathbf{Z}$, but is false over $\mathbf{Z}[\zeta_p]$ because there may be a common divisor which is not principal. Once this is realized it is not hard to alter Kummer's proof so that it holds in full generality. Hilbert does so in Section 172 of [H]. See [La] for another presentation. We give Kummer's proof of the more restricted result because it is relatively short and the main ideas show through more clearly.

Theorem 3.1′. *Let p be a regular prime. Then, the equation*

$$x^p + y^p = z^p$$

has no solution with $x, y, z \in \mathbf{Z}[\zeta_p]$, $\lambda | xyz$, and x, y, z pairwise relatively prime.

PROOF. Assume x, y, z is such a solution. It is no loss of generality to assume that $\lambda | z$. It follows that λ does not divide x or y. Write $z = \lambda^m z_0$ with $(\lambda, z_0) = 1$. Then, x, y, z_0 is a solution to $X^p + Y^p = \lambda^{mp} Z^p$ with $(xyz_0, \lambda) = 1$, x, y, z pairwise relatively prime, and $m \geq 1$. Let u be a unit in $\mathbf{Z}[\zeta_p]$. We will show that there are no solutions x, y, z to

$$X^p + Y^p = u\lambda^{mp} Z^p \tag{$*$}$$

with x, y, z pairwise relatively prime, $(xyz, \lambda) = 1$, and $m \geq 1$. This will prove the theorem.

The strategy is this. Assume such a solution exists. One shows that, in fact, m must be greater than 1. Then one finds a solution of the same type to a similar equation but with m replaced by $m - 1$. This yields a contradiction via "infinite descent."

We need a Lemma.

Lemma 3.2. *Let $v \in \mathbf{Z}[\zeta_p]$ with $(v, \lambda) = 1$. Then, there is a rational integer k such that $\zeta_p^k v \equiv a \pmod{\lambda}$ with $a \in \mathbf{Z}$.*

PROOF. Since the residue class field modulo λ has p elements, we may write $v \equiv m + n\lambda \pmod{\lambda^2}$ where $m, n \in \mathbf{Z}$ and $(m, p) = 1$. Now, $\zeta_p^k = (1+\lambda)^k \equiv 1 + k\lambda \pmod{\lambda^2}$. Thus, $\zeta_p^k v \equiv m + (n + km)\lambda \pmod{\lambda^2}$. Choose k to be a solution of $n + mx \equiv 0 \pmod{p}$ and the Lemma follows.

Now, assume x, y, z is a solution of equation $(*)$ above with $(xyz, \lambda) = 1$ and x, y, z pairwise relatively prime. By Lemma 3.2 we can assume that

$x, y \equiv a, b \pmod{\lambda^2}$ where $a, b \in \mathbf{Z}$. We have,

$$\prod_{i=0}^{p-1}(x + \zeta_p^i y) = u\lambda^{mp}z^p. \qquad (**)$$

It follows that λ must divide at least one term on the left hand side of this equation, and consequently it must divide all the terms. Since $x + y \equiv a + b \pmod{\lambda^2}$, we must have $p \mid a + b$ and so $\lambda^2 \mid x + y$. From equation $(**)$ it now follows that λ^{p+1} divides the left hand side, and so $m > 1$, which is our first goal.

We have $x + \zeta_p^i y = x + y + (\zeta_p^i - 1)y$ and so for $i > 0$, $x + \zeta_p^i y$ is exactly divisible by λ. Thus, passing to ideals, we find $(x+y) = (\lambda)^{p(m-1)+1}C_0$ and $(x + \zeta_p^i y) = (\lambda)C_i$ for $i = 1, 2, \ldots, p-1$, where the C_i for $i = 0, 1, \ldots, p-1$ are pairwise relatively prime principal ideals each of which is prime to (λ). From equation $(**)$ we deduce $C_0 C_1 \ldots C_{p-1} = (z)^p$. It follows that each C_i is a pth power, i.e., $C_i = D_i^p$. Since D_i^p is a principal ideal and p is a regular prime, we deduce that D_i is a principal ideal, i.e., $D_i = (w_i)$ for $w_i \in \mathbf{Z}[\zeta_p]$. Returning to the level of elements, we see that there are units u_0, u_1, u_2 such that

$$x + y = u_0\lambda^{p(m-1)+1}w_0^p, \quad x + \lambda y = u_1\lambda w_1^p, \quad x + \lambda^2 y = u_2\lambda w_2^p.$$

Eliminating x and y from these three equations and dividing the result by u_1, we find units e_2 and e such that

$$w_1^p + e_2 w_2^p = e\lambda^{(m-1)p}w_0^p.$$

Taking congruences modulo λ^p, we see that e_2 is congruent to a pth power modulo λ^p. By Theorem 2.7, $e = f^p$ for some $f \in \mathbf{Z}[\zeta_p]$. Setting $x' = w_1$, $y' = fw_2$, and $z' = w_0$, we see that x', y', z' is a solution to $X^p + Y^p = e\lambda^{(m-1)p}Z^p$ for which $(x'y'z', \lambda) = 1$ and x', y', z' are pairwise relatively prime. We have reached our second goal and so completed the proof of Theorem 3.1$'$.

Kummer went well beyond the case of regular primes in his attempt to prove FLT in general. He produced some explicit cyclotomic units E_i for $i = 2, 4, \ldots, p-3$ and stated that FLT is true if the following conditions hold; h_p^- is divisible by p but not p^2, and for $i = 2, 4, \ldots, p-3$, B_{pi} is not divisible by p^3 and E_i is not a pth power. He then verified on the basis of this result and Theorem 3.1 that FLT holds for all primes less than 100.

Kummer's work was reconsidered by H.S. Vandiver in the 1920's. He found some problems with the proof of the above mentioned result which he was able to fix. He went on to improve upon Kummer's work in several ways. For example, he proved the following theorem [V], [W; Theorem 9.4].

Theorem 3.3. *Suppose B_{pi} is not divisible by p^3 for $i = 2, 4, \dots, p-3$ and that h_p^+ is not divisible by p. Then, the second case of FLT is true for p.*

In Chapter 9 of [W] there are further results of Vandiver which give rational criteria for proving the second case of FLT. See, in particular, Theorem 9.5. The first case of FLT can also be tested by even simpler rational criteria (see [LS]). Thus, it became possible to test FLT computationally. Vandiver and his students, using desk calculators, extended Kummer's verification to all primes less than 620. In 1954, using a computer, Vandiver, D.H. Lehmer, and E. Lehmer verified FLT for all primes up to 2,000. In 1955, J.L. Selfridge, C.A. Nicol, and H.S. Vandiver verified FLT up to 4,001. In 1976, S.Wagstaff showed FLT is true for all $p < 125{,}000$. In 1993 it was shown that FLT is true for all primes up to 4 million [BCEM]. This is the largest bound achieved before FLT was proven to hold for all primes p and so all $n > 2$.

Vandiver (1882-1973) made many contributions to FLT and other parts of number theory. For references to some of his work on FLT see his interesting expository article (and the short follow up article) [V]. According to an interesting obituary notice written by D.H. Lehmer [Lmr], Vandiver never graduated from high school. He spent most of his professional life at the University of Texas. He is the only American mentioned in E. Landau's monumental three volume treatise on number theory, *Vorlesungen über Zahlentheorie*.

Vandiver conjectured that h_p^+ is not divisible by p for all primes p. This conjecture is referred to simply as Vandiver's conjecture. Serge Lang has pointed out that Kummer made the conjecture many years earlier in a letter to Kronecker where he refers to it as a "noch zu beweisenden Satz" (Kummer's *Collected Works*, vol. **1**, page 85). In any case, it has held up well. In [BCEM] it is verified for all primes less than 4 million. Larry Washington has produced a probabilistic argument [W, page 159] which shows that the number of exceptions to Vandiver's conjecture up to a given bound x should be approximately $\frac{1}{2}\log\log x$. For $x = 4{,}000{,}000$ this is approximately 1.361, so the fact that no counter-example has shown up is perhaps not surprising. On the other hand Washington's reasoning rests on certain randomness assumptions which may or may not hold. In any case, the conjecture is true very often! This is good because, as we shall see in the next section, Vandiver's conjecture has very interesting implications.

Section 4. The structure of the p-class group.
A prime is irregular if A_p, the p-part of the class group of $\mathbf{Q}(\zeta_p)$, is nontrivial. Much work has been devoted to understanding the structure of A_p beginning with Kummer. Important contributions have been made by people like Hilbert, Herbrand, Leopoldt, Iwasawa, Ribet, Mazur, and Wiles, among others. In this section we will review some of this work. Among other things we will show that if one accepts the Vandiver conjecture as true, then it is possible to give a completely satisfying description of the structure of A_p.

We begin with a few preliminary remarks. It is convenient to write the group operation in A_p additively. Also, since p is fixed in this discussion, we will write A instead of A_p. Since A is a torsion p-group, we may consider it as a $\mathbf{Z}_p$-module. It is also a module for the Galois group $G_p = \mathrm{Gal}(\mathbf{Q}(\zeta_p)/\mathbf{Q})$. Thus, A is a module over the group ring $\mathbf{Z}_p[G_p]$. As is well known, $(\mathbf{Z}/p\mathbf{Z})^*$ is isomorphic to G_p, where the automorphism corresponding to a, σ_a, takes ζ_p to ζ_p^a. Let ω be the p-adic valued character introduced in Section 2, and for each i such that $0 \leq i < p-1$ define

$$\epsilon_i = \frac{1}{p-1}\sum_a \omega(a)^{-i}\sigma_a \in \mathbf{Z}[G_p].$$

Here as elsewhere in this section, the summation goes from $a = 0$ to $a = p-2$. It is easy to check that these elements constitute a complete set of mutually orthogonal idempotents in the group ring. Define $A_i = \epsilon_i A$. Then,

$$A = \bigoplus_i A_i, \quad \text{and if } x \in A_i, \text{ then } \sigma_a x = \omega(a)^i x.$$

Because of this decomposition, to understand the structure of A, it suffices to understand the structure of each A_i.

The automorphism σ_{-1} is simply complex conjugation. Since $\omega(-1) = -1$, it follows that complex conjugation fixes A_i if i is even, and acts as multiplication by -1 if i is odd. It is not hard to show that the part of A which is fixed by complex conjugation is isomorphic to the p-part of the class group of $\mathbf{Q}(\zeta_p)^+$. Thus, if Vandiver's conjecture is true, $A_i = (0)$ for i even.

A final preliminary comment is that ϵ_0 is a constant times the norm map, and the norm map annihilates the class group. It follows that ϵ_0 simultaneously annihilates and fixes A_0. The conclusion is that $A_0 = (0)$.

Do any other elements of the group ring $\mathbf{Z}[G_p]$ annihilate A? This question is the key to the deeper part of the theory. The answer is that yes, there are other elements beside the norm map which annihilate the class group. The honor of being the first to see this goes (once again) to Kummer. Let l be a rational prime such that $l \equiv 1 \pmod{p}$, $\mathcal{L}$ a prime lying above it in $\mathbf{Q}(\zeta_p)$, and let χ be an appropriately chosen character of $(\mathbf{Z}/l\mathbf{Z})^*$ of order p. Form the Gauss sum

$$G(\chi) = \sum_{x=0}^{l-1} \chi(x)\zeta_l^x.$$

Then, it is easy to show $G(\chi)^p \in \mathbf{Q}(\zeta_p)$. What Kummer did (his notation was different) was to determine the prime decomposition of this element. He showed,

$$(G(\chi)^p) = \mathcal{L}^{\sum a\sigma_a^{-1}}.$$

Kummer also showed that in every ideal class there are ideals all of whose prime factors have absolute degree 1. It follows from these considerations that $\sum a\sigma_a^{-1}$ annihilates the class group of $\mathbf{Q}(\zeta_p)$. Somewhat later, 1890 to be precise, L. Stickelberger generalized all these considerations to the fields $\mathbf{Q}(\zeta_m)$ with m arbitrary. Complete proofs of Stickelberger's theorem can be found in [IR] and [W]. We will deal only with the case $m = p$. Let us state the result we will need.

Theorem 4.1. *Define $\theta = \frac{1}{p}\sum a\sigma_a^{-1}$. Then, $p\theta$ annihilates the class group. Further, suppose $(b, p) = 1$. Then, $(\sigma_b - b)\theta \in \mathbf{Z}_p[G_p]$ and $(\sigma_b - b)\theta$ annihilates the class group.*

Corollary. *$A_0 = (0)$ and $A_1 = (0)$. For $i \geq 2$, $B_{1,\omega^{-i}}$ annihilates A_i.* (Note that this is only interesting for i odd, since the generalized Bernoulli numbers are zero for even characters).

PROOF. We have already shown that $A_0 = (0)$. Now, $p\theta$ acts on A_1 by multiplication by $\sum a\omega(a)^{-1} = \sum a\omega(a)^{p-2} \equiv \sum a^{p-1} \equiv p - 1 \equiv -1 \pmod p$. This shows that $p\theta$ acts on A_1 by multiplication by a p-adic unit. On the other hand, it annihilates A_1 by the theorem. Thus, $A_1 = (0)$.

Assume now that $i \geq 2$. Then, $(\sigma_b - b)\theta$ acts on A_i by multiplication by $(\omega(b)^i - b)B_{1,\omega^{-i}}$. Choose b to be a primitive root modulo p. Then,

$$\omega(b)^i - b \equiv b(b^{i-1} - 1) \not\equiv 0 \pmod p.$$

This factor is therefore a p-adic unit and so $B_{1,\omega^{-i}}$ annihilates A_i as asserted. (It is easy to check using the defining property of ω that $B_{1,\omega^{-i}} \in \mathbf{Z}_p$ for $i > 1$.)

We can now prove the following important theorem of J. Herbrand which appeared (posthumously) in 1932 [He].

Theorem 4.2. *Let i be odd and $3 \leq i \leq p - 2$. Set $j = p - i$. If p does not divide B_j, then $A_i = (0)$.*

PROOF. By the above corollary, $B_{1,\omega^{-i}}$ annihilates A_i. By Proposition 2.3,

$$B_{1,\omega^{-i}} = B_{1,\omega^{p-1-i}} \equiv \frac{B_{p-i}}{p-i} \pmod p.$$

Thus, if p doesn't divide $B_{p-i} = B_j$, it follows that $B_{1,\omega^{-i}}$ is a p-adic unit and so $A_i = (0)$.

Herbrand also proved the converse to this theorem, but under the assumption that Vandiver's conjecture is true. In 1976, Ribet proved the converse is true unconditionally [R].

Theorem 4.3. *Let i be odd and $1 \leq i \leq p-2$. Set $j = p - i$. If p divides B_j, then $A_i \neq 0$.*

Ribet's methods are completely different from those used by Herbrand, which were classical, descending in a direct line from Kummer. Ribet used the rapidly developing arithmetic theory of modular forms and Galois representation theory, the same tools used eventually to successfully attack FLT itself.

Herbrand's theorem and its converse are somewhat qualitative in nature. They provide a rational criterion for determining when A_i is trivial or not. In 1980, Wiles proved the following, more quantitative, result [Wi].

Theorem 4.4. *Let i be odd and $3 \leq i \leq p-2$. Assume that A_i is cyclic. Then $\#(A_i) = p^{m_i}$, where $m_i = \mathrm{ord}_p(B_{1,\omega^{-i}})$.*

Once again, the methods used were the modern ones involving modular forms, modular curves, and representation theory. The result is very nice, but is conditional on the hypothesis of cyclicity. This was removed a few years later in 1984 when B. Mazur and A. Wiles published a proof of the main theorem of Iwasawa theory for abelian extensions of $\mathbf{Q}$. See [MW]. As a corollary of their work, they deduce the following unconditional result.

Theorem 4.5. *Let i be odd and $3 \leq i \leq p-2$. Then $\#(A_i) = p^{m_i}$, where $m_i = \mathrm{ord}_p(B_{1,\omega^{-i}})$.*

This result makes no assertion as to whether A_i is cyclic or not. It gives its size, but not its structure.

The Mazur-Wiles result once again uses sophisticated modern methods involving the theory of modular forms, modular curves, and modular Jacobians. A different approach was found by V.A. Kolyvagin, who invented the method of Euler systems which is in principal much more elementary. The application of these methods to Iwasawa theory is exposited in an appendix by K. Rubin to the new edition of Lang's book on cyclotomic fields [Lg1].

We will end this section by showing how "classical" methods together with the conjecture of Vandiver enables one to give a complete structure theorem for A. In essence, this is due to Herbrand. However, we will make good use of the "Spiegelungsatz" which is due to H.W. Leopoldt [Le] (1958). Special cases were proved independently by Iwasawa. We state such a special case next. The proof involves Kummer theory and class field theory. See Section 10.2 of [W]. If B is a finite abelian p-group, define $\operatorname{rank} B$ to be the dimension over $\mathbf{Z}/p\mathbf{Z}$ of B/pB. This can be shown to be the minimal number of generators for B.

Theorem 4.6. *Suppose j is even, $1 < i, j < p-1$, and $i + j = p$. Then,*

$$\operatorname{rank} A_j \leq \operatorname{rank} A_i \leq \operatorname{rank} A_j + 1.$$

Corollary. *Assuming the Vandiver conjecture, $A_j = (0)$ for j even and A_i is cyclic for i odd.*

It should also be remarked that the first inequality of the Theorem can be viewed as a refinement of Kummer's result, Theorem 2.3, that if $p|h_p^+$, then $p|h_p^-$.

Define

$$A^+ = \{x \in A|\ \sigma_{-1}x = x\} \text{ and } A^- = \{x \in A|\ \sigma_{-1}x = -x\}.$$

Then $A = A^+ + A^-$ and $A^+ = (0)$ if Vandiver's conjecture holds.

Theorem 4.7. *Assume Vandiver's conjecture is true. Then,*

$$A = A^- = A_3 \oplus A_5 \oplus \ldots \oplus A_{p-2}.$$

For i odd, $3 \leq i \leq p-2$, set $m_i = \operatorname{ord}_p B_{1,\omega^{-i}}$. Then,

$$A_i \simeq \mathbf{Z}_p/B_{1,\omega^{-i}}\mathbf{Z}_p \simeq \mathbf{Z}/p^{m_i}\mathbf{Z}.$$

Finally, $m_i > 0$ if and only if $p|B_{p-i}$.

PROOF. Vandiver's conjecture implies $A^+ = (0)$, and so $A = A^-$. By the Corollary to Theorem 4.1, $A_1 = (0)$, so the first assertion follows. By the same Corollary, $B_{1,\omega^{-i}}$ annihilates A_i, so A_i is a $\mathbf{Z}_p/B_{1,\omega^{-i}}\mathbf{Z}_p \simeq \mathbf{Z}/p^{m_i}\mathbf{Z}$ module. By the Corollary to Theorem 4.6 we have that A_i is cyclic. It follows that $\#(A_i) \leq p^{m_i}$. Taking the product over all i, we find $\#(A^-) \leq p^{\sum m_i}$. We claim that this inequality is actually an equality. Consider the class number formula given in Theorem 2.2 P, and recall that $pB_{1,\omega^{-1}}$ is a p-adic unit. It follows that $\operatorname{ord}_p(h_p^-) = m_3 + m_5 + \ldots + m_{p-2}$. Since $\operatorname{ord}_p\#(A^-) = \operatorname{ord}_p(h_p^-)$, we have proven our claim. This shows that $\#(A_i) = p^{m_i}$ for all i. Since A_i is a cyclic $\mathbf{Z}/p^{m_i}\mathbf{Z}$ module, the conclusion is that $A_i \simeq \mathbf{Z}/p^{m_i}\mathbf{Z}$.

The final assertion follows from the congruence

$$B_{1,\omega^{-i}} \equiv B_{p-i}/(p-i) \pmod{p}.$$

The index of irregularity of a prime p is defined to be the number of Bernoulli numbers in the set $\{B_2, B_4, \ldots, B_{p-3}\}$ which are divisible by p.

Call this number r_p. The above theorem shows that when Vandiver's conjecture is true, the rank of A_p is equal to r_p. By [BCEM] we know that Vandiver's conjecture is true for all primes up to 4 million. They also are able to verify that in this range, A_p is an elementary p-group. Thus,

$$A_p \simeq (\mathbf{Z}/p\mathbf{Z})^{r_p}$$

for all primes less than 4 million. They also show a similar result for the p-part of the class group of $\mathbf{Q}(\zeta_{p^n})$.

It is interesting to ask about the properties of the invariant r_p. For example, can it be arbitrarily large? If one assumes that the probability that a given Bernoulli number is divisible by p is $1/p$, the probability (assuming some independencies) that $r_p = k$ is given by

$$\binom{\frac{p-3}{2}}{k}\left(\frac{1}{p}\right)^k\left(1-\frac{1}{p}\right)^{\frac{p-3}{2}-k} \approx \frac{e^{-\frac{1}{2}}}{2^k k!}.$$

(The left hand side approaches the right hand side as $p \to \infty$.) The calculations of [B-C-E-M] show that this is in excellent agreement with the facts for $k = 0, 1, 2, 3, 4, 5$ for the set of prime less than 4 million. The largest value of r_p found was 7 which occurs only once for $p = 3{,}238{,}481$. If one accepts the validity of the probabilistic calculation just given, it would follow that r_p can be arbitrarily large, but no proof of this is known. It is also unknown if the exponent of the group A_p can be arbitrarily large.

Section 5. Suggested readings.

In this brief final section we will point to a few sources for further reading.

Since so much of our story has concerned Kummer, it is perhaps most suitable to begin with the first volume of his collected works [Ku]. The first volume contains his contributions to number theory. The introductory essay by A. Weil is very enlightening, both for its assessment of the significance of Kummer's contributions and as a guide to the papers themselves. These volumes were first published in 1975, and there are two informative reviews to recommend, one by Edwards [Ed4] and the other by B. Mazur [M]. Mazur's review is especially valuable for connecting Kummer's work to modern developments, much of which is due to Mazur himself.

As for FLT itself, there is the interesting, but mostly non-mathematical, book by E.T. Bell [B]. More substantial is the book by Edwards [Ed1]. In spite of its title, the emphasis is not so much FLT as the development of algebraic number theory in the hands of Kummer. In particular, it elaborates Kummer's theory of ideal numbers in much greater detail than any other work available in English. Also by Edwards are the interesting papers [Ed2] and [Ed3], already mentioned earlier, which debunk the story that Kummer thought he had proven FLT but had mistakenly assumed unique factorization in $\mathbf{Z}[\zeta_p]$. The best source for covering the spectrum of work done on FLT and related topics up through 1978 is the book of Ribenboim [Ri]. It is erudite and readable at the same time. Moreover, the bibliography is very extensive and useful.

The area of p-adic analysis is a modern development with origins in Kummer's arithmetic work. A good introduction is the book of N. Koblitz [Ko]. More advanced, and more inclusive in its coverage, is the book of Lang [Lg1]. As has been pointed out earlier, the new edition of this book contains a valuable appendix by Rubin which gives a proof of the main theorem of Iwasawa theory using the methods of Kolyvagin.

For the purposes of finding the proofs of most of the facts related in this paper, the best reference is the book by L. Washington [W]. A second edition of this book is in production and it will contain a number of interesting new results , e.g., a proof of the converse to Herbrand's theorem which uses Kolyvagin's Euler systems and avoids the more sophisticated methods used by Ribet. Of course, it should not be forgotten that these more sophisticated methods are what underlie the magnificent proof found by Wiles of FLT, a proof that eluded mathematicians for hundreds of years!

Appendix A. Kummer congruence and Hilbert's theorem 94.

We will sketch the proof of two important results mentioned in the text. The first is a congruence which, in essence, is due to Kummer. The second is Hilbert's theorem 94, which is not as famous as his theorem 90, but nevertheless of great importance in the history of class field theory. Our proof will use some properties of cohomology of groups.

Theorem A1. *Let p be an odd prime, $n > 0$ an odd integer, and assume that $p-1$ does not divide $n+1$. Then,*

$$B_{1,\omega^n} \equiv \frac{B_{n+1}}{n+1} \pmod{p}.$$

PROOF. It is well known and easily proved that $\omega(a) = \lim_{k\to\infty} a^{p^k}$, where a is any p-adic integer and the limit is taken p-adically. It follows that for k sufficiently large,

$$pB_{1,\omega^n} \equiv \sum_{a=1}^{p-1} a^{1+np^k} \equiv pB_{1+np^k} \pmod{p^2}. \tag{$*$}$$

The last congruence follows from the Corollary to Proposition 15.2.2 of [IR] and the observation that $1+np^k \equiv 1+n \pmod{p-1}$. Then Kummer's congruences [IR; Thm.5, Ch 15] imply

$$\frac{B_{1+np^k}}{1+np^k} \equiv \frac{B_{n+1}}{n+1} \pmod{p}. \tag{$**$}$$

Dividing congruence (*) by p and using congruence (**) yields the result.

The second result we want to prove is Hilbert's Theorem 94. We actually prove a slightly more general result which, given the cohomological tools we shall use, is no harder to prove (For further results using this type of analysis, see [CR]).

Theorem A2. *Let L/K be a cyclic unramified extension of algebraic number fields of odd degree n. Then the order of the kernel of the natural map from Cl_K to Cl_L is divisible by n. In particular, the class number of K is divisible by n. (Here, Cl_K and Cl_L denote the class groups of K and L respectively.)*

PROOF. Let D_L, P_L, and U_L denote the divisors, principal divisors, and units of L respectively, with similar notation for the field K. Let $G = \mathrm{Gal}(L/K)$. Consider the two exact sequences,

$$(0) \to U_L \to L^* \to P_L \to (0), \tag{1}$$

$$(0) \to P_L \to D_L \to Cl_L \to (0). \tag{2}$$

Using Hilbert's Theorem 90 and equation (1), we find $H^1(G, U_L) = P_L^G/\mathrm{im}(P_K)$. From equation (2) we derive

$$(0) \to P_L^G/\mathrm{im}(P_K) \to D_L^G/\mathrm{im}(P_K) \to Cl_L^G.$$

Since L/K is unramified, $D_L^G = \mathrm{im}(D_K)$. Combining these remarks we derive the following exact sequence,

$$(0) \to H^1(G, U_L) \to Cl_K \to Cl_L^G. \tag{3}$$

For any G module M, the Herbrand quotient of M is defined to be

$$h(M) = \frac{\#H^0(G, M)}{\#H^1(G, M)}.$$

Given the hypotheses of the theorem it is well known that $h(U_L) = n^{-1}$. For the proof, which depends on an analysis of the structure of $U_L \otimes \mathbf{Q}$ as a $\mathbf{Q}[G]$-module, see [Lg], [A-T], or the article of J. Tate in [C-F]. It follows immediately that n divides the order of $H^1(G, U_L)$, which by equation (3) is isomorphic to the kernel of the natural map from Cl_K to Cl_L. This completes the proof.

If n is even, it is still possible to derive results of a similar nature by making more restrictive hypotheses. For example, as Hilbert remarks, if $n = 2$ the result remains valid if we assume that every real prime of K splits in L.

Bibliography

[AT] E. Artin and J. Tate, *Class Field Theory*, Benjamin Publishers, New York, 1967.

[B] E.T. Bell, *The Last Problem*, Simon and Schuster, New York, 1961.

[BTW] J. Brillhart, J. Tonascia, and P. Weinberger, On the Fermat quotient, in *Computers in Number Theory*, Academic Press, New York, 1971, 213–222.

[BS] Z. Borevich and I. Shafarevich, *Number Theory*, Academic Press, London and New York, 1966.

[BCEM] J. Buhler, R. Crandall, R. Ernvall, and T. Metsänkyla, Irregular Primes and Cyclotomic Invariants to Four Million, *Math. Comp.*, 61, No. 203 (1993), 151–153.

[CF] J. Cassels and A. Fröhlich, ed., *Algebraic Number Theory*, Academic Press, London and New York, 1967.

[CR] G. Cornell and M. Rosen, Cohomological analysis of the class group extension problem, Proceedings of the Queen's Number theory Conference 1979, in *Queen's papers in Pure and Appl. Math.*, No. 54 (1980), 287–308.

[Ed1] H.M. Edwards, *Fermat's Last Theorem*, GTM 64, Springer Verlag, New York-Berlin-Heidelberg, 1977.

[Ed2] H.M. Edwards, The background of Kummer's proof of Fermat's Last Theorem for regular primes, *Arch. History Exact Sci.* 14 (1975), 219–236.

[Ed3] H.M. Edwards, Postscript to "The background of Kummer's proof...," *Arch. History Exact Sci.* 17 (1977), 381–394.

[Ed4] H.M. Edwards, Review of "Ernst Edward Kummer, Collected Papers, Vol. I," in *Historia Math.*, Vol. 4, No. 4 (1977), 475–478.

[G] C.F. Gauss, Theorie der biquadratische Reste I and II, in *Arithmetischen Untersuchungen*, Chelsea Publishing Co., Bronx, New York, 1965.

[Gr] N. Greenleaf, On Fermat's Equation in $\mathbf{C}(t)$, *Amer. Math. Monthly*, 76 (1969), 808–809.

[H] D. Hilbert, Die theorie der algebraischen Zahlkörper, in *Gesammelte Abhandlungen*, Vol. I, Chelsea Publishing Co., New York, 1965, 63–363.

[He] J. Herbrand, Sur les classes des corps circulaires, *J. Math. Pures Appl.*(9), 11 (1932), 417–441.

[IR] K. Ireland and M. Rosen, *A Classical Introduction to Number Theory*, GTM 84, Springer Verlag, New York-Berlin-Heidelberg, 1982.

[Ko] N. Koblitz, *p-adic Numbers, p-adic Analysis, and Zeta Functions*, GTM 58, Springer Verlag, New York-Berlin-Heidelberg, 1979.

[Ku] E. Kummer, *Collected Papers* (ed. by A. Weil), Springer Verlag, New York-Berlin-Heidelberg, 1975.

[La] E. Landau, *Vorlesungen über Zahlentheorie*, Vol. III, Chelsea Publishing Co., 1947.

[Lg1] S. Lang, *Cyclotomic Fields I and II*, Combined Second Edition, GTM 121, Springer Verlag, New York, 1990.

[Lg2] S. Lang, *Algebraic Number Theory*, GTM 110, Springer Verlag, New York-Berlin-Heidelberg, 1986.

[LS] H. W. Lenstra, Jr. and P. Stevenhagen, Class Field Theory and the First Case of Fermat's Last Theorem, this volume

[Le] H.W. Leopoldt, Zur Struktur der l-Klassengruppe galoisscher Zahlkörper, *J. reine angew. Math.*, 199 (1958), 165–174.

[M] B. Mazur, Book review of "Ernst Edward Kummer, Collected Papers, Vol. I and II," in *Bull. Amer. Math. Soc.*, Vol. 83, No. 5 (1977), 976–986.

[Ri] P. Ribenboim, *13 Lectures on Fermat's Last Theorem*, Springer Verlag, New York-Heidelberg-Berlin, 1979.

[R] K. Ribet, A Modular Construction of Unramified p-extensions of $\mathbf{Q}(\mu_p)$, *Invent. Math.*, 34 (1976), 151–162.

[W] L. Washington, *Introduction to Cyclotomic Fields*, GTM 83, Springer Verlag, New York-Heidelberg-Berlin, 1982.

[Wi] A. Wiles, Modular Curves and Class Groups of $\mathbf{Q}(\zeta_p)$, *Invent. Math.*, 58 (1980), 1–35.

[V] H.S. Vandiver, Fermat's Last Theorem: Its history and the nature of the known results concerning it, *Amer. Math. Monthly*, 53 (1946), 555–578 ; 60 (1953) 164–167.

ON TERNARY EQUATIONS OF FERMAT TYPE AND RELATIONS WITH ELLIPTIC CURVES

GERHARD FREY

§1. CONJECTURES

The main purpose of this chapter is to show how arithmetical properties of elliptic curves E defined over global fields K and corresponding Galois representations are often related to interesting diophantine questions, amongst which the most prominent is without doubt Fermat's Last Theorem, which has now become Wiles' theorem.

Of course, the most important case for us is when K is a number field, or even equal to $\mathbb{Q}$, but many of the conjectured (or proved) assertions make sense for fields of finite type and become more convincing in the case that K is a function field or, in other words, for families of elliptic curves. The reason for this is that the conjectures predict the behaviour of "generic" elliptic curves and it is expected that each "special" property gives rise to interesting diophantine properties of K.

So assume from now on that K is a global field, i.e., that K is either a finite number field or a function field of one variable over a perfect field K_0. For simplicity we always assume that $\mathrm{char}(K_0) \neq 2, 3$. By Σ_K we denote the set of non-archimedean places of K. For divisors $D \in \mathbb{Z}[\Sigma_K]$ we define $\deg(D)$ as the divisor degree in the usual sense if K is a function field, and as the logarithm of its norm in the number field case. For a subset $S_0 \subset \Sigma_K$, let O_{S_0} be the ring of S_0-integers; its units are denoted by $O^*_{S_0}$ and called S_0-units.

The field K has two important numerical invariants. By $g(K)$ we denote the genus of K, which is the usual genus in the function field case, and equal to $\frac{1}{2}\log|\Delta_{K/\mathbb{Q}}|$ in the number field case (where $\Delta_{K/\mathbb{Q}}$ is the discriminant of $K/\mathbb{Q}$). By $d(K)$ we denote the degree of K, which is equal to $[K : \mathbb{Q}]$ if K is a number field. If K is a function field, $d(K)$ is equal to

$$\min_{f \in K \setminus K_0} \deg(f)_\infty = \min_{f \in K \setminus K_0} [K : K_0(f)],$$

where $(f)_\infty$ is the polar divisor of f; so $d(K)$ is the minimal degree of a covering map from the curve corresponding to K to $\mathbb{P}^1_{/K_0}$.

(a) Conjectures about diophantine equations.

One of the most basic conjectures about solutions of ternary diophantine equations comes a little bit disguised. For $x \in K^*$, let $h_K(x)$ denote its height, and

$$\operatorname{supp}(x) = \prod_{\substack{v_\mathfrak{p}(x)\neq 0 \\ \mathfrak{p}\in\Sigma_K}} \mathfrak{p}$$

its divisor of support.

Conjecture 1. *There are constants c and d such that the following holds. Assume that $x \neq 0, 1$, and that $K/K_0(x)$ is separable if K is a function field. Then*

$$h_K(x) \leq c \cdot \deg \operatorname{supp}(x(x-1)) + d.$$

Of course here and in all of the conjectures which will follow, it is of great importance to specify how the constants depend on K. A refinement of Conjecture 1 is

Conjecture 1′. *For all $\varepsilon \in \mathbb{R}_{>0}$, one can take $c = 1+\varepsilon$ and $d = d(\varepsilon, g(K))$ in Conjecture 1. Further, the dependence of d on $g(K)$ is linear, and for K a function field, $\varepsilon = 0$ is allowed.*

In section 2 we shall see that Conjecture 1′ is true for function fields. Now we shall translate Conjecture 1 into the A-B-C-Conjecture due to Masser-Oesterlé for $K = \mathbb{Q}$.

First assume that K is a function field. We can asume that K has a prime $\mathfrak{p}_\infty$ of degree 1. Let $(x)_\infty$ be the polar divisor of x. The Riemann-Roch theorem implies that there is an element $C \in K$ with

$$(C)_\infty | \mathfrak{p}_\infty^{\deg((x)_\infty)+2g(K)-1} \quad \text{and} \quad (x)_\infty | (C)_0,$$

where $(C)_0$ is the zero divisor of C. Take $A = x \cdot C$. Then

$$(A)_\infty = ((x)(C))_\infty | (\mathfrak{p}_\infty)^{\deg((x)_\infty)+2g(K)-1}$$

and

$$(A)_0 | (x)_0 \cdot (C)_0 \cdot (x)_0^{-1} = (x)_0 \cdot D'$$

with $\deg D' \leq 2g(K) - 1$.

Let $S_0(x)$ be the set of primes $\mathfrak{p} \in \Sigma_K$ dividing $\mathfrak{p}_\infty \cdot D'$. Then A and C are relatively prime elements in $O_{S_0(x)}$, and the same is true for $B := A - C$, which is equal to $(x-1)C$, and $\deg(\gcd((A)_0, (B)_0)) \leq 2g - 1$. If K is a number field of genus $g(K)$, we use the theorem of Minkowski to get a corresponding result: There are elements $A, B, C \in O_K$ such that $x = A/C$ and $x - 1 = B/C$ with $\deg(\gcd((A), (B))$ and $\deg(\gcd((A), (C)))$ bounded linearly in $g(K)$. Hence Conjecture 1′ becomes:

A-B-C-Conjecture. *Let O_K be the ring of integers of K (with respect to a fixed place $\mathfrak{p}_\infty$ of K in the function field case). Let A and B be elements in O_K, let $D(A,B)$ be their greatest common divisor in O_K, and assume that $K/K(A/B)$ is separable. Then*

$$h_K(A) \leq (1+\varepsilon)\deg\Big(\prod_{\mathfrak{p}|AB(A-B)} \mathfrak{p}\Big) + d(\varepsilon, \deg(g(K), D(A,B))),$$

and the dependence on $\deg(D(A,B))$ *and* $g(K)$ *should be linear.*

For $K = \mathbb{Q}$, this means

$$|A| \leq \tilde{d}(\varepsilon)\Big(\prod_{p|AB(A-B)} p\Big)^{1+\varepsilon}$$

whenever $\gcd(A,B) = 1$, with $\tilde{d}(\varepsilon) = e^{d(\varepsilon,1)}$ (this is the original form of the conjecture of Masser-Oesterlé).

Remark 1.1. Of course one would be glad to have any bounds c and d at all. But this sharp version has the advantage that an effective version of Faltings' theorem about the finiteness of K-rational points on curves of genus ≥ 2 would follow (see [E]).

Now we fix a finite set S_0 of non-archimedean places of K and let $s_0 := \deg(\prod_{\mathfrak{p}\in S_0} \mathfrak{p})$. We assume that

$$a, b, c \in K^* \quad \text{satisfy} \quad \operatorname{supp}(a)\operatorname{supp}(b)\operatorname{supp}(c) \;\Big|\; \prod_{\mathfrak{p}\in S_0} \mathfrak{p}^3.$$

Let n be an integer prime to $\operatorname{char}(K)$ and define

$$L_{(a,b,c),n}(K) = \left\{ (x,y,z) \in K^3\backslash(0,0,0);\ \begin{array}{r} ax^n - by^n = cz^n, \text{ and} \\ K/K_0(\frac{a}{c}(\frac{x}{z})^n) \text{ is separable} \\ \text{if } K \text{ is a function field} \end{array} \right\} \Big/ \sim$$

where $\sim$ means projective equivalence.

Conjecture 2. ("Asymptotic Fermat Conjecture") *There are numbers $N(g(K), s_0)$, such that for all $n > N(g(K), s_0)$ and all elements $(x,y,z) \in L_{(a,b,c),n}(K)$, either $xyz = 0$ or $\{x/z, y/z\} \subset O^*_{S_0}$.*

Remark 1.2. If K is a number field or if K_0 is finite, it follows from Conjecture 2 that

$$\bigcup_{\substack{n\geq 4 \\ \operatorname{supp}(a)\operatorname{supp}(b)\operatorname{supp}(c) | \prod_{\mathfrak{p}\in S_0} \mathfrak{p}^3}} L_{(a,b,c),n}(K)$$

is a finite set, and for n large enough, $L_{(a,b,c),n}(K)$ consists of triplets with coordinates in $\{0\} \cup \{\text{roots of unity}\}$.

An easy observation is

Proposition 1.1. *Conjecture 1 implies the Asymptotic Fermat conjecture.*

Proof. We assume that $xyz \neq 0$. We take $t_1 = (a/c)(x/z)^n$ and $t_2 = (b/c)(y/z)^n$ and assume that $x \neq 0, 1$. So by Conjecture 1 we get:

$$n\Bigg(\sum_{\substack{v_\mathfrak{p}(x/z)>0 \\ \mathfrak{p}\notin S_0}} v_\mathfrak{p}\left(\frac{x}{z}\right) + \sum_{\substack{v_\mathfrak{p}(y/z)>0 \\ \mathfrak{p}\notin S_0}} v_\mathfrak{p}\left(\frac{y}{z}\right)\Bigg) \leq h_K(t_1) + h_K(t_2)$$
$$\leq 2(1+\varepsilon)s_0 \deg \operatorname{supp}(xyz) + d(g(K), \varepsilon).$$

Hence for n large enough and $v_\mathfrak{p}(x/z) > 0$, it follows that $\mathfrak{p} \in S_0$, and the same is true for z/x and so $x/z \in O^*_{S_0}$, and the analogous equations for $(b/c)(y/z)^n$ yield $(y/z)^n \in O^*_{S_0}$.

Of course it is most interesting to find (a, b, c) such that one can determine exactly the elements in $L_{(a,b,c),n}(K)$; in the last section we shall see how Wiles' theorem about modularity of semi-stable elliptic curves can be used to find such examples if $K = \mathbb{Q}$, and especially $a = b = c = 1$ will lead to Fermat's Last Theorem.

(b) Conjectures about elliptic curves.

Now we shall state conjectures about elliptic curves and corresponding Galois representations which are, as we shall see below, related to Conjectures 1 and 2. Since they seem to be interesting for their own sake, we shall begin by describing a more general context.

So let A/K be an abelian variety with conductor N_A, Faltings height $h_K(A)$ and geometric height $h_{\text{geom}}(A)$. For us it is important that for elliptic curves E/K, the height $h_K(E)$ is closely related to $\deg(\Delta_E)$, where Δ_E is the minimal discriminant divisor of E, and $h_{\text{geom}}(E) \sim h_K(j_E)$ with j_E the absolute invariant of E (see [Si]). It should be noted that, if K is a number field or K_0 is finite, then there are only finitely many abelian varieties over K with fixed dimension and bounded height.

Let n be a natural number and let A_n be the kernel of the multiplication by $n \cdot id_A$ in A. We shall always assume that n is prime to $\operatorname{char}(K)$. The Galois group $G_K = \operatorname{Aut}(K_{\text{sep}}|K)$ acts on A_n, and so we get a representation

$$\rho_{A,n} : G_\mathbb{Q} \to \operatorname{GL}(A_n) \cong \operatorname{GL}(2 \cdot \dim A, \mathbb{Z}/n\mathbb{Z})$$

induced by this action.

In general the image of $\rho_{A,n}$ is "as big as possible" subject to the restrictions coming from the symplectic structure induced by the Weil pairing and the decomposition of A into simple factors.

Definition 1.1. Let H be a finite subgroup of A defined over K and of order prime to $\operatorname{char}(K)$. H is called exceptional if

(i) there is no subgroup H_1 with $0 \neq H_1 < H$ such that A is isomorphic to A/H_1, and
(ii) there is no proper abelian subvariety B of A containing H.

Convention. We say that an abelian variety over K has multiplicity 1 if the (absolute) simple factors of A are pairwise non-isogenous.

Question 1. Is there a number N and a number M such that the existence of an exceptional subscheme H of order $> N$ in an abelian variety of multiplicity 1 implies $h_{\text{geom}}(A) < M$?

Of course this question only makes sense if one specifies how N and M should depend on A and K. A possible guess could be that both depend only on the maximum of the dimensions of the simple factors of A and on the genus of K or, even stronger, on the irrationality degree of K. It may be too early to make more explicit guesses now, and we shall restrict ourselves to an observation depending on deep results of Faltings and of Masser - Wüstholz (see [B]):

> Let K be a number field. Assume that we fix $d \in \mathbb{N}$ and a finite set $S_0 \subset \Sigma_K$ and that we look at all abelian varieties A/K of dimension d whose conductor N_A has support inside of S_0. Then A lies in a finite set of isomorphism classes $\mathcal{A}$ of abelian varieties, and if $A, A' \in \mathcal{A}$ are isogenous then there is a K-rational isogeny $\eta : A \to A'$ with $\deg(\eta)$ bounded by a number depending on K and S_0 only. Therefore it follows immediately that numbers N and M as in the question exist if we restrict ourselves to abelian varieties with simple factors whose dimension and conductors are bounded, and that these numbers depend on S_0 and, in fact, on $g(K)$.

From now on we look at the special case that the factors of A are elliptic curves, and all essential features of the question can be found by looking at elliptic curves E or abelian varieties of dimension 2 which are isogenous to $E_1 \times E_2$ with E_i/K non-isogenous elliptic curves. We begin with $A = E$ an elliptic curve defined over K.

Definition 1.2. E/K is admissible if $\text{char}(K) = 0$ or $K/K_0(j_E)$ is separable.

Let H be an exceptional K-subgroup of E. It follows that H is cyclic and hence our question leads us to:

Conjecture 3. *There are numbers $M(g(K))$ and $N(g(K))$ such that if E/K is an admissible elliptic curve with a separable cyclic K-rational isogeny of degree $N > N(g(K))$, then we have $h_{\text{geom}}(E) < M(g(K))$.*

Remark 1.3. (1) If K is a number field and if Conjecture 3 holds, then there exists a number N such that only elliptic curves with complex multiplication defined over K have cyclic K-isogenies of degree $\geq N$. It seems to be not unreasonable to believe this statement.
(2) In the next section we shall prove that Conjecture 3 is true if K is a function field.

(3) If we sharpen the rationality condition to "H has a K-rational generator," then it is a theorem of Merel that in the number field case there is a number $N(d(K))$ with $N \leq M(d(K))$.

(4) A (much) weaker version of the stated conjectures is that M depends on N as well as on K, and then the corresponding statements are true over number fields, (see [Fr 2]).

As a consequence of (4) one gets:

Proposition 1.2. *Let d be a natural number and p a prime. There exists a number $l_p(d)$ such that for all $n_p \geq l_p(d)$, all number fields K with $[K : \mathbb{Q}] \leq d$ and all elliptic curves E/K, if E/K has a cyclic K-isogeny of degree l^{n_p}, then E has complex multiplication.*

Proof. Corollary 2 in [Fr 2] states that for $p^{l_0} > 120d$, there are only finitely many j-invariants j_E in fields K with $[K : \mathbb{Q}] \leq d$ such that a corresponding elliptic curve E has a K-rational cyclic isogeny of degree p^{l_0}. Let E_i be an elliptic curve over K_i corresponding to j_i. If E_i has no complex multiplication, then there is an l_i such that E_i (and no twist of E_i) has a K_i-rational cyclic isogeny of degree p^{l_i}. Then $l_p(d) = \max(l_i)$ satisfies the assertion of the proposition.

Now we assume that $\dim A = 2$. In this case the existence of exceptional subgroups is closely related to the geometric fundamental groups of curves of genus 2 over K. For a discussion of some aspects of this relation we refer to [Fr 3].

In the following we restrict our attention to the case that A is not K-simple, i.e., A is K-isogenous to $E_1 \times E_2$ with E_i/K being elliptic curves. Let $H < A$ be exceptional. Inside of A we have two non-isogenous elliptic curves E_1', E_2' intersecting in $H' := E_1' \cap E_2'$ (which is not exceptional in our sense) and $A/H' \cong E_1 \times E_2$.

The case that H (or H') contains cyclic subgroups has been discussed above. So we shall assume that $H' = E'_{i,n'}$ for some $n' \in \mathbb{N}$, and so $E_i \cong E_i'$.

Let $\eta : A \to E_1 \times E_2$ be the quotient map and $\hat{\eta}$ its dual map whose kernel $\hat{H}'$ is isomorphic (as G_K-module) to H'. Both $\hat{H}'$ and $\eta(H)$ are exceptional, and since $h_K(A) = h_K(E_1 \times E_2)$, we lose nothing by assuming that $A = E_1 \times E_2$ and that H does not contain a proper cyclic subgroup.

It follows from these assumptions that H is isomorphic to $p_i(H) = E_{i,N}$ (where p_i is the projection to E_i) for $i = 1, 2$ and some $N \in \mathbb{N}$. So H is given as the graph of an element $\alpha \in \mathrm{Iso}_{G_K}(E_{1,N}, E_{2,N})$,

$$H = \Delta_\alpha = \{(P, \alpha P); P \in E_{1,N}\},$$

and the representation $\rho_{E_{1,N}}$ of G_K induced by the action on $E_{1,N}$ is equivalent to $\rho_{E_{2,N}}$. Hence the question about the existence of exceptional subgroups in the special case under consideration is a question about equivalence of representations on torsion elements of elliptic curves raised by Mazur in [Ma 2] for $K = \mathbb{Q}$.

We can give a geometric interpretation (see [K-S]). Let $X(N)$ be the modular curve parametrizing level-N structures of elliptic curves. Its group of automorphisms is equal to $\mathrm{SL}(2,\mathbb{Z}/N)/\pm id$. The triplet (E_1, E_2, α) gives rise to a K-rational point on

$$\alpha_\varepsilon(\mathrm{SL}(2,\mathbb{Z}))\backslash X(N) \times X(N) =: Z_{N,\varepsilon}.$$

Here $\alpha_\varepsilon(\mathrm{SL}(2,\mathbb{Z}/N))/\pm id$ is a diagonal embedding

$$(\mathrm{SL}(2,\mathbb{Z}/N\mathbb{Z}))/\pm id) \hookrightarrow (\mathrm{SL}(2,\mathbb{Z}/N\mathbb{Z}))/\pm id)^2$$

induced by the conjugation with a matrix $g \in \mathrm{GL}(2,\mathbb{Z}/N\mathbb{Z})$ satisfying $\det g = \varepsilon \in (\mathbb{Z}/N\mathbb{Z})^*$. The number ε is determined by $e_{2,N}(\alpha P, \alpha Q) = e_{1,N}(P,Q)^\varepsilon$ for $P, Q \in E_{1,N}$ and $e_{i,N}$ the Weil pairing on points of order N of E_i.

One should hope that geometrical properties of the surface $Z_{N,\varepsilon}$ will yield results for the questions raised above which can now be expressed as follows: Let $Z'_{N,\varepsilon}(K) \subset Z_{N,\varepsilon}(K)$ be the set of K-rational points P corresponding to (E_1, E_2, α) with E_i admissible and E_1 not isogenous to E_2.

Conjecture 4. *There are numbers $N(g(K))$, $M(g(K))$ such that for all $N > N(g(K))$ and $P \in Z'_{N,\varepsilon}(K)$ with corresponding curves E_1, E_2, one has*

$$\max_{i=1,2}(h_{\mathrm{geom}}(E_i)) < M(g(K)).$$

It is not hard to see, but worthwhile to remark, that for K a number field it follows from Conjecture 4 that for N large enough, $Z'_{N,\varepsilon}(K) = \emptyset$. In the conjecture both N and M depend on K. Another possible guess is

Conjecture 4′. *There is an N_0, independent of K, such that for $N \geq N_0$ there is a number $M(g(K), N)$ such that for all $P \in Z'_{N,\varepsilon}(K)$ with corresponding curves E_1, E_2, one has $\max_{i=1,2}(h_{\mathrm{geom}}(E_i)) \leq M(g(K), N)$.*

Remark 1.4. In the number field case, Conjectures 4 and 4' are essentially due to Darmon ([Da]).

Kani and Schanz proved in [K-S] that for $N \geq 13$, the surface $Z_{N,\varepsilon}$ is of general type. Hence if K is a number field, one would expect according to Lang's conjecture that the K-rational points of $Z_{N,\varepsilon}$ are concentrated on curves of genus ≤ 1. This and the explicit knowledge of intermediate curves between $X(N)$ and $X(1)$ motivates

Conjecture 4″. (Kani)[1] *If N is a prime ≥ 23 and $C \subset Z_{N,\varepsilon}$ is a curve of genus ≤ 1, then it is a twisted Hecke correspondence. In particular, $C \cap$*

[1]We use this opportunity to thank E.Kani for many very valuable discussions and hints.

$Z'_{N,\varepsilon} = \emptyset$ (see [K]), and hence Conjecture 4′ follows from Lang's conjecture, if K is a number field and N is a prime ≥ 23.

We shall see in the next section that in the "generic" case (e.g., K a function field over an algebraically closed field of characteristic 0) one can guess how curves of higher genus can be embedded into $Z_{N,\varepsilon}$ to come nearer to Conjecture 4′. The situation becomes much simpler if we fix one of the curves E_i which we denote by E_0 from now on.

Conjecture 5. *Let E_0/K be admissible. There are numbers $N_0(g(K), E_0)$ and $M(g(K), E_0)$ such that for $N \geq N_0(g(K), E_0)$ and (E_0, E) corresponding to a point $P \in Z'_{N,\varepsilon}(K)$, we get have $h_K(E) < M(g(K), E_0)$.*

Again Conjecture 5 is true over function fields. A crucial point is the fact that $\rho_{E_{1,N}} \cong \rho_{E_{2,N}}$ imposes strong arithmetical conditions on E.

Proposition 1.3. *Assume that $H < E_1 \times E_2$ is exceptional over K of order N^2 with $N \geq 5$. Let N_{E_1} be the conductor of E_1.*
(1) If $\mathfrak{p} \in \Sigma_K$ does not divide $N \cdot N_{E_1}$, then E_2 is semi-stable at $\mathfrak{p}$.
(2) If $\mathfrak{p}$ does not divide N_{E_1}, then $\max(-v_{\mathfrak{p}}(j_{E_2}), 0) \equiv 0 \bmod N$.

Proof. If $\mathfrak{p} \nmid N \cdot N_{E_1}$, then $K(E_{1,N}) = K(E_{2,N})$ is unramified at $\mathfrak{p}$, and so E_2 has to be semi-stable at $\mathfrak{p}$ (see [Si]). If $\mathfrak{p} \nmid N_{E_1}$, then $E_{1,N}$ is a finite group scheme at $\mathfrak{p}$, and using the theory of Tate curves (see [Si] again) it follows that if $v_{\mathfrak{p}}(j_{E_2}) < 0$, then $v_{\mathfrak{p}}(j_{E_2}) \equiv 0 \bmod N$.

As a consequence we see that prime divisors of N_{E_2} which don't divide N_{E_1} give a large contribution to the height of E_2, and this should contradict the conjectured relation between the height of elliptic curves and the degree of their conductors which is stated as:

Conjecture 6. (Height Conjecture for Elliptic Curves) *There are constants c and $d(g(K))$ such that for all admissible elliptic curves over K we have*

$$h_K(E) \leq c \deg N_E + d(g(K)).$$

Moreover, for any $\varepsilon > 0$, the numbers $c = \frac{1}{2} + \varepsilon$ and $d = d(g(K), \varepsilon)$ should work, with $\varepsilon = 0$ allowed if K is a function field.

Remark 1.5. If we replace $h_K(E)$ by $\frac{1}{12} \deg \Delta_E$ (i.e., if we take care only of the contribution of the non-archimedean places to the height of E), we get:

Szpiro's Conjecture. *There exist constants c', d' such that*

$$\deg \Delta \leq c' \deg N_E + d'.$$

This conjecture would suffice for many applications.

Here are some consequences of Conjecture 6.

Proposition 1.4. *Assume that Conjecture 6 is true over K with constants c and d. Let δ_K be equal to 1 if K is a function field, and equal to $[K:\mathbb{Q}]+6$ if K is a number field.*
(i) *Let S_0 be a finite set of places of K, $s_0 = \deg \prod_{\mathfrak{p}\in S_0} \mathfrak{p}$. There is a number $N_0(s_0, c, d, d(K))$ such that all admissible elliptic curves E/K which are semi-stable outside of S_0 and have a K-rational cyclic isogeny of degree $N > N_0$ have potential good reduction at all $\mathfrak{p} \in \Sigma_K$.*
(ii) *Let P be a K-rational point of an admissible E with order N prime to $\mathrm{char}(K)$. Then either $N < \max(5\delta_K, 2c+2d)$, or E has good reduction at all non-archimedean places of K, and so $j_E \in K_0$ if K is function field and $h_k(E)$ is bounded (depending on d and $d(K)$) if K is a number field.*
(iii) *Fix an admissible E_0/K. There is a number $N_0(c, d, N_{E_0}, d(K)) =: N_0$, such that for all $N > N_0$ and all admissible elliptic curves E/K with (E_0, E) corresponding to a point $P \in Z_{N,\varepsilon}(K)$, we get*

$$h_{\mathrm{geom}}(E)_{\mathrm{fin}} \leq M(c, d, NN_{E_0}, d(K)),$$

where $h_{\mathrm{geom}}(E)_{\mathrm{fin}}$ is the non-archimedean part of the geometric height of E. If E is semi-stable at divisors of N, we can replace $h_{\mathrm{geom}}(E)_{\mathrm{fin}}$ by $h_K(E)$. If $c = \frac{1}{2} + \varepsilon$, and if d depends only on $g(K)$ and ε, we get Conjecture 5 for pairs (E_0, E) corresponding to $P \in Z_{N,\varepsilon}(K)$ with E admissible and semi-stable at divisors of N.

Proof. (i) Let $N_E = N'_E \prod_{\mathfrak{p}\in S_0} \mathfrak{p}^{n_\mathfrak{p}}$ be the conductor of E and hence of $\eta(E)$, where N'_E is prime to S_0 and the $n_\mathfrak{p}$'s are bounded by $d(K)$. Using the theory of Tate curves, one gets (see [Fr 1]): If $\mathfrak{p}|N'_E$, then $-v_\mathfrak{p}(j_E) - v_\mathfrak{p}(j_{\eta(E)}) \geq N+1$, and so

$$2c \deg N_E + 2d \geq 2h_K(E) \geq \frac{1}{6}(N+1)\deg N'_E + \tilde{d}(d(K))$$

for some $\tilde{d}(d(K)) \in \mathbb{R}$. It follows that for

$$N > 12\Big(\deg(\prod_{\mathfrak{p}\in S_0} \mathfrak{p}^{n_\mathfrak{p}} + d + c) - \tilde{d}(d(K))\Big),$$

we have $\deg(N'_E) = 0$.
(ii) If E has a K-rational point of order $N \geq 5$, then E has semi-stable reduction at all places $\mathfrak{p} \nmid N$. So if $N = p^\alpha N'$ and $N' \geq 5$, then E is semi-stable at all divisors of p. If this condition is not satisfied but $p^\alpha > v_\mathfrak{p}(p)+6$, then again E is semi-stable at $q|p$. Hence E/K is semi-stable at all $\mathfrak{p} \in \Sigma_K$ if

$$N \geq 5\delta_K = \begin{cases} 5 & \text{if } K \text{ is a function field,} \\ 5([K:\mathbb{Q}]+6) & \text{if } K \text{ is a number field.} \end{cases}$$

Hence by using (i) (with η corresponding to $\langle P\rangle$), we get: If

$$N \geq \max(5\delta_K, 24c + 24d),$$

then E has good reduction at all places of K, and the assertion follows.
(iii) Assume that (E_0, E) corresponds to $P \in Z_{N,\varepsilon}(K)$. If $N \geq 5$ it follows from Proposition 1.3 that

$$N_E = N'_E \prod_{\mathfrak{p}|N//v_\mathfrak{p}(j_E)\geq 0//} \mathfrak{p}^{\delta_\mathfrak{p}} \prod_{\mathfrak{p}|N_{E_0}} \mathfrak{p}^{\delta_\mathfrak{p}}$$

with N'_E prime to N_{E_0} and $0 \leq \delta_\mathfrak{p}$ bounded by numbers depending on $d(K)$. Moreover $v_\mathfrak{p}(j_E) \equiv 0 \bmod N$ if $\mathfrak{p}|N'_E$. Hence

$$\frac{N}{12} \deg N'_E \leq c(\deg N'_E + \deg \prod_{\mathfrak{p}|N} \mathfrak{p}^{\delta_\mathfrak{p}}) + d'(N_{E_0}, c, d).$$

Since $\deg \prod_{\mathfrak{p}|N} \mathfrak{p}$ is bounded by a number depending on $d(K)$, it follows that for large N (depending on $N_{E_0}, d(K), c, d$), N'_E is trivial, and so the non-archimedean part of $h_{\text{geom}}(E)$ is bounded. If E is semi-stable at all divisors of N, then $N_E = N'_E \prod_{\mathfrak{p}|N_{E_0}} \mathfrak{p}^{\delta_\mathfrak{p}}$, and hence for large N we have $N_E = \prod_{\mathfrak{p}|N_{E_0}} \mathfrak{p}^{\delta_\mathfrak{p}}$. So in this case $h_K(E)$ is bounded.

Corollary 1.1. *Assume that Conjecture 6 is true over K and that K is a number field or a function field over a finite field K_0. Then for N large enough we get: If E/K is semi-stable at all divisors of N and $\rho_{E,N} \cong \rho_{E_0,N}$, then E is isogenous to E_0. Especially, Conjecture 6 for function fields implies Conjecture 5 for function fields.*

(c) Relations between elliptic curves and Fermat equations.

The considerations in the last section have shown that the arithmetical properties of the representations $\rho_{E,n}$ are closely connected with arithmetical properties of Δ_E. If E has K-rational points of order 2, these properties can be translated into properties of the X-coordinates of such points as was done in various papers by Hellegouarch [He] and the author. The procedure is described in detail in [Fr 1], but since it is a crucial point we shall repeat it for the convenience of the reader.

Take $x \in K^* \backslash \{0, 1\}$, choose a prime $\mathfrak{p}_\infty \in \Sigma_K$ if K is a function field, and let O_K be the ring of integers of K (with respect to $\mathfrak{p}_\infty$ in the function field case). Now we define

$$E_{(x)} : Y^2 = X(X-1)(X-x).$$

Since

$$j_{E_{(x)}} =: j_x = 2^8 \frac{(x^2 - x + 1)^3}{x^2(x-1)^2},$$

it follows that $E_{(x)}$ is admissible if $K/K_0(x)$ is separable, and that

$$h_K(j_{E_{(x)}}) \geq c' + 6h_K(x).$$

$E_{(x)}$ has good reduction at all primes $\mathfrak{p}$ not dividing $2 \cdot \mathrm{supp}(x(x-1))$, and since for $\mathfrak{p} | \mathrm{supp}(x(x-1))$ but $\mathfrak{p} \nmid 2$, we have

$$v_{\mathfrak{p}}(j_{(x)}) = \begin{cases} 2v_{\mathfrak{p}}(x(x-1)) & \text{if } v_{\mathfrak{p}}(x) \geq 0, \\ 2v_{\mathfrak{p}}(x) & \text{if } v_{\mathfrak{p}}(x) < 0, \end{cases}$$

and hence we have bad reduction at all divisors of $\mathrm{supp}(x(x-1))$ which are prime to 2.

Now choose $A, C \in O_K$ such that $x = A/C$ and $\deg(\gcd(A,C)) \leq s_0(g(K))$ (see the discussion of the A-B-C-Conjecture), and take $B = A - C$. Then

$$E_{(A,B)} : Y^2 = X(X-A)(X-B)$$

is a twist of $E_{(x)}$. Define

$$S_0 = \{\mathfrak{p} \in \Sigma_K ; \mathfrak{p} | 2\mathfrak{p}_\infty \cdot \gcd(A,C)\}.$$

Let $\Delta_{(A,B)}$ be the discriminant of $E_{(A,B)}$. Then for $\mathfrak{p} \notin S_0$ we have:

$$v_{\mathfrak{p}}(\Delta_{A,B}) = -v_{\mathfrak{p}}(j_x) = -v_{\mathfrak{p}}\left(2^8 \frac{(A^2+B^2-AB)^3}{A^2B^2C^2}\right) = 2v_{\mathfrak{p}}(ABC).$$

Moreover we get for $\mathfrak{p} \notin S_0$ that $X(X-A)(X-B) \bmod \mathfrak{p}$ has at least two different zeros, and hence $E_{(A,B)}$ has semi-stable reduction modulo $\mathfrak{p}$ for all $\mathfrak{p} \notin S_0$, and has good reduction if $\mathfrak{p} \notin S_0 \cup \mathrm{supp}(ABC)$. Hence

$$N_{(A,B)} := N_{E_{(A,B)}} = \prod_{\mathfrak{p} \in S_0} \mathfrak{p}^{\delta_{\mathfrak{p}}} \prod_{\substack{\mathfrak{p} \nmid 2 \\ \mathfrak{p} \notin S_0 \\ \mathfrak{p} | \mathrm{supp}(x(x-1))}} \mathfrak{p} = \prod_{\mathfrak{p} \in S_0} \mathfrak{p}^{\delta_{\mathfrak{p}}} \prod_{\substack{\mathfrak{p} | (ABC) \\ \mathfrak{p} \notin S_0}} \mathfrak{p}.$$

Now we are ready to state

Proposition 1.5. *Assume that Conjecture 6 holds with constants c and d. Then for all* $\varepsilon \in \mathbb{R}_{>0}$,

$$h_K(x) \leq 2(c+\varepsilon) \deg \mathrm{supp}(x(x-1)) + \tilde{d}(d, \varepsilon, d(K))$$

If K *is a function field, then* $\varepsilon = 0$ *is allowed.*

So the height conjecture for elliptic curves implies the A-B-C-Conjecture and the Asymptotic Fermat Conjecture. We continue to look at the curve $E_{(A,B)}$ and get with $C = A - B$:

Proposition 1.6. *For* $N \in \mathbb{N}$, *define*

$$S_N := S_0 \cup \{\mathfrak{p} \in \mathrm{supp}(ABC); 2v_{\mathfrak{p}}(ABC) \not\equiv 0 \bmod N\}.$$

Then $E_{(A,B),N}$ *is a finite group scheme at all* $\mathfrak{p} \notin S_N$, *and the Artin conductor* $N_{\rho_{(A,B),N}}$ *of* $\rho_{E_{(A,B),N}}$ *is equal to*

$$N_{\rho(A,B),N} = \prod_{\mathfrak{p} \in S_N \cup \mathrm{supp}(N)} \mathfrak{p}^{f_{\mathfrak{p}}}.$$

Corollary 1.2. *Assume that $(x,y,z) \in L_{(a,b,c),N}(K)$, and suppose that $S_0 \subset \Sigma_K$ satisfies $x,y,z \in O_{S_0}$, $\{a,b,c\} \subset O^*_{S_0}$, and the prime divisors of* $\operatorname{supp}(\gcd(x,y,z))$ *are in S_0. Then*

$$N_{\rho_{E_{(ax^N,by^N),N}}} = \prod_{\mathfrak{p}\in S_0\cup\operatorname{supp}(N)} \mathfrak{p}^{f_\mathfrak{p}},$$

and at $\mathfrak{p}|N$, $\mathfrak{p}\notin S_0$, the subgroup scheme $E_{(ax^N,by^N),N}$ is finite.

The disturbing fact is that in our general setting the set S_0 can depend on (x,y,z). But since the height conjecture, and hence the asymptotic Fermat conjecture, are true in the function field case, the interesting application of Corollary 1.2 is to the case that K is a number field. In this case we find a fixed set $S_K \subset \Sigma_K$ such that O_{S_K} is a principal ideal domain and such that $\operatorname{supp}(S) \subset S_K$ (where the degree of $\prod_{\mathfrak{p}\in S_K}\mathfrak{p}$ can be bounded by a number depending on $g(K)$), and hence we can represent elements in $L_{a,b,c,N}(K)$ up to projective equivalence by $(x,y,z) \subset O_{S_K}$ which are relatively prime.

Hence we get

Corollary 1.3. *Assume that K is a number field and S_K is as above. Then for each $(x',y',z') \in L_{(a,b,c),N}(K)$ with $x'y'z' \neq 0$, we can find a projectively equivalent triple (x,y,z) such that*

$$N_{\rho_{E_{(ax^N,by^N),N}}} = \prod_{\mathfrak{p}\in S_K\cup\operatorname{supp}(abcN)} \mathfrak{p}^{f_\mathfrak{p}}$$

and $E_{(ax^N,by^N),N}$ is a finite group scheme at $\mathfrak{p}|N$, but $\mathfrak{p}\nmid abc$.

Conversely assume that we have an elliptic curve E/K, a finite set S_0 of primes containing all divisors of 2, and $N \geq 5$ such that O_{S_0} is a principal ideal domain and such that E_{2N} is finite at $\mathfrak{p}\notin S_0$. It follows that $g(K(E_2))$ is controlled by $\deg\prod_{\mathfrak{p}\in S_0}\mathfrak{p}$, and hence in our context we can assume without great loss of generality that $E_2 \subset E(K)$. Then we find an equation $Y^2 = X(X-A)(X-B)$ for E with $A,B \in O_{S_0}$ relatively prime. (Here we use that E has to have semi-stable reduction outside of S_0.) So for $\mathfrak{p}\notin S_0$, we get

$$v_\mathfrak{p}(A) \equiv v_\mathfrak{p}(B) \equiv v_\mathfrak{p}(A-B) \equiv 0 \bmod N,$$

and hence

$$A = ax^N, \quad B = by^N, \quad \text{and} \quad C = cz^N$$

with $a,b,c \in O^*_{S_0}$ and (x,y,z) relatively prime in O_{S_0}. So $(x,y,z) \in L_{(a,b,c),N}(K)$, and we can summarize:

Proposition 1.7. *The Asymptotic Fermat Conjecture (Conjecture 2) implies Conjecture 5 for even N. That is, for given E_0 there is a number $N_0 = N_0(g(K), E_0)$ such that for $N > N_0$ and $\rho_{E_0,N} \cong \rho_{E,N}$, it follows that E_0 is isogenous to E. Conversely, the Asymptotic Fermat Conjecture for a given set S_0 holds if there are only finitely many odd irreducible two-dimensional representations of G_K into $\mathrm{GL}(2, \mathbb{Z}/N\mathbb{Z})$ (with some $N \in \mathbb{N}$, N large enough) which are unramified outside of S_0 and finite at divisors of N. Especially there are no solutions (x, y, z) in $L_{(a,b,c),N}(K)$ with $xyz \neq 0$ and $(x, y, z) \in O_K$ relatively prime if there are no odd representations*

$$\rho : G_K \to \mathrm{GL}(2, \mathbb{Z}/N\mathbb{Z})$$

satisfying the conditions above with $S_0 = \{\mathfrak{p}; \mathfrak{p}|2abc\}$.

It is obvious that the verification of such non-existence theorems as discussed above for G_K-representations ρ needs strong number theoretical tools. Since in general the image of ρ is not solvable, "classical" methods of class field theory cannot be applied. Looking for other methods, it is quite natural that one gets the idea to use "Langland's philosophy," or over $\mathbb{Q}$, its beautiful concretization given by Serre's conjecture, and hence one is naturally lead to the theory of modular forms.

(d) Dependence on $d(K)$.

Above we have stated conjectures involving constants which should depend on $g(K)$. This seems to be quite natural for the A-B-C-Conjecture and the height conjecture of elliptic curves, but for other conjectures, for instance about the existence of cyclic isogenies, one could hope that the constants depend only on $d(K)$. (Cf. Merel's theorem about the order of torsion points of elliptic curves.) Especially this seems to be convincing if the objects one looks for are parametrized by a family of curves D_n and if one can prove that the irrationality degree of D_n grows with n. For simplicity we restrict the discussion to the case that K is a number field.

We have used already a bound for the irrationality degree of $X_0(N)$ over $\mathbb{Q}$ to get finiteness results for isogenies. Now we shall use a result for this degree over $\mathbb{C}$ of Abramovich[2] to get corresponding results for elliptic curves with isomorphic torsion structures:

Fix E_0/K and N. Consider the set of elliptic curves E over extension fields L of K such that there is a map $\alpha : E_{0,N} \longrightarrow E_N$ so that the triple (E_0, E, α) corresponds to a point $P \in Z_{N,\varepsilon}(L)$. This set is parametrized by a twisted modular curve $X_{\rho_{E_0,N},\varepsilon}$, defined over K, which is isomorphic to $X(N)$ over $\mathbb{C}$. Hence $X_{\rho_{E_0,N},\varepsilon}$ has the same irrationality degree over $\mathbb{C}$ as $X(N)$, which is at least equal to the irrationality degree of $X_0(N)$ over $\mathbb{C}$, and in [A] it is proved that $d(X_0(N) \times \mathbb{C}) \geq N/256$. Hence we get:

[2]I would like to thank B. Edixhoven for this reference

Proposition 1.8. *Let K be a number field and E_0/K an elliptic curve. For every $d \in \mathbb{N}$ there exists a number $N_0 = N_0(d, E_0)$ such that for $N > N_0$, the set*

$$\bigcup_{[L:K]\leq d} \{P \in Z_{N,\varepsilon}(L); (E_0, E, \alpha) \text{ with } E \neq E_0 P \text{ corresponds to a triple}\}$$

is finite. Further, for all primes p, there is a number $l_p = l(p, d, E_0)$ such that for $N = p^{l_p}$, the corresponding set is empty.

In the case of Fermat curves the situation is similar: Let $C_{a,b,c,N}$ be the curve given by

$$aX^N - bY^N = cZ^N.$$

This is a plane projective curve without singularities over K, and a theorem going back to M.Noether[3] and proved (in a generalized version) by Hartshorne in [H] implies that $d(C_{a,b,c,N}) = N$. Hence we get:

Proposition 1.9. *Let K be a number field, let $a, b, c \in K^*$, let $d \in \mathbb{N}$, and let $N \geq 2d$. Then $C_{a,b,c,N}$ has only finitely many points of degree $\leq d$ over K, and hence for any finite set S_0 of primes of K, the set*

$$\bigcup_{\substack{[L:K]\leq N/2 \\ \operatorname{supp}(a)\operatorname{supp}(b)\operatorname{supp}(c) \mid \prod_{\mathfrak{p}\in S_0} \mathfrak{p}^3}} L_{a,b,c,N}(L)$$

is finite. Further, for any extension field L of K and a fixed prime p, there is a number l_p such that every solution of $L_{a,b,c,p^{l_p}}(L)$ has, up to projective equivalence, coordinates in $\{0\} \cup \mu'(L)$, where $\mu'(L)$ are the roots of unity in L with order prime to p.

§2. The generic case

In this section we shall assume that K in a function field in one variable over a perfect field K_0.

Proposition 2.1. *Let E/K be an admissible elliptic curve. Then*

$$h_K(E) \leq \frac{1}{2} \deg N_E + (g(K) - 1),$$

hence Conjecture 6 is true.

Proof. The height conjecture was proved in the case that $\operatorname{char}(K_0) = 0$ by Parshin and in general by Szpiro using the geometry of the elliptic surface $\mathcal{E}$ defined by E over K_0 using the Bogomolov-Miyaoka-Yau inequality between Chern classes of $\mathcal{E}$, and it is an idea and hope of Parshin [P] that this

[3] I would like to thank H.Stichtenoth and W.-D.Geyer for this reference

kind of proof can possibly be "translated" into the frame of arithmetical surfaces over number fields. Here we shall give a very elementary proof which uses only Hurwitz' genus formula for function fields, it can be found in [Fr 1] already, and we repeat it for the convenience of the reader.

First we can assume that $j_E \notin K_0$ and that K_0 is algebraically closed. The next observation is that it is enough to prove the inequality for a quadratic twist of E and so we can assume that E is semi-stable at all $\mathfrak{p} \in \Sigma_K$ with $v_\mathfrak{p}(j_E) < 0$. We have to prove:

$$\deg \Delta_E \leq 6 \deg N_E + 12(g-1).$$

Our assumptions imply:

If $\mathfrak{p}^2 | N_E$ then $v_\mathfrak{p}(j_E) \geq 0$ and $0 < v_\mathfrak{p}(\Delta_E) \leq 10$. If $j_E \equiv 12^3 \bmod \mathfrak{p}$ then $v_\mathfrak{p}(\Delta_E) \leq 9$, and if $j_E \equiv 0 \bmod \mathfrak{p}$ then $v_\mathfrak{p}(\Delta_E) \leq 8$. If $v_\mathfrak{p}(j_E) < 0$ then $v_\mathfrak{p}(\Delta_E) = -v_\mathfrak{p}(j_E)$ and $[K : K_0(j_E)] = \deg \prod_{v_\mathfrak{p}(j_E)<0} \mathfrak{p}^{-v_\mathfrak{p}(j_E)} =: d_\infty$. Let N'_E be the square free part of N_E. Now use Hurwitz genus formula to get:

$$\begin{aligned} 2g(K) - 2 \geq & -2d_\infty + d_\infty - \deg N'_E \\ & + \deg\Big(\prod_{\substack{\mathfrak{p} \nmid N_E \\ v_\mathfrak{p}(j_E - 12^3) > 0}} \mathfrak{p}\Big) + 2\deg\Big(\prod_{\substack{\mathfrak{p} \nmid N_E \\ v_\mathfrak{p}(j_E) > 0}} \mathfrak{p}\Big) + \deg\Big(\prod_{\substack{\mathfrak{p}^2 | N_E \\ v_\mathfrak{p}(\Delta_E) = 10}} \mathfrak{p}\Big). \end{aligned}$$

Define

$$s^{(0)} = \deg\Big(\prod_{\substack{\mathfrak{p} | N_E \\ v_\mathfrak{p}(j_E - 12^3) > 0}} \mathfrak{p}\Big), \quad s^{(1)} = \deg\Big(\prod_{\substack{\mathfrak{p} | N_E \\ v_\mathfrak{p}(j_E) > 0}} \mathfrak{p}\Big), \quad s^{(2)} = \deg\Big(\prod_{\substack{\mathfrak{p} | N_E \\ v_\mathfrak{p}(\Delta_E) = 10}} \mathfrak{p}\Big).$$

Then:

$$2g - 2 \geq -d_\infty - \deg N'_E + \frac{d_\infty - s^{(0)}}{2} + \frac{d_\infty - s^{(1)}}{3} + \frac{1}{3} s^{(2)},$$

or

$$\frac{1}{6} d_\infty \leq \deg N'_E + \frac{s^{(0)}}{2} + \frac{2}{3} s^{(1)} - \frac{1}{3} s^{(2)},$$

and so

$$\begin{aligned} \deg \Delta_E - 12(g-1) \leq & \; 6 \deg N'_E + 10 \deg\Big(\prod_{\substack{\mathfrak{p} | N_E \\ v_\mathfrak{p}(j_E(j_E - 12^3)) = 0}} \mathfrak{p}\Big) \\ & + 12 \deg\Big(\prod_{\substack{\mathfrak{p} | N_E \\ v_\mathfrak{p}(j_E - 12^3) > 0}} \mathfrak{p}\Big) + 12 \deg\Big(\prod_{\substack{\mathfrak{p} | N_E \\ v_\mathfrak{p}(j_E) > 0}} \mathfrak{p}\Big) \\ \leq & \; 6 \deg N_E. \end{aligned}$$

Corollary 2.1. *There are constants* d_0 *and* d_1 *such that for all function fields* K *over perfect fields* K_0 *and for all* $x \in K\setminus\{0,1\}$ *with* $K/K_0(x)$ *separable we get:*

$$h_K(x) \leq \frac{1}{2} \deg \operatorname{supp}(x(x-1)) + d_1 g(K) + d_0,$$

hence Conjecture 1′ *is true, as well as Conjecture 2 and Conjecture 5.*

We can use the height formula to bound the order of K-rational torsion subgroups for elliptic curves defined over K but we can do much better:

Assume that $j_E \notin K_0$ and that E is admissible. Let $\eta : E \to E'$ be a separable cyclic isogeny of order N defined over K. Then (j_E, j'_E) generate a separable subfield in K which is isomorphic to the function field of $X_0(N) \times K_0$, hence $g(K) \geq g(X_0(N)) \sim \frac{N}{12}$ and so $N \leq 12g(K)$, and Conjecture 3 is true. Proposition 1.8 can be used to get: $N \leq 256d(K)$, and so we can replace $g(K)$ by $d(K)$ in Conjecture 3.

What about Conjectures 4 ff?

To give a flavour of the kind of problems which arise we simplify by assuming that $\operatorname{char}(K_0) = 0$ and that K is algebraically closed. A non-constant point $P \in Z_{N,\varepsilon}(K)$ corresponds to a K_0-rational curve C_P on $Z_{N,\varepsilon}$ whose genus is bounded by $g(K)$ and whose irrationality degree is bounded by $d(K)$. Assume that P corresponds to the pair (E_1, E_2) of elliptic curves. Then the absolute invariants j_i of E_i satisfy an equation over K_0 whose degree is a lower bound for the heights of E_1 and E_2 over K, and this degree is the degree of the image curve of C_P under the natural map from $Z_{N,\varepsilon}$ to $\mathbb{P}^1 \times \mathbb{P}^1$ induced by the modular interpretation.

Hence Conjectures 4 and 4′ lead to a question which generalizes in some sense Kani's Conjecture 4″ :

Question 2. Let g (resp. d) be non-negative integers. There exist numbers $N_0(g)$ (resp. $N'_0(d)$) such that for $N \geq N_0(g)$ ($N > N'_0(d)$) and irreducible curves $C \subset Z_{N,\varepsilon}$ with $g(C) \leq g$ (resp. $d(C) \leq d$) it follows that C is either a twist of a Hecke correspondence or the degree of its image in $\mathbb{P}^1 \times \mathbb{P}^1$ is bounded by a constant depending on g and N (resp. d and N).

§3. $K = \mathbb{Q}$

What can we say about our conjectures if $K = \mathbb{Q}$?

There is one essential tool to use: The arithmetic of modular curves and its applications to elliptic curves over $\mathbb{Q}$. There are two major reasons for the relations between modular curves and elliptic curves. Firstly level-n-structures are parametrized by points on these curves and secondly elliptic curves appear as factors of their Jacobians.

The great power of the "modular method" is due to major contributions of many mathematicians. In our context we have to mention especially B. Mazur, J.P. Serre, K. Ribet and above all A. Wiles.

That elliptic curves with cyclic isogenies of degree N correspond to points on $X_0(N)$ was used already in the previous sections. Over $\mathbb{Q}$ Mazur in [Ma 2] succeeded in proving

Theorem 3.1. *Let $E/\mathbb{Q}$ be an elliptic curve with $\mathbb{Q}$-rational cyclic isogeny of degree N. Then $N \leq 163$. Hence Conjecture 3 is true over $\mathbb{Q}$.*

Moreover Mazur gives the entire list of such curves E.

The importance of congruence primes is already emphasized by Mazur in [Ma 1]. These primes play a major role for the study of elliptic curve E which are factors of $J_0(N_E)$, i.e. which are modular elliptic curves; they link arithmetical properties with attached Galois representations.

Here is the definition of these primes corresponding to a modular curve E: Let $\varphi_E : X_0(N_E) \to E$ be a $\mathbb{Q}$-rational non-trivial morphism of minimal degree. Let ω_E be the Néron differential of E. Then $\varphi_E^* \omega_E = c_\varphi f_E(z)\, dz$ with $c_\varphi \in \mathbb{Z}\backslash\{0\}$ and $f_E(z) = q + \sum\limits_{i=2}^{\infty} a_i q^i$ with $f_E \in S_2(N_E)(\mathbb{Z})$ a newform. For primes $l \nmid pN_E$ one has: $Tr(\rho_{E,p}(\sigma_l) \equiv a_l \bmod p$ where σ_l is a Frobenius element at l. We call f_E the cusp form attached to E.

Now assume that K is a number field and $g(z) = q + \sum\limits_{i=2}^{\infty} b_i q^i \in S_2(N, O_K)$ is an eigenform such that there is a prime divisor $\mathfrak{p}$ of p with $b_l \equiv a_l \bmod \mathfrak{p}$ for all $l \nmid pNN_E$. Then we say by abuse of language that g is congruent to f_E modulo p and if $g \neq f_E$, that p is a congruence prime for f_E resp. for E.

The following theorem uses Corollary 1.3 and is a special case of a beautiful result of K. Ribet [Ribet] in which he proved a large part of Serre's conjecture for odd two-dimensional representations of $G_\mathbb{Q}$ (cf. [Se]) in the modular case:

Theorem 3.2. (Ribet) *Let p be an odd prime. Assume that E is a modular elliptic curve which is semi-stable outside of S_0, let f_E be the attached newform in $S_2(N_E)(\mathbb{Z})$. Assume that $\rho_{E,p}$ is irreducible. Let*

$$N_{\rho'_{E,p}} := \prod_{l \in S_0} l^{v_l(N_E)} \prod_{\substack{l \notin S_0 \\ v_l(\Delta_E) \not\equiv 0 \bmod p}} l.$$

Then there is a number $N_0 \mid N'_{\rho_{E,p}}$ and a newform $g \in S_2(N_0)$ congruent to f_E modulo p.

Finally we come to Andrew Wiles' [W] celebrated result. The following version includes the extension by F. Diamond, who showed that Wiles' original everywhere semi-stability condition could be substantially relaxed:

Theorem 3.3. (A. Wiles, F. Diamond) *Assume that E has a twist E' which is semi-stable at 3 and 5. Then E is a modular elliptic curve. Especially it follows that E is modular if $E_2 \subset E(\mathbb{Q})$.*

As a first application we shall give a "modular version" of the height conjecture for elliptic curves and hence of the A-B-C-Conjecture: Assume that E is modular with attached eigenform f_E. Then (cf. [Fr 1]) we have

$$h_{\mathbb{Q}}(E) = \frac{1}{2}\log\deg\varphi_E - \log|c_\varphi| - \frac{1}{2}\log\Big(\frac{1}{2\pi}\int_{X_0(N_E)\otimes\mathbb{C}} |f_E|^2 dv\Big)$$

with $dv = 4\pi\, dx\, dy$.

Lemma 3.1.
(i) *There is a number $k > 0$ (independent of E) such that*

$$\frac{1}{2\pi}\int_{X_0(N_E)\otimes\mathbb{C}} |f_E|^2 dv \geq k.$$

(ii) $\log\Big(\int_{X_0(N_E)\otimes\mathbb{C}} |f_E|^2 dv\Big) \leq O(\log N_E)$.

The (easy) proof of i) can be found in [Fr 1], the much more involved proof of ii) is a result due to Mai-Murty (see [MM]).

Proposition 3.1.
(i) *Assume that E is modular. Then* $2h_{\mathbb{Q}}(E) \leq \log(\deg\varphi_E) + O(1)$.
(ii) *Assume that E is semi-stable. Then* $\log\deg\varphi \leq 2h_{\mathbb{Q}}(E) + O(\log N_E)$.

Proof. Obvious using Lemma 3.1 and for the second part, Theorem 3.3 and the fact that $|c_\varphi|$ is bounded if E is semi-stable (conjecturally $|c_\varphi| = 1$).

Hence the height conjecture for modular elliptic curves E is true with constant c (and computable d) if for such curves $\log(\deg\varphi_E) \leq 2c\log N_E + d'$ and conversely for semi-stable elliptic curves the existence of such constants follows from the height conjecture for elliptic curves.

Corollary 3.1. *The A-B-C-Conjecture for relatively prime numbers A and B holds with constants c, d, if for modular elliptic curves E we have:* $\log\ (\deg\ \varphi_E) \leq c\ \log\ N_E + d + \log|c_\varphi| - \frac{1}{2}\log\ k$, *and its truth (with some constants $\tilde{c}$, $\tilde{d}$) is equivalent with an estimate of* $\deg\varphi_E$ *of this type for all E with E_2 contained in $E(\mathbb{Q})$.*

Remark 3.1. It is not difficult to show that $\log\deg\varphi_E \leq O(N_E\log N_E)$ and hence we get an exponential version of the A-B-C-Conjecture but this is not surprising since by transcendental methods one gets this type of results as well.

There is an obvious relation between φ_E and congruence primes, and more generally, Question 1:

$$\varphi_E \text{ induces a morphism } \psi_E : J_0(N_E) \longrightarrow E.$$

Let $\hat{\psi}_E$ its dual map. Using theorem 1 we can assume that ψ_E does not factor through another elliptic curve and so $\hat{\psi}_E$ is injective and $\ker(\psi_E) = A_E \subset J_0(N_E)$ is an abelian variety. Since $\psi_E \circ \hat{\psi}_E = \deg \psi_E \circ id$ it follows that $\hat{\psi}_E(E) \cap A_E = \hat{\psi}_E(E)_{\deg \varphi_E}$, and so any prime divisior of $\deg \varphi_E$ is a congruence prime. Moreover we find a subvariety B in $A_E/\hat{\psi}_E(E)_{\deg \varphi_E}$ and a monomorphism $\alpha : E_{\deg \varphi_E} \longrightarrow B$ such that

$$\Delta_\alpha = \{(P, \alpha P); (P \in E_{\deg \varphi})\}$$

is an exceptional subgroup of $E \times B$.

So a positive answer to Question 1 would have strong consequences for the height conjecture for elliptic curves, the A-B-C-Conjecture and the Asymptotic Fermat Conjecture (remember that $\dim B \leq O(N_E)$).

Example 3.1. Assume that $E/\mathbb{Q}$ has prime conductor $N_E = l$. Using Theorem 3 and Theorem 2 and checking small primes one gets: $\Delta_E | l^6$, and so Szpiro's conjecture is true. To get the height conjecture one would need more, namely the Hall Conjecture (cf. [Si]) which predicts that

$$\log(g_2(E)) \leq d'(\varepsilon) + (2+\varepsilon)\log|\Delta_E|$$

for all $\varepsilon \in \mathbb{R}_{>0}$.

Let n be a non zero integer and define

$$\delta_n = \begin{cases} 0 & \text{if } v_2(n) \geq 4, \\ 2 & \text{if } 2 \leq v_2(n) \leq 3, \\ 4 & \text{if } v_2(n) = 1, \\ 0 & \text{if } v_2(n) = 0, \end{cases}$$

Now take $A, B \in \mathbb{Z}$ relatively prime and assume that A is even and that $B \equiv 3 \bmod 4$. We use again theorems 3.1,3.2,3.3 and get:

Proposition 3.2. *The curve $E_{(A,B)} : Y^2 = X(X-A)(X-B)$ is modular, and for primes $p \geq 5$ with $\rho_{E_{(A,B)},p}$ irreducible (e.g. $p > 163$) we get: $f_{E_{(A,B)}}$ is congruent modulo p to $g \in S_2(N_{(A,B),p})$ with*

$$N_{(A,B),p} = 2^{\delta_A} \prod_{v_l(2^{-4}AB(A-B)) \not\equiv 0 \bmod p} l.$$

This result implies strong conditions for solutions of equations of Fermat-type. Take $a, b, c \in \mathbb{Z}\backslash\{0\}$ relatively prime. We want to study $L_{(a,b,c),p}(\mathbb{Q})$ for large primes p . We can assume without loss of generality that $v_2(a) \geq v_2(b) + v_2(c)$ and that $p > v_2(a)$.

Define $\tilde{L}_{(a,b,c),p}(\mathbb{Q})$ to be the set of triplets $(x, y, z) \in L_{(a,b,c),p}(\mathbb{Q})$ with $xyz \neq 0$ and assume that $\gcd(x, y, z) = 1$. Now apply Proposition 3.2 with $A = ax^p$, $B = by^p$ to get:

Corollary 3.2. *Let p be larger than 163. $\tilde{L}_{(a,b,c),p}(\mathbb{Q})$ is empty if the genus of*

$$X_0\left(2^{\delta_a}\prod_{v_l(2^{-4}abc)\not\equiv 0 \bmod p} l\right)$$

is equal to 0, and there is no triplet $(x,y,z)\in\tilde{L}_{(a,b,c),p}(\mathbb{Q})$ with $v_2(x)>0$ if the genus of

$$X_0\left(\prod_{v_l(2^{-4}abc)\not\equiv 0 \bmod p} l\right)$$

is equal to 0.

Since $g(X_0(N))=0$ if $N\leq 10$ or$N=12,13,16,18,25$ one can use this corollary together with a careful discussion of small exponents to determine $L_{(a,b,c),p}(\mathbb{Q})$ for suitable a,b,c (cf. [R 2]).

We look only at one special case: $a=b=c=1$ and so $N=2$ and get

Fermat's Last Theorem. *Let n be a natural number ≥ 3 and $x,y,z\in\mathbb{Z}$ with $x^n-y^n=z^n$. Then $xyz=0$.*

Another interesting case is that $f_{E_{(A,B)}}$ (with $A=ax^p, B=by^p, A-B=cz^p$ as above) is congruent modulo p to a newform g_0 belonging to an elliptic curve E_0 with conductor $N_0|N_{(A,B),p}|2^{\delta_a+1}abc$, with $g_0(z)=q+\sum_{i=2}^{\infty}b_iq^i$.

Let l be a prime not dividing pN_0 but dividing ABC.Then there is a quadratic extension K_1 of $\mathbb{Q}_l$ such that $E_{(A,B)}(K_1)$ contains a point of order p. Hence $E_0(K_1)$ contains a point of order p and since it has good reduction at l we get:$l\geq p^{1/2}+1$. So for $l\leq p^{1/2}$ the curve $E_{(A,B)}$ has good reduction and $a_l\equiv b_l \bmod p$. Since $|a_l|\leq 2\sqrt{l}+1$, $|b_l|\leq 2\sqrt{l}+1$ we have $a_l=b_l$ for $p>4\sqrt{l}+2$. Especially $E_0 \bmod l$ has 4 rational points of two-power order .

Hence we get for l prime to N_0 and $l<Min((p-2)^2/16,p^{1/2})$ that $E_0 \bmod l$ has rational points of order 2. For p large enough (depending on N_0 only) it follows that E_0 is isogenous to an elliptic curve E_0' with $E_{0,2}'\subset E_0'(\mathbb{Q})$. Replace E_0 by E_0'. Then E_0 is given by an equation

$$Y^2=X(X-a')(X-b') \quad \text{with} \quad \operatorname{supp}(a'b'(a'-b'))|2^{\delta_a+1}\left(\prod_{l|abc} l\right).$$

There is one case in which we can be sure that g_0 belongs to an elliptic curve:

Proposition 3.3. *Suppose for $N_0=2^{\delta_a+1}\left(\prod_{l|abc} l\right)$ we have $g(X_0(N_0))=1$ and that $X_0(N_0)$ does not have 4 $\mathbb{Q}$-rational points of 2-power order. Then* *there is an effectively computable number* $p_0(N_0)$ *such that for* $p \geq p_0(N_0)$ *we have:* $\tilde{L}_{(a,b,c),p}(\mathbb{Q}) = \emptyset$.

Now there are only very few modular curves of genus 1 and hence the assumption of the proposition is not often satisfied. But it is an observation of Mazur that one can enforce the existence of E_0 by choosing p large:

The situation is as above. We assume that $p > 163$ and apply theorem 3.3 to get a newform $g_0 \in S_2(N_0)(K)$ with Fourier coefficient $b_i \in O_K$ such that for a divisor $\mathfrak{p}$ of p and prime numbers $l \nmid pN_0$ we get: $a_l \equiv b_l \bmod \mathfrak{p}$.

Now the degree of $K/\mathbb{Q}$ can be bounded by $g(X_0(N_0)) \sim N_0/6$, and since the absolute values of the conjugates of b_l are bounded by $2l^{1/2}$ it follows that for $log(p) > O(N_0 log(N_0))$ we have: $a_l = b_l$ for all $l \leq 2g(X_0(N_0))$ and hence $g_0 \in S_2(N_0)(\mathbb{Q})$.

We summarize:

Proposition 3.3. *Assume that a,b,c are relatively prime integers different from 0 and define* N_0 *as above. Then there is an effectively computable number* $p_0(N_0)(< O(N_0 log(N_0)))$ *such that for* $p \geq p_0(N_0)$ *and* $\tilde{L}_{(a,b,c),p}(\mathbb{Q}) \neq \vee$ *we get:* $X_0(N_0)$ *has an elliptic factor* E_0 *with* $E_{0,2} \subset E_0/(\mathbb{Q})$ *and* $\rho_{E_{0,p}} \cong \rho_{E_{(ax^p, by^p)},p}$ *for* $(x, y, z) \in \tilde{L}_{(a,b,c),p}(\mathbb{Q})$.

Corollary. (Mazur) *Let q be a prime. If the equation* $2^4X^p - Y^p = qZ^p$ *has solutions for infinitely many p then* $q = 17$.

Proof. $X_0(q)$ has to have an elliptic factor given by $Y^2 = X(X-a')(X-b')$ and (since it has good reduction outside of q) $a' = 2^4$ and $|b'(a'-b')| = q$.

Finally we observe a consequence of proposition 3.4 which brings our conjectures about solutions of equations of Fermat-type and about exceptional subgroups of $E_1 \times E_2$ nicely together:

Proposition 3.5. *Conjecture 2 is equivalent to Conjecture 5, i.e., for any finite set* $S_0 \in \mathbb{P}$ *the set*

$$\bigcup_{\substack{\text{supp}(abc) | \prod_{l \in S_0} l \\ p \geq 5}} L_{(a,b,c),p}(\mathbb{Q})$$

is finite if and only if the set

$$\left\{ E/\mathbb{Q};\ \begin{array}{c} E_2 \subset E(\mathbb{Q}) \text{ and there is a prime } p \geq 5 \text{ and} \\ \text{a curve } E_0 \text{ with } N_{E_0} | \prod_{l \in S_0} l \text{ and } \rho_{E_{0,p}} \cong \rho_{E,p} \end{array} \right\}$$

is finite.

Proof. One direction is true in general and was proved in Proposition 1.7. Conversely assume that Conjecture 5 is true and that there are infinitely many solutions $(x, y, z) \in \tilde{L}_{(a,b,c),p}(\mathbb{Q})$ for $p \geq 5$, $\text{supp}(abc) | \prod_{l \in S_0} l$.

Hence there are infinitely many non-isogenous elliptic curves $E_{(ax^p,by^p)}$ which are modular with $N_{(ax^p,by^p),p}|2^{n_2}\prod_{l\in S_0} l$ with $n_2\le 5$. By Proposition 3.4 there is one elliptic curve $E/\mathbb{Q}$ of level N_0 dividing $2^{n_2}\prod_{l\in S_0} l$ with $\rho_{E_0,P}\cong\rho_{E_{(ax^p,by^p)},p}$ for infinitely many primes p which contradicts Conjecture 5.

REFERENCES

[A] D.Abramovich, *Subvarieties of Abelian Varieties and Jacobians of Curves*, Thesis Harvard 1991.

[B] J.B. Bost, *Périodes et isogénies des variétés abeliennes sur les corps de nombres*, Sém. Bourbaki **795** (1994–95).

[Da] H. Darmon, *Serre's conjecture*, Sem. on Fermat's Last Theorem, CMS Conf. Proc. AMS, to appear.

[De] P. Deligne, *Démonstration des conjectures de Tate et de Shafarevich (d'après G. Faltings)*, Séminaire Bourbaki 616, Astérisque **121–122** (1985), 25–41.

[E] N.D. Elkies, *ABC implies Mordell*, Int. Math. Res. Not. **7**, 99–109.

[Fa] G. Faltings, *The general case of S. Lang's conjecture*, Princeton University, 1992.

[Fr 1] G. Frey, *Links between solutions of $A-B=C$ and elliptic curves*, Number Theory (Ulm 1987), Lecture Notes in Math. 1380, Springer, 1989, pp. 31–62.

[Fr 2] G. Frey, *Curves with infinitely many points of fixed degree*, Israel J. of Math. **85**, 79–83.

[Fr 3] G. Frey, *On elliptic curves with isomorphic torsion structures and corresponding curves of genus 2*, Conference on Elliptic Curves and Modular Forms (Hong Kong), Intern. Press, 1995, pp. 79–98.

[H] R.Hartshorne, *Generalized divisors and Gorenstein curves and a theorem of Noether*, Journ. Math. Kyoto **26** (1986), 375–385.

[He] Y. Hellegouarch, *Points d'ordre $2p^h$ sur les courbes elliptiques*, Acta Arith. **26** (1975), 253–263.

[K] E.Kani, Letter to Mazur, Sept. 1995.

[K-S] E. Kani and W. Schanz, *Diagonal quotient surfaces* (to appear).

[MM] L. Mai and M.R. Murty, *The Phragmén-Lindelöf theorem and modular elliptic curves*, Contemp. Math. **166** (1994), 335–340.

[Ma 1] B. Mazur, *Modular curves and the Eisenstein ideal*, Publ. Math. IHES **47** (1977), 33–186.

[Ma 2] B. Mazur, *Rational isogenies of prime degree*, Invent. Math. **44** (1978), 129–162.

[P] A.N. Parshin, *The Bogomolov-Miyaoka-Yau inequality for arithmetical surfaces and its applications*, Sém. Théorie Nombres Paris 1986/87, Progr. Math. **75** (1989), 299–312.

[R 1] K. Ribet, *On modular representations of $G(\bar{\mathbb{Q}}\mid\mathbb{Q})$ arising from modular forms*, Journ. Math. **100** (1990), 431–476.

[R 2] K. Ribet, *On the equation $a^p\cdot 2^\alpha+b^p+c^p=0$*, Acta Arithmetica (to appear).

[Se] J.P. Serre, *Sur les représentations modulaires de degré 2 de $G(\bar{\mathbb{Q}}\mid\mathbb{Q})$*, Duke Math. J. **54** (1987), 179–230.

[Si] J.H. Silverman, *The arithmetic of elliptic curves*, Springer, 1986.

[W] A. Wiles, *Modular elliptic curves and Fermat's Last Theorem*, Ann. of Math. **142** (1995), 443–551.

WILES' THEOREM AND THE ARITHMETIC OF ELLIPTIC CURVES

HENRI DARMON

Thanks to the work of Wiles [Wi], completed by Taylor-Wiles [TW] and extended by Diamond [Di], we now know that all elliptic curves over the rationals (having good or semi-stable reduction at 3 and 5) are modular. This breakthrough has far-reaching consequences for the arithmetic of elliptic curves. As Mazur wrote in [Ma3], "It has been abundantly clear for years that one has a much more tenacious hold on the arithmetic of an elliptic curve $E/\mathbf{Q}$ if one supposes that it is [...] parametrized [by a modular curve]." This expository article explores some of the implications of Wiles' theorem for the theory of elliptic curves, with particular emphasis on the Birch and Swinnerton-Dyer conjecture, now the main outstanding problem in the field.

1 Prelude: Plane Conics, Fermat and Gauss

In a volume devoted to Wiles' proof of Fermat's Last Theorem, what better place to begin this discussion than the Diophantine equation

$$C : x^2 + y^2 = 1, \tag{1}$$

which also figured prominently in Diophantus' treatise, and prompted Fermat's famous marginal comment, more than 350 years ago?

The set $C(\mathbf{Q})$ of rational solutions to equation (1) is well understood, thanks to the parametrization

$$(x, y) = \left(\frac{t^2-1}{t^2+1}, \frac{2t}{t^2+1}\right), \tag{2}$$

giving the classification of Pythagorean triples well-known to the ancient Babylonians. The integer solutions are even simpler: there are $N_{\mathbf{Z}} = 4$ integer lattice points $(\pm 1, 0)$ and $(0, \pm 1)$ on the circle of radius 1.

It has become a dominant theme in number theory that curves such as C ought to be studied over various fields, such as the real or complex numbers, the finite fields $\mathbf{F}_p$, and the p-adic fields $\mathbf{Q}_p$, for each prime p.

The solutions to (1) in $\mathbf{R}^2$ describe the locus of points on the circle of radius 1. A natural measure of the size of this solution set is the circumference of the circle: $N_{\mathbf{R}} = 2\pi$.

The set of $\mathbf{F}_p$-valued solutions $C(\mathbf{F}_p)$ is finite, of cardinality N_p. Let $a_p = p - N_p$. Is there a convenient formula for N_p, or equivalently, for a_p? Letting t run over the values $t = 0, 1, 2, \ldots, p-1, \infty \in \mathbf{P}_1(\mathbf{F}_p)$ in the parametrization (2) gives $p+1$ distinct points in $C(\mathbf{F}_p)$, with one important caveat: if $t^2+1=0$ has a solution $t_0 \in \mathbf{F}_p$, then the values $t = \pm t_0$ do not give rise to points over $\mathbf{F}_p$. Hence, if p is odd:

$$a_p = \begin{cases} +1 & \text{if } -1 \text{ is a square mod } p; \\ -1 & \text{if } -1 \text{ is not a square mod } p. \end{cases}$$

The condition which determines the value of a_p might seem subtle to the uninitiated. But much the opposite is true, thanks to the following result which is due to Fermat himself:

Theorem 1.1 (Fermat) *If p is an odd prime,*

$$a_p = \begin{cases} +1 & \text{if } p \equiv 1 \pmod 4, \\ -1 & \text{if } p \equiv 3 \pmod 4, \end{cases}$$

and $a_2 = 0$.

The computational advantage of this formula is obvious. It now suffices to glance at the last two decimal digits of p to determine whether N_p is equal to $p-1$ or $p+1$.

Let

$$L(C/\mathbf{Q}, s) = \prod_p (1 - a_p p^{-s})^{-1}$$

be the "Hasse-Weil zeta-function" associated to C. Thanks to Fermat's theorem 1.1, one has:

Corollary 1.2 *The Hasse-Weil L-function $L(C/\mathbf{Q}, s)$ is equal to a Dirichlet L-series $L(s, \chi)$, where $\chi : (\mathbf{Z}/4\mathbf{Z})^\times \longrightarrow \pm 1$ is the unique non-trivial quadratic Dirichlet character of conductor 4. In particular, $L(C/\mathbf{Q}, s)$ has a functional equation and an analytic continuation to the entire complex plane.*

More precisely (see [Was] chapter 4), setting

$$\Lambda(C/\mathbf{Q}, s) = \left(\frac{4}{\pi}\right)^{s/2} \Gamma\left(\frac{s+1}{2}\right) L(C/\mathbf{Q}, s),$$

we have:

$$\Lambda(C/\mathbf{Q}, s) = \Lambda(C/\mathbf{Q}, 1-s). \tag{3}$$

The special value $L(C/\mathbf{Q},1)$ is given by:

$$L(C/\mathbf{Q},1) = L(1,\chi) = 1 - \frac{1}{3} + \frac{1}{5} - \cdots = \frac{\pi}{4}. \tag{4}$$

The formal equality $L(C/\mathbf{Q},1)$"="$\prod_p \frac{p}{N_p}$, equation (4) can be rewritten in the suggestive form:

$$\prod_p \frac{N_p}{p} \cdot N_{\mathbf{R}} = 2N_{\mathbf{Z}}, \tag{5}$$

a formula which suggests a mysterious link between the solutions to C over the reals, the finite fields $\mathbf{F}_p$, and the integers. The proof that we have sketched, although quite simple, does little to dispel the mystery.

Another example which was also at the center of Fermat's preoccupations is the Fermat-Pell equation

$$H : x^2 - Dy^2 = 1, \tag{6}$$

where D is a positive square free integer. (H is for "hyperbola.") Assume for simplicity that D is congruent to 5 mod 8.

Defining the integers N_p and $a_p = p - N_p$ as before, one finds that for p not dividing $2D$,

$$a_p = \begin{cases} +1 & \text{if } D \text{ is a square mod } p, \\ -1 & \text{if } D \text{ is not a square mod } p. \end{cases}$$

Extend the definition of a_p by setting $a_p = 0$ if $p|2D$. By Gauss's theorem of quadratic reciprocity:

Theorem 1.3 (Gauss) *Let*

$$\chi_D : (\mathbf{Z}/D\mathbf{Z})^\times \longrightarrow \pm 1$$

be the even (non-primitive) Dirichlet character of conductor $2D$ defined by $\chi_D(n) = \left(\frac{n}{D}\right)$. Then $a_p = \chi_D(p)$.

Define the Hasse-Weil L-function $L(H/\mathbf{Q},s) = \prod_p (1 - a_p p^{-s})^{-1}$ as before.

Corollary 1.4 *The function $L(H/\mathbf{Q},s)$ is equal to the Dirichlet L-series $L(s,\chi_D)$, so that it has a functional equation and an analytic continuation to the entire complex plane.*

The precise functional equation, similar to equation (3), can be found in [Was], chapter 4. As before the value $L(H/\mathbf{Q},1)$ can be evaluated in closed form (see [Was], theorem 4.9):

$$L(H/\mathbf{Q},1) = L(1,\chi_D) = \sum_{n=1}^{\infty} \frac{\chi_D(n)}{n} = \frac{3}{2\sqrt{D}} \sum_{a=1}^{D-1} \chi_D(a) \log|1 - \zeta_D^a|, \tag{7}$$

where $\zeta_D = e^{2\pi i/D}$ is a primitive D-th root of unity.

To gain further insight into the arithmetic significance of this special value, one uses the following result of Gauss, which is a primary ingredient in one of his proofs of quadratic reciprocity, and is in fact essentially equivalent to it.

Theorem 1.5 *Every quadratic field is contained in a cyclotomic field generated by roots of unity. More precisely, the quadratic field $\mathbf{Q}(\sqrt{D})$ is contained in $\mathbf{Q}(\zeta_D)$, and the homomorphism of Galois theory*

$$\mathrm{Gal}(\mathbf{Q}(\zeta_D)/\mathbf{Q}) = (\mathbf{Z}/D\mathbf{Z})^\times \longrightarrow \mathrm{Gal}(\mathbf{Q}(\sqrt{D})/\mathbf{Q}) = \pm 1$$

is identified with the Dirichlet character χ_D.

One of the applications of theorem 1.5 is that it gives a natural way of finding units in $\mathbf{Q}(\sqrt{D})$, and thereby solving Pell's equation. Indeed, the cyclotomic field $\mathbf{Q}(\zeta)$ is equipped with certain natural units, the so-called *circular units.* These are algebraic integers of the form $(1-\zeta_D^a)$ if D is not prime, and of the form $\frac{1-\zeta_D^a}{1-\zeta_D}$ if D is prime, with $a \in (\mathbf{Z}/D\mathbf{Z})^\times$. In particular, theorem 1.5 implies that the expression

$$u_D = \prod_{a=1}^{D} (1-\zeta_D^a)^{3\chi_D(a)}$$

is an element of norm 1 in the quadratic field $\mathbf{Q}(\sqrt{D})$, and in fact, in the ring $\mathbf{Z}[\sqrt{D}]$. Hence, formula (7) can be rewritten:

$$L(H/\mathbf{Q},1) = \frac{1}{\sqrt{4D}} \log |x_0 + y_0\sqrt{D}|, \tag{8}$$

where (x_0, y_0) is an integer solution to equation (6). The non-vanishing of $L(1,\chi_D)$, (or, equivalently, by the functional equation, of $L'(0,\chi_D)$) implies that this solution is non-trivial.

Remark: A natural generalization of theorem 1.5, the Kronecker-Weber theorem, states that every abelian extension of the rationals is contained in a cyclotomic field. The norms of circular units always give a subgroup of finite index in the group of units of L.

2 Elliptic Curves and Wiles' Theorem

Let $E/\mathbf{Q}$ be an elliptic curve over the rationals of conductor N, given by the projective equation

$$y^2z + a_1xyz + a_3yz^2 = x^3 + a_2x^2z + a_4xz^2 + a_6z^3. \tag{9}$$

By the Mordell-Weil theorem, the Mordell-Weil group $E(\mathbf{Q})$ is a finitely generated abelian group,

$$E(\mathbf{Q}) \simeq \mathbf{Z}^r \oplus T,$$

where T is the finite torsion subgroup of $E(\mathbf{Q})$. Paraphrasing a remark of Mazur ([Ma3], page 186), there are resonances between the problem of studying integer points on plane conics and rational points on elliptic curves. In the basic trichotomy governing the study of curves over $\mathbf{Q}$, these Diophantine problems correspond to the only classes of curves having Euler characteristic equal to 0. The Euler characteristic $\chi(X)$ depends only on the Riemann surface $X(\mathbf{C})$, which is topologically equivalent to a compact surface of genus g with s points removed. The Euler characteristic is defined by

$$\chi(X) = (2 - 2g) - s.$$

2.1 Wiles' Theorem and $L(E/\mathbf{Q},s)$

If p is a prime of good reduction for E, let N_p be the number of distinct solutions to equation (9) in $\mathbf{P}^2(\mathbf{F}_p)$, and set

$$a_p = p + 1 - N_p.$$

Further, set $a_p = 1$ if $E/\mathbf{Q}_p$ has split multiplicative reduction, $a_p = -1$ if $E/\mathbf{Q}_p$ has non-split multiplicative reduction, and $a_p = 0$ otherwise. Define the Hasse-Weil L-function $L(E/\mathbf{Q}, s)$ by the formula

$$L(E/\mathbf{Q}, s) = \prod_{p \nmid N} (1 - a_p p^{-s} + p^{1-2s})^{-1} \prod_{p|N} (1 - a_p p^{-s})^{-1}.$$

To study the elliptic curve E along the lines of section 1, one needs a better understanding of the coefficients a_p, allowing an analysis of the L-function $L(E/\mathbf{Q}, s)$. This is precisely the content of Wiles' theorem, stated here in a form which is analogous to theorems 1.1 and 1.3.

Theorem 2.1 ([Wi], [TW], [Di]) *Assume E has good or semi-stable reduction at 3 and 5. Then the coefficients a_p are the Fourier coefficients of a modular form f of weight 2 and level N which is an eigenform for all the Hecke operators T_p.*

This result gives has the following elliptic curve analogue of corollary 1.4.

Corollary 2.2 (Hecke) *The L-function $L(E/\mathbf{Q}, s)$ is equal to the L-function $L(f, s)$ attached to the eigenform f. In particular, it has an analytic continuation and a functional equation.*

More precisely, setting

$$\Lambda(E/\mathbf{Q},s) = N^{s/2}(2\pi)^{-s}\Gamma(s)L(E/\mathbf{Q},s),$$

we have

$$\Lambda(E/\mathbf{Q},s) = \int_0^\infty f\left(\frac{iy}{\sqrt{N}}\right) y^s \frac{dy}{y}, \tag{10}$$

and

$$\Lambda(E/\mathbf{Q},s) = w\Lambda(E/\mathbf{Q},2-s), \tag{11}$$

where $w = \pm 1$ can be computed as a product of local signs. For example:

Proposition 2.3 *If $E/\mathbf{Q}$ is a semistable curve, then w is equal to $(-1)^{s+1}$, where s is the number of primes of split multiplicative reduction for $E/\mathbf{Q}$.*

Remark: The statements of Wiles' theorem given in theorem 2.1 and corollary 2.2 bear a strong ressemblance to theorem 1.3 and corollary 1.4 respectively. This is only fitting, as Wiles' theorem is a manifestation of a non-abelian reciprocity law for $\mathbf{GL}_2$, having its roots ultimately in the fundamental quadratic reciprocity law of Gauss.

More germane to the discussion of section 1, Wiles' achievement allows one to make sense of the special values $L(E/\mathbf{Q},s) = L(f,s)$ even when s is outside the domain $\{\text{Real}(s) > \frac{3}{2}\}$ of absolute convergence of the infinite product used to define $L(E/\mathbf{Q},s)$. This is of particular interest for the point $s = 1$, which is related conjecturally to the arithmetic of $E/\mathbf{Q}$ by the Birch and Swinnerton-Dyer conjecture.

Conjecture 2.4 *The Hasse-Weil L-function $L(E/\mathbf{Q},s)$ vanishes to order r, the rank of $E(\mathbf{Q})$, at $s = 1$, and*

$$\lim_{s\to 1}(s-1)^{-r}L(E/\mathbf{Q},s) =$$

$$\#\text{Ш}(E/\mathbf{Q})\left(\det\left(\langle P_i,P_j\rangle\right)_{1\le i,j\le r}\right)\#T^{-2}\left(\int_{E(\mathbf{R})}\omega\right)\prod_p m_p,$$

where $\text{Ш}(E/\mathbf{Q})$ is the (conjecturally finite) Shafarevich-Tate group of $E/\mathbf{Q}$, the points $P_1,\ldots,P_r$ are a basis for $E(\mathbf{Q})$ modulo torsion, $\langle\ ,\ \rangle$ is the Néron-Tate canonical height, ω is the Néron differential on E, and m_p is the number of connected components in the Néron model of $E/\mathbf{Q}_p$.

Motivated by this conjecture, one calls the order of vanishing of $L(E/\mathbf{Q},s)$ at $s = 1$ the *analytic rank* of $E/\mathbf{Q}$, and denotes it r_{an}.

If E is a semistable elliptic curve, then the formula for w given in proposition 2.3 implies that $t + r_{an}$ is always even, where t denotes the number of analytic uniformizations (complex and p-adic) with which $E/\mathbf{Q}$ is endowed. Hence, a corollary of conjecture 2.4 is the following parity conjecture for the rank:

Conjecture 2.5 *If $E/\mathbf{Q}$ is semistable, then the integer $r+t$ is even.*

A great deal of theoretical evidence is available for conjecture 2.4 when the analytic rank r_{an} is equal to 0 or 1. By contrast, very little is known when $\mathrm{ord}_{s=1}L(E/\mathbf{Q},s) > 1$, and so conjecture 2.4, and even conjecture 2.5, remain very mysterious. (Some numerical evidence has been gathered for certain specific elliptic curves, such as the curve of rank 3 and conductor 5077, see [BGZ].)

2.2 Geometric Versions of Wiles' Theorem

To tackle conjecture 2.4 requires an explicit formula for the leading term of $L(E/\mathbf{Q},s)$ at $s=1$. There are such formulae when the analytic rank is 0 or 1. In deriving them, essential use is made of the following geometric version of Wiles' theorem, which may be seen as a direct analogue of theorem 1.5:

Theorem 2.6 *Suppose that E has good or semistable reduction at 3 and 5. Then the elliptic curve E is uniformized by the modular curve $X_0(N)$, i.e., there is a non-constant algebraic map defined over $\mathbf{Q}$:*

$$\phi : X_0(N) \longrightarrow E.$$

Here $X_0(N)$ is the usual modular curve which is a (coarse) moduli space classifying elliptic curves together with a cyclic subgroup of order N. Its complex points can be decribed analytically as a compactification of the quotient

$$Y_0(N)_{/\mathbf{C}} = \mathcal{H}/\Gamma_0(N),$$

where $\Gamma_0(N)$ is the usual congruence subgroup of level N of $\mathbf{SL}_2(\mathbf{Z})$, and $\mathcal{H}$ is the complex upper half plane of complex numbers τ with $\mathrm{Im}(\tau) > 0$.

The pull-back of the Néron differential ω on E is an integer multiple of the differential $2\pi i f(\tau)d\tau = f(q)\frac{dq}{q}$, where f is the modular form given in theorem 2.1 and $q = e^{2\pi i\tau}$:

$$\phi^*\omega = cf(q)\frac{dq}{q}. \qquad (12)$$

The integer c is called the *Manin constant* associated to ϕ. When the degree of ϕ is minimal among all possible maps $X_0(N) \longrightarrow E'$ with E' isogenous to E, it is conjectured that $c=1$.

When theorem 2.6 is satisfied, the elliptic curve E is also uniformized by other arithmetic curves, the Shimura curves associated to indefinite quaterion algebras. Although somewhat less studied than classical modular curves, they are endowed with a similarly rich arithmetic structure. They play an important role in Ribet's fundamental "lowering the level" results (see the article of Edixhoven in this volume). It is also likely that a deeper understanding of the arithmetic of elliptic curves might be achieved by considering the collection of all modular and Shimura curve parametrizations simultaneously. (See for example the remarks in [Ma3].)

Let $N = N^+N^-$ be a factorization of N such that N^- is square-free, is the product of an even number of primes, and satisfies $\gcd(N^+, N^-) = 1$. Let B be the indefinite quaternion algebra which is ramified exactly at the primes dividing N^-, and let R be a maximal order in B. The algebra B is unique up to isomorphism, and any two maximal orders in B are conjugate. (For more on the arithmetic of quaternion algebras over $\mathbf{Q}$, see [Vi].) The Shimura curve X_{1,N^-} is defined as a (coarse) moduli space for abelian surfaces with quaternionic multiplication by R, i.e., abelian surfaces A equipped with a map

$$R \longrightarrow \mathrm{End}(A).$$

The curve X_{N^+,N^-} is a (coarse) moduli space for abelian surfaces with quaternionic multiplication by R, together with a subgroup scheme generically isomorphic to $\mathbf{Z}/N^+\mathbf{Z} \times \mathbf{Z}/N^+\mathbf{Z}$ and stable under the action of R. Shimura showed that the curves X_{1,N^-} and X_{N^+,N^-} have canonical models over $\mathbf{Q}$. Let J_{N^+,N^-} be the Jacobian of X_{N^+,N^-}. By a theorem of Jacquet-Langlands [JL], it is isogenous to a factor of the Jacobian $J_0(N)$ corresponding to the forms of level N which are *new* at the primes dividing N^-, and hence we have:

Theorem 2.7 *Suppose that E has good or semistable reduction at* 3 *and* 5. *Then E is a factor of the Jacobian J_{N^+,N^-}, i.e., there is a non-constant algebraic map ϕ_{N^+,N^-} defined over $\mathbf{Q}$:*

$$\phi_{N^+,N^-} : J_{N^+,N^-} \longrightarrow E.$$

A nice account of the theory of Shimura curves can be found in [Jo] and [Ro].

Remark: The case where $N^- = 1$ corresponds to the case of the usual modular curves. In this case, the algebra B is the matrix algebra $M_2(\mathbf{Q})$, the order R can be chosen to be $M_2(\mathbf{Z})$, and an abelian surface with endomorphisms by R is isomorphic to a product $A = E \times E$, where E is an elliptic curve. The level N structure on A corresponds to a usual level N structure on E, so that the curve $X_{N,1}$ is isomorphic to $X_0(N)$.

In general, there is considerable freedom in choosing the map ϕ_{N^+,N^-}. One rigidifies the situation by requiring that ϕ_{N^+,N^-} be *optimal*, i.e., that its kernel be a (connected) abelian subvariety of J_{N^+,N^-}. This can always be accomplished, if necessary by replacing E by another elliptic curve in the same isogeny class.

Likewise, we will always assume in the next section that the morphism ϕ of theorem 2.6 sends the cusp $i\infty$ to the identity of E, and that the map induced by ϕ on Jacobians is optimal.

3 The Special Values of $L(E/\mathbf{Q},s)$ at $s = 1$

We now review some of the information on the leading term of $L(E/\mathbf{Q}, s)$ at $s = 1$ which can be extracted from the knowledge that E is modular.

3.1 Analytic Rank 0

Theorem 3.1 *There is a rational number M such that*

$$L(E/\mathbf{Q}, 1) = M \int_{E(\mathbf{R})} \omega,$$

where ω is a Néron differential on E.

Proof: Let $0, i\infty$ be the usual cusps in the extended upper half-plane, and let ϕ be the modular parametrization of theorem 2.6. The theorem of Manin-Drinfeld says that the divisor $(i\infty) - (0)$ is torsion in $J_0(N)$, and hence, if ϕ sends $i\infty$ to the point at infinity on E, then $\phi(0)$ is a torsion point in E. By composing ϕ with an isogeny, assume without loss of generality that $\phi(0) = \phi(i\infty)$ is the identity element in $E(\mathbf{Q})$. Then the modular parametrization ϕ induces a map from the interval $[0, i\infty]$ (with the points 0 and $i\infty$ identified) to the connected component $E_0(\mathbf{R})$ of $E(\mathbf{R})$. Let M_0 be the winding number of this map between two circles. By the formula for $L(E/\mathbf{Q}, 1)$ of equation (10),

$$L(E/\mathbf{Q}, 1) = 2\pi i \int_0^{i\infty} f(\tau) d\tau = \frac{1}{c} \int_0^{i\infty} \phi^* \omega = \frac{M_0}{c} \int_{E_0(\mathbf{R})} \omega = M \int_{E(\mathbf{R})} \omega,$$

where $M = \frac{M_0}{c}[E(\mathbf{R}) : E_0(\mathbf{R})]^{-1}$.

The reader should compare theorem 3.1 with equation (4), which also expresses the special value $L(C/\mathbf{Q}, 1)$ as a rational multiple of the period 2π.

While theorem 3.1 gives some evidence for the Birch and Swinnerton-Dyer conjecture, proving that the value of $L(E/\mathbf{Q}, 1)$ is the correct one "up to rational multiples", it does not shed much light on the relation between M and arithmetic quantities associated to E such as the rank of $E/\mathbf{Q}$ and the order of $Ш(E/\mathbf{Q})$.

3.2 Analytic Rank 1: The Gross-Zagier Formula

Assume now that the sign w in the functional equation (11) for $L(E/\mathbf{Q},s)$ is -1, so that the L-function of $E/\mathbf{Q}$ vanishes to odd order. The L-function $L(E/\mathbf{Q},s)$ now has an "automatic zero" at $s=1$, and one might hope for a natural closed form expression for the special value $L'(E/\mathbf{Q},1)$.

Rather surprisingly, no really "natural" closed form expression is known. Instead, a formula can only be written down after choosing an auxiliary quadratic imaginary field K. Let K be such a field, D its discriminant, and let χ be the associated odd Dirichlet character. Let $E^{(D)}$ be the quadratic twist of E, relative to the character χ. Consider the L-series

$$L(E/K,s) := L(E/\mathbf{Q},s)L(E^{(D)}/\mathbf{Q},s).$$

There are (at least) two different ways to show that this L-series has an analytic continuation and a functional equation relating its value at s and $2-s$. Since E and $E^{(D)}$ are both modular, each of the two factors on the right has a functional equation and analytic continuation. Alternately, the functional equation for $L(E/K,s)$ can be obtained by expressing $L(E/K,s)$ as the Rankin convolution of the L-series $L(f,s)$ with the L-series of a theta-function of weight 1 associated to the imaginary quadratic field K, and applying Rankin's method. (See [GZ], chapter IV). If K is an arbitrary quadratic field (not necessarily quadratic imaginary) one has

Proposition 3.2 *The sign w_K in the functional equation for $L(E/K,s)$ can be expressed as a product of local signs*

$$w_K = \prod_v w_v,$$

where $w_v = \pm 1$ depends only on the behaviour of E over the completion K_v. In particular,

1. *If E has good reduction at v, then $w_v = 1$;*
2. *If v is archimedean, then $w_v = -1$;*
3. *If E/K_v has split (resp. non-split) multiplicative reduction at v then $w_v = -1$ (resp. $w_v = 1$).*

Heegner Points:

Just as cyclotomic fields are equipped with certain canonical units (the circular units) whose logarithms express the special values of Dirichlet L-series, so modular curves and Shimura curves are equipped with a certain natural set of algebraic points, the *Heegner points* associated to the imaginary quadratic field K, whose heights express first derivatives of the L-functions attached to cusp forms.

Let A be any elliptic curve which has complex multiplication by the maximal order $\mathcal{O}_K$ of K. There are exactly h such curves, where h is the class number of K. They are all defined over the Hilbert class field H of K and are conjugate to each other under the action of $\mathrm{Gal}(H/K)$.

Assume further that all the primes dividing the conductor N are split in the imaginary quadratic field K. By proposition 3.2, this implies that $w_K = -1$, so that the analytic rank of $E(K)$ is odd.

Under this hypothesis, the complex multiplication curve A has a rational subgroup of order N which is defined over H. This subgroup is not unique, and choosing one amounts to choosing an integral ideal of norm N in the quadratic field K. Choose such a subgroup C of A. The pair (A, C) gives rise to a point α on $X_0(N)$ which is defined over H. It is called a *Heegner point* on $X_0(N)$ (associated to the maximal order $\mathcal{O}_K$). Let $P_H = \phi(\alpha)$ be the image of α on $E(H)$ by the modular parametrization ϕ of theorem 2.6, and let $P_K = \mathrm{trace}_{H/K} P_H$ be its trace to $E(K)$. The point P_K (up to sign) depends only on the quadratic imaginary field K, not on the choice of A and C. Hence, its Néron-Tate height is canonical.

The fundamental theorem of Gross and Zagier expresses the special value of $L'(E/K, 1)$ in terms of the height of P_K.

Theorem 3.3 $L'(E/K,1) = \left(\iint_{E(\mathbf{C})} \omega \wedge i\bar{\omega}\right) \langle P_K, P_K \rangle / c^2 u_K^2 |D|^{\frac{1}{2}}.$

The proof of this beautiful theorem, which is quite involved, is given in [GZ].

Remarks:
1. Theorem 3.3 gives a formula for $L'(E/\mathbf{Q}, 1)L(E^{(D)}/\mathbf{Q}, 1)$, and in this sense does not give a "natural" formula for $L'(E/\mathbf{Q}, 1)$ alone.
2. Theorem 3.3 is also true when $w = 1$. In this case, the twisted L-function $L(E^{(D)}/\mathbf{Q}, s)$ vanishes at $s = 1$, and theorem 3.3 gives a formula for $L(E/\mathbf{Q}, 1)L'(E^{(D)}/\mathbf{Q}, 1)$.

3.3 Some Variants of the Gross-Zagier Formula

The fundamental formula of Gross and Zagier has been extended and generalized in various directions in the last years. Let us mention very briefly a few of these variants:

A. *Shimura curve analogues*: Assume here for simplicity that E is semistable so that N is square-free, and that K is a quadratic imaginary field of discriminant D with $\gcd(N, D) = 1$. Let $N = N^+N^-$ be the factorization of N such that N^+ is the product of all primes which are split in K, and

N^- is the product of the primes which are inert in K. By proposition 3.2, the integer N^- is a product of an even number of prime factors if and only if the sign w_K in the functional equation for $L(E/K,s)$ is -1. Assume that $w_K = -1$. The Gross-Zagier formula given in theorem 3.3 corresponds to the case where $N^+ = N, N^- = 1$. Assume now that $N^- \neq 1$. One can then define the Shimura curve X_{N^+,N^-} as in section 2.2.

The curve X_{N^+,N^-} is equipped with Heegner points defined over the Hilbert class field H of K, which correspond to moduli of quaternionic surfaces with level N^+ structure having complex multiplication by $\mathcal{O}_K$, i.e, quaternionic surfaces A endowed with a map

$$\mathcal{O}_K \longrightarrow \underline{\mathrm{End}}(A),$$

where $\underline{\mathrm{End}}(A)$ denotes the algebraic endomorphisms of A which commute with the quaternionic multiplications. By considering the image in the Mordell-Weil group $E(H)$ of certain degree zero divisors supported on Heegner points in $J_{N^+,N^-}(H)$ by ϕ_{N^+,N^-}, one obtains a Heegner point P_K in $E(K)$, which cannot be obtained from the modular curve parametrization ϕ. One expects that that the height of P_K can be expressed in terms of the derivative $L'(E/K,1)$, in a manner analogous to theorem 3.3. In particular, one expects that P_K is of infinite order in $E(K)$ if and only if $L'(E/K,1) \neq 0$. Nothing as precise has yet been established, but some work in progress of Keating and Kudla supports this expectation.

B. *Perrin Riou's p-adic analogue*: In [PR], a formula is obtained (when all primes dividing N are split in K) relating the first derivative of the two-variable p-adic L-function of E/K to the p-adic height of the Heegner point P_K. The calculations of [PR] are also quite involved, but on a conceptual level they follow those of Gross and Zagier quite closely.

C. *Rubin's p-adic formula*: Let E be an elliptic curve with complex multiplication by $\mathcal{O}_K$. In [Ru], Rubin obtains a formula expressing the derivative of the two-variable p-adic L-function of E/K at a point which lies outside the range of classical interpolation, to the p-adic logarithm in the formal group attached to E over $K \otimes \mathbf{Q}_p$ of a Heegner point in $E(K)$. The proof of this formula uses the theory of elliptic units, as well as the formula of Gross-Zagier and Perrin-Riou's p-adic analogue, in an essential way. A striking feature of Rubin's formula is that it allows one to recover a rational point in $E(K)$ as the formal group exponential evaluated on an expression involving the first derivative of a p-adic L-function, in much the same way that, if χ is an even Dirichlet character, exponentiating $L'(0,\chi)$ yields a unit in the real quadratic field cut out by χ.

D. *Formulae for* $L(E/K,1)$ *when* E *is a Tate curve*: Suppose that E has a prime p of multiplicative reduction which is inert in K, and suppose that all other primes dividing N are split in K. Then the sign in the functional equation for $L(E/K,s)$ is 1 by prop. 3.2, and one expects no Heegner point construction yielding a point on $E(K)$. However, there are Heegner points $P_n \in E(H_n)$, where H_n is the ring class field of H of conductor p^n, constructed from elliptic curves with level N structure having complex multiplication by the orders of conductor p^n in $\mathcal{O}_K$. The precise construction is explained in [BD2], where it is shown that these points are trace-compatible, and that $\text{trace}_{H_1/H}(P_1) = 0$. Assume to simplify the exposition that K has unit group $\mathcal{O}_K^\times = \pm 1$ and class number 1, so that $H = K$, and that the group of connected components of the Néron model of E/K at the prime p is trivial. The prime p is totally ramified in H_n/K; let p_n be the unique prime of H_n over p, and let Φ_n be the group of connected components of E/K_n at the prime p_n; one has

$$\begin{aligned}\Phi_n &= \mathbf{Z}/(p+1)\mathbf{Z} \times \mathbf{Z}/p^{n-1}\mathbf{Z},\\ \Phi_\infty &:= \lim_{\leftarrow} \Phi_n = \mathbf{Z}/(p+1)\mathbf{Z} \times \mathbf{Z}_p,\end{aligned}$$

where the inverse limit is taken with respect to the norm maps.

The main formula of [BD2] relates the image $\bar{P}_n$ of P_n in the group Φ_n to the special value $L(E/K,1)$. The norm-compatible system of points P_n gives rise to a canonical Heegner element $P_\infty \in \lim_{\leftarrow} E(H_n)$, and hence to an element $\bar{P}_\infty$ in Φ_∞. As a corollary to the main result of [BD2] one obtains:

Theorem 3.4 *The element $\bar{P}_\infty$ is non-torsion if and only if $L(E/K,1) \neq 0$.*

The calculations involved in the proof of theorem 3.4 are considerably simpler than those of [GZ] needed to prove theorem 3.3. The main ingredients in this proof are a formula of Gross for the special value $L(E/K,1)$ (generalized somewhat in [Dag]) and a moduli description due to Edixhoven for the specialization map to the group of connected components of $J_0(N)$. For more details, see [BD2].

A precursor of theorem 3.4 for Eisenstein quotients can be found in Mazur's article [Ma2].

E. *p-adic analytic construction of Heegner points from derivatives of p-adic L-functions*: Assume for simplicity that E is semi-stable, and that, as before, $E/\mathbf{Q}$ has a prime p of multiplicative reduction which is inert in K, so that it is equipped with the analytic Tate parametrization

$$\Phi_{Tate} : K_p^\times \longrightarrow E(K_p),$$

where $K_p := K \otimes \mathbf{Q}_p$. Assume now that $L(E/K,s)$ has sign -1 in its functional equation. Let H_∞ be the compositum of all the ring class fields of K of conductor p^n, whose Galois group $G_\infty = \mathrm{Gal}(H_\infty/K)$ is canonically isomorphic to an extension of the class group $\Delta = \mathrm{Gal}(H/K)$ by the group $(K_p^\times)_1$ of elements in K_p of norm 1. By a generalization of the work of Gross [Gr2] explained in [Dag], there exists an element $\mathcal{L}$ in the completed integral group ring $\mathbf{Z}[\![G_\infty]\!] := \lim_{\leftarrow} \mathbf{Z}[G_n]$ such that

$$|\chi(\mathcal{L})|^2 = \mathcal{L}(E/K,\chi,1) \Big/ \iint_{E(\mathbf{C})} \omega \wedge \bar{\omega} \prod_{\ell | N^-} m_\ell \sqrt{D}, \qquad (13)$$

for all finite order characters $\chi : G_\infty \longrightarrow \mathbf{C}^\times$. The element $\mathcal{L}$ plays the role of the p-adic L-function associated to the anti-cyclotomic $\mathbf{Z}_p$-extension in this setting. (It really might be more accurate to view it as a*square root* of the p-adic L-function.)

Note that if χ_{triv} denotes the trivial character, then $\chi_{triv}(\mathcal{L}) = 0$, since $L(E/K,1) = 0$. Hence $\mathcal{L}$ belongs to the augmentation ideal I in the completed group ring $\mathbf{Z}[\![G_\infty]\!]$. Let $\mathcal{L}'$ be the natural projection of $\mathcal{L}$ in $I/I^2 = G_\infty$. One shows (see [BD3]) that $\mathcal{L}'$ belongs to $(K_p^\times)_1 \subset G_\infty$. The element $\mathcal{L}'$ in $K_p^\times$ should be viewed as the first derivative of the p-adic L-function of E/K (in the anticyclotomic direction, at the trivial character).

Let P_K be the Heegner point on $E(K)$ coming from the Shimura curve parametrization ϕ_{N^+,N^-} that was introduced in paragraph A of this section, and let $\bar{P}_K$ be its Galois conjugate. The following theorem is the main result of [BD3]:

Theorem 3.5 *Let w_p be a local sign which is -1 if $E/\mathbf{Q}_p$ has split multiplicative reduction, and 1 if $E/\mathbf{Q}_p$ has non-split multiplicative reduction. Then*

$$\Phi_{Tate}(\mathcal{L}') = \pm(P_K + w_p \bar{P}_K).$$

Note that, since $p|N^-$, the curve X_{N^+,N^-} is never a classical modular curve. Like the formula of Rubin described in paragraph C, theorem 3.5 allows one to recover a global point in $E(K)$ from the first derivative of a p-adic L-function.

The main ingredients in the proof of theorem 3.5 are the explicit construction of $\mathcal{L}$ given in [Gr2] and [Dag] and the Cerednik-Drinfeld theory of p-adic uniformization of the Shimura curve X_{N^+,N^-} [Cer], [Dr], [BC]. The details of the proof are given in [BD3].

Remarks:
1. The formulas described in paragraphs D and E were inspired by some fundamental ideas of Mazur, Tate, and Teitelbaum on p-adic analogues of the Birch and Swinnerton-Dyer conjecture. The connection with this circle of ideas is explained in [BD1].
2. There are many other generalizations of the Gross-Zagier formula which were not mentionned here because they are not directly relevant to modular

elliptic curves: for example, the work of Nekovar [Ne] and Zhang [Zh] extending the work of Gross-Zagier and Kolyvagin to modular forms of higher weight, replacing Heegner points by higher-dimensional cycles on Kuga-Sato varieties.
3. In connection with the results described in paragraphs C and E, one should also mention an intriguing result of Ulmer [Ul], who constructs global points on certain universal elliptic curves over the function fields of modular curves in characteristic p. Some of the results described above (and, in particular, the formula of paragraph E) should extend to the function field setting; this extension has some tantalizing similarities, as well as differences, with Ulmer's constructions.

4 The Birch and Swinnerton-Dyer Conjecture

4.1 Analytic Rank 0

For modular elliptic curves of analytic rank 0, one has the following theorem.

Theorem 4.1 *If $L(E/\mathbf{Q},1) \neq 0$, then $E(\mathbf{Q})$ is finite, and so is $Ш(E/\mathbf{Q})$.*

There are now several ways of proving this theorem. We will review the different strategies, giving only the briefest indication of the details of the proofs.

4.1.1 Kolyvagin's proof

It can be divided into three steps.

Step 1 (Non-vanishing lemma): Choose an auxiliary imaginary quadratic field $K/\mathbf{Q}$ such that

1. All primes dividing N are split in K.
2. Under assumption 1, the sign w_K is -1 and the L-function $L(E/K,s)$ necessarily vanishes at $s=1$. In addition, one requires that the L-function $L(E/K,s)$ has only a simple zero, that is, $L'(E^{(D)}/\mathbf{Q},1) \neq 0$.

The existence of such a quadratic field K follows from the theorems of Bump-Friedberg-Hoffstein [BFH] and Murty-Murty [MM] on non-vanishing of first derivatives of twists of automorphic L-series.

Step 2 (Gross-Zagier formula): Invoking the Gross-Zagier formula (theorem 3.3), one concludes that the Heegner point $P_K \in E(K)$ is of infinite order. In particular the rank of $E(K)$ is at least 1.

Step 3 (Kolyvagin's descent): In [Ko1], Kolyvagin proves the following theorem:

Theorem 4.2 *If the Heegner point P_K is of infinite order, then $E(K)$ has rank* 1 *and* $Ш(E/K)$ *is finite.*

Crucial to the proof of theorem 4.2 is the fact that the Heegner point P_K does not come alone. Namely, for each abelian extension L/K such that the Galois group $\mathrm{Gal}(L/\mathbf{Q})$ is dihedral, satisfying $\gcd(\mathrm{Disc}(L/K), ND) = 1$, there is a Heegner point P_L in $E(L)$ and this system of points is norm-compatible in the sense that, if $L_1 \subset L_2$, then

$$\mathrm{trace}_{L_2/L_1} P_{L_2} = \ell(L_2/L_1) P_{L_1},$$

where $\ell(L_2/L_1) \in \mathbf{Z}[\mathrm{Gal}(L_1/K)]$ is an element whose definition involves the local Euler factors in $L(E/K, s)$ at the primes dividing $\mathrm{Disc}(L_2/L_1)$. Kummer theory allows one to construct Galois cohomology classes $c_L \in H^1(L, T_p(E))$ from the points P_L, where $T_p(E)$ is the p-adic Tate module of E. These classes satisfy the same trace-compatibility properties as the P_L. Kolyvagin calls such a system of cohomology classes an *Euler System* [Ko2], and shows that if the "initial" class c_K is non-zero, the rank of $E(K)$ is less than or equal to 1 and $Ш(E/K)$ is finite.

We will not go into the details of Kolyvagin's ingenious argument, referring the reader instead to [Ko1] and [Gr3] for more details.

4.1.2 A variant

The following variant of Kolyvagin's basic strategy avoids the non-vanishing result of Bump-Friedberg-Hoffstein and Murty-Murty, as well as the formula of Gross and Zagier. It only works, however, for elliptic curves having a prime p of multiplicative reduction, and does not prove the finiteness of $Ш(E/\mathbf{Q})$, but only of the p-primary part of $Ш(E/\mathbf{Q})$.

Step 1 (Non-vanishing lemma): Choose now an auxiliary imaginary quadratic field $K/\mathbf{Q}$ such that

1. The prime p is inert in K, and all the other primes dividing N are split in K.

2. By proposition 3.2, the L-function $L(E/K, s)$ has sign $w_K = 1$ in its functional equation. One requires also that $L(E^{(D)}/\mathbf{Q}, 1) \neq 0$, so that $L(E/K, 1) \neq 0$.

The existence of such a quadratic field K follows from a theorem of Waldspurger [Wal] on non-vanishing of the values of twists of automorphic L-series.

Step 2 (A variant of the Gross-Zagier formula): Invoking theorem 3.4, one finds that the element P_∞ has non-trivial image in Φ_∞.

Step 3 (A variant of Kolyvagin's descent): In [BD2], the following theorem is proved:

Theorem 4.3 *If the image $\bar{P}_\infty$ of P_∞ in Φ_∞ is non-torsion, then $E(K)$ has rank* 0 *and* $Ш(E/K) \otimes \mathbf{Z}_p$ *is finite.*

This theorem is proved by a minor adaptation of Kolyvagin's argument. The entire system of points P_n is now used to construct a cohomology class $c_K \in H^1(K, T_p(E))$, which is part of an Euler system. The non-vanishing of $\bar{P}_\infty$ translates into the non-triviality of the class c_K, and in fact of its image in a certain quotient (the "singular part") of the local cohomology group $H^1(K_p, T_p(E))$. Such a non-triviality is used to uniformly bound the p^n Selmer group of E/K, following the ideas of Kolyvagin.

The details of the argument are explained in [BD2].

4.1.3 Kato's proof

Recently Kato [Ka2] has discovered a wholly original proof of theorem 4.1 which does not require the choice of an auxiliary imaginary quadratic field and does not use Heegner points.

Kato's argument constructs cohomology classes $c_L \in H^1(L, T_p(E))$, where L is a cyclotomic extension of the rationals with discriminant prime to N. These classes are constructed from certain elements introduced by Beilinson, belonging to the K_2 of modular function fields. Defined via explicit modular units (Siegel units), these classes yield elements in $H^1(L, T_p(J_0(N)))$ which are mapped to $H^1(L, T_p(E))$ via the map ϕ of theorem 2.6. (In particular, theorem 2.6 is also crucial to Kato's construction.)

Kato's classes c_L obey norm-compatibility properties similar to those of Kolyvagin, and hence deserve to be viewed as an Euler system [Ka1]. The most difficult part of Kato's argument, given in [Ka2], is to relate the basic class $c_{\mathbf{Q}} \in H^1(\mathbf{Q}, T_p(E))$ (or rather, its localization in a certain quotient — the "singular part" — of the local cohomology group $H^1(\mathbf{Q}_p, T_p(E))$) to the special value $L(E/\mathbf{Q}, 1)$.

4.2 Analytic Rank 1

In the case of analytic rank 1, there is:

Theorem 4.4 *Suppose that $L(E/\mathbf{Q}, 1) = 0$, but that $L'(E/\mathbf{Q}, 1) \neq 0$. Then $E(\mathbf{Q})$ has rank* 1, *and* $Ш(E/\mathbf{Q})$ *is finite.*

This theorem lies somewhat deeper than theorem 4.1. Presently, the only proof follows the basic strategy of Kolyvagin based on the Gross-Zagier formula.

Step 1 (Non-vanishing lemma): Choose an auxiliary imaginary quadratic field $K/\mathbf{Q}$ such that:

1. All primes dividing N are split in K.
2. The Hasse-Weil L-function $L(E/K,s)$ has a simple zero at $s=1$, so that $L(E^{(D)}/\mathbf{Q},1)\neq 0$.

The existence of such a quadratic field K follows from the same theorem of Waldspurger [Wal] on non-vanishing of values of twists of automorphic L-series used in step 1 of section 4.1.2.

Step 2 (Gross-Zagier formula): Invoking the Gross-Zagier formula (theorem 3.3), one finds that the Heegner point $P_K\in E(K)$ is of infinite order. In particular, the rank of $E(K)$ is at least 1. More precisely, by analyzing the action of complex conjugation on P_K, one finds that P_K (up to torsion) actually belongs to $E(\mathbf{Q})$ in this case, so that the rank of $E(\mathbf{Q})$ is at least 1.

Step 3 (Kolyvagin's descent): By theorem 4.2, one concludes that $E(K)$ has rank 1 and finite Shafarevich-Tate group. Hence the rank of $E(\mathbf{Q})$ is exactly 1, its Shafarevich-Tate group $Ш(E/\mathbf{Q})$ is finite, and, as a by-product, $E^{(D)}(\mathbf{Q})$ and $Ш(E^{(D)}/\mathbf{Q})$ are also finite.

Remark: When the sign in the functional equation for $L(E/\mathbf{Q},s)$ is -1, the class $c_{\mathbf{Q}}$ constructed by Kato gives rise to a natural element in the pro-p Selmer group of $E/\mathbf{Q}$, defined as the inverse limit $\lim_{\leftarrow}\mathrm{Sel}(\mathbf{Q},E_{p^n})$. One might expect that this class is non-zero if and only if $L'(E/\mathbf{Q},1)\neq 0$. A proof of this would show that

$$L'(E/\mathbf{Q},1)\neq 0\Rightarrow rank(E(\mathbf{Q}))\leq 1,$$

which represents a part of theorem 4.4. The reverse inequality seems harder to obtain with Kato's methods.

References

[BC] J-F. Boutot, H. Carayol, *Uniformisation p-adique des courbes de Shimura: les théorèmes de Cerednik et de Drinfeld*, Astérisque **196-197** (1991), 45-158.

[BD1] M. Bertolini and H. Darmon, *Heegner points on Mumford-Tate curves*, CICMA preprint; Invent. Math., to appear.

[BD2] M. Bertolini and H. Darmon (with an appendix by B. Edixhoven) *A rigid analytic Gross-Zagier formula and arithmetic applications*, CICMA preprint; submitted.

[BD3] M. Bertolini and H. Darmon, *Heegner points, p-adic L-functions and the Cerednik-Drinfeld uniformization*, CICMA preprint; submitted.

[BFH] D. Bump, S. Friedberg, and J. Hoffstein, *Non-vanishing theorems for L-functions of modular forms and their derivatives*, Annals of Math. (1991).

[BGZ] J. Buhler, B. Gross, and D. Zagier, *On the conjecture of Birch and Swinnerton-Dyer for an elliptic curve of rank* 3, Math. Comp. **44** (1985), 473–481.

[Cer] I.V. Cerednik, *Uniformization of algebraic curves by discrete arithmetic subgroups of* $\mathbf{PGL}_2(k_w)$ *with compact quotient* (in Russian) Math. Sbornik **100** (1976), 59-88. Transl. in Math. USSR Sb. **29** (1976), 55-78.

[Dag] H. Daghigh, McGill PhD. Thesis, in progress.

[Di] F. Diamond, *On deformation rings and Hecke rings*, to appear in Annals of Math.

[Dr] V.G. Drinfeld, *Coverings of p-adic symmetric regions*, (in Russian), Funkts. Anal. Prilozn. **10** (1976), 29-40. Transl. in Funct. Anal. Appl. **10** (1976), 107-115.

[Gr1] B.H. Gross, *Heegner points on* $X_0(N)$, in Modular Forms, R.A. Rankin ed., p. 87-107, Ellis Horwood Ltd., 1984.

[Gr2] B.H. Gross, *Heights and the special values of L-series*, CMS conference proceedings, Volume 7 (1987).

[Gr3] B.H. Gross, *Kolyvagin's work on modular elliptic curves*, in L-functions and arithmetic (Durham, 1989) 235-256, LMS Lecture Notes **153** Cambridge University Press, Cambridge 1991.

[GZ] B.H. Gross, D.B. Zagier, *Heegner points and derivatives of L-series*, Inv. Math. (1986).

[JL] H. Jacquet, R.P. Langlands, *Automorphic forms on* $\mathbf{GL}(2)$, Springer Lecture Notes, v. 114, (1970).

[Jo] B. Jordan, *On the Diophantine arithmetic of Shimura curves*, Harvard PhD thesis, 1981.

[Ka1] K. Kato, *Euler systems, Iwasawa theory, and Selmer groups*, preprint.

[Ka2] K. Kato, *p-adic Hodge theory and special values of zeta functions of elliptic cusp forms*, in preparation.

[Ko1] V.A. Kolyvagin, *Finiteness of* $E(\mathbf{Q})$ *and* $\text{Ш}(E/\mathbf{Q})$ *for a subclass of Weil curves*, (Russian) Izv. Akad. Nauk. Ser. Mat. **52** (1988) no 6., 1154-1180; translation in Math USSR Izv. **33** no. 3 (1989), 473-499.

[Ko2] V.A. Kolyvagin, *Euler Systems*, The Grothendieck festschrift, Vol. II, 435–483, Progr. in Math. **87**, Birkhaüser, 1990.

[Ma1] B. Mazur, P. Swinnerton-Dyer, *Arithmetic of Weil curves*, Invent. Math. **35** (1974), 1-61.

[Ma2] B. Mazur, *On the arithmetic of special values of L-functions*, Invent. Math. **55** no. 3 (1979), 207–240.

[Ma3] B. Mazur, *Modular curves and arithmetic*, Proceedings of the Int. Congress of Math., (1983), Warszawa, pp. 185-209.

[MTT] B. Mazur, J. Tate, and J. Teitelbaum, *On p-adic analogues of the conjectures of Birch and Swinnerton-Dyer*, Invent. Math. **84** (1986), 1-48.

[MM] Murty, M.R., and Murty, V.K., *Mean values of derivatives of modular L-series*, Annals of Math. **133** (1991), 447-475.

[Ne] J. Nekovar, *On the p-adic heights of Heegner cycles*, Math. Ann. **302** (1995), 609-686.

[PR] B. Perrin-Riou, *Points de Heegner et dérivées de fonctions L p-adiques*, Invent. Math. **89** no. 3 (1987), 455-510.

[Ro] D. Roberts, Shimura curves analogous to $X_0(N)$, Harvard PhD. Thesis, 1989.

[Ru] K. Rubin, *p-adic L-functions and rational points on elliptic curves with complex multiplication*, Invent. Math. **107** (1992), 323-350.

[TW] R. Taylor and A. Wiles, *Ring theoretic properties of certain Hecke algebras*, Annals of Math. **141**, No. 3 (1995), 553-572.

[Ul] D. Ulmer, *A construction of local points on elliptic curves over modular curves*, Inter. Math. Res. Notices **7** (1995), 349-363.

[Vi] M-F. Vignéras, Arithmétique des algèbres de quaternions, Lecture Notes in Math. 800, Springer Verlag.

[Wal] J-L. Waldspurger, *Sur les valeurs de certaines fonctions L automorphes en leur centre de symmétrie*, Compos. Math. **54** no. 2 (1985), 173–242.

[Was] L. Washington, Introduction to Cyclotomic Fields, GTM 83, Springer-Verlag, 1982.

[Wi] A. Wiles, *Modular elliptic curves and Fermat's last theorem*, Annals of Math. **141**, No. 3 (1995), 443-551.

[Zh] S. Zhang, *Heights of Heegner cycles and derivatives of L-series*, preprint.

Index

A-augmentation, 258, 267
A-B-C-Conjecture, 2, 529, 544
A-modules, exact sequence of, 320
A-representation, 314
Abelian group, 101
Abelian varieties, 530
  modular, 94–95
Abelianization, 248
Absolute irreducibility, 318–320
  relaxing condition for, 324–325
Absolute logarithmic height, 35
Additive groups, 127
Adelic representations, 168–170
Admissible representations, 165
Affine scheme, 381
Albanese functoriality maps, 368
Algebra, 121
Algebra representations, group representations versus, 251–252
Algebraic integers, 513
Analytic rank, 554
Antipode, 126
Archimedean case, 176–177
Archimedean representation theory, 165–167
Arithmetic theory of elliptic curves, 17–40
Arthur-Selberg, trace formula of, 194–196
Artin conductor, 212
Artin L-function, 179
Artin motives, 73, 96
Artin symbol, 494
Artinian object, 397
Artin's conjecture, 181
  base change and, 197–204
Associativity, 122
Asymptotic Fermat conjecture, 529–530, 539
Atkin-Lehner involution, 226
Atkin-Lehner notation, 331
Augmentation, 126
Augmentation ideal, 126
Augmentation ideal sheaf, 138
Automorphic cuspidal representations, 164, 169
  of weight one, 171
Automorphic induction, 186–188
Automorphic representations
  defined, 169
  of weight one, 164–174
Automorphism group of E, 19

Automorphisms
continuous, 5
diamond, 68
Frobenius, *see* Frobenius automorphism
Auxiliary set of primes, 425
Azumaya Algebra, 255

Bad multiplicative reduction, elliptic curves with, 116
Base change, 129
Artin's conjecture and, 197–204
proof of, 196–197
theory of, 190
Base change lifting, 192–193
Base change theory, 192–197
Bernoulli numbers, 510
generalized, 509
Biquadratic reciprocity, 508
Birch and Swinnerton-Dyer conjecture, 37, 554, 557, 562, 563–566
Bogomolov-Miyaoka-Yau inequality, 540
Brauer group, 33, 102
Brauer-Nesbitt Theorem, 391
Bump-Friedberg-Hoffstein theorem, 563

Canonical height, 35–36
Canonical pairing, non-degenerate, 118
Carayol's Lemma, 218, 219
Carayol's reductions, 237–239
Cartan subgroup, 223
Cartesian diagram, 268
Cartier duality, 145
Cartier map, 337
Cartier-Nishi duality theorem, 388
Casimir operator, 173
Cassels' global duality theorem, 33
Category, group objects in, 122–125
Central character, 167, 199
Central function, 257
Cerednik-Drinfeld theory, 562
Change-of-basis matrix, 57
Character-functions, characterizing, 257
Character group schemes, 130
Characters, 129, 168
fundamental, 149, 214, 376–377
representations and, 252–254
universal, 263
Chebotarev density theorem, 93, 249
Circular units, 552
Class field theory, first case of Fermat's Last Theorem and, 499–502
Class number, 509
relative, 509
Classification theorem, 149–152, 443
CM (complex multiplication), 19, 37
Coboundaries, 101–102
Coboundary conditions, 103
Cocommutative Hopf algebras, 126
Cocycle conditions, 103
Cocycles, 101–102
difference, 285
method of obtaining, 109
Coefficient-ring, 4, 249
Coefficient-ring homomorphism, 249
Cohen-Macaulay ring, 329
Cohomological interpretation of Zariski tangent A-modules, 284–287
Cokernels, 125
Collection of local conditions, 113–114
Commutative algebra, 363–365
Commutative p-group schemes, Raynaud's results on, 146–153
Commutative triangle, 364
Commutativity, 123
Compactification of quotient, 555
Complete intersection rings, 13
Complete intersections, 343
criteria for, 343–355
Complete rings, local, 314
Completed tensor product, 265
Complex conjugation, 517
Complex multiplication (CM), 19, 37
Comultiplication, 126

Conductor, 3, 31–32, 168, 173
  nebentypus character of, 426
Congruence groups, 467
Congruence ideal, 343
Congruence modules, 14, 366–370
Congruence primes, 543
Connected components, 139
Connected-étale exact sequence over henselian local ring, 138–142
Connectedness, 391
Constant on orbits morphisms, 135
Constant S-schemes, 130
Continuity proposition, 280
Continuous automorphisms, 5
Continuous functors, 267
Continuous Kähler differentials, 274–276
Contragredient representation on dual space, 74
Contravariant set functor, 121
Converse theorem to Hecke theory, 187–188
Corank, 337
Correspondence, 61
Counit, 126
Crossed homomorphism, 102
Cubic reciprocity, 508
Cup products, 106–107
Cusp, 18, 45
Cusp forms, 76
  Maass, 173
  of weight 2, 85–88
Cuspidal automorphic representations, 169
Cuspidal representations, 161
Cyclotomic character, 5
Cyclotomic fields, Kummer's work on, 508–513
Cyclotomic units, 509

D-module, finite, 407
Deck-transformation group, 434
Decomposition subgroups, 5
Deformation, 245
Deformation conditions, 289–291
Deformation functors
  flat, 373–418
  universal, 394
Deformation problem, reduced tangent space of, 438
Deformation rings, 375
  universal, 12
Deformation theory
  of Galois representations, 243–309
  local Galois cohomology and, 397–406
Deformation types, 11, 421
Deformations, 4, 259, 313, 425–427
  flat, 324
  of galois representations, classifying, 108
  of group representations, 257–259
  infinitesimal, 109
  minimal, 423
  modular, *see* Modular deformations
  ordinary, 304, 323–324
  restrictions on, 323–324
Degeneracy maps, 368
Degree of representation, 250
Deligne theorem, 211–215
Descent, 35
  of group representations, 254–256
Determinant, fixed, 110
  global Galois deformation problem with, 294
Determinant conditions, 291–292
Deuring's theorem, 38
Dévissage, 147
Diagonalizable group schemes, 128
Diamond automorphisms, 68
Diamond operators, 8, 77, 211
Dieudonné modules, 337, 406–407
Difference cocycles, 285
Differential forms, 85–88, 234–237
Dihedral case, 159
Diophantine equations, 549
  conjectures about, 528–530
Dirichlet character, 169, 552
Dirichlet L-series, 550, 551
Discrete series representation, 177
Discriminant, minimal, 3, 31
Dual Hopf algebra, 144
Dual isogeny, 19
Dual Selmer group, 439

Dual space, contragredient
representation on, 74
Duality theory, 33

Eichler-Shimura-Igusa theorem,
429
Eichler-Shimura theory, 7, 8
"Eisenstein" primes, 432
Eisenstein series, 24, 26
Elliptic curves, 2, 17
arithmetic theory of, 17–40
with bad multiplicative
reduction, 116
conjectures about, 530–536
elliptic functions and, 24–26
explicit families of, with galois
representations mod n,
449–461
formal groups of, 26–27
galois representations mod 5 and,
471–473
height conjecture for, 534
isogenous, 23
isomorphic, 25
modular, *see* Modular elliptic
curves
over finite fields, 22–24
over local fields, 27–29
relations between Fermat
equations and, 536–539
relations with, 527–553
with same galois representations
mod N, 450–454
supersingular, 24
Weierstrass parameterization of,
52
Wiles' theorem and, 552–557
zeta functions of, 23
Elliptic functions, 24, 52–54
elliptic curves and, 24–26
Elliptic modular surfaces
of level N, 450–449
twists of, 449–450
Elliptic regulator, 36
Endomorphism ring of E, 19
Epsilon conjecture, 1, 9
Euler characteristics, 108, 553
Euler product, 185
Euler System, 564

Extensions, 108, 288

Faltings' construction, 263
Faltings' height, 530
Faltings' isogeny theorem, 32
Faltings' theorem, 34, 470, 529
Fermat conjecture, asymptotic,
529–530, 539
Fermat equations, relations between
elliptic curves and, 536–539
Fermat-Pell equation, 551
Fermat type, ternary equations of,
527–553
Fermat's Last Theorem (FLT), 1,
527, 546
first case of, 499–500
class field theory and, 499–502
overview of proof of, 1–15
for polynomials, 507
proof of, 10
for regular primes, 513–516
remarks of history on, 505–523
second case of, 506
Shimura-Taniyama-Weil
conjecture and, 220
suggested readings on, 521–522
Fiber products, 269
representability and, 267
Filtration, 28
Finite D-module, 407
Finite étale S-group schemes,
136–138
Finite fields, elliptic curves over,
22–24
Finite flat condition, 308
Finite flat group schemes, 121–153
Fontaine's approach to, 406–412
passage to quotient by, 132–135
techniques, 379
Finite Honda systems, 410
Finiteness at p, 214–215
Fitting ideals, 344, 345
Fixed determinant, 110
Flat deformation functors, 373–418
Flat deformations, 324
applications to, 413–418
Flat group schemes, finite, 121–153
Flat representations, 116, 375–393,
398

Flatness, 383
FLT, *see* Fermat's Last Theorem
Fontaine-Laffaille modules, 413
Fontaine-Laffaille theory, 116
Fontaine's approach to finite flat group schemes, 406–412
Forms of level N, 83
Formal group law, 27
Formal groups of elliptic curves, 26–27
Freeness of Hecke algebra, 434–436
Frey, Gerhart, 1
Frobenius automorphism, 6
  Hecke correspondences and, 69–73
Frobenius conjugacy class, 157
Frobenius element, 38, 71, 94, 247
Frobenius morphism, 22
Frobenius-semilinearity, 407
Functoriality, 193
  established examples of, 186–190
Functoriality maps, Albanese, 368
Functors
  continuous, 267
  defining, 394–397
  flat deformation, 373–418
  representability and, 267–284
  smooth morphisms of, 278
  universal deformation, 394
Fundamental characters, 149, 214, 376–377
Fundamental groups, 137
Fundamental lemma, 197

G-module, 101
Galois cohomology, 101–120
  local, deformation theory and, 397–406
Galois deformation conditions, 294–296
Galois deformation problem, global, 294
Galois-equivariant pairing, 333
Galois extension, 42
Galois groups, 246
  local, 5
Galois-invariant vector, 441
Galois representations, 2, 3–7, 212, 246–251
  arising naturally, 250–251
  associated to newforms, 7–8
  attached to E, 20–21
  classification of, by j-invariant of E, 493–498
  classifying deformations of, 108
  deformation theory of, 243–309
  determinants of, 4
  global, 5–6
  local, 476–479
  local behavior of, 475
  local properties of, 5–6
  mod 3, 231–233
  mod 4, 456–457
  mod 5
    elliptic curves and, 471–473
    modularity of, 463–473
  mod n, explicit families of elliptic curves with, 447–461
  mod p, 157–158, 220
  modular, 2, 7–9
  modularity of, 8
  in number theory, 6
  ordinary, 11
  p-adic, 6–7
  remarkable, 7
  residual representations of, 4
  semistable, 11
  unramified, 6
Gauss's theorem, 551
General linear groups, 127
General power reciprocity law, 501
Generalized Bernoulli numbers, 509
Generalized Selmer groups, 111–113, 365
Geometric height, 530
Geometric versions of Wiles' theorem, 555–557
Global duality theorem, 33
Global field, 527
Global Galois deformation problem, 294
  with fixed determinant, 294
Global galois representations, 5–6
Global L-series, 32
Global minimal Weierstrass equation, 31
Good reduction, 29
Gorenstein condition, 349

Gorenstein property, 12, 327–330
  importance of, for Hecke algebra, 327–341
Goursat's Lemma, 253
Gross-Zagier formula, 558–559, 563, 566
  variants of, 559–563, 565
Grossencharacters, 169
Grothendieck's semistable reduction theorem, 74
Grothendieck's theorem, 135, 277
Group, 101
Group functors, 124
Group law, 18
Group-like elements, 130
Group objects, 123
  in category, 122–125
Group representations, *see also* Representations
  algebra representations versus, 251–252
  deformations of, 257–259
  descent of, 254–256
Group schemes, 121, 125–132, 381
  diagonalizable, 128
Gunderson's theorem, 500

Haar measure, 81
Hall Conjecture, 545
Hasse invariant, 209
Hasse-Weil conjecture, 73–75
Hasse-Weil L-function, 32, 550, 551, 566
  analytic rank 0, 557
  analytic rank 1, 558–559
  special values of, 557–563
  Wiles' theorem and, 553–555
Hasse-Weil zeta-function, 550
Hasse's theorem, 23
Hecke algebras, 8, 89–94, 330–331, 359–361, 482–483
  freeness of, 434–436
  importance of Gorenstein property for, 327–341
Hecke congruence subgroup, 168
Hecke corollary, 553
Hecke correspondences, 61–73
  Frobenius automorphism and, 69–73
  moduli interpretation of, 63–66
  on upper half-plane, 66–67
Hecke eigenform, 80
Hecke-Jacquet-Langlands L-function, 179
Hecke L-series, 38
Hecke operators, 7, 78–81, 211, 429
Hecke rings, 375, 427–430
  universal deformation rings and, 421–444
Hecke theory, Converse theorem to, 187–188
Heegner points, 558–559
  p-adic analytic construction of, 561–562
Height conjecture for elliptic curves, 534
Heights, 35, 530
Henselian local ring, connected-étale exact sequence over, 138–142
Hensel's lemma, 26
Herbrand quotient, 523
Herbrand's theorem, 518–519
Hermite-Minkowski theorem, 117
Hilbert class field, 38
Hilbert space, 165
Hilbert symbol, 102
Hilbert's theorem 90, 104, 523
Hilbert's theorem 94, 523
Hochschild-Serre spectral sequence, 105
Holomorphic form, 173
Hom, 103–104
Homomorphisms
  coefficient-ring, 249
  crossed, 102
  k-algebra, 141
  lifting, to matrix groups, 317–318
Homothetic lattices, 25
Honda systems, 438
  finite, 410
Hopf algebras, 121, 125–126
  dual, 144
Hurwitz' genus formula, 42, 235, 541

I-ordinary cohomology group, 305

I-ordinary cohomology submodule, 306
I-ordinary representation, 304–305
Icosahedral case, 159
Ideal numbers, 505
Ideals, 508
Idele class group, 169
Index of H in G, 135
Inertia group, 5, 101, 376
Infinitesimal deformations, 109
Inflation-restriction proposition, 105
Inflation-restriction sequence, 117
Integral points, 39–40
Intersection rings, complete, 13
Intersections, complete, *see* Complete intersections
Invariant subspace, 166
Inverses, 123
Irreducibility, 161, 257
  absolute, *see* Absolute irreducibility
  supersingular case and, 377
Irreducibility theorem, 464
  proof of, 470
Irregular primes, 509
Isogenies, 19
  kernels of, 121
Isogenous elliptic curves, 23
Isogeny theorem, 423
Isomorphic elliptic curves, 25
Isomorphism classes, 46
Iwasawa decomposition, 166
Iwasawa theory, 519

J-function, 50
j-invariants, 17
  classification of galois representations by, 491–498
  modular, 454–455
j-minimal curves, 493
Jacobian variety, 71
Jacobi's formula, 26
Jacquet-Langlands correspondence, 221, 487
Jacquet-Langlands theorem, 556
Jordan-Hölder factors, 391–392

K-algebra homomorphisms, 141
K-finite vectors, 165
k-representation, 313
k-vector space, 351
Kähler differentials, 210, 236
  continuous, 274–276
Kamienny, Mazur, Merel theorem, 34
Kani's conjecture, 542
Kato's proof, 565
Katz's definition of modular forms, 209
Kernel of Norm, 104
Kernels, 124–125
  of isogenies, 121
Klein 4-group, 198
Kodaira-Spencer isomorphism, 217, 234
Kodaira symbols, 492
Kolyvagin's descent, 564, 566
  variant of, 565
Kolyvagin's proof, 563–564
Koszul complexes, 346
Kronecker-Weber theorem, 113, 552
Krull-Schmidt-Akizuki theorem, 338
Krull topology, 4
Kuga-Sato varieties, 563
Kummer congruence, 522
Kummer generators, 105
Kummer sequence, 29
Kummer's lemma, 512
Kummer's work on cyclotomic fields, 508–513

L-functions, 73–99, 192
L-groups, 183
L-morphism, 185
L-series, 558
  Rankin convolution of, 558
Langlands class, 171, 174, 183, 184
Langlands functoriality conjecture, 175–176
  statement of, 185–186
  theory and results, 182–191
Langlands L-factors, 184
Langlands parameter, 175, 182
Langlands program, 175–191
  Shimura-Taniyama-Weil Conjecture and, 190–191

Langlands reciprocity conjecture (LRC), 164, 179–182
Langlands theory, 14
Langlands-Tunnell form, 230–239
Langlands-Tunnell theorem, 158–159, 230–231
 proof of, 192–204
 reformulation of, 179–180
Lang's conjecture, 533–534
Laplace-Beltrami operator, 173
Lattices, homothetic, 25
Law of composition, 122
Left actions, 136
Leopoldt Conjecture, 262
Lie algebras, 406
Lifting homomorphisms to matrix groups, 317–318
Liftings, 4, 216
 strictly equivalent, 258
LLC (local Langlands conjecture), 175
Local behavior of galois representations, 475
Local complete rings, 314
Local conditions, collection of, 113–114
Local duality theorem, 33
Local fields, elliptic curves over, 27–29
Local Galois cohomology, deformation theory and, 397–406
Local galois groups, 5
Local Galois representations, 476–479
Local Langlands conjecture (LLC), 175
Local Langlands correspondence, 175, 176–179
Local Tate duality, 107–108
Local terms, computation of, 439–442
LRC (Langlands reciprocity conjecture), 164, 179–182

Maass cusp form, 173
Magmas, 122
Manin constant, 555
Manin-Drinfeld theorem, 557
Masser-Oesterle A-B-C conjecture, 2, 529, 544
Matrix algebra, 556
Matrix groups, lifting homomorphisms to, 317–318
Mayer-Vietoris property, 270, 277
Mazur's corollary, 547
Mazur's modular lifting conjecture, 158
Mazur's result, 225–227
Mazur's theorem, 34
Merel's theorem, 532, 539
Meromorphic function on punctured disk, 49
Minimal case, 14
Minimal deformations, 423
Minimal discriminant, 3, 31
Minimal ramification conditions, 300
Minimal Weierstrass equation, 27
Minimality condition, 422
Minimally ramified liftings, 480–481
"Mittag-Leffler" argument, 431
Modified Tate cohomology group, 101
Modular abelian varieties, 94–95
Modular curves, 41–60, 466–470
 classical theory of, 468
 of level N, 450–449
 as quotients of upper half-plane, 58–60
 twists of, 449–450, 469
Modular deformation ring, universal, 12
Modular deformations, Wiles's "main conjecture" and, 357–370
Modular elliptic curves, 95, 543
 explicit families of, 454–461
Modular forms, 2, 76–78
 Katz's definition of, 209
 lifting, 216
Modular functions, 49–51
Modular galois representations, 7–9
Modular j-invariants, 454–455
Modular lifting conjecture, Mazur's, 158
Modularity, 155–204

of galois representations mod 5, 463–473
Modularity Conjecture, 1, 9–10
semistable, 11
Modularity theorem, 464
proof of, 470–471
Moduli interpretation of Hecke correspondences, 63–66
Moduli scheme parametrizing triples, 210
Mordell Conjecture, 470
Mordell-Weil group, 553
Mordell-Weil theorem, 29
proof of, 36
Multiplication maps, 18
Multiplicative groups, 127
Murty-Murty theorem, 563

N-division point representation, 250
Nagell-Lutz Theorem, 460
Nakayama's Lemma, 138, 308
Nearly representable element, 277
Nebentypus character of conductor, 426
Néron differential, 554
Néron model, 380
Néron-Ogg-Shafarevich criterion, 28
Néron property, 225, 226
Néron-Tate canonical height, 35–36, 554
Newforms, 7, 173, 428
galois representations associated to, 7–8
of level N, 83
Node, 18
Noetherian rings, 121, 313
Non-minimal case, 14
Non-split reduction, 27
Non-vanishing lemma, 564
Norm, 101
kernel of, 104
Normalized basis, 55–57
Number theory, galois representations in, 6

Octahedral case, 159, 202–204
Old forms of level N, 83
Ordinary deformations, 304, 323–324
Ordinary galois representations, 11
Ordinary representation, 304–305

P-adic analogue, Perrin-Riou's, 560
p-adic analytic construction of Heegner points, 561–562
p-adic case, 177–179
p-adic galois representations, 6–7
p-adic L-function, square root of, 562
p-adic numbers, 510
p-adic representation theory, 167–168
p-class group, structure of, 517–521
p-finiteness condition, 246
Pell's equation, 552
Perrin-Riou's p-adic analogue, 560
Petersson inner product, 81–83
Picard functor, 225
Poitou-Tate proposition, 119
Polynomials, Fermat's Last Theorem for, 507
Potential good reduction, 29
Primes
auxiliary set of, 425
congruence, 543
"Eisenstein," 434
irregular, 509
regular, *see* Regular primes
Primitive forms, 7
of level N, 83
Pro-finite flat condition, 308
Pro-representable hull, 278
Profinite group, 313
Projective limits, 320–323
Prolongations, 146
unicity of, 153
Pseudo-characters, 257
Pseudo-representation, 257
Punctured disk, meromorphic function on, 49
Pythagorean triples, 549

Quantum groups, 126
Quasi-period of lattice L, 26
Quaternionic surfaces, 560
Quotient

Quotient (*continued*)
compactification of, 555
Herbrand, 523
passage to, by finite flat group schemes, 132–135
unramified, 179
of upper half-plane, 57–58
modular curves as, 58–60
Quotient ring, 89

Ramakrishna's theorem, 396–397, 403–406
proof of, 374
Ramakrishna's theory, 292–294
Ramification conditions, minimal, 300
Rank, analytic, 556
Rankin convolution of L-series, 558
Rankin-Selberg L-function, 201
Rankin-Selberg method, 97
Rankin-Selberg products, 189–190
Rational torsion, 34
Raynaud F-module scheme, 148
Raynaud's results on commutative p-group schemes, 146–153
Reduced tangent space of deformation problem, 438
Reduction of E, 27
Regular primes, 509
Fermat's Last Theorem for, 513–516
Relative class number, 509
Relative Zariski tangent A-modules, 286
Relative Zariski tangent space, 274–276
Relatively representable subfunctors, 278–279
Representability
fiber products and, 267
functors and, 267–284
strong near, 282
weak near, 281
Representations, *see also* Group representations
adelic, 168–170
admissible, 165
characters and, 252–254
residual, 4, 258, 259
Residual representations, 4, 258, 259
Restrictions on deformations, 323–324
Ribet, Ken, 1
Ribet's theorem, 9, 227-230
Riemann-Roch formula, 217
Riemann zeta function, 26
Right invariance, 132
Ring, 121
Rosati involution, 333
Rubin's p-adic formula, 560

S-group schemes, 125
Satake isomorphism, 194, 196
Scheme of left cosets of H in G, 135
Schemes, 121
Schlessinger's Criteria, 262–263
Schlessinger's representability theorem, 276–278
Schur-type theorems, 254–256
Schur's Lemma, 252
Selmer groups, 13, 30, 365–366, 436–439
dual, 439
generalized, 111–113
Semisimple ring, 91
Semistability, 3, 376
Semistable galois representations, 11
Semistable Modularity Conjecture, 11
Semistable reduction, 455–456
Separable closure, 101
Serre, Jean-Pierre, 1
Serre duality, 217
Serre's conjectures, 8–9, 209–239
cases for, 222–224
statement and results, 209–222
Serre's theorem, 34
Shafarevich-Tate group, 30, 554
Shapiro's lemma, 434
Shimura curves, 221, 556
analogues, 561-562
Shimura-Taniyama conjecture, 97–99
Shimura-Taniyama-Weil conjecture
Fermat's Last Theorem and, 220
Langlands program and, 190–191

Siegel's theorems, 39
Singular cubics, 18
Smooth morphisms of functors, 278
Snake Lemma, 110
Space of new forms of level N, 83
Space of old forms of level N, 83
Split reduction, 27
Square root of p-adic L-function, 562
Stickelberger's theorem, 506, 518
Strictly equivalent liftings, 258
Strictly free actions, 134
Strong Artin conjecture, 164, 179
Strong finiteness, 286
Strong multiplicity one, 174
Strong near representability, 282
Subfunctors, relatively representable, 278–279
Supersingular case, 373
  irreducibility and, 377
Supersingular elliptic curves, 24
Supersingular points, 233
Sylow subgroup, 148
Symmetric square lifting, 188–189
Szpiro's conjecture, 3, 534

Tame ramification group, 213
Tame ramification theory, 492
Taniyama-Weil conjecture, *see* Shimura-Taniyama-Weil conjecture
Tannakian approach, 248
Tate cohomology group, modified, 101
Tate curves, 34–35, 210
Tate duality, local, 33, 107–108
Tate module, 6, 20, 357
Tate period, 423
Tate-Poitou exact sequence, 439
Tate's local duality theorem, 33, 107–108
Taylor, Richard, 1
Taylor-Wiles-Faltings criterion, 430–432, 485
Teichmüller lift, 216
Ternary equations of Fermat type, 527–553
Tetrahedral case, 159, 198–202
Topological generator, 104
Topological ring, 4
Torsion points, 20
Torsion subgroups, 457–460
Trace formula, 192
  of Arthur-Selberg, 194–196
Trace of Frobenius, 23
Translation-by-Q map, 18
Triple of integers, 2
Twisted regular representation, 196
Twists
  of elliptic modular surfaces, 449–450
  of modular curves, 449–450, 469

Unicity of prolongations, 153
Unique factorization, failure of, 508
Unit elements, 122
Universal characters, 263
Universal coefficient-ring, 261
Universal deformation, 12, 259, 426–427
Universal deformation functors, 394
Universal deformation rings, 12, 259, 313, 362, 426, 481–482
  explicit construction of, 313–325
  Hecke rings and, 421–444
  structure of, 436–442
Universal deformation space, 259
Universal modular deformation, 12
Universal modular deformation ring, 12
Unramified galois representations, 6
Unramified induced representation, 178
Unramified quotient, 179
Unramifiedness criterion, 390
Upper half-plane
  Hecke correspondences on, 66–67
  quotient of, *see* Quotient of upper half-plane

Vandiver's conjecture, 506, 516, 520

Waldspurger's theorem, 566
Weak finiteness, 286
Weak near representability, 281
Weber function on E, 37
Wedderburn's Theorem, 385

Weierstrass $\wp$-function, 24
Weierstrass σ-function, 25
Weierstrass equation, 2, 17
  global minimal, 31
  minimal, 27
Weierstrass model, 450
Weierstrass parameterization of elliptic curves, 52
Weil-Châtelet group, 30
Weil form, 183
Weil group, 175, 176
Weil pairing, 21–22, 157
Wild ramification group, 214
Wiles, Andrew, 2
Wiles-Lenstra criterion, 488
Wiles' "main conjecture," modular deformations and, 357–370
Wiles' numerical criterion, 13
Wiles' results, extension of, 475–488
Wiles' theorem, 10, 527, 554
  elliptic curves and, 552–557
  geometric versions of, 555–557
  Hasse-Weil L-function and, 553–555
  proof of, 10–15
Witt vectors, 262

Yoneda's pairing, 330

Zariski-Nagata theorem, 389
Zariski tangent A-modules, 273
  cohomological interpretation of, 284–287
  relative, 286
Zariski tangent space, 270–272
  relative, 274–276
Zeta functions of elliptic curves, 23